HOMOLOGICAL THEORY OF REPRESENTATIONS

Modern developments in representation theory rely heavily on homological methods. This book for advanced graduate students and researchers introduces these methods from their foundations up and discusses several landmark results that illustrate their power and beauty.

The categorical foundations include abelian and derived categories, with an emphasis on localisation, spectra, and purity. The representation theoretic focus is on module categories of Artin algebras, with discussions of the representation theory of finite groups and finite quivers. Also covered are Gorenstein and quasi-hereditary algebras, including Schur algebras, which model polynomial representations of general linear groups, and the Morita theory of derived categories via tilting objects. The final part is devoted to a systematic introduction to the theory of purity for locally finitely presented categories, covering pure-injectives, definable subcategories, and Ziegler spectra.

With its clear, detailed exposition of important topics in modern representation theory, many of which have been unavailable in one volume until now, this book deserves a place in every representation theorist's library.

Henning Krause is Professor of Mathematics at Bielefeld University. He works in the area of representation theory of finite dimensional algebras, with a particular interest in homological structures. His previous publications include the *Handbook of Tilting Theory* (Cambridge, 2007). Professor Krause is Fellow of the American Mathematical Society.

CAMBRIDGE STUDIES IN ADVANCED MATHEMATICS

All the titles listed below can be obtained from good booksellers or from Cambridge University Press.
For a complete series listing, visit www.cambridge.org/mathematics.

Already Published
155 G. Pisier *Martingales in Banach Spaces*
156 C. T. C. Wall *Differential Topology*
157 J. C. Robinson, J. L. Rodrigo & W. Sadowski *The Three-Dimensional Navier–Stokes Equations*
158 D. Huybrechts *Lectures on K3 Surfaces*
159 H. Matsumoto & S. Taniguchi *Stochastic Analysis*
160 A. Borodin & G. Olshanski *Representations of the Infinite Symmetric Group*
161 P. Webb *Finite Group Representations for the Pure Mathematician*
162 C. J. Bishop & Y. Peres *Fractals in Probability and Analysis*
163 A. Bovier *Gaussian Processes on Trees*
164 P. Schneider *Galois Representations and (ϕ, Γ)-Modules*
165 P. Gille & T. Szamuely *Central Simple Algebras and Galois Cohomology (2nd Edition)*
166 D. Li & H. Queffelec *Introduction to Banach Spaces, I*
167 D. Li & H. Queffelec *Introduction to Banach Spaces, II*
168 J. Carlson, S. Müller-Stach & C. Peters *Period Mappings and Period Domains (2nd Edition)*
169 J. M. Landsberg *Geometry and Complexity Theory*
170 J. S. Milne *Algebraic Groups*
171 J. Gough & J. Kupsch *Quantum Fields and Processes*
172 T. Ceccherini-Silberstein, F. Scarabotti & F. Tolli *Discrete Harmonic Analysis*
173 P. Garrett *Modern Analysis of Automorphic Forms by Example, I*
174 P. Garrett *Modern Analysis of Automorphic Forms by Example, II*
175 G. Navarro *Character Theory and the McKay Conjecture*
176 P. Fleig, H. P. A. Gustafsson, A. Kleinschmidt & D. Persson *Eisenstein Series and Automorphic Representations*
177 E. Peterson *Formal Geometry and Bordism Operators*
178 A. Ogus *Lectures on Logarithmic Algebraic Geometry*
179 N. Nikolski *Hardy Spaces*
180 D.-C. Cisinski *Higher Categories and Homotopical Algebra*
181 A. Agrachev, D. Barilari & U. Boscain *A Comprehensive Introduction to Sub-Riemannian Geometry*
182 N. Nikolski *Toeplitz Matrices and Operators*
183 A. Yekutieli *Derived Categories*
184 C. Demeter *Fourier Restriction, Decoupling and Applications*
185 D. Barnes & C. Roitzheim *Foundations of Stable Homotopy Theory*
186 V. Vasyunin & A. Volberg *The Bellman Function Technique in Harmonic Analysis*
187 M. Geck & G. Malle *The Character Theory of Finite Groups of Lie Type*
188 B. Richter *Category Theory for Homotopy Theory*
189 R. Willett & G. Yu *Higher Index Theory*
190 A. Bobrowski *Generators of Markov Chains*
191 D. Cao, S. Peng & S. Yan *Singularly Perturbed Methods for Nonlinear Elliptic Problems*
192 E. Kowalski *An Introduction to Probabilistic Number Theory*
193 V. Gorin *Lectures on Random Lozenge Tilings*
194 E. Riehl & D. Verity *Elements of ∞-Category Theory*

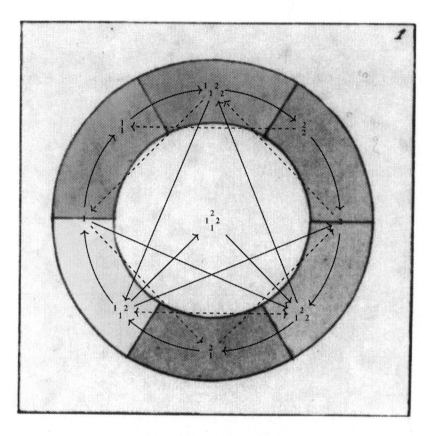

This illustration combines Goethe's Farbkreis [J. W. von Goethe, Zur Farbenlehre, Erster Band, Nebst einem Hefte mit sechzehn Kupfertafeln, Tübingen, 1810] with the Auslander–Reiten quiver of a Gorenstein algebra of dimension one (Figure 6.1).

Homological Theory of Representations

HENNING KRAUSE

Universität Bielefeld, Germany

CAMBRIDGE
UNIVERSITY PRESS

CAMBRIDGE
UNIVERSITY PRESS

University Printing House, Cambridge CB2 8BS, United Kingdom

One Liberty Plaza, 20th Floor, New York, NY 10006, USA

477 Williamstown Road, Port Melbourne, VIC 3207, Australia

314–321, 3rd Floor, Plot 3, Splendor Forum, Jasola District Centre,
New Delhi – 110025, India

103 Penang Road, #05–06/07, Visioncrest Commercial, Singapore 238467

Cambridge University Press is part of the University of Cambridge.

It furthers the University's mission by disseminating knowledge in the pursuit of
education, learning, and research at the highest international levels of excellence.

www.cambridge.org
Information on this title: www.cambridge.org/9781108838894
DOI: 10.1017/9781108979108

First published 2022

A catalogue record for this publication is available from the British Library.

ISBN 978-1-108-83889-4 Hardback

Cambridge University Press has no responsibility for the persistence or accuracy
of URLs for external or third-party internet websites referred to in this publication
and does not guarantee that any content on such websites is, or will remain,
accurate or appropriate.

Contents

Introduction		*page* x
Conventions and Notations		xv
Glossary		xvii
Standard Functors and Isomorphisms		xxxiii
PART ONE ABELIAN AND DERIVED CATEGORIES		1
1	**Localisation**	3
	1.1 Localisation	3
	1.2 Calculus of Fractions	10
	Notes	12
2	**Abelian Categories**	14
	2.1 Exact Categories	15
	2.2 Localisation of Additive and Abelian Categories	28
	2.3 Module Categories and Their Localisations	42
	2.4 Commutative Noetherian Rings	47
	2.5 Grothendieck Categories	55
	Notes	70
3	**Triangulated Categories**	72
	3.1 Triangulated Categories	73
	3.2 Localisation of Triangulated Categories	77
	3.3 Frobenius Categories	83
	3.4 Brown Representability	89
	Notes	100
4	**Derived Categories**	101
	4.1 Derived Categories	102
	4.2 Resolutions and Extensions	110

	4.3	Resolutions and Derived Functors	122
	4.4	Examples of Derived Categories	133
	Notes		144
5	**Derived Categories of Representations**		146
	5.1	Examples Related to the Projective Line	146
	5.2	Derived Categories of Finitely Presented Modules	164
	Notes		172

PART TWO   ORTHOGONAL DECOMPOSITIONS   173

6	**Gorenstein Algebras, Approximations, Serre Duality**		175
	6.1	Approximations	176
	6.2	Gorenstein Rings	179
	6.3	Serre Duality	189
	6.4	The Derived Nakayama Functor	195
	6.5	Examples	203
	Notes		205
7	**Tilting in Exact Categories**		207
	7.1	Cotorsion Pairs	208
	7.2	Tilting in Exact Categories	215
	Notes		226
8	**Polynomial Representations**		228
	8.1	Quasi-hereditary Algebras	231
	8.2	Symmetric Tensors	241
	8.3	Polynomial Representations	250
	8.4	Cauchy Decompositions	264
	8.5	Schur and Weyl Modules and Functors	272
	8.6	Schur Algebras	284
	Notes		291

PART THREE   DERIVED EQUIVALENCES   295

9	**Derived Equivalences**		297
	9.1	Differential Graded Algebras	298
	9.2	Derived Equivalences	309
	9.3	Finite Global Dimension	318
	Notes		328

10	**Examples of Derived Equivalences**	329
	10.1 Coherent Sheaves on Projective Space	329
	10.2 Koszul Duality	330
	10.3 The BGG Correspondence	332
	10.4 Koszul Duality for the Beilinson Algebra	333
	10.5 Weighted Projective Lines	334
	10.6 Gentle Algebras	336
	PART FOUR PURITY	339
11	**Locally Finitely Presented Categories**	341
	11.1 Locally Finitely Presented Categories	342
	11.2 Grothendieck Categories	356
	11.3 Gröbner Categories	367
	Notes	376
12	**Purity**	377
	12.1 Purity	378
	12.2 Definable Subcategories	384
	12.3 Indecomposable Pure-Injective Objects	392
	12.4 Pure-Injective Modules	400
	Notes	412
13	**Endofiniteness**	414
	13.1 Endofinite Objects and Subadditive Functions	415
	13.2 Endofinite Modules	426
	Notes	433
14	**Krull–Gabriel Dimension**	435
	14.1 The Krull–Gabriel Filtration	435
	14.2 Examples of Krull–Gabriel Filtrations	444
	Notes	456
	References	457
	Notation	469
	Index	477

Introduction

This book is devoted to representations of associative algebras and their homological theory. There are two basic approaches. The passage from representations to chain complexes of representations leads to the study of *derived categories*. The other approach identifies representations with appropriate functors; this leads to the study of *functor categories*. We offer an introduction to both approaches and present results which illustrate their beauty and importance.

History. The appearance of the book *Homological Algebra* by Cartan and Eilenberg in 1956 established the subject [46]. In the following year Grothendieck published his seminal paper *Sur quelques points d'algèbre homologique* which inspired a whole generation [94]. Two students from the circle around Grothendieck then developed the foundations for the subject of this book. There is the thesis *Des catégories abéliennes* from 1960 by Peter Gabriel [79] and the thesis *Des catégories dérivées des catégories abéliennes* from 1967 by Jean-Louis Verdier [199]. Substantial parts of this book are devoted to explaining their work so that it can be applied to the study of representations.

Topics. We focus on representation theoretic results, most of which originate from the 1980s and early 1990s. A major part of the book is devoted to derived categories, and the notion of tilting plays an important role. Orthogonal decompositions provide another organisational principle for several results. The final part is about purity and involves the use of functor categories. The context for most results is the category of modules over a ring. When appropriate we work more generally with abelian categories, or we restrict to certain classes of modules or rings. For instance, of particular interest from the representation theory perspective are modules of finite length over Artin algebras.

The following is a list of topics and results which are treated in this book, beyond the foundational material discussed further below.

- *Gorenstein algebras.* The module category of a Gorenstein algebra admits an orthogonal decomposition into the subcategory of modules of finite projective dimension and the subcategory of Gorenstein projective modules. For Artin algebras the corresponding bounded derived category of modules of finite projective dimension and the stable category of Gorenstein projective modules admit Serre functors.
- *Tilting modules.* For Artin algebras there is a bijective correspondence between equivalence classes of tilting modules and covariantly finite coresolving subcategories.
- *Characteristic tilting modules.* For every quasi-hereditary algebra there is a canonical tilting module; its indecomposable direct summands are precisely the indecomposable modules which have a standard and a costandard filtration.
- *Schur algebras.* Polynomial representations of general linear groups identify with modules over Schur algebras. Every Schur algebra is quasi-hereditary and the characteristic tilting modules are given by tensor products of exterior powers.
- *Happel's theorem.* A tilting object of an exact category induces a triangle equivalence between its bounded derived category and the category of perfect complexes over the endomorphism algebra of the tilting object.
- *Happel's functor.* The bounded derived category of modules over an Artin algebra embeds into the stable category of graded modules over the corresponding trivial extension algebra, and equality holds if and only if the algebra has finite global dimension.
- *Rickard's theorem.* Two algebras have equivalent derived categories if and only if one admits a tilting complex with endomorphism algebra isomorphic to the other algebra.
- *Global dimension.* Tilting preserves finite global dimension. If the bounded derived category of an abelian category of finite global dimension admits a tilting object, then its endomorphism ring is of finite global dimension.
- *Gröbner categories.* Representations of the category of finite sets in some locally noetherian Grothendieck category form again a locally noetherian Grothendieck category. This is a vast generalisation of Hilbert's basis theorem and has several applications.
- *Definable subcategories.* The definable subcategories of a module category (that is, subcategories closed under filtered colimits, products and pure submodules) are in bijective correspondence to Ziegler closed subsets of indecomposable pure-injective modules.
- *Injective cohomology representations.* For a finite group, every injective module over its cohomology ring can be realised as the cohomology of

a representation. Such a representation is essentially unique and Σ-pure-injective; therefore it decomposes uniquely into indecomposable representations corresponding to homogeneous prime ideals of the cohomology ring.

– *Endofinite modules.* Modules of finite length over their endomorphism ring decompose uniquely into indecomposables, and the isomorphism classes of indecomposables are in bijective correspondence to irreducible subadditive functions on finitely presented modules.

– *Finite representation type.* A ring is of finite representation type if and only if every module is endofinite.

– *Krull–Gabriel filtrations.* Pure-injective objects are classified via Krull–Gabriel filtrations. Examples include modules over Dedekind domains, quasi-coherent sheaves on the projective line, and representations of the Kronecker quiver.

Foundations. Several chapters of this book are devoted to basic concepts and foundational results. Let us mention some of these topics.

– *Localisation.* Localisation is a process of adding formal inverses to an algebraic structure; it is used throughout the book. The localisation of additive categories amounts to annihilating appropriate subcategories. For abelian and triangulated categories the morphisms of a localised category can be described via a calculus of fractions.

– *Abelian categories.* Abelian categories generalise module categories. Of particular interest are Grothendieck categories, which are precisely the localisations of module categories. Objects in these categories admit injective envelopes; so one can do homological algebra.

– *Triangulated and derived categories.* The derived category of an abelian category provides the proper context for studying derived functors. An important ingredient is the construction of resolutions. Triangulated categories form the appropriate categorical framework. Useful tools include Verdier localisation and Brown representability for cohomological functors.

– *Locally finitely presented categories.* These are cocomplete additive categories such that every object is a filtered colimit of finitely presented objects; in fact they are determined by their full subcategories of finitely presented objects. For locally finitely presented Grothendieck categories there is a well-developed theory of injective objects. Pure-injective objects are studied via an embedding into a Grothendieck category that is locally finitely presented.

There are further fundamental concepts that appear throughout this work. We mention the notion of a finitely presented (or coherent) functor. In fact, functor categories play an important role, not only because each representation (of an algebra, a quiver, or a group) may be viewed as a functor. The basic idea is to identify an object X of an additive category with the corresponding representable functor $\mathrm{Hom}(-, X)$, often restricted to some appropriate generating subcategory. In this way categories of representations are presented as categories of functors. This idea goes back to Gabriel [79] and Auslander [7], but continues to be useful, also in the study of triangulated categories [150].

Another key concept that pervades this book is the notion of a spectrum. Indecomposable representations are often viewed as points of some space. The analogue in commutative algebra is the Zariski spectrum, but the spectrum of indecomposable injective objects of a Grothendieck category is the more general concept which is used throughout.

Prerequisites. The exposition is demanding in terms of the background and mathematical experience expected of the reader. We assume a basic knowledge of representation theory and homological algebra, including the appropriate categorical language.

Basic concepts and facts that are used throughout the book are arranged in a glossary which also serves to fix notation. Some topics from the glossary are explained in more detail in later chapters.

For unexplained terminology and further details, the following books are recommended: Cartan and Eilenberg [46], Mac Lane [141] (homological algebra), Schubert [183] (categories), Lam [136] and Stenström [197] (rings and modules).

Organisation. This book does not attempt to give a complete and systematic introduction to the homological theory of representations. It is rather motivated by a series of representation theoretic results (cf. the above list) for which we provide proper foundations and complete proofs. The choice of these results is based on personal taste and is by no means systematic.

The material is organised into 14 chapters and each is devoted to a particular topic. We have tried to keep the chapters as independent of each other as possible. This causes some repetition, but it will help the reader who is only interested in a particular topic.

Each chapter ends with notes and historical comments. This compensates for the fact that the body of the text gives no credits for definitions and theorems. We make no attempt to discuss the early history of the subject. For instance, the development of 'modern algebra' by Emmy Noether and her school inspired many concepts and results which are presented in this volume; for a detailed account we refer to Corry [55].

A final warning seems appropriate. We try to present concepts and results in their natural generality, even though readers may only be interested in some special cases. For instance, we treat derived categories of *exact categories*, and it is obvious that these are abelian in most applications. Or we study purity for *locally finitely presented categories*, despite the fact that module categories are the most interesting examples. Our motivation for generality is twofold. Concepts and arguments often become more transparent by identifying the ingredients that are essential. And we believe in potential applications beyond those which are obvious and well known.

Acknowledgements. I have been fortunate to meet several excellent teachers: Herbert Kupisch and Josef Waschbüsch (Berlin), Sheila Brenner and Michael Butler (Liverpool), Dieter Happel and Claus Michael Ringel (Bielefeld). Later on, Maurice Auslander, Bill Crawley-Boevey, and Helmut Lenzing provided much inspiration. I am very grateful to all of them; their style and taste has guided me throughout the work on this book. Also, I learned a lot from my coauthors. It is a special pleasure to thank Dave Benson and Srikanth Iyengar for great fun through collaboration.

Many students as well as colleagues contributed specifically to this book with numerous suggestions and friendly criticism. I would like to record my thanks to those people. In particular, the combinatorially most challenging exposition of Schur algebras benefited greatly from comments by Darij Grinberg, and Sondre Kvamme provided thoughtful comments on many chapters. The discussions with Andrew Hubery and the advice from Dieter Vossieck on all aspects of this work are very much appreciated.

It remains to thank the staff at Cambridge University Press for their efficient work, and in particular Tom Harris for his enthusiastic support.

Conventions and Notations

Categories

We follow the von Neumann–Bernays–Gödel set theory and distinguish between *sets* and *classes*. All categories are assumed to be *locally small* in the sense that the objects form a class and for each pair of objects the morphisms between them form a set. When a category is abelian or exact, we assume in addition that for each pair of objects the extensions (in the sense of Yoneda) form a set.

We denote by Set the category of *sets* and by Ab the category of *abelian groups*. The *cardinality* of a set X is denoted by card X.

Morphisms are composed from right to left. For the composite $X \xrightarrow{\alpha} Y \xrightarrow{\beta} Z$ we write $\beta\alpha$.

Functors $\mathcal{C} \to \mathcal{D}$ are by convention covariant. Replacing one of the categories by its opposite category identifies contravariant functors $\mathcal{C} \to \mathcal{D}$ with covariant functors $\mathcal{C}^{\mathrm{op}} \to \mathcal{D}$ or $\mathcal{C} \to \mathcal{D}^{\mathrm{op}}$.

Rings and Modules

All rings are associative and have a unit.

For a ring Λ we consider the category Mod Λ of *right Λ-modules* but drop the adjective 'right'. Left Λ-modules are identified with modules over the *opposite ring* Λ^{op}. The full subcategory of finitely presented Λ-modules is denoted by mod Λ, and proj Λ denotes the full subcategory of finitely generated projective Λ-modules.

When Λ and Γ are k-algebras over a commutative ring k, then Γ-Λ-*bimodules* $_\Gamma M_\Lambda$ are identified with modules over the algebra $\Gamma^{\mathrm{op}} \otimes_k \Lambda$.

Numbers

We denote by $\mathbb{Z}$ the set of integers and write

$$\mathbb{N} = \{0, 1, 2, 3, \ldots\}$$

for the set of non-negative integers.

Glossary

Category. A *category* $\mathcal{C}$ is given by a class of *objects* $\mathrm{Ob}\,\mathcal{C}$ and a class of *morphisms* $\mathrm{Mor}\,\mathcal{C}$, together with an associative unital composition. For objects $X, Y \in \mathcal{C}$ let $\mathrm{Hom}_{\mathcal{C}}(X, Y)$ denote the set of morphisms $X \to Y$, and $\mathrm{id}_X \colon X \to X$ the identity morphism. We write $\mathrm{End}_{\mathcal{C}}(X)$ for the set of endomorphisms of X. Sometimes we simplify the notation and write $\mathrm{Hom}(X, Y)$ or $\mathcal{C}(X, Y)$. The composition is given by a map

$$\mathrm{Hom}_{\mathcal{C}}(Z, Y) \times \mathrm{Hom}_{\mathcal{C}}(X, Y) \longrightarrow \mathrm{Hom}_{\mathcal{C}}(X, Z), \quad (\psi, \phi) \mapsto \psi\phi$$

for each triple of objects $X, Y, Z \in \mathcal{C}$.

The category $\mathcal{C}$ is *small* if $\mathrm{Ob}\,\mathcal{C}$ is a set, and $\mathcal{C}$ is *essentially small* if the isomorphism classes of objects in $\mathcal{C}$ form a set. The *opposite category* of $\mathcal{C}$ is denoted by $\mathcal{C}^{\mathrm{op}}$.

Morphisms. A morphism $X \to Y$ in a category $\mathcal{C}$ is a *monomorphism* if the induced map $\mathrm{Hom}(C, X) \to \mathrm{Hom}(C, Y)$ is injective for all $C \in \mathcal{C}$ (notation: $X \rightarrowtail Y$), an *epimorphism* if the map $\mathrm{Hom}(Y, C) \to \mathrm{Hom}(X, C)$ is injective for all $C \in \mathcal{C}$ (notation: $X \twoheadrightarrow Y$), and an *isomorphism* if the map $\mathrm{Hom}(C, X) \to \mathrm{Hom}(C, Y)$ is bijective for all $C \in \mathcal{C}$ (notation: $X \xrightarrow{\sim} Y$).

The morphisms in $\mathcal{C}$ form the *category of morphisms* $\mathcal{C}^2$; this identifies with the category of functors $\mathbf{2} \to \mathcal{C}$ where $\mathbf{2}$ denotes the category given by two objects which are connected by one morphism.

Functor. A *functor* $F \colon \mathcal{C} \to \mathcal{D}$ is given by a map on objects $\mathrm{Ob}\,\mathcal{C} \to \mathrm{Ob}\,\mathcal{D}$, together with maps on morphisms

$$F_{X,Y} \colon \mathrm{Hom}_{\mathcal{C}}(X, Y) \longrightarrow \mathrm{Hom}_{\mathcal{D}}(FX, FY)$$

for all objects $X, Y \in \mathcal{C}$, preserving the composition of morphisms. The identity functor is denoted by $\mathrm{id}_{\mathcal{C}} \colon \mathcal{C} \to \mathcal{C}$.

The functor F is *faithful* if all $F_{X,Y}$ are injective and *full* if all $F_{X,Y}$ are surjective. The notation $\mathcal{C} \rightarrowtail \mathcal{D}$ is used when F is fully faithful.

We write $\mathcal{H}om(\mathcal{C}, \mathcal{D})$ or $\mathrm{Fun}(\mathcal{C}, \mathcal{D})$ for the 'category' of functors $\mathcal{C} \to \mathcal{D}$. The morphisms between two functors are the natural transformations, but we do not require that they form a set.

Essential image. The *essential image* of a functor $F\colon \mathcal{C} \to \mathcal{D}$ is the full subcategory of $\mathcal{D}$ given by $\mathrm{Im}\, F = \{Y \in \mathcal{D} \mid Y \cong F(X) \text{ for some } X \in \mathcal{C}\}$. The functor F is *essentially surjective* if $\mathrm{Im}\, F = \mathcal{D}$.

Quotient functor. A *quotient functor* is a functor $F\colon \mathcal{C} \to \mathcal{D}$ such that for some class $S \subseteq \mathrm{Mor}\,\mathcal{C}$ of morphisms the functor F inverts all morphisms in S (so $F\phi$ is invertible for all $\phi \in S$) and every functor $F'\colon \mathcal{C} \to \mathcal{D}'$ factors uniquely through F provided that F' inverts all morphisms in S. In this case we set $\mathcal{C}[S^{-1}] = \mathcal{D}$ and the notation $\mathcal{C} \twoheadrightarrow \mathcal{D}$ is used for F.

Equivalence. A functor $F\colon \mathcal{C} \to \mathcal{D}$ is an *equivalence* if there is a functor $G\colon \mathcal{D} \to \mathcal{C}$, together with natural isomorphisms $\mathrm{id}_\mathcal{C} \xrightarrow{\sim} GF$ and $FG \xrightarrow{\sim} \mathrm{id}_\mathcal{D}$. An equivalent condition is that F is fully faithful and every object $Y \in \mathcal{D}$ is isomorphic to FX for some object $X \in \mathcal{C}$.[1] Notation: $\mathcal{C} \xrightarrow{\sim} \mathcal{D}$.

Adjoint. A pair (F, G) of functors $F\colon \mathcal{C} \to \mathcal{D}$ and $G\colon \mathcal{D} \to \mathcal{C}$ is *adjoint* if there are natural bijections:

$$\mathrm{Hom}_\mathcal{D}(FX, Y) \cong \mathrm{Hom}_\mathcal{C}(X, GY) \qquad (X \in \mathcal{C}, Y \in \mathcal{D}).$$

Then the notation $\mathcal{C} \rightleftarrows \mathcal{D}$ is used and there are two natural morphisms:

$$\eta_X\colon X \longrightarrow GF(X) \quad \text{(unit)} \qquad\qquad \varepsilon_Y\colon FG(Y) \longrightarrow Y \quad \text{(counit)}.$$

The composite

$$\mathrm{Hom}_\mathcal{C}(X', X) \xrightarrow{\ F\ } \mathrm{Hom}_\mathcal{D}(F(X'), F(X)) \xrightarrow{\sim} \mathrm{Hom}_\mathcal{C}(X', GF(X))$$

is given by composition with the unit η_X, and the composite

$$\mathrm{Hom}_\mathcal{D}(Y, Y') \xrightarrow{\ G\ } \mathrm{Hom}_\mathcal{C}(G(Y), G(Y')) \xrightarrow{\sim} \mathrm{Hom}_\mathcal{D}(FG(Y), Y')$$

is given by composition with the counit ε_Y. Moreover, the following conditions are equivalent.[2]

(1) The functor G is fully faithful.

---

[1] [197, Proposition IV.1.1]
[2] Proposition 1.1.3

(2) The counit ε_Y is an isomorphism for every object $Y \in \mathcal{D}$.

(3) The functor F is the composite $\mathcal{C} \twoheadrightarrow \mathcal{C}[S^{-1}] \xrightarrow{\sim} \mathcal{D}$ of a quotient functor for some class $S \subseteq \mathrm{Mor}\,\mathcal{C}$ and an equivalence.

This is expressed by the following diagram.

$$\mathcal{C} \underset{G}{\overset{F}{\rightleftarrows}} \mathcal{D}$$

Sometimes we denote by F_λ the left adjoint of F, and by F_ρ the right adjoint of F.

Localisation functor. A functor $L\colon \mathcal{C} \to \mathcal{C}$ is called a *localisation functor* if there exists a morphism $\eta\colon \mathrm{id}_\mathcal{C} \to L$ such that $L\eta\colon L \to L^2$ is an isomorphism and $L\eta = \eta L$.

Any localisation functor gives a pair (F, G) of adjoint functors such that F is (up to an equivalence) a quotient functor and G is fully faithful (by taking the inclusion $G\colon \mathcal{D} \to \mathcal{C}$ for $\mathcal{D} = \{X \in \mathcal{C} \mid \eta_X \text{ is invertible}\}$ and $FX = LX$ for every object $X \in \mathcal{C}$). Conversely, any pair (F, G) of adjoint functors such that F is a quotient functor or G is fully faithful gives a localisation functor (L, η) (by taking $L = GF$ and for η the unit).[3]

Limit and colimit. Let $\mathcal{C}$ be a category. A *diagram* of *type* $\mathcal{J}$ is a functor $\mathcal{J} \to \mathcal{C}$ where the category $\mathcal{J}$ is essentially small. Let $\mathcal{C}^{\mathcal{J}}$ denote the category of such diagrams. The *diagonal functor* $\Delta\colon \mathcal{C} \to \mathcal{C}^{\mathcal{J}}$ takes an object to the constant functor. For $F \in \mathcal{C}^{\mathcal{J}}$ the *limit* $\lim F$ (also written $\lim_{i \in \mathcal{J}} F(i)$) is given by a natural bijection

$$\mathrm{Hom}_\mathcal{C}(X, \lim F) \cong \mathrm{Hom}_{\mathcal{C}^{\mathcal{J}}}(\Delta X, F) \qquad (X \in \mathcal{C})$$

provided it exists in $\mathcal{C}$. Thus the limit is the right adjoint of the diagonal functor and the counit provides canonical morphisms $\lim F \to F(i)$ for all $i \in \mathcal{J}$. Analogously, the *colimit* $\mathrm{colim}\,F$ is given by

$$\mathrm{Hom}_\mathcal{C}(\mathrm{colim}\,F, X) \cong \mathrm{Hom}_{\mathcal{C}^{\mathcal{J}}}(F, \Delta X) \qquad (X \in \mathcal{C})$$

and comes with canonical morphisms $F(i) \to \mathrm{colim}\,F$ for all $i \in \mathcal{J}$.

Filtered category. A category $\mathcal{J}$ is *filtered* if

(Fil1) the category is non-empty,

(Fil2) given objects i, i' there is an object j with morphisms $i \to j \leftarrow i'$, and

---

[3] Proposition 1.1.5

(Fil3) given morphisms $\alpha, \alpha' : i \to j$ there is a morphism $\beta : j \to k$ such that
$\beta\alpha = \beta\alpha'$.

A partially ordered set $(I, \leq)$ can be viewed as a category: the objects are
the elements of I and there is a unique morphism $i \to j$ whenever $i \leq j$. This
category is filtered if and only if $(I, \leq)$ is non-empty and *directed*, that is, for
each pair of elements i, i' there is an element j such that $i, i' \leq j$.

Filtered colimit. A *filtered colimit* is the colimit of a functor $F : \mathfrak{I} \to \mathcal{C}$ such
that the category $\mathfrak{I}$ is filtered. When $\mathfrak{I}$ is given by a directed partially ordered
set, this colimit is also called the *directed colimit* (or confusingly *direct limit*).

Let $\mathfrak{I}$ be an essentially small filtered category. A fully faithful functor $\phi : \mathfrak{J} \to$
$\mathfrak{I}$ is *cofinal* if for every object $i \in \mathfrak{I}$ there exists a morphism $i \to \phi(j)$ for
some $j \in \mathfrak{J}$. In that case $\mathfrak{J}$ is filtered and any functor $F : \mathfrak{I} \to \mathcal{C}$ induces an
isomorphism $\operatorname{colim}(F \circ \phi) \xrightarrow{\sim} \operatorname{colim} F$.[4]

For each essentially small filtered category $\mathfrak{I}$ there exists a functor $\phi : \mathfrak{J} \to \mathfrak{I}$
such that $\mathfrak{J}$ is the category corresponding to a directed partially ordered set and
any functor $F : \mathfrak{I} \to \mathcal{C}$ induces an isomorphism $\operatorname{colim}(F \circ \phi) \xrightarrow{\sim} \operatorname{colim} F$.[5]

Additive category. A category $\mathcal{A}$ is *additive* if

(Ad1) for every finite family of objects $X_1, \ldots, X_r$ in $\mathcal{A}$ there exists a product
$X_1 \times \cdots \times X_r$ in $\mathcal{A}$,

(Ad2) each morphism set $\operatorname{Hom}_A(X, Y)$ is an abelian group, and

(Ad3) the composition maps $\operatorname{Hom}_A(Y, Z) \times \operatorname{Hom}_A(X, Y) \to \operatorname{Hom}_A(X, Z)$
are biadditive.

If $\mathcal{A}$ is an additive category, then finite coproducts also exist. Moreover,
finite products and coproducts in $\mathcal{A}$ coincide. For objects X, Y in $\mathcal{A}$ the group
structure on $\operatorname{Hom}_A(X, Y)$ is determined by the following commuting diagram
for any pair $\phi, \psi \in \operatorname{Hom}_A(X, Y)$:

$$
\begin{array}{ccc}
X \times X & \xrightarrow{\phi \times \psi} & Y \times Y \\
{\scriptstyle \Delta}\uparrow & & \downarrow{\scriptstyle \nabla} \\
X & \xrightarrow{\phi + \psi} & Y
\end{array}
$$

---

[4] Lemma 11.1.5
[5] [98, Proposition 8.1.6]

Direct sum. Given a finite number of objects $X_1, \ldots, X_r$ of an additive category $\mathcal{A}$, there exists a *direct sum* $X_1 \oplus \cdots \oplus X_r$, which is by definition an object X together with morphisms $\iota_i \colon X_i \to X$ and $\pi_i \colon X \to X_i$ for $1 \leq i \leq r$ such that $\sum_{i=1}^{r} \iota_i \pi_i = \mathrm{id}_X$, $\pi_i \iota_i = \mathrm{id}_{X_i}$, and $\pi_j \iota_i = 0$ for all $i \neq j$. Note that the morphisms ι_i and π_i induce isomorphisms

$$\coprod_{i=1}^{r} X_i \cong \bigoplus_{i=1}^{r} X_i \cong \prod_{i=1}^{r} X_i.$$

An object X' is a *direct summand* of X if $X = X' \oplus X''$ for some object X''.

For a class of objects $\mathcal{X} \subseteq \mathcal{A}$, let add $\mathcal{X}$ denote the smallest full subcategory of $\mathcal{A}$ containing $\mathcal{X}$ and closed under finite direct sums and direct summands.

Decomposition. Let $\mathcal{A}$ be an additive category and $(\mathcal{A}_i)_{i \in I}$ a family of full additive subcategories. We have an *orthogonal decomposition*

$$\mathcal{A} = \coprod_{i \in I} \mathcal{A}_i$$

of $\mathcal{A}$ if $\mathcal{A} = \sum_i \mathcal{A}_i$ (so each object in $\mathcal{A}$ can be written as a coproduct $\coprod_i X_i$ with $X_i \in \mathcal{A}_i$ for all i), and $\mathrm{Hom}_{\mathcal{A}}(X_i, X_j) = 0$ for all $X_i \in \mathcal{A}_i$, $X_j \in \mathcal{A}_j$, and $i \neq j$. The additive category $\mathcal{A}$ is *connected* if it admits no proper decomposition $\mathcal{A} = \mathcal{A}_1 \amalg \mathcal{A}_2$.

A *direct decomposition*

$$\mathcal{A} = \bigvee_{i \in I} \mathcal{A}_i$$

means that $\mathcal{A} = \sum_i \mathcal{A}_i$ and $\mathcal{A}_j \cap \sum_{i \neq j} \mathcal{A}_i = 0$ for all j.

Ideal. Let $\mathcal{A}$ be an additive category. An *ideal* $\mathfrak{I}$ in $\mathcal{A}$ is given by subgroups

$$\mathfrak{I}(X, Y) \subseteq \mathrm{Hom}_{\mathcal{A}}(X, Y) \qquad (X, Y \in \mathcal{A})$$

such that any composite $X \xrightarrow{\phi} Y \xrightarrow{\psi} Z$ of morphisms in $\mathcal{A}$ belongs to $\mathfrak{I}$ if ϕ or ψ belongs to $\mathfrak{I}$.

Additive functor. A functor $F \colon \mathcal{A} \to \mathcal{B}$ between additive categories is *additive* if it preserves finite products. An equivalent condition is that the induced map $\mathrm{Hom}_{\mathcal{A}}(X, Y) \to \mathrm{Hom}_{\mathcal{B}}(FX, FY)$ is additive for all objects $X, Y \in \mathcal{A}$.

Let $F \colon \mathcal{A} \to \mathcal{B}$ be an additive functor. Then the morphisms that are annihilated by F form an ideal in $\mathcal{A}$. The *kernel* of F is the full subcategory of $\mathcal{A}$ given by $\mathrm{Ker}\, F = \{X \in \mathcal{A} \mid F(X) = 0\}$.

Kernel and cokernel. Let $\phi\colon X \to Y$ be a morphism in an additive category. The *kernel* of ϕ consists of an object $\operatorname{Ker}\phi$ and a morphism $\phi'\colon \operatorname{Ker}\phi \to X$ such that $\phi\phi' = 0$ and every morphism $\alpha\colon X' \to X$ satisfying $\phi\alpha = 0$ factors uniquely through ϕ'. If the kernel exists, then $\operatorname{Ker}\phi$ and ϕ' are unique up to isomorphism and ϕ' is a monomorphism. The *cokernel* $Y \to \operatorname{Coker}\phi$ of ϕ is defined dually.

Abelian category. An additive category $\mathcal{A}$ is *abelian* if for every morphism $\phi\colon X \to Y$ there is a kernel and a cokernel, and if the canonical factorisation

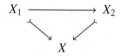

of ϕ induces an isomorphism $\bar{\phi}$.

Image. Let $\phi\colon X \to Y$ be a morphism in an abelian category. The *image* of ϕ is the kernel of the canonical morphism $\phi''\colon Y \to \operatorname{Coker}\phi$ and we write $\operatorname{Im}\phi = \operatorname{Ker}\phi''$.

Subobject. Let $\mathcal{A}$ be any category. We say that two monomorphisms $X_1 \rightarrowtail X$ and $X_2 \rightarrowtail X$ are *equivalent*, if there exists an isomorphism $X_1 \to X_2$ making the following diagram commutative.

$$X_1 \longrightarrow X_2$$
$$X$$

An equivalence class of monomorphisms into X is called a *subobject* of X. Given subobjects $X_1 \rightarrowtail X$ and $X_2 \rightarrowtail X$, we write $X_1 \subseteq X_2$ if there is a morphism $X_1 \to X_2$ making the above diagram commutative; this yields a partial order.

A morphism $\phi\colon X \to Y$ in an abelian category yields subobjects $\operatorname{Ker}\phi \subseteq X$ and $\operatorname{Im}\phi \subseteq Y$. The *quotient* of X with respect to a subobject $X' \subseteq X$ is $X/X' = \operatorname{Coker}(X' \to X)$. For a family of subobjects $(X_i)_{i\in I}$ of an object X one has

$$\sum_{i\in I} X_i = \operatorname{Im}\left(\coprod_{i\in I} X_i \to X\right) \quad \text{and} \quad \bigcap_{i\in I} X_i = \operatorname{Ker}\left(X \to \prod_{i\in I} X/X_i\right),$$

assuming that these (co)products exist.

Cocomplete category. A category $\mathcal{A}$ is *cocomplete* if every functor $\mathcal{I} \to \mathcal{A}$ from an essentially small category $\mathcal{I}$ admits a colimit. When $\mathcal{A}$ is additive, an equivalent condition is that for every family $(X_i)_{i \in I}$ of objects indexed by a set there is a coproduct $\coprod_{i \in I} X_i$ in $\mathcal{A}$, and every morphism in $\mathcal{A}$ admits a cokernel.

For an object $X \in \mathcal{A}$, we write $X^{(I)}$ or $X[I]$ for $\coprod_{i \in I} X_i$ when $X_i = X$ for all i. For a class of objects $\mathcal{X} \subseteq \mathcal{A}$, let Add $\mathcal{X}$ denote the smallest full subcategory of $\mathcal{A}$ containing $\mathcal{X}$ and closed under all coproducts and direct summands.

Locally finitely presented category. Let $\mathcal{A}$ be a cocomplete additive category. An object $X \in \mathcal{A}$ is *finitely presented* if the functor $\mathrm{Hom}(X; -)$ preserves filtered colimits. Thus for every filtered colimit $\mathrm{colim}_i \, Y_i$ in $\mathcal{A}$ the canonical map

$$\mathrm{colim}_i \, \mathrm{Hom}_{\mathcal{A}}(X, Y_i) \longrightarrow \mathrm{Hom}_{\mathcal{A}}(X, \mathrm{colim}_i \, Y_i)$$

is bijective. Let fp $\mathcal{A}$ denote the full subcategory of finitely presented objects; it is an additive category and closed under cokernels. The category $\mathcal{A}$ is called *locally finitely presented* if fp $\mathcal{A}$ is essentially small and every object in $\mathcal{A}$ is a filtered colimit of finitely presented objects.

Grothendieck category. A category $\mathcal{A}$ is a *Grothendieck category* if

(Gr1) the category $\mathcal{A}$ is abelian,

(Gr2) for every set of objects $(X_i)_{i \in I}$ there is a coproduct $\coprod_{i \in I} X_i$ in $\mathcal{A}$,

(Gr3) there is a *generator* G, that is, for every object X the canonical morphism $\coprod_{\phi \in \mathrm{Hom}_{\mathcal{A}}(G, X)} G \to X$ is an epimorphism, and

(Gr4) for every directed set of subobjects $(X_i)_{i \in I}$ of an object X and $Y \subseteq X$ one has

$$\left(\sum_{i \in I} X_i \right) \cap Y = \sum_{i \in I} (X_i \cap Y). \tag{AB5}$$

A condition equivalent to (AB5) says that every filtered colimit of exact sequences is exact.

In a Grothendieck category the subobjects $X' \subseteq X$ of any given object X form a set.

The category Ab of abelian groups is the prototype of a Grothendieck category.[6]

Simple object. An object X in an abelian category is *simple* if $X \neq 0$ and if $X' \subseteq X$ implies $X' = 0$ or $X' = X$. An object is *semisimple* if it admits a decomposition as a coproduct of simple objects.

---

[6] Corollary 2.5.3

Finite length. Let $\mathcal{A}$ be an abelian category. An object X has *finite length* if it has a finite composition series

$$0 = X_0 \subseteq X_1 \subseteq \cdots \subseteq X_n = X,$$

that is, each X_i/X_{i-1} is simple. In this case the length n of a composition series is an invariant of X (Jordan–Hölder theorem); it is called the *length* of X and is denoted by $\ell(X)$. When X is a vector space over a field we write $\mathrm{rank}(X)$ for its length. For example, X is simple if and only if $\ell(X) = 1$. Note that X has finite length if and only if X is both *artinian* (i.e. satisfies the descending chain condition on subobjects) and *noetherian* (i.e. satisfies the ascending chain condition on subobjects).

Length category. An abelian category is called a *length category* if all objects have finite length and the isomorphism classes of objects form a set.

Indecomposable object. An object X is called *indecomposable* if $X \neq 0$ and if $X = X_1 \oplus X_2$ implies $X_1 = 0$ or $X_2 = 0$.

A ring is called *local* if all non-invertible elements form a proper ideal. Thus an object is indecomposable if its endomorphism ring is local.

A finite length object admits a decomposition into a finite direct sum of indecomposable objects having local endomorphism rings. Such a decomposition is unique up to isomorphism (Krull–Remak–Schmidt theorem).

Socle. Let X be an object of an abelian category. The *socle* $\mathrm{soc}(X)$ is the sum of all simple subobjects of X. We set $\mathrm{soc}^0(X) = 0$ and $\mathrm{soc}^{n+1}(X)$ is given by $\mathrm{soc}^{n+1}(X)/\mathrm{soc}^n(X) = \mathrm{soc}(X/\mathrm{soc}^n(X))$ for all $n \geq 0$. The *height* $\mathrm{ht}(X)$ is the smallest $n \geq 0$ such that $\mathrm{soc}^n(X) = X$.

Radical. Let X be an object of an abelian category. The *radical* $\mathrm{rad}(X)$ of X is the intersection of all maximal subobjects of X. We set $\mathrm{rad}^0(X) = X$ and $\mathrm{rad}^{n+1}(X) = \mathrm{rad}(\mathrm{rad}^n X)$ for all $n \geq 0$. The *Loewy length* of X is the smallest $n \geq 0$ such that $\mathrm{rad}^n(X) = 0$. The *top* of X is the quotient $\mathrm{top}(X) = X/\mathrm{rad}(X)$.

For a ring Λ let $J(\Lambda) = \mathrm{rad}\,\Lambda$ denote the *Jacobson radical*, that is, the intersection of all maximal right ideals. Note that $J(\Lambda) = J(\Lambda^{\mathrm{op}})$. For a Λ-module X we have $XJ(\Lambda) \subseteq \mathrm{rad}\,X$ and equality when $\Lambda/J(\Lambda)$ is semisimple.

Let $\mathcal{A}$ be an additive category $\mathcal{A}$. Then $\mathrm{Rad}_{\mathcal{A}}$ denotes the ideal satisfying

$$\mathrm{Rad}_{\mathcal{A}}(X, X) = J(\mathrm{End}_{\mathcal{A}}(X))$$

for every object $X \in \mathcal{A}$. The morphisms belonging to $\mathrm{Rad}_{\mathcal{A}}$ are called *radical*.

Exact sequence. Given an abelian category $\mathcal{A}$, a finite or infinite sequence of morphisms

$$\cdots \xrightarrow{\phi_{n-2}} X_{n-1} \xrightarrow{\phi_{n-1}} X_n \xrightarrow{\phi_n} X_{n+1} \xrightarrow{\phi_{n+1}} \cdots$$

in $\mathcal{A}$ is *exact* if $\operatorname{Im} \phi_i = \operatorname{Ker} \phi_{i+1}$ for all i. An exact sequence of the form $0 \to X \to Y \to Z \to 0$ is called *short exact*.

An additive functor $\mathcal{A} \to \mathcal{B}$ between abelian categories is *exact* if it sends each exact sequence in $\mathcal{A}$ to an exact sequence in $\mathcal{B}$.

Extension group. Let $\mathcal{A}$ be an abelian category, or more generally an exact category. For a pair of objects X, Y and $n \geq 1$, let $\operatorname{Ext}^n_{\mathcal{A}}(X, Y)$ denote the group of *n-extensions* in the sense of Yoneda, i.e. equivalence classes of exact sequences

$$0 \longrightarrow Y \longrightarrow E_n \longrightarrow \cdots \longrightarrow E_2 \longrightarrow E_1 \longrightarrow X \longrightarrow 0,$$

assuming they form a set. Set $\operatorname{Ext}^0_{\mathcal{A}}(X, Y) = \operatorname{Hom}_{\mathcal{A}}(X, Y)$ and $\operatorname{Ext}^n_{\mathcal{A}}(X, Y) = 0$ for $n < 0$. Splicing together exact sequences yields the composition maps

$$\operatorname{Ext}^m_{\mathcal{A}}(Y, Z) \times \operatorname{Ext}^n_{\mathcal{A}}(X, Y) \longrightarrow \operatorname{Ext}^{m+n}_{\mathcal{A}}(X, Z)$$

for all $m, n \in \mathbb{Z}$.

For each exact sequence $\xi \colon 0 \to X' \to X \to X'' \to 0$ in $\mathcal{A}$ and $n \geq 0$, composition with ξ yields a *connecting morphism*

$$\operatorname{Ext}^n_{\mathcal{A}}(X', Y) \longrightarrow \operatorname{Ext}^{n+1}_{\mathcal{A}}(X'', Y)$$

and these fit into a *long exact sequence*:[7]

$$0 \longrightarrow \operatorname{Hom}_{\mathcal{A}}(X'', -) \longrightarrow \operatorname{Hom}_{\mathcal{A}}(X, -) \longrightarrow \operatorname{Hom}_{\mathcal{A}}(X', -) \rightharpoondown$$
$$\hookrightarrow \operatorname{Ext}^1_{\mathcal{A}}(X'', -) \longrightarrow \operatorname{Ext}^1_{\mathcal{A}}(X, -) \longrightarrow \operatorname{Ext}^1_{\mathcal{A}}(X', -) \longrightarrow \cdots$$

Idempotent complete category. Let $\mathcal{A}$ be an additive category. Then $\mathcal{A}$ is *idempotent complete* if every idempotent endomorphism in $\mathcal{A}$ admits a kernel. Note that $X = \operatorname{Ker} e \oplus \operatorname{Ker}(\operatorname{id}_X - e)$ for $e^2 = e \in \operatorname{End}_{\mathcal{A}}(X)$.

Let $\mathcal{A}$ be idempotent complete and let $\mathcal{C} \subseteq \mathcal{A}$ be a full additive subcategory. Then $\mathcal{C}$ is idempotent complete if and only if $\mathcal{C}$ is closed under direct summands. For any object X in $\mathcal{A}$, the functor $\operatorname{Hom}_{\mathcal{A}}(X, -)$ induces an equivalence

$$\operatorname{add} X \xrightarrow{\sim} \operatorname{proj} \operatorname{End}_{\mathcal{A}}(X).$$

---

[7] Corollary 4.2.12

Krull–Schmidt category. Let $\mathcal{A}$ be an additive category. Then $\mathcal{A}$ is *Krull–Schmidt* if every object decomposes into a finite direct sum of objects having local endomorphism rings. Such a decomposition is essentially unique. An equivalent condition is that $\mathcal{A}$ is idempotent complete and every object has a semiperfect endomorphism ring.[8]

A ring Λ with Jacobson radical $J(\Lambda)$ is *semiperfect* if $\Lambda/J(\Lambda)$ is semisimple and idempotents can be lifted modulo $J(\Lambda)$.

Extension closed subcategory. Let $\mathcal{A}$ be an abelian category, or more generally an exact category. A full additive subcategory $\mathcal{C} \subseteq \mathcal{A}$ is *extension closed* if for every exact sequence $0 \to X' \to X \to X'' \to 0$ in $\mathcal{A}$ the object X is in $\mathcal{C}$ when X' and X'' are in $\mathcal{C}$.

Given a class of objects $\mathcal{X} \subseteq \mathcal{A}$, we write $\mathrm{Filt}(\mathcal{X})$ for the smallest extension closed subcategory of $\mathcal{A}$ that contains $\mathcal{X}$. Note that an object $X \in \mathcal{A}$ belongs to $\mathrm{Filt}(\mathcal{X})$ if and only if there exists a finite chain

$$0 = X_0 \subseteq X_1 \subseteq \cdots \subseteq X_n = X$$

such that X_i/X_{i-1} is in $\mathcal{X}$ for $1 \leq i \leq n$.

Exact category. Let $\mathcal{A}$ be an additive category. A sequence

$$0 \longrightarrow X \xrightarrow{\alpha} Y \xrightarrow{\beta} Z \longrightarrow 0$$

of morphisms in $\mathcal{A}$ is *exact* if α is a kernel of β and β is a cokernel of α. An *exact category* is a pair $(\mathcal{A}, \mathcal{E})$ consisting of an additive category $\mathcal{A}$ and a class $\mathcal{E}$ of exact sequences in $\mathcal{A}$ (called *admissible* and given by an *admissible monomorphism* followed by an *admissible epimorphism*) which is closed under isomorphisms and satisfies the following axioms.

(Ex1) The identity morphism of each object is an admissible monomorphism and an admissible epimorphism.

(Ex2) The composite of two admissible monomorphisms is an admissible monomorphism, and the composite of two admissible epimorphisms is an admissible epimorphism.

(Ex3) Each pair of morphisms $X' \xleftarrow{\phi} X \xrightarrow{\alpha} Y$ with α an admissible monomorphism can be completed to a pushout diagram

$$
\begin{array}{ccc}
X & \xrightarrow{\alpha} & Y \\
{\scriptstyle \phi}\downarrow & & \downarrow \\
X' & \xrightarrow{\alpha'} & Y'
\end{array}
$$

---

[8] [131, Corollary 4.4]

such that α' is an admissible monomorphism. And each pair of morphisms $Y \xrightarrow{\beta} Z \xleftarrow{\psi} Z'$ with β an admissible epimorphism can be completed to a pullback diagram

$$
\begin{array}{ccc}
Y' & \xrightarrow{\beta'} & Z' \\
\downarrow & & \downarrow{\psi} \\
Y & \xrightarrow{\beta} & Z
\end{array}
$$

such that β' is an admissible epimorphism.

For example, an abelian category endowed with all short exact sequences is an exact category. Any extension closed subcategory $\mathcal{A}$ of an abelian category $\mathcal{B}$ is exact by taking for $\mathcal{E}$ all short exact sequences from $\mathcal{B}$. Conversely, any essentially small exact category arises as an extension closed subcategory of an abelian category.[9]

Thick subcategory. Let $\mathcal{A}$ be an exact category. A full additive subcategory $\mathcal{C} \subseteq \mathcal{A}$ is *thick* if it is closed under direct summands and satisfies the following *two out of three property*: an exact sequence $0 \to X \to Y \to Z \to 0$ lies in $\mathcal{C}$ if two of X, Y, Z are in $\mathcal{C}$.

Given a class of objects $\mathcal{X} \subseteq \mathcal{A}$, we write $\mathrm{Thick}(\mathcal{X})$ for the smallest thick subcategory of $\mathcal{A}$ that contains $\mathcal{X}$.

Serre subcategory. Let $\mathcal{A}$ be an abelian category. A full additive subcategory $\mathcal{C} \subseteq \mathcal{A}$ is a *Serre subcategory* provided that $\mathcal{C}$ is closed under taking subobjects, quotients and extensions. This means that for every exact sequence $0 \to X' \to X \to X'' \to 0$ in $\mathcal{A}$, the object X is in $\mathcal{C}$ if and only if X' and X'' are in $\mathcal{C}$.

If $\mathcal{A}$ is a length category, then a Serre subcategory $\mathcal{C} \subseteq \mathcal{A}$ is determined by the simple objects of $\mathcal{A}$ that are contained in $\mathcal{C}$.

Torsion pair. Let $\mathcal{A}$ be an exact category. A pair $(\mathcal{T}, \mathcal{F})$ of full additive subcategories is a *torsion pair* for $\mathcal{A}$, if

$$
\mathcal{T} = \{X \in \mathcal{A} \mid \mathrm{Hom}(X, Y) = 0 \text{ for all } Y \in \mathcal{F}\},
$$
$$
\mathcal{F} = \{Y \in \mathcal{A} \mid \mathrm{Hom}(X, Y) = 0 \text{ for all } X \in \mathcal{T}\},
$$

and each object $X \in \mathcal{A}$ fits into an exact sequence $\xi_X \colon 0 \to X' \to X \to X'' \to 0$ with $X' \in \mathcal{T}$ and $X'' \in \mathcal{F}$. In that case $X \mapsto X'$ provides a right adjoint for the inclusion $\mathcal{T} \to \mathcal{A}$, and $X \mapsto X''$ provides a left adjoint for the inclusion $\mathcal{F} \to \mathcal{A}$.

---

[9] Proposition 2.3.7

A torsion pair $(\mathcal{T}, \mathcal{F})$ is *split* if the sequence ξ_X is split exact for all $X \in \mathcal{A}$. An equivalent condition is that $\mathcal{A} = \mathcal{T} \vee \mathcal{F}$.

Projective object. Let $\mathcal{A}$ be an exact category. An object P in $\mathcal{A}$ is *projective* if every admissible epimorphism $X \to Y$ in $\mathcal{A}$ induces a surjective map $\mathrm{Hom}_{\mathcal{A}}(P, X) \to \mathrm{Hom}_{\mathcal{A}}(P, Y)$. The full subcategory of projective objects in $\mathcal{A}$ is denoted by $\mathrm{Proj}\,\mathcal{A}$.

The category $\mathcal{A}$ has *enough projective objects* if for every object $X \in \mathcal{A}$ there is an admissible epimorphism $P \to X$ such that P is projective. A projective object P is a *projective generator* if $\mathrm{Hom}_{\mathcal{A}}(P, X) = 0$ implies $X = 0$ for every object $X \in \mathcal{A}$.

Suppose $\mathrm{Proj}\,\mathcal{A} = \mathrm{add}\,P$ for some object P and set $\Lambda = \mathrm{End}_{\mathcal{A}}(P)$. If $\mathcal{A}$ has enough projectives, then the functor

$$\mathrm{Hom}_{\mathcal{A}}(P, -) \colon \mathcal{A} \longrightarrow \mathrm{mod}\,\Lambda$$

is fully faithful and exact. This functor is an equivalence if $\mathcal{A}$ is abelian.[10]

Injective object. Let $\mathcal{A}$ be an exact category. An object I in $\mathcal{A}$ is *injective* if every admissible monomorphism $X \to Y$ induces a surjective map $\mathrm{Hom}_{\mathcal{A}}(Y, I) \to \mathrm{Hom}_{\mathcal{A}}(X, I)$. The full subcategory of injective objects in $\mathcal{A}$ is denoted by $\mathrm{Inj}\,\mathcal{A}$.

The category $\mathcal{A}$ has *enough injective objects* if for every object $X \in \mathcal{A}$ there is an admissible monomorphism $X \to I$ such that I is injective. An injective object I is an *injective cogenerator* if $\mathrm{Hom}_{\mathcal{A}}(X, I) = 0$ implies $X = 0$ for every object $X \in \mathcal{A}$.

Injective envelope. Let $\mathcal{A}$ be an abelian category. A monomorphism $\phi \colon X \to Y$ is *essential* if any morphism $\alpha \colon Y \to Y'$ is a monomorphism provided that the composite $\alpha\phi$ is a monomorphism. This condition can be rephrased as follows: if $U \subseteq Y$ is a subobject with $U \cap \mathrm{Im}\,\phi = 0$, then $U = 0$. A monomorphism $\phi \colon X \to I$ is an *injective envelope* of X if I is injective and ϕ is essential. The injective object given by an injective envelope of X is denoted by $E(X)$.

Every object in a Grothendieck category admits an injective envelope.[11]

Homological dimension. Let $\mathcal{A}$ be an exact category. The *projective dimension* of an object X is by definition

$$\mathrm{proj.dim}\,X = \inf\{n \geq 0 \mid \mathrm{Ext}_{\mathcal{A}}^{n+1}(X, -) = 0\}.$$

---

[10] Lemma 2.1.14
[11] Corollary 2.5.4

Set proj.dim $X = \infty$ if such a number n does not exist. An application of the long exact sequence for $\text{Ext}^*(-, -)$ shows that $\text{Ext}_{\mathcal{A}}^{n+1}(X, -) = 0$ if and only if $\text{Ext}_{\mathcal{A}}^{n}(X, -)$ is right exact.

The category $\mathcal{A}$ has enough projective objects if every object X fits into an exact sequence

$$0 \longrightarrow \Omega \longrightarrow P \longrightarrow X \longrightarrow 0$$

such that P is projective. Then Ω is called a *syzygy* of X. We set $\Omega^0 X = X$ and $\Omega^{n+1}X = \Omega(\Omega^n X)$ for $n \geq 0$. Also, proj.dim $X \leq n$ if and only if there exists an exact sequence

$$0 \longrightarrow P_n \longrightarrow \cdots \longrightarrow P_1 \longrightarrow P_0 \longrightarrow X \longrightarrow 0$$

such that all P_i are projective objects.

The *injective dimension* of X is by definition

$$\text{inj.dim } X = \inf\{n \geq 0 \mid \text{Ext}_{\mathcal{A}}^{n+1}(-, X) = 0\}.$$

The *global dimension* gl.dim $\mathcal{A}$ of $\mathcal{A}$ is defined as the smallest integer $n \geq 0$ such that $\text{Ext}_{\mathcal{A}}^{n+1}(-, -) = 0$. Note that the global dimension of $\mathcal{A}$ is equal to $\sup\{\text{proj.dim } X \mid X \in \mathcal{A}\}$ and $\sup\{\text{inj.dim } X \mid X \in \mathcal{A}\}$.

Diagram lemmas. For any abelian category there are several statements about commutative diagrams with exactness properties which are known as diagram lemmas: the *five lemma*[12], the *snake lemma*[13], and the *horseshoe lemma*[14].

Split exact category. An exact category $\mathcal{A}$ is called *split exact* provided that $\text{Ext}_{\mathcal{A}}^1(-, -) = 0$. This means that every admissible exact sequence splits.

A ring Λ is *semisimple* if its module category Mod Λ is split exact. An equivalent condition is that every module is semisimple. Another equivalent condition is that Λ is isomorphic to a finite product of matrix rings $\prod_i M_{n_i}(K_i)$ where the n_i are natural numbers and the K_i are division rings (Wedderburn–Artin theorem).

Hereditary category. An exact category $\mathcal{A}$ is called *hereditary* provided that $\text{Ext}_{\mathcal{A}}^2(-, -) = 0$.

A ring Λ is *right hereditary* if its module category Mod Λ is hereditary.

---

[12] [141, Lemma I.3.3]
[13] [141, Lemma II.5.2]
[14] [46, Proposition I.2.5]

Frobenius category. An exact category $\mathcal{A}$ is called *Frobenius* provided that there are enough projective and enough injective objects, and if projective and injective objects coincide.

A ring Λ is *quasi-Frobenius* if its module category $\operatorname{Mod} \Lambda$ is Frobenius.

Grothendieck group. Let $\mathcal{A}$ be an essentially small abelian category, or more generally an exact category. Denote by $F(\mathcal{A})$ the free abelian group generated by the isomorphism classes of objects in $\mathcal{A}$. Let $F_0(\mathcal{A})$ be the subgroup generated by $[X] - [Y] + [Z]$ for all exact sequences $0 \to X \to Y \to Z \to 0$ in $\mathcal{A}$. The *Grothendieck group* $K_0(\mathcal{A})$ of $\mathcal{A}$ is by definition the factor group $F(\mathcal{A})/F_0(\mathcal{A})$.

If $\mathcal{A}$ is a length category, then $K_0(\mathcal{A})$ is a free abelian group and the isomorphism classes of simple objects in $\mathcal{A}$ form a basis (Jordan–Hölder theorem).

For a ring Λ, the *Grothendieck group* $K_0(\Lambda) = K_0(\operatorname{proj} \Lambda)$ is the quotient of the free abelian group generated by the isomorphism classes of finitely generated projective Λ-modules, modulo the relations given by split short exact sequences.

For Λ semiperfect, $K_0(\Lambda)$ is a free abelian group and the isomorphism classes of indecomposable projective Λ-modules form a basis (Krull–Remak–Schmidt theorem).

For Λ right coherent, $\operatorname{mod} \Lambda$ is abelian and the embedding $\operatorname{proj} \Lambda \to \operatorname{mod} \Lambda$ induces a homomorphism $K_0(\operatorname{proj} \Lambda) \to K_0(\operatorname{mod} \Lambda)$. This is an isomorphism when the global dimension is finite, since every Λ-module has a finite projective resolution.

Centre. Let $\mathcal{A}$ be an additive category. The *centre* $Z(\mathcal{A})$ is given as the ring of all natural transformations $\operatorname{id}_{\mathcal{A}} \to \operatorname{id}_{\mathcal{A}}$. This ring is commutative; it acts on $\operatorname{Hom}_{\mathcal{A}}(X, Y)$ for all objects X, Y and the composition maps are bilinear. Given a commutative ring k, the category $\mathcal{A}$ is *k-linear* if k acts on all morphisms in $\mathcal{A}$ via a ring homomorphism $k \to Z(\mathcal{A})$.

Let Λ be a ring. Then the centre of its module category identifies with the centre $Z(\Lambda) = \{x \in \Lambda \mid xy = yx \text{ for all } y \in \Lambda\}$. The structure of a *k-algebra* is given by a ring homomorphism $k \to Z(\Lambda)$.

Let Λ be a k-algebra over a commutative ring k. Then Λ is a *noetherian algebra* if Λ is noetherian as a k-module, and Λ is an *Artin algebra* if Λ is of finite length as a k-module. Clearly, a noetherian algebra is a noetherian ring, and an Artin algebra is an artinian ring.

Matlis duality. Let k be a commutative ring. An injective envelope $E = E(\coprod_S S)$ of the coproduct of a representative set of simple k-modules provides a minimal injective cogenerator. The *Matlis duality* for k-modules is given by the assignment $X \mapsto DX := \operatorname{Hom}_k(X, E)$. There is a natural isomorphism

$$\operatorname{Hom}_k(X, DY) \cong \operatorname{Hom}_k(X \otimes_k Y, E) \cong \operatorname{Hom}_k(Y, DX)$$

for all k-modules X, Y. Matlis duality is faithful and exact; it maps finite length modules to finite length modules. The natural map $X \to D^2 X$ is an isomorphism when X has finite length.

For the ring $\mathbb{Z}$ of integers, Matlis duality is given by $D = \operatorname{Hom}_{\mathbb{Z}}(-, \mathbb{Q}/\mathbb{Z})$.

Lattice. Let $L = (L, \leq)$ be a partially ordered set. Then L is a *lattice* provided that

(L1) for each pair $x, y \in L$ the supremum $x \vee y := \sup(x, y)$ and the infimum $x \wedge y := \inf(x, y)$ exist in L, and

(L2) the supremum $1 := \sup L$ and the infimum $0 := \inf L$ exist in L.

A lattice L is *modular* if $a \leq b$ implies for each $x \in L$

$$a \vee (x \wedge b) = (a \vee x) \wedge b.$$

The subobjects of an object X in an abelian category form a modular lattice which is denoted by $\mathbf{L}(X)$. For any pair of subobjects U, V the *Noether isomorphisms*

$$U/(U \cap V) \xrightarrow{\sim} (U + V)/V \qquad \text{and} \qquad V/(U \cap V) \xrightarrow{\sim} (U + V)/U$$

yield a commutative diagram with exact rows and columns.

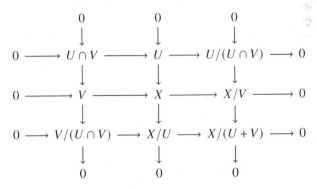

Quiver. A *quiver* is a quadruple $\Gamma = (\Gamma_0, \Gamma_1, s, t)$ consisting of a set Γ_0 of *vertices*, a set Γ_1 of *arrows*, and two maps $s, t \colon \Gamma_1 \to \Gamma_0$. An arrow $\alpha \in \Gamma_1$ *starts* at $s(\alpha)$ and *terminates* at $t(\alpha)$. A *path* $\alpha = \alpha_1 \cdots \alpha_n$ of *length n* in Q is a sequence of arrows $\circ \xrightarrow{\alpha_n} \cdots \xrightarrow{\alpha_1} \circ$ satisfying $s(\alpha_i) = t(\alpha_{i+1})$ for all i. We set $s(\alpha) = s(\alpha_n)$ and $t(\alpha) = t(\alpha_1)$.

A *representation* of Γ in a category $\mathcal{C}$ is a collection

$$(X_i, X_\alpha)_{i \in \Gamma_0, \alpha \in \Gamma_1}$$

of objects X_i and morphisms $X_\alpha \colon X_{s(\alpha)} \to X_{t(\alpha)}$ in $\mathcal{C}$. Thus a quiver is a representation of the *Kronecker quiver* $\cdot \rightrightarrows \cdot$ in the category of sets.

Standard Functors and Isomorphisms

Tensor functors. Fix a pair of rings Λ, Γ. A bimodule ${}_\Lambda M_\Gamma$ yields an adjoint pair of functors

$$- \otimes_\Lambda M : \operatorname{Mod}\Lambda \longrightarrow \operatorname{Mod}\Gamma \quad \text{and} \quad \operatorname{Hom}_\Gamma(M, -) : \operatorname{Mod}\Gamma \longrightarrow \operatorname{Mod}\Lambda.$$

An additive functor $F : \operatorname{Mod}\Lambda \to \operatorname{Mod}\Gamma$ is of the form $F = - \otimes_\Lambda M$ for some bimodule ${}_\Lambda M_\Gamma$ if and only if F preserves all coproducts and cokernels. In that case $M = F(\Lambda)$ with Λ acting via $\Lambda = \operatorname{End}_\Lambda(\Lambda) \to \operatorname{End}_\Gamma(F(\Lambda))$.

Tensor-hom adjunction. Fix a pair of rings Λ, Γ and modules $(X_\Lambda, Y_\Gamma, {}_\Lambda M_\Gamma)$. Then there is a natural isomorphism

$$\operatorname{Hom}_\Lambda(X, \operatorname{Hom}_\Gamma(M, Y)) \xrightarrow{\;\sim\;} \operatorname{Hom}_\Gamma(X \otimes_\Lambda M, Y)$$

given by

$$\phi \longmapsto (x \otimes m \mapsto \phi(x)(m)).$$

Modules over algebras. Fix an algebra Λ over a commutative ring k and modules $(X_\Lambda, {}_\Lambda Y, Z_k)$. Then there are natural isomorphisms

$$\operatorname{Hom}_\Lambda(X, \operatorname{Hom}_k(Y, Z)) \cong \operatorname{Hom}_k(X \otimes_\Lambda Y, Z) \cong \operatorname{Hom}_\Lambda(Y, \operatorname{Hom}_k(X, Z)).$$

Finitely generated projective modules. Fix a pair of rings Λ, Γ and modules $(X_\Lambda, Y_\Gamma, {}_\Lambda M_\Gamma)$. Then there is a natural homomorphism

$$X \otimes_\Lambda \operatorname{Hom}_\Gamma(Y, M) \longrightarrow \operatorname{Hom}_\Gamma(Y, X \otimes_\Lambda M)$$

given by

$$x \otimes \phi \longmapsto (y \mapsto x \otimes \phi(y)),$$

which is invertible if X or Y is finitely generated projective.

Duality. Fix a pair of rings Λ, Γ and modules $(X_\Lambda, {}_\Gamma Y, {}_\Gamma M_\Lambda)$. Then there is a natural homomorphism

$$X \otimes_\Lambda \mathrm{Hom}_\Gamma(M, Y) \longrightarrow \mathrm{Hom}_\Gamma(\mathrm{Hom}_\Lambda(X, M), Y)$$

given by

$$x \otimes \phi \longmapsto (\psi \mapsto \phi(\psi(x))),$$

which is invertible if X is finitely generated projective or if X is finitely presented and Y is injective.

Change of rings. Let $\phi: \Lambda \to \Gamma$ be a ring homomorphism, which yields canonical bimodules ${}_\Lambda \Gamma_\Gamma$ and ${}_\Gamma \Gamma_\Lambda$. Then the functor $\mathrm{Mod}\,\Gamma \to \mathrm{Mod}\,\Lambda$ given by *restriction of scalars*

$$\phi^* := \mathrm{Hom}_\Gamma(\Gamma, -) \cong - \otimes_\Gamma \Gamma =: \phi^!$$

admits a left adjoint $\phi_!$ (*extension of scalars*) and a right adjoint ϕ_*

$$\mathrm{Mod}\,\Gamma \; \begin{array}{c} \xleftarrow{\;\;\phi_!\;\;} \\ \xrightarrow{\;\;\phi^*=\phi^!\;\;} \\ \xleftarrow{\;\;\phi_*\;\;} \end{array} \; \mathrm{Mod}\,\Lambda$$

which are given by

$$\phi_! := - \otimes_\Lambda \Gamma \qquad \text{and} \qquad \phi_* := \mathrm{Hom}_\Lambda(\Gamma, -).$$

Change of categories. Let $f: \mathcal{C} \to \mathcal{D}$ be an additive functor between additive categories. Then the functor $f^*: \mathrm{Mod}\,\mathcal{D} \to \mathrm{Mod}\,\mathcal{C}$ given by $Y \mapsto Y \circ f$ admits a left adjoint $f_!$ and a right adjoint f_*

$$\mathrm{Mod}\,\mathcal{D} \; \begin{array}{c} \xleftarrow{\;\;f_!\;\;} \\ \xrightarrow{\;\;f^*=f^!\;\;} \\ \xleftarrow{\;\;f_*\;\;} \end{array} \; \mathrm{Mod}\,\mathcal{C}$$

which for $X \in \mathrm{Mod}\,\mathcal{C}$ with presentation

$$\coprod_j \mathrm{Hom}_\mathcal{C}(-, C_j) \longrightarrow \coprod_i \mathrm{Hom}_\mathcal{C}(-, C_i) \longrightarrow X \longrightarrow 0$$

are given by the presentation

$$\coprod_j \mathrm{Hom}_\mathcal{D}(-, f(C_j)) \longrightarrow \coprod_i \mathrm{Hom}_\mathcal{D}(-, f(C_i)) \longrightarrow f_!(X) \longrightarrow 0$$

and

$$f_*(X)(D) = \mathrm{Hom}(\mathrm{Hom}_\mathcal{D}(f-, D), X) \qquad (D \in \mathcal{D}).$$

PART ONE

ABELIAN AND DERIVED CATEGORIES

1

Localisation

Contents

	1.1	**Localisation**	**3**
		Localisation of Categories	3
		Local Objects	4
		Adjoint Functors	5
		Localisation Functors	7
		Localisation of Adjoints	9
		Localisation and Coproducts	9
	1.2	**Calculus of Fractions**	**10**
		Calculus of Fractions	10
		Calculus of Fractions for Subcategories	12
	Notes		**12**

A localisation of a category is obtained by formally inverting a specific class of morphisms. Forming localisations is one of the standard techniques in algebra; it is used throughout this book. The calculus of fractions helps to describe the morphisms of a localised category.

1.1 Localisation

We introduce the concept of localisation for categories. A localisation is obtained by formally inverting a specific class of morphisms.

Localisation of Categories

Let $\mathcal{C}$ be a category and let $S \subseteq \operatorname{Mor} \mathcal{C}$ be a class of morphisms in $\mathcal{C}$. The *localisation* of $\mathcal{C}$ with respect to S is a category $\mathcal{C}[S^{-1}]$ together with a functor $Q \colon \mathcal{C} \to \mathcal{C}[S^{-1}]$ satisfying the following.

(L1) For every $\sigma \in S$, the morphism $Q\sigma$ is invertible.

(L2) For every functor $F \colon \mathcal{C} \to \mathcal{D}$ such that $F\sigma$ is invertible for all $\sigma \in S$, there exists a unique functor $\bar{F} \colon \mathcal{C}[S^{-1}] \to \mathcal{D}$ such that $F = \bar{F} \circ Q$.

The localisation solves a universal problem and is therefore unique, up to a unique isomorphism. We sketch the construction of $Q \colon \mathcal{C} \to \mathcal{C}[S^{-1}]$. At this stage, we ignore set-theoretic issues, that is, the morphisms between two objects of $\mathcal{C}[S^{-1}]$ need not form a set. However, later on we pay attention and formulate criteria such that $\mathcal{C}[S^{-1}]$ is locally small. We put $\mathrm{Ob}\,\mathcal{C}[S^{-1}] = \mathrm{Ob}\,\mathcal{C}$. To define the morphisms of $\mathcal{C}[S^{-1}]$, consider the quiver with class of vertices $\mathrm{Ob}\,\mathcal{C}$ and class of arrows the disjoint union $(\mathrm{Mor}\,\mathcal{C}) \sqcup S^-$, where

$$S^- = \{\sigma^- \colon Y \to X \mid S \ni \sigma \colon X \to Y\}.$$

Let $\mathcal{P}$ be the category of paths in this quiver, that is, finite sequences of composable arrows, together with the obvious composition given by concatenation and denoted by $\circ_{\mathcal{P}}$. We define $\mathrm{Mor}\,\mathcal{C}[S^{-1}]$ as the quotient of $\mathcal{P}$ modulo the following relations:

(1) $\beta \circ_{\mathcal{P}} \alpha = \beta \circ \alpha$ for all composable morphisms $\alpha, \beta \in \mathrm{Mor}\,\mathcal{C}$,

(2) $\mathrm{id}_{\mathcal{P}}\,X = \mathrm{id}_{\mathcal{C}}\,X$ for all $X \in \mathrm{Ob}\,\mathcal{C}$,

(3) $\sigma^- \circ_{\mathcal{P}} \sigma = \mathrm{id}_{\mathcal{P}}\,X$ and $\sigma \circ_{\mathcal{P}} \sigma^- = \mathrm{id}_{\mathcal{P}}\,Y$ for all $\sigma \colon X \to Y$ in S.

The composition of morphisms in $\mathcal{P}$ induces the composition in $\mathcal{C}[S^{-1}]$. The functor Q is the identity on objects and on $\mathrm{Mor}\,\mathcal{C}$ the composite

$$\mathrm{Mor}\,\mathcal{C} \xrightarrow{\mathrm{inc}} (\mathrm{Mor}\,\mathcal{C}) \sqcup S^- \xrightarrow{\mathrm{inc}} \mathcal{P} \xrightarrow{\mathrm{can}} \mathrm{Mor}\,\mathcal{C}[S^{-1}].$$

The following is a more precise formulation of the properties of the canonical functor $Q \colon \mathcal{C} \to \mathcal{C}[S^{-1}]$.

Lemma 1.1.1. *For any category $\mathcal{D}$, the functor*

$$\mathcal{H}om(\mathcal{C}[S^{-1}], \mathcal{D}) \longrightarrow \mathcal{H}om(\mathcal{C}, \mathcal{D}), \quad F \mapsto F \circ Q,$$

is fully faithful and identifies $\mathcal{H}om(\mathcal{C}[S^{-1}], \mathcal{D})$ with the full subcategory of functors in $\mathcal{H}om(\mathcal{C}, \mathcal{D})$ that make all morphisms in S invertible. $\square$

Local Objects

Let $\mathcal{C}$ be a category and $S \subseteq \mathrm{Mor}\,\mathcal{C}$. An object Y in $\mathcal{C}$ is called *S-local* (or *S-closed*, or *S-orthogonal*) if the map $\mathrm{Hom}_{\mathcal{C}}(\sigma, Y)$ is bijective for all $\sigma \in S$. We denote by $S^{\perp}$ the full subcategory of S-local objects in $\mathcal{C}$.

Lemma 1.1.2. *An object Y in $\mathcal{C}$ is S-local if and only if the canonical map*

$$p_{X,Y} \colon \operatorname{Hom}_{\mathcal{C}}(X,Y) \longrightarrow \operatorname{Hom}_{\mathcal{C}[S^{-1}]}(X,Y)$$

is bijective for all $X \in \mathcal{C}$.

Proof If Y is S-local, then $\operatorname{Hom}_{\mathcal{C}}(-,Y) \colon \mathcal{C}^{\mathrm{op}} \to$ Set induces a functor $\operatorname{Hom}_{\mathcal{C}}(-,Y) \colon \mathcal{C}[S^{-1}]^{\mathrm{op}} \to$ Set. Yoneda's lemma yields a morphism

$$\operatorname{Hom}_{\mathcal{C}[S^{-1}]}(-,Y) \longrightarrow \operatorname{Hom}_{\mathcal{C}}(-,Y)$$

corresponding to id_Y, and it is straightforward to check that this is an inverse for the canonical morphism $\operatorname{Hom}_{\mathcal{C}}(-,Y) \to \operatorname{Hom}_{\mathcal{C}[S^{-1}]}(-,Y)$.

Now assume that $p_{X,Y}$ is bijective for all $X \in \mathcal{C}$. Then $\operatorname{Hom}_{\mathcal{C}}(\sigma, Y)$ is bijective for all $\sigma \in S$ since $\operatorname{Hom}_{\mathcal{C}[S^{-1}]}(\sigma, Y)$ is bijective. $\qquad\square$

Adjoint Functors

Let $F \colon \mathcal{C} \to \mathcal{D}$ and $G \colon \mathcal{D} \to \mathcal{C}$ be a pair of functors and assume that F is left adjoint to G. We set

$$S = S(F) = \{\sigma \in \operatorname{Mor} \mathcal{C} \mid F\sigma \text{ is invertible}\}$$

and obtain the following diagram

with $\qquad F = \bar{F} \circ Q.$

Proposition 1.1.3. *The following statements are equivalent.*

(1) *The functor G is fully faithful.*
(2) *The counit $FG(X) \to X$ is invertible for every object $X \in \mathcal{D}$.*
(3) *The functor F induces an equivalence $\bar{F} \colon \mathcal{C}[S^{-1}] \xrightarrow{\sim} \mathcal{D}$.*

Moreover, in that case G induces an equivalence $\mathcal{D} \xrightarrow{\sim} S^{\perp}$ with quasi-inverse $S^{\perp} \hookrightarrow \mathcal{C} \xrightarrow{F} \mathcal{D}$.

Proof We denote by $\eta \colon \mathrm{id}_{\mathcal{C}} \to GF$ the unit and by $\varepsilon \colon FG \to \mathrm{id}_{\mathcal{D}}$ the counit of the adjunction. Note that the composite $F \xrightarrow{F\eta} FGF \xrightarrow{\varepsilon F} F$ equals id_F, and $G \xrightarrow{\eta G} GFG \xrightarrow{G\varepsilon} G$ equals id_G; this characterises the fact that (F, G) is an adjoint pair.

(1) $\Leftrightarrow$ (2): The counit $\varepsilon_X\colon FG(X) \to X$ induces for all $Y \in \mathcal{D}$ a natural map

$$\mathrm{Hom}_{\mathcal{D}}(X, Y) \to \mathrm{Hom}_{\mathcal{C}}(GX, GY) \xrightarrow{\sim} \mathrm{Hom}_{\mathcal{D}}(FG(X), Y),$$

which is a bijection if and only if ε_X is an isomorphism, by Yoneda's lemma.

(2) $\Rightarrow$ (3): We claim that QG is a quasi-inverse of $\bar{F}$. Clearly, $\bar{F}(QG) = FG \cong \mathrm{id}_{\mathcal{D}}$. On the other hand, $F\eta$ is invertible, since εF is invertible. Thus $Q\eta\colon Q \to QGF$ is invertible, and therefore

$$(QG)\bar{F}Q = QGF \cong Q \cong \mathrm{id}_{\mathcal{C}[S^{-1}]} Q.$$

Then the defining property of Q implies $(QG)\bar{F} \cong \mathrm{id}_{\mathcal{C}[S^{-1}]}$.

(3) $\Rightarrow$ (2): If $\bar{F}$ is an equivalence, then composition with F induces a fully faithful functor $\mathcal{Hom}(\mathcal{D}, \mathcal{X}) \to \mathcal{Hom}(\mathcal{C}, \mathcal{X})$ for any category $\mathcal{X}$, by Lemma 1.1.1. For $\mathcal{X} = \mathcal{D}$, this implies that there is $\eta'\colon \mathrm{id}_{\mathcal{D}} \to FG$ such that $F\eta = \eta'F$. We claim that $(\mathrm{id}_{\mathcal{D}}, FG)$ is an adjoint pair with unit η' and counit ε. Clearly, then FG is an equivalence and ε is an isomorphism.

From the fact that $F \xrightarrow{F\eta} FGF \xrightarrow{\varepsilon F} F$ equals id_F it follows that $(\varepsilon \circ \eta')F = \varepsilon F \circ \eta'F = \mathrm{id}_F$, and therefore $\varepsilon \eta' = \mathrm{id}_{\mathrm{id}_{\mathcal{D}}}$. On the other hand, the fact that $G \xrightarrow{\eta G} GFG \xrightarrow{G\varepsilon} G$ equals id_G implies by applying F that $FG\varepsilon \circ \eta'FG = FG\varepsilon \circ F\eta G = \mathrm{id}_{FG}$. Thus $(\mathrm{id}_{\mathcal{D}}, FG)$ is an adjoint pair.

Now suppose that the equivalent conditions hold. In order to show that G induces an equivalence $\mathcal{D} \xrightarrow{\sim} S^{\perp}$, we need to show that the essential image of G equals $S^{\perp}$. The inclusion $\mathrm{Im}\, G \subseteq S^{\perp}$ is clear. If $X \in S^{\perp}$, then $\mathrm{Hom}_{\mathcal{C}}(\eta_X, X)$ is bijective since $\eta_X \in S$. This gives an inverse of η_X, so $X \cong GF(X)$. $\quad\square$

Example 1.1.4. Let $\mathcal{C}$ be an additive category and consider the category $\mathrm{mod}\,\mathcal{C}$ of functors $F\colon \mathcal{C}^{\mathrm{op}} \to \mathrm{Ab}$ that fit into an exact sequence

$$\mathrm{Hom}_{\mathcal{C}}(-, X) \longrightarrow \mathrm{Hom}_{\mathcal{C}}(-, Y) \longrightarrow F \longrightarrow 0.$$

Then the *Yoneda functor*

$$\mathcal{C} \longrightarrow \mathrm{mod}\,\mathcal{C}, \quad X \mapsto h_X := \mathrm{Hom}_{\mathcal{C}}(-, X)$$

admits a left adjoint if and only if every morphism in $\mathcal{C}$ admits a cokernel. The left adjoint sends $F = \mathrm{Coker}\, h_\phi$ in $\mathrm{mod}\,\mathcal{C}$ (given by a morphism ϕ in $\mathcal{C}$) to $\mathrm{Coker}\,\phi$.

Proof Suppose that $\mathcal{C}$ has cokernels. For $C \in \mathcal{C}$ we have

$$\mathrm{Hom}(\mathrm{Coker}\, h_\phi, h_C) \cong \mathrm{Ker}\,\mathrm{Hom}(h_\phi, h_C)$$
$$\cong \mathrm{Ker}\,\mathrm{Hom}_{\mathcal{C}}(\phi, C)$$
$$\cong \mathrm{Hom}_{\mathcal{C}}(\mathrm{Coker}\,\phi, C).$$

This follows from Yoneda's lemma and yields the adjointness. The converse follows from the fact that a left adjoint preserves cokernels. □

We introduce the following terminology. A diagram of additive functors

$$\mathcal{C}' \underset{E_\rho}{\overset{E}{\rightleftarrows}} \mathcal{C} \underset{F_\rho}{\overset{F}{\rightleftarrows}} \mathcal{C}''$$

is called a *localisation sequence* if

(LS1) (E, E_ρ) and (F, F_ρ) are adjoint pairs,
(LS2) E and F_ρ are fully faithful,
(LS3) Im $E =$ Ker F (equivalently, $EE_\rho(X) \overset{\sim}{\to} X$ if and only if $F(X) = 0$).

The dual notion is called a *colocalisation sequence* and is given by a diagram of additive functors

$$\mathcal{C}' \underset{E}{\overset{E_\lambda}{\rightleftarrows}} \mathcal{C} \underset{F}{\overset{F_\lambda}{\rightleftarrows}} \mathcal{C}''$$

satisfying the dual properties.

The above Example 1.1.4 gives rise to a localisation sequence

$$\text{Ker } F \rightleftarrows \text{mod } \mathcal{C} \underset{F_\rho}{\overset{F}{\rightleftarrows}} \mathcal{C}$$

provided that $\mathcal{C}$ is abelian. In that case the functor F is exact and the right adjoint of the inclusion Ker $F \to$ mod $\mathcal{C}$ sends an object X to the kernel of the unit $X \to F_\rho F(X)$.

Localisation Functors

Suppose that the canonical functor $\mathcal{C} \to \mathcal{C}[S^{-1}]$ corresponding to a class of morphisms $S \subseteq \text{Mor } \mathcal{C}$ admits a right adjoint. Then the above Proposition 1.1.3 suggests we think of localisation as an endofunctor $\mathcal{C} \to \mathcal{C}$. The following definition makes this idea precise. Moreover, we see that both ways of thinking about localisation are equivalent.

A functor $L\colon \mathcal{C} \to \mathcal{C}$ is called a *localisation functor* if there exists a morphism $\eta\colon \text{id}_\mathcal{C} \to L$ such that $L\eta\colon L \to L^2$ is an isomorphism and $L\eta = \eta L$. Note that we only require the existence of η; the actual morphism is not part of the definition of L. However, we will see that η is determined by L, up to a unique isomorphism $L \to L$.

Proposition 1.1.5. *Let $L\colon \mathcal{C} \to \mathcal{C}$ be a functor and $\eta\colon \text{id}_\mathcal{C} \to L$ a morphism. Then the following are equivalent.*

(1) $L\eta\colon L \to L^2$ *is an isomorphism and* $L\eta = \eta L$.
(2) *There exists a functor* $F\colon \mathcal{C} \to \mathcal{D}$ *and a fully faithful right adjoint* $G\colon \mathcal{D} \to \mathcal{C}$ *such that* $L = G \circ F$ *and* $\eta\colon \mathrm{id}_{\mathcal{C}} \to G \circ F$ *is the unit of the adjunction.*

Proof (1) $\Rightarrow$ (2): Let $\mathcal{D}$ denote the essential image of L, that is, the full subcategory of $\mathcal{C}$ consisting of objects isomorphic to LX for some $X \in \mathcal{C}$. Note that $X \in \mathcal{D}$ if and only if η_X is invertible. In this case let $\theta_X\colon LX \to X$ denote the inverse of η_X. Define $F\colon \mathcal{C} \to \mathcal{D}$ by $FX = LX$ and let $G\colon \mathcal{D} \to \mathcal{C}$ be the inclusion. We claim that F and G form an adjoint pair. To this end, one checks that the maps

$$\mathrm{Hom}_{\mathcal{D}}(FX, Y) \longrightarrow \mathrm{Hom}_{\mathcal{C}}(X, GY), \quad \alpha \mapsto G\alpha \circ \eta_X,$$

and

$$\mathrm{Hom}_{\mathcal{C}}(X, GY) \longrightarrow \mathrm{Hom}_{\mathcal{D}}(FX, Y), \quad \beta \mapsto \theta_Y \circ F\beta,$$

are mutually inverse bijections. Consider a pair of morphisms $\alpha\colon FX \to Y$ and $\beta\colon X \to GY$. This yields a pair of commutative squares

$$
\begin{array}{ccc}
FX & \xrightarrow{\ \alpha\ } & Y \\
{\scriptstyle \eta_{FX}}\downarrow & & \downarrow{\scriptstyle \eta_Y} \\
GF(FX) & \xrightarrow{GF(\alpha)} & GF(Y)
\end{array}
\qquad
\begin{array}{ccc}
X & \xrightarrow{\ \beta\ } & GY \\
{\scriptstyle \eta_X}\downarrow & & \downarrow{\scriptstyle \eta_{GY}} \\
GF(X) & \xrightarrow{GF(\beta)} & GF(GY)
\end{array}
$$

giving the desired identities

$$\alpha = \theta_Y \circ \eta_Y \circ \alpha = \theta_Y \circ GF(\alpha) \circ \eta_{FX} = \theta_Y \circ FG(\alpha) \circ F\eta_X$$

and

$$\beta = \theta_{GY} \circ \eta_{GY} \circ \beta = \theta_{GY} \circ GF(\beta) \circ \eta_X = G\theta_Y \circ GF(\beta) \circ \eta_X.$$

(2) $\Rightarrow$ (1): Let $\varepsilon\colon FG \to \mathrm{id}_{\mathcal{D}}$ denote the counit. Then it is well known that the composites

$$F \xrightarrow{\ F\eta\ } FGF \xrightarrow{\ \varepsilon F\ } F \quad \text{and} \quad G \xrightarrow{\ \eta G\ } GFG \xrightarrow{\ G\varepsilon\ } G$$

are identity morphisms. We know from Proposition 1.1.3 that ε is invertible because G is fully faithful. Therefore $L\eta = GF\eta$ is invertible. Moreover, we have

$$L\eta = GF\eta = (G\varepsilon F)^{-1} = \eta GF = \eta L. \qquad \square$$

Localisation of Adjoints

Localising a pair of adjoint functors yields an adjoint pair of functors between the localised categories.

Lemma 1.1.6. *Let (F, G) be an adjoint pair of functors $\mathcal{C} \rightleftarrows \mathcal{D}$. If $S \subseteq \mathrm{Mor}\,\mathcal{C}$ and $T \subseteq \mathrm{Mor}\,\mathcal{D}$ are classes of morphisms such that $F(S) \subseteq T$ and $G(T) \subseteq S$, then (F, G) induces an adjoint pair of functors $(\bar{F}, \bar{G})$ such that the following diagram commutes.*

$$
\begin{array}{ccc}
\mathcal{C} & \underset{\overleftarrow{G}}{\overset{F}{\rightrightarrows}} & \mathcal{D} \\
\downarrow & & \downarrow \\
\mathcal{C}[S^{-1}] & \underset{\bar{G}}{\overset{\bar{F}}{\rightrightarrows}} & \mathcal{D}[T^{-1}]
\end{array}
$$

Proof The functors F and G induce a pair of functors $\bar{F} \colon \mathcal{C}[S^{-1}] \to \mathcal{D}[T^{-1}]$ and $\bar{G} \colon \mathcal{D}[T^{-1}] \to \mathcal{C}[S^{-1}]$. We have by definition a natural isomorphism

$$\alpha \colon \mathrm{Hom}_{\mathcal{D}}(F-, -) \xrightarrow{\sim} \mathrm{Hom}_{\mathcal{C}}(-, G-)$$

of functors $\mathcal{C}^{\mathrm{op}} \times \mathcal{D} \to \mathrm{Set}$. These functors invert morphisms in S and T. Thus α induces a natural isomorphism

$$\mathrm{Hom}_{\mathcal{D}[T^{-1}]}(\bar{F}-, -) \xrightarrow{\sim} \mathrm{Hom}_{\mathcal{C}[S^{-1}]}(-, \bar{G}-)$$

of functors $\mathcal{C}^{\mathrm{op}}[S^{-1}] \times \mathcal{D}[T^{-1}] \to \mathrm{Set}$. It follows that $(\bar{F}, \bar{G})$ is an adjoint pair. $\qquad\square$

There is a useful consequence which is obtained by setting $T = \varnothing$.

Lemma 1.1.7. *Consider a composite $\mathcal{C} \twoheadrightarrow \mathcal{C}[S^{-1}] \to \mathcal{D}$ of functors and suppose there exists a right adjoint. Then $\mathcal{C}[S^{-1}] \to \mathcal{D}$ admits a right adjoint.* $\qquad\square$

Localisation and Coproducts

Let $\mathcal{C}$ be a category and $S \subseteq \mathrm{Mor}\,\mathcal{C}$. We provide a criterion for the canonical functor $\mathcal{C} \to \mathcal{C}[S^{-1}]$ to preserve coproducts.

Lemma 1.1.8. *Let $\mathcal{C}$ be a category which admits coproducts and let $S \subseteq \mathrm{Mor}\,\mathcal{C}$ be a class of morphisms. If $\coprod_i \sigma_i$ belongs to S for every family $(\sigma_i)_{i \in I}$ in S, then the category $\mathcal{C}[S^{-1}]$ admits coproducts and the canonical functor $\mathcal{C} \to \mathcal{C}[S^{-1}]$ preserves coproducts.*

Proof Let $(X_i)_{i \in I}$ be a family of objects in $\mathcal{C}[S^{-1}]$. Then the coproduct is obtained by applying the left adjoint of the diagonal functor $\Delta \colon \mathcal{C} \to \prod_{i \in I} \mathcal{C}$. The assumption on S means that we can apply Lemma 1.1.6. Thus the diagonal functor $\Delta \colon \mathcal{C}[S^{-1}] \to \prod_{i \in I} \mathcal{C}[S^{-1}]$ admits a left adjoint which provides the coproduct $\coprod_{i \in I} X_i$ in $\mathcal{C}[S^{-1}]$. □

1.2 Calculus of Fractions

We introduce the calculus of fractions; this helps to describe explicitly the morphisms of a localised category.

Calculus of Fractions

Let $\mathcal{C}$ be a category and $S \subseteq \operatorname{Mor} \mathcal{C}$. There is an explicit description of the localisation $\mathcal{C}[S^{-1}]$ provided that the class S admits a *calculus of left fractions*, that is, the following conditions are satisfied.

(LF1) The identity morphism of each object is in S. The composite of two morphisms in S is again in S.

(LF2) Each pair of morphisms $X' \xleftarrow{\sigma} X \to Y$ with $\sigma \in S$ can be completed to a commutative diagram

$$
\begin{array}{ccc}
X & \longrightarrow & Y \\
\sigma \downarrow & & \downarrow \tau \\
X' & \longrightarrow & Y'
\end{array}
$$

such that $\tau \in S$.

(LF3) Let $\alpha, \beta \colon X \to Y$ be morphisms in $\mathcal{C}$. If there is $\sigma \colon X' \to X$ in S such that $\alpha\sigma = \beta\sigma$, then there is $\tau \colon Y \to Y'$ in S such that $\tau\alpha = \tau\beta$.

The class S admits a *calculus of right fractions* if it admits a calculus of left fractions in the opposite category $\mathcal{C}^{\mathrm{op}}$.

Now assume that S admits a calculus of left fractions. Then one obtains a new category $S^{-1}\mathcal{C}$ as follows. The objects are those of $\mathcal{C}$. Given objects X and Y, we call a pair (α, σ) of morphisms

$$
X \xrightarrow{\ \alpha\ } Y' \xleftarrow{\ \sigma\ } Y
$$

in $\mathcal{C}$ with σ in S a *left fraction*. The morphisms $X \to Y$ in $S^{-1}\mathcal{C}$ are equivalence classes $[\alpha, \sigma]$ of such left fractions, where (α_1, σ_1) and (α_2, σ_2) are *equivalent* if there exists a commutative diagram

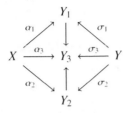

with σ_3 in S. The composite of $[\alpha, \sigma]$ and $[\beta, \tau]$ is by definition $[\beta'\alpha, \sigma'\tau]$ where σ' and β' are obtained from condition (LF2) as in the following commutative diagram.

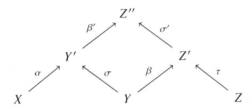

The canonical functor $P\colon \mathcal{C} \to S^{-1}\mathcal{C}$ is the identity on objects and sends a morphism $\alpha\colon X \to Y$ to $[\alpha, \mathrm{id}_Y]$.

Lemma 1.2.1. *Let S admit a calculus of left fractions. The functor $F\colon S^{-1}\mathcal{C} \to \mathcal{C}[S^{-1}]$ which is the identity on objects and takes a morphism $[\alpha, \sigma]$ to $(Q\sigma)^{-1} \circ Q\alpha$ is an isomorphism.*

Proof   The functor P inverts all morphisms in S and factors therefore through $Q\colon \mathcal{C} \to \mathcal{C}[S^{-1}]$ via a functor $G\colon \mathcal{C}[S^{-1}] \to S^{-1}\mathcal{C}$. It is straightforward to check that $F \circ G = \mathrm{id}$ and $G \circ F = \mathrm{id}$.   □

From now on, we identify $S^{-1}\mathcal{C}$ with $\mathcal{C}[S^{-1}]$ whenever S admits a calculus of left fractions.

A category $\mathcal{J}$ is called *filtered* if it is non-empty, for each pair of objects i, i' there is an object j with morphisms $i \to j \leftarrow i'$, and for each pair of morphisms $\alpha, \alpha'\colon i \to j$ there is a morphism $\beta\colon j \to k$ such that $\beta\alpha = \beta\alpha'$.

Lemma 1.2.2. *Let S admit a calculus of left fractions and fix objects X, Y in $\mathcal{C}$. The morphisms $\sigma\colon Y \to Y'$ in S form a filtered category, and taking σ to $\mathrm{Hom}_{\mathcal{C}}(X, Y')$ gives a bijection*

$$\mathop{\mathrm{colim}}_{\sigma\colon Y \to Y'} \mathrm{Hom}_{\mathcal{C}}(X, Y') \xrightarrow{\sim} \mathrm{Hom}_{\mathcal{C}[S^{-1}]}(X, Y).$$

This map sends a morphism α in $\mathrm{Hom}_{\mathcal{C}}(X, Y')$ to $[\alpha, \sigma]$.

Proof Straightforward.   □

Examples for classes of morphisms with a calculus of fractions arise from pairs of adjoint functors (F, G) by taking left fractions of the form

$$X \xrightarrow{\alpha} GF(Y) \xleftarrow{\eta_Y} Y.$$

Example 1.2.3. Let $F\colon \mathcal{C} \to \mathcal{D}$ be a functor with a fully faithful right adjoint. Then $S = \{\sigma \in \mathrm{Mor}\,\mathcal{C} \mid F\sigma \text{ is invertible}\}$ admits a calculus of left fractions.

Another class of examples arises from localising a ring. A ring may be viewed as a category with one object, by viewing the elements as morphisms.

Example 1.2.4. Let A be a ring. Then a subset $S \subseteq A$ admits a calculus of right fractions if the following holds.

(1) If $s, t \in S$, then $st \in S$. Also $1_A \in S$.
(2) For $a \in A$ and $s \in S$ there are $b \in A$ and $t \in S$ such that $at = sb$.
(3) If $sa = 0$ for $a \in A$ and $s \in S$, then there is $t \in S$ such that $at = 0$.

In this case $AS^{-1} = A[S^{-1}]$ is a ring and $A \to A[S^{-1}]$ is the universal homomorphism that makes all elements in S invertible.

Calculus of Fractions for Subcategories

Let $\mathcal{C}$ be a category and $S \subseteq \mathrm{Mor}\,\mathcal{C}$. A full subcategory $\mathcal{D}$ of $\mathcal{C}$ is *left cofinal* with respect to S if for every morphism $\sigma\colon X \to Y$ in S with X in $\mathcal{D}$ there is a morphism $\tau\colon Y \to Z$ with $\tau \circ \sigma$ in $S \cap \mathcal{D}$.

Lemma 1.2.5. *Let S admit a calculus of left fractions and $\mathcal{D} \subseteq \mathcal{C}$ be left cofinal with respect to S. Then $S \cap \mathcal{D}$ admits a calculus of left fractions and the induced functor $\mathcal{D}[(S \cap \mathcal{D})^{-1}] \to \mathcal{C}[S^{-1}]$ is fully faithful.*

Proof It is straightforward to check (LF1)–(LF3) for $S \cap \mathcal{D}$. Now let X, Y be objects in $\mathcal{D}$. We need to show that the induced map

$$f\colon \mathrm{Hom}_{\mathcal{D}[(S \cap \mathcal{D})^{-1}]}(X, Y) \longrightarrow \mathrm{Hom}_{\mathcal{C}[S^{-1}]}(X, Y)$$

is bijective. The map sends the equivalence class of a fraction to the equivalence class of the same fraction. If $[\alpha, \sigma]$ belongs to $\mathrm{Hom}_{\mathcal{C}[S^{-1}]}(X, Y)$ and τ is a morphism with $\tau \circ \sigma$ in $S \cap \mathcal{D}$, then $[\tau \circ \alpha, \tau \circ \sigma]$ belongs to $\mathrm{Hom}_{\mathcal{D}[(S \cap \mathcal{D})^{-1}]}(X, Y)$ and f sends it to $[\alpha, \sigma]$. Thus f is surjective. A similar argument shows that f is injective.

For an alternative proof using filtered colimits, combine Lemma 1.2.2 and Lemma 11.1.5.   □

Notes

The standard reference for localisation and the calculus of fractions is the book of Gabriel and Zisman [85]. The localisation of a category generalises the concept for rings. For instance, rings of functions are localised in order to study the local properties of a geometric object. The localisation of non-commutative rings was pioneered by Ore in 1931, who introduced the 'Ore condition' [151]. For a survey about localisation in algebra and topology, see [166].

2

Abelian Categories

Contents

	2.1	**Exact Categories**	**15**
		Additive and Abelian Categories	15
		Finitely Presented Functors	16
		Exact Categories	21
		Projective and Injective Objects	23
		Projective Covers and Injective Envelopes	25
		Stable Categories	27
	2.2	**Localisation of Additive and Abelian Categories**	**28**
		Additive Categories	29
		Abelian Categories	30
		Localisation and Adjoints	33
		Categories with Injective Envelopes	35
		Categories with Enough Projectives or Injectives	38
		Pullbacks of Abelian Categories	41
	2.3	**Module Categories and Their Localisations**	**42**
		Effaceable and Left Exact Functors	42
		Epimorphisms of Rings	45
		Universal Localisation	46
	2.4	**Commutative Noetherian Rings**	**47**
		Support of Modules	48
		Injective Modules	50
		Artinian Modules	52
		Graded Rings and Modules	53
	2.5	**Grothendieck Categories**	**55**
		The Embedding Theorem	56
		Injective Envelopes	57
		Decompositions into Indecomposables	58
		Locally Presentable Categories	60
		Localisation of Grothendieck Categories	64
		Coherent Functors	67
	Notes		**70**

This chapter is devoted to some of the foundations which are used throughout this book. The main theme is the theory of localisation for additive and abelian categories. We begin with a brief introduction into additive and exact categories. Then we describe specific constructions for localising additive and abelian categories. We provide many examples. For instance, we study localisations of module categories, and it is shown that Grothendieck categories are precisely the abelian categories arising from localising a module category. Also, for the category of modules over a commutative noetherian ring the localising subcategories are classified in terms of support.

2.1 Exact Categories

We introduce the notion of an exact category and begin with the more fundamental notions of additive and abelian categories. An exact category is by definition an additive category together with an extra structure given by a distinguished class of short exact sequences. Extreme cases arise either from additive categories by taking all split exact sequences as distinguished sequences, or from abelian categories by taking any possible short exact sequence as a distinguished sequence. Categories of finitely presented functors are a useful tool.

Additive and Abelian Categories

A category $\mathcal{A}$ is *additive* if it admits finite products, including the product indexed over the empty set, for each pair of objects X, Y the set $\mathrm{Hom}_\mathcal{A}(X, Y)$ is an abelian group, and the composition maps

$$\mathrm{Hom}_\mathcal{A}(Y, Z) \times \mathrm{Hom}_\mathcal{A}(X, Y) \longrightarrow \mathrm{Hom}_\mathcal{A}(X, Z)$$

sending a pair (ψ, ϕ) to the composite $\psi \circ \phi$ are biadditive.

Lemma 2.1.1. *In an additive category finite coproducts also exist. Moreover, finite products and coproducts coincide.*

Proof For a pair of objects X, Y the product $X \times Y$ together with the morphisms $(\mathrm{id}_X, 0) \colon X \to X \times Y$ and $(0, \mathrm{id}_Y) \colon Y \to X \times Y$ represents the coproduct of X and Y, since any pair of morphisms $\phi \colon X \to A$ and $\psi \colon Y \to A$ induces the morphism

$$X \times Y \xrightarrow{\phi \times \psi} A \times A \xrightarrow{\nabla} A$$

where ∇ denotes the sum of both projections $A \times A \to A$. $\square$

We write $X \oplus Y$ for the (co)product of objects X, Y in $\mathcal{A}$ and note that the group structure on $\mathrm{Hom}_{\mathcal{A}}(X, Y)$ is determined by the following commuting diagram for any pair $\phi, \psi \in \mathrm{Hom}_{\mathcal{A}}(X, Y)$:

$$
\begin{array}{ccc}
X \oplus X & \xrightarrow{\ \phi \oplus \psi\ } & Y \oplus Y \\
\Big\uparrow{\scriptstyle \Delta} & & \Big\downarrow{\scriptstyle \nabla} \\
X & \xrightarrow{\ \phi + \psi\ } & Y
\end{array}
\tag{2.1.2}
$$

A functor $F \colon \mathcal{A} \to \mathcal{B}$ between additive categories is *additive* if it preserves finite products. An equivalent condition is that the induced map

$$\mathrm{Hom}_{\mathcal{A}}(X, Y) \longrightarrow \mathrm{Hom}_{\mathcal{B}}(FX, FY)$$

is additive for every pair of objects X, Y in $\mathcal{A}$.

The *kernel* $\mathrm{Ker}\, F$ of an additive functor $F \colon \mathcal{A} \to \mathcal{B}$ is the full subcategory of objects X in $\mathcal{A}$ such that $F(X) = 0$.

An additive category $\mathcal{A}$ is *abelian* if every morphism $\phi \colon X \to Y$ has a kernel and a cokernel, and if the canonical factorisation

$$
\begin{array}{ccccccc}
\mathrm{Ker}\,\phi & \xrightarrow{\ \phi'\ } & X & \xrightarrow{\ \phi\ } & Y & \xrightarrow{\ \phi''\ } & \mathrm{Coker}\,\phi \\
& & \Big\downarrow & & \Big\uparrow & & \\
& & \mathrm{Coker}\,\phi' & \xrightarrow{\ \bar{\phi}\ } & \mathrm{Ker}\,\phi'' & &
\end{array}
$$

of ϕ induces an isomorphism $\bar{\phi}$.

Remark 2.1.3. An additive category may be characterised as follows. It is a category with finite products and coproducts (including the (co)product indexed over the empty set) such that products and coproducts coincide, and the monoid structure on $\mathrm{Hom}(X, Y)$ given by (2.1.2) yields a group structure for all objects X, Y.

Example 2.1.4. (1) Let $\mathcal{A}$ be an additive category and X an object. Set $\Lambda = \mathrm{End}_{\mathcal{A}}(X)$. Then $\mathrm{Hom}_{\mathcal{A}}(X, -) \colon \mathcal{A} \to \mathrm{Mod}\,\Lambda$ induces a fully faithful functor $\mathrm{add}\, X \to \mathrm{proj}\,\Lambda$. This functor is an equivalence if $\mathcal{A}$ is idempotent complete.

(2) The category of modules over an associative ring is an abelian category.

Finitely Presented Functors

Let $\mathcal{C}$ be an additive category. We consider additive functors $\mathcal{C}^{\mathrm{op}} \to \mathrm{Ab}$. Morphisms between such functors are the natural transformations. This gives a category which is denoted by $\mathrm{Mod}\,\mathcal{C}$. For F and G in $\mathrm{Mod}\,\mathcal{C}$, we write

$\text{Hom}_{\mathcal{C}}(F, G)$ for the class of morphisms $F \to G$. Note that $\text{Hom}_{\mathcal{C}}(F, G)$ is a set when $\mathcal{C}$ is essentially small.

For each object X in $\mathcal{C}$ there is the *representable functor*

$$h_X = \text{Hom}_{\mathcal{C}}(-, X) \colon \mathcal{C}^{\text{op}} \longrightarrow \text{Ab}.$$

An important tool is *Yoneda's lemma*.

Lemma 2.1.5 (Yoneda). *For objects F in* $\text{Mod}\,\mathcal{C}$ *and X in $\mathcal{C}$, the map*

$$\text{Hom}_{\mathcal{C}}(h_X, F) \longrightarrow F(X), \quad \phi \mapsto \phi_X(\text{id}_X)$$

is an isomorphism of abelian groups.

Proof The inverse map sends $x \in F(X)$ to $\psi \colon h_X \to F$ given by $\psi_C(\alpha) = F(\alpha)(x)$ for $C \in \mathcal{C}$ and $\alpha \in \text{Hom}_{\mathcal{C}}(C, X)$. $\square$

It follows from this lemma that the *Yoneda functor*

$$\mathcal{C} \longrightarrow \text{Mod}\,\mathcal{C}, \quad X \mapsto h_X$$

is fully faithful. Also, one sees for $F, G \in \text{Mod}\,\mathcal{C}$ that $\text{Hom}_{\mathcal{C}}(F, G)$ is a set when there is an epimorphism $h_X \to F$ for some object $X \in \mathcal{C}$.

(Co)kernels and (co)products in $\text{Mod}\,\mathcal{C}$ are computed *pointwise*, and it follows that $\text{Mod}\,\mathcal{C}$ is an abelian category which has set-indexed products and coproducts. A sequence $F \to G \to H$ of morphisms in $\text{Mod}\,\mathcal{C}$ is exact if and only if the sequence $F(X) \to G(X) \to H(X)$ is exact for all X in $\mathcal{C}$.

Let $\text{mod}\,\mathcal{C}$ denote the category of finitely presented functors $F \colon \mathcal{C}^{\text{op}} \to \text{Ab}$, where a functor F is *finitely presented* if it fits into an exact sequence

$$\text{Hom}_{\mathcal{C}}(-, X) \longrightarrow \text{Hom}_{\mathcal{C}}(-, Y) \longrightarrow F \longrightarrow 0.$$

The morphisms in $\text{mod}\,\mathcal{C}$ are given by the natural transformations.

A morphism $X \to Y$ in $\mathcal{C}$ is a *weak kernel* of a morphism $Y \to Z$ if the induced sequence

$$\text{Hom}_{\mathcal{C}}(-, X) \longrightarrow \text{Hom}_{\mathcal{C}}(-, Y) \longrightarrow \text{Hom}_{\mathcal{C}}(-, Z)$$

is exact. Let $\mathcal{D}$ be an abelian category. Then an additive functor $F \colon \mathcal{C} \to \mathcal{D}$ is *weakly left exact* if it sends each weak kernel sequence $X \to Y \to Z$ in $\mathcal{C}$ to an exact sequence $F(X) \to F(Y) \to F(Z)$ in $\mathcal{D}$.

Lemma 2.1.6. *The category* $\text{mod}\,\mathcal{C}$ *is additive and all morphisms in* $\text{mod}\,\mathcal{C}$ *have cokernels. The category is abelian if and only if all morphisms in $\mathcal{C}$ have weak kernels.*

Proof   We fix a pair of functors with finite presentations

$$\operatorname{Hom}(-, X_i) \longrightarrow \operatorname{Hom}(-, Y_i) \longrightarrow F_i \longrightarrow 0 \qquad (i = 1, 2).$$

A morphism $\phi \colon F_1 \to F_2$ gives rise to a commutative diagram

$$
\begin{array}{ccccccc}
\operatorname{Hom}(-, X_1) & \longrightarrow & \operatorname{Hom}(-, Y_1) & \longrightarrow & F_1 & \longrightarrow & 0 \\
\downarrow & & \downarrow & & \downarrow{\scriptstyle\phi} & & \\
\operatorname{Hom}(-, X_2) & \longrightarrow & \operatorname{Hom}(-, Y_2) & \longrightarrow & F_2 & \longrightarrow & 0
\end{array}
$$

in mod $\mathcal{C}$. We obtain presentations

$$\operatorname{Hom}(-, X_1 \oplus X_2) \longrightarrow \operatorname{Hom}(-, Y_1 \oplus Y_2) \longrightarrow F_1 \oplus F_2 \longrightarrow 0$$

and

$$\operatorname{Hom}(-, X_2 \oplus Y_1) \longrightarrow \operatorname{Hom}(-, Y_2) \longrightarrow \operatorname{Coker} \phi \longrightarrow 0.$$

It follows that mod $\mathcal{C}$ is an additive category with cokernels.

Now suppose that $\mathcal{C}$ has weak kernels. Choose weak kernel sequences

$$Y_0 \longrightarrow X_2 \oplus Y_1 \longrightarrow Y_2 \qquad \text{and} \qquad X_0 \longrightarrow X_1 \oplus Y_0 \longrightarrow Y_1.$$

This gives rise to a commutative diagram

$$
\begin{array}{ccccccc}
\operatorname{Hom}(-, X_0) & \longrightarrow & \operatorname{Hom}(-, Y_0) & \longrightarrow & \operatorname{Ker} \phi & \longrightarrow & 0 \\
\downarrow & & \downarrow & & \downarrow & & \\
\operatorname{Hom}(-, X_1) & \longrightarrow & \operatorname{Hom}(-, Y_1) & \longrightarrow & F_1 & \longrightarrow & 0
\end{array}
$$

in mod $\mathcal{C}$. Thus mod $\mathcal{C}$ has kernels and it follows that mod $\mathcal{C}$ is abelian.

Finally, suppose mod $\mathcal{C}$ is abelian and fix a morphism $Y \to Z$ in $\mathcal{C}$. Let F denote the kernel of the induced morphism $\operatorname{Hom}(-, Y) \to \operatorname{Hom}(-, Z)$ in mod $\mathcal{C}$. Then there is an epimorphism $\operatorname{Hom}(-, X) \to F$, and the composite $\operatorname{Hom}(-, X) \to F \to \operatorname{Hom}(-, Y)$ induces a weak kernel $X \to Y$ for the morphism $Y \to Z$.   □

If $\mathcal{D}$ is an additive category with cokernels, then every additive functor $F \colon \mathcal{C} \to \mathcal{D}$ extends essentially uniquely to a right exact functor $\bar{F} \colon \operatorname{mod} \mathcal{C} \to \mathcal{D}$ such that $\bar{F}(h_X) = F(X)$ for all $X \in \mathcal{C}$. To be precise, set $\bar{F}(\operatorname{Coker} h_\phi) = \operatorname{Coker} F(\phi)$ for an object $\operatorname{Coker} h_\phi$ in mod $\mathcal{C}$ given by a morphism ϕ in $\mathcal{C}$. This universal property of the Yoneda functor $h \colon \mathcal{C} \to \operatorname{mod} \mathcal{C}$ can be reformulated as follows.

Lemma 2.1.7. *For any additive category* $\mathcal{D}$ *with cokernels, composition with the Yoneda functor induces a functor*

$$\mathcal{H}om(\mathrm{mod}\,\mathcal{C}, \mathcal{D}) \longrightarrow \mathcal{H}om(\mathcal{C}, \mathcal{D}), \qquad F \mapsto F \circ h$$

that yields an equivalence when restricted to the full subcategory of right exact functors in $\mathcal{H}om(\mathrm{mod}\,\mathcal{C}, \mathcal{D})$ *and the full subcategory of additive functors in* $\mathcal{H}om(\mathcal{C}, \mathcal{D})$. $\qquad\square$

The above lemma has an analogue for functors $\mathrm{mod}\,\mathcal{C} \to \mathcal{D}$ that are exact. Thus we suppose that all morphisms in $\mathcal{C}$ have weak kernels so that $\mathrm{mod}\,\mathcal{C}$ is abelian.

Lemma 2.1.8. *For any abelian category* $\mathcal{D}$, *composition with the Yoneda functor induces a functor*

$$\mathcal{H}om(\mathrm{mod}\,\mathcal{C}, \mathcal{D}) \longrightarrow \mathcal{H}om(\mathcal{C}, \mathcal{D}), \qquad F \mapsto F \circ h$$

that yields an equivalence when restricted to the full subcategory of exact functors in $\mathcal{H}om(\mathrm{mod}\,\mathcal{C}, \mathcal{D})$ *and the full subcategory of additive functors in* $\mathcal{H}om(\mathcal{C}, \mathcal{D})$ *that are weakly left exact.*

Proof Fix an additive functor $F: \mathcal{C} \to \mathcal{D}$ and its right exact extension $\bar{F}: \mathrm{mod}\,\mathcal{C} \to \mathcal{D}$ satisfying $\bar{F}(h_X) = F(X)$ for all $X \in \mathcal{C}$. We claim that $\bar{F}$ is exact if and only if F is weakly left exact. One direction is clear. So suppose that F is weakly left exact. Choose an exact sequence $0 \to X_1 \to X_2 \to X_3 \to 0$ in $\mathrm{mod}\,\mathcal{C}$. Then we may choose presentations

$$h_{X_{i2}} \longrightarrow h_{X_{i1}} \longrightarrow h_{X_{i0}} \longrightarrow X_i \longrightarrow 0 \qquad (i = 1, 2, 3)$$

that induce a commutative diagram

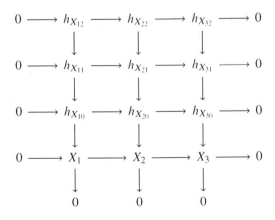

such that each sequence

$$0 \longrightarrow h_{X_{1j}} \longrightarrow h_{X_{2j}} \longrightarrow h_{X_{3j}} \longrightarrow 0 \qquad (j = 0, 1, 2)$$

is split exact. It follows that each sequence

$$F(X_{i2}) \longrightarrow F(X_{i1}) \longrightarrow F(X_{i0}) \longrightarrow \bar{F}(X_i) \longrightarrow 0 \qquad (i = 1, 2, 3)$$

is exact since F is weakly left exact and $\bar{F}$ is right exact. Thus the snake lemma implies that $0 \to \bar{F}(X_1) \to \bar{F}(X_2) \to \bar{F}(X_3) \to 0$ is exact. $\qquad \square$

Remark 2.1.9. Let $\mathcal{C}$ be an additive category with kernels. Then left exact functors and weakly left exact functors $F : \mathcal{C} \to \mathcal{D}$ agree.

We end our discussion of finitely presented functors with an equivalent description. Let $\mathcal{C}$ be an additive category and denote by $\mathcal{C}^2$ the *category of morphisms* in $\mathcal{C}$. The objects are morphisms $x : X_1 \to X_0$ in $\mathcal{C}$, and for an object $y : Y_1 \to Y_0$ the morphisms $\phi : x \to y$ are given by pairs of morphisms (ϕ_0, ϕ_1) making the following square commutative.

$$
\begin{array}{ccc}
X_1 & \xrightarrow{\;x\;} & X_0 \\
\downarrow{\scriptstyle \phi_1} & & \downarrow{\scriptstyle \phi_0} \\
Y_1 & \xrightarrow{\;y\;} & Y_0
\end{array}
$$

Such a morphism ϕ is called *null-homotopic* if there is a morphism $\rho : X_0 \to Y_1$ satisfying $y \circ \rho = \phi_0$. Let us denote by $\mathcal{C}^2/\mathrm{htp}$ the category which is obtained from $\mathcal{C}^2$ by identifying parallel morphisms ϕ and ψ if $\phi - \psi$ is null-homotopic.

Lemma 2.1.10. *Taking an object $x : X_1 \to X_0$ in $\mathcal{C}^2$ to the functor F_x with presentation*

$$\mathrm{Hom}_{\mathcal{C}}(-, X_1) \longrightarrow \mathrm{Hom}_{\mathcal{C}}(-, X_0) \longrightarrow F_x \longrightarrow 0$$

yields an equivalence $\mathcal{C}^2/\mathrm{htp} \xrightarrow{\sim} \mathrm{mod}\,\mathcal{C}$. $\qquad \square$

We give an application. Let $f : \mathcal{C} \to \mathcal{D}$ be an additive functor between additive categories. We write $f_! : \mathrm{mod}\,\mathcal{C} \to \mathrm{mod}\,\mathcal{D}$ for the right exact functor sending h_X to $h_{f(X)}$ for each $X \in \mathcal{C}$.

Lemma 2.1.11. $f_! : \mathrm{mod}\,\mathcal{C} \to \mathrm{mod}\,\mathcal{D}$ *is fully faithful if and only if $f : \mathcal{C} \to \mathcal{D}$ is fully faithful.*

Proof Clearly, f is fully faithful if and only if the induced functor $\mathcal{C}^2/\mathrm{htp} \to \mathcal{D}^2/\mathrm{htp}$ is fully faithful. $\qquad \square$

Exact Categories

Let $\mathcal{A}$ be an additive category. A sequence

$$0 \longrightarrow X \xrightarrow{\alpha} Y \xrightarrow{\beta} Z \longrightarrow 0$$

of morphisms in $\mathcal{A}$ is *exact* if α is a kernel of β and β is a cokernel of α. An *exact category* is a pair $(\mathcal{A}, \mathcal{E})$ consisting of an additive category $\mathcal{A}$ and a class $\mathcal{E}$ of exact sequences in $\mathcal{A}$ (called *admissible* and given by an *admissible monomorphism* followed by an *admissible epimorphism*) which is closed under isomorphisms and satisfies the following axioms.

(Ex1) The identity morphism of each object is an admissible monomorphism and an admissible epimorphism.

(Ex2) The composite of two admissible monomorphisms is an admissible monomorphism, and the composite of two admissible epimorphisms is an admissible epimorphism.

(Ex3) Each pair of morphisms $X' \xleftarrow{\phi} X \xrightarrow{\alpha} Y$ with α an admissible monomorphism can be completed to a pushout diagram

$$\begin{array}{ccc} X & \xrightarrow{\alpha} & Y \\ {\scriptstyle \phi}\downarrow & & \downarrow \\ X' & \xrightarrow{\alpha'} & Y' \end{array}$$

such that α' is an admissible monomorphism. And each pair of morphisms $Y \xrightarrow{\beta} Z \xleftarrow{\psi} Z'$ with β an admissible epimorphism can be completed to a pullback diagram

$$\begin{array}{ccc} Y' & \xrightarrow{\beta'} & Z' \\ \downarrow & & \downarrow{\scriptstyle \psi} \\ Y & \xrightarrow{\beta} & Z \end{array}$$

such that β' is an admissible epimorphism.

Observe that in (Ex3) the morphism ϕ induces an isomorphism $\operatorname{Coker}\alpha \xrightarrow{\sim} \operatorname{Coker}\alpha'$, while ψ induces an isomorphism $\operatorname{Ker}\beta' \xrightarrow{\sim} \operatorname{Ker}\beta$.

A pair of admissible exact sequences ξ and ξ' is called *equivalent* if there is a commutative diagram of the following form.

$$\begin{array}{ccccccccc} \xi: & 0 & \longrightarrow & X & \xrightarrow{\alpha} & Y & \xrightarrow{\beta} & Z & \longrightarrow & 0 \\ & & & \| & & \downarrow{\scriptstyle \phi} & & \| & & \\ \xi': & 0 & \longrightarrow & X & \xrightarrow{\alpha'} & Y' & \xrightarrow{\beta'} & Z & \longrightarrow & 0 \end{array}$$

In this case ϕ is an isomorphism. We write $\mathrm{Ext}^1_{\mathcal{A}}(Z, X)$ for the set of equivalence classes of such *extensions* and note that it is an abelian group via the *Baer sum*, which is given by the following diagram.

$$
\begin{array}{ccccccccc}
\xi_1 \oplus \xi_2: & 0 & \longrightarrow & X \oplus X & \longrightarrow & Y_1 \oplus Y_2 & \longrightarrow & Z \oplus Z & \longrightarrow & 0 \\
& & & \| & & \uparrow & & \Delta\uparrow & & \\
& 0 & \longrightarrow & X \oplus X & \longrightarrow & Y' & \longrightarrow & Z & \longrightarrow & 0 \\
& & & \nabla\downarrow & & \downarrow & & \| & & \\
\xi_1 + \xi_2: & 0 & \longrightarrow & X & \longrightarrow & Y & \longrightarrow & Z & \longrightarrow & 0
\end{array}
$$

We obtain a functor

$$\mathrm{Ext}^1_{\mathcal{A}}(-,-)\colon \mathcal{A}^{\mathrm{op}} \times \mathcal{A} \longrightarrow \mathrm{Ab}$$

which is given on morphisms by taking pullbacks (in the first argument) and pushouts (in the second argument).

Given exact categories $\mathcal{A}$ and $\mathcal{B}$, a functor $\mathcal{A} \to \mathcal{B}$ is *exact* if it is additive and takes admissible exact sequences in $\mathcal{A}$ to admissible exact sequences in $\mathcal{B}$. A *full exact subcategory* of an exact category $\mathcal{A}$ is a full additive subcategory $\mathcal{B} \subseteq \mathcal{A}$ that is *extension closed*, which means that for an admissible exact sequence in $\mathcal{A}$ with end terms in $\mathcal{B}$ the middle term is also in $\mathcal{B}$.

Example 2.1.12. (1) An additive category endowed with all split exact sequences is an exact category.

(2) An abelian category endowed with all short exact sequences is an exact category. Conversely, an exact category is an abelian category, if each morphism ϕ admits a factorisation $\phi = \phi''\phi'$ such that ϕ' is an admissible epimorphism and ϕ'' is an admissible monomorphism.

(3) Let $\mathcal{B}$ be an exact category and let $\mathcal{A} \subseteq \mathcal{B}$ be a full exact subcategory. Then $\mathcal{A}$ becomes an exact category by taking as admissible exact sequences those which are admissible in $\mathcal{B}$.

(4) Any essentially small exact category $\mathcal{A}$ can be embedded into an abelian category $\mathcal{B}$ such that it identifies with a full extension closed subcategory. For instance, take for $\mathcal{B}$ the category of left exact functors $F\colon \mathcal{A}^{\mathrm{op}} \to \mathrm{Ab}$; see Proposition 2.3.7. This yields an alternative definition for essentially small categories: an exact category is a full extension closed subcategory $\mathcal{A} \subseteq \mathcal{B}$ of an abelian category $\mathcal{B}$, endowed with all sequences which are short exact in $\mathcal{B}$.

(5) Let $\mathcal{C}$ be an additive category. Then $\mathrm{mod}\,\mathcal{C}$ is an exact category, if one chooses as admissible exact sequences the sequences that are pointwise exact.

Projective and Injective Objects

Let $\mathcal{A}$ be an exact category. An object P in $\mathcal{A}$ is *projective* if every admissible epimorphism $X \to Y$ induces a surjective map $\operatorname{Hom}_{\mathcal{A}}(P, X) \to \operatorname{Hom}_{\mathcal{A}}(P, Y)$. Dually, an object I is *injective* if every admissible monomorphism $X \to Y$ induces a surjective map $\operatorname{Hom}_{\mathcal{A}}(Y, I) \to \operatorname{Hom}_{\mathcal{A}}(X, I)$.

An exact category $\mathcal{A}$ has *enough projective objects* if every object X in $\mathcal{A}$ admits an admissible epimorphism $P \to X$ such that P is projective, and $\mathcal{A}$ has *enough injective objects* if every object X in $\mathcal{A}$ admits an admissible monomorphism $X \to I$ such that I is injective.

Example 2.1.13. (1) The category of modules over a ring Λ has enough projective objects, because every free module is projective. We write $\operatorname{Proj}\Lambda$ for the full subcategory of projective Λ-modules.

(2) The category of modules over a ring Λ has enough injective objects, and we write $\operatorname{Inj}\Lambda$ for the full subcategory of injective Λ-modules. More generally, any Grothendieck category has enough injective objects; cf. Corollary 2.5.4.

(3) Let $\mathcal{C}$ be an additive category and view $\operatorname{mod}\mathcal{C}$ as an exact category, with exact structure given by all pointwise exact sequences. Then each representable functor $\operatorname{Hom}_{\mathcal{C}}(-, X)$ is a projective object in $\operatorname{mod}\mathcal{C}$ by Yoneda's lemma.

Let $\mathcal{A}$ be an exact category and write $\mathcal{C} := \operatorname{Proj}\mathcal{A}$ for the full subcategory of projective objects in $\mathcal{A}$. Suppose that $\mathcal{A}$ has enough projective objects. Then every object $X \in \mathcal{A}$ admits a *projective presentation*

$$ P_1 \longrightarrow P_0 \longrightarrow X \longrightarrow 0, $$

that is, an exact sequence such that each P_i is projective. This yields an exact sequence

$$ \operatorname{Hom}_{\mathcal{C}}(-, P_1) \longrightarrow \operatorname{Hom}_{\mathcal{C}}(-, P_0) \longrightarrow \operatorname{Hom}_{\mathcal{A}}(-, X)|_{\mathcal{C}} \longrightarrow 0 $$

and therefore the functor

$$ F : \mathcal{A} \longrightarrow \operatorname{mod}\mathcal{C}, \quad X \mapsto \operatorname{Hom}_{\mathcal{A}}(-, X)|_{\mathcal{C}} $$

is well defined.

Lemma 2.1.14. *The functor F is fully faithful; it is an equivalence when $\mathcal{A}$ is abelian.*

Proof For the first assertion fix objects X, Y in $\mathcal{A}$ and choose projective presentations

$$ P_1 \xrightarrow{p} P_0 \to X \to 0 \qquad \text{and} \qquad Q_1 \xrightarrow{q} Q_0 \to Y \to 0. $$

Then the morphisms $X \to Y$ in $\mathcal{A}$ correspond to equivalence classes of commutative squares in $\mathcal{C}$

$$
\begin{array}{ccc}
P_1 & \xrightarrow{\;p\;} & P_0 \\
\downarrow & & \downarrow \\
Q_1 & \xrightarrow{\;q\;} & Q_0
\end{array}
$$

which in turn correspond to morphisms $\mathrm{Hom}_{\mathcal{A}}(-, X)|_{\mathcal{C}} \to \mathrm{Hom}_{\mathcal{A}}(-, Y)|_{\mathcal{C}}$, by Lemma 2.1.10.

Now suppose that $\mathcal{A}$ is abelian. Then the inclusion $\mathcal{C} \to \mathcal{A}$ extends to a quasi-inverse $\mathrm{mod}\,\mathcal{C} \to \mathcal{A}$ for F. □

We obtain the following correspondence; it provides a useful principle when dealing with abelian categories having enough projectives.

Proposition 2.1.15. *The assignments $\mathcal{C} \mapsto \mathrm{mod}\,\mathcal{C}$ and $\mathcal{A} \mapsto \mathrm{Proj}\,\mathcal{A}$ induce (up to equivalence) mutually inverse bijections between*

- *additive categories that are idempotent complete such that each morphism admits a weak kernel, and*
- *abelian categories with enough projective objects.* □

An application of this correspondence is the following criterion.

Corollary 2.1.16. *Let $\mathcal{A}$ be an abelian category with enough projective objects. Then a right exact functor $F\colon \mathcal{A} \to \mathcal{B}$ between abelian categories is exact if and only if for each exact sequence $X_2 \to X_1 \to X_0$ in $\mathcal{A}$ with each $X_i \in \mathrm{Proj}\,\mathcal{A}$ the sequence $FX_2 \to FX_1 \to FX_0$ is exact.*

Proof Set $\mathcal{C} = \mathrm{Proj}\,\mathcal{A}$ and identify $\mathcal{A} = \mathrm{mod}\,\mathcal{C}$. Then apply Lemma 2.1.8. □

There is a dual version of the above proposition for abelian categories with enough injective objects.

Proposition 2.1.17. *The assignments $\mathcal{C} \mapsto (\mathrm{mod}(\mathcal{C}^{\mathrm{op}}))^{\mathrm{op}}$ and $\mathcal{A} \mapsto \mathrm{Inj}\,\mathcal{A}$ induce (up to equivalence) mutually inverse bijections between*

- *additive categories that are idempotent complete such that each morphism admits a weak cokernel, and*
- *abelian categories with enough injective objects.* □

We end our discussion of projectives and injectives with a basic fact that will be used throughout without further reference.

Lemma 2.1.18. *The left adjoint of an exact functor takes projective objects to projective objects. Dually, the right adjoint of an exact functor takes injective objects to injective objects.* □

Projective Covers and Injective Envelopes

Let $\mathcal{A}$ be an abelian category. An epimorphism $\phi\colon X \to Y$ is *essential* if any morphism $\alpha\colon X' \to X$ is an epimorphism provided that the composite $\phi\alpha$ is an epimorphism. This condition can be rephrased as follows: if $U \subseteq X$ is a subobject with $U + \operatorname{Ker}\phi = X$, then $U = X$. An epimorphism $\phi\colon P \to X$ is a *projective cover* of X if P is projective and ϕ is essential.

There are the following dual notions. A monomorphism $\phi\colon X \to Y$ is *essential* if any morphism $\alpha\colon Y \to Y'$ is a monomorphism provided that the composite $\alpha\phi$ is a monomorphism. This condition can be rephrased as follows: if $U \subseteq Y$ is a subobject with $U \cap \operatorname{Im}\phi = 0$, then $U = 0$. A monomorphism $\phi\colon X \to I$ is an *injective envelope* of X if I is injective and ϕ is essential.

We collect some basic properties of projective covers and injective envelopes. In most cases we provide only one formulation (say, about injective envelopes) and leave the dual result (about projective covers) to the reader.

Lemma 2.1.19. *Let I be an injective object. Then the following are equivalent for a monomorphism $\phi\colon X \to I$.*

(1) *The morphism ϕ is an injective envelope of X.*
(2) *Every endomorphism $\alpha\colon I \to I$ satisfying $\alpha\phi = \phi$ is an isomorphism.*

Proof (1) $\Rightarrow$ (2): Let $\alpha\colon I \to I$ be an endomorphism satisfying $\alpha\phi = \phi$. Then α is a monomorphism since ϕ is essential. Thus there exists $\alpha'\colon I \to I$ satisfying $\alpha'\alpha = \operatorname{id}_I$ since I is injective. It follows that $\alpha'\phi = \phi$ and therefore α' is a monomorphism. On the other hand, α' is an epimorphism. Thus α' and α are isomorphisms.

(2) $\Rightarrow$ (1): Let $\alpha\colon I \to I'$ be a morphism such that $\alpha\phi$ is a monomorphism. Then ϕ factors through $\alpha\phi$ via a morphism $\alpha'\colon I' \to I$ since I is injective. The composite $\alpha'\alpha$ is an isomorphism and therefore α is a monomorphism. Thus ϕ is essential. □

We write $E(X) = I$ when $X \to I$ is an injective envelope. The following statement justifies this notation.

Lemma 2.1.20. *Let $\phi\colon X \to I$ and $\phi'\colon X \to I'$ be injective envelopes of an object X. Then there is an isomorphism $\alpha\colon I \to I'$ such that $\phi' = \alpha\phi$.* □

There is a close relation between projective covers and radical morphisms. We establish this in two steps: first for modules, and then for general abelian categories.

Lemma 2.1.21. *Let Λ be a ring and $P_1 \xrightarrow{\phi} P_0 \xrightarrow{\psi} X \to 0$ an exact sequence of Λ-modules such that each P_i is finitely generated projective. Then the following are equivalent.*

(1) ψ *is essential.*
(2) $\operatorname{Im} \phi \subseteq \operatorname{rad} P_0$.
(3) $\phi \in \operatorname{Rad}(P_1, P_0)$.

Proof (1) $\Rightarrow$ (2): Set $U = \operatorname{Im} \phi$. Suppose that ψ is essential and let $V \subseteq P_0$ be a maximal subobject not containing U. Then $U + V = P_0$ and therefore $V = P_0$. This is a contradiction and therefore U is contained in every maximal subobject. Thus $U \subseteq \operatorname{rad} P_0$.

(2) $\Rightarrow$ (1): Suppose that $U \subseteq \operatorname{rad} P_0$ and let $V \subseteq P_0$ be a subobject with $U + V = P_0$. If $V \neq P_0$, then there is a maximal subobject $V' \subseteq P_0$ containing V since P_0 is finitely generated. Thus $P_0 = U + V \subseteq V'$. This is a contradiction and therefore $V = P_0$. It follows that ψ is essential.

(2) $\Leftrightarrow$ (3): When $P_1 = \Lambda$ we have the identification $\operatorname{Rad}(\Lambda, P_0) = \operatorname{rad} P_0$ via $\lambda \mapsto \lambda(1)$. In particular, for $\lambda \colon \Lambda \to P_0$ we have $\lambda \in \operatorname{Rad}(\Lambda, P_0)$ if and only if $\operatorname{Im} \lambda \subseteq \operatorname{rad} P_0$. More generally, for $\lambda \colon \Lambda^n \to P_0$ we have $\lambda \in \operatorname{Rad}(\Lambda^n, P_0)$ if and only if $\operatorname{Im} \lambda \subseteq \operatorname{rad} P_0$.

For the implication (3) $\Rightarrow$ (2) choose an epimorphism $\pi \colon \Lambda^n \to P_1$. Then $\phi\pi$ is a radical morphism and therefore $\operatorname{Im} \phi = \operatorname{Im} \pi\phi \subseteq \operatorname{rad} P_0$.

For the implication (2) $\Rightarrow$ (3) choose an epimorphism $\Lambda^n \to U$. Then $\lambda \colon \Lambda^n \to U \rightarrowtail P_0$ is a radical morphism, and therefore $\phi \in \operatorname{Rad}(P_1, P_0)$ since ϕ factors through λ. $\qquad\square$

Proposition 2.1.22. *Let $P_1 \xrightarrow{\phi} P_0 \xrightarrow{\psi} X \to 0$ be an exact sequence in an abelian category such that each P_i is projective. Then ψ is a projective cover if and only if $\phi \in \operatorname{Rad}(P_1, P_0)$.*

Proof Let $\mathcal{A}$ denote the abelian category and $\mathcal{C}$ the smallest full additive subcategory which is closed under cokernels and contains $P = P_0 \oplus P_1$. Set $\Lambda = \operatorname{End}(P)$. The functor $H = \operatorname{Hom}(P, -) \colon \mathcal{A} \to \operatorname{Mod} \Lambda$ restricts to an equivalence $\mathcal{C} \xrightarrow{\sim} \operatorname{mod} \Lambda$. It follows from the dual of Lemma 2.1.19 that ψ is a projective cover if and only if $H\psi$ is a projective cover. On the other hand, $\operatorname{Rad}(P_1, P_0) \xrightarrow{\sim} \operatorname{Rad}(HP_1, HP_0)$ via H. Thus the assertion follows from Lemma 2.1.21. $\qquad\square$

We record the dual characterisation of injective envelopes via the radical.

Proposition 2.1.23. *Let* $0 \to X \xrightarrow{\phi} I^0 \xrightarrow{\psi} I^1$ *be an exact sequence in an abelian category such that each I^i is injective. Then ϕ is an injective envelope if and only if $\psi \in \mathrm{Rad}(I^0, I^1)$.* $\qquad\square$

We say that a morphism $\phi \colon X \to Y$ in an additive category admits a *minimal decomposition* if ϕ can be written as a direct sum

$$X = X' \oplus X'' \xrightarrow{\phi' \oplus \phi''} Y' \oplus Y'' = Y$$

such that ϕ' is an isomorphism and ϕ'' is a radical morphism.

An abelian category has *injective envelopes* if every object admits an injective envelope. Dually, an abelian category has *projective covers* if every object admits a projective cover.

Corollary 2.1.24. *Let $\mathcal{A}$ be an abelian category with enough injective objects. Then $\mathcal{A}$ has injective envelopes if and only if all morphisms in $\mathrm{Inj}\,\mathcal{A}$ admit minimal decompositions.*

Proof Suppose first that $\mathcal{A}$ has injective envelopes. Let $\phi \colon X \to Y$ be a morphism in $\mathrm{Inj}\,\mathcal{A}$. Choose a decomposition $X = X' \oplus X''$ such that $X'' = E(\mathrm{Ker}\,\phi)$. Let $\phi' \colon X' \xrightarrow{\sim} Y' = \phi(X')$ be the restriction $\phi|_{X'}$. Then ϕ' is a direct summand of ϕ. Thus we get a decomposition $\phi = \phi' \oplus \phi''$ and ϕ'' is radical by Proposition 2.1.23.

For the converse let $A \in \mathcal{A}$ and choose an exact sequence $0 \to A \to X \xrightarrow{\phi} Y$ with $\phi \in \mathrm{Inj}\,\mathcal{A}$. Decomposing $\phi = \phi' \oplus \phi''$ yields an injective envelope $A \to X''$, again by Proposition 2.1.23. $\qquad\square$

Example 2.1.25. Let $\mathcal{A}$ be a Krull–Schmidt category. Then every morphism $\phi \colon X \to Y$ in $\mathcal{A}$ admits a minimal decomposition.

To see this, choose decompositions $X = \bigoplus_i X_i$ and $Y = \bigoplus_j Y_j$ into indecomposables. Then $\phi = (\phi_{ij})$ belongs to $\mathrm{Rad}(X, Y)$ if and only if $\phi_{ij} \in \mathrm{Rad}(X_i, Y_j)$ for all i, j. Suppose $\phi_{i_0 j_0}$ is not radical. Then $\phi_{i_0 j_0}$ is an isomorphism and we may decompose $X = X_{i_0} \oplus \bar{X}$ and $Y = Y_{j_0} \oplus \bar{Y}$ such that $\phi = \phi_{i_0 j_0} \oplus \bar{\phi}$. Removing successively summands ϕ_{ij} that are not radical we obtain the decomposition $\phi = \phi' \oplus \phi''$ as required.

Stable Categories

Let $\mathcal{A}$ be an exact category and suppose that $\mathcal{A}$ has enough injective objects. Thus for each object $X \in \mathcal{A}$ we can choose an exact sequence $0 \to X \to I_X \to X' \to 0$ such that I_X is injective.

The *injectively stable category* $\mathrm{St}\,\mathcal{A}$ has by definition the same objects as $\mathcal{A}$ while the morphisms for objects X, Y are given by the quotient

$$\mathrm{Hom}_{\mathrm{St}\,\mathcal{A}}(X, Y) = \mathrm{Hom}_{\mathcal{A}}(X, Y)/\{\phi \mid \phi \text{ factors through an injective object}\}.$$

Lemma 2.1.26. *The assignment* $X \mapsto \mathrm{Ext}^1(-, X)$ *induces a fully faithful functor* $\mathrm{St}\,\mathcal{A} \to \mathrm{mod}\,\mathcal{A}$.

Proof For each object $X \in \mathcal{A}$ the sequence $0 \to X \to I_X \to X' \to 0$ induces a presentation

$$0 \to \mathrm{Hom}(-, X) \to \mathrm{Hom}(-, I_X) \to \mathrm{Hom}(-, X') \to \mathrm{Ext}^1(-, X) \longrightarrow 0.$$

Given a morphism $\phi \colon X \to Y$ in $\mathcal{A}$, we have $\mathrm{Ext}^1(-, \phi) = 0$ if and only if ϕ factors through $X \to I_X$. On the other hand, given a morphism of functors, $\psi \colon \mathrm{Ext}^1(-, X) \to \mathrm{Ext}^1(-, Y)$, we use Yoneda's lemma and obtain from the above presentation a morphism $\mathrm{Hom}(-, X) \to \mathrm{Hom}(-, Y)$ which corresponds to a morphism $\bar{\psi} \colon X \to Y$ in $\mathcal{A}$. Clearly, $\mathrm{Ext}^1(-, \bar{\psi}) = \psi$. $\qquad\square$

We call a pair of objects X, Y in $\mathcal{A}$ *stably equivalent* if the equivalent conditions in the following lemma are satisfied.

Lemma 2.1.27. *For objects* $X, Y \in \mathcal{A}$ *the following are equivalent.*

(1) $\mathrm{Ext}^1(-, X) \cong \mathrm{Ext}^1(-, Y)$ *in* $\mathrm{mod}\,\mathcal{A}$.

(2) $X \cong Y$ *in* $\mathrm{St}\,\mathcal{A}$.

(3) $X \oplus I \cong Y \oplus J$ *in* $\mathcal{A}$ *for some injective objects* $I, J \in \mathcal{A}$.

Proof (1) $\Leftrightarrow$ (2): See Lemma 2.1.26.

(2) $\Rightarrow$ (3): Let $\phi \colon X \to Y$ be a morphism in $\mathcal{A}$ that becomes invertible in $\mathrm{St}\,\mathcal{A}$. Adding $X \to I_X$ yields a split monomorphism $X \to Y \oplus I_X$, so $X \oplus I \cong Y \oplus I_X$ for some object I. We have $I = 0$ in $\mathrm{St}\,\mathcal{A}$, so I is injective.

(3) $\Rightarrow$ (2): Clear. $\qquad\square$

2.2 Localisation of Additive and Abelian Categories

There are specific constructions for localising additive and abelian categories. In both cases the localisation amounts to annihilating a class of objects. Also, the additional categorical structure is preserved. This means the localisation provides an additive functor $\mathcal{A} \to \mathcal{A}[S^{-1}]$ when $\mathcal{A}$ is additive and an exact functor when $\mathcal{A}$ is abelian.

Additive Categories

Let $\mathcal{A}$ be an additive category. When $F\colon \mathcal{A} \to \mathcal{B}$ is an additive functor, then the class $S = \{\sigma \in \mathrm{Mor}\,\mathcal{A} \mid F\sigma \text{ is invertible}\}$ contains the identities and is closed under finite direct sums. The following criterion shows that this is sufficient for $\mathcal{A}[S^{-1}]$ to be an additive category.

Lemma 2.2.1. *Let $\mathcal{A}$ be an additive category and $S \subseteq \mathrm{Mor}\,\mathcal{A}$ a class of morphisms. Suppose that S contains the identity morphism of each object and that $\sigma, \tau \in S$ implies $\sigma \oplus \tau \in S$. Then $\mathcal{A}[S^{-1}]$ is an additive category and the canonical functor $\mathcal{A} \to \mathcal{A}[S^{-1}]$ is additive.*

Proof We use the characterisation of an additive category from Remark 2.1.3. Also, we make a number of additional observations.

(1) Finite coproducts in a category $\mathcal{C}$ are given by a left adjoint of the diagonal $\Delta\colon \mathcal{C} \to \mathcal{C}^n$ for any $n \geq 0$. Dually, finite products are given by a right adjoint.

(2) If $\mathcal{C}_i$ and $S_i \subseteq \mathrm{Mor}\,\mathcal{C}_i$ are categories with classes of morphisms, then

$$\left(\prod_i \mathcal{C}_i\right)\left[\left(\prod_i S_i\right)^{-1}\right] \xrightarrow{\sim} \prod_i \mathcal{C}_i[S_i^{-1}].$$

(3) Let (F, G) be an adjoint pair of functors $\mathcal{C} \rightleftarrows \mathcal{D}$. If $S \subseteq \mathrm{Mor}\,\mathcal{C}$ and $T \subseteq \mathrm{Mor}\,\mathcal{D}$ are classes of morphisms such that $F(S) \subseteq T$ and $G(T) \subseteq S$, then (F, G) induces an adjoint pair of functors $\mathcal{C}[S^{-1}] \rightleftarrows \mathcal{D}[T^{-1}]$ (Lemma 1.1.6).

Now it follows that $\mathcal{A}[S^{-1}]$ is a category with finite products and coproducts, and the canonical functor $\mathcal{A} \to \mathcal{A}[S^{-1}]$ preserves these (co)products. Moreover, in $\mathcal{A}[S^{-1}]$ the monoid structure on $\mathrm{Hom}(X, Y)$ given by (2.1.2) yields a group structure for all objects X, Y. $\square$

Let $\mathcal{A}$ be an additive category and let $\mathcal{C} \subseteq \mathcal{A}$ be a full additive subcategory. The *additive quotient category* $\mathcal{A}/\mathcal{C}$ of $\mathcal{A}$ with respect to $\mathcal{C}$ has the same objects as $\mathcal{A}$ while the morphisms for objects X, Y are defined by the quotient

$$\mathrm{Hom}_{\mathcal{A}/\mathcal{C}}(X, Y) = \mathrm{Hom}_{\mathcal{A}}(X, Y)/\{\phi \mid \phi \text{ factors through an object in } \mathcal{C}\}.$$

For a morphism ϕ in $\mathcal{A}$ we write $\bar{\phi}$ for the corresponding morphism in $\mathcal{A}/\mathcal{C}$.

Lemma 2.2.2. *Let $\mathcal{A}$ be an additive category and let $\mathcal{C} \subseteq \mathcal{A}$ be a full additive subcategory. Set*

$$S = S(\mathcal{C}) = \{\sigma \in \mathrm{Mor}\,\mathcal{A} \mid \bar{\sigma} \text{ is invertible in } \mathcal{A}/\mathcal{C}\}.$$

Then the canonical functor $\mathcal{A} \to \mathcal{A}/\mathcal{C}$ induces an isomorphism $\mathcal{A}[S^{-1}] \xrightarrow{\sim} \mathcal{A}/\mathcal{C}$.

Proof   Consider the canonical functors $P: \mathcal{A} \to \mathcal{A}/\mathcal{C}$ and $Q: \mathcal{A} \to \mathcal{A}[S^{-1}]$. Clearly, P factors through Q via a functor $\bar{P}$. Now observe for morphisms α, β in $\mathcal{A}$ that $\bar{\alpha} = \bar{\beta}$ implies $Q\alpha = Q\beta$, since Q is additive by Lemma 2.2.1. Thus Q factors through P via a functor $\bar{Q}$. It follows that $\bar{P}\bar{Q} = \mathrm{id}$ and $\bar{Q}\bar{P} = \mathrm{id}$, since P and Q provide solutions of some universal problems.   □

Abelian Categories

Let $\mathcal{A}$ be an abelian category. A full additive subcategory $\mathcal{C} \subseteq \mathcal{A}$ is a *Serre subcategory* provided that $\mathcal{C}$ is closed under taking subobjects, quotients and extensions. This means that for every exact sequence $0 \to X' \to X \to X'' \to 0$ in $\mathcal{A}$, the object X is in $\mathcal{C}$ if and only if X' and X'' are in $\mathcal{C}$.

Example 2.2.3. The kernel of an exact functor $\mathcal{A} \to \mathcal{B}$ between abelian categories is a Serre subcategory of $\mathcal{A}$.

Fix a Serre subcategory $\mathcal{C}$ of $\mathcal{A}$. We set

$$S(\mathcal{C}) = \{\sigma \in \mathrm{Mor}\,\mathcal{A} \mid \mathrm{Ker}\,\sigma, \mathrm{Coker}\,\sigma \in \mathcal{C}\}$$

and

$$\mathcal{C}^{\perp} = \{Y \in \mathcal{A} \mid \mathrm{Hom}_{\mathcal{A}}(X, Y) = 0 = \mathrm{Ext}^1_{\mathcal{A}}(X, Y) \text{ for all } X \in \mathcal{C}\}.$$

The *abelian quotient category* $\mathcal{A}/\mathcal{C}$ of $\mathcal{A}$ with respect to $\mathcal{C}$ has the same objects while the morphisms for objects X, Y are defined as follows. There is for each pair of subobjects $X' \subseteq X$ and $Y' \subseteq Y$ an induced map $\mathrm{Hom}_{\mathcal{A}}(X, Y) \to \mathrm{Hom}_{\mathcal{A}}(X', Y/Y')$. The pairs (X', Y') such that both X/X' and Y' lie in $\mathcal{C}$ form a directed set, and one obtains a directed system of abelian groups $\mathrm{Hom}_{\mathcal{A}}(X', Y/Y')$. Then one defines

$$\mathrm{Hom}_{\mathcal{A}/\mathcal{C}}(X, Y) = \operatorname*{colim}_{(X', Y')} \mathrm{Hom}_{\mathcal{A}}(X', Y/Y').$$

The composition of morphisms in $\mathcal{A}$ induces the composition in $\mathcal{A}/\mathcal{C}$.

Lemma 2.2.4. *For a Serre subcategory $\mathcal{C} \subseteq \mathcal{A}$ the following holds.*

(1) *$S(\mathcal{C})$ admits a calculus of left and right fractions.*

(2) *An object in $\mathcal{A}$ is $S(\mathcal{C})$-local if and only if it is in $\mathcal{C}^{\perp}$.*

(3) *The canonical functor $\mathcal{A} \to \mathcal{A}/\mathcal{C}$ induces an isomorphism $\mathcal{A}[S(\mathcal{C})^{-1}] \xrightarrow{\sim} \mathcal{A}/\mathcal{C}$.*

Proof (1) and (2) are straightforward. For (3) we apply Lemma 1.2.2. Given objects X, Y in $\mathcal{A}$ we have

$$\mathrm{Hom}_{\mathcal{A}/\mathcal{C}}(X,Y) = \underset{(X',Y')}{\mathrm{colim}}\, \mathrm{Hom}_{\mathcal{A}}(X',Y/Y')$$

$$\cong \underset{(\sigma,\tau)}{\mathrm{colim}}\, \mathrm{Hom}_{\mathcal{A}}(\bar{X},\bar{Y})$$

$$\cong \mathrm{Hom}_{\mathcal{A}[S^{-1}]}(X,Y)$$

where $\sigma\colon \bar{X} \to X$ and $\tau\colon Y \to \bar{Y}$ run through all morphisms in $S(\mathcal{C})$. $\square$

A consequence is the following useful observation describing the morphisms in $\mathcal{A}/\mathcal{C}$. For each morphism $\phi\colon X \to Y$ in $\mathcal{A}/\mathcal{C}$ we have a commutative square

$$\begin{array}{ccc} X' & \longrightarrow & Y/Y' \\ \downarrow & & \uparrow \\ X & \xrightarrow{\ \phi\ } & Y \end{array}$$

such that the other three morphisms are in the image of $\mathcal{A} \to \mathcal{A}/\mathcal{C}$ and the vertical morphisms are isomorphisms in $\mathcal{A}/\mathcal{C}$, since X/X' and Y' lie in $\mathcal{C}$. There is an analogue for exact sequences in $\mathcal{A}/\mathcal{C}$; see Lemma 14.1.9.

The following provides another useful fact about the morphisms in $\mathcal{A}/\mathcal{C}$.

Lemma 2.2.5. *Let $\mathcal{C} \subseteq \mathcal{A}$ be a Serre subcategory and $Y \in \mathcal{A}$. Then the canonical map*

$$\mathrm{Hom}_{\mathcal{A}}(X,Y) \longrightarrow \mathrm{Hom}_{\mathcal{A}/\mathcal{C}}(X,Y)$$

is a bijection for all $X \in \mathcal{A}$ if and only if $Y \in \mathcal{C}^{\perp}$.

Proof This follows from Lemma 1.1.2 and Lemma 2.2.4. $\square$

Proposition 2.2.6. *Let $\mathcal{A}$ be an abelian category and $\mathcal{C} \subseteq \mathcal{A}$ a Serre subcategory. Then the following holds.*

(1) *The category $\mathcal{A}/\mathcal{C}$ is abelian and the canonical functor $Q\colon \mathcal{A} \to \mathcal{A}/\mathcal{C}$ is an exact functor that annihilates $\mathcal{C}$.*

(2) *If $\mathcal{B}$ is an abelian category and $F\colon \mathcal{A} \to \mathcal{B}$ is an exact functor that annihilates $\mathcal{C}$, then there exists a unique exact functor $\bar{F}\colon \mathcal{A}/\mathcal{C} \to \mathcal{B}$ such that $F = \bar{F} \circ Q$.*

Proof (1) We apply Lemma 2.2.4. Thus $\mathcal{A}/\mathcal{C} = \mathcal{A}[S^{-1}]$ for $S = S(\mathcal{C})$, and S admits a calculus of left and right fractions. The category $\mathcal{A}/\mathcal{C}$ is additive by Lemma 2.2.1. A morphism $X \to Y$ in $\mathcal{A}/\mathcal{C}$ is up to an isomorphism of the form

$Q\phi$ for some $\phi\colon X \to Y$ in $\mathcal{A}$. Choosing a cokernel $\psi\colon Y \to Z$ yields for each $A \in \mathcal{A}$ an exact sequence

$$0 \to \mathrm{Hom}_{\mathcal{A}}(Z, A) \to \mathrm{Hom}_{\mathcal{A}}(Y, A) \to \mathrm{Hom}_{\mathcal{A}}(X, A)$$

and therefore an exact sequence

$$0 \to \operatorname*{colim}_{A \to A'} \mathrm{Hom}_{\mathcal{A}}(Z, A') \to \operatorname*{colim}_{A \to A'} \mathrm{Hom}_{\mathcal{A}}(Y, A') \to \operatorname*{colim}_{A \to A'} \mathrm{Hom}_{\mathcal{A}}(X, A')$$

where $A \to A'$ runs through all morphisms in S starting at A. Thus the sequence

$$0 \to \mathrm{Hom}_{\mathcal{A}/\mathcal{C}}(Z, A) \to \mathrm{Hom}_{\mathcal{A}/\mathcal{C}}(Y, A) \to \mathrm{Hom}_{\mathcal{A}/\mathcal{C}}(X, A)$$

is exact by Lemma 1.2.2, and it follows that $Q\psi$ is a cokernel of $Q\phi$. The dual argument shows that each morphism in $\mathcal{A}/\mathcal{C}$ admits a kernel. Clearly, Q preserves kernels and cokernels; so the property of $\mathcal{A}$ to be abelian carries over to $\mathcal{A}/\mathcal{C}$.

(2) If $F\colon \mathcal{A} \to \mathcal{B}$ is an exact functor and $F|_{\mathcal{C}} = 0$, then F inverts all morphisms in S. Thus F factors through $Q\colon \mathcal{A} \to \mathcal{A}/\mathcal{C}$ via a unique functor $\bar{F}\colon \mathcal{A}/\mathcal{C} \to \mathcal{B}$. The functor $\bar{F}$ is exact, because any exact sequence in $\mathcal{A}/\mathcal{C}$ is up to isomorphism the image of an exact sequence in $\mathcal{A}$.   □

Remark 2.2.7. (1) The properties (1)–(2) in Proposition 2.2.6 provide a universal property that determines the canonical functor $\mathcal{A} \to \mathcal{A}/\mathcal{C}$ up to a unique isomorphism.

(2) The canonical functor $\mathcal{A} \to \mathcal{A}/\mathcal{C}$ preserves all coproducts in $\mathcal{A}$ if and only if $\mathcal{C}$ is closed under coproducts; see Lemma 1.1.8.

Next we describe all Serre subcategories of a quotient $\mathcal{A}/\mathcal{C}$.

Proposition 2.2.8. *Let $\mathcal{C} \subseteq \mathcal{B} \subseteq \mathcal{A}$ be Serre subcategories of an abelian category $\mathcal{A}$. Then $\mathcal{B}/\mathcal{C}$ identifies with a Serre subcategory of $\mathcal{A}/\mathcal{C}$, and every Serre subcategory of $\mathcal{A}/\mathcal{C}$ is of this form. Moreover, the canonical functor $\mathcal{A} \to \mathcal{A}/\mathcal{C}$ induces an isomorphism $\mathcal{A}/\mathcal{B} \xrightarrow{\sim} (\mathcal{A}/\mathcal{C})/(\mathcal{B}/\mathcal{C})$.*

We capture the situation in the following commutative diagram.

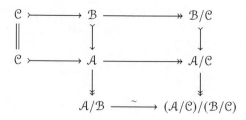

Proof The inclusion $\mathcal{B} \to \mathcal{A}$ induces a fully faithful functor $\mathcal{B}/\mathcal{C} \to \mathcal{A}/\mathcal{C}$ since $\mathcal{B}$ is left and right cofinal with respect to $S(\mathcal{C})$; see Lemma 1.2.5. It is easily checked that $\mathcal{B}/\mathcal{C}$ yields a Serre subcategory of $\mathcal{A}/\mathcal{C}$. If $\mathcal{D} \subseteq \mathcal{A}/\mathcal{C}$ is a Serre subcategory, set $\mathcal{B} := Q^{-1}(\mathcal{D})$. Then $\mathcal{B}/\mathcal{C} \xrightarrow{\sim} \mathcal{D}$. The final assertion is clear, since the kernel of the composite $\mathcal{A} \to \mathcal{A}/\mathcal{C} \to (\mathcal{A}/\mathcal{C})/(\mathcal{B}/\mathcal{C})$ equals $\mathcal{B}$. $\qquad\square$

Remark 2.2.9. The above correspondence $\mathcal{B} \mapsto \mathcal{B}/\mathcal{C}$ between Serre subcategories is inclusion preserving, and $\mathcal{B}/\mathcal{B}' \xrightarrow{\sim} (\mathcal{B}/\mathcal{C})/(\mathcal{B}'/\mathcal{C})$ for $\mathcal{B}' \subseteq \mathcal{B}$.

Localisation and Adjoints

Let $\mathcal{A}$ be an abelian category. We consider the situation that the canonical functor $\mathcal{A} \to \mathcal{A}/\mathcal{C}$ given by a Serre subcategory $\mathcal{C}$ admits a right adjoint.

Lemma 2.2.10. *Let $\mathcal{A}$ be an abelian category and $\mathcal{C} \subseteq \mathcal{A}$ a Serre subcategory. Suppose the canonical functor $Q \colon \mathcal{A} \to \mathcal{A}/\mathcal{C}$ admits a right adjoint $Q_\rho \colon \mathcal{A}/\mathcal{C} \to \mathcal{A}$. Then the following holds.*

(1) *The functor Q_ρ is fully faithful and induces an equivalence*

$$\mathcal{A}/\mathcal{C} \xrightarrow{\sim} \mathcal{C}^{\perp} \qquad \text{with quasi-inverse} \qquad \mathcal{C}^{\perp} \hookrightarrow \mathcal{A} \xrightarrow{Q} \mathcal{A}/\mathcal{C}.$$

(2) *The adjunction yields for X in $\mathcal{A}$ a natural exact sequence*

$$0 \longrightarrow X' \longrightarrow X \xrightarrow{\eta} Q_\rho Q(X) \longrightarrow X'' \longrightarrow 0$$

with X' and X'' in $\mathcal{C}$.

(3) *The assignment $X \mapsto X'$ gives a right adjoint of the inclusion $\mathcal{C} \to \mathcal{A}$.*

Proof (1) This follows from Proposition 1.1.3 and Lemma 2.2.4.

(2) This follows from the fact that $Q(\eta)$ is invertible.

(3) The map $\operatorname{Hom}_{\mathcal{A}}(C, X') \to \operatorname{Hom}_{\mathcal{A}}(C, X)$ is bijective for $C \in \mathcal{C}$ since $Q_\rho Q(X)$ is in $\mathcal{C}^{\perp}$. $\qquad\square$

We capture the situation in the following diagram

$$\mathcal{C} \underset{I_\rho}{\overset{I}{\rightleftarrows}} \mathcal{A} \underset{Q_\rho}{\overset{Q}{\rightleftarrows}} \mathcal{A}/\mathcal{C}$$

which is a localisation sequence. Each object $X \in \mathcal{A}$ fits into a functorial exact sequence

$$0 \longrightarrow II_\rho(X) \longrightarrow X \longrightarrow Q_\rho Q(X).$$

A Serre subcategory $\mathcal{C} \subseteq \mathcal{A}$ is called *localising* if the canonical functor

$Q: \mathcal{A} \to \mathcal{A}/\mathcal{C}$ admits a right adjoint. Note that in this case $\mathcal{C}$ is closed under all coproducts which exist in $\mathcal{A}$, since Q preserves coproducts.

Proposition 2.2.11. *Let (F, G) be an adjoint pair of functors*

$$\mathcal{A} \underset{G}{\overset{F}{\rightleftarrows}} \mathcal{B}$$

between abelian categories such that F is exact and set $\mathcal{C} = \mathrm{Ker}\, F$. Then G is fully faithful if and only if F induces an equivalence $\mathcal{A}/\mathcal{C} \xrightarrow{\sim} \mathcal{B}$.

Proof Let $S = \{\sigma \in \mathrm{Mor}\,\mathcal{A} \mid F\sigma \text{ is invertible}\}$. Then G is fully faithful if and only if F induces an equivalence $\mathcal{A}[S^{-1}] \xrightarrow{\sim} \mathcal{B}$, by Proposition 1.1.3. It remains to observe that $\mathcal{A}[S^{-1}] \xrightarrow{\sim} \mathcal{A}/\mathcal{C}$, by Lemma 2.2.4.

Let us give a more direct proof for one implication. So suppose that G is fully faithful. Then it is easily checked that the counit $\varepsilon_X: FG(X) \to X$ is an isomorphism for all $X \in \mathcal{B}$; see Proposition 1.1.3. We show that F satisfies, up to an isomorphism, the universal property of the canonical functor $\mathcal{A} \to \mathcal{A}/\mathcal{C}$; see Remark 2.2.7. Clearly, F is exact and annihilates $\mathcal{C}$. Now let $H: \mathcal{A} \to \mathcal{A}'$ be an exact functor between abelian categories that annihilates $\mathcal{C}$. Set $\bar{H} = H \circ G$. We claim that $\bar{H}$ is exact, that $H \cong \bar{H} \circ F$, and that $\bar{H}$ is unique with these properties. For the exactness, choose an exact sequence $0 \to X \to Y \to Z \to 0$ in $\mathcal{B}$ which yields an exact sequence

$$0 \to GX \to GY \to GZ \to X' \to 0$$

in $\mathcal{A}$ since G is left exact. We have $FX' = 0$ since $F \circ G \cong \mathrm{id}$, so $X' \in \mathcal{C}$, and therefore $HX' = 0$. Thus $\bar{H}$ is exact. Let $X \in \mathcal{A}$. Then F maps the unit $\eta_X: X \to GF(X)$ to an isomorphism, since the counit ε_{FX} is an inverse. Thus $\mathrm{Ker}\, \eta_X$ and $\mathrm{Coker}\, \eta_X$ are in $\mathcal{C}$. It follows that $H\eta$ yields an isomorphism $H \xrightarrow{\sim} \bar{H} \circ F$. If $\tilde{H}: \mathcal{B} \to \mathcal{A}'$ is another functor such that $H \cong \tilde{H} \circ F$, then one composes this isomorphism with G. Thus $\bar{H} = H \circ G \cong \tilde{H} \circ F \circ G \cong \tilde{H}$. $\square$

Remark 2.2.12. There are dual versions of Lemma 2.2.10 and Proposition 2.2.11 for abelian categories where the canonical functor $\mathcal{A} \to \mathcal{A}/\mathcal{C}$ admits a left adjoint.

Example 2.2.13. Let $\mathcal{A}$ be an abelian category and $i_*: \mathcal{A}' \to \mathcal{A}$ the inclusion of a Serre subcategory. Set $\mathcal{A}'' = \mathcal{A}/\mathcal{A}'$ and suppose that the canonical functor $j^*: \mathcal{A} \to \mathcal{A}''$ admits both adjoints. Then one obtains a *recollement* of abelian categories.

$$\mathcal{A}' \underset{i^!}{\overset{i^*}{\underset{\longleftarrow}{\overset{\longleftarrow}{\longrightarrow}}}} {\scriptstyle i_* = i_!} \mathcal{A} \underset{j_*}{\overset{j_!}{\underset{\longleftarrow}{\overset{\longleftarrow}{\longrightarrow}}}} {\scriptstyle j' = j^*} \mathcal{A}''$$

For an object X in $\mathcal{A}$, there are natural exact sequences relating the left and the right halves of the diagram.

$$j_!j^!(X) \longrightarrow X \longrightarrow i_*i^*(X) \longrightarrow 0 \qquad 0 \longrightarrow i_!i^!(X) \longrightarrow X \longrightarrow j_*j^*(X)$$

Each recollement of abelian categories is, up to equivalence, of the above form. A prototypical example arises from the category $\mathrm{Sh}(X)$ of sheaves on a topological space X and the inclusion $i \colon V \to X$ of a closed subset plus the inclusion $j \colon U \to X$ for $U = X \setminus V$.

$$\mathrm{Sh}(V) \overset{i^*}{\underset{i^!}{\underset{i_* = i_!}{\rightleftarrows}}} \mathrm{Sh}(X) \overset{j_!}{\underset{j_*}{\underset{j' = j^*}{\rightleftarrows}}} \mathrm{Sh}(U)$$

which involves the following functors:

$i^*, j^* = $ restriction $i^! = $ sections with support

$i_*, j_* = $ direct image $j_! = $ extension by zero.

This example explains the notation.

Categories with Injective Envelopes

Recall that an abelian category has *injective envelopes* if every object admits an injective envelope.

Proposition 2.2.14. *Let $\mathcal{A}$ be an abelian category with injective envelopes and let $\mathcal{C} \subseteq \mathcal{A}$ be a Serre subcategory. Then the inclusion $\mathcal{C} \to \mathcal{A}$ admits a right adjoint if and only if the canonical functor $\mathcal{A} \to \mathcal{A}/\mathcal{C}$ admits a right adjoint. In that case $\mathcal{C}$ and $\mathcal{A}/\mathcal{C}$ are categories with injective envelopes. Moreover, both right adjoints induce a sequence of functors*

$$\mathrm{Inj}(\mathcal{A}/\mathcal{C}) \rightarrowtail \mathrm{Inj}\,\mathcal{A} \twoheadrightarrow \mathrm{Inj}\,\mathcal{C}$$

that induces an equivalence

$$(\mathrm{Inj}\,\mathcal{A})/\mathrm{Inj}(\mathcal{A}/\mathcal{C}) \xrightarrow{\sim} \mathrm{Inj}\,\mathcal{C}.$$

Proof If the functor $\mathcal{A} \to \mathcal{A}/\mathcal{C}$ admits a right adjoint, then the inclusion $\mathcal{C} \to \mathcal{A}$ admits a right adjoint, by Lemma 2.2.10. For the other implication, suppose that $\mathcal{C} \to \mathcal{A}$ admits a right adjoint, sending $X \in \mathcal{A}$ to the maximal subobject $tX \subseteq X$ that belongs to $\mathcal{C}$. Choose an injective envelope $X/tX \to I$. Then I belongs to $\mathcal{C}^{\perp}$ because there are no non-zero subobjects in $\mathcal{C}$. We form

the following pullback

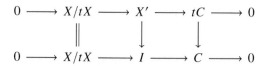

and also X' belongs to $\mathcal{C}^{\perp}$. Then $X \mapsto X'$ yields a right adjoint of the canonical functor $\mathcal{A} \to \mathcal{A}/\mathcal{C}$, since the kernel and cokernel of the morphism $X \to X'$ belong to $\mathcal{C}$ by construction.

Now suppose that both adjoints exist. It is convenient to identify $\mathcal{C}^{\perp} = \mathcal{A}/\mathcal{C}$. If $X \to I$ is an injective envelope in $\mathcal{A}$, then it is easily checked that $tX \to tI$ is an injective envelope in $\mathcal{C}$. In particular, t induces a functor $\operatorname{Inj}\mathcal{A} \to \operatorname{Inj}\mathcal{C}$ that is surjective on isoclasses of objects and full. In fact, for $X \in \operatorname{Inj}\mathcal{C}$ we have $X \cong tE(X)$. Also, any morphism $\phi\colon tX \to tY$ can be extended to a morphism $\tilde{\phi}\colon X \to Y$ since Y is injective, and $t\tilde{\phi} = \phi$. We claim that

$$\operatorname{Ker} t \cap \operatorname{Inj}\mathcal{A} = \mathcal{C}^{\perp} \cap \operatorname{Inj}\mathcal{A} = \operatorname{Inj}(\mathcal{C}^{\perp}).$$

The first equality is clear. Also, an object in $\mathcal{C}^{\perp} \cap \operatorname{Inj}\mathcal{A}$ is injective in $\mathcal{C}^{\perp}$, since the inclusion $\mathcal{C}^{\perp} \to \mathcal{A}$ is left exact. Given an object $X \in \mathcal{C}^{\perp}$, then its injective envelope $E(X)$ is also in $\mathcal{C}^{\perp}$, since $tE(X) = 0$. This yields the second equality and shows that $\mathcal{A}/\mathcal{C}$ has injective envelopes. Morover, it follows that t induces an equivalence between the additive quotient $(\operatorname{Inj}\mathcal{A})/\operatorname{Inj}(\mathcal{C}^{\perp})$ and $\operatorname{Inj}\mathcal{C}$. $\quad\square$

Corollary 2.2.15. *Let $\mathcal{A}$ be an abelian category with injective envelopes and let $\mathcal{C} \subseteq \mathcal{A}$ be a localising subcategory. Then we have $\mathcal{C}^{\perp} \cap \operatorname{Inj}\mathcal{A} = \operatorname{Inj}(\mathcal{C}^{\perp})$ and $\mathcal{C}^{\perp} \subseteq \mathcal{A}$ is closed under injective envelopes.* $\quad\square$

Grothendieck categories form an important class of abelian categories with injective envelopes. Thus we can apply the above proposition.

Proposition 2.2.16. *Let $\mathcal{A}$ be a Grothendieck category and $\mathcal{C} \subseteq \mathcal{A}$ a Serre subcategory that is closed under coproducts. Then $\mathcal{C}$ and the quotient $\mathcal{A}/\mathcal{C}$ are Grothendieck categories. Moreover, the canonical functors $\mathcal{C} \to \mathcal{A}$ and $\mathcal{A} \to \mathcal{A}/\mathcal{C}$ admit right adjoints.*

Proof Let $G \in \mathcal{A}$ be a generator of $\mathcal{A}$. The right adjoints are constructed as follows. Fix an object $X \in \mathcal{A}$. Observe that the subobjects of X form a set which has its cardinality bounded by 2^{α}, where $\alpha = \operatorname{card} \operatorname{Hom}(G, X)$. The subobjects $C \subseteq X$ with $C \in \mathcal{C}$ form a directed subset and we set $tX := \operatorname{colim}_{C \subseteq X} C$; this is the largest subobject of X belonging to $\mathcal{C}$. Then $X \mapsto tX$ yields a right adjoint of the inclusion $\mathcal{C} \to \mathcal{A}$. The right adjoint of $\mathcal{A} \to \mathcal{A}/\mathcal{C}$ then exists by Proposition 2.2.14.

The object G is also a generator of $\mathcal{A}/\mathcal{C}$, and the coproduct of all quotients of G that belong to $\mathcal{C}$ is a generator for $\mathcal{C}$. It is straightforward to check that the condition (AB5) holds for $\mathcal{C}$ and $\mathcal{A}/\mathcal{C}$. □

Corollary 2.2.17. *A Serre subcategory of a Grothendieck category is localising if and only if it is closed under coproducts.* □

Let $\mathcal{A}$ be a Grothendieck category. We denote by $\operatorname{Sp}\mathcal{A}$ a representative set of the isomorphism classes of indecomposable injective objects in $\mathcal{A}$ (the *spectrum* of $\mathcal{A}$). Note that $\operatorname{Sp}\mathcal{A}$ is a set, because $\mathcal{A}$ has a generator G and each object in $\operatorname{Sp}\mathcal{A}$ is the injective envelope of G/U for some subobject $U \subseteq G$.

Corollary 2.2.18. *Let $\mathcal{A}$ be a Grothendieck category and $\mathcal{C} \subseteq \mathcal{A}$ a localising subcategory. Every injective object $X \in \mathcal{A}$ admits a canonical decomposition $X = X' \oplus X''$ satisfying $tX' = tX$ and $X'' \in \mathcal{C}^{\perp}$. In particular, there is a canonical bijection*

$$\operatorname{Sp}\mathcal{C} \sqcup \operatorname{Sp}\mathcal{A}/\mathcal{C} \xrightarrow{\sim} \operatorname{Sp}\mathcal{A}.$$

Proof Let $X \in \mathcal{A}$ be injective. Then the injective envelope $X' = E(tX)$ is a direct summand of X and $X'' = X/X'$ belongs to $\mathcal{C}^{\perp}$. The map $\operatorname{Sp}\mathcal{C} \sqcup \operatorname{Sp}\mathcal{A}/\mathcal{C} \to \operatorname{Sp}\mathcal{A}$ sends $X \in \operatorname{Sp}\mathcal{C}$ to $E(X)$ and $X \in \operatorname{Sp}\mathcal{A}/\mathcal{C}$ to its image under $\mathcal{A}/\mathcal{C} \xrightarrow{\sim} \mathcal{C}^{\perp} \hookrightarrow \mathcal{A}$. □

Example 2.2.19. (1) Let $\mathcal{A}$ be a length category and denote by $S(\mathcal{A})$ a representative set of the isomorphism classes of simple objects. Then the maps

$$\mathcal{A} \supseteq \mathcal{C} \longmapsto \mathcal{C} \cap S(\mathcal{A}) \qquad \text{and} \qquad S(\mathcal{A}) \supseteq \mathcal{S} \longmapsto \operatorname{Filt}(\mathcal{S}) \subseteq \mathcal{A}$$

give mutually inverse and inclusion preserving bijections between the Serre subcategories of $\mathcal{A}$ and the subsets of $S(\mathcal{A})$.

(2) Let Λ be a *semiprimary ring*. Thus the Jacobson radical $J(\Lambda)$ is nilpotent and $\Lambda/J(\Lambda)$ is semisimple. Denote by $S(\Lambda)$ a representative set of the isomorphism classes of simple Λ-modules. Every Λ-module has a finite filtration with semisimple factors. It follows that the map

$$\operatorname{Mod}\Lambda \supseteq \mathcal{C} \longmapsto \mathcal{C} \cap S(\Lambda)$$

gives an inclusion preserving bijection between the localising subcategories of $\operatorname{Mod}\Lambda$ and the subsets of $S(\Lambda)$.

(3) Let $\mathcal{A}$ be a Grothendieck category that is *locally noetherian*. This means that every object is the directed union of its noetherian subobjects. Let $\operatorname{noeth}\mathcal{A}$ denote the full subcategory of noetherian objects in $\mathcal{A}$. Then the map

$$\mathcal{A} \supseteq \mathcal{C} \longmapsto \mathcal{C} \cap \operatorname{noeth}\mathcal{A}$$

gives an inclusion preserving bijection between the localising subcategories of $\mathcal{A}$ and the Serre subcategories of $\mathrm{noeth}\,\mathcal{A}$.

Categories with Enough Projectives or Injectives

Recall that an abelian category $\mathcal{A}$ has enough projective objects if and only if the inclusion $\mathcal{C} := \mathrm{Proj}\,\mathcal{A} \hookrightarrow \mathcal{A}$ induces an equivalence $\mathrm{mod}\,\mathcal{C} \xrightarrow{\sim} \mathcal{A}$; see Proposition 2.1.15. In this case the localisation theory for $\mathcal{A}$ is determined by certain subcategories of $\mathcal{C}$.

Let $\mathcal{C}$ be a category and $\mathcal{X} \subseteq \mathcal{C}$ a full subcategory. Given an object $C \in \mathcal{C}$, a morphism $X \to C$ with $X \in \mathcal{X}$ is called a *right $\mathcal{X}$-approximation* of C if the induced map $\mathrm{Hom}_{\mathcal{C}}(X', X) \to \mathrm{Hom}_{\mathcal{C}}(X', C)$ is surjective for every object $X' \in \mathcal{X}$. The subcategory $\mathcal{X}$ is *contravariantly finite* if every object $C \in \mathcal{C}$ admits a right $\mathcal{X}$-approximation.

Let $\mathcal{C}$ be an additive category. We denote by $\mathrm{Mod}\,\mathcal{C}$ the category of additive functors $\mathcal{C}^{\mathrm{op}} \to \mathrm{Ab}$. An additive functor $f\colon \mathcal{C} \to \mathcal{D}$ induces an adjoint pair $(f_!, f^*)$

$$
\begin{array}{ccccc}
\mathcal{C} & \hookrightarrow & \mathrm{mod}\,\mathcal{C} & \hookrightarrow & \mathrm{Mod}\,\mathcal{C} \\
\downarrow{\scriptstyle f} & & \downarrow{\scriptstyle f_!} & & {\scriptstyle f_!}\downarrow\uparrow{\scriptstyle f^*} \\
\mathcal{D} & \hookrightarrow & \mathrm{mod}\,\mathcal{D} & \hookrightarrow & \mathrm{Mod}\,\mathcal{D}
\end{array}
$$

where f^* is given by $Y \mapsto Y \circ f$ and $f_!$ is given by $X \mapsto f_!(X)$ via presentations

$$\mathrm{Hom}_{\mathcal{C}}(-, C_1) \longrightarrow \mathrm{Hom}_{\mathcal{C}}(-, C_0) \longrightarrow X \longrightarrow 0$$

and

$$\mathrm{Hom}_{\mathcal{D}}(-, f(C_1)) \longrightarrow \mathrm{Hom}_{\mathcal{D}}(-, f(C_0)) \longrightarrow f_!(X) \longrightarrow 0.$$

The following proposition describes the localisation of an abelian category with enough projective objects.

Proposition 2.2.20. *Let $\mathcal{C}$ be an additive category such that $\mathrm{mod}\,\mathcal{C}$ is abelian. If $\mathcal{D} \subseteq \mathcal{C}$ is a contravariantly finite subcategory, then the sequence of additive functors*

$$\mathcal{D} \overset{i}{\rightarrowtail} \mathcal{C} \overset{p}{\twoheadrightarrow} \mathcal{C}/\mathcal{D}$$

induces a diagram of functors between abelian categories

$$
\mathrm{mod}(\mathcal{C}/\mathcal{D}) \underset{p^*}{\overset{p_!}{\rightleftarrows}} \mathrm{mod}\,\mathcal{C} \underset{i^*}{\overset{i_!}{\rightleftarrows}} \mathrm{mod}\,\mathcal{D}
$$

which is a colocalisation sequence. The functors i^* *and* p^* *are exact and induce equivalences*

$$\mathrm{mod}(\mathcal{C}/\mathcal{D}) \xrightarrow{\sim} \mathrm{Ker}\, i^* \quad and \quad (\mathrm{mod}\, \mathcal{C})/(\mathrm{Ker}\, i^*) \xrightarrow{\sim} \mathrm{mod}\, \mathcal{D}.$$

Proof For any additive functor f the assignment $F \mapsto f^*(F)$ is exact, but we need to show that it maps finitely presented functors to finitely presented functors when f is one of i or p. It suffices to show this when F is representable. In the first case, let $F = \mathrm{Hom}_{\mathcal{C}}(-, C)$ and choose a presentation $D_1 \to D_0 \to C$ with $D_i \in \mathcal{D}$, using that $\mathcal{D}$ is contravariantly finite and that $\mathcal{C}$ has weak kernels. Thus $D_0 \to C$ is a right $\mathcal{D}$-approximation of C, and $D_1 \to D_0$ is given by a right $\mathcal{D}$-approximation of a weak kernel of $D_0 \to C$. This yields a presentation

$$\mathrm{Hom}_{\mathcal{D}}(-, D_1) \longrightarrow \mathrm{Hom}_{\mathcal{D}}(-, D_0) \longrightarrow \mathrm{Hom}_{\mathcal{C}}(-, C)|_{\mathcal{D}} \longrightarrow 0$$

in $\mathrm{mod}\, \mathcal{D}$. Now let $F = \mathrm{Hom}_{\mathcal{C}/\mathcal{D}}(-, C)$. This yields in $\mathrm{mod}\, \mathcal{C}$ a presentation

$$\mathrm{Hom}_{\mathcal{C}}(-, D_0) \longrightarrow \mathrm{Hom}_{\mathcal{C}}(-, C) \longrightarrow \mathrm{Hom}_{\mathcal{C}/\mathcal{D}}(-, C) \longrightarrow 0.$$

The equivalence $\mathrm{mod}(\mathcal{C}/\mathcal{D}) \xrightarrow{\sim} \mathrm{Ker}\, i^*$ is clear, since additive functors $\mathcal{C} \to$ Ab vanishing on $\mathcal{D}$ identify with additive functors $\mathcal{C}/\mathcal{D} \to$ Ab. The second equivalence follows from the fact that $i^* i_! \cong \mathrm{id}$; see Proposition 2.2.11. $\square$

Example 2.2.21. Let $\mathcal{C}$ be an exact category and $(\mathcal{T}, \mathcal{F})$ a torsion pair for $\mathcal{C}$. Then the subcategory $\mathcal{T} \subseteq \mathcal{C}$ is contravariantly finite. If the torsion pair is split, so $\mathcal{C} = \mathcal{T} \vee \mathcal{F}$, then we have an equivalence $\mathcal{F} \xrightarrow{\sim} \mathcal{C}/\mathcal{T}$.

We have the following converse of Proposition 2.2.20, showing that any colocalisation sequence of abelian categories

$$\mathcal{A}' \xleftarrow{\longleftarrow}_{\longrightarrow} \mathcal{A} \xleftarrow{\longleftarrow}_{\longrightarrow} \mathcal{A}''$$

is of the above form, provided that every object in $\mathcal{A}$ admits a projective cover.

Proposition 2.2.22. *Let* $\mathcal{A}$ *be an abelian category with projective covers and let* $\mathcal{A}' \subseteq \mathcal{A}$ *be a Serre subcategory. Suppose that the canonical functors* $\mathcal{A}' \to \mathcal{A}$ *and* $\mathcal{A} \to \mathcal{A}'' := \mathcal{A}/\mathcal{A}'$ *admit left adjoints. Set* $\mathcal{C} := \mathrm{Proj}\, \mathcal{A}$, $\mathcal{C}' := \mathrm{Proj}\, \mathcal{A}'$, *and* $\mathcal{C}'' := \mathrm{Proj}\, \mathcal{A}''$. *Then the left adjoints restrict to functors*

$$\mathcal{C}'' \xrightarrow{\ i\ } \mathcal{C} \xrightarrow{\ p\ } \mathcal{C}'$$

which induce the following commutative diagram.

$$\begin{array}{ccccc}
\mathcal{A}' & \rightarrowtail & \mathcal{A} & \twoheadrightarrow & \mathcal{A}'' \\
\downarrow\wr & & \downarrow\wr & & \downarrow\wr \\
\mathrm{mod}\, \mathcal{C}' & \xrightarrow{\ p^*\ } & \mathrm{mod}\, \mathcal{C} & \xrightarrow{\ i^*\ } & \mathrm{mod}\, \mathcal{C}''
\end{array}$$

Proof We have an equivalence $\mathcal{A} \xrightarrow{\sim} \text{mod}\,\mathcal{C}$ by Proposition 2.1.15 since $\mathcal{A}$ has enough projectives. Now apply the dual of Proposition 2.2.14 which shows that $\mathcal{A}'$ and $\mathcal{A}''$ have enough projectives.   □

Example 2.2.23. Let A be a ring and $e = e^2$ an idempotent in A. Multiplication of A-modules by e identifies with $\text{Hom}_A(eA, -)$ and yields an exact functor $\text{Mod}\,A \to \text{Mod}\,eAe$ which has a fully faithful left adjoint given by $-\otimes_{eAe} eA$.

$$\text{Mod}\,A \xleftarrow[\text{Hom}_A(eA,-)]{-\otimes_{eAe}eA} \text{Mod}\,eAe$$

The kernel of $\text{Hom}_A(eA, -)$ identifies with $\text{Mod}\,A/AeA$. On the other hand, multiplication by e identifies with $-\otimes_A Ae$ and the corresponding functor $\text{Mod}\,A \to \text{Mod}\,eAe$ has a fully faithful right adjoint given by $\text{Hom}_{eAe}(Ae, -)$.

$$\text{Mod}\,A \xleftarrow[\text{Hom}_{eAe}(Ae,-)]{-\otimes_A Ae} \text{Mod}\,eAe$$

Multiplication by an idempotent can be viewed as evaluation or restriction. Thus the following example generalises the previous one.

Example 2.2.24. Let $\mathcal{C}$ be an essentially small additive category and fix an object $X \in \mathcal{C}$. Set $\mathcal{D} = \text{add}\,X$ and let $i\colon \mathcal{D} \to \mathcal{C}$ denote the inclusion. Then the evaluation $F \mapsto F(X)$ induces a functor

$$i^*\colon \text{Mod}\,\mathcal{C} \longrightarrow \text{Mod}\,\mathcal{D} = \text{Mod}\,\text{End}(X)$$

which gives rise to the following recollement

$$\text{Mod}\,\mathcal{C}/\mathcal{D} \xrightleftharpoons[p_*]{\overset{p_!}{\longleftarrow}\ \ p^*\ \longrightarrow} \text{Mod}\,\mathcal{C} \xrightleftharpoons[i_*]{\overset{i_!}{\longleftarrow}\ \ i^*\ \longrightarrow} \text{Mod}\,\text{End}(X)$$

where $p\colon \mathcal{C} \to \mathcal{C}/\mathcal{D}$ denotes the canonical functor.

Remark 2.2.25. There are dual versions of Proposition 2.2.20 and Proposition 2.2.22 for abelian categories with enough injective objects. For instance, let $\mathcal{A}$ be an abelian category with enough injective objects and let $\mathcal{A}' \subseteq \mathcal{A}$ be a Serre subcategory that is localising. Set $\mathcal{C} = \text{Inj}\,\mathcal{A}$ and $\mathcal{C}'' = \text{Inj}(\mathcal{A}/\mathcal{A}')$. Then $\mathcal{A} \xrightarrow{\sim} (\text{mod}\,\mathcal{C}^{\text{op}})^{\text{op}}$ and $\mathcal{C}/\mathcal{C}'' \xrightarrow{\sim} \text{Inj}\,\mathcal{A}'$.

Pullbacks of Abelian Categories

Each diagram of abelian categories and exact functors

$$
\begin{array}{c}
\mathcal{A}_2 \\
\downarrow F_2 \\
\mathcal{A}_1 \xrightarrow{\;F_1\;} \mathcal{A}
\end{array}
$$

can be completed to a commutative diagram

$$
\begin{array}{ccc}
\mathcal{A}_1 \times_{\mathcal{A}} \mathcal{A}_2 & \xrightarrow{\;P_2\;} & \mathcal{A}_2 \\
\downarrow{P_1} & & \downarrow{F_2} \\
\mathcal{A}_1 & \xrightarrow{\;F_1\;} & \mathcal{A}
\end{array}
$$

as follows. The objects of $\mathcal{A}_1 \times_{\mathcal{A}} \mathcal{A}_2$ are given by triples (X_1, X_2, μ), where $X_i \in \mathcal{A}_i$ are objects, and $\mu \colon F_1(X_1) \xrightarrow{\sim} F_2(X_2)$ is an isomorphism. A morphism from (X_1, X_2, μ) to (Y_1, Y_2, ν) is a pair (ϕ_1, ϕ_2) of morphisms $\phi_i \colon X_i \to Y_i$ such that $\nu F_1(\phi_1) = F_2(\phi_2)\mu$. The composition of morphisms is given by the formula

$$
(\psi_1, \psi_2) \circ (\phi_1, \phi_2) = (\psi_1 \circ \phi_1, \psi_2 \circ \phi_2).
$$

It is straightforward to check that $\mathcal{A}_1 \times_{\mathcal{A}} \mathcal{A}_2$ is an abelian category and that the canonical functors $P_i \colon \mathcal{A}_1 \times_{\mathcal{A}} \mathcal{A}_2 \to \mathcal{A}_i$ given by $P_i(X_1, X_2, \mu) = X_i$ are exact.

Proposition 2.2.26. *Let $\mathcal{C}$ be a category and $E_i \colon \mathcal{C} \to \mathcal{A}_i$ functors such that $F_1 E_1 \cong F_2 E_2$. Then there exists, up to isomorphism, a unique functor $E \colon \mathcal{C} \to \mathcal{A}_1 \times_{\mathcal{A}} \mathcal{A}_2$ such that $P_i E \cong E_i$ for $i = 1, 2$.*

Proof Let $\tau \colon F_1 E_1 \xrightarrow{\sim} F_2 E_2$ be a natural isomorphism. Then one defines $E \colon \mathcal{C} \to \mathcal{A}_1 \times_{\mathcal{A}} \mathcal{A}_2$ by $E(X) = (E_1(X), E_2(X), \tau_X)$. $\square$

The proposition justifies the notation $\mathcal{A}_1 \times_{\mathcal{A}} \mathcal{A}_2$ and we call the category a *pullback* (strictly speaking, a *2-pullback*); it is unique, up to equivalence.

The following lemma describes a property of pullbacks *of* abelian categories which is the analogue of a property of a pullback *in* an abelian category.

Lemma 2.2.27. *Let $F_i \colon \mathcal{A}_i \to \mathcal{A}$ be exact functors and suppose that F_1 induces an equivalence $\mathcal{A}_1/\mathrm{Ker}\, F_1 \xrightarrow{\sim} \mathcal{A}$. Then P_1 restricts to an equivalence $\mathrm{Ker}\, P_2 \xrightarrow{\sim} \mathrm{Ker}\, F_1$ and P_2 induces an equivalence $(\mathcal{A}_1 \times_{\mathcal{A}} \mathcal{A}_2)/\mathrm{Ker}\, P_2 \xrightarrow{\sim} \mathcal{A}_2$.*

The following diagram illustrates the assertion of the lemma.

$$
\begin{array}{ccccc}
\operatorname{Ker} P_2 & \rightarrowtail & \mathcal{A}_1 \times_{\mathcal{A}} \mathcal{A}_2 & \xrightarrow{\ P_2\ } & \mathcal{A}_2 \\
\Big\downarrow{\wr} & & P_1 \Big\downarrow & & \Big\downarrow{F_2} \\
\operatorname{Ker} F_1 & \rightarrowtail & \mathcal{A}_1 & \xrightarrow[\ F_1\ ]{} & \mathcal{A}
\end{array}
$$

Proof We provide for both functors a quasi-inverse. For $\operatorname{Ker} P_2 \to \operatorname{Ker} F_1$ the quasi-inverse $\operatorname{Ker} F_1 \to \operatorname{Ker} P_2$ is given by $X \mapsto (X, 0, 0)$. Now choose a quasi-inverse $G_1 \colon \mathcal{A} \to \mathcal{A}_1/\operatorname{Ker} F_1$ for $\bar{F}_1 \colon \mathcal{A}_1/\operatorname{Ker} F_1 \xrightarrow{\sim} \mathcal{A}$ together with an isomorphism $\tau \colon \bar{F}_1 G_1 \xrightarrow{\sim} \mathrm{id}$. Then the quasi-inverse $\mathcal{A}_2 \to (\mathcal{A}_1 \times_{\mathcal{A}} \mathcal{A}_2)/\operatorname{Ker} P_2$ is given by $X \mapsto (G_1 F_2(X), X, \tau_{F_2(X)})$. $\qquad\square$

2.3 Module Categories and Their Localisations

For several classes of abelian categories we describe specific Serre subcategories and the corresponding localisations. We begin with categories of functors and the interplay between effaceable and left exact functors. Then we consider module categories and see the connection with the localisation of a ring.

Effaceable and Left Exact Functors

Let $\mathcal{A}$ be an abelian category. Fix $F \in \operatorname{mod} \mathcal{A}$ given by a presentation

$$
0 \longrightarrow \operatorname{Hom}_{\mathcal{A}}(-, X_2) \longrightarrow \operatorname{Hom}_{\mathcal{A}}(-, X_1) \longrightarrow \operatorname{Hom}_{\mathcal{A}}(-, X_0) \longrightarrow F \longrightarrow 0
$$

$$(2.3.1)$$

coming from an exact sequence $0 \to X_2 \to X_1 \to X_0$ in $\mathcal{A}$.

Lemma 2.3.2. *For* $G \in \operatorname{mod} \mathcal{A}$ *we have* $\operatorname{Ext}^i(F, G) \cong H^i G(X)$ *where* $G(X)$ *is the complex*

$$
\cdots \longrightarrow 0 \longrightarrow G(X_0) \longrightarrow G(X_1) \longrightarrow G(X_2) \longrightarrow 0 \longrightarrow \cdots
$$

Proof This is clear since (2.3.1) provides a projective resolution of F. $\qquad\square$

The functor F is called *effaceable* if $X_1 \to X_0$ is an epimorphism. This definition does not depend on the presentation of F, since an equivalent condition is that $\operatorname{Hom}(F, G) = 0$ for each representable functor $G = \operatorname{Hom}_{\mathcal{A}}(-, X)$. Let eff $\mathcal{A}$ denote the full subcategory of effaceable functors.

Proposition 2.3.3. *Let* $\mathcal{A}$ *be an abelian category. The functor* $\operatorname{mod} \mathcal{A} \to \mathcal{A}$ *that* *sends* Coker Hom$_A$$(-, \phi)$ *(given by a morphism ϕ in A) to* Coker ϕ *provides an exact left adjoint of the Yoneda functor $A \to$* mod A *and induces an equivalence*

$$(\text{mod}\, A)/(\text{eff}\, A) \xrightarrow{\sim} A.$$

Proof   For the adjointness, see Example 1.1.4. The exactness of the left adjoint follows from Lemma 2.1.8. Now the equivalence is a consequence of Proposition 2.2.11.   □

Remark 2.3.4.   (1) The inclusion eff $A \hookrightarrow$ mod A admits a right adjoint that sends F with presentation (2.3.1) to F' with presentation

$$0 \longrightarrow \text{Hom}_A(-, X_2) \longrightarrow \text{Hom}_A(-, X_1) \longrightarrow \text{Hom}_A(-, X) \longrightarrow F' \longrightarrow 0$$

where $X = \text{Coker}(X_2 \to X_1)$.

(2) There is an equivalence $(\text{eff}\, A)^{\text{op}} \xrightarrow{\sim} \text{eff}(A^{\text{op}})$ given by

$$F \longmapsto F^\vee \qquad \text{with} \qquad F^\vee(X) = \text{Ext}^2(F, \text{Hom}_A(-, X)).$$

When F is given by (2.3.1), then $F^\vee$ has a presentation

$$0 \longrightarrow \text{Hom}_A(X_0, -) \longrightarrow \text{Hom}_A(X_1, -) \longrightarrow \text{Hom}_A(X_2, -) \longrightarrow F^\vee \longrightarrow 0$$

and we have $F^{\vee\vee} \cong F$.

We give an alternative description of the equivalence in Proposition 2.3.3 when $A = \text{mod}\,\Lambda$ is the module category of a ring. Let $\underline{\text{mod}}\,\Lambda$ denote the *projectively stable category* which is obtained from mod Λ by setting for Λ-modules X and Y

$$\underline{\text{Hom}}_\Lambda(X, Y) = \text{Hom}_\Lambda(X, Y)/\{\phi \mid \phi \text{ factors through a projective module}\}.$$

Proposition 2.3.5.   *Let Λ be a right coherent ring so that* mod Λ *is abelian. Then the sequence of additive functors* proj $\Lambda \rightarrowtail$ mod $\Lambda \twoheadrightarrow \underline{\text{mod}}\,\Lambda$ *induces a sequence of exact functors*

$$\text{mod}(\underline{\text{mod}}\,\Lambda) \rightarrowtail \text{mod}(\text{mod}\,\Lambda) \xrightarrow{\ \pi\ } \text{mod}(\text{proj}\,\Lambda) = \text{mod}\,\Lambda$$

and an equivalence

$$\text{mod}(\underline{\text{mod}}\,\Lambda) \xrightarrow{\sim} \text{Ker}\,\pi = \text{eff}(\text{mod}\,\Lambda).$$

Proof   The subcategory proj $\Lambda \subseteq$ mod Λ is contravariantly finite. Now apply Proposition 2.2.20.   □

Now let A be an exact category and let Mod A denote the category of additive functors $A^{\text{op}} \to$ Ab. A functor $F \in \text{Mod}\,A$ is *locally effaceable* if for each object C in A and $x \in F(C)$ there exists an admissible epimorphism

$\phi: B \to C$ such that $F(\phi)(x) = 0$. We write Eff $\mathcal{A}$ for the full subcategory of locally effaceable functors.

Lemma 2.3.6. *When $\mathcal{A}$ is abelian we have* eff $\mathcal{A}$ = Eff $\mathcal{A}$ ∩ mod $\mathcal{A}$.

Proof Let $F \in$ mod $\mathcal{A}$ be given by a presentation (2.3.1). Suppose first that $F \in$ eff $\mathcal{A}$. An element $x \in F(C)$ is given by a morphism $C \to X_0$, and forming the pullback with $X_1 \to X_0$ yields an epimorphism $\phi: B \to C$ such that $F(\phi)(x) = 0$. Thus $F \in$ Eff $\mathcal{A}$.

Now let $F \in$ Eff $\mathcal{A}$. Choose $C = X_0$ and take for $x \in F(C)$ the element given by id: $X_0 \to X_0$. This yields an epimorphism $\phi: B \to C$ that factors through $X_1 \to X_0$. Thus the morphism $X_1 \to X_0$ is an epimorphism and therefore $F \in$ eff $\mathcal{A}$.   □

We denote by Lex $\mathcal{A}$ the category of additive functors $F: \mathcal{A}^{\mathrm{op}} \to$ Ab that are *left exact*, that is, each exact sequence $0 \to X \to Y \to Z \to 0$ in $\mathcal{A}$ induces an exact sequence $0 \to FZ \to FY \to FX$ of abelian groups.

Proposition 2.3.7. *Let $\mathcal{A}$ be an essentially small exact category.*

(1) *The inclusion* Lex $\mathcal{A}$ → Mod $\mathcal{A}$ *admits an exact left adjoint* Mod $\mathcal{A}$ → Lex $\mathcal{A}$ *that induces an equivalence*

$$(\mathrm{Mod}\,\mathcal{A})/(\mathrm{Eff}\,\mathcal{A}) \xrightarrow{\sim} \mathrm{Lex}\,\mathcal{A}.$$

(2) *The category* Lex $\mathcal{A}$ *is a Grothendieck category.*
(3) *The Yoneda functor* $\mathcal{A} \to$ Lex $\mathcal{A}$ *that takes X to* $\mathrm{Hom}_{\mathcal{A}}(-, X)$ *is exact and identifies $\mathcal{A}$ with a full extension closed subcategory of* Lex $\mathcal{A}$.

Proof Using (Ex1) and (Ex2) one shows that Eff $\mathcal{A}$ is a Serre subcategory of Mod $\mathcal{A}$ and closed under coproducts. From (Ex3) it follows that $(\mathrm{Eff}\,\mathcal{A})^{\perp} =$ Lex $\mathcal{A}$. Thus the canonical functor Mod $\mathcal{A} \to \frac{\mathrm{Mod}\,\mathcal{A}}{\mathrm{Eff}\,\mathcal{A}}$ admits a fully faithful right adjoint, which identifies $\frac{\mathrm{Mod}\,\mathcal{A}}{\mathrm{Eff}\,\mathcal{A}}$ with Lex $\mathcal{A}$; see Lemma 2.2.10 and Proposition 2.2.16. In particular, Lex $\mathcal{A}$ is a Grothendieck category.

Now let $\xi: 0 \to \mathrm{Hom}_{\mathcal{A}}(-, X) \xrightarrow{\alpha} E \xrightarrow{\beta} \mathrm{Hom}_{\mathcal{A}}(-, Z) \to 0$ be an exact sequence in Lex $\mathcal{A}$. Then Coker β is locally effaceable, and there exists an admissible epimorphism $V \to Z$ inducing the following commutative diagram with exact rows.

$$
\begin{array}{ccccccccc}
0 & \longrightarrow & \mathrm{Hom}_{\mathcal{A}}(-, U) & \longrightarrow & \mathrm{Hom}_{\mathcal{A}}(-, V) & \longrightarrow & \mathrm{Hom}_{\mathcal{A}}(-, Z) & \longrightarrow & 0 \\
 & & \downarrow & & \downarrow & & \| & & \\
0 & \longrightarrow & \mathrm{Hom}_{\mathcal{A}}(-, X) & \longrightarrow & E & \longrightarrow & \mathrm{Hom}_{\mathcal{A}}(-, Z) & \longrightarrow & 0
\end{array}
$$

Apply condition (Ex3) by forming the following pushout.

$$\begin{array}{ccccccccc}
0 & \longrightarrow & U & \longrightarrow & V & \longrightarrow & Z & \longrightarrow & 0 \\
 & & \downarrow & & \downarrow & & \| & & \\
0 & \longrightarrow & X & \longrightarrow & Y & \longrightarrow & Z & \longrightarrow & 0
\end{array}$$

Then the bottom row identifies with ξ, and therefore the image of the Yoneda functor $\mathcal{A} \to \mathrm{Lex}\,\mathcal{A}$ is extension closed. $\qquad\square$

The injective objects in $\mathrm{Lex}\,\mathcal{A}$ admit the following explicit description. The functors $I_X = \mathrm{Hom}_{\mathbb{Z}}(\mathrm{Hom}_{\mathcal{A}}(X, -), \mathbb{Q}/\mathbb{Z})$ (with $X \in \mathcal{A}$) form a set of injective cogenerators for the abelian category $\mathrm{Mod}\,\mathcal{A}$. Thus the direct summands of products $\prod_{\alpha} I_{X_{\alpha}}$ are precisely the injective objects in $\mathrm{Mod}\,\mathcal{A}$. Moreover,

$$\mathrm{Inj}(\mathrm{Lex}\,\mathcal{A}) = \{F \in \mathrm{Inj}(\mathrm{Mod}\,\mathcal{A}) \mid F \text{ is exact}\}.$$

Epimorphisms of Rings

A ring homomorphism $\phi \colon A \to B$ is by definition an *epimorphism of rings* if for any pair of homomorphisms $\psi, \psi' \colon B \to C$ we have that $\psi\phi = \psi'\phi$ implies $\psi = \psi'$. An equivalent condition is that restriction of scalars $\phi^* \colon \mathrm{Mod}\,B \to \mathrm{Mod}\,A$ is fully faithful [197, Proposition XI.1.2]. In fact, we have an adjoint pair $(\phi_!, \phi^*)$ with counit $X \otimes_A B \to X$ given by scalar multiplication for any B-module X. Then ϕ^* is fully faithful if and only if the counit is an isomorphism for all X if and only if $B \otimes_A B \xrightarrow{\sim} B$. It follows that the adjoint pair $(\phi_!, \phi^*)$ gives rise to a localisation functor $\phi^* \circ \phi_! \colon \mathrm{Mod}\,A \to \mathrm{Mod}\,A$ when ϕ is an epimorphism, cf. Proposition 1.1.5.

Proposition 2.3.8. *Let* $L \colon \mathrm{Mod}\,A \to \mathrm{Mod}\,A$ *be a localisation functor. Then the following are equivalent.*

(1) *The functor L is, up to an equivalence, of the form $\phi^* \circ \phi_!$ for some ring epimorphism $\phi \colon A \to B$.*
(2) *The subcategory $\mathrm{Im}\,L$ is closed under all coproducts and cokernels.*

Proof (1) $\Rightarrow$ (2): An epimorphism $\phi \colon A \to B$ yields an adjoint pair $(\phi_!, \phi^*)$, and we have $\mathrm{Im}\,\phi^* = \mathrm{Im}\,L$ for $L = \phi^* \circ \phi_!$. Clearly, ϕ^* is right exact and preserves coproducts.

(2) $\Rightarrow$ (1): Recall from Proposition 1.1.5 that a localisation functor L can be written as the composite $L = G \circ F$ given by an adjoint pair (F, G) such that F is a quotient functor and G is fully faithful. Let $\mathcal{B} = \{X \in \mathrm{Mod}\,A \mid X \xrightarrow{\sim} L(X)\}$ be the localised category. It is of the form $S^{\perp}$ for a class S of morphisms in $\mathrm{Mod}\,A$, so closed under all limits in $\mathrm{Mod}\,A$; see Proposition 1.1.3. Also,

$\mathcal{B} = \operatorname{Im} L$ is closed under colimits and it follows that $\mathcal{B}$ is abelian. The inclusion $G\colon \mathcal{B} \to \operatorname{Mod} A$ is exact, and therefore F takes projectives to projectives. It follows that FA is a projective generator of $\mathcal{B}$. Also, $\operatorname{Hom}_{\mathcal{B}}(FA, -)$ preserves coproducts since G preserves coproducts. Set $B = \operatorname{End}_{\mathcal{B}}(FA)$. It follows that $\operatorname{Hom}_{\mathcal{B}}(FA, -)\colon \mathcal{B} \to \operatorname{Mod} B$ is an equivalence. Let $\phi\colon A \to B$ denote the homomorphism that is induced by F. Then the composite

$$\operatorname{Mod} A \xrightarrow{\quad F \quad} \mathcal{B} \xrightarrow{\operatorname{Hom}(FA,-)} \operatorname{Mod} B$$

is isomorphic to $\phi_! = - \otimes_B B$, and therefore $L \cong \phi^* \circ \phi_!$. $\square$

Examples of ring epimorphisms arise from localising a ring by universally inverting a set of fixed elements.

Universal Localisation

Let A be a ring and Σ a set of morphisms between finitely generated projective A-modules. The *universal localisation* of A with respect to Σ is a ring A_Σ together with a ring homomorphism $q\colon A \to A_\Sigma$ satisfying the following:

(UL1) For every $\sigma \in \Sigma$, the morphism $\sigma \otimes_A A_\Sigma$ is invertible.

(UL2) For every ring homomorphism $f\colon A \to B$ such that $\sigma \otimes_A B$ is invertible for all $\sigma \in \Sigma$, there exists a unique ring homomorphisms $\bar{f}\colon A_\Sigma \to B$ such that $f = \bar{f}q$.

The universal localisation solves a universal problem and is therefore unique. In particular, a universal localisation is an epimorphism of rings.

Any element $x \in A$ can be viewed as a morphism $\lambda_x\colon A \to A$ (left multiplication by x). Thus the universal localisation generalises the localisation of A with respect to a subset $S \subseteq A$, because we have $A[S^{-1}] = A_\Sigma$ for $\Sigma = \{\lambda_x \mid x \in S\}$.

We sketch the construction of A_Σ. Set $\mathcal{C} = \operatorname{proj} A$ so that $\Sigma \subseteq \operatorname{Mor} \mathcal{C}$. We may assume that Σ contains the identity morphism of each object and that $\sigma, \tau \in \Sigma$ implies $\sigma \oplus \tau \in \Sigma$. Then $\mathcal{C}[\Sigma^{-1}]$ is an additive category and the canonical functor $\mathcal{C} \to \mathcal{C}[\Sigma^{-1}]$ is additive, by Lemma 2.2.1. Set $A_\Sigma = \operatorname{End}_{\mathcal{C}[\Sigma^{-1}]}(A)$. The functor $\operatorname{Hom}_{\mathcal{C}[\Sigma^{-1}]}(A, -)$ makes the following diagram commutative

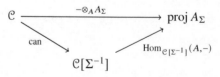

and identifies the idempotent completion of $\mathcal{C}[\Sigma^{-1}]$ with $\operatorname{proj} A_\Sigma$.

There is an alternative construction of A_Σ. Set $\mathcal{A} = \mathrm{Mod}\, A$ and consider the full subcategory $\mathcal{A}' \subseteq \mathcal{A}$ of A-modules X such that $\mathrm{Hom}_A(\sigma, X)$ is invertible for all $\sigma \in \Sigma$. It is easily checked that $\mathcal{A}'$ is closed under taking (co)kernels, (co)products, and extensions. Moreover, the inclusion $\mathcal{A}' \to \mathcal{A}$ admits a left adjoint $F\colon \mathcal{A} \to \mathcal{A}'$ (for instance by [84, Satz 8.5] or [1, Theorem 1.39]) which takes A to a projective generator of $\mathcal{A}'$. Set $A_\Sigma = \mathrm{End}_A(FA)$. Then we obtain an equivalence

$$\mathrm{Hom}_A(FA, -)\colon \mathcal{A}' \xrightarrow{\;\sim\;} \mathrm{Mod}\, A_\Sigma.$$

The inverse is given by the canonical functor $\mathrm{Mod}\, A_\Sigma \to \mathrm{Mod}\, A$, via restriction of scalars along the morphism $A \to A_\Sigma$ induced by F. Now set $\bar{\Sigma} = \{\sigma \in \mathrm{Mor}\,\mathcal{A} \mid \sigma \otimes_A A_\Sigma$ is invertible$\}$. Then it follows from Proposition 1.1.3 that the following diagram commutes

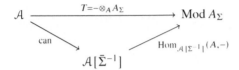

which equals the 'completion' of the above diagram for $\mathrm{proj}\, A$. Note that $\bar{T} = \mathrm{Hom}_{\mathcal{A}[\bar{\Sigma}^{-1}]}(A, -)$ is an equivalence.

In general, the universal localisation A_Σ is not a flat A-module.

Example 2.3.9. Let Σ be a set of morphisms between finitely generated projective A-modules such that $\mathrm{proj.dim}\,\mathrm{Coker}\,\sigma \le 1$ for all $\sigma \in \Sigma$. Then $\mathrm{Mod}\, A_\Sigma$ identifies with $\mathcal{C}^\perp$ where $\mathcal{C} = \{\mathrm{Ker}\,\sigma, \mathrm{Coker}\,\sigma \mid \sigma \in \Sigma\}$ and

$$\mathcal{C}^\perp = \{X \in \mathrm{Mod}\, A \mid \mathrm{Hom}_A(C, X) = 0 = \mathrm{Ext}^1_A(C, X) \text{ for all } C \in \mathcal{C}\}.$$

2.4 Commutative Noetherian Rings

We consider modules over commutative rings. There is a notion of support for modules which yields a classification of Serre subcategories for the category of noetherian modules. This extends to a classification of localising subcategories for the category of all modules provided the ring is noetherian. Also, we discuss injective and artinian modules.

Let A be a commutative ring. For the main results of this section we need to assume that A is noetherian.

Support of Modules

Let A be a commutative ring. The *spectrum* Spec A of A is the set of *prime ideals* $\mathfrak{p} \subseteq A$. A subset of Spec A is *Zariski closed* if it is of the form

$$\mathcal{V}(\mathfrak{a}) = \{\mathfrak{p} \in \text{Spec } A \mid \mathfrak{a} \subseteq \mathfrak{p}\}$$

for some ideal $\mathfrak{a}$ of A. A subset $\mathcal{V}$ of Spec A is *specialisation closed* if for any pair $\mathfrak{p} \subseteq \mathfrak{q}$ of prime ideals, $\mathfrak{p} \in \mathcal{V}$ implies $\mathfrak{q} \in \mathcal{V}$. For $\mathfrak{p} \in \text{Spec } A$ set $S = A \setminus \mathfrak{p}$ and denote by $A_{\mathfrak{p}} = A[S^{-1}]$ the localisation. Note that $X \mapsto X_{\mathfrak{p}} := X \otimes_A A_{\mathfrak{p}}$ yields an exact functor Mod $A \to$ Mod $A_{\mathfrak{p}}$. The *support* of an A-module X is the subset

$$\text{Supp } X = \{\mathfrak{p} \in \text{Spec } A \mid X_{\mathfrak{p}} \neq 0\}.$$

Observe that this is a specialisation closed subset of Spec A.

Lemma 2.4.1. *We have* Supp $A/\mathfrak{a} = \mathcal{V}(\mathfrak{a})$ *for each ideal $\mathfrak{a}$ of A.*

Proof Fix $\mathfrak{p} \in \text{Spec } A$ and let $S = A \setminus \mathfrak{p}$. Recall that for any A-module X, an element x/s in $S^{-1}X = X_{\mathfrak{p}}$ is zero if and only if there exists $t \in S$ such that $tx = 0$. Thus we have $(A/\mathfrak{a})_{\mathfrak{p}} = 0$ if and only if there exists $t \in S$ with $t(1 + \mathfrak{a}) = t + \mathfrak{a} = 0$ if and only if $\mathfrak{a} \not\subseteq \mathfrak{p}$. □

Lemma 2.4.2. *Let* $0 \to X' \to X \to X'' \to 0$ *be an exact sequence of A-modules. Then* Supp $X =$ Supp $X' \cup$ Supp X''.

Proof The sequence $0 \to X'_{\mathfrak{p}} \to X_{\mathfrak{p}} \to X''_{\mathfrak{p}} \to 0$ is exact for each $\mathfrak{p}$ in Spec A. □

Lemma 2.4.3. *Let $X = \sum_i X_i$ be an A-module, written as a sum of submodules X_i. Then* Supp $X = \bigcup_i$ Supp X_i.

Proof The assertion is clear if the sum $\sum_i X_i$ is direct, since

$$\bigoplus_i (X_i)_{\mathfrak{p}} = \left(\bigoplus_i X_i\right)_{\mathfrak{p}}.$$

As $X_i \subseteq X$ for all i one gets $\bigcup_i$ Supp $X_i \subseteq$ Supp X, from Lemma 2.4.2. On the other hand, $X = \sum_i X_i$ is a factor of $\bigoplus_i X_i$, so Supp $X \subseteq \bigcup_i$ Supp X_i. □

We write Ann X for the ideal of elements in A that annihilate X; it is the kernel of the natural homomorphism $A \to \text{End}_A(X)$.

Lemma 2.4.4. *We have* Supp $X \subseteq \mathcal{V}(\text{Ann } X)$, *with equality when X is in* mod A.

Proof   Write $X = \sum_i X_i$ as a sum of cyclic modules $X_i \cong A/\mathfrak{a}_i$. Then

$$\operatorname{Supp} X = \bigcup_i \operatorname{Supp} X_i = \bigcup_i \mathcal{V}(\mathfrak{a}_i) \subseteq \mathcal{V}\left(\bigcap_i \mathfrak{a}_i\right) = \mathcal{V}(\operatorname{Ann} X),$$

and equality holds if the sum is finite.   □

Lemma 2.4.5. *Let $X \neq 0$ be an A-module. If $\mathfrak{p}$ is maximal in the set of ideals which annihilate a non-zero element of X, then $\mathfrak{p}$ is prime.*

Proof   Suppose $0 \neq x \in X$ and $\mathfrak{p}x = 0$. Let $a, b \in A$ with $ab \in \mathfrak{p}$ and $a \notin \mathfrak{p}$. Then $(\mathfrak{p}, b)$ annihilates $ax \neq 0$, so the maximality of $\mathfrak{p}$ implies $b \in \mathfrak{p}$. Thus $\mathfrak{p}$ is prime.   □

Lemma 2.4.6. *Let $X \neq 0$ be a noetherian A-module. There exists a submodule of X which is isomorphic to $A/\mathfrak{p}$ for some prime ideal $\mathfrak{p}$.*

Proof   The ring $\bar{A} = A/(\operatorname{Ann} X)$ is noetherian. Thus the set of ideals of $\bar{A}$ annihilating a non-zero element has a maximal element. Now apply Lemma 2.4.5.   □

Lemma 2.4.7. *For each noetherian A-module X there exists a finite filtration*

$$0 = X_0 \subseteq X_1 \subseteq \cdots \subseteq X_n = X$$

such that each factor X_i/X_{i-1} is isomorphic to $A/\mathfrak{p}_i$ for some prime ideal $\mathfrak{p}_i$. In that case we have $\operatorname{Supp} X = \bigcup_i \mathcal{V}(\mathfrak{p}_i)$.

Proof   Repeated application of Lemma 2.4.6 yields a chain of submodules $0 = X_0 \subseteq X_1 \subseteq X_2 \subseteq \cdots$ of X such that each X_i/X_{i-1} is isomorphic to $A/\mathfrak{p}_i$ for some $\mathfrak{p}_i$. This chain stabilises since X is noetherian, and therefore $\bigcup_i X_i = X$.

The last assertion follows from Lemma 2.4.2 and Lemma 2.4.1.   □

For a class $\mathcal{C} \subseteq \operatorname{Mod} A$ we set

$$\operatorname{Supp} \mathcal{C} = \bigcup_{X \in \mathcal{C}} \operatorname{Supp} X.$$

Proposition 2.4.8. *Let A be a commutative noetherian ring. Then the assignment $\mathcal{C} \mapsto \operatorname{Supp} \mathcal{C}$ induces a bijection between*

   – *the set of Serre subcategories of $\operatorname{mod} A$, and*
   – *the set of specialisation closed subsets of $\operatorname{Spec} A$.*

Its inverse takes $\mathcal{V} \subseteq \operatorname{Spec} A$ to $\{X \in \operatorname{mod} A \mid \operatorname{Supp} X \subseteq \mathcal{V}\}$.

Proof Both maps are well defined by Lemma 2.4.2 and Lemma 2.4.4. If $\mathcal{V} \subseteq \operatorname{Spec} A$ is a specialisation closed subset, let $\mathcal{C}_\mathcal{V}$ denote the smallest Serre subcategory containing $\{A/\mathfrak{p} \mid \mathfrak{p} \in \mathcal{V}\}$. Then we have $\operatorname{Supp} \mathcal{C}_\mathcal{V} = \mathcal{V}$, by Lemma 2.4.1 and Lemma 2.4.2. Now let $\mathcal{C}$ be a Serre subcategory of mod A. Then

$$\operatorname{Supp} \mathcal{C} = \{\mathfrak{p} \in \operatorname{Spec} A \mid A/\mathfrak{p} \in \mathcal{C}\}$$

by Lemma 2.4.7. It follows that $\mathcal{C} = \mathcal{C}_\mathcal{V}$ for each Serre subcategory $\mathcal{C}$, where $\mathcal{V} = \operatorname{Supp} \mathcal{C}$. Thus $\operatorname{Supp} \mathcal{C}_1 = \operatorname{Supp} \mathcal{C}_2$ implies $\mathcal{C}_1 = \mathcal{C}_2$ for each pair $\mathcal{C}_1, \mathcal{C}_2$ of Serre subcategories. □

Corollary 2.4.9. *Let X and Y be in* mod A. *Then* $\operatorname{Supp} Y \subseteq \operatorname{Supp} X$ *if and only if Y belongs to the smallest Serre subcategory containing X.*

Proof With $\mathcal{C}$ denoting the smallest Serre subcategory containing X, there is an equality $\operatorname{Supp} \mathcal{C} = \operatorname{Supp} X$ by Lemma 2.4.2. Now apply Proposition 2.4.8. □

Corollary 2.4.10. *The assignment $\mathcal{C} \mapsto \operatorname{Supp} \mathcal{C}$ induces a bijection between*

- *the set of localising subcategories of* Mod A, *and*
- *the set of specialisation closed subsets of* Spec A.

Proof The proof is essentially the same as that of Proposition 2.4.8 if we observe that any A-module X is the sum $X = \sum_i X_i$ of its finitely generated submodules; see also Example 2.2.19. Note that X belongs to a localising subcategory $\mathcal{C}$ if and only if all X_i belong to $\mathcal{C}$. In addition, we use that $\operatorname{Supp} X = \bigcup_i \operatorname{Supp} X_i$; see Lemma 2.4.3. □

Injective Modules

Let A be a commutative noetherian ring. For an A-module X we say that $\mathfrak{p} \in \operatorname{Spec} A$ is *associated* to X if $A/\mathfrak{p}$ is isomorphic to a submodule of X. The set of associated primes is denoted by $\operatorname{Ass} X$.

Lemma 2.4.11. *We have* $\operatorname{Supp} X = \bigcup_{\mathfrak{p} \in \operatorname{Ass} X} \mathcal{V}(\mathfrak{p})$ *for each A-module X.*

Proof We have $\mathcal{V}(\mathfrak{p}) \subseteq \operatorname{Supp} X$ when $A/\mathfrak{p} \subseteq X$, by Lemma 2.4.1 and Lemma 2.4.2. For the other direction, let $\mathfrak{p} \in \operatorname{Supp} X$, and we need to show that $\mathfrak{p} \in \operatorname{Ass} X$ when $\mathfrak{p}$ is minimal in $\operatorname{Supp} X$. We may assume that X is finitely generated, and as in Lemma 2.4.7 we have submodules

$$0 = X_0 \subseteq X_1 \subseteq \cdots \subseteq X_n = X$$

such that each factor X_i/X_{i-1} is isomorphic to $A/\mathfrak{p}_i$ for some prime ideal $\mathfrak{p}_i$.

Choose $\mathfrak{p} = \mathfrak{p}_i$ to be minimal in $\{\mathfrak{p}_1, \dots, \mathfrak{p}_n\}$, and let i be minimal such that $\mathfrak{p} = \mathfrak{p}_i$. Pick $x \in X_i \setminus X_{i-1}$. Then $\mathfrak{p}_1 \cdots \mathfrak{p}_i \subseteq \operatorname{Ann} Ax \subseteq \mathfrak{p}$. Also $\mathfrak{p}_j \nsubseteq \mathfrak{p}$ for $j < i$ and therefore $\mathfrak{p}_1 \cdots \mathfrak{p}_{i-1} \nsubseteq \mathfrak{p}$. Pick $a \in \mathfrak{p}_1 \cdots \mathfrak{p}_{i-1} \setminus \mathfrak{p}$. Then $\operatorname{Ann} Aax = \mathfrak{p}$, and therefore $\mathfrak{p} \in \operatorname{Ass} X$. □

Lemma 2.4.12. *Let* $\mathfrak{p} \in \operatorname{Spec} A$. *Then* $\operatorname{Ass} A/\mathfrak{p} = \{\mathfrak{p}\}$.

Proof We have $A/\mathfrak{q} \cong X \subseteq A/\mathfrak{p}$ if and only if $\mathfrak{q}$ equals the ideal annihilating $a + \mathfrak{p}$ for some $a \in A$. Then $b \in \mathfrak{q}$ if and only if $ab \in \mathfrak{p}$ if and only if $b \in \mathfrak{p}$, since $\mathfrak{p}$ is prime. □

Recall that for an A-module X, $E(X)$ denotes an injective envelope.

Lemma 2.4.13. *We have* $\operatorname{Ass} E(X) = \operatorname{Ass} X$ *for every* A-*module* X.

Proof Clearly, $\operatorname{Ass} X \subseteq \operatorname{Ass} E(X)$. If $A/\mathfrak{p} \cong X' \subseteq E(X)$ for some $\mathfrak{p} \in \operatorname{Spec} A$, then $X' \cap X \neq 0$, and we have $A/\mathfrak{q} \cong X'' \subseteq X' \cap X$ for some $\mathfrak{q} \in \operatorname{Spec} A$, by Lemma 2.4.6. This implies $\mathfrak{p} = \mathfrak{q}$, by Lemma 2.4.12. □

Corollary 2.4.14. *Let* X *be an* A-*module. Then* $\operatorname{Supp} X = \operatorname{Supp} E(X)$. *Therefore localising subcategories of* $\operatorname{Mod} A$ *are closed under injective envelopes.*

Proof We have $\operatorname{Ass} E(X) = \operatorname{Ass} X$ by Lemma 2.4.13, and then Lemma 2.4.11 implies that $\operatorname{Supp} E(X) = \operatorname{Supp} X$. If $\mathcal{C} \subseteq \operatorname{Mod} A$ is localising and $X \in \mathcal{C}$, then $E(X) \in \mathcal{C}$ by Corollary 2.4.10. □

Corollary 2.4.15. *The assignments* $\mathfrak{p} \mapsto E(A/\mathfrak{p})$ *and* $X \mapsto \operatorname{Ass} X$ *yield mutually inverse bijections between* $\operatorname{Spec} A$ *and* $\operatorname{Sp}(\operatorname{Mod} A)$.

Proof We have $\operatorname{Ass}(E(A/\mathfrak{p})) = \{\mathfrak{p}\}$ by Lemma 2.4.12 and Lemma 2.4.13. On the other hand, if X is indecomposable injective, then $\operatorname{Ass} X \neq \varnothing$ by Lemma 2.4.6. Clearly, $X \cong E(A/\mathfrak{p})$ when $\mathfrak{p} \in \operatorname{Ass} X$. □

For a subset $\mathcal{U} \subseteq \operatorname{Spec} A$ we set

$$\operatorname{Inj}_{\mathcal{U}} A = \{X \in \operatorname{Inj} A \mid \operatorname{Ass} X \subseteq \mathcal{U}\}.$$

Corollary 2.4.16. *Let* $\mathcal{V} \subseteq \operatorname{Spec} A$ *be specialisation closed and set* $\mathcal{W} = \operatorname{Spec} A \setminus \mathcal{V}$. *Then we have for* $\operatorname{Inj} A$ *a split torsion pair* $(\operatorname{Inj}_{\mathcal{V}} A, \operatorname{Inj}_{\mathcal{W}} A)$.

Proof Consider the localising subcategory

$$\mathcal{C} = \{X \in \operatorname{Mod} A \mid \operatorname{Supp} X \subseteq \mathcal{V}\};$$

see Corollary 2.4.10. Because $\mathcal{C}$ is closed under injective envelopes by Corollary 2.4.14, we have

$$\operatorname{Inj}_{\mathcal{V}} A = \mathcal{C} \cap \operatorname{Inj} A \qquad \text{and} \qquad \operatorname{Inj}_{\mathcal{W}} A = \mathcal{C}^{\perp} \cap \operatorname{Inj} A. \qquad \square$$

Localising subcategories of module categories over non-commutative rings are usually not closed under injective envelopes.

Example 2.4.17. Let k be a field and $\Lambda = \left[\begin{smallmatrix} k & 0 \\ k & k \end{smallmatrix}\right]$. Consider the simple Λ-module $S = e\Lambda$ where $e = \left[\begin{smallmatrix} 1 & 0 \\ 0 & 0 \end{smallmatrix}\right]$. The localising subcategory generated by S consists of all direct sums of copies of S since $\mathrm{Ext}^1_\Lambda(S,S) = 0$; so it does not contain $E(S) = \mathrm{Hom}_k(\Lambda e, k)$.

Artinian Modules

Let A be a commutative ring and let $\mathfrak{a}$ be an ideal. We set $\mathrm{gr}(A)_n = \mathfrak{a}^n/\mathfrak{a}^{n+1}$ for $n \in \mathbb{Z}$, where $\mathfrak{a}^n = A$ for all $n \leq 0$. The *associated graded ring*

$$\mathrm{gr}(A) = \bigoplus_{n \in \mathbb{Z}} \mathrm{gr}(A)_n$$

is $\mathbb{Z}$-graded with multiplication induced by that in A.

Lemma 2.4.18. *If the ideal $\mathfrak{a}$ is finitely generated over A, then $\mathrm{gr}(A)$ is a finitely generated $A/\mathfrak{a}$-algebra.*

Proof Let $x_1, \ldots, x_n$ generate $\mathfrak{a}$. Then $\mathrm{gr}(A) = (A/\mathfrak{a})[\bar{x}_i, \ldots \bar{x}_n]$, where $\bar{x}_i = x_i + \mathfrak{a}^2$, and $\mathrm{gr}(A)$ is a quotient of the polynomial ring $(A/\mathfrak{a})[X_i, \ldots X_n]$ as a graded ring. $\square$

For an A-module X and $m \in \mathbb{Z}$ let X_m denote the submodule of elements annihilated by $\mathfrak{a}^m$. We set $\mathrm{gr}^{\mathfrak{a}}(X)_m = X_{-m+1}/X_{-m}$ and obtain a graded $\mathrm{gr}(A)$-module

$$\mathrm{gr}^{\mathfrak{a}}(X) = \bigoplus_{n \in \mathbb{Z}} \mathrm{gr}^{\mathfrak{a}}(X)_n.$$

The assignment $(x, a) \mapsto xa$ yields an $A/\mathfrak{a}$-bilinear map

$$\mathrm{gr}^{\mathfrak{a}}(X)_m \times \mathrm{gr}(A)_n \longrightarrow \mathrm{gr}^{\mathfrak{a}}(X)_{m+n},$$

which induces a homomorphism

$$\mu_X \colon \mathrm{gr}^{\mathfrak{a}}(X) \longrightarrow \mathrm{Hom}_{A/\mathfrak{a}}(\mathrm{gr}(A), X_1)$$

of graded $\mathrm{gr}(A)$-modules since $\mathrm{gr}^{\mathfrak{a}}(X)_0 = X_1$.

For each submodule $U \subseteq X$ let $\mathfrak{a}_U$ denote the graded ideal of $\mathrm{gr}(A)$ consisting in degree n of elements $a \in \mathrm{gr}(A)_n$ such that $xa = 0$ for all $x \in ((U \cap X_{n+1}) + X_n)/X_n$.

Lemma 2.4.19. *Let A be a commutative noetherian ring and let $\mathfrak{m}$ be a maximal ideal. Then the injective envelope $E(A/\mathfrak{m})$ is artinian over A.*

Proof Set $X = E(A/\mathfrak{m})$. We consider $\mathrm{gr}(A)$ for $\mathfrak{a} = \mathfrak{m}$ and $\mathrm{gr}^{\mathfrak{m}}(X)$. First observe that $X = \bigcup_{n \geq 0} X_n$. To see this, let $U \subseteq X$ be a finitely generated submodule. Then we have $\mathrm{Supp}\, U \subseteq \mathrm{Supp}\, X = \{\mathfrak{m}\}$ by Corollary 2.4.14. Thus U admits a finite filtration with factors isomorphic to $A/\mathfrak{m}$ by Lemma 2.4.7. This means U is annihilated by $\mathfrak{m}^n$ for some $n \geq 0$, so $U \subseteq X_n$.

Our first observation implies that μ_X is an isomorphism. For submodules U, V of X, it follows that $\mathfrak{m}_U = \mathfrak{m}_V$ implies

$$(U \cap X_{n+1}) + X_n = (V \cap X_{n+1}) + X_n$$

for all n. Thus $U \cap X_{n+1} = V \cap X_{n+1}$ for all n by induction, and therefore $U = V$. Clearly, $U \subseteq V$ implies $\mathfrak{m}_V \subseteq \mathfrak{m}_U$. Thus X is artinian, because $\mathrm{gr}(A)$ is noetherian by Lemma 2.4.18. □

Proposition 2.4.20. *For a module X over a commutative noetherian ring the following are equivalent.*

(1) *The module X is artinian.*
(2) *The module X is a union of finite length submodules and the socle of X has finite length.*
(3) *The socle of X has finite length and all prime ideals in $\mathrm{Supp}\, X$ are maximal.*

Proof (1) $\Rightarrow$ (2): The module X is a union of its finitely generated submodules, which are both artinian and noetherian, and therefore of finite length. A semisimple artinian module has finite length. Thus $\mathrm{soc}\, X$ has finite length.

(2) $\Rightarrow$ (3): This follows from Lemma 2.4.3, since the support of a finite length module consists of prime ideals which are maximal.

(3) $\Rightarrow$ (1): We have $\mathrm{Supp}\, E(X) = \mathrm{Supp}\, X$ by Corollary 2.4.14. Then Lemma 2.4.19 implies that $E(X)$ is artinian. Thus X is artinian. □

Graded Rings and Modules

The preceding results about modules over commutative noetherian rings generalise to graded modules over graded rings. We sketch the appropriate setting.

Fix an abelian *grading group* G and let A be a G-*graded ring*. Thus A is a ring together with a decomposition of the underlying abelian group

$$A = \bigoplus_{g \in G} A_g$$

such that the multiplication satisfies $A_g A_h \subseteq A_{g+h}$ for all $g, h \in G$. An element in A is called *homogeneous* of *degree* g if it belongs to A_g for some $g \in G$.

We consider graded A-modules and homogeneous ideals of A. An A-module M is *G-graded* if the underlying abelian group admits a decomposition

$$M = \bigoplus_{g \in G} M_g$$

such that the multiplication satisfies $M_g A_h \subseteq M_{g+h}$ for all $g, h \in G$. We write GrMod A for the category of graded A-modules (with degree zero morphisms) and grmod A for the full subcategory of finitely presented modules. Later on we will consider the full subcategory grproj A of finitely generated projective modules and the projectively stable category $\overline{\text{grmod}}\, A$.

Now suppose that G is endowed with a symmetric bilinear form

$$(-,-) \colon G \times G \longrightarrow \mathbb{Z}/2.$$

A typical example is $G = \mathbb{Z}$ with $\mathbb{Z} \times \mathbb{Z} \to \mathbb{Z}/2$ the multiplication map modulo two. We say that A is *G-graded commutative* when $xy = (-1)^{(g,h)} yx$ for all homogeneous $x \in A_g$, $y \in A_h$. A homogeneous element in A is *even* if it belongs to A_g for some $g \in G$ satisfying $(g, h) = 0$ for all $h \in G$.

Let us fix such a G-graded commutative ring A. Note that all homogeneous ideals are automatically two-sided. The graded localisation of A at a multiplicative set consisting of even (and therefore central) homogeneous elements is the obvious one and enjoys the usual properties; in particular, it is again a G-graded commutative ring. Similarly, one localises any graded A-module at such a multiplicative set. For instance, when $\mathfrak{p}$ is a homogeneous prime ideal of A and M is a graded A-module, then $M_{\mathfrak{p}}$ is the localisation of M with respect to the multiplicative set of even homogeneous elements in $A \setminus \mathfrak{p}$.

Suppose now that A is *noetherian* as a G-graded ring, that is, the ascending chain condition holds for homogeneous ideals of A. Then all results of this section carry over to the category of graded A-modules. However, it is necessary to twist. For any graded A-module M and $g \in G$, the *twisted module* $M(g)$ is the A-module M with the new grading defined by $M(g)_h = M_{g+h}$ for each $h \in G$. For instance, in Lemma 2.4.6 one shows that each graded non-zero module has a submodule of the form $(A/\mathfrak{p})(g)$ for some homogeneous prime ideal $\mathfrak{p}$ and some $g \in G$. This affects all subsequent statements. The following is then the analogue of Proposition 2.4.8.

Proposition 2.4.21. *The assignment* $\mathcal{C} \mapsto \operatorname{Supp} \mathcal{C}$ *induces a bijection between*

- *the set of Serre subcategories of* grmod A *that are closed under twists, and*
- *the set of specialisation closed sets of homogeneous prime ideals of A.* $\square$

Example 2.4.22. Let k be a field and $\mathbb{X} = \mathbb{P}^1_k$ the projective line with homogeneous coordinate ring $S = k[x_0, x_1]$. Then a theorem of Serre [188] provides the following localisation sequence

$$\text{GrMod}_0\, S \ \underset{\longleftarrow}{\overset{\longrightarrow}{\rightarrowtail}}\ \text{GrMod}\, S \ \underset{\Gamma_*(\mathbb{X},-)}{\overset{\longrightarrow}{\longleftarrow}}\ \text{Qcoh}\, \mathbb{X}$$

where $\text{GrMod}_0\, S$ denotes the category of torsion modules. Note that $\text{GrMod}_0\, S$ is the localising subcategory corresponding to the category $\text{grmod}_0\, S$ of finite length modules. These are precisely the modules with support only containing the unique maximal homogeneous ideal of positive degree elements. The fact that the subcategory $\text{GrMod}_0\, S$ is not closed under products leads to an example showing that products in $\text{Qcoh}\, \mathbb{X}$ need not be exact.

For each $n \geq 0$, we have a canonical map

$$\pi_n \colon \mathcal{O}(-n) \otimes_k \text{Hom}_{\mathbb{X}}(\mathcal{O}(-n), \mathcal{O}) \longrightarrow \mathcal{O}$$

which is an epimorphism in $\text{Qcoh}\, \mathbb{X}$. We claim that the product

$$\pi \colon \prod_{n \geq 0}\left(\mathcal{O}(-n) \otimes_k \text{Hom}_{\mathbb{X}}(\mathcal{O}(-n), \mathcal{O})\right) \longrightarrow \prod_{n \geq 0} \mathcal{O}$$

is not an epimorphism. Taking graded global sections gives for each $n \geq 0$ the multiplication map

$$\Gamma_*(\mathbb{X}, \pi_n) \colon S(-n) \otimes_k S_n \longrightarrow S$$

which is a morphism of graded S-modules with cokernel of finite length. However, the cokernel of

$$\Gamma_*(\mathbb{X}, \pi) = \prod_{n \geq 0} \Gamma_*(\mathbb{X}, \pi_n)$$

is not a torsion module. The left adjoint of $\Gamma_*(\mathbb{X}, -)$ is exact and takes $\Gamma_*(\mathbb{X}, \pi)$ to π. It follows that the cokernel of π is non-zero, because the left adjoint of $\Gamma_*(\mathbb{X}, -)$ annihilates exactly those S-modules which are torsion.

2.5 Grothendieck Categories

We study the basic properties of Grothendieck categories. It is shown that an abelian category is a Grothendieck category if and only if it is the localisation of a module category. From this we deduce that objects in a Grothendieck category admit injective envelopes. Also, it follows that any Grothendieck category is a locally presentable category. This means that every object is an α-filtered colimit of α-presentable objects for some regular cardinal α. Finally, we characterise the coherent functors for any locally presentable category.

The Embedding Theorem

Let $\mathcal{A}$ be an abelian category and suppose that $\mathcal{A}$ admits arbitrary coproducts. We fix an object $C \in \mathcal{A}$ and set $\Lambda = \mathrm{End}(C)$. Then the functor

$$H : \mathcal{A} \longrightarrow \mathrm{Mod}\,\Lambda, \quad X \mapsto \mathrm{Hom}(C, X)$$

admits a left adjoint $T \colon \mathrm{Mod}\,\Lambda \to \mathcal{A}$. We obtain this by first extending the equivalence $\mathrm{add}\,\Lambda \to \mathrm{add}\,C$ to a functor $\tilde{T} \colon \mathrm{Add}\,\Lambda \to \mathrm{Add}\,C$ preserving coproducts. Then extend $\tilde{T}$ to a right exact functor $\mathrm{Mod}\,\Lambda \to \mathcal{A}$.

Recall that C is a generator for $\mathcal{A}$ if for every object $X \in \mathcal{A}$ the canonical morphism $\coprod_{\phi \in \mathrm{Hom}(C,X)} C \to X$ is an epimorphism.

Lemma 2.5.1. *Suppose that filtered colimits in $\mathcal{A}$ are exact and that C is a generator. If $\phi \colon X \to H(Y)$ is a monomorphism in $\mathrm{Mod}\,\Lambda$, then the adjoint morphism $\psi \colon T(X) \to Y$ is a monomorphism.*

Proof Suppose $K = \mathrm{Ker}\,\psi \neq 0$. Choose an epimorphism $\Lambda^{(I)} \to X$ which yields an epimorphism $\pi \colon T(\Lambda^{(I)}) \to T(X)$. Write $\Lambda^{(I)} = \bigcup_{J \subseteq I} \Lambda^{J}$ as filtered colimit, where $J \subseteq I$ runs through all finite subsets. This implies $T(\Lambda^{(I)}) = \bigcup_{J \subseteq I} T(\Lambda^{J})$ and therefore

$$\bigcup_{J \subseteq I} \left(\pi^{-1}(K) \cap T(\Lambda^{J}) \right) = \pi^{-1}(K) \neq 0.$$

Thus we obtain a non-zero morphism

$$\tau \colon T(\Lambda) \to \pi^{-1}(K) \cap T(\Lambda^{J}) \hookrightarrow T(\Lambda^{J}) \to T(X)$$

such that $\psi\tau = 0$, since $C = T(\Lambda)$ is a generator. Note that $\tau = T(\sigma)$ for some $\sigma \colon \Lambda \to X$ which yields the following commutative diagram.

We have $\phi\sigma = 0$ and this implies $\sigma = 0$ since ϕ is a monomorphism. This is a contradiction since $T(\sigma) \neq 0$, and therefore $\mathrm{Ker}\,\psi = 0$.   □

The following is known as the *Gabriel–Popescu theorem*.

Theorem 2.5.2 (Gabriel–Popescu). *Let $\mathcal{A}$ be a category such that filtered colimits are exact. Given C, H, and T as above, the following are equivalent.*

(1) *C is a generator for $\mathcal{A}$.*
(2) *H is fully faithful.*
(3) *T is exact and induces an equivalence $(\mathrm{Mod}\,\Lambda)/(\mathrm{Ker}\,T) \xrightarrow{\sim} \mathcal{A}$.*

Proof (1) ⇔ (2): Clearly, C is a generator when H is faithful. For the converse suppose that C is a generator. For $X \in \mathcal{A}$ consider the counit $\varepsilon_X : TH(X) \to X$. Then we need to show that this is invertible for all $X \in \mathcal{A}$; see Proposition 1.1.3. Each morphism $C \to X$ factors through ε_X since e_C is invertible, and therefore ε_X is an epimorphism. On the other hand, ε_X is adjoint to id: $H(X) \to H(X)$ and therefore a monomorphism by Lemma 2.5.1.

(1) & (2) ⇒ (3): We show that T is exact. Then it follows from Proposition 2.2.11 that T induces an equivalence $(\operatorname{Mod}\Lambda)/(\operatorname{Ker} T) \xrightarrow{\sim} \mathcal{A}$.

For the exactness of T we apply the criterion from Corollary 2.1.16. Thus we need to show that for each exact sequence $X \to Y \to Z$ of projective Λ-modules, the sequence $T(X) \to T(Y) \to T(Z)$ is exact. To show this, it suffices to prove that for each exact sequence $0 \to X \to Y \to Z \to 0$ of Λ-modules, the sequence $0 \to T(X) \to T(Y) \to T(Z) \to 0$ is exact provided that Y is projective. Moreover, it suffices to show that $T(X) \to T(Y)$ is a monomorphism since T is right exact. We may assume that $Y = \Lambda^{(I)}$ is free and write this as the filtered colimit $Y = \operatorname{colim} Y_J$, where $Y_J = \Lambda^J$ and $J \subseteq I$ runs through all finite subsets. Then $X \to Y$ is the filtered colimit of monomorphisms $X_J \to Y_J$, where $X_J = X \cap Y_J$. The morphism $T(X_J) \to T(Y_J) = C^J$ is adjoint to $X_J \to Y_J = H(C^J)$ and therefore a monomorphism by Lemma 2.5.1. It remains to note that T preserves colimits and that filtered colimits in $\mathcal{A}$ are exact. Thus $T(X) \to T(Y)$ is a monomorphism since it identifies with the filtered colimit of monomorphisms $T(X_J) \to T(Y_J)$.

(3) ⇒ (2): See Proposition 2.2.11. □

Corollary 2.5.3. *An abelian category is a Grothendieck category if and only if it is the localisation of a module category, so of the form* $(\operatorname{Mod}\Lambda)/\mathcal{C}$ *for some ring* Λ *and a localising subcategory* $\mathcal{C} \subseteq \operatorname{Mod}\Lambda$.

Proof Combine Theorem 2.5.2 with Proposition 2.2.16. □

Injective Envelopes

We are now able to establish injective envelopes in Grothendieck categories.

Corollary 2.5.4. *A Grothendieck category admits arbitrary products, and every object admits an injective envelope.*

Proof Fix a Grothendieck category $\mathcal{A}$. We apply the above Theorem 2.5.2 and identify $\mathcal{A}$ with $(\operatorname{Ker} T)^{\perp} \subseteq \operatorname{Mod}\Lambda$. The category $\operatorname{Mod}\Lambda$ has arbitrary products, and $(\operatorname{Ker} T)^{\perp}$ is closed under products. From this the first assertion follows. The existence of injective envelopes in $\mathcal{A}$ follows from Corollary 2.2.15, once we have shown that $\operatorname{Mod}\Lambda$ has injective envelopes.

We proceed in two steps. Set $\mathcal{A} = \mathrm{Mod}\,\Lambda$ and fix an object $X \in \mathcal{A}$.

(1) *The object X admits an embedding into an injective object.* It suffices to find an injective cogenerator, say E, because then $X \to \prod_{\phi \in \mathrm{Hom}(X,E)} E$ is a monomorphism.

If $\Lambda = \mathbb{Z}$, then

$$\mathbb{Q}/\mathbb{Z} \cong \coprod_{p \text{ prime}} \mathbb{Z}_{p^\infty} \cong E\left(\coprod_{p \text{ prime}} \mathbb{Z}/(p) \right)$$

is an injective cogenerator. This can be shown using the notion of a divisible module. For an arbitrary ring Λ, we use restriction of scalars via the canonical homomorphism $\mathbb{Z} \to \Lambda$. So $\mathrm{Hom}_{\mathbb{Z}}(\Lambda, \mathbb{Q}/\mathbb{Z})$ is an injective cogenerator since

$$\mathrm{Hom}_{\Lambda}(-, \mathrm{Hom}_{\mathbb{Z}}(\Lambda, \mathbb{Q}/\mathbb{Z})) \cong \mathrm{Hom}_{\mathbb{Z}}(-, \mathbb{Q}/\mathbb{Z})$$

by adjunction.

(2) *The object X admits an essential embedding into an injective object.* Let $\phi\colon X \to E$ be a monomorphism such that E is injective. Consider the partially ordered set of subobjects $\{E' \subseteq E \mid \mathrm{Im}\,\phi \hookrightarrow E' \text{ essential}\}$. Using the fact that filtered colimits are exact, it follows that this has a maximal element by Zorn's lemma, say E_0. It is easily checked that $X \to E_0$ is an injective envelope. In fact, choose a maximal subobject $E'' \subseteq E$ such that $E'' \cap E_0 = 0$, using again Zorn's lemma. Then the composite $E_0 \hookrightarrow E \twoheadrightarrow E/E''$ is an essential monomorphism and therefore an isomorphism by the maximality of E_0. Thus the inclusion $E_0 \hookrightarrow E$ is split and E_0 is injective. $\square$

Corollary 2.5.5. *A Grothendieck category admits an injective cogenerator.*

Proof Fix a generator C and choose $E = \prod_{C' \subseteq C} E(C/C')$ where $C' \subseteq C$ runs through all subjects. It follows that any non-zero morphism $C \to X$ can be extended to a non-zero morphism $X \to E$. $\square$

Decompositions into Indecomposables

We provide a brief discussion about decompositions of objects into indecomposable objects. In particular, we include a result about the uniqueness of such decompositions into indecomposable objects with local endomorphism rings.

Recall that an object X is *indecomposable* if $X \neq 0$ and if $X = X_1 \oplus X_2$ implies $X_1 = 0$ or $X_2 = 0$.

A non-zero object X is called *uniform* provided any two non-zero subobjects intersect non-trivially. Clearly, X is uniform if and only if its injective envelope $E(X)$ is indecomposable. An object X is called *super-decomposable* if X has no indecomposable direct summands. Note that $E(X)$ is super-decomposable if and only if X has no uniform subobjects. This is clear since a direct summand E of $E(X)$ is the injective envelope of the intersection $E \cap X$.

Example 2.5.6. Let $\Lambda = k\langle x, y \rangle$ be the free algebra on two generators. Then the Λ-module $E(\Lambda)$ is super-decomposable.

To see this, observe that if $a \in \Lambda$, then $ax\Lambda \cap ay\Lambda = 0$. Thus Λ has no uniform right ideals, and hence $E(\Lambda)$ is super-decomposable.

A ring is called *local* if all non-invertible elements form a proper ideal. Thus an object is indecomposable if its endomorphism ring is local.

Lemma 2.5.7. *If X is an indecomposable injective object in a Grothendieck category, then* $\mathrm{End}(X)$ *is a local ring.*

Proof We need to show that if ϕ and ψ in $\mathrm{End}(X)$ are non-invertible, then $\phi + \psi$ is non-invertible. If ϕ or ψ is a monomorphism, then it splits. Thus we need to show that $\mathrm{Ker}\,\phi \neq 0$ and $\mathrm{Ker}\,\psi \neq 0$ implies $\mathrm{Ker}(\phi + \psi) \neq 0$. But this is clear, since X is the injective envelope of any non-zero subobject. Thus

$$0 \neq (\mathrm{Ker}\,\phi) \cap (\mathrm{Ker}\,\psi) \subseteq \mathrm{Ker}(\phi + \psi). \qquad \square$$

The following is known as *Krull–Remak–Schmidt–Azumaya theorem*.

Theorem 2.5.8 (Krull–Remak–Schmidt–Azumaya). *Let X be an object in a Grothendieck category with decompositions $X = \coprod_{i \in I} X_i$ and $X = \coprod_{j \in J} Y_j$ such that $\mathrm{End}(X_i)$ is a local ring for all i and Y_j is indecomposable for all j. Then there is a bijection $\sigma \colon I \xrightarrow{\sim} J$ such that $X_i \cong Y_{\sigma(i)}$ for all $i \in I$.*

Proof See for example [156, Section 4.8]. $\qquad \square$

The appropriate tool for studying decompositions of objects in a Grothendieck category is its spectral category. Let $\mathcal{A}$ be a Grothendieck category and denote by Ess the class of essential monomorphisms in $\mathcal{A}$. This class admits a calculus of right fractions and is closed under coproducts. We obtain the canonical functor

$$P \colon \mathcal{A} \longrightarrow \mathcal{A}[\mathrm{Ess}^{-1}]$$

and call $\mathcal{A}[\mathrm{Ess}^{-1}]$ the *spectral category* of $\mathcal{A}$. It is not difficult to show that this is again a Grothendieck category which is split exact [82, Satz 1.3].

We have the following explicit description of the spectral category.

Proposition 2.5.9. *The canonical functor $\mathcal{A} \to \mathcal{A}[\mathrm{Ess}^{-1}]$ restricted to $\mathrm{Inj}\,\mathcal{A}$ induces an equivalence $(\mathrm{Inj}\,\mathcal{A})/\mathrm{Rad}(\mathrm{Inj}\,\mathcal{A}) \xrightarrow{\sim} \mathcal{A}[\mathrm{Ess}^{-1}]$.*

The assertion says that P induces for $X, Y \in \operatorname{Inj} \mathcal{A}$ an isomorphism

$$\operatorname{Hom}_{\mathcal{A}}(X, Y)/\operatorname{Rad}_{\mathcal{A}}(X, Y) \xrightarrow{\sim} \operatorname{Hom}_{\mathcal{A}[\operatorname{Ess}^{-1}]}(X, Y).$$

Proof The functor P identifies each object X with its injective envelope $E(X)$. Thus the restriction $P|_{\operatorname{Inj} \mathcal{A}}$ is essentially surjective. This restriction is also surjective on morphisms, because each morphism in $\mathcal{A}[\operatorname{Ess}^{-1}]$ is given by a right fraction $X \xleftarrow{\sigma} X' \xrightarrow{\alpha} Y$ (Lemma 1.2.1). Indeed, α extends to a morphisms $\bar{\alpha}: X \to Y$ when Y is injective, and then the right fraction equals $P(\bar{\alpha})$. Finally, we apply Proposition 2.1.23 and see that P annihilates a morphism ϕ in $\operatorname{Inj} \mathcal{A}$ if and only if ϕ is radical, since P is left exact.   □

Locally Presentable Categories

A cardinal α is called *regular* if α is not the sum of fewer than α cardinals, all smaller than α. For example, $\aleph_0$ is regular because the sum of finitely many finite cardinals is finite. Also, the successor κ^+ of every infinite cardinal κ is regular. In particular, there are arbitrarily large regular cardinals.

Let α be a regular cardinal. A category $\mathcal{J}$ is called *α-filtered* if

(Fil1) the category is non-empty,
(Fil2) for each family $(x_i)_{i \in I}$ of fewer than α objects there is an object x with morphisms $x_i \to x$ for all i, and
(Fil3) for each family $(\phi_i: x \to y)_{i \in I}$ of fewer than α morphisms there exists a morphism $\psi: y \to z$ such that $\psi \phi_i = \psi \phi_j$ for all i, j.

An *α-filtered colimit* is the colimit of a functor $\mathcal{J} \to \mathcal{C}$ such that the category $\mathcal{J}$ is α-filtered. An *α-small colimit* is the colimit of a functor $\mathcal{J} \to \mathcal{C}$ such that the category $\mathcal{J}$ has fewer than α morphisms.

We record a characteristic property of α-filtered categories; it is well known when $\alpha = \aleph_0$ and says that α-filtered colimits in the category of sets commute with α-small limits.

Lemma 2.5.10. *For a regular cardinal α let $F: \mathcal{J} \times \mathcal{J} \to \operatorname{Set}$ be a functor such that $\mathcal{J}$ is α-filtered and $\mathcal{J}$ is α-small. Then the canonical map*

$$\operatorname{colim}_i \lim_j F(i, j) \longrightarrow \lim_j \operatorname{colim}_i F(i, j)$$

is bijective.

Proof Adapt the proof of the case $\alpha = \aleph_0$; see [142, Section IX.2].   □

Now fix an additive category $\mathcal{A}$ and suppose that $\mathcal{A}$ is cocomplete. An object

$X \in \mathcal{A}$ is called *α-presentable* if $\mathrm{Hom}(X, -)$ preserves α-filtered colimits, that is, for every α-filtered colimit $\mathrm{colim}_{i \in \mathcal{I}} Y_i$ in $\mathcal{A}$ the canonical map

$$\mathrm{colim}_i \mathrm{Hom}(X, Y_i) \longrightarrow \mathrm{Hom}(X, \mathrm{colim}_i Y_i)$$

is bijective. Let $\mathcal{A}^\alpha$ denote the full subcategory of α-presentable objects.

Lemma 2.5.11. *The α-presentable objects are closed under taking α-small colimits.*

Proof Let $\mathrm{colim}_{i \in \mathcal{I}} X_i$ be an α-small colimit of α-presentable objects X_i. For an α-filtered colimit $\mathrm{colim}_{j \in \mathcal{I}} Y_j$ we compute

$$\mathrm{colim}_j \mathrm{Hom}(\mathrm{colim}_i X_i, Y_j) \cong \mathrm{colim}_j \lim_i \mathrm{Hom}(X_i, Y_j)$$

$$\cong \lim_i \mathrm{colim}_j \mathrm{Hom}(X_i, Y_j)$$

$$\cong \mathrm{Hom}(\mathrm{colim}_i X_i, \mathrm{colim}_j Y_j)$$

where the second isomorphism follows from the fact that α-small limits commute with α-filtered colimits in the category of sets, by Lemma 2.5.10. □

Lemma 2.5.12. *Let $\alpha \le \beta$ be regular cardinals. Then any colimit of α-presentable objects can be written canonically as a β-filtered colimit of β-presentable objects, which are β-small colimits of α-presentable objects.*

Proof Let $X \colon \mathcal{I} \to \mathcal{A}$ be a functor such that $X(i)$ is α-presentable for each $i \in \mathcal{I}$. Consider the set $\binom{\mathcal{I}}{\beta}$ of all subcategories $\mathcal{J} \subseteq \mathcal{I}$ having fewer than β morphisms. This set is partially ordered by inclusion and can be viewed as a category, which is β-filtered. For each $\mathcal{J} \subseteq \mathcal{I}$ set $X(\mathcal{J}) = \mathrm{colim} X|_{\mathcal{J}}$; this induces a functor $X_\beta \colon \binom{\mathcal{I}}{\beta} \to \mathcal{A}$. Then it is straightforward to check that the morphisms $X(\mathcal{J}) \to \mathrm{colim} X$ induce an isomorphism $\phi \colon \mathrm{colim} X_\beta \overset{\sim}{\to} \mathrm{colim} X$. In fact, for each $i \in \mathcal{I}$ there is a canonical morphism $X(i) \to \mathrm{colim} X_\beta$. These morphisms are compatible and induce the inverse of ϕ. It remains to observe that each $X(\mathcal{J})$ is β-presentable by Lemma 2.5.11. □

A cocomplete category $\mathcal{A}$ is called *locally α-presentable* if the category $\mathcal{A}^\alpha$ is essentially small and each object is an α-filtered colimit of α-presentable objects. The category is *locally presentable* if it is locally α-presentable for some regular cardinal α.

Lemma 2.5.13. *Let $\mathcal{A}$ be a locally presentable category. Then*

$$\mathcal{A} = \bigcup_\alpha \mathcal{A}^\alpha$$

where α runs through all regular cardinals.

If $\mathcal{A}$ is locally α-presentable, then $\mathcal{A}$ is locally β-presentable for all $\beta \geq \alpha$. Moreover, $\mathcal{A}^\beta$ equals the closure of $\mathcal{A}^\alpha$ under β-small colimits.

Proof Let $X \in \mathcal{A}$ be the α-filtered colimit of α-presentable objects, given by a functor $\mathfrak{I} \to \mathcal{A}$. Choose a regular cardinal $\beta \geq \alpha$ such that $\mathfrak{I}$ has fewer than β morphisms. Then X is β-presentable by Lemma 2.5.11.

Let $\mathcal{A}$ be locally α-presentable. Then every object is a β-filtered colimit of β-presentable objects, by Lemma 2.5.12. In fact, we can choose β-presentable objects that are β-small colimits of α-presentable objects. In particular, every β-presentable object is of this form. □

Next we consider more specifically the category $\mathcal{A} = \mathrm{Mod}\,\Lambda$ for a ring Λ.

Lemma 2.5.14. *Let Λ be a ring, α a regular cardinal, and $n \geq 0$ an integer. If a Λ-module X admits a free presentation*

$$\Lambda^{(\alpha_{n+1})} \longrightarrow \Lambda^{(\alpha_n)} \longrightarrow \cdots \longrightarrow \Lambda^{(\alpha_0)} \longrightarrow X \longrightarrow 0$$

with $\alpha_p < \alpha$ for $0 \leq p \leq n+1$, then $\mathrm{Ext}_\Lambda^n(X, -)$ preserves α-filtered colimits.

Proof We view the presentation of X as a complex and have

$$\mathrm{Ext}_\Lambda^n(X, -) \cong H^n \mathrm{Hom}_\Lambda(\Lambda^{(\alpha_p)}, -).$$

For an α-filtered colimit $\mathrm{colim}_{i \in \mathfrak{I}} Y_i$ of Λ-modules we compute

$$\begin{aligned}
\mathrm{colim}_i \mathrm{Ext}_\Lambda^n(X, Y_i) &\cong \mathrm{colim}_i H^n \mathrm{Hom}_\Lambda(\Lambda^{(\alpha_p)}, Y_i) \\
&\cong H^n \mathrm{colim}_i \mathrm{Hom}_\Lambda(\Lambda^{(\alpha_p)}, Y_i) \\
&\cong H^n \mathrm{Hom}_\Lambda(\Lambda^{(\alpha_p)}, \mathrm{colim}_i Y_i) \\
&\cong \mathrm{Ext}_\Lambda^n(X, \mathrm{colim}_i Y_i).
\end{aligned}$$

The second isomorphism follows from the fact that taking α-filtered colimits is exact, and the third isomorphism uses that $\Lambda^{(\alpha_p)}$ is α-presentable for each $p \leq n+1$, by Lemma 2.5.11. □

Lemma 2.5.15. *Let Λ be a ring. For every family of Λ-modules $(X_i)_{i \in I}$ and every $n \geq 0$ we have a canonical isomorphism*

$$\mathrm{Ext}_\Lambda^n \left(\coprod_i X_i, - \right) \xrightarrow{\sim} \prod_i \mathrm{Ext}_\Lambda^n(X_i, -).$$

Proof Choose a projective resolution $p(X_i) \to X_i$ for each i. Because taking (co)products of modules is exact, we obtain for every Λ-module Y

$$\operatorname{Ext}_\Lambda^n \left(\coprod_i X_i, Y \right) \cong H^n \operatorname{Hom} \left(\coprod_i p(X_i), Y \right)$$

$$\cong H^n \prod_i \operatorname{Hom}(p(X_i), Y)$$

$$\cong \prod_i \operatorname{Ext}_\Lambda^n(X_i, Y). \qquad \square$$

Proposition 2.5.16. *Any Grothendieck category is locally presentable.*

Proof Fix a Grothendieck category $\mathcal{A}$ with generator C and set $\Lambda = \operatorname{End}(C)$. We deduce the assertion from Theorem 2.5.2. Let $T \colon \operatorname{Mod}\Lambda \to \mathcal{A}$ be the exact left adjoint of the full faithful functor $\operatorname{Hom}(C, -)$. Then $\operatorname{Hom}(C, -)$ identifies $\mathcal{A}$ with $(\operatorname{Ker} T)^\perp$ by Lemma 2.2.10. Now choose a generator K of $\operatorname{Ker} T$. It is not difficult to check that $(\operatorname{Ker} T)^\perp = K^\perp$, since any exact sequence $0 \to X' \to K^{(\alpha)} \to X \to 0$ in $\operatorname{Mod}\Lambda$ (α any cardinal) yields an exact sequence

$$
\begin{array}{c}
0 \longrightarrow \operatorname{Hom}(X, -) \longrightarrow \operatorname{Hom}(K^{(\alpha)}, -) \longrightarrow \operatorname{Hom}(X', -) \\
\longrightarrow \operatorname{Ext}^1(X, -) \longrightarrow \operatorname{Ext}^1(K^{(\alpha)}, -) \longrightarrow \operatorname{Ext}^1(X', -) \longrightarrow \cdots
\end{array}
$$

and keeping in mind that

$$\operatorname{Ext}^i(K^{(\alpha)}, -) \cong \operatorname{Ext}^i(K, -)^\alpha \qquad (i \geq 0)$$

by Lemma 2.5.15. Now choose a free presentation

$$\Lambda^{(\alpha_2)} \longrightarrow \Lambda^{(\alpha_1)} \longrightarrow \Lambda^{(\alpha_0)} \longrightarrow K \longrightarrow 0$$

and a regular cardinal α such that $\alpha_i < \alpha$ for all i. Then it follows from Lemma 2.5.14 that $\operatorname{Hom}(K, -)$ and $\operatorname{Ext}^1(K, -)$ preserve α-filtered colimits. Thus the functor $\operatorname{Hom}(C, -)$ preserves α-filtered colimits, because it identifies with the inclusion $\mathcal{A} \hookrightarrow \operatorname{Mod}\Lambda$. Then the lemma below implies that T maps α-presentable objects to α-presentable objects.

Any Λ-module is a filtered colimit of finitely presented modules (Proposition 11.1.9), and therefore an α-filtered colimit of α-presentable modules by Lemma 2.5.12. Applying the functor T it follows that any object in $\mathcal{A}$ is an α-filtered colimit of α-presentable objects. $\qquad \square$

Lemma 2.5.17. *Let (F, G) be an adjoint pair of functors and α a regular cardinal. If G preserves α-filtered colimits, then F maps α-presentable objects to α-presentable objects.*

Proof For an α-presentable object X and an α-filtered colimit $\mathrm{colim}_{i \in \mathcal{I}} Y_i$ we have

$$\mathrm{colim}_i \mathrm{Hom}(FX, Y_i) \cong \mathrm{colim}_i \mathrm{Hom}(X, GY_i)$$

$$\cong \mathrm{Hom}(X, \mathrm{colim}_i GY_i)$$

$$\cong \mathrm{Hom}(X, G(\mathrm{colim}_i Y_i))$$

$$\cong \mathrm{Hom}(FX, \mathrm{colim}_i Y_i). \qquad \square$$

Remark 2.5.18. Let $\mathcal{C}$ be an essentially small additive category and fix a regular cardinal α. When $\mathcal{C}$ has α-small colimits we write

$$\mathrm{Ind}_\alpha \mathcal{C} := \mathrm{Lex}_\alpha(\mathcal{C}^{\mathrm{op}}, \mathrm{Ab})$$

for the category of left exact functors $\mathcal{C}^{\mathrm{op}} \to \mathrm{Ab}$ preserving α-small products. This category is locally α-presentable with

$$\mathcal{C} \xrightarrow{\sim} (\mathrm{Ind}_\alpha \mathcal{C})^\alpha.$$

Conversely, for any locally α-presentable additive category $\mathcal{A}$ the assignment $X \mapsto \mathrm{Hom}_{\mathcal{A}}(-, X)|_{\mathcal{A}^\alpha}$ induces an equivalence

$$\mathcal{A} \xrightarrow{\sim} \mathrm{Ind}_\alpha(\mathcal{A}^\alpha).$$

This generalises (with similar proofs) a correspondence for locally finitely presented categories, which is the case $\alpha = \aleph_0$ (Theorem 11.1.15). A consequence is the fact that a locally presentable category is complete, because the subcategory $\mathrm{Ind}_\alpha \mathcal{C} \subseteq \mathrm{Mod}\,\mathcal{C}$ is closed under limits.

Remark 2.5.19. Let $\mathcal{A}^2$ denote the category of morphisms in $\mathcal{A}$. If $\mathcal{A}$ is locally α-presentable, then $\mathcal{A}^2$ is locally α-presentable and $(\mathcal{A}^\alpha)^2 \xrightarrow{\sim} (\mathcal{A}^2)^\alpha$. This means that each morphism in $\mathcal{A}$ can be written as an α-filtered colimit of morphisms in $\mathcal{A}^\alpha$.

Localisation of Grothendieck Categories

In the following we sketch the localisation theory for Grothendieck categories, using the fact that any Grothendieck category $\mathcal{A}$ admits a filtration $\mathcal{A} = \bigcup_\alpha \mathcal{A}^\alpha$. In fact, we will see that $\mathcal{A}^\alpha$ is abelian when α is sufficiently large.

Lemma 2.5.20. *Let $\mathcal{A}$ be a locally α-presentable Grothendieck category. Then $\mathcal{A}^\alpha$ is abelian if and only if $\mathcal{A}^\alpha$ is closed under kernels. Moreover, in this case the inclusion $\mathcal{A}^\alpha \to \mathcal{A}$ is exact and $\mathcal{A}^\alpha$ is an extension closed subcategory.*

Proof We use the fact that $\mathcal{A}^\alpha$ is closed under cokernels. Thus when $\mathcal{A}^\alpha$ is closed under kernels, then $\mathcal{A}^\alpha$ is abelian and the inclusion $\mathcal{A}^\alpha \to \mathcal{A}$ is exact. Conversely, suppose that $\mathcal{A}^\alpha$ is abelian. Given an exact sequence $0 \to X \xrightarrow{\phi} Y \xrightarrow{\psi} Z$ in $\mathcal{A}^\alpha$, we need to show that it is also exact in $\mathcal{A}$. Let colim X_i be the kernel of ψ in $\mathcal{A}$, written as α-filtered colimit of objects in $\mathcal{A}^\alpha$. Each $X_i \to Y$ factors through ϕ, so colim $X_i \to Y$ factors through ϕ. Thus ϕ is a kernel in $\mathcal{A}$.

In order to show that $\mathcal{A}^\alpha$ is extension closed, let $\eta \colon 0 \to X \to Y \to Z \to 0$ be an exact sequence in $\mathcal{A}$ with $X, Z \in \mathcal{A}^\alpha$. Write $Y = $ colim Y_i as α-filtered colimit of objects in $\mathcal{A}^\alpha$. Then η is the colimit of exact sequences $0 \to X_i \to Y_i \to Z$, and for some index i_0 the induced morphisms $\phi \colon X_{i_0} \to X$ and $Y_{i_0} \to Z$ are epimorphisms. It follows that Y is isomorphic to the cokernel of Ker $\phi \to Y_{i_0}$ and therefore in $\mathcal{A}^\alpha$. $\square$

Proposition 2.5.21. *Let $\mathcal{A}$ be a Grothendieck category and α a regular cardinal. Suppose that $\mathcal{A}$ is locally α-presentable and that $\mathcal{A}^\alpha$ is abelian. For a localising subcategory $\mathcal{B} \subseteq \mathcal{A}$ such that $\mathcal{B} \cap \mathcal{A}^\alpha$ generates $\mathcal{B}$, the following holds.*

(1) *$\mathcal{B}$ and $\mathcal{A}/\mathcal{B}$ are locally α-presentable Grothendieck categories.*
(2) *$\mathcal{B}^\alpha = \mathcal{B} \cap \mathcal{A}^\alpha$ and the quotient functor $\mathcal{A} \to \mathcal{A}/\mathcal{B}$ induces an equivalence*

$$\mathcal{A}^\alpha / \mathcal{B}^\alpha \xrightarrow{\sim} (\mathcal{A}/\mathcal{B})^\alpha.$$

(3) *The inclusion $\mathcal{B} \to \mathcal{A}$ induces a localisation sequence.*

$$
\begin{array}{ccccc}
\mathcal{B}^\alpha & \rightarrowtail & \mathcal{A}^\alpha & \twoheadrightarrow & \mathcal{A}^\alpha/\mathcal{B}^\alpha \\
\downarrow & & \downarrow & & \downarrow \\
\mathcal{B} & \rightleftarrows & \mathcal{A} & \rightleftarrows & \mathcal{A}/\mathcal{B}
\end{array}
$$

Proof The proof amounts to identifying the sequence $\mathcal{B} \rightarrowtail \mathcal{A} \twoheadrightarrow \mathcal{A}/\mathcal{B}$ with the sequence $\mathrm{Ind}_\alpha(\mathcal{B}^\alpha) \to \mathrm{Ind}_\alpha(\mathcal{A}^\alpha) \to \mathrm{Ind}_\alpha(\mathcal{A}^\alpha/\mathcal{B}^\alpha)$ which is induced by $\mathcal{B}^\alpha \rightarrowtail \mathcal{A}^\alpha \twoheadrightarrow \mathcal{A}^\alpha/\mathcal{B}^\alpha$. Proposition 11.1.31 gives the details when $\alpha = \aleph_0$, and the general case is similar. $\square$

Let $\mathcal{C}$ be an essentially small additive category and fix a regular cardinal α. We write

$$\mathrm{mod}_\alpha\, \mathcal{C} := (\mathrm{Mod}\, \mathcal{C})^\alpha \qquad \text{and} \qquad \mathrm{proj}_\alpha\, \mathcal{C} := \mathrm{Proj}\, \mathcal{C} \cap \mathrm{mod}_\alpha\, \mathcal{C},$$

where $\mathrm{Proj}\, \mathcal{C}$ denotes the full subcategory of projective objects in $\mathrm{Mod}\, \mathcal{C}$. It is easily checked that $X \in \mathrm{Mod}\, \mathcal{C}$ belongs to $\mathrm{mod}_\alpha\, \mathcal{C}$ if and only if there is a presentation

$$\coprod_{i \in I} \mathrm{Hom}_{\mathcal{C}}(-, C_i) \longrightarrow \coprod_{j \in J} \mathrm{Hom}_{\mathcal{C}}(-, D_j) \longrightarrow X \longrightarrow 0$$

satisfying card I, card $J < \alpha$; see Lemma 2.5.12.

The next lemma shows that $\mathrm{mod}_\alpha\, \mathcal{C}$ is abelian when α is sufficiently large.

Lemma 2.5.22. *The following conditions are equivalent.*

(1) *The kernel of each morphism in* $\mathrm{mod}\,\mathcal{C}$ *belongs to* $\mathrm{mod}_\alpha\, \mathcal{C}$.
(2) *The category* $\mathrm{proj}_\alpha\, \mathcal{C}$ *has weak kernels.*
(3) *The category* $\mathrm{mod}_\alpha\, \mathcal{C}$ *is abelian.*

Proof (1) $\Rightarrow$ (2): We apply Lemma 2.5.11. The objects in $\mathrm{proj}_\alpha\, \mathcal{C}$ are precisely the direct summands of coproducts $X = \coprod_{i \in I} \mathrm{Hom}_\mathcal{C}(-, X_i)$ with card $I < \alpha$. Clearly, X is the filtered colimit of subobjects $\coprod_{i \in J} \mathrm{Hom}_\mathcal{C}(-, X_i)$ with card $J <$ $\aleph_0$. This colimit is α-small, and it follows that any morphism $X \to Y$ in $\mathrm{proj}_\alpha\, \mathcal{C}$ is an α-small filtered colimit of morphisms $X_\lambda \to Y_\lambda$ in $\mathrm{proj}\,\mathcal{C} \subseteq \mathrm{mod}\,\mathcal{C}$. Thus

$$\mathrm{Ker}(X \to Y) = \operatorname*{colim}_\lambda \mathrm{Ker}(X_\lambda \to Y_\lambda)$$

belongs to $\mathrm{mod}_\alpha\, \mathcal{C}$. It remains to observe that each object in $\mathrm{mod}_\alpha\, \mathcal{C}$ is the quotient of an object in $\mathrm{proj}_\alpha\, \mathcal{C}$.

(2) $\Rightarrow$ (3): That $\mathrm{mod}_\alpha\, \mathcal{C}$ is abelian follows from Lemma 2.1.6 since each object in $\mathrm{mod}_\alpha\, \mathcal{C}$ is the cokernel of a morphism in $\mathrm{proj}_\alpha\, \mathcal{C}$, and therefore

$$\mathrm{mod}_\alpha\, \mathcal{C} \xrightarrow{\sim} \mathrm{mod}(\mathrm{proj}_\alpha\, \mathcal{C}).$$

(3) $\Rightarrow$ (1): This is clear, since $\mathrm{mod}\,\mathcal{C} \subseteq \mathrm{mod}_\alpha\, \mathcal{C}$.   □

Corollary 2.5.23. *Let $\mathcal{A}$ be a Grothendieck category. There exists a regular cardinal α_0 such that for all regular $\alpha \geq \alpha_0$ the category $\mathcal{A}^\alpha$ is abelian and an extension closed subcategory of $\mathcal{A}$ with exact inclusion $\mathcal{A}^\alpha \to \mathcal{A}$.*

Proof We apply Theorem 2.5.2 and write $\mathcal{A}$ as the quotient $(\mathrm{Mod}\,\Lambda)/\mathcal{C}$ for some ring Λ and a localising subcategory $\mathcal{C} \subseteq \mathrm{Mod}\,\Lambda$. Choose α such that $\mathrm{mod}_\alpha\, \Lambda$ is abelian and $\mathcal{C} \cap \mathrm{mod}_\alpha\, \Lambda$ generates $\mathcal{C}$. Then the assertion follows from Proposition 2.5.21. More precisely, for $\alpha \geq \alpha_0$ we have $(\mathrm{mod}_\alpha\, \Lambda)/\mathcal{C}^\alpha \xrightarrow{\sim} \mathcal{A}^\alpha$. Thus $\mathcal{A}^\alpha$ is abelian and the inclusion $\mathcal{A}^\alpha \to \mathcal{A}$ is exact. Also, $\mathcal{A}^\alpha$ is extension closed by Lemma 2.5.20.   □

When $\mathcal{C}$ has α-small colimits, then the Yoneda functor $\mathcal{C} \to \mathrm{mod}_\alpha\, \mathcal{C}$ admits a left adjoint; it is the α-small colimit preserving functor $\mathrm{mod}_\alpha\, \mathcal{C} \to \mathcal{C}$ taking each representable functor $\mathrm{Hom}_\mathcal{C}(-, X)$ to X. The special case $\alpha = \aleph_0$ is Example 1.1.4. Let $\mathrm{eff}_\alpha\, \mathcal{C}$ denote the full subcategory of $\mathrm{mod}_\alpha\, \mathcal{C}$ consisting of the objects annihilated by this left adjoint, and set $\mathrm{Eff}_\alpha\, \mathcal{C} := \mathrm{Ind}_\alpha(\mathrm{eff}_\alpha\, \mathcal{C})$. Then the following is an analogue of Proposition 2.3.3.

Proposition 2.5.24. *Let* $\mathcal{C}$ *be an essentially small abelian category with α-small coproducts and suppose that* $\mathrm{Ind}_\alpha\,\mathcal{C}$ *is a Grothendieck category. Then the inclusion* $\mathrm{Ind}_\alpha\,\mathcal{C} \to \mathrm{Mod}\,\mathcal{C}$ *induces a localisation sequence of abelian categories*

$$\mathrm{Eff}_\alpha\,\mathcal{C} \rightleftarrows \mathrm{Mod}\,\mathcal{C} \rightleftarrows \mathrm{Ind}_\alpha\,\mathcal{C}$$

which restricts to the localisation sequence

$$\mathrm{eff}_\alpha\,\mathcal{C} \rightleftarrows \mathrm{mod}_\alpha\,\mathcal{C} \rightleftarrows \mathcal{C}.$$

Proof The inclusion $\mathrm{Ind}_\alpha\,\mathcal{C} \to \mathrm{Mod}\,\mathcal{C}$ has a left adjoint; it is the colimit preserving functor which is the identity on the representable functors. This left adjoint is exact by an analogue of Theorem 2.5.2, and it sends α-presentable objects to α-presentable objects, since the right adjoint preserves α-filtered colimits; see Lemma 2.5.17. This yields the left adjoint of the Yoneda functor $\mathcal{C} \to \mathrm{mod}_\alpha\,\mathcal{C}$. The rest then follows from Proposition 2.5.21. $\square$

The following immediate consequence provides a canonical presentation of a Grothendieck category as the quotient of a module category.

Corollary 2.5.25. *Let* $\mathcal{A}$ *be a locally α-presentable Grothendieck category such that* $\mathcal{C} = \mathcal{A}^\alpha$ *is abelian. Then*

$$(\mathrm{Mod}\,\mathcal{C})/(\mathrm{Eff}_\alpha\,\mathcal{C}) \xrightarrow{\sim} \mathcal{A}. \qquad \square$$

Coherent Functors

Let $\mathcal{A}$ be a cocomplete additive category. We call a functor $F: \mathcal{A} \to \mathrm{Ab}$ *coherent* if there is an exact sequence

$$\mathrm{Hom}_\mathcal{A}(Y, -) \longrightarrow \mathrm{Hom}_\mathcal{A}(X, -) \longrightarrow F \longrightarrow 0.$$

More precisely, we say for a regular cardinal α that F is *α-coherent* if X and Y are α-presentable objects. Note that every coherent functor is α-coherent for some regular cardinal α when $\mathcal{A}$ is locally presentable, thanks to Lemma 2.5.13.

Recall that a locally presentable category is complete and cocomplete; see Remark 2.5.18.

Theorem 2.5.26. *Let* $\mathcal{A}$ *be a locally α-presentable category. Then a functor* $F: \mathcal{A} \to \mathrm{Ab}$ *is α-coherent if and only if F preserves products and α-filtered colimits.*

The proof requires some preparations. In particular, we need a characterisation of finitely presented functors in terms of a tensor product.

Let $\mathcal{C}$ be an essentially small additive category. Recall that there exists a *tensor product*

$$\mathrm{Mod}(\mathcal{C}) \times \mathrm{Mod}(\mathcal{C}^{\mathrm{op}}) \longrightarrow \mathrm{Ab}, \qquad (X, Y) \longmapsto X \otimes_{\mathcal{C}} Y,$$

where the tensor functors $X \otimes_{\mathcal{C}} -$ and $- \otimes_{\mathcal{C}} Y$ are determined by the fact that they preserve colimits and that for $C \in \mathcal{C}$ there are natural isomorphisms

$$X \otimes_{\mathcal{C}} \mathrm{Hom}_{\mathcal{C}}(C, -) \cong X(C) \quad \text{and} \quad \mathrm{Hom}_{\mathcal{C}}(-, C) \otimes_{\mathcal{C}} Y \cong Y(C).$$

Recall that $Y \in \mathrm{Mod}(\mathcal{C}^{\mathrm{op}})$ is finitely presented if there is a presentation

$$\mathrm{Hom}_{\mathcal{C}}(D, -) \longrightarrow \mathrm{Hom}_{\mathcal{C}}(C, -) \longrightarrow Y \longrightarrow 0.$$

Proposition 2.5.27. *A functor $Y \in \mathrm{Mod}(\mathcal{C}^{\mathrm{op}})$ is finitely presented if and only if the functor $- \otimes_{\mathcal{C}} Y$ preserves all products.*

Proof Let $Y \in \mathrm{Mod}(\mathcal{C}^{\mathrm{op}})$. We choose a family of objects $(X_i)_{i \in I}$ in $\mathrm{Mod}\,\mathcal{C}$ and consider the canonical map

$$\alpha_Y : \left(\prod_i X_i \right) \otimes_{\mathcal{C}} Y \longrightarrow \prod_i (X_i \otimes_{\mathcal{C}} Y).$$

When $Y = \mathrm{Hom}_{\mathcal{C}}(C, -)$ for some $C \in \mathcal{C}$, then α_Y is a bijection. Now choose an exact sequence $Y_1 \to Y_0 \to Y \to 0$ in $\mathrm{Mod}(\mathcal{C}^{\mathrm{op}})$ and consider the following commutative diagram with exact rows.

$$
\begin{array}{ccccccc}
(\prod_i X_i) \otimes_{\mathcal{C}} Y_1 & \longrightarrow & (\prod_i X_i) \otimes_{\mathcal{C}} Y_0 & \longrightarrow & (\prod_i X_i) \otimes_{\mathcal{C}} Y & \longrightarrow & 0 \\
\downarrow{\scriptstyle \alpha_{Y_1}} & & \downarrow{\scriptstyle \alpha_{Y_0}} & & \downarrow{\scriptstyle \alpha_Y} & & \\
\prod_i (X_i \otimes_{\mathcal{C}} Y_1) & \longrightarrow & \prod_i (X_i \otimes_{\mathcal{C}} Y_0) & \longrightarrow & \prod_i (X_i \otimes_{\mathcal{C}} Y) & \longrightarrow & 0
\end{array}
$$

When $Y_t = \mathrm{Hom}_{\mathcal{C}}(C_t, -)$ for $C_0, C_1 \in \mathcal{C}$, then all vertical maps are bijective. Thus $- \otimes_{\mathcal{C}} Y$ preserves all products.

It is convenient to set

$$h_C = \mathrm{Hom}_{\mathcal{C}}(-, C) \qquad (C \in \mathcal{C})$$

and for any family of objects $(C_i)_{i \in I}$ in $\mathcal{C}$ we consider the canonical map

$$\beta_Y : \left(\prod_i h_{C_i} \right) \otimes_{\mathcal{C}} Y \longrightarrow \prod_i (h_{C_i} \otimes_{\mathcal{C}} Y).$$

Suppose that β_Y is surjective. We claim that Y is finitely generated. To this end consider the product of representable functors

$$\prod_{C \in \mathcal{C}} h_C^{Y(C)}$$

so that the canonical map

$$\beta: \left(\prod_{C \in \mathcal{C}} h_C^{Y(C)} \right) \otimes_\mathcal{C} Y \longrightarrow \prod_{C \in \mathcal{C}} (h_C \otimes_\mathcal{C} Y)^{Y(C)} = \prod_{C \in \mathcal{C}} Y(C)^{Y(C)}$$

is surjective. For any finite subset

$$I \subseteq \bigsqcup_{C \in \mathcal{C}} Y(C)$$

there is by Yoneda's lemma an induced morphism $\bigsqcup_{i \in I} h_{C_i} \to Y$ and we denote by Y_I its image. Then $Y = \operatorname{colim} Y_I$ and therefore

$$\operatorname*{colim}_I \left(\prod_{C \in \mathcal{C}} h_C^{Y(C)} \right) \otimes_\mathcal{C} Y_I \xrightarrow{\sim} \left(\prod_{C \in \mathcal{C}} h_C^{Y(C)} \right) \otimes_\mathcal{C} Y.$$

It follows that for some finite set I_0 there is an element

$$x \in \left(\prod_{C \in \mathcal{C}} h_C^{Y(C)} \right) \otimes_\mathcal{C} Y_{I_0}$$

such that $\beta(x) = \operatorname{id}_Y$, and therefore $Y = Y_{I_0}$ is finitely generated.

Now choose an exact sequence $0 \to Y_1 \to Y_0 \to Y \to 0$ in $\operatorname{Mod}(\mathcal{C}^{\mathrm{op}})$ and consider the following commutative diagram with exact rows.

$$
\begin{array}{ccccccc}
(\prod_i h_{C_i}) \otimes_\mathcal{C} Y_1 & \longrightarrow & (\prod_i h_{C_i}) \otimes_\mathcal{C} Y_0 & \longrightarrow & (\prod_i h_{C_i}) \otimes_\mathcal{C} Y & \longrightarrow & 0 \\
\downarrow{\scriptstyle\beta_{Y_1}} & & \downarrow{\scriptstyle\beta_{Y_0}} & & \downarrow{\scriptstyle\beta_Y} & & \\
0 \longrightarrow \prod_i(h_{C_i} \otimes_\mathcal{C} Y_1) & \longrightarrow & \prod_i(h_{C_i} \otimes_\mathcal{C} Y_0) & \longrightarrow & \prod_i(h_{C_i} \otimes_\mathcal{C} Y) & \longrightarrow & 0
\end{array}
$$

Suppose that β_Y is bijective. Then Y is finitely generated and we may choose $Y_0 = \operatorname{Hom}_\mathcal{C}(C, -)$ for some $C \in \mathcal{C}$. Thus β_{Y_0} is bijective and it follows that β_{Y_1} is surjective. Then Y_1 is finitely generated, and we conclude that Y is finitely presented. $\qquad\square$

Proof of Theorem 2.5.26 Suppose first that F is α-coherent. A representable functor $\operatorname{Hom}_A(X, -)$ preserves products and α-filtered colimits provided that X is α-presentable. Clearly, this property is preserved when one passes to the cokernel of a morphism $\operatorname{Hom}_A(Y, -) \to \operatorname{Hom}_A(X, -)$ where X and Y are α-presentable.

Now suppose that F preserves products and α-filtered colimits. Let $\mathcal{C}$ denote the full subcategory of α-presentable objects in A. We set $G = F|_\mathcal{C}$ and note that

$$F(X) \cong \operatorname{Hom}_A(-, X)|_\mathcal{C} \otimes_\mathcal{C} G \qquad (X \in A) \qquad (2.5.28)$$

since F preserves α-filtered colimits and every object in A is an α-filtered colimit of objects in $\mathcal{C}$.

The assumption on F to preserve products implies that any family of objects $(C_i)_{i \in I}$ in $\mathcal{C}$ induces an isomorphism

$$\left(\prod_i \mathrm{Hom}_{\mathcal{C}}(-, X_i) \right) \otimes_{\mathcal{C}} G \xrightarrow{\sim} \prod_i \left(\mathrm{Hom}_{\mathcal{C}}(-, X_i) \otimes_{\mathcal{C}} G \right).$$

We conclude from Proposition 2.5.27 that G has a presentation

$$\mathrm{Hom}_{\mathcal{C}}(Y, -) \longrightarrow \mathrm{Hom}_{\mathcal{C}}(X, -) \longrightarrow G \longrightarrow 0$$

with $X, Y \in \mathcal{C}$. Combining this presentation with the isomorphism (2.5.28) gives a presentation

$$\mathrm{Hom}_A(Y, -) \longrightarrow \mathrm{Hom}_A(Y, -) \longrightarrow F \longrightarrow 0$$

of F. Thus F is α-coherent. □

Notes

We follow Gabriel [79] and recall that abelian categories were introduced by Buchsbaum and Grothendieck in order to generalise the homological methods of Cartan and Eilenberg [46]. The localisation theory for abelian categories is developed in Gabriel's thesis [79], following Grothendieck's fundamental work [94]. In particular, [79] contains the description of Serre and localising subcategories for commutative noetherian rings. Also, the idea of presenting a Grothendieck category as a category of left exact functors is from [79]. Exact categories were introduced by Heller under the name 'abelian category' [107]; we follow expositions by Keller and Quillen [120, 165].

Projective and injective objects are important ingredients of homological algebra. We focus on injective objects because Grothendieck categories always have enough injectives, but not necessarily enough projectives. The study of injective modules goes back to the work of Baer [21]; the notion of an injective envelope was introduced by Eckmann and Schopf [70]. In [71] Eilenberg proposes an axiomatic description of minimal resolutions. Our treatment of projective covers and injective envelopes in terms of minimal decompositions of morphisms follows closely [131].

Finitely presented (or coherent) functors were studied in a famous article by Auslander [7]. Closely related is the correspondence between additive categories with weak kernels and abelian categories having enough projective objects, which is due to Freyd [75]. The notion of an effaceable functor goes back to Grothendieck [94]. The presentation of an abelian category as the quotient of the category of finitely presented functors modulo the subcategory of effaceable functors is also known as 'Auslander's formula' [139].

The notion of a recollement was introduced by Beilinson, Bernšteĭn and Deligne [26] in their study of perverse sheaves; it describes a diagram of six additive functors and makes sense equally for abelian as for triangulated categories. Universal localisations of (not necessarily commutative) rings were introduced by Cohn [54] and Schofield [182]; see also [33].

Grothendieck categories were introduced by Grothendieck in his Tôhoku paper [94] as an appropriate setting for homological algebra. While coproducts and filtered colimits are exact in Grothendieck categories, taking products need not be exact. The example of sheaves on the projective line over a field was suggested by Keller. The embedding theorem for Grothendieck categories is due to Popescu and Gabriel [159]. The Krull–Remak–Schmidt–Azumaya theorem is Azumaya's generalisation of the uniqueness result for decompositions of finite length modules into indecomposables [19]. Gabriel and Oberst introduced the spectral category of a Grothendieck category [82]; it provides a general context for the study of direct sum decompositions.

Locally presentable categories were introduced and studied by Gabriel and Ulmer [84]; for a modern account see [1]. The characterisation of coherent functors on locally presentable categories is taken from [128]; it generalises the characterisation of functors preserving products and filtered colimits for module categories by Crawley-Boevey [59]. The crucial ingredient of its proof is Lenzing's theorem which characterises finitely presented modules via their tensor functors [138].

3

Triangulated Categories

Contents

3.1	**Triangulated Categories**		**73**
	The Axioms		73
	Exact Functors		74
	Cohomological Functors		75
	Uniqueness of Exact Triangles		76
	Triangulated and Thick Subcategories		76
	Dévissage		77
3.2	**Localisation of Triangulated Categories**		**77**
	Verdier Localisation		77
	Localisation of Subcategories		80
	Localisation and Adjoints		81
3.3	**Frobenius Categories**		**83**
	Stable Categories of Frobenius Categories		83
	Frobenius Pairs		88
3.4	**Brown Representability**		**89**
	Homotopy (Co)limits		89
	A Brown Representability Theorem		92
	Compact Objects		96
	Compact Generators		98
	Notes		**100**

In this chapter we introduce triangulated categories. These provide the appropriate framework for studying derived functors and derived categories. A triangulated category is an additive category together with a suspension functor and a distinguished class of triangles. Important examples are stable categories of Frobenius categories. A basic tool is the localisation theory for triangulated categories. Another useful result is Brown's representability theorem for cohomological functors which requires the existence of generators satisfying certain finiteness conditions.

3.1 Triangulated Categories

Triangulated categories are defined via a set of four axioms. Then we discuss some of the basic properties of triangulated categories.

The Axioms

A *suspended category* is a pair $(\mathcal{T}, \Sigma)$ consisting of an additive category $\mathcal{T}$ and an equivalence $\Sigma \colon \mathcal{T} \xrightarrow{\sim} \mathcal{T}$ which we call a *suspension* or *shift*. A *triangle* in $(\mathcal{T}, \Sigma)$ is a sequence (α, β, γ) of morphisms

$$X \xrightarrow{\ \alpha\ } Y \xrightarrow{\ \beta\ } Z \xrightarrow{\ \gamma\ } \Sigma X$$

and a morphism between triangles (α, β, γ) and $(\alpha', \beta', \gamma')$ is given by a triple (ϕ_1, ϕ_2, ϕ_3) of morphisms in $\mathcal{T}$ making the following diagram commutative.

$$
\begin{array}{ccccccc}
X & \xrightarrow{\ \alpha\ } & Y & \xrightarrow{\ \beta\ } & Z & \xrightarrow{\ \gamma\ } & \Sigma X \\
\downarrow{\scriptstyle \phi_1} & & \downarrow{\scriptstyle \phi_2} & & \downarrow{\scriptstyle \phi_3} & & \downarrow{\scriptstyle \Sigma\phi_1} \\
X' & \xrightarrow{\ \alpha'\ } & Y' & \xrightarrow{\ \beta'\ } & Z' & \xrightarrow{\ \gamma'\ } & \Sigma X'
\end{array}
$$

A *triangulated category* is a triple $(\mathcal{T}, \Sigma, \mathcal{E})$ consisting of a suspended category $(\mathcal{T}, \Sigma)$ and a class $\mathcal{E}$ of distinguished triangles in $(\mathcal{T}, \Sigma)$ (called *exact triangles*) satisfying the following conditions.

(Tr1) A triangle isomorphic to an exact triangle is exact. For each object X, the triangle $0 \to X \xrightarrow{\text{id}} X \to \Sigma 0$ is exact. Each morphism α fits into an exact triangle (α, β, γ).

(Tr2) A triangle (α, β, γ) is exact if and only if $(\beta, \gamma, -\Sigma\alpha)$ is exact.

(Tr3) Given two exact triangles (α, β, γ) and $(\alpha', \beta', \gamma')$, each pair of morphisms ϕ_1 and ϕ_2 satisfying $\phi_2\alpha = \alpha'\phi_1$ can be completed to a morphism

$$
\begin{array}{ccccccc}
X & \xrightarrow{\ \alpha\ } & Y & \xrightarrow{\ \beta\ } & Z & \xrightarrow{\ \gamma\ } & \Sigma X \\
\downarrow{\scriptstyle \phi_1} & & \downarrow{\scriptstyle \phi_2} & & \downarrow{\scriptstyle \phi_3} & & \downarrow{\scriptstyle \Sigma\phi_1} \\
X' & \xrightarrow{\ \alpha'\ } & Y' & \xrightarrow{\ \beta'\ } & Z' & \xrightarrow{\ \gamma'\ } & \Sigma X'
\end{array}
$$

of triangles.

(Tr4) Given exact triangles $(\alpha_1, \alpha_2, \alpha_3)$, $(\beta_1, \beta_2, \beta_3)$, and $(\gamma_1, \gamma_2, \gamma_3)$ with $\gamma_1 = \beta_1\alpha_1$, there exists an exact triangle $(\delta_1, \delta_2, \delta_3)$ making the following diagram commutative.

$$
\begin{array}{ccccccc}
X & \xrightarrow{\alpha_1} & Y & \xrightarrow{\alpha_2} & U & \xrightarrow{\alpha_3} & \Sigma X \\
\| & & \downarrow{\beta_1} & & \downarrow{\delta_1} & & \| \\
X & \xrightarrow{\gamma_1} & Z & \xrightarrow{\gamma_2} & V & \xrightarrow{\gamma_3} & \Sigma X \\
& & \downarrow{\beta_2} & & \downarrow{\delta_2} & & \downarrow{\Sigma\alpha_1} \\
& & W & = & W & \xrightarrow{\beta_3} & \Sigma Y \\
& & \downarrow{\beta_3} & & \downarrow{\delta_3} & & \\
& & \Sigma Y & \xrightarrow{\Sigma\alpha_2} & \Sigma U & &
\end{array}
$$

The axiom (Tr4) is also known as the *octahedral axiom*, because the objects and morphisms of the diagram can be arranged to produce the skeleton of an octahedron, four of whose faces are exact triangles, so of the form

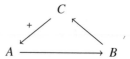

corresponding to an exact triangle $A \to B \to C \to \Sigma A$.

Given a triangulated category $(\mathcal{T}, \Sigma, \mathcal{E})$, we simplify the notation and identify $\mathcal{T} = (\mathcal{T}, \Sigma, \mathcal{E})$.

Exact Functors

An *exact functor* (or *triangle functor*) $\mathcal{T} \to \mathcal{U}$ between triangulated categories is a pair (F, η) consisting of an additive functor $F \colon \mathcal{T} \to \mathcal{U}$ and a natural isomorphism $\eta \colon F \circ \Sigma_{\mathcal{T}} \xrightarrow{\sim} \Sigma_{\mathcal{U}} \circ F$ such that for every exact triangle $X \xrightarrow{\alpha} Y \xrightarrow{\beta} Z \xrightarrow{\gamma} \Sigma_{\mathcal{T}} X$ in $\mathcal{T}$ the triangle

$$
FX \xrightarrow{F\alpha} FY \xrightarrow{F\beta} FZ \xrightarrow{\eta_X \circ F\gamma} \Sigma_{\mathcal{U}}(FX)
$$

is exact in $\mathcal{U}$. In the following we simplify the notation and identify $F = (F, \eta)$.

An exact functor $F \colon \mathcal{T} \to \mathcal{U}$ is called a *triangle equivalence* if F is an equivalence of categories. The terminology is justified by the following observation, because then a quasi-inverse is again exact.

Lemma 3.1.1. *Let (F, G) be an adjoint pair of functors between triangulated categories. Then F is exact if and only if G is exact.* □

Cohomological Functors

Let $\mathcal{T}$ be a triangulated category. An additive functor $F: \mathcal{T} \to \mathcal{A}$ into an abelian category $\mathcal{A}$ is called *cohomological* if it sends each exact triangle $X \to Y \to Z \to \Sigma X$ in $\mathcal{T}$ to an exact sequence $FX \to FY \to FZ$ in $\mathcal{A}$.

Lemma 3.1.2. *For each object X in $\mathcal{T}$, the representable functors*

$$\mathrm{Hom}_{\mathcal{T}}(X, -): \mathcal{T} \longrightarrow \mathrm{Ab} \quad and \quad \mathrm{Hom}_{\mathcal{T}}(-, X): \mathcal{T}^{\mathrm{op}} \longrightarrow \mathrm{Ab}$$

into the category Ab *of abelian groups are cohomological functors.*

Proof We show that $\mathrm{Hom}_{\mathcal{T}}(X, -)$ is cohomological. For $\mathrm{Hom}_{\mathcal{T}}(-, X)$ the proof is dual.

Fix an exact triangle $U \xrightarrow{\alpha} V \xrightarrow{\beta} W \xrightarrow{\gamma} \Sigma U$. We need to show the exactness of the induced sequence

$$\mathrm{Hom}_{\mathcal{T}}(X, U) \longrightarrow \mathrm{Hom}_{\mathcal{T}}(X, V) \longrightarrow \mathrm{Hom}_{\mathcal{T}}(X, W).$$

To this end fix a morphism $\phi: X \to V$ and consider the following diagram.

$$
\begin{array}{ccccccc}
X & \xrightarrow{\mathrm{id}} & X & \longrightarrow & 0 & \longrightarrow & \Sigma X \\
& & \downarrow{\phi} & & & & \\
U & \xrightarrow{\alpha} & V & \xrightarrow{\beta} & W & \xrightarrow{\gamma} & \Sigma U
\end{array}
$$

If ϕ factors through α, then (Tr3) implies the existence of a morphism $0 \to W$ making the diagram commutative. Thus $\beta \circ \phi = 0$. Now assume $\beta \circ \phi = 0$. Applying (Tr2) and (Tr3), we find a morphism $X \to U$ making the diagram commutative. Thus ϕ factors through α. □

We discuss some consequences. For example, we see that in any exact triangle $X \xrightarrow{\alpha} Y \xrightarrow{\beta} Z \xrightarrow{\gamma} \Sigma X$ the morphism α is a weak kernel of β. Also, the Yoneda functor $\mathcal{T} \to \mathrm{mod}\,\mathcal{T}$ is a universal cohomological functor.

Proposition 3.1.3. *The category* $\mathrm{mod}\,\mathcal{T}$ *is abelian and the Yoneda functor* $\mathcal{T} \to \mathrm{mod}\,\mathcal{T}$ *is cohomological. Any cohomological functor* $\mathcal{T} \to \mathcal{A}$ *factors uniquely (up to a unique isomorphism) through the Yoneda functor via an exact functor* $\mathrm{mod}\,\mathcal{T} \to \mathcal{A}$.

Proof Every morphism in $\mathcal{T}$ admits a weak kernel by Lemma 3.1.2. Therefore the category $\mathrm{mod}\,\mathcal{T}$ is abelian by Lemma 2.1.6. Moreover, Lemma 3.1.2 implies that the Yoneda functor is cohomological. Given a cohomological functor $F: \mathcal{T} \to \mathcal{A}$, the functor $\mathrm{mod}\,\mathcal{T} \to \mathcal{A}$ takes $\mathrm{Coker}\,\mathrm{Hom}_{\mathcal{T}}(-, \phi)$ (given by a morphism ϕ in $\mathcal{T}$) to $\mathrm{Coker}\,F(\phi)$. This functor is exact and essentially unique; see Lemma 2.1.8. □

Proposition 3.1.4. *A functor* $\mathcal{T}^{\mathrm{op}} \to \mathrm{Ab}$ *is cohomological if and only if it is a filtered colimit of representable functors.*

Proof One direction is clear, since filtered colimits in Ab are exact and representable functors are cohomological. Now fix an additive functor $F \colon \mathcal{T}^{\mathrm{op}} \to \mathrm{Ab}$. Let $\mathcal{T}/F$ denote the category consisting of pairs (X, f) with $X \in \mathcal{T}$ and $f \in F(X)$. A morphism $(X, f) \to (X', f')$ is given by a morphism $\alpha \colon X \to X'$ in $\mathcal{T}$ such that $F(\alpha)(f') = f$. We write $\mathrm{Add}(\mathcal{T}^{\mathrm{op}}, \mathrm{Ab})$ for the category of additive functors $\mathcal{T}^{\mathrm{op}} \to \mathrm{Ab}$. Then F equals the colimit of the functor

$$\mathcal{T}/F \longrightarrow \mathrm{Add}(\mathcal{T}^{\mathrm{op}}, \mathrm{Ab}), \quad (X, f) \mapsto \mathrm{Hom}_{\mathcal{T}}(-, X)$$

(Lemma 11.1.8). It is easily checked that $\mathcal{T}/F$ is filtered when F is cohomological. $\square$

Uniqueness of Exact Triangles

Let $\mathcal{T}$ be a triangulated category. Given a morphism $\alpha \colon X \to Y$ in $\mathcal{T}$ and two exact triangles $\Delta = (\alpha, \beta, \gamma)$ and $\Delta' = (\alpha, \beta', \gamma')$ which complete α, there exists a comparison morphism $(\mathrm{id}_X, \mathrm{id}_Y, \phi)$ between Δ and Δ', by (Tr3). The morphism ϕ is an isomorphism, by the following lemma, but it need not be unique.

Lemma 3.1.5. *Let* (ϕ_1, ϕ_2, ϕ_3) *be a morphism between exact triangles. If two of* ϕ_1, ϕ_2, ϕ_3 *are isomorphisms, then the third is also an isomorphism.*

Proof Use Lemma 3.1.2 and apply the five lemma. $\square$

The third object Z in an exact triangle $X \xrightarrow{\alpha} Y \to Z \to \Sigma X$ is called the *cone* of α and is denoted by $\mathrm{Cone}\,\alpha$, despite the fact that it is not unique. Later on we will see specific constructions which justify this terminology.

Triangulated and Thick Subcategories

Let $\mathcal{T}$ be a triangulated category. A full subcategory $\mathcal{S}$ is a *triangulated subcategory* if $\mathcal{S}$ is non-empty and the following conditions hold.

(TS1) $\Sigma^n X \in \mathcal{S}$ for all $X \in \mathcal{S}$ and $n \in \mathbb{Z}$.
(TS2) Let $X \to Y \to Z \to \Sigma X$ be an exact triangle in $\mathcal{T}$. Then $X, Y \in \mathcal{S}$ implies $Z \in \mathcal{S}$.

A triangulated subcategory $\mathcal{S}$ is *thick* if in addition the following condition holds.

(TS3) Every direct summand of an object in $\mathcal{S}$ belongs to $\mathcal{S}$, that is, a decomposition $X = X' \oplus X''$ for $X \in \mathcal{S}$ implies $X' \in \mathcal{S}$.

Note that a triangulated subcategory $\mathcal{S}$ inherits a canonical triangulated structure from $\mathcal{T}$.

Example 3.1.6. The kernel of an exact functor $\mathcal{T} \to \mathcal{U}$ between triangulated categories is a thick subcategory of $\mathcal{T}$.

Example 3.1.7. An object X in $\mathcal{T}$ is *homologically finite* if for every object Y in $\mathcal{T}$ we have $\mathrm{Hom}_{\mathcal{T}}(X, \Sigma^n Y) = 0$ for almost all $n \in \mathbb{Z}$. The homologically finite objects form a thick subcategory of $\mathcal{T}$.

Dévissage

For a triangulated category $\mathcal{T}$ and a class of objects $\mathcal{C} \subseteq \mathcal{T}$ let $\mathrm{Thick}(\mathcal{C})$ denote the smallest thick subcategory of $\mathcal{T}$ that contains $\mathcal{C}$.

Lemma 3.1.8. *Let $F : \mathcal{T} \to \mathcal{U}$ be an exact functor between triangulated categories and let $\mathcal{C} \subseteq \mathcal{T}$ be a class of objects in $\mathcal{T}$. If the induced map*

$$\mathrm{Hom}_{\mathcal{T}}(X, \Sigma^n Y) \to \mathrm{Hom}_{\mathcal{U}}(FX, \Sigma^n FY)$$

is bijective for all $X, Y \in \mathcal{C}$ and $n \in \mathbb{Z}$, then F restricted to $\mathrm{Thick}(\mathcal{C})$ is fully faithful.

Proof   Use Lemma 3.1.2 and apply the five lemma.   □

3.2 Localisation of Triangulated Categories

We introduce the localisation of a triangulated category with respect to a triangulated subcategory. Localising amounts to annihilating a class of objects, and the triangulated structure is preserved.

Verdier Localisation

Let $\mathcal{T}$ be a triangulated category and fix a triangulated subcategory $\mathcal{S}$. Set

$$S(\mathcal{S}) = \{\sigma \in \mathrm{Mor}\,\mathcal{T} \mid \mathrm{Cone}\,\sigma \in \mathcal{S}\}.$$

Also, we set

$$\mathcal{S}^{\perp} = \{Y \in \mathcal{T} \mid \mathrm{Hom}_{\mathcal{T}}(X, Y) = 0 \text{ for all } X \in \mathcal{S}\}$$

and

$$^{\perp}\mathcal{S} = \{X \in \mathcal{T} \mid \mathrm{Hom}_{\mathcal{T}}(X, Y) = 0 \text{ for all } Y \in \mathcal{S}\}.$$

Lemma 3.2.1. *For a triangulated subcategory* $\mathcal{S} \subseteq \mathcal{T}$ *the following holds.*

(1) $S(\mathcal{S})$ *admits a calculus of left and right fractions.*
(2) *An object in* $\mathcal{T}$ *is* $S(\mathcal{S})$*-local if and only if it is in* $\mathcal{S}^{\perp}$.

Proof   Set $S = S(\mathcal{S})$.

(1) We check for S the conditions (LF1)–(LF3) to admit a calculus of left fractions. The proof that S admits a calculus of right fractions is dual.

(LF1) The class S contains the identity morphisms by (Tr1) and the composite of two morphisms in S by (Tr4).

(LF2) Fix a pair of morphisms $X' \xleftarrow{\sigma} X \xrightarrow{\alpha} Y$ in $\mathcal{T}$ with $\sigma \in S$. Completing the composite $\Sigma^{-1}(\mathrm{Cone}\,\sigma) \to X \to Y$ to an exact triangle and applying (Tr3) yields a commutative diagram

$$
\begin{array}{ccc}
X & \longrightarrow & Y \\
\sigma \downarrow & & \downarrow \tau \\
X' & \longrightarrow & Y'
\end{array}
$$

with $\mathrm{Cone}\,\sigma \cong \mathrm{Cone}\,\tau$. Thus $\tau \in S$.

(LF3) Let $\alpha, \beta \colon X \to Y$ be morphisms in $\mathcal{T}$ and suppose there is $\sigma \colon X' \to X$ in S such that $\alpha\sigma = \beta\sigma$. Complete σ to an exact triangle $X' \xrightarrow{\sigma} X \xrightarrow{\phi} \mathrm{Cone}\,\sigma \to \Sigma X'$. Then $\alpha - \beta$ factors through ϕ via a morphism $\psi \colon \mathrm{Cone}\,\sigma \to Y$. Now complete ψ to an exact triangle $\mathrm{Cone}\,\sigma \xrightarrow{\psi} Y \xrightarrow{\tau} Y' \to \Sigma(\mathrm{Cone}\,\sigma)$. Then $\tau\alpha = \tau\beta$ and $\tau \in S$.

(2) Fix $Y \in \mathcal{T}$ and suppose that $\mathrm{Hom}_{\mathcal{T}}(X, Y) = 0$ for all $X \in \mathcal{S}$. Then every $\sigma \in S$ induces a bijection $\mathrm{Hom}_{\mathcal{T}}(\sigma, Y)$ because $\mathrm{Hom}_{\mathcal{T}}(-, Y)$ is cohomological. Thus Y is S-local.

Now suppose that Y is S-local. If X belongs to $\mathcal{S}$, then the morphism $\sigma \colon X \to 0$ belongs to S and therefore induces a bijection $\mathrm{Hom}_{\mathcal{T}}(\sigma, Y)$. Thus Y belongs to $\mathcal{S}^{\perp}$. $\square$

The *Verdier localisation* of $\mathcal{T}$ with respect to $\mathcal{S}$ is by definition the localisation

$$\mathcal{T}/\mathcal{S} = \mathcal{T}[S(\mathcal{S})^{-1}]$$

together with the canonical functor $\mathcal{T} \to \mathcal{T}/\mathcal{S}$.

Proposition 3.2.2. *Let* $\mathcal{T}$ *be a triangulated category and* $\mathcal{S}$ *a triangulated subcategory. Then the following holds.*

(1) *The category $\mathcal{T}/\mathcal{S}$ carries a unique triangulated structure such that the canonical functor $Q: \mathcal{T} \to \mathcal{T}/\mathcal{S}$ is exact and annihilates $\mathcal{S}$.*

(2) *If $\mathcal{U}$ is a triangulated category and $F: \mathcal{T} \to \mathcal{U}$ is an exact functor that annihilates $\mathcal{S}$, then there exists a unique exact functor $\bar{F}: \mathcal{T}/\mathcal{S} \to \mathcal{U}$ such that $F = \bar{F} \circ Q$.*

Proof (1) We apply Lemma 3.2.1. Thus $S(\mathcal{S})$ admits a calculus of left and right fractions. The category $\mathcal{T}/\mathcal{S}$ is additive by Lemma 2.2.1. The class $S(\mathcal{S})$ is invariant under the suspension Σ. Thus Σ induces an equivalence $\mathcal{T}/\mathcal{S} \xrightarrow{\sim} \mathcal{T}/\mathcal{S}$. A triangle in $\mathcal{T}/\mathcal{S}$ is by definition exact if it is isomorphic to the image under Q of an exact triangle in $\mathcal{T}$. It is straightforward to check the conditions (Tr1)–(Tr4), and the functor Q is exact by construction. Clearly, $Q|_{\mathcal{S}} = 0$.

(2) If $F: \mathcal{T} \to \mathcal{U}$ is an exact functor and $F|_{\mathcal{S}} = 0$, then F inverts all morphisms in $S(\mathcal{S})$. Thus F factors through $Q: \mathcal{T} \to \mathcal{T}/\mathcal{S}$ via a unique functor $\bar{F}: \mathcal{T}/\mathcal{S} \to \mathcal{U}$. The functor $\bar{F}$ is exact, because any exact triangle in $\mathcal{T}/\mathcal{S}$ is up to isomorphism the image under Q of an exact triangle in $\mathcal{T}$. □

Remark 3.2.3. (1) The properties (1)–(2) in Proposition 3.2.2 provide a universal property that determines the canonical functor $\mathcal{T} \to \mathcal{T}/\mathcal{S}$ up to a unique isomorphism.

(2) The canonical functor $Q: \mathcal{T} \to \mathcal{T}/\mathcal{S}$ annihilates a morphism α in $\mathcal{T}$ if and only if α factors through an object in $\mathcal{S}$. In particular, $QX = 0$ for an object X in $\mathcal{T}$ if and only if X is a direct summand of an object in $\mathcal{S}$. Thus $\operatorname{Ker} Q = \operatorname{Thick}(\mathcal{S})$.

(3) A cohomological functor $H: \mathcal{T} \to \mathcal{A}$ factors through $\mathcal{T} \to \mathcal{T}/\mathcal{S}$ via a unique cohomological functor $\mathcal{T}/\mathcal{S} \to \mathcal{A}$ if and only if $H|_{\mathcal{S}} = 0$.

(4) The canonical functor $\mathcal{T} \to \mathcal{T}/\mathcal{S}$ preserves all coproducts in $\mathcal{T}$ if and only if $\mathcal{S}$ is closed under coproducts; see Lemma 1.1.8.

The following provides a useful fact about the morphisms in $\mathcal{T}/\mathcal{S}$.

Lemma 3.2.4. *Let $\mathcal{S} \subseteq \mathcal{T}$ be a triangulated subcategory and $X \in \mathcal{T}$. Then the canonical map*

$$\operatorname{Hom}_{\mathcal{T}}(X', X) \longrightarrow \operatorname{Hom}_{\mathcal{T}/\mathcal{S}}(X', X)$$

is a bijection for all $X' \in \mathcal{T}$ if and only if $X \in \mathcal{S}^{\perp}$. Analogously,

$$\operatorname{Hom}_{\mathcal{T}}(X, X') \longrightarrow \operatorname{Hom}_{\mathcal{T}/\mathcal{S}}(X, X')$$

is a bijection for all $X' \in \mathcal{T}$ if and only if $X \in {}^{\perp}\mathcal{S}$.

Proof This follows from Lemma 1.1.2 and Lemma 3.2.1. □

Localisation of Subcategories

We consider a Verdier localisation and its triangulated subcategories. The following lemma provides a useful criterion.

Lemma 3.2.5. *Let $\mathcal{U}, \mathcal{V} \subseteq \mathcal{T}$ be triangulated subcategories of a triangulated category $\mathcal{T}$. Suppose that one of the following conditions holds.*

(1) *Every morphism $\mathcal{V} \ni V \to U \in \mathcal{U}$ factors through an object in $\mathcal{U} \cap \mathcal{V}$.*
(2) *Every morphism $\mathcal{U} \ni U \to V \in \mathcal{V}$ factors through an object in $\mathcal{U} \cap \mathcal{V}$.*

Then the induced functor $\mathcal{U}/(\mathcal{U} \cap \mathcal{V}) \to \mathcal{T}/\mathcal{V}$ is fully faithful.

We capture the situation in the following commutative diagram

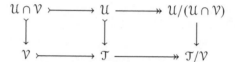

and provide a criterion for the functor on the right to be fully faithful.

Proof   Suppose (1) holds; the other case is dual. We claim that $\mathcal{U}$ is left cofinal with respect to $S(\mathcal{V})$. Then the inclusion $\mathcal{U} \to \mathcal{T}$ induces a fully faithful functor $\mathcal{U}/(\mathcal{U} \cap \mathcal{V}) \to \mathcal{T}/\mathcal{V}$ by Lemma 1.2.5, since $S(\mathcal{U} \cap \mathcal{V}) = S(\mathcal{V}) \cap \mathcal{U}$.

To prove the claim choose a morphism $U \to Y$ in $S(\mathcal{V})$ with $U \in \mathcal{U}$. This yields an exact triangle $V \to U \to Y \to \Sigma V$. The first morphism factors through an object $X \in \mathcal{U} \cap \mathcal{V}$. Applying the octahedral axiom yields a commutative diagram

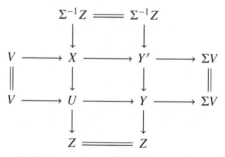

with exact rows and columns. Then $Y \to Z$ is the desired morphism with $Z \in \mathcal{U}$. □

Next we describe all triangulated subcategories of a Verdier localisation.

Proposition 3.2.6. *Let $\mathcal{V} \subseteq \mathcal{U} \subseteq \mathcal{T}$ be triangulated subcategories of a triangulated category $\mathcal{T}$. Then $\mathcal{U}/\mathcal{V}$ identifies with a triangulated subcategory of*

$\mathcal{T}/\mathcal{V}$, *and every triangulated subcategory of* $\mathcal{T}/\mathcal{V}$ *is of this form. Moreover, the canonical functor* $\mathcal{T} \to \mathcal{T}/\mathcal{V}$ *induces an isomorphism* $\mathcal{T}/\mathcal{U} \xrightarrow{\sim} (\mathcal{T}/\mathcal{V})/(\mathcal{U}/\mathcal{V})$.

We capture the situation in the following commutative diagram.

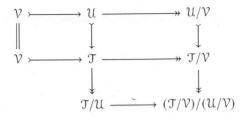

Proof The inclusion $\mathcal{U} \to \mathcal{T}$ induces a fully faithful functor $\mathcal{U}/\mathcal{V} \to \mathcal{T}/\mathcal{V}$ by the above Lemma 3.2.5. It is easily checked that $\mathcal{U}/\mathcal{V}$ yields a triangulated subcategory of $\mathcal{T}/\mathcal{V}$. If $\mathcal{W} \subseteq \mathcal{T}/\mathcal{V}$ is a triangulated subcategory, set $\mathcal{U} := Q^{-1}(\mathcal{W})$. Then $\mathcal{U}/\mathcal{V} \xrightarrow{\sim} \mathcal{W}$. The final assertion is clear, since the kernel of the composite $\mathcal{T} \to \mathcal{T}/\mathcal{V} \to (\mathcal{T}/\mathcal{V})/(\mathcal{U}/\mathcal{V})$ equals $\mathcal{U}$. □

Localisation and Adjoints

Let $\mathcal{T}$ be a triangulated category and $\mathcal{S} \subseteq \mathcal{T}$ a triangulated subcategory. Suppose that the canonical functor $Q\colon \mathcal{T} \to \mathcal{T}/\mathcal{S}$ admits a right adjoint $Q_\rho\colon \mathcal{T}/\mathcal{S} \to \mathcal{T}$. Then Q_ρ is fully faithful and induces an equivalence

$$\mathcal{T}/\mathcal{S} \xrightarrow{\sim} \mathcal{S}^\perp \qquad \text{with quasi-inverse} \qquad \mathcal{S}^\perp \hookrightarrow \mathcal{T} \xrightarrow{Q} \mathcal{T}/\mathcal{S}.$$

This follows from Proposition 1.1.3 and Lemma 3.2.1. The unit of the adjunction yields for X in $\mathcal{T}$ an exact triangle

$$X' \longrightarrow X \xrightarrow{\eta} Q_\rho Q(X) \longrightarrow \Sigma X'$$

with X' a direct summand of an object in $\mathcal{S}$ since $Q(\eta)$ is invertible.

Lemma 3.2.7. *When* $\mathcal{S} \subseteq \mathcal{T}$ *is thick and* Q *admits a right adjoint, then the assignment* $X \mapsto X'$ *provides a right adjoint of the inclusion* $\mathcal{S} \to \mathcal{T}$.

Proof The map $\mathrm{Hom}_{\mathcal{T}}(-, X') \to \mathrm{Hom}_{\mathcal{T}}(-, X)$ is bijective when restricted to $\mathcal{S}$ since $Q_\rho Q(X)$ and $\Sigma^{-1}Q_\rho Q(X)$ are in $\mathcal{S}^\perp$. □

The following proposition expresses the symmetry which arises from localising a triangulated category with respect to a thick subcategory.

Proposition 3.2.8. *Let* $\mathcal{S} \subseteq \mathcal{T}$ *be a thick subcategory. Then the following are equivalent.*

(1) *The inclusion* $S \to \mathcal{T}$ *admits a right adjoint.*
(2) *For each* $X \in \mathcal{T}$ *there exists an exact triangle* $X' \to X \to X'' \to \Sigma X'$ *with* $X' \in S$ *and* $X'' \in S^\perp$.
(3) *The canonical functor* $\mathcal{T} \to \mathcal{T}/S$ *admits a right adjoint.*
(4) *The composite* $S^\perp \hookrightarrow \mathcal{T} \twoheadrightarrow \mathcal{T}/S$ *is a triangle equivalence.*

In that case the right adjoint $\mathcal{T} \to S$ *induces a triangle equivalence*

$$\mathcal{T}/(S^\perp) \xrightarrow{\sim} S \qquad and \qquad {}^\perp(S^\perp) = S.$$

Proof   (1) $\Rightarrow$ (2): Suppose that the inclusion $I\colon S \to \mathcal{T}$ admits a right adjoint $I_\rho\colon \mathcal{T} \to S$, and consider for X in $\mathcal{T}$ the exact triangle

$$\Sigma^{-1}X'' \longrightarrow II_\rho(X) \longrightarrow X \longrightarrow X''$$

given by the counit of the adjunction. Then we have $II_\rho(X) \in S$ and $X'' \in S^\perp$.

(2) $\Rightarrow$ (3): Suppose there is for X in $\mathcal{T}$ an exact triangle $X' \to X \to X'' \to \Sigma X'$ with $X' \in S$ and $X'' \in S^\perp$. The assignment $X \mapsto X''$ provides a left adjoint for the inclusion $S^\perp \to \mathcal{T}$, say $F\colon \mathcal{T} \to S^\perp$. The kernel of F equals ${}^\perp(S^\perp) = S$, and F induces an equivalence $\mathcal{T}/S \xrightarrow{\sim} S^\perp$. Composing this with the inclusion $S^\perp \to \mathcal{T}$ provides the desired right adjoint of $\mathcal{T} \to \mathcal{T}/S$.

(3) $\Rightarrow$ (4): Combine Proposition 1.1.3 and Lemma 3.2.1.

(4) $\Rightarrow$ (1): A quasi-inverse of $S^\perp \xrightarrow{\sim} \mathcal{T}/S$ composed with the inclusion $S^\perp \to \mathcal{T}$ provides a right adjoint of $\mathcal{T} \to \mathcal{T}/S$. Then the inclusion $S \to \mathcal{T}$ admits a right adjoint, by Lemma 3.2.7.

This completes the first part of the proof. We have already seen that a right adjoint $I_\rho\colon \mathcal{T} \to S$ of the inclusion $S \to \mathcal{T}$ arises from a localisation, by Proposition 1.1.3, and its kernel equals $S^\perp$. Thus I_ρ induces a triangle equivalence $\mathcal{T}/(S^\perp) \xrightarrow{\sim} S$. $\square$

We capture the situation in the following diagram

$$S \underset{I_\rho}{\overset{I}{\rightleftarrows}} \mathcal{T} \underset{Q_\rho}{\overset{Q}{\rightleftarrows}} \mathcal{T}/S$$

which is a localisation sequence. The adjunctions yield for each object $X \in \mathcal{T}$ an exact triangle

$$II_\rho(X) \longrightarrow X \longrightarrow Q_\rho Q(X) \longrightarrow \Sigma II_\rho(X).$$

The following proposition complements Proposition 3.2.8.

Proposition 3.2.9. *Let* (F, G) *be an adjoint pair of functors*

$$\mathcal{T} \underset{G}{\overset{F}{\rightleftarrows}} \mathcal{U}$$

between triangulated categories such that F is exact and set $\mathcal{S} = \operatorname{Ker} F$. Then G is fully faithful if and only if F induces a triangle equivalence $\mathcal{T}/\mathcal{S} \xrightarrow{\sim} \mathcal{U}$.

Proof Let $S = \{\sigma \in \operatorname{Mor} \mathcal{T} \mid F\sigma \text{ is invertible}\}$. Then G is fully faithful if and only if F induces an equivalence $\mathcal{T}[S^{-1}] \xrightarrow{\sim} \mathcal{U}$, by Proposition 1.1.3. It remains to observe that $\mathcal{T}[S^{-1}] = \mathcal{T}/\mathcal{S}$, since $S = S(\mathcal{S})$. Here we use that F is exact. $\square$

We note the symmetry for triangulated categories which differs from that for abelian categories. For an abelian category $\mathcal{A}$ and a Serre subcategory $\mathcal{C} \subseteq \mathcal{A}$, a right adjoint of $\mathcal{A} \to \mathcal{A}/\mathcal{C}$ implies the existence of a right adjoint of $\mathcal{C} \hookrightarrow \mathcal{A}$ (Lemma 2.2.10), but the converse is not true without further assumptions. Also, a right adjoint of $\mathcal{A} \to \mathcal{A}/\mathcal{C}$ need not be exact.

3.3 Frobenius Categories

Stable categories of Frobenius categories provide important examples of triangulated categories. The exact structure of a Frobenius category induces a canonical triangulated structure of the stable category. In particular, there are canonical choices of exact triangles and morphisms between such triangles. With these choices the formation of cones becomes functorial.

Stable Categories of Frobenius Categories

An exact category $\mathcal{A}$ is a *Frobenius category* if there are enough projective and enough injective objects, and if projective and injective objects coincide. Let $\mathcal{P}$ denote the full subcategory of projective objects. The *stable category* $\operatorname{St}\mathcal{A}$ is by definition the additive quotient $\mathcal{A}/\mathcal{P}$. For objects X, Y we set

$$\underline{\operatorname{Hom}}_{\mathcal{A}}(X,Y) = \operatorname{Hom}_{\operatorname{St}\mathcal{A}}(X,Y).$$

Let $\mathcal{A}$ be a Frobenius category and fix for each object X an admissible monomorphism $X \to IX$ such that IX is an injective object. The *cone* of a morphism $\phi \colon X \to Y$ is obtained by forming the following pushout diagram

$$
\begin{array}{ccccccccc}
0 & \longrightarrow & X & \longrightarrow & IX & \longrightarrow & \Sigma X & \longrightarrow & 0 \\
& & \phi \downarrow & & \downarrow & & \| & & \\
0 & \longrightarrow & Y & \xrightarrow{\phi'} & \operatorname{Cone} \phi & \xrightarrow{\phi''} & \Sigma X & \longrightarrow & 0
\end{array}
$$

and we call (ϕ, ϕ', ϕ'') a *cone sequence* induced by ϕ. Note that this diagram depends on the choice of $X \to IX$, but it is unique up to an isomorphism when one passes to the stable category of $\mathcal{A}$. In particular, a morphism in $\mathcal{A}$ has a projective cone if and only if its image under $\mathcal{A} \to \mathrm{St}\,\mathcal{A}$ is invertible in $\mathrm{St}\,\mathcal{A}$.

Now let $\xi\colon 0 \to X \to Y \to Z \to 0$ be an admissible exact sequence in $\mathcal{A}$. We consider the induced commutative diagram

$$
\begin{array}{ccccccccc}
0 & \longrightarrow & X & \xrightarrow{\alpha} & Y & \xrightarrow{\beta} & Z & \longrightarrow & 0 \\
& & \parallel & & \downarrow & & \downarrow{\scriptstyle\gamma} & & \\
0 & \longrightarrow & X & \longrightarrow & IX & \longrightarrow & \Sigma X & \longrightarrow & 0
\end{array}
$$

and call (α, β, γ) a *standard triangle* induced by ξ. Again, the triangle is unique up to isomorphism in $\mathrm{St}\,\mathcal{A}$.

Let us compare cone sequences and standard triangles by taking them into the stable category $\mathrm{St}\,\mathcal{A}$.

Lemma 3.3.1. *In $\mathrm{St}\,\mathcal{A}$, a triangle (α, β, γ) is isomorphic to a cone sequence induced by a morphism in $\mathcal{A}$ if and only if (α, β, γ) is isomorphic to a standard triangle induced by an admissible exact sequence in $\mathcal{A}$.*

Proof   Given a morphism $\phi\colon X \to Y$ in $\mathcal{A}$, the pushout defining the cone sequence (ϕ, ϕ', ϕ'') yields an admissible exact sequence $0 \to X \to Y \oplus IX \to \mathrm{Cone}\,\phi \to 0$. On the other hand, an admissible exact sequence $0 \to X \to Y \to Z \to 0$ yields the following pushout diagram

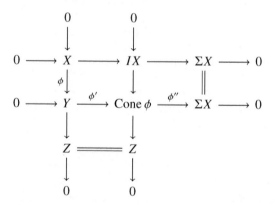

and it is clear that $\mathrm{Cone}\,\phi \to Z$ is an isomorphism in $\mathrm{St}\,\mathcal{A}$. $\square$

Proposition 3.3.2. *Let $\mathcal{A}$ be a Frobenius category. Then the assignment $X \mapsto \Sigma X$ induces an equivalence $\mathrm{St}\,\mathcal{A} \xrightarrow{\sim} \mathrm{St}\,\mathcal{A}$, and the category $\mathrm{St}\,\mathcal{A}$ together with all triangles isomorphic to the image of a standard triangle in $\mathcal{A}$ is a triangulated category.*

A triangulated category that is triangle equivalent to the stable category of a Frobenius category is called *algebraic*. In fact, all specific triangulated categories arising in this book are algebraic. Further descriptions are provided in Proposition 9.1.5 and Proposition 9.1.15.

The proof of Proposition 3.3.2 requires some preparation. For each $X \in \mathcal{A}$ fix an exact sequence

$$\omega_X : \quad 0 \longrightarrow X \xrightarrow{\ x\ } IX \xrightarrow{\ \bar{x}\ } \Sigma X \longrightarrow 0.$$

Lemma 3.3.3. *Multiplication by ω_X induces a natural isomorphism*

$$\underline{\mathrm{Hom}}(-, \Sigma X) \xrightarrow{\ \sim\ } \mathrm{Ext}^1(-, X).$$

A standard triangle (α, β, γ) corresponding to an exact sequence $\xi : 0 \to X \xrightarrow{\alpha} Y \xrightarrow{\beta} Z \to 0$ in $\mathcal{A}$ induces an exact sequence of functors

$$\mathrm{Hom}(-, X) \xrightarrow{(-,\alpha)} \mathrm{Hom}(-, Y) \xrightarrow{(-,\beta)} \mathrm{Hom}(-, Z) \xrightarrow{(-,\gamma)} \underline{\mathrm{Hom}}(-, \Sigma X)$$

which is functorial in X and Z. Moreover, we have $\omega_X \cdot \gamma = \xi$.

Proof The cokernel of $\mathrm{Hom}(-, IX) \to \mathrm{Hom}(-, \Sigma X)$ equals $\underline{\mathrm{Hom}}(-, \Sigma X)$ which is therefore isomorphic to $\mathrm{Ext}^1(-, X)$. Thus for $Z \in \mathcal{A}$ the isomorphism

$$\underline{\mathrm{Hom}}(Z, \Sigma X) \xrightarrow{\ \sim\ } \mathrm{Ext}^1(Z, X)$$

maps ϕ to $\omega_X \cdot \phi$. The identity $\omega_X \cdot \gamma = \xi$ follows from the definition of a standard triangle, and then the exact sequence of functors is clear. □

Proof of Proposition 3.3.2 The first assertion is easily checked. For the verification of the axioms of a triangulated category we use Lemma 3.3.1 and standard properties of exact categories.

(Tr1) The class of exact triangles is closed under isomorphisms by definition. The standard triangle given by the exact sequence $0 \to 0 \to X \xrightarrow{\mathrm{id}} X \to 0$ equals $0 \to X \xrightarrow{\mathrm{id}} X \to 0$. From the definition of a cone sequence (ϕ, ϕ', ϕ'') it is clear that each morphism ϕ fits into an exact triangle.

(Tr2) Fix a standard triangle (α, β, γ) given by the following commutative diagram with exact rows.

$$
\begin{array}{ccccccccc}
0 & \longrightarrow & X & \xrightarrow{\ \alpha\ } & Y & \xrightarrow{\ \beta\ } & Z & \longrightarrow & 0 \\
 & & \| & & \downarrow{\scriptstyle \alpha'} & & \downarrow{\scriptstyle \gamma} & & \\
0 & \longrightarrow & X & \xrightarrow{\ x\ } & IX & \xrightarrow{\ \bar{x}\ } & \Sigma X & \longrightarrow & 0
\end{array}
$$

Then consider the following diagram with exact rows.

$$
\begin{array}{ccccccccc}
0 & \longrightarrow & Y & \xrightarrow{\left[\begin{smallmatrix}\beta\\\alpha'\end{smallmatrix}\right]} & Z \oplus IX & \xrightarrow{\left[\gamma\ -\bar{x}\right]} & \Sigma X & \longrightarrow & 0 \\
& & \| & & \downarrow{\scriptstyle\left[\beta'\ I\alpha\right]} & & \downarrow{\scriptstyle -\Sigma\alpha} & & \\
0 & \longrightarrow & Y & \xrightarrow{\ y\ } & IY & \xrightarrow{\ \bar{y}\ } & \Sigma Y & \longrightarrow & 0
\end{array}
$$

From the identity

$$(y - I\alpha\alpha')\alpha = y\alpha - I\alpha x = 0$$

we obtain $\beta' \colon Z \to IY$ satisfying $y - I\alpha\alpha' = \beta'\beta$; so the left hand square commutes. For the commutativity of the other square we compute

$$\bar{y}\beta'\beta = \bar{y}y - \bar{y}I\alpha\alpha' = -\Sigma\alpha\bar{x}\alpha' = -\Sigma\alpha\gamma\beta.$$

Thus $\bar{y}\beta' = -\Sigma\alpha\gamma$ since β is an epimorphism. Now the diagram yields a standard triangle which is isomorphic to $(\beta, \gamma, -\Sigma\alpha)$.

   (Tr3) Fix exact triangles (α, β, γ) and $(\alpha', \beta', \gamma')$ with a pair of morphisms ϕ_1 and ϕ_2 satisfying $\phi_2\alpha = \alpha'\phi_1$. We may assume them to be standard triangles and that the equality $\phi_2\alpha = \alpha'\phi_1$ holds in $\mathcal{A}$, by adding to α an injective summand if necessary. This yields the following commutative diagram with exact rows.

$$
\begin{array}{ccccccccc}
\xi\colon & 0 & \longrightarrow & X & \xrightarrow{\alpha} & Y & \xrightarrow{\beta} & Z & \longrightarrow & 0 \\
& & & \downarrow{\scriptstyle\phi_1} & & \downarrow{\scriptstyle\phi_2} & & \downarrow{\scriptstyle\phi_3} & & \\
\xi'\colon & 0 & \longrightarrow & X' & \xrightarrow{\alpha'} & Y' & \xrightarrow{\beta'} & Z' & \longrightarrow & 0
\end{array}
$$

We need to show that $\Sigma\phi_1\gamma = \gamma'\phi_3$. Clearly, this follows from a commutative diagram of the following form.

$$
\begin{array}{ccccccc}
\mathrm{Hom}(-, X) & \longrightarrow & \mathrm{Hom}(-, Y) & \longrightarrow & \mathrm{Hom}(-, Z) & \longrightarrow & \underline{\mathrm{Hom}}(-, \Sigma X) \\
\downarrow{\scriptstyle(-,\phi_1)} & & \downarrow{\scriptstyle(-,\phi_2)} & & \downarrow{\scriptstyle(-,\phi_3)} & & \downarrow{\scriptstyle(-,\Sigma\phi_1)} \\
\mathrm{Hom}(-, X') & \longrightarrow & \mathrm{Hom}(-, Y') & \longrightarrow & \mathrm{Hom}(-, Z') & \longrightarrow & \underline{\mathrm{Hom}}(-, \Sigma X')
\end{array}
$$

We obtain this from Lemma 3.3.3, since the horizontal sequence is functorial, and using that $\phi_1\xi = \xi'\phi_3$.

   (Tr4) Fix exact triangles $\alpha = (\alpha_1, \alpha_2, \alpha_3)$, $\beta = (\beta_1, \beta_2, \beta_3)$, and $\gamma = (\gamma_1, \gamma_2, \gamma_3)$ with $\gamma_1 = \beta_1\alpha_1$. We may assume them to be standard and that the equality $\gamma_1 = \beta_1\alpha_1$ holds in $\mathcal{A}$. Then we obtain in $\mathcal{A}$ a commutative diagram with exact rows.

$$
\begin{array}{ccccccccc}
0 & \longrightarrow & X & \xrightarrow{\alpha_1} & Y & \xrightarrow{\alpha_2} & U & \longrightarrow & 0 \\
 & & \| & & \downarrow{\beta_1} & & \downarrow{\delta_1} & & \\
0 & \longrightarrow & X & \xrightarrow{\gamma_1} & Z & \xrightarrow{\gamma_2} & V & \longrightarrow & 0
\end{array}
$$

From this we obtain a standard triangle $(\delta_1, \delta_2, \delta_3)$ making the following diagram commutative.

$$
\begin{array}{ccccccc}
X & \xrightarrow{\alpha_1} & Y & \xrightarrow{\alpha_2} & U & \xrightarrow{\alpha_3} & \Sigma X \\
\| & & \downarrow{\beta_1} & & \downarrow{\delta_1} & & \| \\
X & \xrightarrow{\gamma_1} & Z & \xrightarrow{\gamma_2} & V & \xrightarrow{\gamma_3} & \Sigma X \\
 & & \downarrow{\beta_2} & & \downarrow{\delta_2} & & \\
 & & W & = & W & & \\
 & & \downarrow{\beta_3} & & \downarrow{\delta_3} & & \\
 & & \Sigma Y & \xrightarrow{\Sigma\alpha_2} & \Sigma U & &
\end{array}
$$

It remains to check the identity $\beta_3\delta_2 = \Sigma\alpha_1\gamma_3$ in $\underline{\mathrm{Hom}}(V, \Sigma Y)$. This follows from the following commutative diagram

$$
\begin{array}{ccccc}
\underline{\mathrm{Hom}}(V, \Sigma X) & \xrightarrow{(V,\Sigma\alpha_1)} & \underline{\mathrm{Hom}}(V, \Sigma Y) & \xleftarrow{(\delta_2,\Sigma Y)} & \underline{\mathrm{Hom}}(W, \Sigma Y) \\
\downarrow{\iota} & & \downarrow{\iota} & & \downarrow{\iota} \\
\mathrm{Ext}^1(V, X) & \longrightarrow & \mathrm{Ext}^1(V, Y) & \longleftarrow & \mathrm{Ext}^1(W, Y)
\end{array}
$$

and the description of the third morphism in a standard triangle given in Lemma 3.3.1, since the maps in the bottom row send the extensions corresponding to β and γ to the same extension

$$
0 \longrightarrow Y \xrightarrow{\left|\begin{smallmatrix}\alpha_2\\\beta_1\end{smallmatrix}\right|} U \oplus Z \xrightarrow{\left|\begin{smallmatrix}-\delta_1 & \gamma_2\end{smallmatrix}\right|} V \longrightarrow 0. \qquad \square
$$

Example 3.3.4. For a ring Λ the following conditions are equivalent (Theorem 13.2.13).

(1) Projective and injective Λ-modules coincide.
(2) The category $\mathrm{Mod}\,\Lambda$ of Λ-modules is a Frobenius category.
(3) The ring Λ is right artinian and $\mathrm{mod}\,\Lambda$ is a Frobenius category.
(4) The ring Λ is right noetherian and the module Λ_Λ is injective.

A ring satisfying these equivalent conditions is called *quasi-Frobenius*. This notion is symmetric, so Λ is quasi-Frobenius if and only if Λ^{op} is quasi-Frobenius. A ring Λ is called *right self-injective* if the module Λ_Λ is injective, and Λ is *self-injective* if it is both right and left self-injective. Thus for noetherian rings the concepts 'quasi-Frobenius' and 'self-injective' coincide. For example, the group algebra kG of a finite group G over a field k is quasi-Frobenius and self-injective.

We write $\mathrm{StMod}\,\Lambda = \mathrm{St}(\mathrm{Mod}\,\Lambda)$ when Λ is quasi-Frobenius.

Frobenius Pairs

A *Frobenius pair* $(\mathcal{A}, \mathcal{A}_0)$ is a Frobenius category $\mathcal{A}$ together with a full additive subcategory $\mathcal{A}_0 \subseteq \mathcal{A}$ such that $\mathcal{A}_0$ contains all projective objects of $\mathcal{A}$ and the *two out of three property* holds: for an admissible exact sequence in $\mathcal{A}$ with two terms in $\mathcal{A}_0$, the third term is also in $\mathcal{A}_0$.

We observe that for a fixed Frobenius category $\mathcal{A}$ the Frobenius pairs $(\mathcal{A}, \mathcal{A}_0)$ correspond bijectively to triangulated subcategories of $\mathrm{St}\,\mathcal{A}$. The assignment sends $\mathcal{A}_0$ to its stable category $\mathrm{St}\,\mathcal{A}_0$, where $\mathcal{A}_0$ is viewed as a Frobenius category having the same projective and injective objects as $\mathcal{A}$.

Let $(\mathcal{A}, \mathcal{A}_0)$ be a Frobenius pair and set

$$S = \{\phi \in \mathrm{Mor}\,\mathcal{A} \mid \mathrm{Cone}\,\phi \in \mathcal{A}_0\}.$$

The *derived category* $\mathbf{D}(\mathcal{A}, \mathcal{A}_0)$ of $(\mathcal{A}, \mathcal{A}_0)$ is obtained by formally inverting all morphisms in S. Thus one defines

$$\mathbf{D}(\mathcal{A}, \mathcal{A}_0) = \mathcal{A}[S^{-1}].$$

For a morphism ϕ in $\mathcal{A}$ we write $\bar{\phi}$ for the corresponding morphism in $\mathrm{St}\,\mathcal{A}$.

Proposition 3.3.5. *For a Frobenius pair* $(\mathcal{A}, \mathcal{A}_0)$ *the following holds.*

(1) *The class* $\bar{S} = \{\bar{\phi} \mid \phi \in S\} \subseteq \mathrm{Mor}\,\mathrm{St}\,\mathcal{A}$ *admits a calculus of left and right fractions, and the canonical functor* $\mathcal{A} \to \mathbf{D}(\mathcal{A}, \mathcal{A}_0)$ *induces an equivalence*

$$(\mathrm{St}\,\mathcal{A})[\bar{S}^{-1}] = \mathrm{St}\,\mathcal{A}/\mathrm{St}\,\mathcal{A}_0 \xrightarrow{\sim} \mathbf{D}(\mathcal{A}, \mathcal{A}_0).$$

(2) *The assignment* $X \mapsto \Sigma X$ *induces an equivalence* $\mathbf{D}(\mathcal{A}, \mathcal{A}_0) \xrightarrow{\sim} \mathbf{D}(\mathcal{A}, \mathcal{A}_0)$, *and the category* $\mathbf{D}(\mathcal{A}, \mathcal{A}_0)$ *together with all triangles isomorphic to the localisation of a cone sequence* (ϕ, ϕ', ϕ'') *in* $\mathcal{A}$ *is a triangulated category.*

Proof The stable category $\mathrm{St}\,\mathcal{A}$ is the localisation of $\mathcal{A}$ with respect to the class of morphisms ϕ in $\mathcal{A}$ such that $\mathrm{Cone}\,\phi$ is projective, by Lemma 2.2.2. Thus $(\mathrm{St}\,\mathcal{A})[\bar{S}^{-1}]$ identifies with $\mathcal{A}[S^{-1}]$. Next observe that $\mathrm{St}\,\mathcal{A}_0$ is a triangulated subcategory of $\mathrm{St}\,\mathcal{A}$. It follows from Lemma 3.2.1 that $\bar{S}$ admits a calculus of left and right fractions, and the localisation $(\mathrm{St}\,\mathcal{A})[\bar{S}^{-1}]$ equals the Verdier localisation $\mathrm{St}\,\mathcal{A}/\mathrm{St}\,\mathcal{A}_0$. Now the triangulated structure of $\mathbf{D}(\mathcal{A},\mathcal{A}_0)$ is induced by that of $\mathrm{St}\,\mathcal{A}$, using Proposition 3.2.2 and Proposition 3.3.2. □

The class $S \subseteq \mathrm{Mor}\,\mathcal{A}$ admits a calculus of left fractions if and only if $\mathcal{A}_0 = \mathcal{A}$. For instance, (LF3) fails for a pair $\alpha, \beta \colon X \to Y$ where $\alpha = 0$ and β is an epimorphism with projective X and $Y \in \mathcal{A} \setminus \mathcal{A}_0$.

The construction of the derived category $\mathbf{D}(\mathcal{A},\mathcal{A}_0)$ yields the following universal property.

Corollary 3.3.6. *Let* $(\mathcal{A},\mathcal{A}_0)$ *be a Frobenius pair. If* $\mathcal{T}$ *is a triangulated category and* $F \colon \mathrm{St}\,\mathcal{A} \to \mathcal{T}$ *is an exact functor such that* $F|_{\mathcal{A}_0} = 0$, *then there exists a unique exact functor* $\bar{F} \colon \mathbf{D}(\mathcal{A},\mathcal{A}_0) \to \mathcal{T}$ *making the following diagram commutative:*

$$
\begin{array}{ccc}
\mathcal{A} & \longrightarrow & \mathbf{D}(\mathcal{A},\mathcal{A}_0) \\
\downarrow & & \downarrow{\scriptstyle \bar{F}} \\
\mathrm{St}\,\mathcal{A} & \xrightarrow{\ F\ } & \mathcal{T}
\end{array}
$$

Proof Combine Proposition 3.2.2 and Proposition 3.3.5. □

3.4 Brown Representability

In this section we study triangulated categories that admit arbitrary coproducts. An important aspect in this context is the representability of cohomological functors. We discuss two versions of Brown's representability theorem. In each case the category needs to be generated by objects satisfying certain finiteness conditions. The most natural condition is 'compactness', which means that the functor $\mathrm{Hom}(X, -)$ preserves all coproducts. The construction of representing objects is fairly explicit and involves homotopy colimits.

Homotopy (Co)limits

Let $\mathcal{T}$ be a triangulated category and suppose that countable coproducts exist in $\mathcal{T}$. Let

$$
X_0 \xrightarrow{\ \phi_0\ } X_1 \xrightarrow{\ \phi_1\ } X_2 \xrightarrow{\ \phi_2\ } \cdots
$$

be a sequence of morphisms in $\mathcal{T}$. A *homotopy colimit* of this sequence is by definition an object X that occurs in an exact triangle

$$
\Sigma^{-1}X \longrightarrow \coprod_{n \geq 0} X_n \xrightarrow{\ \mathrm{id}-\phi\ } \coprod_{n \geq 0} X_n \xrightarrow{\ \mu\ } X.
$$

Here, the nth component of the morphism $\mathrm{id} - \phi$ is the composite

$$X_n \xrightarrow{\begin{vmatrix} \mathrm{id} \\ -\phi_n \end{vmatrix}} X_n \oplus X_{n+1} \xrightarrow{\ \mathrm{inc}\ } \coprod_{n \geq 0} X_n.$$

We write $\mathrm{hocolim}_n X_n$ for X; this comes with canonical morphisms

$$\mu_i \colon X_i \longrightarrow \operatorname*{hocolim}_n X_n \qquad\qquad (i \geq 0).$$

Note that a homotopy colimit is unique up to a non-unique isomorphism. In some cases the obstruction for uniqueness is controlled by phantom morphisms; see Lemma 5.2.5.

Lemma 3.4.1. *Let* $(\alpha_n \colon X_n \to Y)_{n \geq 0}$ *be a sequence of morphisms in* $\mathcal{T}$ *such that* $\alpha_n = \alpha_{n+1} \phi_n$ *for all* n. *Then there exists a (usually non-unique) morphism* $\bar{\alpha} \colon \mathrm{hocolim}_n X_n \to Y$ *such that* $\alpha_n = \bar{\alpha} \mu_n$ *for all* n.

Proof The α_n yield a morphism $\alpha \colon \coprod_{n \geq 0} X_n \to Y$ satisfying $\alpha(\mathrm{id} - \phi) = 0$. Thus α factors through $\mathrm{Cone}(\mathrm{id} - \phi) = \mathrm{hocolim}_n X_n$. $\qquad\square$

The dual construction requires the existence of countable products in $\mathcal{T}$ and yields the *homotopy limit* of a sequence

$$\cdots \xrightarrow{\phi_2} X_2 \xrightarrow{\phi_1} X_1 \xrightarrow{\phi_0} X_0$$

which is by definition an object X occurring in an exact triangle

$$X \longrightarrow \prod_{n \geq 0} X_n \xrightarrow{\ \mathrm{id} - \phi\ } \prod_{n \geq 0} X_n \longrightarrow \Sigma X.$$

Again, this is unique up to a non-unique isomorphism and we write $\mathrm{holim}_n X_n$.

Remark 3.4.2. Given sequences $X_0 \to X_1 \to X_2 \to \cdots$ and $Y_0 \to Y_1 \to Y_2 \to \cdots$ of morphisms in $\mathcal{T}$, we have

$$\Big(\operatorname*{hocolim}_n X_n\Big) \oplus \Big(\operatorname*{hocolim}_n Y_n\Big) \cong \operatorname*{hocolim}_n (X_n \oplus Y_n).$$

Let us compute the functor $\mathrm{Hom}_{\mathcal{T}}(-, \mathrm{hocolim}_n X_n)$. To this end observe that a sequence

$$A_0 \xrightarrow{\phi_0} A_1 \xrightarrow{\phi_1} A_2 \xrightarrow{\phi_2} \cdots$$

of maps between abelian groups induces an exact sequence

$$0 \longrightarrow \coprod_{n \geq 0} A_n \xrightarrow{\ \mathrm{id} - \phi\ } \coprod_{n \geq 0} A_n \longrightarrow \mathrm{colim}_n A_n \longrightarrow 0$$

because it identifies with the colimit of the exact sequences

$$0 \longrightarrow \coprod_{i=0}^{n-1} A_i \xrightarrow{\ \mathrm{id} - \phi\ } \coprod_{i=0}^{n} A_i \longrightarrow A_n \longrightarrow 0.$$

Lemma 3.4.3. *Let C be an object in $\mathcal{T}$ such that $\mathrm{Hom}_{\mathcal{T}}(C, -)$ preserves all coproducts. Then any sequence $X_0 \to X_1 \to X_2 \to \cdots$ in $\mathcal{T}$ induces an isomorphism*

$$\operatorname*{colim}_{n} \mathrm{Hom}_{\mathcal{T}}(C, X_n) \xrightarrow{\sim} \mathrm{Hom}_{\mathcal{T}}(C, \operatorname*{hocolim}_{n} X_n).$$

Proof The above observation gives an exact sequence

$$0 \to \coprod_n \mathrm{Hom}_{\mathcal{T}}(C, X_n) \to \coprod_n \mathrm{Hom}_{\mathcal{T}}(C, X_n) \to \operatorname{colim}_n \mathrm{Hom}_{\mathcal{T}}(C, X_n) \to 0.$$

Now apply $\mathrm{Hom}_{\mathcal{T}}(C, -)$ to the defining triangle for $\operatorname{hocolim}_n X_n$. Comparing both sequences yields the assertion, since

$$\coprod_n \mathrm{Hom}_{\mathcal{T}}(C, X_n) \cong \mathrm{Hom}_{\mathcal{T}}\left(C, \coprod_n X_n\right). \qquad \square$$

Example 3.4.4. Let $\phi \colon X \to X$ be an idempotent morphism in $\mathcal{T}$. Consider the following sequences:

$$X \xrightarrow{\phi} X \xrightarrow{\phi} X \xrightarrow{\phi} \cdots$$

$$X \xrightarrow{\mathrm{id} - \phi} X \xrightarrow{\mathrm{id} - \phi} X \xrightarrow{\mathrm{id} - \phi} \cdots$$

Write X' for a homotopy colimit of the first sequence and X'' for a homotopy colimit of the second sequence. Then we have $X \cong X' \oplus X''$ with $X' = \mathrm{Ker}(\mathrm{id} - \phi)$ and $X'' = \mathrm{Ker}\,\phi$. In particular, a triangulated category with countable coproducts is idempotent complete.

Proof The object $X' \oplus X''$ is isomorphic to the homotopy colimit of the sequence

$$X \oplus X \xrightarrow{\left|\begin{smallmatrix} \phi & 0 \\ 0 & 1-\phi \end{smallmatrix}\right|} X \oplus X \xrightarrow{\left|\begin{smallmatrix} \phi & 0 \\ 0 & 1-\phi \end{smallmatrix}\right|} X \oplus X \xrightarrow{\left|\begin{smallmatrix} \phi & 0 \\ 0 & 1-\phi \end{smallmatrix}\right|} \cdots$$

by Remark 3.4.2. Now consider the following commutative diagram

$$
\begin{array}{ccccccc}
X \oplus X & \xrightarrow{\left|\begin{smallmatrix} \phi & 0 \\ 0 & 1-\phi \end{smallmatrix}\right|} & X \oplus X & \xrightarrow{\left|\begin{smallmatrix} \phi & 0 \\ 0 & 1-\phi \end{smallmatrix}\right|} & X \oplus X & \xrightarrow{\left|\begin{smallmatrix} \phi & 0 \\ 0 & 1-\phi \end{smallmatrix}\right|} & \cdots \\
\downarrow{\scriptstyle \alpha} & & \downarrow{\scriptstyle \alpha} & & \downarrow{\scriptstyle \alpha} & & \\
X \oplus X & \xrightarrow{\left|\begin{smallmatrix} 1 & 0 \\ 0 & 0 \end{smallmatrix}\right|} & X \oplus X & \xrightarrow{\left|\begin{smallmatrix} 1 & 0 \\ 0 & 0 \end{smallmatrix}\right|} & X \oplus X & \xrightarrow{\left|\begin{smallmatrix} 1 & 0 \\ 0 & 0 \end{smallmatrix}\right|} & \cdots
\end{array}
$$

with α given by $\left[\begin{smallmatrix} \phi & 1-\phi \\ 1-\phi & \phi \end{smallmatrix}\right]$. Observe that α is an isomorphism, since $\alpha^2 = \mathrm{id}$. The homotopy colimit of the bottom row is X, again using Remark 3.4.2, and therefore $X \cong X' \oplus X''$. $\qquad \square$

A Brown Representability Theorem

Let $\mathcal{T}$ be a triangulated category with arbitrary coproducts. A triangulated subcategory $\mathcal{S} \subseteq \mathcal{T}$ is called *localising* if it is closed under all coproducts. Given a class $\mathcal{X} \subseteq \mathcal{T}$ of objects we denote by $\mathrm{Loc}(\mathcal{X})$ the smallest localising subcategory of $\mathcal{T}$ that contains $\mathcal{X}$.

A set $\mathcal{S}$ of objects in $\mathcal{T}$ is called *perfectly generating* if $\mathrm{Loc}(\mathcal{S}) = \mathcal{T}$ and the following holds

(PG) Given a countable family of morphisms $X_i \to Y_i$ in $\mathcal{T}$ such that the map $\mathrm{Hom}_{\mathcal{T}}(S, X_i) \to \mathrm{Hom}_{\mathcal{T}}(S, Y_i)$ is surjective for all i and $S \in \mathcal{S}$, the induced map

$$\mathrm{Hom}_{\mathcal{T}}\left(S, \coprod_i X_i\right) \longrightarrow \mathrm{Hom}_{\mathcal{T}}\left(S, \coprod_i Y_i\right)$$

is surjective.

The condition $\mathrm{Loc}(\mathcal{S}) = \mathcal{T}$ can be reformulated, saying that $\mathrm{Hom}_{\mathcal{T}}(\Sigma^n S, X) = 0$ for all $S \in \mathcal{S}$ and $n \in \mathbb{Z}$ implies $X = 0$; see Corollary 3.4.8. The triangulated category $\mathcal{T}$ is called *perfectly generated* if $\mathcal{T}$ admits a perfectly generating set.

We have the following *Brown representability theorem* for a perfectly generated triangulated category.

Theorem 3.4.5 (Brown). *Let $\mathcal{T}$ be a perfectly generated triangulated category. Then a functor $F \colon \mathcal{T}^{\mathrm{op}} \to \mathrm{Ab}$ is cohomological and sends all coproducts in $\mathcal{T}$ to products if and only if $F \cong \mathrm{Hom}_{\mathcal{T}}(-, X)$ for some object X in $\mathcal{T}$.*

The proof employs the category $\mathrm{mod}\,\mathcal{T}$ of finitely presented functors on $\mathcal{T}$. The following lemma explains the basic facts which are needed; it is independent of the triangulated structure of $\mathcal{T}$. In particular, the crucial condition (PG) is explained.

Lemma 3.4.6. *Let $\mathcal{T}$ be an additive category with arbitrary coproducts and weak kernels. Let $\mathcal{S}_0$ be a set of objects in $\mathcal{T}$, and denote by $\mathcal{S}$ the full subcategory of all coproducts of objects in $\mathcal{S}$.*

(1) *The category $\mathrm{mod}\,\mathcal{T}$ is abelian and has arbitrary coproducts. Moreover, the Yoneda functor $\mathcal{T} \to \mathrm{mod}\,\mathcal{T}$ preserves all coproducts.*

(2) *The category $\mathcal{S}$ has weak kernels and $\mathrm{mod}\,\mathcal{S}$ is an abelian category.*

(3) *The assignment $F \mapsto F|_{\mathcal{S}}$ induces an exact functor $\mathrm{mod}\,\mathcal{T} \to \mathrm{mod}\,\mathcal{S}$.*

(4) *The functor $\mathcal{T} \to \mathrm{mod}\,\mathcal{S}$ sending X to $\mathrm{Hom}_{\mathcal{T}}(-, X)|_{\mathcal{S}}$ preserves countable coproducts if and only if condition (PG) holds.*

Proof First observe that for every X in $\mathcal{T}$, there exists an *approximation* $X' \to X$ such that $X' \in \mathcal{S}$ and $\mathrm{Hom}_{\mathcal{T}}(T, X') \to \mathrm{Hom}_{\mathcal{T}}(T, X)$ is surjective for all $T \in \mathcal{S}$. Take $X' = \coprod_{S \in \mathcal{S}_0} \coprod_{\alpha \in \mathrm{Hom}_{\mathcal{T}}(S, X)} S$ and the canonical morphism $X' \to X$.

(1) The category $\mathrm{mod}\,\mathcal{T}$ is abelian since every morphism in $\mathcal{T}$ has a weak kernel; see Lemma 2.1.6.

Let $(F_i)_{i \in I}$ be a family of functors in $\mathrm{mod}\,\mathcal{T}$ with presentations

$$\mathrm{Hom}_{\mathcal{T}}(-, X_i) \longrightarrow \mathrm{Hom}_{\mathcal{T}}(-, Y_i) \longrightarrow F_i \longrightarrow 0.$$

Then the coproduct $\coprod_i F$ is given by the presentation

$$\mathrm{Hom}_{\mathcal{T}}\left(-, \coprod_i X_i\right) \longrightarrow \mathrm{Hom}_{\mathcal{T}}\left(-, \coprod_i Y_i\right) \longrightarrow \coprod_i F_i \longrightarrow 0.$$

To see this we need to check that

$$\mathrm{Hom}\left(\coprod_i F_i, G\right) \cong \prod_i \mathrm{Hom}(F_i, G)$$

for each $G \in \mathrm{mod}\,\mathcal{T}$. This reduces to the case that $G = \mathrm{Hom}_{\mathcal{T}}(-, Z)$ is representable, and then it follows from Yoneda's lemma. In particular, the coproduct is *not* computed pointwise in Ab.

(2) To prove that $\mathrm{mod}\,\mathcal{S}$ is abelian, it is sufficient to show that every morphism in $\mathcal{S}$ has a weak kernel. In order to obtain a weak kernel of a morphism $Y \to Z$ in $\mathcal{S}$, take the composite of a weak kernel $X \to Y$ in $\mathcal{T}$ and an approximation $X' \to X$.

(3) It follows from Proposition 2.2.20 that restriction to $\mathcal{S}$ yields a functor $\mathrm{mod}\,\mathcal{T} \to \mathrm{mod}\,\mathcal{S}$. Clearly, restriction is exact.

(4) We denote by $i \colon \mathcal{S} \to \mathcal{T}$ the inclusion and write $i^* \colon \mathrm{mod}\,\mathcal{T} \to \mathrm{mod}\,\mathcal{S}$ for the restriction functor. Then i^* induces an equivalence $(\mathrm{mod}\,\mathcal{T})/(\mathrm{Ker}\,i^*) \xrightarrow{\sim} \mathrm{mod}\,\mathcal{S}$; see again Proposition 2.2.20.

Thus the functor $\mathcal{T} \to \mathrm{mod}\,\mathcal{S}$ preserves countable coproducts if and only if i^* preserves countable coproducts, and this happens if and only if if $\mathrm{Ker}\,i^*$ is closed under countable coproducts; see Remark 2.2.7.

Now observe that $\mathrm{Ker}\,i^*$ being closed under countable coproducts is a reformulation of the condition (PG). □

Proof of Theorem 3.4.5 Fix a perfectly generating set $\mathcal{S}_0$ and denote by S the coproduct of all suspensions of objects in $\mathcal{S}_0$. It is easily checked that $\{S\}$ is perfectly generating. Taking coproducts and suspensions does not affect the condition (PG). Also, $\mathrm{Loc}(S) = \mathrm{Loc}(\mathcal{S}_0)$ because a triangulated subcategory closed under countable coproducts is closed under direct summands; see Example 3.4.4.

We construct inductively a sequence

$$X_0 \xrightarrow{\phi_0} X_1 \xrightarrow{\phi_1} X_2 \xrightarrow{\phi_2} \cdots$$

of morphisms in $\mathcal{T}$ and elements π_i in FX_i as follows. Set $X_0 = 0$ and $\pi_0 = 0$. Let $X_1 = S^{[FS]}$ be the coproduct of copies of S indexed by the elements in FS, and let π_1 be the element corresponding to id_{FS} in $FX_1 \cong (FS)^{FS}$. Suppose we have already constructed ϕ_{i-1} and π_i for some $i > 0$. Let

$$K_i = \{\alpha \in \mathrm{Hom}_{\mathcal{T}}(S, X_i) \mid (F\alpha)\pi_i = 0\}$$

and complete the canonical morphism $\chi_i : S^{[K_i]} \to X_i$ to an exact triangle

$$S^{[K_i]} \xrightarrow{\chi_i} X_i \xrightarrow{\phi_i} X_{i+1} \longrightarrow \Sigma S^{[K_i]}.$$

Now choose an element π_{i+1} in FX_{i+1} such that $(F\phi_i)\pi_{i+1} = \pi_i$. This is possible since $(F\chi_i)\pi_i = 0$ and F is cohomological.

Let $\mathcal{S}$ denotes the full subcategory of all coproducts of copies of S in $\mathcal{T}$. We identify each π_i via Yoneda's lemma with a morphism $\mathrm{Hom}_{\mathcal{T}}(-, X_i) \to F$ and obtain in mod $\mathcal{S}$ the following commutative diagram with split exact rows, where $\psi_i = \mathrm{Hom}_{\mathcal{T}}(-, \phi_i)|_{\mathcal{S}}$.

$$
\begin{array}{ccccccccc}
0 & \longrightarrow & \mathrm{Ker}\,\pi_i|_{\mathcal{S}} & \longrightarrow & \mathrm{Hom}_{\mathcal{T}}(-, X_i)|_{\mathcal{S}} & \xrightarrow{\pi_i} & F|_{\mathcal{S}} & \longrightarrow & 0 \\
& & \downarrow{\scriptstyle 0} & & \downarrow{\scriptstyle \psi_i} & & \| & & \\
0 & \longrightarrow & \mathrm{Ker}\,\pi_{i+1}|_{\mathcal{S}} & \longrightarrow & \mathrm{Hom}_{\mathcal{T}}(-, X_{i+1})|_{\mathcal{S}} & \xrightarrow{\pi_{i+1}} & F|_{\mathcal{S}} & \longrightarrow & 0
\end{array}
$$

We wish to compute the colimit of the sequence $(\psi_i)_{i \geq 0}$. Taking coproducts yields the following commutative diagram with exact rows

$$
\begin{array}{ccccccccc}
0 & \to & \coprod_{i \geq 0} \mathrm{Ker}\,\pi_i|_{\mathcal{S}} & \to & \coprod_{i \geq 0} \mathrm{Hom}_{\mathcal{T}}(-, X_i)|_{\mathcal{S}} & \to & \coprod_{i \geq 0} F|_{\mathcal{S}} & \to & 0 \\
& & \downarrow{\scriptstyle \mathrm{id}-0} & & \downarrow{\scriptstyle \mathrm{id}-\psi} & & \downarrow{\scriptstyle \mathrm{id}-\mathrm{id}} & & \\
0 & \to & \coprod_{i \geq 0} \mathrm{Ker}\,\pi_i|_{\mathcal{S}} & \to & \coprod_{i \geq 0} \mathrm{Hom}_{\mathcal{T}}(-, X_i)|_{\mathcal{S}} & \to & \coprod_{i \geq 0} F|_{\mathcal{S}} & \to & 0
\end{array}
$$

and then the snake lemma yields the following exact sequence.

$$0 \longrightarrow \coprod_{i \geq 0} \mathrm{Hom}_{\mathcal{T}}(-, X_i)|_{\mathcal{S}} \xrightarrow{\mathrm{id}-\psi} \coprod_{i \geq 0} \mathrm{Hom}_{\mathcal{T}}(-, X_i)|_{\mathcal{S}} \longrightarrow F|_{\mathcal{S}} \longrightarrow 0.$$

$$(3.4.7)$$

Next consider the exact triangle

$$\Sigma^{-1} X \longrightarrow \coprod_{i \geq 0} X_i \xrightarrow{\mathrm{id}-\phi} \coprod_{i \geq 0} X_i \longrightarrow X$$

and observe that

$$(\pi_i) \in \prod_{i \geq 0} FX_i \cong F\left(\coprod_{i \geq 0} X_i\right)$$

induces a morphism

$$\pi \colon \operatorname{Hom}_{\mathcal{T}}(-, X) \longrightarrow F$$

by Yoneda's lemma. We have an isomorphism

$$\coprod_{i \geq 0} \operatorname{Hom}_{\mathcal{T}}(-, X_i)|_{\mathcal{S}} \cong \operatorname{Hom}_{\mathcal{T}}\left(-, \coprod_{i \geq 0} X_i\right)\Big|_{\mathcal{S}}$$

because of the reformulation of condition (PG) in Lemma 3.4.6, and we obtain in mod $\mathcal{S}$ the following exact sequence:

$$\coprod_{i \geq 0} \operatorname{Hom}_{\mathcal{T}}(-, X_i)|_{\mathcal{S}} \xrightarrow{\mathrm{id}-\psi} \coprod_{i \geq 0} \operatorname{Hom}_{\mathcal{T}}(-, X_i)|_{\mathcal{S}} \longrightarrow \operatorname{Hom}_{\mathcal{T}}(-, X)|_{\mathcal{S}}$$

$$\longrightarrow \coprod_{i \geq 0} \operatorname{Hom}_{\mathcal{T}}(-, \Sigma X_i)|_{\mathcal{S}} \xrightarrow{\mathrm{id}-\Sigma\psi} \coprod_{i \geq 0} \operatorname{Hom}_{\mathcal{T}}(-, \Sigma X_i)|_{\mathcal{S}}.$$

A comparison with the exact sequence (3.4.7) shows that

$$\pi|_{\mathcal{S}} \colon \operatorname{Hom}_{\mathcal{T}}(-, X)|_{\mathcal{S}} \longrightarrow F|_{\mathcal{S}}$$

is an isomorphism since $\mathrm{id} - \Sigma\psi$ is a monomorphism. Here one uses that $\Sigma S \cong S$.

Finally, observe that the objects Y in $\mathcal{T}$ such that π_Y is an isomorphism form a localising subcategory of $\mathcal{T}$. We conclude that π is an isomorphism, since $\operatorname{Loc}(\mathcal{S}_0) = \mathcal{T}$. □

We collect several consequences of the Brown representability theorem. For instance, the following provides a useful reformulation of the definition of a perfectly generating set.

Corollary 3.4.8. *Let $\mathcal{T}$ be a triangulated category with arbitrary coproducts and let $\mathcal{S}_0$ be a set of objects satisfying* (PG). *Then* $\operatorname{Loc}(\mathcal{S}_0) = \mathcal{T}$ *if and only if* $\operatorname{Hom}_{\mathcal{T}}(\Sigma^n S, X) = 0$ *for all $S \in \mathcal{S}_0$ and $n \in \mathbb{Z}$ implies $X = 0$ for each $X \in \mathcal{T}$.*

Proof Suppose that $\operatorname{Loc}(\mathcal{S}_0) = \mathcal{T}$ holds. Let $X \in \mathcal{T}$ satisfy $\operatorname{Hom}_{\mathcal{T}}(\Sigma^n S, X) = 0$ for all $S \in \mathcal{S}_0$ and $n \in \mathbb{Z}$. The objects $U \in \mathcal{T}$ satisfying $\operatorname{Hom}_{\mathcal{T}}(\Sigma^n U, X) = 0$ for all $n \in \mathbb{Z}$ form a localising subcategory of $\mathcal{T}$ containing $\mathcal{S}_0$. Thus $X = 0$.

For the other implication fix an object $X \in \mathcal{T}$. The above proof yields for $F = \operatorname{Hom}_{\mathcal{T}}(-, X)$ an object $X' \in \operatorname{Loc}(\mathcal{S}_0)$ and a morphism $\pi \colon X' \to X$ which restricts to an isomorphism

$$\operatorname{Hom}_{\mathcal{T}}(-, X')|_{\mathcal{S}} \xrightarrow{\sim} \operatorname{Hom}_{\mathcal{T}}(-, X)|_{\mathcal{S}}.$$

The condition on $\mathcal{S}_0$ implies $\operatorname{Cone} \pi = 0$. Thus $X' \cong X$, so $\operatorname{Loc}(\mathcal{S}_0) = \mathcal{T}$. □

Corollary 3.4.9. *Let $\mathcal{S}$ be a perfectly generating set for $\mathcal{T}$. Then every object in $\mathcal{T}$ can be written as a homotopy colimit of a sequence*

$$X_0 \xrightarrow{\ \phi_0\ } X_1 \xrightarrow{\ \phi_1\ } X_2 \xrightarrow{\ \phi_2\ } \cdots$$

of morphisms in $\mathcal{T}$ such that $X_0 = 0$ and the cone of each ϕ_i is a coproduct of suspensions of objects in $\mathcal{S}$.

Proof Let $X \in \mathcal{T}$ and consider the functor $F = \mathrm{Hom}_{\mathcal{T}}(-, X)$. Then the construction of the representing object in the above proof yields X as a homotopy colimit of a sequence having the desired properties. □

Corollary 3.4.10. *A perfectly generated triangulated category has arbitrary products.*

Proof Given a family of objects X_i, the product $\prod_i X_i$ is the object representing the functor $\prod_i \mathrm{Hom}(-, X_i)$. □

Corollary 3.4.11. *Let $\mathcal{T}$ be a perfectly generated triangulated category. Then an exact functor $\mathcal{T} \to \mathcal{U}$ between triangulated categories preserves all coproducts if and only if it has a right adjoint.*

Proof Let $F \colon \mathcal{T} \to \mathcal{U}$ be an exact functor. If F preserves all coproducts, then one defines the right adjoint G by sending an object X in $\mathcal{U}$ to the object in $\mathcal{T}$ representing $\mathrm{Hom}_{\mathcal{U}}(F-, X)$. Thus

$$\mathrm{Hom}_{\mathcal{U}}(F-, X) \cong \mathrm{Hom}_{\mathcal{T}}(-, GX).$$

Conversely, given a right adjoint of F, it is automatic that F preserves all coproducts. □

Remark 3.4.12. There is the dual concept of a *perfectly cogenerating* set for a triangulated category. The dual Brown representability theorem for a perfectly cogenerated triangulated category $\mathcal{T}$ characterises the representable functors $\mathrm{Hom}_{\mathcal{T}}(X, -)$ as the cohomological and product preserving functors $\mathcal{T} \to \mathrm{Ab}$.

Compact Objects

Let $\mathcal{T}$ be a triangulated category. An object X in $\mathcal{T}$ is called *compact* (or *small*) if for any morphism $\phi \colon X \to \coprod_{i \in I} Y_i$ in $\mathcal{T}$ there is a finite set $J \subseteq I$ such that ϕ factors through $\coprod_{i \in J} Y_i$. It is easily checked that X is compact if and only if the canonical map

$$\coprod_{i \in I} \mathrm{Hom}_{\mathcal{T}}(X, Y_i) \longrightarrow \mathrm{Hom}_{\mathcal{T}}\left(X, \coprod_{i \in I} Y_i\right)$$

is bijective for all coproducts $\coprod_{i \in I} Y_i$ in $\mathcal{T}$. It follows that the compact objects form a thick subcategory of $\mathcal{T}$.

We wish to describe all compact objects of a triangulated category. To this end we make the following definition.

For classes $\mathcal{U}$ and $\mathcal{V}$ of objects in a triangulated category $\mathcal{T}$ we denote by $\mathcal{U} * \mathcal{V}$ the class of objects $X \in \mathcal{T}$ that fit into an exact triangle $U \to X \to V \to \Sigma U$ such that $U \in \mathcal{U}$ and $V \in \mathcal{V}$. The octahedral axiom implies that the operation $*$ is associative. For a class $\mathcal{X}$ the objects of $\mathcal{X} * \mathcal{X} * \cdots * \mathcal{X}$ (n factors) are called *extensions of length n* of objects in $\mathcal{X}$.

Let $\mathcal{C} \subseteq \mathcal{T}$ be a class of objects and suppose that $\mathcal{C}$ is closed under all suspensions. We write $\coprod \mathcal{C}$ for the class of all coproducts of objects in $\mathcal{C}$.

Proposition 3.4.13. *Let $X \in \mathcal{T}$ be an object that is a direct summand of an extension of objects in $\coprod \mathcal{C}$. If X and all objects in $\mathcal{C}$ are compact, then X is a direct summand of an extension of objects in $\mathcal{C}$.*

Proof Let $X \to Y$ be a split monomorphism such that Y is an extension of objects in $\coprod \mathcal{C}$. Then the assertion follows from the lemma below by choosing $Y' = 0$. More precisely, complete the morphism $X' \to X$ in this lemma to an exact triangle $X' \to X \to X'' \to \Sigma X'$. The choice for Y' implies that the morphism $X \to Y$ factors through $X \to X''$. In particular, $X \to X''$ is a split monomorphism, so X is a direct summand of an extension of objects in $\mathcal{C}$. $\quad\square$

Lemma 3.4.14. *Let X and all objects in $\mathcal{C}$ be compact. Also, let $Y' \to Y$ be a morphism such that its cone is an extension of objects in $\coprod \mathcal{C}$. Then each morphism $X \to Y$ fits into a commutative square*

$$
\begin{array}{ccc}
X' & \longrightarrow & X \\
\downarrow & & \downarrow \\
Y' & \longrightarrow & Y
\end{array}
$$

such that the cone of $X' \to X$ is an extension of objects in $\mathcal{C}$.

Proof Complete $\psi : Y' \to Y$ to an exact triangle $Y' \to Y \to Y'' \to \Sigma Y'$. We use induction on the length l of Y''. If $l = 1$, then $Y'' \in \coprod \mathcal{C}$ and the composite $X \to Y \to Y''$ factors through a summand X'' of Y'' that lies in $\mathcal{C}$ since X is compact. We complete $X \to X''$ to an exact triangle $X' \to X \to X'' \to \Sigma X'$ and $X' \to X$ factors through $Y' \to Y$ by construction. Now let $l > 1$ and write Y'' as an extension $Y''_0 \to Y'' \to Y''_1 \to \Sigma Y''_0$ of objects having smaller length than l. Using the octahedral axiom we obtain the following morphism of exact triangles

$$
\begin{array}{ccccccc}
Y' & \xrightarrow{\psi_0} & Y_0 & \longrightarrow & Y_0'' & \longrightarrow & \Sigma Y' \\
\Big\| & & \Big\downarrow{\psi_1} & & \Big\downarrow & & \Big\| \\
Y' & \xrightarrow{\psi} & Y & \longrightarrow & Y'' & \longrightarrow & \Sigma Y'
\end{array}
$$

where ψ admits a factorisation $\psi = \psi_1 \psi_0$ with $\mathrm{Cone}\,\psi_i = Y_i''$. By induction we have a pair of commutative squares

$$
\begin{array}{ccccc}
X' & \xrightarrow{\phi_0} & X_0 & \xrightarrow{\phi_1} & X \\
\Big\downarrow & & \Big\downarrow & & \Big\downarrow \\
Y' & \xrightarrow{\psi_0} & Y_0 & \xrightarrow{\psi_1} & Y
\end{array}
$$

such that the cone of each ϕ_i is an extension of objects in $\mathcal{C}$. Then the same holds for the cone of $\phi_1 \phi_0$ by the octahedral axiom. □

Compact Generators

Let $\mathcal{T}$ be a triangulated category that admits arbitrary coproducts. A set $\mathcal{C}$ of compact objects is called *compactly generating* if $\mathcal{T}$ has no proper localising subcategory containing $\mathcal{C}$. In this case $\mathcal{T}$ is called *compactly generated*.

Proposition 3.4.15. *Let $\mathcal{T}$ be a compactly generated triangulated category and $\mathcal{C}$ a generating set of compact objects. Then $\mathcal{C}$ is a perfectly generating set for $\mathcal{T}$ and the full subcategory of compact objects equals* Thick($\mathcal{C}$).

Proof The first assertion follows easily from the fact that for any family of maps $\phi_i \colon A_i \to B_i$ between abelian groups we have

$$
\prod_i \phi_i \text{ is an epimorphism} \iff \text{each } \phi_i \text{ is an epimorphism}
$$

$$
\iff \coprod_i \phi_i \text{ is an epimorphism.}
$$

Clearly, the compact objects form a thick subcategory of $\mathcal{T}$. It follows from Corollary 3.4.9 that each object $X \in \mathcal{T}$ can be written as the homotopy colimit hocolim X_n of objects that are extensions of coproducts of suspension of objects in $\mathcal{C}$. If X is compact, then Lemma 3.4.3 implies that id_X factors through the canonical morphism $X_n \to X$ for some n. We conclude from Proposition 3.4.13 that X belongs to Thick($\mathcal{C}$). □

The following *Brown representability theorem* is an immediate consequence of Theorem 3.4.5. In fact, all corollaries of Theorem 3.4.5 apply to compactly generated triangulated categories as well. In particular, the definition of 'compactly generated' may be reformulated: a set $\mathcal{C}$ of compact objects generates if $\mathrm{Hom}(\Sigma^n C, X) = 0$ for all $C \in \mathcal{C}$ and $n \in \mathbb{Z}$ implies $X = 0$; see Corollary 3.4.8.

Theorem 3.4.16 (Brown). *Let $\mathcal{T}$ be a compactly generated triangulated category. Then a functor $F : \mathcal{T}^{\mathrm{op}} \to \mathrm{Ab}$ is cohomological and sends all coproducts in $\mathcal{T}$ to products if and only if $F \cong \mathrm{Hom}_{\mathcal{T}}(-, X)$ for some object X in $\mathcal{T}$.* □

There is also a version of Brown representability for functors preserving products, keeping in mind that arbitrary products exist in a compactly generated triangulated category, by Corollary 3.4.10.

Theorem 3.4.17. *Let $\mathcal{T}$ be a compactly generated triangulated category. Then a functor $F : \mathcal{T} \to \mathrm{Ab}$ is cohomological and preserves all products in $\mathcal{T}$ if and only if $F \cong \mathrm{Hom}_{\mathcal{T}}(X, -)$ for some object X in $\mathcal{T}$.*

Proof Let $\mathcal{C}$ be a set of compact generators for $\mathcal{T}$. We claim that $\mathcal{T}^{\mathrm{op}}$ is also perfectly generated. Then the assertion follows from Theorem 3.4.5. For $C \in \mathcal{C}$ let C^* denote the object in $\mathcal{T}$ that represents $\mathrm{Hom}_{\mathbb{Z}}(\mathrm{Hom}_{\mathcal{T}}(C, -), \mathbb{Q}/\mathbb{Z})$. Then it is straightforward to check that $\{C^* \mid C \in \mathcal{C}\}$ perfectly generates $\mathcal{T}^{\mathrm{op}}$, using the equivalent description from Corollary 3.4.8. □

We end our discussion of compact objects with a lemma that addresses the question when a right adjoint functor preserves coproducts.

Lemma 3.4.18. *Let $F : \mathcal{T} \to \mathcal{U}$ be an exact functor between triangulated categories that admit arbitrary coproducts, and suppose there exists a right adjoint G. If G preserves all coproducts, then F preserves compactness. The converse holds when $\mathcal{T}$ is compactly generated.*

Proof Fix objects $X \in \mathcal{T}$ and $\coprod_{i \in I} Y_i \in \mathcal{U}$, and suppose that X is compact. We consider the following commutative diagram.

$$
\begin{array}{ccc}
\coprod_i \mathrm{Hom}_{\mathcal{U}}(FX, Y_i) & \xrightarrow{\ \ \ \alpha\ \ \ } & \mathrm{Hom}_{\mathcal{U}}(FX, \coprod_i Y_i) \\
\downarrow{\wr} & & \downarrow{\wr} \\
\coprod_i \mathrm{Hom}_{\mathcal{T}}(X, GY_i) \xrightarrow{\ \sim\ } \mathrm{Hom}_{\mathcal{T}}(X, \coprod_i GY_i) & \xrightarrow{\ \beta\ } & \mathrm{Hom}_{\mathcal{T}}(X, G(\coprod_i Y_i))
\end{array}
$$

Suppose that G preserves coproducts. Then β is an isomorphism, and therefore α is an isomorphism. Thus FX is compact. The converse requires that the compact objects of $\mathcal{T}$ are generating. □

An application of Brown representability provides a description of the localisation with respect to a localising subcategory generated by compact objects.

Example 3.4.19. Let $\mathcal{T}$ be a triangulated category that admits arbitrary co-products. Then a localising subcategory $\mathcal{S} \subseteq \mathcal{T}$ generated by a set of compact objects in $\mathcal{T}$ fits into a localisation sequence

$$\mathcal{S} \rightleftarrows \mathcal{T} \rightleftarrows \mathcal{T}/\mathcal{S}$$

because the inclusion $\mathcal{S} \to \mathcal{T}$ admits a right adjoint; see Corollary 3.4.11 and Proposition 3.2.8. In fact, the right adjoint $\mathcal{T} \to \mathcal{S}$ preserves all coproducts by Lemma 3.4.18. Applying Brown representability once more (assuming that $\mathcal{T}$ is perfectly generated) we obtain the following recollement.

$$\mathcal{S} \rightleftarrows \mathcal{T} \rightleftarrows \mathcal{T}/\mathcal{S}$$

Notes

Triangulated categories and derived categories were introduced simultaneously in 1963 by Verdier in his thesis, and most of the basic properties can be found in his work [199]. For a modern exposition we refer to Neeman's book [150]. A similar notion of a 'stable category' was defined by Puppe, but without the octahedral axiom [164]. There is no example known of a 'pre-triangulated category' (so all axioms except (Tr4) are required), which is not triangulated.

The study of Frobenius categories and their stable categories was initiated by Heller [108]; for Frobenius pairs see [181]. The terminology reflects the properties of modules for quasi-Frobenius and self-injective rings [40, 73].

In algebraic topology the Brown representability theorem for cohomology theories is due to Brown [42]. An analogue for compactly generated triangulated categories was established by Keller [121] and Neeman [148]. The method of describing the compact objects in such categories as the direct summands of extensions of compact generators goes back to Ravenel [167]. More general representability theorems for cohomological functors are due to Franke [74] and Neeman [150]; for the dual version see [149]. The formulation in terms of perfect generators, which is presented here, uses categories of finitely presented functors and is taken from [127].

4

Derived Categories

Contents

4.1	**Derived Categories**		**102**
	Categories of Complexes		102
	The Mapping Cone		104
	Cohomology		104
	The Derived Category of an Exact Category		106
	Bounded Derived Categories		109
	Grothendieck Groups		109
4.2	**Resolutions and Extensions**		**110**
	Truncations		111
	Resolutions		112
	Extension Groups		116
	Exact Subcategories		118
4.3	**Resolutions and Derived Functors**		**122**
	Homotopy Injective and Projective Resolutions		122
	Grothendieck Categories		125
	Derived Functors		127
	Derived Functors Between Module Categories		129
	Homotopically Minimal Complexes		131
4.4	**Examples of Derived Categories**		**133**
	Quotient Categories		133
	Serre Subcategories		134
	Homological Epimorphisms		137
	Thick Subcategories		138
	Hereditary Categories		139
	Frobenius Categories		142
	A Proper Class of Extensions		144
Notes			**144**

In this chapter we introduce the derived category of an abelian category, and more generally the derived category of an exact category. For instance, we see that derived categories provide a natural context for describing the functors $\mathrm{Ext}^n(-, -)$. Also, we discuss the existence of resolutions because they yield an efficient method of making computations in derived categories, including the construction of derived functors. Important examples are module categories. In fact, we establish the existence of homotopy injective resolutions for complexes in any Grothendieck category. Then we analyse derived categories of exact subcategories and quotient categories. There are two classes of exact categories which deserve special attention: hereditary categories and Frobenius categories.

4.1 Derived Categories

Derived categories are introduced and their triangulated structure is explained. The objects are cochain complexes. The morphisms are somewhat delicate because the construction of a derived category involves the localisation with respect to a class of morphisms. These are the quasi-isomorphisms which induce an isomorphism when passing to the cohomology of a cochain complex.

Categories of Complexes

Let $\mathcal{A}$ be an additive category. A *cochain complex* (or simply a *complex*) in $\mathcal{A}$ is a sequence of morphisms

$$\cdots \longrightarrow X^{n-1} \xrightarrow{\ d^{n-1}\ } X^n \xrightarrow{\ d^n\ } X^{n+1} \longrightarrow \cdots$$

such that $d^n \circ d^{n-1} = 0$ for all $n \in \mathbb{Z}$. We think of a complex X as a graded object with *differential d* and refer to n as the *degree*.

We denote by $\mathbf{C}(\mathcal{A})$ the category of complexes, where a morphism $\phi \colon X \to Y$ between complexes consists of morphisms $\phi^n \colon X^n \to Y^n$ with $d_Y^n \circ \phi^n = \phi^{n+1} \circ d_X^n$ for all $n \in \mathbb{Z}$.

A morphism $\phi \colon X \to Y$ is *null-homotopic* if there are morphisms $\rho^n \colon X^n \to Y^{n-1}$ such that $\phi^n = d_Y^{n-1} \circ \rho^n + \rho^{n+1} \circ d_X^n$ for all $n \in \mathbb{Z}$. A complex is *contractible* if its identity morphism is null-homotopic. For example a complex of the form

$$\cdots \longrightarrow 0 \longrightarrow X \xrightarrow{\ \mathrm{id}\ } X \longrightarrow 0 \longrightarrow \cdots$$

is contractible.

The null-homotopic morphisms form an *ideal* $\mathfrak{I}$ in $\mathbf{C}(\mathcal{A})$, that is, for each pair X, Y of complexes there is a subgroup

$$\mathfrak{I}(X,Y) \subseteq \mathrm{Hom}_{\mathbf{C}(\mathcal{A})}(X,Y)$$

such that any composite $\psi \circ \phi$ of morphisms in $\mathbf{C}(\mathcal{A})$ belongs to $\mathfrak{I}$ if ϕ or

ψ belongs to $\mathfrak{I}$. The *homotopy category* $\mathbf{K}(\mathcal{A})$ is the quotient of $\mathbf{C}(\mathcal{A})$ with respect to this ideal. Thus

$$\text{Hom}_{\mathbf{K}(\mathcal{A})}(X,Y) = \text{Hom}_{\mathbf{C}(\mathcal{A})}(X,Y)/\mathfrak{I}(X,Y)$$

for every pair of complexes X, Y. One calls X and Y *homotopy equivalent* if they are isomorphic in $\mathbf{K}(\mathcal{A})$.

Given a complex X in $\mathcal{A}$, we denote by ΣX or $X[1]$ the *shifted complex* with

$$(\Sigma X)^n = X^{n+1} \quad \text{and} \quad d^n_{\Sigma X} = -d^{n+1}_X.$$

The category $\mathbf{C}(\mathcal{A})$ becomes an exact category if one takes as admissible exact sequences $0 \to X \to Y \to Z \to 0$ those where in each degree n the sequence $0 \to X^n \to Y^n \to Z^n \to 0$ is split exact in $\mathcal{A}$.

Lemma 4.1.1. *The exact category $\mathbf{C}(\mathcal{A})$ is a Frobenius category. The projective and injective objects are precisely the contractible complexes, and $\mathbf{K}(\mathcal{A})$ identifies with the stable category of $\mathbf{C}(\mathcal{A})$.*

Proof We view $\prod_{n \in \mathbb{Z}} \mathcal{A}$ as an exact category, with componentwise split exact structure. Then the functor $u : \mathbf{C}(\mathcal{A}) \to \prod_{n \in \mathbb{Z}} \mathcal{A}$ that takes X to $(X^n)_{n \in \mathbb{Z}}$ is exact; it admits a left adjoint u_λ and right adjoint u_ρ. The object $u_\rho u(X)$ is the direct sum of the complexes

$$\cdots \longrightarrow 0 \longrightarrow X^n \overset{\text{id}}{\longrightarrow} X^n \longrightarrow 0 \longrightarrow \cdots$$

concentrated in degrees $n-1$ and n. We have $u_\rho u(X) = u_\lambda u(\Sigma X)$ and this complex is projective, injective, and contractible. Here we use that each object in $\prod_{n \in \mathbb{Z}} \mathcal{A}$ is projective and injective, and that adjoints of an exact functor preserve projectivity and injectivity (Lemma 2.1.18). A morphism $X \to Y$ is null-homotopic if and only if it factors through the unit $X \to u_\rho u(X)$. It remains to observe that the unit and counit yield an admissible exact sequence

$$0 \longrightarrow X \longrightarrow u_\rho u(X) \longrightarrow \Sigma X \longrightarrow 0 \qquad (4.1.2)$$

which in degree n is of the form

$$0 \longrightarrow X^n \xrightarrow{\left|\begin{smallmatrix}\text{id}\\d^n\end{smallmatrix}\right|} X^n \oplus X^{n+1} \xrightarrow{\left|\begin{smallmatrix}-d^n & \text{id}\end{smallmatrix}\right|} X^{n+1} \longrightarrow 0. \qquad \square$$

A consequence is the useful fact that the homotopy category $\mathbf{K}(\mathcal{A})$ carries a triangulated structure; see Proposition 3.3.2.

Note that an additive functor $\mathcal{A} \to \mathcal{B}$ induces exact functors $\mathbf{C}(\mathcal{A}) \to \mathbf{C}(\mathcal{B})$ and $\mathbf{K}(\mathcal{A}) \to \mathbf{K}(\mathcal{B})$.

The Mapping Cone

Given a morphism $\phi\colon X \to Y$ of complexes in an additive category $\mathcal{A}$, the *mapping cone* is the complex Z with $Z^n = Y^n \oplus X^{n+1}$ and differential $\begin{bmatrix} d_Y^n & \phi^{n+1} \\ 0 & -d_X^{n+1} \end{bmatrix}$. The mapping cone fits into a *mapping cone sequence*

$$X \xrightarrow{\phi} Y \longrightarrow Z \longrightarrow \Sigma X$$

which is defined in degree n by the following sequence:

$$X^n \xrightarrow{\phi^n} Y^n \xrightarrow{\begin{bmatrix} \mathrm{id} \\ 0 \end{bmatrix}} Y^n \oplus X^{n+1} \xrightarrow{\begin{bmatrix} 0 & -\mathrm{id} \end{bmatrix}} X^{n+1}.$$

This mapping cone sequence equals the cone sequence (ϕ, ϕ', ϕ'') in $\mathbf{C}(\mathcal{A})$ with respect to the admissible monomorphism $X \to u_\rho u(X)$; in particular, it yields an exact triangle in $\mathbf{K}(\mathcal{A})$.

Cohomology

Let $\mathcal{A}$ be an abelian category. Then one defines for a complex X and each $n \in \mathbb{Z}$ the *cohomology*

$$H^n X = \operatorname{Ker} d^n / \operatorname{Im} d^{n-1}$$

of *degree n*. Sometimes we write $Z^n X$ for $\operatorname{Ker} d^n$ and call the elements *cocycles* of degree n. Note that $H^n X = H^0(\Sigma^n X)$. Two morphisms $\phi, \psi\colon X \to Y$ induce the same morphism $H^n \phi = H^n \psi$, if $\phi - \psi$ is null-homotopic.

Proposition 4.1.3. *Let $\mathcal{A}$ be an abelian category. An exact triangle*

$$X \xrightarrow{\alpha} Y \xrightarrow{\beta} Z \xrightarrow{\gamma} \Sigma X$$

in $\mathbf{K}(\mathcal{A})$ induces the following long exact sequence:

$$\cdots \longrightarrow H^{n-1} Z \xrightarrow{H^{n-1}\gamma} H^n X \xrightarrow{H^n\alpha} H^n Y \xrightarrow{H^n\beta} H^n Z \xrightarrow{H^n\gamma} H^{n+1} X \longrightarrow \cdots$$

Proof We may assume that the triangle comes from an admissible exact sequence $0 \to X \to Y \to Z \to 0$ in $\mathbf{C}(\mathcal{A})$. Then we obtain in $\mathbf{C}(\mathcal{A})$ a commutative diagram with exact rows

$$
\begin{array}{ccccccccc}
0 & \longrightarrow & X & \longrightarrow & Y & \longrightarrow & Z & \longrightarrow & 0 \\
& & \downarrow{\scriptstyle d_X} & & \downarrow{\scriptstyle d_Y} & & \downarrow{\scriptstyle d_Z} & & \\
0 & \longrightarrow & \Sigma X & \longrightarrow & \Sigma Y & \longrightarrow & \Sigma Z & \longrightarrow & 0
\end{array}
$$

which induces the following commutative diagram with exact rows.

$$
\begin{array}{ccccccc}
\operatorname{Coker} d_{\Sigma^{-1}X} & \longrightarrow & \operatorname{Coker} d_{\Sigma^{-1}Y} & \longrightarrow & \operatorname{Coker} d_{\Sigma^{-1}Z} & \longrightarrow & 0 \\
\downarrow{\scriptstyle d_X} & & \downarrow{\scriptstyle d_Y} & & \downarrow{\scriptstyle d_Z} & & \\
0 & \longrightarrow & \operatorname{Ker} d_{\Sigma X} & \longrightarrow & \operatorname{Ker} d_{\Sigma Y} & \longrightarrow & \operatorname{Ker} d_{\Sigma Z}
\end{array}
$$

Now observe that the differentials induce for each $n \in \mathbb{Z}$ an exact sequence

$$0 \to \operatorname{Ker} d^n / \operatorname{Im} d^{n-1} \to \operatorname{Coker} d^{n-1} \to \operatorname{Ker} d^{n+1} \to \operatorname{Ker} d^{n+1} / \operatorname{Im} d^n \to 0.$$

It remains to apply the snake lemma. □

A morphism $\phi \colon X \to Y$ between complexes induces in each degree n a morphism $H^n \phi \colon H^n X \to H^n Y$, and ϕ is a *quasi-isomorphism* if $H^n \phi$ is an isomorphism for all $n \in \mathbb{Z}$. A complex X is *acyclic* if $H^n X = 0$ for all $n \in \mathbb{Z}$.

Lemma 4.1.4. *Let $\mathcal{A}$ be an abelian category. A morphism between complexes is a quasi-isomorphism if and only if its mapping cone is acyclic.*

Proof Apply Proposition 4.1.3. □

Example 4.1.5. Let X, Y be complexes and suppose X is concentrated in degree zero. Then $\operatorname{Hom}_A(X, Y)$ may be considered as complex and

$$H^n \operatorname{Hom}_A(X, Y) \cong \operatorname{Hom}_{\mathbf{K}(\mathcal{A})}(X, \Sigma^n Y)$$

because $\operatorname{Ker} d^n$ identifies with $\operatorname{Hom}_{\mathbf{C}(\mathcal{A})}(X, \Sigma^n Y)$ and $\operatorname{Im} d^{n-1}$ identifies with the ideal of null-homotopic maps $X \to \Sigma^n Y$.

Example 4.1.6. A complex X is called *split* if there are morphisms $\rho^n \colon X^n \to X^{n-1}$ such that $d^n = d^n \rho^{n+1} d^n$ for all $n \in \mathbb{Z}$. An equivalent condition is that each d^n is the composite $X^n \twoheadrightarrow \operatorname{Im} d^n \rightarrowtail X^{n+1}$ of a split epimorphism and a split monomorphism. In this case the morphisms $d^{n-1} \rho^n + \rho^{n+1} d^n$ yield an endomorphism $\phi \colon X \to X$, which induces quasi-isomorphisms $\operatorname{Ker} \phi \to X$ and $X \to \operatorname{Coker} \phi$. Moreover, $\operatorname{Ker} \phi$ and $\operatorname{Coker} \phi$ identify with the complex

$$\cdots \xrightarrow{0} H^{n-1}X \xrightarrow{0} H^n X \xrightarrow{0} H^{n+1}X \xrightarrow{0} \cdots .$$

Proof The morphisms ρ^n provide decompositions $X^n = \operatorname{Ker} d^n \oplus U^n$ and $\operatorname{Ker} d^n = \operatorname{Im} d^{n-1} \oplus V^n$ for each $n \in \mathbb{Z}$. Then $\rho^{n+1} d^n$ is projecting onto U^n and $d^{n-1} \rho^n$ onto $\operatorname{Im} d^{n-1}$. It follows that $\operatorname{Ker} \phi^n$ and $\operatorname{Coker} \phi^n$ identify with $V^n \cong H^n X$. □

The Derived Category of an Exact Category

Let $\mathcal{A}$ be an exact category. A complex X in $\mathcal{A}$ is called *acyclic* if for each $n \in \mathbb{Z}$ there is an admissible exact sequence

$$\eta_n : \qquad 0 \longrightarrow Z^n \xrightarrow{\ \alpha^n\ } X^n \xrightarrow{\ \beta^n\ } Z^{n+1} \longrightarrow 0$$

in $\mathcal{A}$ such that $d_X^n = \alpha^{n+1} \circ \beta^n$. This definition generalises the definition for abelian categories. In fact, there is no reasonable definition of $H^n(X)$ when $\mathcal{A}$ is not abelian, but let us write $H^n(X) = 0$ when d_X^n admits a kernel and d_X^{n-1} can be written as the composite

$$X^{n-1} \twoheadrightarrow \operatorname{Ker} d_X^n \rightarrowtail X^n$$

of an admissible epimorphism and an admissible monomorphism in $\mathcal{A}$.

Lemma 4.1.7. *The mapping cone of a morphism between acyclic complexes is acyclic.*

Proof This is clear when $\mathcal{A}$ is abelian, thanks to Proposition 4.1.3. The general case requires a calculation which is straightforward. □

Lemma 4.1.8. *The following are equivalent for an exact category $\mathcal{A}$.*

(1) *Each contractible complex in $\mathcal{A}$ is acyclic.*
(2) *The category $\mathcal{A}$ is idempotent complete.*
(3) *The class of acyclic complexes is closed under isomorphisms in $\mathbf{K}(\mathcal{A})$.*

Proof (1) $\Rightarrow$ (2): Each idempotent $\phi \in \operatorname{End}_A(X)$ gives rise to a contractible complex

$$\cdots \longrightarrow X \xrightarrow{\ 1-\phi\ } X \xrightarrow{\ \phi\ } X \xrightarrow{\ 1-\phi\ } X \xrightarrow{\ \phi\ } X \longrightarrow \cdots$$

The fact that this complex is acyclic provides a kernel of ϕ in $\mathcal{A}$.

(2) $\Rightarrow$ (1): Consider a complex X that is contractible. Then the monomorphism $X \to u_\rho u(X)$ in (4.1.2) splits. Note that $u_\rho u(X)$ is acyclic. A small calculation shows that acyclic complexes are closed under direct summands since $\mathcal{A}$ is idempotent complete. Thus X is acyclic.

(1) $\Rightarrow$ (3): Let $X \to Y$ be an isomorphism in $\mathbf{K}(\mathcal{A})$. Then the cone is contractible and therefore acyclic. If one of X or Y is acyclic, then so also is the other, since acyclic complexes form a triangulated subcategory by the above lemma.

(3) $\Rightarrow$ (1): A contractible complex X is isomorphic in $\mathbf{K}(\mathcal{A})$ to the zero complex, which is acyclic. Thus X is acyclic. □

We denote by $\mathbf{Ac}(\mathcal{A})$ the full subcategory of complexes in $\mathbf{C}(\mathcal{A})$ that are isomorphic to an acyclic complex in $\mathbf{K}(\mathcal{A})$. A morphism of complexes is a *quasi-isomorphism* if its cone is in $\mathbf{Ac}(\mathcal{A})$, and we write Qis for the class of all quasi-isomorphisms in $\mathbf{C}(\mathcal{A})$.

Lemma 4.1.9. $(\mathbf{C}(\mathcal{A}), \mathbf{Ac}(\mathcal{A}))$ *is a Frobenius pair.*

Proof The category $\mathbf{C}(\mathcal{A})$ is a Frobenius category by Lemma 4.1.1. The subcategory $\mathbf{Ac}(\mathcal{A})$ contains all contractible complexes and has the two out of three property by Lemma 4.1.7. □

The *derived category* $\mathbf{D}(\mathcal{A})$ of $\mathcal{A}$ is obtained from $\mathbf{C}(\mathcal{A})$ by formally inverting all quasi-isomorphisms. Thus one defines

$$\mathbf{D}(\mathcal{A}) = \mathbf{C}(\mathcal{A})[\text{Qis}^{-1}]$$

and this is precisely the derived category of the Frobenius pair $(\mathbf{C}(\mathcal{A}), \mathbf{Ac}(\mathcal{A}))$.

Viewing $\mathbf{Ac}(\mathcal{A})$ as a full subcategory of $\mathbf{K}(\mathcal{A})$, the canonical functor $\mathbf{C}(\mathcal{A}) \to \mathbf{D}(\mathcal{A})$ induces a triangle equivalence

$$\mathbf{K}(\mathcal{A})/\mathbf{Ac}(\mathcal{A}) \xrightarrow{\sim} \mathbf{D}(\mathcal{A}).$$

This follows from Proposition 3.3.5. In particular, we can apply the description of morphisms for a Verdier quotient as follows.

For a pair of complexes X, Y, we denote by Qis$/X$ the category of quasi-isomorphisms $X' \to X$ in $\mathbf{K}(\mathcal{A})$, and dually by $Y/$Qis the category of quasi-isomorphisms $Y \to Y'$.

Lemma 4.1.10. *The categories* Qis$/X$ *and* $Y/$Qis *are filtered, and we have natural isomorphisms*

$$\operatorname*{colim}_{X' \to X} \operatorname{Hom}_{\mathbf{K}(\mathcal{A})}(X', Y) \xrightarrow{\sim} \operatorname{Hom}_{\mathbf{D}(\mathcal{A})}(X, Y) \xleftarrow{\sim} \operatorname*{colim}_{Y \to Y'} \operatorname{Hom}_{\mathbf{K}(\mathcal{A})}(X, Y'),$$

where $X' \to X$ *runs through* Qis$/X$, *and* $Y \to Y'$ *runs through* $Y/$Qis.

Proof The quasi-isomorphisms are by definition the morphisms having their cone in $\mathbf{Ac}(\mathcal{A})$, which is a triangulated subcategory of $\mathbf{K}(\mathcal{A})$. Thus the assertion follows from Lemma 1.2.2, because the quasi-isomorphisms in $\mathbf{K}(\mathcal{A})$ admit a calculus of left and right fractions by Lemma 3.2.1. □

There is a canonical functor $\mathcal{A} \to \mathbf{D}(\mathcal{A})$ that takes an object X in $\mathcal{A}$ to the corresponding complex $\bar{X}$ concentrated in degree zero. On the other hand, there is also the functor $H^0 \colon \mathbf{D}(\mathcal{A}) \to \mathcal{A}$ when $\mathcal{A}$ is abelian. Clearly, $H^0 \bar{X} = X$ for all $X \in \mathcal{A}$.

Lemma 4.1.11. *For X, Y in $\mathcal{A}$ the assignment $\phi \mapsto \bar{\phi}$ gives a bijection*

$$\mathrm{Hom}_{\mathcal{A}}(X, Y) \xrightarrow{\sim} \mathrm{Hom}_{\mathbf{D}(\mathcal{A})}(\bar{X}, \bar{Y}).$$

Proof We give the argument when $\mathcal{A}$ is abelian. We have already seen that $H^0 \bar{\phi} = \phi$. Thus the map is injective. A morphism $\psi \colon \bar{X} \to \bar{Y}$ is given by a diagram $\bar{X} \xrightarrow{\alpha} Z \xleftarrow{\sigma} \bar{Y}$ in $\mathbf{K}(\mathcal{A})$ such that σ is a quasi-isomorphism. Then for $\phi \colon X \xrightarrow{H^0 \alpha} H^0 Z \xrightarrow{(H^0 \sigma)^{-1}} Y$ we have $\bar{\phi} = \psi$. $\qquad \square$

Lemma 4.1.12. *Every admissible exact sequence $0 \to X \to Y \to Z \to 0$ in $\mathcal{A}$ induces an exact triangle $\bar{X} \to \bar{Y} \to \bar{Z} \to \Sigma \bar{X}$ in $\mathbf{D}(\mathcal{A})$.*

Proof The morphism $\alpha \colon X \to Y$ yields an exact triangle $\bar{X} \to \bar{Y} \to \mathrm{Cone}\, \bar{\phi} \to \Sigma \bar{X}$ in $\mathbf{K}(\mathcal{A})$. Now observe that $Y \to Z$ induces a morphism $\mathrm{Cone}\, \bar{\phi} \to \bar{Z}$ which is a quasi-isomorphism. $\qquad \square$

Lemma 4.1.13. *For an exact category $\mathcal{A}$ the following are equivalent.*

(1) *Every admissible exact sequence in $\mathcal{A}$ is split exact.*
(2) *Every acyclic complex is contractible.*
(3) *The canonical functor $\mathbf{K}(\mathcal{A}) \to \mathbf{D}(\mathcal{A})$ is an equivalence.*

Proof The implications (1) $\Rightarrow$ (2) $\Rightarrow$ (3) are clear.

(3) $\Rightarrow$ (1): Let $\eta \colon 0 \to X^0 \to X^1 \to X^2 \to 0$ be an admissible exact sequence. We view this as an acyclic complex and denote it by X. Then we have $X = 0$ in $\mathbf{D}(\mathcal{A})$, and therefore also $X = 0$ in $\mathbf{K}(\mathcal{A})$. Thus X is contractible, and therefore η is split exact. $\qquad \square$

Let $F \colon \mathcal{A} \to \mathcal{B}$ be an additive functor. Then F induces an exact functor $\mathbf{K}(\mathcal{A}) \to \mathbf{K}(\mathcal{B})$ by applying F componentwise. There is no obvious way to obtain from this an exact functor $\mathbf{D}(\mathcal{A}) \to \mathbf{D}(\mathcal{B})$, except when F is exact.

Lemma 4.1.14. *An exact functor $\mathcal{A} \to \mathcal{B}$ induces an exact functor $\mathbf{D}(\mathcal{A}) \to \mathbf{D}(\mathcal{B})$.*

Proof The composite $\mathbf{K}(\mathcal{A}) \to \mathbf{K}(\mathcal{B}) \to \mathbf{D}(\mathcal{B})$ annihilates $\mathbf{Ac}(\mathcal{A})$. Thus the assertion follows from Proposition 3.2.2. $\qquad \square$

Now suppose that the category $\mathcal{A}$ admits set-indexed products (or coproducts). Then we say that $\mathcal{A}$ has *exact (co)products* if any (co)product of exact sequences is again exact.

Lemma 4.1.15. *Let $\mathcal{A}$ be an exact category with exact (co)products. Then the derived category $\mathbf{D}(\mathcal{A})$ admits (co)products, which are computed by taking (co)products componentwise in $\mathcal{A}$.*

Proof The category $\mathbf{K}(\mathcal{A})$ inherits set-indexed (co)products from $\mathcal{A}$. The assumption on $\mathcal{A}$ implies that the class of quasi-isomorphisms is closed under (co)products. Now apply Lemma 1.1.8. □

Bounded Derived Categories

Let $\mathcal{A}$ be an additive category. Consider the following full subcategories of $\mathbf{C}(\mathcal{A})$ consisting of *bounded* complexes:

$$\mathbf{C}^b(\mathcal{A}) = \{X \in \mathbf{C}(\mathcal{A}) \mid X^n = 0 \text{ for } |n| \gg 0\} \qquad \text{(bounded)}$$
$$\mathbf{C}^+(\mathcal{A}) = \{X \in \mathbf{C}(\mathcal{A}) \mid X^n = 0 \text{ for } n \ll 0\} \qquad \text{(bounded below)}$$
$$\mathbf{C}^-(\mathcal{A}) = \{X \in \mathbf{C}(\mathcal{A}) \mid X^n = 0 \text{ for } n \gg 0\} \qquad \text{(bounded above)}.$$

For $* \in \{b, +, -\}$, let the homotopy category $\mathbf{K}^*(\mathcal{A})$ be the quotient of $\mathbf{C}^*(\mathcal{A})$ modulo null-homotopic morphisms. When $\mathcal{A}$ is exact let the derived category $\mathbf{D}^*(\mathcal{A})$ be the localisation of $\mathbf{C}^*(\mathcal{A})$ with respect to all quasi-isomorphisms.

Lemma 4.1.16. *For each $* \in \{b, +, -\}$, the inclusion $\mathbf{C}^*(\mathcal{A}) \to \mathbf{C}(\mathcal{A})$ induces fully faithful functors $\mathbf{K}^*(\mathcal{A}) \to \mathbf{K}(\mathcal{A})$ and $\mathbf{D}^*(\mathcal{A}) \to \mathbf{D}(\mathcal{A})$.*

Proof The assertion for $\mathbf{K}^*(\mathcal{A}) \to \mathbf{K}(\mathcal{A})$ is obvious. Now observe that the inclusion $\mathbf{K}^*(\mathcal{A}) \to \mathbf{K}(\mathcal{A})$ is cofinal with respect to the class of quasi-isomorphisms. For example, let $\phi: X \to Y$ be a quasi-isomorphism and $X \in \mathbf{K}^+(\mathcal{A})$ with $X^n = 0$ for $n < 0$. Consider the following truncation, which exists since Cone ϕ is isomorphic to an acyclic complex.

$$
\begin{array}{ccccccccc}
Y & \cdots \to & Y^{-2} & \longrightarrow & Y^{-1} & \longrightarrow & Y^0 & \to & Y^1 \to \cdots \\
& & \downarrow & & \downarrow{\scriptstyle\text{can}} & & \downarrow{\scriptstyle\text{id}} & & \downarrow{\scriptstyle\text{id}} \\
\tau_{\geq -1}Y & \cdots \longrightarrow & 0 & \longrightarrow & \operatorname{Coker} d^{-2} & \longrightarrow & Y^0 & \to & Y^1 \to \cdots
\end{array}
$$

Then $Y \to \tau_{\geq -1}Y$ is a quasi-isomorphism; so the inclusion $\mathbf{K}^+(\mathcal{A}) \to \mathbf{K}(\mathcal{A})$ is left cofinal. Thus the assertion for $\mathbf{D}^*(\mathcal{A}) \to \mathbf{D}(\mathcal{A})$ follows from Lemma 1.2.5. □

Grothendieck Groups

Let $\mathcal{A}$ be an essentially small exact category. Its *Grothendieck group* $K_0(\mathcal{A})$ is defined as the factor group $F(\mathcal{A})/F_0(\mathcal{A})$ given by the free abelian group $F(\mathcal{A})$ generated by the isomorphism classes $[X]$ of objects $X \in \mathcal{A}$, modulo the subgroup $F_0(\mathcal{A})$ generated by $[X] - [Y] + [Z]$ for all exact sequences $0 \to X \to Y \to Z \to 0$ in $\mathcal{A}$.

For an essentially small triangulated category $\mathcal{T}$ we have the following analogue. Denote by $F(\mathcal{T})$ the free abelian group generated by the isomorphism classes $[X]$ of objects $X \in \mathcal{T}$. Let $F_0(\mathcal{T})$ be the subgroup generated by $[X] - [Y] + [Z]$ for all exact triangles $X \to Y \to Z \to \Sigma X$ in $\mathcal{T}$. The *Grothendieck group* $K_0(\mathcal{T})$ of $\mathcal{T}$ is by definition the factor group $F(\mathcal{T})/F_0(\mathcal{T})$.

The embedding $\mathcal{A} \to \mathbf{D}^b(\mathcal{A})$ taking $X \in \mathcal{A}$ to the complex $\bar{X}$ concentrated in degree zero yields a homomorphism

$$\eta_{\mathcal{A}} \colon K_0(\mathcal{A}) \longrightarrow K_0(\mathbf{D}^b(\mathcal{A})),$$

which is well defined by Lemma 4.1.12.

Lemma 4.1.17. *Let $\mathcal{A}$ be an essentially small exact category. Then the assignment $[X] \mapsto \sum_{i \in \mathbb{Z}} (-1)^i [X^i]$ induces an isomorphism*

$$K_0(\mathbf{D}^b(\mathcal{A})) \xrightarrow{\sim} K_0(\mathcal{A}).$$

Proof The assignment $[X] \mapsto \chi(X) := \sum_{i \in \mathbb{Z}} (-1)^i [X^i]$ yields a well-defined map $K_0(\mathbf{K}^b(\mathcal{A})) \to K_0(\mathcal{A})$ since exact triangles in $\mathbf{K}(\mathcal{A})$ come from degreewise split exact sequences of complexes. For an acyclic complex X an induction on the number of integers i with $X^i \neq 0$ shows that $\chi(X) = 0$. Thus $\chi(X) = \chi(Y)$ for any pair X, Y of quasi-isomorphic complexes. It follows that χ is well defined on $K_0(\mathbf{D}^b(\mathcal{A}))$. Clearly, the map $\eta_{\mathcal{A}}$ is a right inverse, so $\chi \circ \eta_{\mathcal{A}} = \mathrm{id}$. Moreover, $\eta_{\mathcal{A}}$ is surjective since we can build any bounded complex via truncations and shifts from complexes of the form $\bar{X}$ with $X \in \mathcal{A}$. □

We have the following analogue for abelian categories.

Lemma 4.1.18. *Let $\mathcal{A}$ be an essentially small abelian category. Then the assignment $[X] \mapsto \sum_{i \in \mathbb{Z}} (-1)^i [H^i X]$ induces an isomorphism*

$$K_0(\mathbf{D}^b(\mathcal{A})) \xrightarrow{\sim} K_0(\mathcal{A}).$$

Proof The map is well defined, given that exact triangles in $\mathbf{D}(\mathcal{A})$ induce exact sequences when taking cohomology, by Proposition 4.1.3. In fact, an induction shows for any complex X that

$$\sum_{i \in \mathbb{Z}} (-1)^i [H^i X] = \sum_{i \in \mathbb{Z}} (-1)^i [X^i].$$

Thus the assertion follows from the previous lemma. □

4.2 Resolutions and Extensions

In this section we explain that derived categories provide a natural context for describing the functors $\mathrm{Ext}^n(-, -)$. In fact there are two possible approaches.

One may use the group of extensions in the sense of Yoneda, or one views $\mathrm{Ext}^n(X, -)$ as a derived functor which is computed via injective resolutions. In any case, resolutions provide a useful tool for working with derived categories. We construct such resolutions using projective or injective objects. This requires some machinery and we discuss the methods that are needed.

Truncations

Let $\mathcal{A}$ be an additive category. For a complex X and $n \in \mathbb{Z}$ there are various possible truncations. We begin with the following morphism

$$
\begin{array}{ccccccccc}
X & \cdots \longrightarrow & X^{n-1} & \longrightarrow & X^n & \longrightarrow & X^{n+1} & \longrightarrow & X^{n+2} \longrightarrow \cdots \\
& & \downarrow{\scriptstyle\mathrm{id}} & & \downarrow{\scriptstyle\mathrm{id}} & & \downarrow & & \downarrow \\
\sigma_{\leq n} X & \cdots \longrightarrow & X^{n-1} & \longrightarrow & X^n & \longrightarrow & 0 & \longrightarrow & 0 \longrightarrow \cdots
\end{array}
$$

and call this *brutal truncation*. There is also the following dual construction:

$$
\begin{array}{ccccccccc}
\sigma_{\geq n} X & \cdots \longrightarrow & 0 & \longrightarrow & 0 & \longrightarrow & X^n & \longrightarrow & X^{n+1} \longrightarrow \cdots \\
& & \downarrow & & \downarrow & & \downarrow{\scriptstyle\mathrm{id}} & & \downarrow{\scriptstyle\mathrm{id}} \\
X & \cdots \longrightarrow & X^{n-2} & \longrightarrow & X^{n-1} & \longrightarrow & X^n & \longrightarrow & X^{n+1} \longrightarrow \cdots
\end{array}
$$

Now suppose that $\mathcal{A}$ is an abelian category. The morphism

$$
\begin{array}{ccccccccc}
\tau_{\leq n} X & \cdots \longrightarrow & X^{n-2} & \longrightarrow & X^{n-1} & \longrightarrow & \mathrm{Ker}\, d^n & \longrightarrow & 0 \longrightarrow \cdots \\
& & \downarrow{\scriptstyle\mathrm{id}} & & \downarrow{\scriptstyle\mathrm{id}} & & \downarrow{\scriptstyle\mathrm{inc}} & & \downarrow \\
X & \cdots \longrightarrow & X^{n-2} & \longrightarrow & X^{n-1} & \longrightarrow & X^n & \longrightarrow & X^{n+1} \longrightarrow \cdots
\end{array}
$$

is called *exact* or *soft truncation*. This morphism induces isomorphisms

$$H^i(\tau_{\leq n} X) \xrightarrow{\sim} H^i(X) \quad \text{for all} \quad i \leq n.$$

The dual construction

$$
\begin{array}{ccccccccc}
X & \cdots \longrightarrow & X^{n-1} & \longrightarrow & X^n & \longrightarrow & X^{n+1} & \longrightarrow & X^{n+2} \longrightarrow \cdots \\
& & \downarrow & & \downarrow{\scriptstyle\mathrm{can}} & & \downarrow{\scriptstyle\mathrm{id}} & & \downarrow{\scriptstyle\mathrm{id}} \\
\tau_{\geq n} X & \cdots \longrightarrow & 0 & \longrightarrow & \mathrm{Coker}\, d^{n-1} & \longrightarrow & X^{n+1} & \longrightarrow & X^{n+2} \longrightarrow \cdots
\end{array}
$$

induces isomorphisms

$$H^i(X) \xrightarrow{\sim} H^i(\tau_{\geq n} X) \quad \text{for all} \quad i \geq n.$$

Using the exact truncations we obtain equivalences

$$\mathbf{D}^-(\mathcal{A}) \xrightarrow{\sim} \{X \in \mathbf{D}(\mathcal{A}) \mid H^n X = 0 \text{ for } n \gg 0\}$$

and

$$\mathbf{D}^+(\mathcal{A}) \overset{\sim}{\longrightarrow} \{X \in \mathbf{D}(\mathcal{A}) \mid H^n X = 0 \text{ for } n \ll 0\}.$$

Lemma 4.2.1. *Let $\mathcal{A}$ be an abelian category. Then each object $X \in \mathbf{D}^b(\mathcal{A})$ belongs to the thick subcategory generated by the objects $H^n X$, viewed as complexes concentrated in degree zero.*

Proof We may assume that $H^n X = 0$ for $n \notin [0, d]$. Then the truncations induce a finite filtration

$$X \cong \tau_{\geq 0} X \twoheadrightarrow \tau_{\geq 1} X \twoheadrightarrow \cdots \twoheadrightarrow \tau_{\geq d} X \twoheadrightarrow \tau_{\geq d+1} X \cong 0$$

such that each subquotient has its cohomology concentrated in a single degree. Thus there are exact triangles

$$\Sigma^{-n}(H^n X) \longrightarrow \tau_{\geq n} X \longrightarrow \tau_{\geq n+1} X \longrightarrow \Sigma^{-n+1}(H^n X)$$

and from this the assertion follows.   □

Example 4.2.2. Let $\mathcal{A}$ be an additive category and suppose that $\mathcal{A}$ has countable (co)products. Any complex X induces a sequence

$$\sigma_{\geq 0} X \longrightarrow \sigma_{\geq -1} X \longrightarrow \sigma_{\geq -2} X \longrightarrow \cdots .$$

It follows that X is the homotopy colimit of its truncations $\sigma_{\geq n} X \to X$ because we have an induced exact triangle

$$\Sigma^{-1} X \longrightarrow \coprod_{n \leq 0} \sigma_{\geq n} X \longrightarrow \coprod_{n \leq 0} \sigma_{\geq n} X \longrightarrow X. \tag{4.2.3}$$

In particular, a complex in $\mathbf{K}^-(\mathcal{A})$ can be built as a homotopy colimit from complexes in $\mathbf{K}^b(\mathcal{A})$. Analogously, the truncations $X \to \sigma_{\leq n} X$ yield the sequence

$$\cdots \longrightarrow \sigma_{\leq 2} X \longrightarrow \sigma_{\leq 1} X \longrightarrow \sigma_{\leq 0} X$$

and an exact triangle

$$X \longrightarrow \prod_{n \geq 0} \sigma_{\leq n} X \longrightarrow \prod_{n \geq 0} \sigma_{\leq n} X \longrightarrow \Sigma X. \tag{4.2.4}$$

In particular, a complex in $\mathbf{K}^+(\mathcal{A})$ can be built as a homotopy limit from complexes in $\mathbf{K}^b(\mathcal{A})$.

Resolutions

Let $\mathcal{A}$ be an exact category. We collect some basic facts about resolutions. For an object $X \in \mathcal{A}$ an *injective resolution* is a complex

$$iX: \qquad \cdots \longrightarrow 0 \longrightarrow I^0 \longrightarrow I^1 \longrightarrow I^2 \longrightarrow \cdots$$

of injective objects in $\mathcal{A}$ together with a quasi-isomorphism $X \to iX$. A *projective resolution* $pX \to X$ is defined dually.

Lemma 4.2.5. *Let X, Y be complexes in $\mathcal{A}$. Suppose that each Y^n is injective and $Y^n = 0$ for $n \ll 0$. If X is acyclic, then $\mathrm{Hom}_{\mathbf{K}(\mathcal{A})}(X, Y) = 0$. Therefore the canonical map*

$$\mathrm{Hom}_{\mathbf{K}(\mathcal{A})}(X, Y) \longrightarrow \mathrm{Hom}_{\mathbf{D}(\mathcal{A})}(X, Y)$$

is bijective.

Proof Let X be acyclic and fix a morphism $\phi \colon X \to Y$. We claim that ϕ is null-homotopic and construct morphisms $\rho^n \colon X^n \to Y^{n-1}$ inductively as follows. Suppose that $Y^n = 0$ for all $n < n_0$. Then set $\rho^n = 0$ for all $n \le n_0$. For $n > n_0$, suppose that ρ^n has been constructed such that $(\phi^n - d_Y^{n-1}\rho^n)d_X^{n-1} = 0$. Then $\phi^n - d_Y^{n-1}\rho^n$ factors through $X^n \twoheadrightarrow \mathrm{Ker}\, d_X^{n+1}$ since X is acyclic and can be extended to a morphism $\rho^{n+1} \colon X^{n+1} \to Y^n$ since Y^n is injective. Then we have $\phi^n = d_X^{n-1}\rho^n + \rho^{n+1}d_X^n$ by construction, and $(\phi^{n+1} - d_Y^n\rho^{n+1})d_X^n = 0$; so we can proceed. Thus ϕ is null-homotopic. The second assertion of the lemma then follows from Lemma 3.2.4. $\square$

Lemma 4.2.6. *Let Y be a complex in $\mathcal{A}$ such that each Y^n is injective. Then an injective resolution $A \to iA$ of an object $A \in \mathcal{A}$ induces an isomorphism*

$$\mathrm{Hom}_{\mathbf{K}(\mathcal{A})}(iA, Y) \xrightarrow{\sim} \mathrm{Hom}_{\mathbf{K}(\mathcal{A})}(A, Y).$$

Proof We complete the morphism $A \to iA$ to an exact triangle

$$aA \longrightarrow A \longrightarrow iA \longrightarrow \Sigma(aA).$$

Then the truncation $Y \to \sigma_{\ge -1}Y$ induces the first isomorphism below

$$\mathrm{Hom}_{\mathbf{K}(\mathcal{A})}(aA, Y) \cong \mathrm{Hom}_{\mathbf{K}(\mathcal{A})}(aA, \sigma_{\ge -1}Y) \cong 0$$

since aA is acyclic and concentrated in non-negative degrees, while the second isomorphism follows from Lemma 4.2.5. It remains to apply $\mathrm{Hom}_{\mathbf{K}(\mathcal{A})}(-, Y)$ to the above triangle. $\square$

Next we fix a subcategory $\mathcal{B} \subseteq \mathcal{A}$ such that each object $X \in \mathcal{A}$ admits an admissible monomorphism $X \to Y$ with $Y \in \mathcal{B}$.

Lemma 4.2.7. *Each object $X \in \mathbf{K}^+(\mathcal{A})$ admits a quasi-isomorphism $X \to Y$ such that $X^n \to Y^n$ is an admissible monomorphism with $Y^n \in \mathcal{B}$ for all $n \in \mathbb{Z}$.*

Proof We construct the morphisms $X^n \to Y^n$ inductively by giving a factorisation $X^n \rightarrowtail C^n \to X^{n+1}$ of d_X^n such that the composite $C^n \to X^{n+1} \xrightarrow{d_X^{n+1}} X^{n+2}$ is zero, and then choosing an admissible monomorphism $C^n \rightarrowtail Y^n$. To begin suppose that $X^n = 0$ for all $n \le n_0$. Set $C^n = 0 = Y^n$ for all $n \le n_0$. For $n \ge n_0$, we form the pushout of the diagram $X^{n+1} \leftarrow C^n \rightarrowtail Y^n$ and obtain the following diagram.

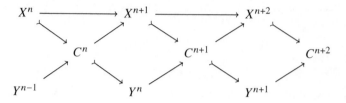

The fact that $C^n \rightarrow X^{n+1} \xrightarrow{d_X^{n+1}} X^{n+2}$ is zero yields a morphism $C^{n+1} \rightarrow X^{n+2}$ giving the factorisation of d_X^{n+1} such that the composite $C^{n+1} \rightarrow X^{n+2} \xrightarrow{d_X^{n+2}} X^{n+3}$ is zero; so we can proceed. The mapping cone of $X \rightarrow Y$ is acyclic since for each n we have an admissible exact sequence

$$0 \longrightarrow C^n \longrightarrow Y^n \oplus X^{n+1} \longrightarrow C^{n+1} \longrightarrow 0. \qquad \square$$

Proposition 4.2.8. *Let A be an exact category.*

(1) *The inclusion* $\operatorname{Inj} A \rightarrow A$ *induces a fully faithful and exact functor*

$$\mathbf{K}^+(\operatorname{Inj} A) \longrightarrow \mathbf{D}^+(A).$$

   If A has enough injective objects then this is a triangle equivalence.

(2) *The inclusion* $\operatorname{Proj} A \rightarrow A$ *induces a fully faithful and exact functor*

$$\mathbf{K}^-(\operatorname{Proj} A) \longrightarrow \mathbf{D}^-(A).$$

   If A has enough projective objects then this is a triangle equivalence.

Proof We prove (1), and (2) is dual. The functor $\mathbf{K}^+(\operatorname{Inj} A) \rightarrow \mathbf{D}^+(A)$ is fully faithful by Lemma 4.2.5. On the other hand, Lemma 4.2.7 implies that every complex in $\mathbf{D}^+(A)$ is isomorphic to a complex of injectives when A has enough injective objects. $\qquad \square$

For a full additive subcategory $\mathcal{C} \subseteq A$ we set

$$\mathbf{K}^{+,b}(\mathcal{C}) = \{X \in \mathbf{K}^+(\mathcal{C}) \mid H^n X = 0 \text{ for } |n| \gg 0\}$$

and

$$\mathbf{K}^{-,b}(\mathcal{C}) = \{X \in \mathbf{K}^-(\mathcal{C}) \mid H^n X = 0 \text{ for } |n| \gg 0\},$$

where the condition $H^n X = 0$ means that d_X^{n-1} can be written as the composite

$$X^{n-1} \twoheadrightarrow \operatorname{Ker} d_X^n \rightarrowtail X^n$$

of an admissible epimorphism and an admissible monomorphism in $\mathcal{A}$. In particular, the subcategories $\mathbf{K}^{+,b}(\mathcal{C})$ and $\mathbf{K}^{-,b}(\mathcal{C})$ depend on the ambient category $\mathcal{A}$, even though it is not part of the notation.

Corollary 4.2.9. *Suppose that $\mathcal{A}$ has enough injective objects. Then the equivalence $\mathbf{K}^+(\mathrm{Inj}\,\mathcal{A}) \xrightarrow{\sim} \mathbf{D}^+(\mathcal{A})$ restricts to an equivalence*

$$\mathbf{K}^{+,b}(\mathrm{Inj}\,\mathcal{A}) \xrightarrow{\sim} \mathbf{D}^b(\mathcal{A}).$$

This equivalence restricts to

$$\mathbf{K}^b(\mathrm{Inj}\,\mathcal{A}) \xrightarrow{\sim} \mathbf{D}^b(\mathcal{A})$$

when every object in $\mathcal{A}$ has finite injective dimension.

When $\mathcal{A}$ has enough projective objects, we have an analogous equivalence

$$\mathbf{K}^{-,b}(\mathrm{Proj}\,\mathcal{A}) \xrightarrow{\sim} \mathbf{D}^b(\mathcal{A}).$$

This equivalence restricts to

$$\mathbf{K}^b(\mathrm{Proj}\,\mathcal{A}) \xrightarrow{\sim} \mathbf{D}^b(\mathcal{A})$$

when every object in $\mathcal{A}$ has finite projective dimension.

Proof Suppose X in $\mathbf{K}^+(\mathrm{Inj}\,\mathcal{A})$ satisfies $H^n X = 0$ for almost all n. Then for $n \gg 0$ the differential d_X^n admits a kernel and $\tau_{\leq n} X \to X$ is a quasi-isomorphism. Clearly, $\tau_{\leq n} X$ is mapped into $\mathbf{D}^b(\mathcal{A})$. On the other hand, each object in $\mathcal{A}$ is in the image of $\mathbf{K}^{+,b}(\mathrm{Inj}\,\mathcal{A}) \to \mathbf{D}^b(\mathcal{A})$, because it identifies with an injective resolution. Thus each bounded complex is in the image. If every object in $\mathcal{A}$ has finite injective dimension, then the objects of $\mathcal{A}$ are in the image of $\mathbf{K}^b(\mathrm{Inj}\,\mathcal{A}) \to \mathbf{D}^b(\mathcal{A})$.

The arguments for $\mathbf{D}^-(\mathcal{A})$ are dual. $\qquad\qquad\qquad\qquad\qquad\qquad\square$

Example 4.2.10. Let $\mathcal{A}$ be an abelian category. Then we have an equivalence

$$(\mathrm{mod}\,\mathcal{A})/(\mathrm{eff}\,\mathcal{A}) \xrightarrow{\sim} \mathcal{A}$$

by Proposition 2.3.3. On the other hand, the Yoneda functor $\mathcal{A} \to \mathrm{mod}\,\mathcal{A}$ induces a triangle equivalence $\mathbf{K}^b(\mathcal{A}) \xrightarrow{\sim} \mathbf{D}^b(\mathrm{mod}\,\mathcal{A})$ by Corollary 4.2.9, since $\mathrm{gl.dim}(\mathrm{mod}\,\mathcal{A}) \leq 2$. This yields the following commutative diagram.

$$
\begin{array}{ccc}
\mathrm{Ac}^b(\mathcal{A}) & \rightarrowtail \mathbf{K}^b(\mathcal{A}) \twoheadrightarrow & \mathbf{D}^b(\mathcal{A}) \\
\downarrow\wr & \downarrow\wr & \| \\
\mathrm{Thick}(\mathrm{eff}\,\mathcal{A}) & \rightarrowtail \mathbf{D}^b(\mathrm{mod}\,\mathcal{A}) \twoheadrightarrow & \mathbf{D}^b(\mathcal{A})
\end{array}
$$

In particular, the triangulated category $\mathbf{Ac}^b(\mathcal{A})$ is generated by the acyclic complexes of the form

$$\cdots \longrightarrow 0 \longrightarrow X^{n-1} \longrightarrow X^n \longrightarrow X^{n+1} \longrightarrow 0 \longrightarrow \cdots .$$

Extension Groups

Let $\mathcal{A}$ be an exact category. For a pair of objects X, Y and $n \geq 1$, let $\mathrm{Ext}^n_{\mathcal{A}}(X, Y)$ denote the abelian group of *n-extensions* in the sense of Yoneda, i.e. equivalence classes of exact sequences

$$\xi: \quad 0 \longrightarrow Y \longrightarrow E_n \longrightarrow \cdots \longrightarrow E_2 \longrightarrow E_1 \longrightarrow X \longrightarrow 0,$$

where such sequences ξ and ξ' are *equivalent* if there is an exact sequence ξ'' that fits into the following commutative diagram.

$$
\begin{array}{ccccccccccccc}
\xi: & 0 & \to & Y & \to & E_n & \to & \cdots & \to & E_2 & \to & E_1 & \to & X & \to & 0 \\
& & & \| & & \downarrow & & & & \downarrow & & \downarrow & & \| & & \\
\xi'': & 0 & \to & Y & \to & E_n'' & \to & \cdots & \to & E_2'' & \to & E_1'' & \to & X & \to & 0 \\
& & & \| & & \uparrow & & & & \uparrow & & \uparrow & & \| & & \\
\xi': & 0 & \to & Y & \to & E_n' & \to & \cdots & \to & E_2' & \to & E_1' & \to & X & \to & 0
\end{array}
$$

Note that $\xi = 0$ if and only if we can choose ξ' such that $Y \to E_n'$ is a split monomorphism or $E_1' \to X$ is a split epimorphism.

From the definition it is not clear that one obtains an equivalence relation; however this follows from the proposition below, using the calculus of fractions.

We identify objects in $\mathcal{A}$ with complexes concentrated in degree zero and show the following.

Proposition 4.2.11. *For all objects X, Y in $\mathcal{A}$ and $n \in \mathbb{Z}$, there is a natural isomorphism*

$$\mathrm{Ext}^n_{\mathcal{A}}(X, Y) \xrightarrow{\sim} \mathrm{Hom}_{\mathbf{D}(\mathcal{A})}(X, \Sigma^n Y), \quad \xi \mapsto \bar{\xi}$$

which is compatible with the Yoneda composition.

Proof This is clear for $n = 0$ by Lemma 4.1.11. So it suffices to consider the case $n \geq 1$. Given $\xi \in \mathrm{Ext}^n_{\mathcal{A}}(X, Y)$, we consider the complex

$$E_\xi: \quad \cdots \longrightarrow 0 \longrightarrow E_n \longrightarrow \cdots \longrightarrow E_2 \longrightarrow E_1 \longrightarrow X \longrightarrow 0 \longrightarrow \cdots$$

with X in degree zero. There is a canonical morphism $\xi_0 \colon X \to E_\xi$ (given by id_X in degree zero) and a quasi-isomorphism $\xi_1 \colon \Sigma^n Y \to E_\xi$ (given by

$Y \to E_n$ in degree $-n$). Then we set $\bar{\xi} = \xi_1^{-1} \circ \xi_0$, which identifies with the left fraction (ξ_0, ξ_1).

Now fix a pair of exact sequences ξ and ξ' of length n, connecting X and Y. If they are equivalent (as in the above definition) then we obtain a commutative diagram

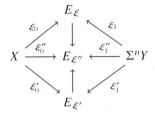

and therefore $\bar{\xi} = \bar{\xi}'$ by Lemma 1.2.1. Conversely, if $\bar{\xi} = \bar{\xi}'$ then one obtains a commutative diagram

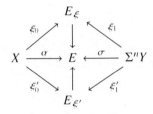

and we will see that ξ and ξ' are equivalent, because we can turn the left fraction (α, σ) into an extension.

Let us construct the inverse map, which sends a morphism $\phi \colon X \to \Sigma^n Y$ to an extension $\bar{\phi}$. The morphism ϕ is given by a left fraction, so a pair of morphisms $X \xrightarrow{\alpha} E \xleftarrow{\sigma} \Sigma^n Y$ between complexes such that σ is a quasi-isomorphism. We consider the truncation $\tau_{\geq -n} \tau_{\leq 0} E$

$$\cdots \longrightarrow 0 \longrightarrow \operatorname{Coker} d^{-n-1} \longrightarrow E^{-n+1} \longrightarrow \cdots$$
$$\cdots \longrightarrow E^{-2} \longrightarrow E^{-1} \longrightarrow \operatorname{Ker} d^0 \longrightarrow 0 \longrightarrow \cdots$$

which is quasi-isomorphic to E. Then σ yields an exact sequence

$$\tilde{\sigma} \colon \quad 0 \longrightarrow Y \longrightarrow \operatorname{Coker} d^{-n-1} \longrightarrow E^{-n+1} \longrightarrow \cdots$$
$$\cdots \longrightarrow E^{-2} \longrightarrow E^{-1} \longrightarrow \operatorname{Ker} d^0 \longrightarrow 0$$

and α induces a morphism $\tilde{\alpha} \colon X \to \operatorname{Ker} d^0$. The pullback of $\tilde{\sigma}$ via $\tilde{\alpha}$ is by definition the element $\bar{\phi}$ in $\operatorname{Ext}_{\mathcal{A}}^n(X, Y)$.

It is straightforward to check the naturality of the bijection; also the Baer sum is preserved. For the Yoneda composition, it suffices to consider a degree one element $\xi \in \operatorname{Ext}^1_{\mathcal{A}}(X', X)$ corresponding to $\bar{\xi} \in \operatorname{Hom}_{\mathbf{D}(\mathcal{A})}(X', \Sigma X)$. Then the connecting morphism $\operatorname{Ext}^n_{\mathcal{A}}(X, Y) \to \operatorname{Ext}^{n+1}_{\mathcal{A}}(X', Y)$ corresponds to the morphism $\operatorname{Hom}_{\mathbf{D}(\mathcal{A})}(X, \Sigma^n Y) \to \operatorname{Hom}_{\mathbf{D}(\mathcal{A})}(X', \Sigma^{n+1} Y)$ given by composition with $\bar{\xi}$. This yields the compatibility. $\qquad \square$

We give a second proof that uses injective resolutions.

Second proof of Proposition 4.2.11 Suppose that Y admits an injective resolution $Y \to iY$. Then we obtain the following isomorphisms

$$\operatorname{Ext}^n_{\mathcal{A}}(X, Y) \cong H^n \operatorname{Hom}_{\mathcal{A}}(X, iY)$$
$$\cong \operatorname{Hom}_{\mathbf{K}(\mathcal{A})}(X, \Sigma^n iY)$$
$$\cong \operatorname{Hom}_{\mathbf{D}(\mathcal{A})}(X, \Sigma^n iY)$$
$$\cong \operatorname{Hom}_{\mathbf{D}(\mathcal{A})}(X, \Sigma^n Y)$$

where the first is based on the description of $\operatorname{Ext}^n(X, -)$ as a derived functor, the second is from Example 4.1.5, the third is from Lemma 4.2.5, and the last is clear. $\qquad \square$

Corollary 4.2.12. *For each exact sequence $\xi \colon 0 \to A' \to A \to A'' \to 0$ in $\mathcal{A}$ and $n \geq 0$, composition with ξ yields a connecting morphism*

$$\operatorname{Ext}^n_{\mathcal{A}}(A', B) \longrightarrow \operatorname{Ext}^{n+1}_{\mathcal{A}}(A'', B)$$

and these fit into a long exact sequence:

$$0 \longrightarrow \operatorname{Hom}_{\mathcal{A}}(A'', -) \longrightarrow \operatorname{Hom}_{\mathcal{A}}(A, -) \longrightarrow \operatorname{Hom}_{\mathcal{A}}(A', -)$$
$$\longrightarrow \operatorname{Ext}^1_{\mathcal{A}}(A'', -) \longrightarrow \operatorname{Ext}^1_{\mathcal{A}}(A, -) \longrightarrow \operatorname{Ext}^1_{\mathcal{A}}(A', -) \longrightarrow \cdots$$

Proof The exact sequence $0 \to A' \to A \to A'' \to 0$ in $\mathcal{A}$ yields an exact triangle $\bar{A}' \to \bar{A} \to \bar{A}'' \to \Sigma \bar{A}'$ in $\mathbf{D}(\mathcal{A})$; see Lemma 4.1.12. Now apply to this triangle for any $X \in \mathcal{A}$ the cohomological functor $\operatorname{Hom}_{\mathbf{D}(\mathcal{A})}(-, \bar{X})$ and use Proposition 4.2.11. $\qquad \square$

Exact Subcategories

Let $\mathcal{A}$ be an exact category and $\mathcal{B} \subseteq \mathcal{A}$ a full exact subcategory. The inclusion induces an exact functor $\mathbf{D}(\mathcal{B}) \to \mathbf{D}(\mathcal{A})$ and we provide criteria for this to be fully faithful.

Lemma 4.2.13. *Let $\mathcal{B} \subseteq \mathcal{A}$ be a full exact subcategory. Then the induced functor $\mathbf{D}^b(\mathcal{B}) \to \mathbf{D}^b(\mathcal{A})$ is fully faithful if and only if the map*

$$\operatorname{Ext}^n_{\mathcal{B}}(X, Y) \longrightarrow \operatorname{Ext}^n_{\mathcal{A}}(X, Y)$$

is bijective for all $X, Y \in \mathcal{B}$ *and* $n \in \mathbb{Z}$.

Proof Identify $\mathrm{Ext}^n_{\mathcal{A}}(X,Y) = \mathrm{Hom}_{\mathbf{D}(\mathcal{A})}(X, \Sigma^n Y)$ using Proposition 4.2.11, and then apply dévissage (Lemma 3.1.8). □

We call $\mathcal{B}$ *left cofinal* in $\mathcal{A}$ if for every admissible monomorphism $\alpha\colon X \to Y$ in $\mathcal{A}$ with X in $\mathcal{B}$ there exists an admissible monomorphism $X \to Y'$ in $\mathcal{B}$ that factors through α. Dually, $\mathcal{B}$ is *right cofinal* in $\mathcal{A}$ if for every admissible epimorphism $\beta\colon Y \to Z$ in $\mathcal{A}$ with Z in $\mathcal{B}$ there exists an admissible epimorphism $Y' \to Z$ in $\mathcal{B}$ that factors through β.

Remark 4.2.14. If $\mathcal{B}$ is left cofinal in $\mathcal{A}$ and idempotent complete, then for every exact sequence $0 \to X \to Y \to Z \to 0$ in $\mathcal{A}$ with $X, Y \in \mathcal{B}$, we have $Z \in \mathcal{B}$.

Proof The property of $\mathcal{B}$ to be left cofinal yields a commutative diagram

$$
\begin{array}{ccccccccc}
0 & \longrightarrow & X & \longrightarrow & Y & \longrightarrow & Z & \longrightarrow & 0 \\
 & & \| & & \downarrow & & \downarrow & & \\
0 & \longrightarrow & X & \longrightarrow & Y' & \longrightarrow & Z' & \longrightarrow & 0
\end{array}
$$

whith exact rows and the bottom one in $\mathcal{B}$. This induces an exact sequence $0 \to Y \to Y' \oplus Z \to Z' \to 0$, and therefore $Y' \oplus Z$ belongs to $\mathcal{B}$. □

Proposition 4.2.15. *Let* $\mathcal{B} \subseteq \mathcal{A}$ *be left or right cofinal. Then the induced functor* $\mathbf{D}^b(\mathcal{B}) \to \mathbf{D}^b(\mathcal{A})$ *is full and faithful.*

Proof Suppose $\mathcal{B} \subseteq \mathcal{A}$ is left cofinal; the other case is dual. We apply Lemma 4.2.13 and need to show that the map

$$\alpha_{X,Y}\colon \mathrm{Ext}^n_{\mathcal{B}}(X,Y) \longrightarrow \mathrm{Ext}^n_{\mathcal{A}}(X,Y)$$

is bijective for all $X, Y \in \mathcal{B}$ and $n \in \mathbb{Z}$. For surjectivity, pick an extension

$$\xi\colon \qquad 0 \longrightarrow Y \longrightarrow A_n \longrightarrow \cdots \longrightarrow A_2 \longrightarrow A_1 \longrightarrow X \longrightarrow 0$$

with all A_i in $\mathcal{A}$. We use induction on n and write $\xi = \xi_1 \circ \xi_{n-1}$ as the composite of extensions of degree 1 and $n-1$ respectively, with ξ_1 given by the exact sequence $\xi_1\colon 0 \to Y \to A_n \to \bar{A}_n \to 0$. The property of $\mathcal{B}$ to be left cofinal yields a commutative diagram

$$
\begin{array}{ccccccccc}
\xi_1\colon & & 0 & \longrightarrow & Y & \longrightarrow & A_n & \longrightarrow & \bar{A}_n & \longrightarrow & 0 \\
 & & & & \| & & \downarrow & & \downarrow{\scriptstyle\phi} & & \\
\xi_1'\colon & & 0 & \longrightarrow & Y & \longrightarrow & B_n & \longrightarrow & \bar{B}_n & \longrightarrow & 0
\end{array}
$$

with exact rows and the bottom one in $\mathcal{B}$. Thus $\xi_1 = \xi_1' \circ \phi$, and therefore $\xi = \xi_1' \circ \phi \circ \xi_{n-1}$ with $\xi_{n-1}' = \phi \circ \xi_{n-1}$ in $\mathrm{Ext}_{\mathcal{A}}^{n-1}(X, \bar{B}_n)$. Then ξ_{n-1}' is in the image of $\mathbf{D}^b(\mathcal{B}) \to \mathbf{D}^b(\mathcal{A})$ by the induction hypothesis, and this yields surjectivity.

To show injectivity, let $\xi \in \mathrm{Ext}_{\mathcal{B}}^n(X, Y)$ be an element such that $\alpha_{X,Y}(\xi) = 0$. This means there is a commutative diagram with exact rows in $\mathcal{A}$

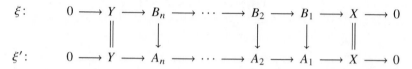

such that the morphism $A_1 \to X$ is a split epimorphism. We claim that we can choose ξ' such that all A_n are in $\mathcal{B}$. This implies $\xi = 0$ in $\mathrm{Ext}_{\mathcal{B}}^n(X, Y)$. As before, we use induction on n and write $\xi' = \xi_1' \circ \xi_{n-1}'$ as the composite of extensions of degree 1 and $n - 1$ respectively. So we can replace ξ' by an extension in such a way that the object A_n is replaced by an object in $\mathcal{B}$. Thus $\xi' = \xi_1'' \circ \xi_{n-1}''$, and the claim holds for ξ_{n-1}'' by the induction hypothesis. $\qquad \square$

Remark 4.2.16. The lemma can be strengthened as follows. Suppose $\mathcal{B} \subseteq \mathcal{A}$ is left cofinal. Then $\mathbf{K}^+(\mathcal{B}) \subseteq \mathbf{K}^+(\mathcal{A})$ is left cofinal with respect to the class of quasi-isomorphisms. In particular, the induced functor $\mathbf{D}^+(\mathcal{B}) \to \mathbf{D}^+(\mathcal{A})$ is full and faithful. This assertion follows from [199, Proposition III.2.3.1], which discusses the dual result (when $\mathcal{A}$ is abelian), because the conditions dual to (E1)–(E3) in [199, Section III.2.2.1] are satisfied. The property of $\mathbf{D}^+(\mathcal{B}) \to \mathbf{D}^+(\mathcal{A})$ then follows from Lemma 1.2.5.

The following lemma provides for an exact category $\mathcal{A}$ a method to compute the extension groups $\mathrm{Ext}_{\mathcal{A}}^n(X, Y)$ for any pair of objects X, Y in $\mathcal{A}$, since $\mathrm{Lex}\,\mathcal{A}$ has enough injective objects; see Proposition 2.3.7 and Proposition 4.2.11.

Lemma 4.2.17. *Let $\mathcal{A}$ be an essentially small exact category. Then the Yoneda functor $\mathcal{A} \to \mathrm{Lex}\,\mathcal{A}$ induces a fully faithful exact functor $\mathbf{D}^b(\mathcal{A}) \to \mathbf{D}^b(\mathrm{Lex}\,\mathcal{A})$.*

Proof The Yoneda functor $\mathcal{A} \to \mathrm{Lex}\,\mathcal{A}$ identifies $\mathcal{A}$ with a full extension closed subcategory of $\mathrm{Lex}\,\mathcal{A}$ by Proposition 2.3.7. We claim that the inclusion $\mathcal{A} \to \mathrm{Lex}\,\mathcal{A}$ is right cofinal. Then the assertion follows from Proposition 4.2.15. To show the cofinality, fix an epimorphism $\phi \colon F \to \mathrm{Hom}_{\mathcal{A}}(-, Z)$ in $\mathrm{Lex}\,\mathcal{A}$. Then the cokernel $C = \mathrm{Coker}\,\phi$ in $\mathrm{Mod}\,\mathcal{A}$ is locally effaceable. Thus there exists an admissible epimorphism $Y \to Z$ in $\mathcal{A}$ such that $CZ \to CY$ annihilates the image of id_Z. This implies that $\mathrm{Hom}_{\mathcal{A}}(-, Y) \to \mathrm{Hom}_{\mathcal{A}}(-, Z)$ factors through ϕ. $\qquad \square$

Example 4.2.18. Let Λ be a right coherent ring. Then $\mathrm{mod}\,\Lambda$ is a right cofinal subcategory of Mod Λ. In fact, the canonical functor $\mathbf{D}^b(\text{mod }\Lambda) \to \mathbf{D}(\text{Mod }A)$ induces a triangle equivalence

$$\mathbf{D}^b(\text{mod }\Lambda) \xrightarrow{\sim} \left\{X \in \mathbf{D}(\text{Mod }A) \mid \coprod_{n\in\mathbb{Z}} H^n X \in \text{mod }\Lambda\right\}.$$

More generally, the inclusion mod $\Lambda \to$ Mod Λ induces a fully faithful functor $\mathbf{D}^-(\text{mod }\Lambda) \to \mathbf{D}(\text{Mod }A)$ by Proposition 4.2.8.

A Grothendieck category $\mathcal{A}$ is *locally finitely presented* if every object in $\mathcal{A}$ is a filtered colimit of finitely presented objects. Here, $X \in \mathcal{A}$ is *finitely presented* if the functor $\text{Hom}_{\mathcal{A}}(X, -)$ preserves filtered colimits. Let fp $\mathcal{A}$ denote the full subcategory of finitely presented objects.

Proposition 4.2.19. *Let $\mathcal{A}$ be a locally finitely presented Grothendieck category and suppose* fp $\mathcal{A}$ *is abelian. Then* fp $\mathcal{A}$ *is an extension closed subcategory and the inclusion* fp $\mathcal{A} \to \mathcal{A}$ *induces a fully faithful functor* $\mathbf{D}^b(\text{fp }\mathcal{A}) \to \mathbf{D}^b(\mathcal{A})$.

Proof We begin with the following observation. If a finitely presented object is written as a filtered colimit $X = \text{colim }X_i$ of objects in $\mathcal{A}$, then for some index i_0 the canonical morphism $X_{i_0} \to X$ is a split epimorphism.

Now let $\eta\colon 0 \to X \to Y \to Z \to 0$ be an exact sequence in $\mathcal{A}$ with $X, Z \in \text{fp }\mathcal{A}$. Write $Y = \text{colim }Y_i$ as a filtered colimit of finitely presented objects. This yields exact sequences $\eta_i\colon 0 \to X_i \to Y_i \to Z$ and we have $\text{colim }\eta_i = \eta$. Thus $\alpha\colon X_{i_0} \to X$ and $Y_{i_0} \to Z$ are epimorphisms for some index i_0. It follows that Y is isomorphic to the cokernel of Ker $\alpha \to Y_{i_0}$ and therefore finitely presented.

The second assertion follows from Proposition 4.2.15 since fp $\mathcal{A}$ is right cofinal as a subcategory of $\mathcal{A}$. To see this, fix an exact sequence $0 \to X \to Y \to Z \to 0$ in $\mathcal{A}$ with $Z \in \text{fp }\mathcal{A}$. As before, we write this as a filtered colimit of exact sequences $0 \to X_i \to Y_i \to Z$ in fp $\mathcal{A}$ and $Y_{i_0} \to Z$ is an epimorphism for some index i_0. Clearly, $Y_{i_0} \to Z$ is admissible in fp $\mathcal{A}$ and factors through $Y \to Z$. $\qquad\square$

Example 4.2.20. Let A be a commutative noetherian ring and $\mathcal{C} \subseteq \text{mod }A$ a Serre subcategory. Then the induced functor $\mathbf{D}^b(\mathcal{C}) \to \mathbf{D}^b(\text{mod }A)$ is fully faithful.

Proof We consider the localising subcategory of Mod A which is generated by $\mathcal{C}$. This is closed under injective envelopes by Corollary 2.4.14. Thus $\mathcal{C} \subseteq \text{mod }A$ is left cofinal, because a monomorphism $\alpha\colon X \to Y$ in mod A with $X \in \mathcal{C}$ yields a morphism $Y \to E(X)$ which factors through a finitely generated submodule $Y' \subseteq E(X)$. We have $Y' \in \mathcal{C}$ and $X \to Y'$ factors through α. Now the assertion follows from Proposition 4.2.15. $\qquad\square$

4.3 Resolutions and Derived Functors

In this section we introduce homotopy injective and projective resolutions of complexes. The construction of such resolutions requires some work. Of particular interest are complexes of modules and we obtain a more explicit description of the derived category of a module category. The construction of resolutions for Grothendieck categories is rather involved, but it can be reduced to the case of a module category. In the second part of this section we use these resolutions to construct derived functors.

Homotopy Injective and Projective Resolutions

Let $\mathcal{A}$ be an exact category. We explain how to construct resolutions for complexes in $\mathcal{A}$ and begin with the relevant definitions.

A complex I in $\mathcal{A}$ is called *K-injective* (or *homotopy injective*) if we have $\mathrm{Hom}_{\mathbf{K}(\mathcal{A})}(X, I) = 0$ for each acyclic complex X. An equivalent condition is that the canonical map

$$\mathrm{Hom}_{\mathbf{K}(\mathcal{A})}(X, I) \longrightarrow \mathrm{Hom}_{\mathbf{D}(\mathcal{A})}(X, I)$$

is bijective for each complex X; see Lemma 3.2.4. A *K-injective resolution* of a complex X is a quasi-isomorphism $X \to I$ such that I is K-injective. Analogously, a complex P is called *K-projective* (or *homotopy projective*) if $\mathrm{Hom}_{\mathbf{K}(\mathcal{A})}(P, X) = 0$ for each acyclic complex X. A *K-projective resolution* of a complex X is a quasi-isomorphism $P \to X$ such that P is K-projective.

The K-injective complexes form a thick subcategory of $\mathbf{K}(\mathcal{A})$ which we denote by $\mathbf{K}_{\mathrm{inj}}(\mathcal{A})$; the category of K-projective complexes is denoted by $\mathbf{K}_{\mathrm{proj}}(\mathcal{A})$.

The following proposition collects the basic properties of K-injective resolutions.

Proposition 4.3.1. *For an exact category $\mathcal{A}$ the following are equivalent.*

(1) *Every object $X \in \mathbf{K}(\mathcal{A})$ admits a K-injective resolution $X \to \mathbf{i}(X)$.*
(2) *The canonical functor $\mathbf{K}(\mathcal{A}) \to \mathbf{D}(\mathcal{A})$ admits a right adjoint.*

In this case, the assignment $X \mapsto \mathbf{i}(X)$ induces a left adjoint for the inclusion $\mathbf{K}_{\mathrm{inj}}(\mathcal{A}) \to \mathbf{K}(\mathcal{A})$ and the resolution $X \to \mathbf{i}(X)$ equals the unit. This yields the following localisation sequence of exact functors

$$\mathbf{Ac}(\mathcal{A}) \begin{array}{c} \longrightarrow \\[-6pt] \longleftarrow \end{array} \mathbf{K}(\mathcal{A}) \begin{array}{c} \xrightarrow{\ \mathbf{i}\ } \\[-6pt] \xleftarrow[\mathrm{inc}]{} \end{array} \mathbf{K}_{\mathrm{inj}}(\mathcal{A})$$

and therefore the canonical functor $\mathbf{K}(\mathcal{A}) \to \mathbf{D}(\mathcal{A})$ restricts to a triangle *equivalence* $\mathbf{K}_{inj}(\mathcal{A}) \xrightarrow{\sim} \mathbf{D}(\mathcal{A})$. *Moreover, the assignment* $X \mapsto \mathbf{i}(X)$ *induces a quasi-inverse* $\mathbf{D}(\mathcal{A}) \xrightarrow{\sim} \mathbf{K}_{inj}(\mathcal{A})$ *and this yields the following diagram.*

$$\mathbf{K}(\mathcal{A}) \underset{\mathbf{i}}{\overset{\text{can}}{\rightleftarrows}} \mathbf{D}(\mathcal{A})$$

Proof Suppose first that each $X \in \mathbf{K}(\mathcal{A})$ admits a K-injective resolution $X \to \mathbf{i}(X)$. We complete this to an exact triangle

$$\mathbf{a}(X) \longrightarrow X \longrightarrow \mathbf{i}(X) \longrightarrow \Sigma\mathbf{a}(X).$$

Clearly, $\mathbf{a}(X)$ is acyclic, and therefore $\mathrm{Hom}_{\mathbf{K}(\mathcal{A})}(\mathbf{a}(X), Y) = 0$ for all $Y \in \mathbf{K}_{inj}(\mathcal{A})$. Thus $X \to \mathbf{i}(X)$ induces for all $Y \in \mathbf{K}_{inj}(\mathcal{A})$ a bijection

$$\mathrm{Hom}_{\mathbf{K}(\mathcal{A})}(\mathbf{i}(X), Y) \xrightarrow{\sim} \mathrm{Hom}_{\mathbf{K}(\mathcal{A})}(X, Y).$$

This means that $X \mapsto \mathbf{i}(X)$ provides a left adjoint for the inclusion $\mathbf{K}_{inj}(\mathcal{A}) \to \mathbf{K}(\mathcal{A})$. Also, we have $\mathbf{Ac}(\mathcal{A})^{\perp} = \mathbf{K}_{inj}(\mathcal{A})$ by definition. Now we apply the general theory from Proposition 3.2.8. Thus $\mathbf{K}(\mathcal{A}) \to \mathbf{D}(\mathcal{A})$ restricts to a triangle equivalence $\mathbf{K}_{inj}(\mathcal{A}) \xrightarrow{\sim} \mathbf{D}(\mathcal{A})$. Moreover, the functor $\mathbf{i}$ annihilates $\mathbf{Ac}(\mathcal{A})$ and induces therefore an equivalence $\mathbf{D}(\mathcal{A}) \xrightarrow{\sim} \mathbf{K}_{inj}(\mathcal{A})$. Then the composite with the inclusion $\mathbf{K}_{inj}(\mathcal{A}) \to \mathbf{K}(\mathcal{A})$ provides a right adjoint for the canonical functor $Q \colon \mathbf{K}(\mathcal{A}) \to \mathbf{D}(\mathcal{A})$.

Now suppose that Q admits a right adjoint $Q_\rho \colon \mathbf{D}(\mathcal{A}) \to \mathbf{K}(\mathcal{A})$. Then the unit $X \to Q_\rho Q(X)$ is a K-injective resolution since $\mathbf{K}_{inj}(\mathcal{A}) = \mathbf{Ac}(\mathcal{A})^{\perp}$, again by Proposition 3.2.8. $\qquad\square$

The above proposition suggests the notation $X \to \mathbf{i}(X)$ for a K-injective resolution and $\mathbf{p}(X) \to X$ for a K-projective resolution of a complex X.

Corollary 4.3.2. *Suppose all complexes in* $\mathbf{K}(\mathcal{A})$ *admit a K-injective resolution. Then the morphisms between two objects in* $\mathbf{D}(\mathcal{A})$ *form a set.* $\qquad\square$

Our aim is to construct K-injective resolutions via homotopy limits. We begin with a remark about limits of abelian groups. For any sequence

$$\cdots \xrightarrow{\phi_3} A_2 \xrightarrow{\phi_2} A_1 \xrightarrow{\phi_1} A_0$$

of maps between abelian groups, the limit and its first derived functor are given by the exact sequence

$$0 \longrightarrow \lim_n A_n \longrightarrow \prod_{n \geq 0} A_n \xrightarrow{\mathrm{id} - \phi} \prod_{n \geq 0} A_n \longrightarrow \lim_n^1 A_n \longrightarrow 0.$$

Note that $\lim_n^1 A_n = 0$ when $A_{n+1} \xrightarrow{\sim} A_n$ for $n \gg 0$.

We say that an abelian category has *exact products* if any product of exact sequences is again exact. For example, products in Ab are exact. Therefore any module category has exact products. However, there are many examples showing that products in Grothendieck categories need not be exact (Example 2.4.22).

Lemma 4.3.3. *Let $\mathcal{A}$ be an abelian category with countable products that are exact and consider in $\mathbf{K}(\mathcal{A})$ a sequence of morphisms $\cdots \to X_2 \to X_1 \to X_0$. Then a compatible sequence of morphisms $X \to X_n$ induces a quasi-isomorphism $X \xrightarrow{\sim} \operatorname{holim}_n X_n$ provided that $H^i(X) \xrightarrow{\sim} H^i(X_n)$ for $n \gg 0$ and each integer i.*

Proof For each $i \in \mathbb{Z}$ the maps $H^i(X_{n+1}) \to H^i(X_n)$ eventually become invertible and therefore induce an exact sequence

$$0 \longrightarrow \lim_n H^i(X_n) \longrightarrow \prod_{n \geq 0} H^i(X_n) \longrightarrow \prod_{n \geq 0} H^i(X_n) \longrightarrow 0.$$

Because products in $\mathcal{A}$ are exact, this sequence identifies with the exact sequence

$$0 \longrightarrow H^i(X) \longrightarrow H^i(\prod_{n \geq 0} X_n) \longrightarrow H^i(\prod_{n \geq 0} X_n) \longrightarrow 0,$$

and applying H^i to the triangle defining $\operatorname{holim}_n X_n$ yields an isomorphism

$$H^i(\operatorname*{holim}_n X_n) \cong \lim_n H^i(X_n).$$

The morphisms $X \to X_n$ induce a morphism $X \to \operatorname{holim}_n X_n$, and it follows that this is a quasi-isomorphism. $\square$

Proposition 4.3.4. *Let $\mathcal{A}$ be an abelian category with enough injective objects, and suppose that $\mathcal{A}$ has countable products that are exact. Then every object in $\mathbf{K}(\mathcal{A})$ admits a K-injective resolution.*

Proof First observe that $\mathbf{K}_{\mathrm{inj}}(\mathcal{A})$ contains $\mathbf{K}^+(\operatorname{Inj}\mathcal{A})$, by Lemma 4.2.5. Also $\mathbf{K}_{\mathrm{inj}}(\mathcal{A})$ is closed under homotopy limits.

Let $X \in \mathbf{K}(\mathcal{A})$. For each integer $n \leq 0$ let $X \to \tau_{\geq n} X$ denote the truncation inducing an isomorphism $H^i(X) \xrightarrow{\sim} H^i(\tau_{\geq n} X)$ for all $i \geq n$. Choose a quasi-isomorphism $\tau_{\geq n} X \to Y_n$ with $Y_n \in \mathbf{K}^+(\operatorname{Inj}\mathcal{A})$, which exists by Lemma 4.2.7. The composite $\tau_{\geq n-1} X \to \tau_{\geq n} X \to Y_n$ extends to a morphism $Y_{n-1} \to Y_n$, since $\operatorname{Hom}_{\mathbf{K}(\mathcal{A})}(\alpha, Y_n)$ is a bijection for every quasi-isomorphism α by Lemma 4.2.5. Now set $Y = \operatorname{holim}_n Y_n$. It follows from Lemma 4.3.3 that the sequence of morphisms $X \to Y_n$ induces a quasi-isomorphism $X \to Y$. It remains to note that $Y \in \mathbf{K}_{\mathrm{inj}}(\mathcal{A})$ by our first observation. $\square$

Remark 4.3.5. The proof of Proposition 4.3.4 shows that $\mathbf{K}_{\mathrm{inj}}(\mathcal{A})$ equals the smallest triangulated subcategory of $\mathbf{K}(\mathcal{A})$ that is closed under countable products and contains $\mathbf{K}^b(\operatorname{Inj}\mathcal{A})$.

There are dual versions of Proposition 4.3.1 and Proposition 4.3.4 for abelian categories with enough projective objects. In particular, for any ring Λ we have K-injective and K-projective resolutions for complexes of Λ-modules.

Corollary 4.3.6. *For a ring Λ every complex in $\mathbf{K}(\mathrm{Mod}\,\Lambda)$ has a K-injective and a K-projective resolution. Thus we have triangle equivalences*

$$\mathbf{K}_{\mathrm{inj}}(\mathrm{Mod}\,\Lambda) \xrightarrow{\sim} \mathbf{D}(\mathrm{Mod}\,\Lambda) \qquad and \qquad \mathbf{K}_{\mathrm{proj}}(\mathrm{Mod}\,\Lambda) \xrightarrow{\sim} \mathbf{D}(\mathrm{Mod}\,\Lambda).$$

In particular, the morphisms between two objects in $\mathbf{D}(\mathrm{Mod}\,\Lambda)$ form a set.   □

Grothendieck Categories

Let $\mathcal{A}$ be a Grothendieck category. Then there is an analogue of Proposition 4.3.4, and its proof amounts to showing that the canonical functor $\mathbf{K}(\mathcal{A}) \to \mathbf{D}(\mathcal{A})$ admits a right adjoint, using Brown's representability theorem. In order to be able to apply this theorem, we need to show that the morphisms between any two objects in $\mathbf{D}(\mathcal{A})$ form a set. This fact uses the property of $\mathcal{A}$ to be locally presentable; see Proposition 2.5.16.

Lemma 4.3.7. *Let $\alpha > \aleph_0$ be a regular cardinal and $\mathcal{A}$ a locally α-presentable Grothendieck category such that $\mathcal{A}^\alpha$ is abelian. Then a morphism $X \to Y$ in $\mathbf{C}(\mathcal{A})$ with $X \in \mathbf{C}(\mathcal{A}^\alpha)$ and $Y \in \mathbf{Ac}(\mathcal{A})$ factors through an object in $\mathbf{Ac}(\mathcal{A}^\alpha)$.*

Proof The assumption $\alpha > \aleph_0$ is used so that $\mathcal{A}^\alpha$ is closed under countable colimits. First we consider the case that X belongs to $\mathbf{C}^-(\mathcal{A}^\alpha)$. Set $X = X_0$ and assume $H^p X = 0$ for $p > 0$. We construct inductively a sequence

$$X = X_0 \longrightarrow X_{-1} \longrightarrow X_{-2} \longrightarrow \cdots \longrightarrow Y$$

of factorisations of $X \to Y$ such that $H^p X_n = 0$ for $p > n$. This yields a factorisation $X \to \mathrm{colim}_{n \le 0} X_n \to Y$ such that $\mathrm{colim}_{n \le 0} X_n$ belongs to $\mathbf{Ac}^-(\mathcal{A}^\alpha)$. The morphism $\phi \colon X \to X_{-1}$ giving the factorisation in each step is constructed as follows. Set $\phi^p = \mathrm{id}$ for all $p \ne -1$. Now form the pullback

$$\begin{array}{ccc} V & \longrightarrow & Z^0(X) \\ \downarrow & & \downarrow \\ Y^{-1} & \longrightarrow & Z^0(Y) \end{array}$$

and write $V = \mathrm{colim}\, V_i$ as α-filtered colimit of objects in $\mathcal{A}^\alpha$. Since Y is acyclic, the morphism $V \to Z^0(X)$ is an epimorphism. Using that X^{-1} and $Z^0(X)$ belong to $\mathcal{A}^\alpha$, there is an index j such that $X^{-1} \to V$ factors through the canonical morphism $V_j \to V$ and the composite $V_j \to V \to Z^0(X)$ is an epimorphism. We set $X_{-1}^{-1} = V_j$ and have constructed the desired factorisation.

Now consider an arbitrary $X \in \mathbf{C}(\mathcal{A}^\alpha)$ and observe that $X = \mathrm{colim}_{n \geq 0}\, \tau_{\leq n} X$. The first part of the proof yields a factorisation $\tau_{\leq 0} X \to X_0 \to Y$ with $X_0 \in \mathbf{Ac}^-(\mathcal{A}^\alpha)$. Then form the pushout

$$
\begin{array}{ccc}
\tau_{\leq 0} X & \longrightarrow & \tau_{\leq 1} X \\
\downarrow & & \downarrow \\
X_0 & \longrightarrow & X_{\leq 1}
\end{array}
$$

and continue with a factorisation $X_{\leq 1} \to X_1 \to Y$. This yields a compatible sequence of factorisations $\tau_{\leq n} X \to X_n \to Y$ and we take its colimit with $\mathrm{colim}_{n \geq 0} X_n$ in $\mathbf{Ac}(\mathcal{A}^\alpha)$.   □

Let $\mathcal{A}$ be a Grothendieck category. We recall from Corollary 2.5.23 that there exists a regular cardinal $\alpha_0 > \aleph_0$ such that for all regular $\alpha \geq \alpha_0$ the category $\mathcal{A}^\alpha$ is abelian.

Proposition 4.3.8. *For every regular cardinal $\alpha \geq \alpha_0$ the inclusion $\mathcal{A}^\alpha \to \mathcal{A}$ induces a fully faithful functor $\mathbf{D}(\mathcal{A}^\alpha) \to \mathbf{D}(\mathcal{A})$. Therefore*

$$
\mathbf{D}(\mathcal{A}) = \bigcup_{\alpha \geq \alpha_0} \mathbf{D}(\mathcal{A}^\alpha)
$$

and the morphisms between two objects in $\mathbf{D}(\mathcal{A})$ form a set.

Proof We apply Lemma 4.3.7. Then it follows from Lemma 3.2.5 that the inclusion $\mathbf{K}(\mathcal{A}^\alpha) \to \mathbf{K}(\mathcal{A})$ induces a fully faithful functor $\mathbf{D}(\mathcal{A}^\alpha) \to \mathbf{D}(\mathcal{A})$. It remains to observe that each complex belongs to $\mathcal{A}^\alpha$ for some $\alpha \geq \alpha_0$ by Lemma 2.5.13.   □

From the fact that morphisms in $\mathbf{D}(\mathcal{A})$ form a set we can deduce the existence of K-injective resolutions.

Theorem 4.3.9. *Let $\mathcal{A}$ be a Grothendieck category. Then every complex in $\mathbf{K}(\mathcal{A})$ admits a K-injective resolution.*

Proof Choose a generator $C \in \mathcal{A}$ and set $\Lambda = \mathrm{End}(C)$. Then the functor $\mathrm{Hom}(C, -) \colon \mathcal{A} \to \mathrm{Mod}\,\Lambda$ is fully faithful and admits an exact left adjoint by Theorem 2.5.2. The left adjoint induces an exact and coproduct preserving functor $\mathbf{D}(\mathrm{Mod}\,\Lambda) \to \mathbf{D}(\mathcal{A})$. This functor admits a right adjoint by Brown representability (Corollary 3.4.11 and Proposition 3.4.15) since $\mathbf{D}(\mathrm{Mod}\,\Lambda)$ is compactly generated. In fact, the ring Λ viewed as a complex concentrated in degree zero is a compact generator since $\mathrm{Hom}(\Lambda, X) = H^0(X)$ for each complex X. Also, we used that morphisms in $\mathbf{D}(\mathcal{A})$ form a set, by Proposition 4.3.8.

Now consider the following commutative square of canonical and exact functors.

$$\begin{array}{ccc} \mathbf{K}(\operatorname{Mod}\Lambda) & \longrightarrow & \mathbf{D}(\operatorname{Mod}\Lambda) \\ \downarrow & & \downarrow \\ \mathbf{K}(\mathcal{A}) & \longrightarrow & \mathbf{D}(\mathcal{A}) \end{array}$$

The functor $\mathbf{K}(\operatorname{Mod}\Lambda) \to \mathbf{D}(\operatorname{Mod}\Lambda)$ admits a right adjoint by Proposition 4.3.1, since complexes in $\mathbf{K}(\operatorname{Mod}\Lambda)$ admit K-injective resolutions by Corollary 4.3.6. Thus the composite $\mathbf{K}(\operatorname{Mod}\Lambda) \to \mathbf{D}(\mathcal{A})$ admits a right adjoint, and then it follows from Lemma 1.1.7 that $\mathbf{K}(\mathcal{A}) \to \mathbf{D}(\mathcal{A})$ admits a right adjoint. Applying Proposition 4.3.1 once more the assertion follows. □

Derived Functors

Homotopy injective and projective resolutions are used to construct derived functors. We begin with a general definition of right derived functors, following Deligne. Left derived functors are defined dually.

Let $\mathcal{A}$ be an exact category and $F: \mathbf{K}(\mathcal{A}) \to \mathcal{T}$ an exact functor into a triangulated category $\mathcal{T}$. Recall that for an object $X \in \mathbf{K}(\mathcal{A})$ the quasi-isomorphisms $X \to X'$ in $\mathbf{K}(\mathcal{A})$ form a filtered category, which we denote by X/Qis; see Lemma 4.1.10. Now consider the functor $\mathcal{T}^{\mathrm{op}} \to \operatorname{Ab}$ given by the filtered colimit

$$\operatorname*{colim}_{X \to X'} \operatorname{Hom}_{\mathcal{T}}(-, FX')$$

where $X \to X'$ runs through the objects in X/Qis. Suppose that this functor is representable for each object X, and denote by $RF(X)$ a representing object corresponding to X. This assignment extends to morphisms in $\mathbf{K}(\mathcal{A})$, since any morphism $X \to Y$ in $\mathbf{K}(\mathcal{A})$ induces a morphism of functors

$$\operatorname*{colim}_{X \to X'} \operatorname{Hom}_{\mathcal{T}}(-, FX') \longrightarrow \operatorname*{colim}_{Y \to Y'} \operatorname{Hom}_{\mathcal{T}}(-, FY').$$

Thus we obtain a functor $\mathbf{K}(\mathcal{A}) \to \mathcal{T}$, and it is straightforward to check that this is exact, because F is exact. Also, a quasi-isomorphism $\sigma: X \to Y$ induces an isomorphism $RF(X) \xrightarrow{\sim} RF(Y)$, since precomposition with σ yields a cofinal functor $Y/\operatorname{Qis} \to X/\operatorname{Qis}$. Then the universal property of the canonical functor $Q: \mathbf{K}(\mathcal{A}) \to \mathbf{D}(\mathcal{A})$ yields an exact functor $\mathbf{R}F: \mathbf{D}(\mathcal{A}) \to \mathcal{T}$, which is by definition the *right derived functor* of F. In fact, the functor $\mathbf{R}F$ together with the canonical morphism $\eta: F \to \mathbf{R}F \circ Q$ enjoys the following universal property.

Proposition 4.3.10. *Let $\mathcal{A}$ be an exact category and $F \colon \mathbf{K}(\mathcal{A}) \to \mathcal{T}$ an exact functor into a triangulated category $\mathcal{T}$. For any exact functor $F' \colon \mathbf{D}(\mathcal{A}) \to \mathcal{T}$ together with a morphism $\eta' \colon F \to F' \circ Q$, there exists a unique morphism $\theta \colon \mathbf{R}F \to F'$ such that $\eta'_X = \theta_X \circ \eta_X$ for all $X \in \mathbf{K}(\mathcal{A})$.*

Proof For an object $X \in \mathbf{K}(\mathcal{A})$ the morphism $\theta_X \colon \mathbf{R}F(X) \to F'X$ is the unique one making the following square commutative

$$
\begin{array}{ccc}
\mathrm{Hom}_{\mathcal{T}}(-, \mathbf{R}F(X)) & \xrightarrow{\ \sim\ } & \underset{X \to X'}{\mathrm{colim}}\, \mathrm{Hom}_{\mathcal{T}}(-, FX') \\[2mm]
\Big\downarrow{\scriptstyle \theta} & & \Big\downarrow{\scriptstyle \eta'} \\[2mm]
\mathrm{Hom}_{\mathcal{T}}(-, F'X) & \xrightarrow{\ \sim\ } & \underset{X \to X'}{\mathrm{colim}}\, \mathrm{Hom}_{\mathcal{T}}(-, (F' \circ Q)X')
\end{array}
$$

where one uses that $X = QX \xrightarrow{\sim} QX'$ for each $X \to X'$ in X/Qis. □

In general, the right derived functor $\mathbf{R}F$ does not exist, because the above functor $\mathcal{T}^{\mathrm{op}} \to \mathrm{Ab}$ need not be representable for each object X in $\mathbf{K}(\mathcal{A})$. However, we have an explicit description in terms of K-injective resolutions.

Proposition 4.3.11. *Let $\mathcal{A}$ be an exact category and suppose that every object in $\mathbf{K}(\mathcal{A})$ admits a K-injective resolution $X \to \mathbf{i}X$. Then for any exact functor $F \colon \mathbf{K}(\mathcal{A}) \to \mathcal{T}$ the right derived functor is given by*

$$
\mathbf{R}F \colon \mathbf{D}(\mathcal{A}) \longrightarrow \mathcal{T}, \qquad X \mapsto F(\mathbf{i}X).
$$

Proof Fix a K-injective resolution $\xi \colon X \to \mathbf{i}X$. Then every $\sigma \colon X \to X'$ in X/Qis admits a morphism $\sigma \to \xi$, and therefore

$$
\underset{X \to X'}{\mathrm{colim}}\, \mathrm{Hom}_{\mathcal{T}}(-, FX') \xrightarrow{\sim} \mathrm{Hom}_{\mathcal{T}}(-, F(\mathbf{i}X)). \qquad \square
$$

Now let $F \colon \mathcal{A} \to \mathcal{B}$ be an additive functor between exact categories. Then F induces a functor $\mathbf{K}(\mathcal{A}) \to \mathbf{K}(\mathcal{B})$ by applying F in each degree; we denote this functor again by F. The above proposition justifies the following definition. The *right derived functor* $\mathbf{R}F \colon \mathbf{D}(\mathcal{A}) \to \mathbf{D}(\mathcal{B})$ of F sends a complex X to $F(\mathbf{i}X)$, and the *left derived functor* $\mathbf{L}F \colon \mathbf{D}(\mathcal{A}) \to \mathbf{D}(\mathcal{B})$ sends a complex X to $F(\mathbf{p}X)$, provided such resolutions exist.

Example 4.3.12. Let $\mathcal{A}$ be a Grothendieck category. Then $\mathbf{D}(\mathcal{A})$ admits coproducts, which are computed componentwise, cf. Lemma 4.1.15. On the other hand, for a family of complexes $(X_i)_{i \in I}$ its product in $\mathbf{D}(\mathcal{A})$ is computed in $\mathbf{K}(\mathcal{A})$ via K-injective resolutions and is represented by

$$
\prod_{i \in I} \mathbf{i}X_i.
$$

One may think of this as the right derived functor of

$$\mathcal{A}^I \longrightarrow \mathcal{A}, \quad (X_i)_{i \in I} \mapsto \prod_{i \in I} X_i.$$

Derived Functors Between Module Categories

We illustrate the construction of derived functors by taking functors between chain complexes of modules.

Let Λ and Γ be a pair of rings and let M be a complex of Λ-Γ-bimodules. There are two functors of complexes associated with M. Given a complex X of Λ-modules, the complex $X \otimes_\Lambda M$ of Γ-modules is defined by

$$(X \otimes_\Lambda M)^n = \bigoplus_{p+q=n} X^p \otimes_\Lambda M^q \qquad (n \in \mathbb{Z})$$

with differential given by

$$d^n(x \otimes m) = d_X(x) \otimes m + (-1)^p x \otimes d_M(m) \qquad (x \in X^p, \ m \in M^q).$$

Given a complex Y of Γ-modules, the complex $\mathrm{Hom}_\Gamma(M,Y)$ of Λ-modules is defined by

$$\mathrm{Hom}_\Gamma(M,Y)^n = \prod_{-p+q=n} \mathrm{Hom}_\Gamma(M^p, Y^q) \qquad (n \in \mathbb{Z})$$

with differential given by

$$d^n(\phi) = d_Y \circ \phi - (-1)^n \phi \circ d_M \qquad (\phi \in \mathrm{Hom}_\Gamma(M,Y)^n).$$

Lemma 4.3.13. *We have a natural isomorphism of complexes*

$$\mathrm{Hom}_\Gamma(X \otimes_\Lambda M, Y) \cong \mathrm{Hom}_\Lambda(X, \mathrm{Hom}_\Gamma(M,Y)).$$

Proof For $n \in \mathbb{Z}$ we have

$$\mathrm{Hom}_\Gamma(X \otimes_\Lambda M, Y)^n \cong \prod_{-p-q+r=n} \mathrm{Hom}_\Gamma(X^p \otimes_\Lambda M^q, Y^r)$$

$$\cong \prod_{-p-q+r=n} \mathrm{Hom}_\Lambda(X^p, \mathrm{Hom}_\Gamma(M^q, Y^r))$$

$$\cong \mathrm{Hom}_\Lambda(X, \mathrm{Hom}_\Gamma(M,Y))^n.$$

The second isomorphism is given by the usual tensor-hom adjunction; it provides a morphism of complexes thanks to the sign rules for the differentials on each side. $\qquad \square$

Lemma 4.3.14. *Let X, Y be complexes of Λ-modules. Then for $n \in \mathbb{Z}$ we have*

$$H^n \mathrm{Hom}_\Lambda(X, Y) \cong \mathrm{Hom}_{\mathbf{K}(\Lambda)}(X, \Sigma^n Y).$$

Proof For $\operatorname{Hom}_\Lambda(X, Y)$ the group $\operatorname{Ker} d^n$ identifies with $\operatorname{Hom}_{\mathbf{C}(\Lambda)}(X, \Sigma^n Y)$ and $\operatorname{Im} d^{n-1}$ identifies with the ideal of null-homotopic morphisms $X \to \Sigma^n Y$.

$\square$

We set

$$X \otimes_\Lambda^L M = \mathbf{p}X \otimes_\Lambda M \qquad \text{and} \qquad \operatorname{RHom}_\Gamma(M, Y) = \operatorname{Hom}_\Gamma(M, \mathbf{i}Y).$$

Proposition 4.3.15. *Let Λ and Γ be a pair of rings and let M be a complex of Λ-Γ-bimodules. Then we have a pair of adjoint functors*

$$\mathbf{K}(\operatorname{Mod}\Lambda) \underset{\operatorname{Hom}_\Gamma(M,-)}{\overset{-\otimes_\Lambda M}{\rightleftarrows}} \mathbf{K}(\operatorname{Mod}\Gamma)$$

which induces a pair of adjoint functors

$$\mathbf{D}(\operatorname{Mod}\Lambda) \underset{\operatorname{RHom}_\Gamma(M,-)}{\overset{-\otimes_\Lambda^L M}{\rightleftarrows}} \mathbf{D}(\operatorname{Mod}\Gamma).$$

Proof Set $T = - \otimes_\Lambda M$ and $H = \operatorname{Hom}_\Gamma(M, -)$. Combining Lemma 4.3.13 and Lemma 4.3.14 it follows that (T, H) yields an adjoint pair of functors $\mathbf{K}(\operatorname{Mod}\Lambda) \rightleftarrows \mathbf{K}(\operatorname{Mod}\Gamma)$.

The derived functors $\mathbf{L}T = - \otimes_\Lambda^L M$ and $\mathbf{R}H = \operatorname{RHom}_\Gamma(M, -)$ are composed from three pairs

$$\mathbf{D}(\operatorname{Mod}\Lambda) \underset{\operatorname{can}}{\overset{\mathbf{p}}{\rightleftarrows}} \mathbf{K}(\operatorname{Mod}\Lambda) \underset{H}{\overset{T}{\rightleftarrows}} \mathbf{K}(\operatorname{Mod}\Gamma) \underset{\mathbf{i}}{\overset{\operatorname{can}}{\rightleftarrows}} \mathbf{D}(\operatorname{Mod}\Gamma)$$

of adjoint functors; see Proposition 4.3.1. Thus $(\mathbf{L}T, \mathbf{R}H)$ is an adjoint pair. $\square$

Example 4.3.16. Let Λ and Γ be a pair of rings and let ${}_\Lambda M_\Gamma$ be a bimodule. Then the pair of adjoint functors

$$\operatorname{Mod}\Lambda \underset{\operatorname{Hom}_\Gamma(M,-)}{\overset{-\otimes_\Lambda M}{\rightleftarrows}} \operatorname{Mod}\Gamma$$

induces a pair of adjoint functors

$$\mathbf{D}(\operatorname{Mod}\Lambda) \underset{\operatorname{RHom}_\Gamma(M,-)}{\overset{-\otimes_\Lambda^L M}{\rightleftarrows}} \mathbf{D}(\operatorname{Mod}\Gamma).$$

For a Λ-module X choose a projective resolution $pX \to X$. Then we have for $n \geq 0$

$$\operatorname{Ext}_\Lambda^n(X, -) = H^n \operatorname{RHom}_\Lambda(X, -) = H^n \operatorname{Hom}_\Lambda(pX, -)$$

and

$$\operatorname{Tor}_n^\Lambda(X, -) = H^n(X \otimes_\Lambda^L -) = H^n(pX \otimes_\Lambda -).$$

Lemma 4.3.17. *Let* X_Λ, $_\Lambda Y_\Gamma$, *and* I_Γ *be modules. Suppose that* I_Γ *is injective. Then we have for* $n \geq 0$ *a natural isomorphism*

$$\mathrm{Ext}^n_\Lambda(X, \mathrm{Hom}_\Gamma(Y, I)) \cong \mathrm{Hom}_\Gamma(\mathrm{Tor}^\Lambda_n(X, Y), I).$$

Proof Using tensor-hom adjunction we compute

$$\mathrm{Ext}^n_\Lambda(X, \mathrm{Hom}_\Gamma(Y, I)) \cong H^n \mathrm{Hom}_\Lambda(pX, \mathrm{Hom}_\Gamma(Y, I))$$
$$\cong H^n \mathrm{Hom}_\Gamma(pX \otimes_\Lambda Y, I)$$
$$\cong \mathrm{Hom}_\Gamma(H^n(pX \otimes_\Lambda Y), I)$$
$$\cong \mathrm{Hom}_\Gamma(\mathrm{Tor}^\Lambda_n(X, Y), I). \qquad \square$$

Homotopically Minimal Complexes

For any object X in a Frobenius category it is natural to ask for a subobject X' that is maximal among all injective subobjects, giving a decomposition $X = X' \oplus X''$ such that X'' is 'injective-free'. Moreover, one may ask in which sense such a decomposition is unique.

We provide a positive answer for the category $\mathbf{C}(\mathcal{A})$ of complexes when $\mathcal{A}$ is an additive category such that its morphisms admit minimal decompositions. Recall that $\phi = \phi' \oplus \phi''$ is a *minimal decomposition* if ϕ' is an isomorphism and ϕ'' is a radical morphism. Examples of additive categories with this property are full subcategories of injective objects of abelian categories with injective envelopes (Corollary 2.1.24) or Krull–Schmidt categories (Example 2.1.25).

Let $\mathcal{A}$ be an additive category. We call a complex $X \in \mathbf{C}(\mathcal{A})$ *homotopically minimal*, if the canonical functor $\mathbf{C}(\mathcal{A}) \to \mathbf{K}(\mathcal{A})$ sends each non-invertible endomorphism of X to a non-invertible endomorphism. Our aim is to establish decompositions $X = X' \oplus X''$ such that X' is contractible and X'' is homotopically minimal.

Let $\mathcal{A}$ be an additive category and suppose that its morphisms admit minimal decompositions. Given a complex $X \in \mathbf{C}(\mathcal{A})$, we construct for each $n \in \mathbb{Z}$ a new complex $X(n)$ as follows. For $i = n, n + 1$ let $X^i = U^i \oplus V^i$ be a decomposition such that the differential decomposes as $d^n_X = d' \oplus d''$ with $d' : U^n \to U^{n+1}$ an isomorphism and $d'' : V^n \to V^{n+1}$ a radical morphism. We set $U^p = 0$ otherwise and obtain a contractible subcomplex $U \subseteq X$. This gives a decomposition $X = U \oplus V$ and we put $X(n) = U$.

Proposition 4.3.18. *Let $\mathcal{A}$ be an additive category and suppose that morphisms in $\mathcal{A}$ admit minimal decompositions. Then the following are equivalent for a complex $X \in \mathbf{C}(\mathcal{A})$.*

(1) *The complex X is homotopically minimal.*

(2) *The complex X has no non-zero direct summand which is contractible.*

(3) *The differential $X^n \to X^{n+1}$ is a radical morphism for all $n \in \mathbb{Z}$.*

Proof (1) $\Rightarrow$ (2): Let $X = X' \oplus X''$ and suppose X' is contractible. The idempotent morphism $\varepsilon \colon X \to X$ with $\operatorname{Ker} \varepsilon = X' = \operatorname{Coker} \varepsilon$ induces an isomorphism in $\mathbf{K}(\mathcal{A})$. Thus (1) implies $X' = 0$.

(2) $\Rightarrow$ (3): Fix $n \in \mathbb{Z}$. Then we have a decomposition $X = X(n) \oplus V$ such that $X(n)$ is contractible. Our assumption implies $X(n) = 0$, and we conclude that the morphism $X^n \to X^{n+1}$ is a radical morphism.

(3) $\Rightarrow$ (1): We may assume that $\mathcal{A}$ identifies with the category of injective objects of an additive category $\bar{\mathcal{A}}$ with kernels such that each object $X \in \bar{\mathcal{A}}$ admits an injective copresentation $0 \to X \xrightarrow{\phi} I^0 \xrightarrow{\psi} I^1$, by taking $\bar{\mathcal{A}} = (\operatorname{mod}(\mathcal{A}^{\mathrm{op}}))^{\mathrm{op}}$. In that case ϕ is essential if and only if $\psi \in \operatorname{Rad}(I^0, I^1)$, cf. Proposition 2.1.23.

Now let $\phi, \psi \colon X \to X$ be a pair of morphisms such that $\psi \circ \phi$ and $\phi \circ \psi$ are chain homotopic to the identity id_X. Thus we have a family of morphisms $\rho^n \colon X^n \to X^{n-1}$ such that

$$\mathrm{id}_{X^n} = (\psi \circ \phi)^n + \delta^{n-1} \circ \rho^n + \rho^{n+1} \circ \delta^n.$$

We claim that $\operatorname{Ker} \phi = 0$. In fact, we show that $K = \operatorname{Ker}(\psi \circ \phi) = 0$. Let $L^n = K^n \cap Z^n X$. Then the restriction $(\delta^{n-1} \circ \rho^n)|_{L^n}$ is a monomorphism, and therefore $\rho^n(L^n) \cap Z^{n-1} X = 0$. The inclusion $Z^{n-1} X \to X^{n-1}$ is essential, and it follows that $L^n = 0$. The same assumption on $Z^n X \to X^n$ implies $K^n = 0$. Thus ϕ is a monomorphism, and in fact a split monomorphism in each degree. Dually, ϕ is an epimorphism, and it follows that ϕ is an isomorphism. $\square$

Corollary 4.3.19. *Let $\mathcal{A}$ be an additive category and suppose that morphisms in $\mathcal{A}$ admit minimal decompositions. Then every complex $X \in \mathbf{C}(\mathcal{A})$ admits a decomposition $X = X' \oplus X''$ such that X' is contractible and X'' is homotopically minimal. Given a second decomposition $X = Y' \oplus Y''$ such that Y' is contractible and Y'' is homotopically minimal, then the canonical morphism $X'' \rightarrowtail X \twoheadrightarrow Y''$ is an isomorphism.*

Proof Take $X' = \coprod_{n \in \mathbb{Z}} X(n)$. This complex is contractible and the canonical morphism $\iota \colon \coprod_{n \in \mathbb{Z}} X(n) \to X$ is a split monomorphism in each degree. Thus ι has a left inverse and we obtain a decomposition $X = X' \oplus X''$. The construction of each $X(n)$ shows that the differentials of X'' are radical morphisms. Thus X'' is homotopically minimal, by Proposition 4.3.18.

Now let $X = Y' \oplus Y''$ be a second decomposition such that Y' is contractible and Y'' is homotopically minimal. The canonical morphism $\phi \colon X'' \rightarrowtail X \twoheadrightarrow Y''$

induces an isomorphism in $\mathbf{K}(\mathcal{A})$ since X' and Y' are contractible. Thus ϕ is an isomorphism in $\mathbf{C}(\mathcal{A})$, since X'' and Y'' are homotopically minimal. □

We provide an application. Let $\mathcal{A}$ be a Grothendieck category and suppose that every injective object is *discrete*, that is, the injective envelope of a coproduct of indecomposable injective objects. For example, a locally noetherian Grothendieck category has this property.

We consider the canonical functor

$$P \colon \mathcal{A} \longrightarrow \mathcal{A}[\mathrm{Ess}^{-1}] \cong \prod_{E \in \mathrm{Sp}\,\mathcal{A}} \mathrm{Mod}\,\Delta_E$$

into the spectral category of $\mathcal{A}$, which identifies with a product of categories of vector spaces given by divison rings $\Delta_E = \mathrm{End}(E)/J(\mathrm{End}(E))$ for each indecomposable injective object E, since the restriction $P|_{\mathrm{Inj}\,\mathcal{A}}$ induces an equivalence $(\mathrm{Inj}\,\mathcal{A})/\mathrm{Rad}(\mathrm{Inj}\,\mathcal{A}) \xrightarrow{\sim} \mathcal{A}[\mathrm{Ess}^{-1}]$ (cf. Proposition 2.5.9). The derived functor

$$\mathbf{R}P \colon \mathbf{D}(\mathcal{A}) \longrightarrow \prod_{E \in \mathrm{Sp}\,\mathcal{A}} \mathbf{D}(\mathrm{Mod}\,\Delta_E)$$

provides a notion of *support* by defining for $X \in \mathbf{D}(\mathcal{A})$

$$\mathrm{Supp}(X) = \{E \in \mathrm{Sp}\,\mathcal{A} \mid \mathbf{R}P(X)_E \neq 0\}.$$

Lemma 4.3.20. *Let* $X \in \mathbf{D}(\mathcal{A})$. *Then* $X \neq 0$ *implies* $\mathrm{Supp}(X) \neq \varnothing$.

Proof We have $\mathbf{R}P(X) = P(\mathbf{i}X)$, and we may assume that $\mathbf{i}X$ is homotopically minimal. Then P annihilates each differential of $\mathbf{i}X$ by Proposition 4.3.18. It follows that $E \in \mathrm{Supp}(X)$ if and only if E arises as a direct summand of $(\mathbf{i}X)^n$ for some $n \in \mathbb{Z}$. □

4.4 Examples of Derived Categories

We consider examples of derived categories and provide explicit descriptions. For instance, for an abelian category we study the passage to Serre subcategories and their quotient categories. There are two classes of exact categories which deserve special attention: hereditary categories and Frobenius categories.

Quotient Categories

Let $\mathcal{A}$ be an abelian category and $\mathcal{C} \subseteq \mathcal{A}$ a Serre subcategory. We set

$$\mathbf{D}_{\mathcal{C}}(\mathcal{A}) = \{X \in \mathbf{D}(\mathcal{A}) \mid H^n(X) \in \mathcal{C} \text{ for all } n \in \mathbb{Z}\} \subseteq \mathbf{D}(\mathcal{A})$$

and obtain a thick subcategory; it is the kernel of the functor $\mathbf{D}(\mathcal{A}) \to \mathbf{D}(\mathcal{A}/\mathcal{C})$.

Lemma 4.4.1. *The quotient functor* $\mathcal{A} \to \mathcal{A}/\mathcal{C}$ *induces a triangle equivalence.*

$$\mathbf{D}(\mathcal{A})/\mathbf{D}_{\mathcal{C}}(\mathcal{A}) \xrightarrow{\sim} \mathbf{D}(\mathcal{A}/\mathcal{C}).$$

Proof The functor $\mathbf{C}(\mathcal{A}) \to \mathbf{C}(\mathcal{A}/\mathcal{C})$ induces an equivalence

$$\mathbf{C}(\mathcal{A})/\mathbf{C}(\mathcal{C}) \xrightarrow{\sim} \mathbf{C}(\mathcal{A}/\mathcal{C}),$$

and a quasi-inverse yields a triangle equivalence

$$F: \mathbf{D}(\mathcal{A}/\mathcal{C}) \xrightarrow{\sim} \mathbf{C}(\mathcal{A})/\mathbf{C}(\mathcal{C})[\mathrm{Qis}^{-1}].$$

The composite $\mathbf{C}(\mathcal{A}) \twoheadrightarrow \mathbf{D}(\mathcal{A}) \twoheadrightarrow \mathbf{D}(\mathcal{A})/\mathbf{D}_{\mathcal{C}}(\mathcal{A})$ factors through the composite

$$\mathbf{C}(\mathcal{A}) \twoheadrightarrow \mathbf{C}(\mathcal{A})/\mathbf{C}(\mathcal{C}) \twoheadrightarrow \mathbf{C}(\mathcal{A})/\mathbf{C}(\mathcal{C})[\mathrm{Qis}^{-1}],$$

and this yields a functor $\mathbf{C}(\mathcal{A})/\mathbf{C}(\mathcal{C})[\mathrm{Qis}^{-1}] \to \mathbf{D}(\mathcal{A})/\mathbf{D}_{\mathcal{C}}(\mathcal{A})$. Composing this with F is a quasi-inverse for $\mathbf{D}(\mathcal{A})/\mathbf{D}_{\mathcal{C}}(\mathcal{A}) \to \mathbf{D}(\mathcal{A}/\mathcal{C})$.   □

Serre Subcategories

Let $\mathcal{A}$ be an abelian category and $\mathcal{C} \subseteq \mathcal{A}$ a Serre subcategory. The inclusion $\mathcal{C} \to \mathcal{A}$ induces a functor $\mathbf{D}^+(\mathcal{C}) \to \mathbf{D}^+(\mathcal{A})$, and we provide a criterion for this to be fully faithful when there are enough injective objects.

Proposition 4.4.2. *Let* $\mathcal{C} \subseteq \mathcal{A}$ *be a Serre subcategory and suppose that* $\mathcal{A}$ *has enough injective objects. Suppose also that the canonical functor* $Q \colon \mathcal{A} \to \mathcal{A}/\mathcal{C}$ *preserves injectivity and admits a right adjoint* $Q_\rho \colon \mathcal{A}/\mathcal{C} \to \mathcal{A}$. *Then the following are equivalent.*

(1) *The unit* $X \to Q_\rho Q(X)$ *is an epimorphism for every injective* $X \in \mathcal{A}$.
(2) *The functor* $\mathbf{D}^+(\mathcal{C}) \to \mathbf{D}^+(\mathcal{A})$ *is fully faithful.*

Proof It is convenient to use the following notation

$$\mathcal{C} = \mathcal{A}' \underset{i^!}{\overset{i_!}{\rightleftarrows}} \mathcal{A} \underset{j_*}{\overset{j^*}{\rightleftarrows}} \mathcal{A}'' = \mathcal{A}/\mathcal{C} \tag{4.4.3}$$

with $Q = j^*$ and $Q_\rho = j_*$.

(1) $\Rightarrow$ (2): Let $\mathcal{J}$ denote the full subcategory of injective objects in $\mathcal{A}$; the categories $\mathcal{J}'$ and $\mathcal{J}''$ are defined analogously. We view $\mathcal{A}'$ and $\mathcal{A}''$ as full subcategories of $\mathcal{A}$ via $i_!$ and j_*, respectively, and write $\mathrm{Filt}(\mathcal{J}', \mathcal{J}'')$ for the smallest extension closed subcategory of $\mathcal{A}$ containing $\mathcal{J}'$ and $\mathcal{J}''$. This contains $\mathcal{J}$ since each injective object X fits into an exact sequence

$$0 \longrightarrow i_! i^!(X) \longrightarrow X \longrightarrow j_* j^*(X) \longrightarrow 0.$$

Note that the diagram (4.4.3) restricts to

$$\mathfrak{I}' \underset{i^!}{\overset{i_!}{\rightleftarrows}} \mathrm{Filt}(\mathfrak{I}',\mathfrak{I}'') \underset{j_*}{\overset{j^*}{\rightleftarrows}} \mathfrak{I}'' \qquad (4.4.4)$$

and all functors in this diagram are exact. The only functor for which this is not obvious is $i^!$. In that case exactness follows from the snake lemma because the unit $X \to j_*j^*(X)$ is an epimorphism for every X in $\mathrm{Filt}(\mathfrak{I}',\mathfrak{I}'')$. Thus the diagram (4.4.4) induces the following diagram.

$$\mathbf{D}^+(\mathfrak{I}') \underset{i^!}{\overset{i_!}{\rightleftarrows}} \mathbf{D}^+(\mathrm{Filt}(\mathfrak{I}',\mathfrak{I}'')) \underset{j_*}{\overset{j^*}{\rightleftarrows}} \mathbf{D}^+(\mathfrak{I}'')$$

We claim that this diagram is equivalent to

$$\mathbf{D}^+(\mathcal{A}') \underset{i^!}{\overset{i_!}{\rightleftarrows}} \mathbf{D}^+(\mathcal{A}) \underset{j_*}{\overset{j^*}{\rightleftarrows}} \mathbf{D}^+(\mathcal{A}'') \qquad (4.4.5)$$

via triangle equivalences induced by the inclusions

$$f' \colon \mathfrak{I}' \to \mathcal{A}' \qquad f'' \colon \mathfrak{I}'' \to \mathcal{A}'' \qquad f \colon \mathrm{Filt}(\mathfrak{I}',\mathfrak{I}'') \to \mathcal{A}.$$

This is clear for f' and f'', since $\mathcal{A}'$ and $\mathcal{A}''$ have enough injective objects. For f it suffices to note that the inclusion $\mathfrak{I} \to \mathrm{Filt}(\mathfrak{I}',\mathfrak{I}'')$ yields a triangle equivalence $\mathbf{D}^+(\mathfrak{I}) \overset{\sim}{\to} \mathbf{D}^+(\mathrm{Filt}(\mathfrak{I}',\mathfrak{I}''))$, since $\mathfrak{I}$ equals the full subcategory of injective objects of the exact category $\mathrm{Filt}(\mathfrak{I}',\mathfrak{I}'')$; see Proposition 4.2.8.

(2) $\Rightarrow$ (1): Suppose the functor $\mathbf{D}^+(\mathcal{A}') \to \mathbf{D}^+(\mathcal{A})$ is fully faithful. Then we have a diagram of the form (4.4.5). Given an injective object X in $\mathcal{A}$, there is an exact triangle

$$i_!i^!(X) \longrightarrow X \longrightarrow j_*j^*(X) \longrightarrow \Sigma i_!i^!(X)$$

in $\mathbf{D}^+(\mathcal{A})$. This uses the fact that for complexes of injectives the derived functors of $i^!$ and j_* are defined degreewise via $i^!$ and j_*, respectively. Taking cohomology, we obtain an exact sequence

$$\cdots \longrightarrow 0 \longrightarrow i_!i^!(X) \longrightarrow X \longrightarrow j_*j^*(X) \longrightarrow 0 \longrightarrow \cdots$$

in $\mathcal{A}$. It follows that the unit $X \to j_*j^*(X)$ is an epimorphism. $\square$

Example 4.4.6. Let $\mathcal{A}$ be an abelian category with enough injective objects and suppose that every object $X \in \mathcal{C}$ admits a monomorphism $X \to Y$ in $\mathcal{C}$ such that Y is injective in $\mathcal{A}$. This implies easily that $Q \colon \mathcal{A} \to \mathcal{A}/\mathcal{C}$ preserves injectivity and that the unit $X \to Q_\rho Q(X)$ is a *split* epimorphism when X is injective. Note that $\mathcal{C} \subseteq \mathcal{A}$ is left cofinal in this case; cf. Proposition 4.2.15.

Example 4.4.7. There is a dual version of Proposition 4.4.2 for categories with enough projective objects. Let A be a ring and $\mathcal{A} = \text{Mod}\, A$ the category of A-modules. The assumptions in Proposition 4.4.2 are satisfied when $\mathcal{C}$ equals the category of A-modules that are annihilated by an idempotent $e^2 = e \in A$ such that eAe is semisimple, since $\mathcal{A}/\mathcal{C}$ identifies with $\text{Mod}\, eAe$; see Example 2.2.23. Note that in this case $Q\colon \text{Mod}\, A \to \text{Mod}\, eAe$ is given by $X \mapsto Xe = \text{Hom}_A(eA, X)$ and the counit equals the multiplication map $Xe \otimes_{eAe} eA \to X$. This counit is a monomorphism for every projective A-module X if and only if the multiplication map $Ae \otimes_{eAe} eA \to AeA$ is bijective. It is not difficult to check that this holds if and only if AeA is a projective A-module.

$\bullet$

Given a ring Λ and an ideal $I \subseteq \Lambda$, it is easily seen that $\text{Mod}\, \Lambda/I$ identifies with a Serre subcategory of $\text{Mod}\, \Lambda$ when I is idempotent. The following provides a criterion for when the induced functor $\mathbf{D}^b(\text{Mod}\, \Lambda/I) \to \mathbf{D}^b(\text{Mod}\, \Lambda)$ is fully faithful.

Lemma 4.4.8. *Let Λ be a ring and $I \subseteq \Lambda$ an idempotent ideal such that I is a projective Λ-module. Then $\text{Ext}^p_{\Lambda/I}(X, Y) \xrightarrow{\sim} \text{Ext}^p_\Lambda(X, Y)$ for all Λ/I-modules X, Y and $p \geq 0$.*

Proof Set $\bar{\Lambda} = \Lambda/I$. We use induction on $p \geq 0$. The assertion is clear for $p = 0, 1$ since $\text{Mod}\, \bar{\Lambda}$ identifies with a Serre subcategory of $\text{Mod}\, \Lambda$. Now let $p > 1$ and fix a pair of $\bar{\Lambda}$-modules X, Y. Choose an exact sequence $0 \to X' \to P \to X \to 0$ of $\bar{\Lambda}$-modules such that P is projective. This induces the following commutative diagram with exact rows.

$$
\begin{array}{ccccccc}
\text{Ext}^{p-1}_{\bar{\Lambda}}(P, Y) & \longrightarrow & \text{Ext}^{p-1}_{\bar{\Lambda}}(X', Y) & \longrightarrow & \text{Ext}^p_{\bar{\Lambda}}(X, Y) & \longrightarrow & \text{Ext}^p_{\bar{\Lambda}}(P, Y) \\
\downarrow & & \downarrow{\scriptstyle \alpha'} & & \downarrow{\scriptstyle \alpha} & & \downarrow \\
\text{Ext}^{p-1}_{\Lambda}(P, Y) & \longrightarrow & \text{Ext}^{p-1}_{\Lambda}(X', Y) & \longrightarrow & \text{Ext}^p_{\Lambda}(X, Y) & \longrightarrow & \text{Ext}^p_{\Lambda}(P, Y)
\end{array}
$$

Observe that $\text{proj.dim}_\Lambda P \leq 1$ since I is projective. Thus the last term in each row is zero. Also, the first term in each row is zero by the induction hypothesis. It follows that α is invertible, since α' is invertible by the induction hypothesis. $\qquad\square$

Homological Epimorphisms

Let $\phi\colon \Lambda \to \Gamma$ be a ring homomorphism. Then the bimodule $_\Lambda\Gamma_\Gamma$ gives rise to an adjoint pair of functors:

$$\mathbf{D}(\mathrm{Mod}\,\Lambda) \underset{\mathrm{RHom}_\Gamma(\Gamma,-)}{\overset{-\otimes^L_\Lambda \Gamma}{\rightleftarrows}} \mathbf{D}(\mathrm{Mod}\,\Gamma).$$

The homomorphism ϕ is called a *homological epimorphism* provided the equivalent conditions of the following proposition are satisfied.

Proposition 4.4.9. *For a ring homomorphism $\phi\colon \Lambda \to \Gamma$ the following are equivalent.*

(1) *Restriction via ϕ induces a fully faithful functor $\mathbf{D}(\mathrm{Mod}\,\Gamma) \to \mathbf{D}(\mathrm{Mod}\,\Lambda)$.*
(2) $\mathrm{Ext}^p_\Gamma(X,Y) \xrightarrow{\sim} \mathrm{Ext}^p_\Lambda(X,Y)$ *for all Γ-modules X,Y and $p \geq 0$.*
(3) $\Gamma \otimes_\Lambda \Gamma \xrightarrow{\sim} \Gamma$ *and $\mathrm{Tor}^\Lambda_p(\Gamma,\Gamma) = 0$ for all $p > 0$.*

Proof (1) $\Rightarrow$ (2): Clear.

(2) $\Rightarrow$ (3): We write $D = \mathrm{Hom}_\mathbb{Z}(-,\mathbb{Q}/\mathbb{Z})$ for the Matlis duality between left and right Λ-modules. Deriving the adjunction $\mathrm{Hom}_\Lambda(\Gamma,DX) \xrightarrow{\sim} D(\Gamma \otimes_\Lambda X)$ for any left Λ-module X yields an isomorphism

$$\mathrm{Ext}^p_\Lambda(\Gamma,DX) \xrightarrow{\sim} D\,\mathrm{Tor}^\Lambda_p(\Gamma,X) \qquad \text{for all } p \geq 0;$$

see Lemma 4.3.17. Then the isomorphism

$$D\Gamma \xrightarrow{\sim} \mathrm{Hom}_\Gamma(\Gamma,D\Gamma) \xrightarrow{\sim} \mathrm{Hom}_\Lambda(\Gamma,D\Gamma)$$

identifies with the Matlis dual of the multiplication map $\Gamma \otimes_\Lambda \Gamma \to \Gamma$, which is therefore an isomorphism. Analogously,

$$\mathrm{Ext}^p_\Gamma(\Gamma,D\Gamma) \xrightarrow{\sim} \mathrm{Ext}^p_\Lambda(\Gamma,D\Gamma) \xrightarrow{\sim} D\,\mathrm{Tor}^\Lambda_p(\Gamma,\Gamma)$$

implies $\mathrm{Tor}^\Lambda_p(\Gamma,\Gamma) = 0$ for all $p > 0$.

(3) $\Rightarrow$ (1): The functor $\mathbf{D}(\mathrm{Mod}\,\Gamma) \to \mathbf{D}(\mathrm{Mod}\,\Lambda)$ is fully faithful if and only if the counit $\varepsilon_X\colon X \otimes^L_\Lambda \Gamma \to X$ is an isomorphism for all $X \in \mathbf{D}(\mathrm{Mod}\,\Gamma)$. The condition (3) says that ε_Γ is an isomorphism. In fact, the objects X such that $\varepsilon_{\Sigma^n X}$ is an isomorphism for all $n \in \mathbb{Z}$ form a triangulated subcategory of $\mathbf{D}(\mathrm{Mod}\,\Gamma)$ that is closed under all coproducts. It remains to note that $\mathbf{D}(\mathrm{Mod}\,\Gamma)$ is generated by Γ, so there is no proper triangulated subcategory that is closed under coproducts and contains Γ, by Corollary 4.3.6. $\qquad\square$

Corollary 4.4.10. *Let $I \subseteq \Lambda$ be an idempotent ideal such that I is projective when viewed as a left or right Λ-module. Then $\Lambda \twoheadrightarrow \Lambda/I$ is a homological epimorphism.*

Proof Because I is idempotent we have $I \otimes_\Lambda \Lambda/I \cong I/I^2 = 0$. Using that I_Λ is projective it follows that $\mathrm{Tor}^\Lambda_*(I, \Lambda/I) = 0$. Then the exact sequence $0 \to I \to \Lambda \to \Lambda/I \to 0$ induces an isomorphism $\mathrm{Tor}^\Lambda_*(\Lambda/I, \Lambda/I) \cong \Lambda/I$. □

Thick Subcategories

Let $\mathcal{A}$ be an exact category. A full additive subcategory $\mathcal{C} \subseteq \mathcal{A}$ is *thick* if it is closed under direct summands and satisfies the following *two out of three property*: an exact sequence $0 \to X \to Y \to Z \to 0$ lies in $\mathcal{C}$ if two of X, Y, Z are in $\mathcal{C}$. Note that this is different from a Serre subcategory of an abelian category.

Example 4.4.11. For each $X \in \mathcal{A}$ the full subcategories

$$\{A \in \mathcal{A} \mid \mathrm{Ext}^n(X, A) = 0 \text{ for all } n \geq 0\}$$

and

$$\{A \in \mathcal{A} \mid \mathrm{Ext}^n(A, X) = 0 \text{ for all } n \geq 0\}$$

are thick. This follows from the long exact sequences given by $\mathrm{Ext}^n(X, -)$ and $\mathrm{Ext}^n(-, X)$ (Corollary 4.2.12).

Example 4.4.12. An object X in $\mathcal{A}$ is *homologically finite* if for every object Y in $\mathcal{A}$ we have $\mathrm{Ext}^n(X, Y) = 0$ for almost all $n \geq 0$. The homologically finite objects form a thick subcategory of $\mathcal{A}$.

We consider the canonical embedding $d \colon \mathcal{A} \to \mathbf{D}^b(\mathcal{A})$. The assignment

$$\mathcal{C} \longmapsto d^{-1}(\mathcal{C})$$

induces a map

$$\{\text{thick subcategories of } \mathbf{D}^b(\mathcal{A})\} \longrightarrow \{\text{thick subcategories of } \mathcal{A}\}$$

which in some interesting cases is a bijection.

Example 4.4.13. Let $\mathcal{A}$ be a Frobenius category and denote by $\mathcal{P}$ the full subcategory of projective (and injective) objects in $\mathcal{A}$.

(1) The canonical functor $s \colon \mathcal{A} \to \mathrm{St}\,\mathcal{A}$ induces via $\mathcal{C} \mapsto s^{-1}(\mathcal{C})$ a bijection between the thick subcategories of $\mathrm{St}\,\mathcal{A}$ and the thick subcategories of $\mathcal{A}$ containing $\mathcal{P}$.

(2) The canonical functor $d \colon \mathcal{A} \to \mathbf{D}^b(\mathcal{A})$ induces via $\mathcal{C} \mapsto d^{-1}(\mathcal{C})$ a bijection between the thick subcategories of $\mathbf{D}^b(\mathcal{A})$ containing $\mathbf{D}^b(\mathcal{P})$ and the thick subcategories of $\mathcal{A}$ containing $\mathcal{P}$.

Hereditary Categories

An abelian category $\mathcal{A}$ is *hereditary* provided that the functor $\mathrm{Ext}^2_{\mathcal{A}}(-,-)$ vanishes. In this case, there is an explicit description of all objects and morphisms in $\mathbf{D}(\mathcal{A})$. We say that a complex X is *quasi-isomorphic to its cohomology* if there is a quasi-isomorphism between X and

$$\cdots \xrightarrow{0} H^{n-1}X \xrightarrow{0} H^n X \xrightarrow{0} H^{n+1}X \xrightarrow{0} \cdots$$

This means there are isomorphisms

$$\coprod_{n\in\mathbb{Z}} \Sigma^{-n}(H^n X) \cong X \cong \prod_{n\in\mathbb{Z}} \Sigma^{-n}(H^n X) \qquad (4.4.14)$$

in $\mathbf{D}(\mathcal{A})$.

Proposition 4.4.15. *An abelian category $\mathcal{A}$ is hereditary if and only if every object in the derived category $\mathbf{D}(\mathcal{A})$ is quasi-isomorphic to its cohomology.*

Proof We first note that a morphism θ in $\mathbf{C}(\mathcal{A})$ of the form

$$
\begin{array}{ccccccccc}
\cdots & \longrightarrow & 0 & \longrightarrow & A & \xrightarrow{f} & B & \longrightarrow & 0 & \longrightarrow & \cdots \\
& & \downarrow & & \downarrow & & \downarrow & & \downarrow & & \\
\cdots & \longrightarrow & X^{-1} & \xrightarrow{d^{-1}} & X^0 & \xrightarrow{d^0} & X^1 & \xrightarrow{d^1} & X^2 & \longrightarrow & \cdots
\end{array}
$$

yields the following morphism between four-term exact sequences in $\mathcal{A}$:

$$
\begin{array}{ccccccccc}
0 & \longrightarrow & \mathrm{Ker}\, f & \longrightarrow & A & \xrightarrow{f} & B & \longrightarrow & \mathrm{Coker}\, f & \longrightarrow & 0 \\
& & \downarrow & & \downarrow & & \downarrow & & \downarrow & & \\
0 & \longrightarrow & H^0 & \longrightarrow & X^0/\mathrm{Im}\, d^{-1} & \longrightarrow & \mathrm{Ker}\, d^1 & \longrightarrow & H^1 & \longrightarrow & 0
\end{array}
$$

In particular, if θ is a quasi-isomorphism, then these two exact sequences represent the same element of $\mathrm{Ext}^2_{\mathcal{A}}(H^1, H^0)$.

Assume first that every object in $\mathbf{D}(\mathcal{A})$ is quasi-isomorphic to its cohomology, and take any morphism $f\colon A \to B$ in $\mathcal{A}$, say with kernel A' and cokernel B'. Then the assumption yields quasi-isomorphisms in $\mathbf{C}(\mathcal{A})$ of the form

$$
\begin{array}{ccccccccc}
\cdots & \longrightarrow & 0 & \longrightarrow & A & \xrightarrow{f} & B & \longrightarrow & 0 & \longrightarrow & \cdots \\
& & \downarrow & & \downarrow & & \downarrow & & \downarrow & & \\
\cdots & \longrightarrow & X^{-1} & \xrightarrow{d^{-1}} & X^0 & \xrightarrow{d^0} & X^1 & \xrightarrow{d^1} & X^2 & \longrightarrow & \cdots \\
& & \uparrow & & \uparrow & & \uparrow & & \uparrow & & \\
\cdots & \longrightarrow & 0 & \longrightarrow & A' & \xrightarrow{0} & B' & \longrightarrow & 0 & \longrightarrow & \cdots
\end{array}
$$

and hence the induced four-term exact sequences all yield the same element in $\operatorname{Ext}^2_{\mathcal{A}}(B', A')$, necessarily the zero element. It follows that $\operatorname{Ext}^2_{\mathcal{A}}(B, A) = 0$ for all objects A, B in $\mathcal{A}$, so that $\mathcal{A}$ is hereditary.

Conversely, let $\mathcal{A}$ be hereditary and take a complex X in $\mathbf{C}(\mathcal{A})$. Note that the vanishing of $\operatorname{Ext}^2_{\mathcal{A}}(H^n X, -)$ implies the existence of a commutative diagram

$$
\begin{array}{ccccccccc}
0 & \longrightarrow & X^{n-1} & \longrightarrow & E^n & \longrightarrow & H^n X & \longrightarrow & 0 \\
& & \downarrow & & \downarrow & & \| & & \\
0 & \longrightarrow & \operatorname{Im} d^{n-1} & \longrightarrow & \operatorname{Ker} d^n & \longrightarrow & H^n X & \longrightarrow & 0
\end{array}
$$

with exact rows. We obtain the following commutative diagram

$$
\begin{array}{ccccccccc}
\cdots \longrightarrow & 0 & \longrightarrow & 0 & \longrightarrow & H^n X & \longrightarrow & 0 & \longrightarrow \cdots \\
& \uparrow & & \uparrow & & \uparrow & & \uparrow & \\
\cdots \longrightarrow & 0 & \longrightarrow & X^{n-1} & \longrightarrow & E^n & \longrightarrow & 0 & \longrightarrow \cdots \\
& \downarrow & & \| & & \downarrow & & \downarrow & \\
\cdots \longrightarrow X^{n-2} \longrightarrow & X^{n-1} & \longrightarrow & X^n & \longrightarrow & X^{n+1} & \longrightarrow \cdots
\end{array}
$$

and the vertical morphisms induce cohomology isomorphism in degree n. Thus we have in $\mathbf{D}(\mathcal{A})$ the required isomorphisms (4.4.14). $\qquad\square$

Remark 4.4.16. There is another useful characterisation. An abelian category is hereditary if and only if, for every morphism $\phi\colon X \to Y$, there exists a commuting square

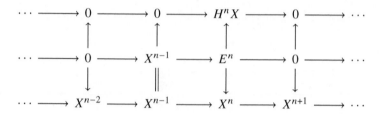

which is a pullback and pushout. In particular, this yields a short exact sequence

$$
0 \longrightarrow X \longrightarrow \operatorname{Im} \phi \oplus E \longrightarrow Y \longrightarrow 0.
$$

Moreover, every thick subcategory of an hereditary category is closed under kernels and cokernels of morphisms.

Proof Suppose that $\mathcal{A}$ is hereditary. A morphism $\phi\colon X \to Y$ induces an exact sequence $0 \to \operatorname{Im}\phi \to Y \to Z \to 0$. The vanishing of $\operatorname{Ext}^2_{\mathcal{A}}(Z, -)$ implies that $X \twoheadrightarrow \operatorname{Im}\phi$ induces a surjective map

$$
\operatorname{Ext}^1_{\mathcal{A}}(Z, X) \longrightarrow \operatorname{Ext}^1_{\mathcal{A}}(Z, \operatorname{Im}\phi).
$$

This yields a commutative diagram

$$
\begin{array}{ccccccccc}
0 & \longrightarrow & X & \longrightarrow & E & \longrightarrow & Z & \longrightarrow & 0 \\
 & & \downarrow & & \downarrow & & \| & & \\
0 & \longrightarrow & \operatorname{Im}\phi & \longrightarrow & Y & \longrightarrow & Z & \longrightarrow & 0
\end{array}
$$

with exact rows, giving the desired pullback and pushout. The argument can be reversed; so the proof of the other direction is similar. $\qquad\square$

For an additive category $\mathcal{C}$ and a family of full additive subcategories $(\mathcal{C}_i)_{i\in I}$ we write

$$
\mathcal{C} = \bigvee_{i \in I} \mathcal{C}_i
$$

if $\mathcal{C} = \sum_i \mathcal{C}_i$ (so each object in $\mathcal{C}$ can be written as $\coprod_i X_i$ with $X_i \in \mathcal{C}_i$ for all i) and $\mathcal{C}_j \cap \sum_{i \neq j} \mathcal{C}_i = 0$ for all j. Thus we have

$$
\mathbf{D}(\mathcal{A}) = \bigvee_{n\in\mathbb{Z}} \Sigma^n\mathcal{A} \quad \text{where} \quad \Sigma^n\mathcal{A} = \{X \in \mathbf{D}(\mathcal{A}) \mid H^i(X) = 0 \text{ for } i \neq -n\}
$$

when $\mathcal{A}$ is hereditary. The bijection $\operatorname{Ext}_{\mathcal{A}}^{n-m}(A, B) \xrightarrow{\sim} \operatorname{Hom}_{\mathbf{D}(\mathcal{A})}(\Sigma^m A, \Sigma^n B)$ for any pair of objects A, B in $\mathcal{A}$ yields a description of all morphisms in $\mathbf{D}(\mathcal{A})$. Each morphism $X \to Y$ in $\mathbf{D}(\mathcal{A})$ corresponds to a family of elements in $\operatorname{Hom}_{\mathcal{A}}(H^n X, H^n Y)$ and a family of elements in $\operatorname{Ext}_{\mathcal{A}}^1(H^n X, H^{n-1}Y)$, with $n \in \mathbb{Z}$. Thus we have non-zero maps $\Sigma^i \mathcal{A} \to \Sigma^j \mathcal{A}$ only if $j - i \in \{0, 1\}$.

For an hereditary category $\mathcal{A}$ there is a close connection between thick subcategories of $\mathcal{A}$ and $\mathbf{D}^b(\mathcal{A})$. This is a consequence of Proposition 4.4.15.

Proposition 4.4.17. *Let $\mathcal{A}$ be an hereditary abelian category. The canonical functor $d \colon \mathcal{A} \to \mathbf{D}^b(\mathcal{A})$ induces via $\mathcal{D} \mapsto d^{-1}(\mathcal{D})$ a bijection between the thick subcategories of $\mathbf{D}^b(\mathcal{A})$ and the thick subcategories of $\mathcal{A}$. The inverse map sends a thick subcategory $\mathcal{C} \subseteq \mathcal{A}$ to $\mathbf{D}^b(\mathcal{C})$.*

Proof If $\mathcal{C} \subseteq \mathcal{A}$ is thick, then the inclusion induces a fully faithful functor $\mathbf{D}^b(\mathcal{C}) \to \mathbf{D}^b(\mathcal{A})$ by Lemma 4.2.13, and its essential image is a thick subcategory of $\mathbf{D}^b(\mathcal{A})$. The subcategory $\mathcal{C}$ is closed under kernels and cokernels of morphisms; see Remark 4.4.16. Thus

$$
H^0(\mathbf{D}^b(\mathcal{C})) = \mathcal{C} = d^{-1}(\mathbf{D}^b(\mathcal{C})).
$$

Now let $\mathcal{D} \subseteq \mathbf{D}^b(\mathcal{A})$ be thick and set $\mathcal{C} = H^0(\mathcal{D}) = d^{-1}(\mathcal{D})$. This is a thick subcategory of $\mathcal{A}$, and we have $\mathbf{D}^b(\mathcal{C}) = \mathcal{D}$ by Proposition 4.4.15. $\qquad\square$

Frobenius Categories

Let $\mathcal{A}$ be a Frobenius category and denote by $\mathcal{P}$ the full subcategory of projective (and injective) objects in $\mathcal{A}$. We provide two useful descriptions of the stable category $\operatorname{St}\mathcal{A}$.

Recall that a complex X in $\mathcal{A}$ is acyclic if for each $n \in \mathbb{Z}$ there is an admissible exact sequence

$$0 \longrightarrow Z^n \xrightarrow{\alpha^n} X^n \xrightarrow{\beta^n} Z^{n+1} \longrightarrow 0$$

in $\mathcal{A}$ such that $d_X^n = \alpha^{n+1} \circ \beta^n$. We set

$$Z^n(X) = \operatorname{Ker}(X^n \xrightarrow{d^n} X^{n+1})$$

and denote $\mathbf{K}(\mathcal{P}) \cap \mathbf{Ac}(\mathcal{A})$ by $\mathbf{K}_{\mathrm{ac}}(\mathcal{P})$.

For any object A in $\mathcal{A}$ there is a *complete resolution*, which is by definition an acyclic complex X of projectives such that $Z^0(X) = A$.

Proposition 4.4.18. *For a Frobenius category $\mathcal{A}$ the composite*

$$F \colon \mathcal{A} \rightarrowtail \mathbf{D}^b(\mathcal{A}) \longrightarrow\!\!\!\rightarrow \mathbf{D}^b(\mathcal{A})/\mathbf{D}^b(\mathcal{P})$$

induces a triangle equivalence

$$\operatorname{St}\mathcal{A} \xrightarrow{\;\sim\;} \mathbf{D}^b(\mathcal{A})/\mathbf{D}^b(\mathcal{P}).$$

Moreover, we have a triangle equivalence

$$Z^0 \colon \mathbf{K}_{\mathrm{ac}}(\mathcal{P}) \xrightarrow{\;\sim\;} \operatorname{St}\mathcal{A}.$$

Proof The functor F is exact: it takes an exact sequence $0 \to X' \to X \to X'' \to 0$ in $\mathcal{A}$ to an exact triangle $F(X') \to F(X) \to F(X'') \to F(X')[1]$. Also, F annihilates all projective objects and therefore yields an exact functor $\bar{F} \colon \operatorname{St}\mathcal{A} \to \mathbf{D}^b(\mathcal{A})/\mathbf{D}^b(\mathcal{P})$. The suspension in $\operatorname{St}\mathcal{A}$ takes X to ΣX, and

$$F(\Sigma X) \cong F(X)[1].$$

We construct a quasi-inverse for $\bar{F}$ as follows.

Consider the category $\mathbf{K}(\mathcal{P})$ and identify the subcategories

$$\mathbf{K}^b(\mathcal{P}) \xrightarrow{\;\sim\;} \mathbf{D}^b(\mathcal{P}) \quad \text{and} \quad \mathbf{K}^{-,b}(\mathcal{P}) \xrightarrow{\;\sim\;} \mathbf{D}^b(\mathcal{A}).$$

For a complex X and $n \in \mathbb{Z}$ we use the following truncation:

$$
\begin{array}{ccccccccc}
X & \cdots \longrightarrow & X^{n-1} & \longrightarrow & X^n & \longrightarrow & X^{n+1} & \longrightarrow & X^{n+2} & \longrightarrow \cdots \\
\Big\downarrow & & \Big\downarrow{\scriptstyle\mathrm{id}} & & \Big\downarrow{\scriptstyle\mathrm{id}} & & \Big\downarrow & & \Big\downarrow & \\
\sigma_{\le n}X & \cdots \longrightarrow & X^{n-1} & \longrightarrow & X^n & \longrightarrow & 0 & \longrightarrow & 0 & \longrightarrow \cdots
\end{array}
$$

Now fix a complex X in $\mathbf{K}^{-,b}(\mathcal{P})$ and choose $n \in \mathbb{Z}$ such that $H^i(X) = 0$ for all $i \le n$. Note that the cone of $X \to \sigma_{\le n} X$ belongs to $\mathbf{K}^b(\mathcal{P})$. Thus $X \cong \sigma_{\le n} X$ in $\mathbf{D}^b(\mathcal{A})/\mathbf{D}^b(\mathcal{P})$ and the assignment

$$X \longmapsto \Sigma^{-n} \operatorname{Coker}(X^{n-1} \to X^n)$$

yields a functor $G: \mathbf{D}^b(\mathcal{A})/\mathbf{D}^b(\mathcal{P}) \to \operatorname{St}\mathcal{A}$ which does not depend on n. It is not difficult to check that $G \circ \bar{F} \cong \operatorname{id}$ and $\bar{F} \circ G \cong \operatorname{id}$.

For the second equivalence observe that Z^0 induces for a pair X, Y in $\mathbf{K}_{\mathrm{ac}}(\mathcal{P})$ a bijection

$$\operatorname{Hom}_{\mathbf{K}(\mathcal{P})}(X, Y) \xrightarrow{\sim} \operatorname{Hom}_{\operatorname{St}\mathcal{A}}(Z^0(X), Z^0(Y)).$$

For instance, when $\psi: Z^0(X) \to Z^0(Y)$ is a morphism in $\mathcal{A}$, then this is easily extended to a morphism $\phi: X \to Y$ such that $Z^0(\phi) = \psi$, using the projectivity of the components X^n for negative degrees and the injectivity of the components Y^n for non-negative degrees. On the other hand, if $\phi: X \to Y$ is a morphism such that $Z^0(\phi)$ factors through a projective object, then this yields inductively morphisms $X^n \to Y^{n-1}$ showing that ϕ is null-homotopic. For any object A in $\mathcal{A}$ there is a complete resolution X such that $Z^0(X) = A$. It remains to observe that the functor Z^0 is exact. Thus Z^0 is a triangle equivalence. $\square$

For an object $X \in \mathcal{A}$ we choose a projective resolution $pX \to X$ and an injective resolution $X \to iX$. Completing the canonical morphism $pX \to iX$ to an exact triangle

$$pX \longrightarrow iX \longrightarrow tX \longrightarrow \Sigma(pX)$$

yields a complete resolution tX satisfying $Z^0(tX) \cong X$. One defines the *Tate cohomology* for $X, Y \in \mathcal{A}$ by

$$\widehat{\operatorname{Ext}}^n_{\mathcal{A}}(X, Y) := H^n \operatorname{Hom}_{\mathcal{A}}(X, tY) \qquad (n \in \mathbb{Z}).$$

Also we set

$$\underline{\operatorname{Hom}}_{\mathcal{A}}(X, Y) := \operatorname{Hom}_{\operatorname{St}\mathcal{A}}(X, Y).$$

The following lemma shows some symmetry of the Tate cohomology. In particular, it could be defined equally well via a complete resolution in the first argument.

Lemma 4.4.19. *For $X, Y \in \mathcal{A}$ and $n \in \mathbb{Z}$ there are natural isomorphisms*

$$\widehat{\operatorname{Ext}}^n_{\mathcal{A}}(X, Y) \cong \operatorname{Hom}_{\mathbf{K}(\mathcal{A})}(tX, \Sigma^n(tY)) \cong \underline{\operatorname{Hom}}_{\mathcal{A}}(X, \Sigma^n Y).$$

Moreover, there is a natural homomorphism

$$\operatorname{Ext}^n_{\mathcal{A}}(X, Y) \longrightarrow \widehat{\operatorname{Ext}}^n_{\mathcal{A}}(X, Y)$$

which is an isomorphism for $n > 0$ and identifies for $n = 0$ with the canonical map $\mathrm{Hom}_{\mathcal{A}}(X, Y) \twoheadrightarrow \underline{\mathrm{Hom}}_{\mathcal{A}}(X, Y)$.

Proof We have

$$
\begin{aligned}
H^n \mathrm{Hom}_{\mathcal{A}}(X, tY) &\cong \mathrm{Hom}_{\mathbf{K}(\mathcal{A})}(X, \Sigma^n(tY)) \\
&\cong \mathrm{Hom}_{\mathbf{K}(\mathcal{A})}(iX, \Sigma^n(tY)) \\
&\cong \mathrm{Hom}_{\mathbf{K}(\mathcal{A})}(tX, \Sigma^n(tY)) \\
&\cong \underline{\mathrm{Hom}}_{\mathcal{A}}(X, \Sigma^n Y).
\end{aligned}
$$

The first isomorphism is by Example 4.1.5, the second by Lemma 4.2.6, the third follows from the triangle defining tX since $\mathrm{Hom}_{\mathbf{K}(\mathcal{A})}(pX, \Sigma^n(tY)) = 0$ by Lemma 4.2.5, and the last is induced by Z^0 as in the above proposition.

The morphism $\tau \colon iY \to tY$ induces the map $\mathrm{Ext}_{\mathcal{A}}^n(X, Y) \to \widehat{\mathrm{Ext}}_{\mathcal{A}}^n(X, Y)$, and it is an isomorphism for $n > 0$ since τ^p equals the identity for $p \geq 0$. $\qquad \square$

A Proper Class of Extensions

We give an example of an abelian category $\mathcal{A}$ and an object X in $\mathcal{A}$ such that

$$
\mathrm{Ext}_{\mathcal{A}}^1(X, X) \cong \mathrm{Hom}_{\mathbf{D}(\mathcal{A})}(X, \Sigma X)
$$

is not a set but a proper class. In particular, we see that the construction of the derived category $\mathbf{D}(\mathcal{A})$ yields a 'category' where morphisms between given objects do not always form a set.

Fix a category $\mathcal{C}$ and a class I. We define a new category $\mathcal{C}\langle I \rangle$. The objects are families $(X, \phi_i)_{i \in I}$ consisting of an object $X \in \mathcal{C}$ and a family of endomorphisms $\phi_i \colon X \to X$. A morphism $(X, \phi_i) \to (Y, \psi_i)$ is given by a morphism $\alpha \colon X \to Y$ in $\mathcal{C}$ such that $\alpha\phi_i = \psi_i\alpha$ for all $i \in I$.

Now suppose that $\mathcal{C}$ is abelian. Then $\mathcal{A} = \mathcal{C}\langle I \rangle$ is an abelian category. For $0 \neq X \in \mathcal{C}$ we consider the 'trivial object' $\bar{X} = (X, \phi_i)_{i \in I}$ with $\phi_i = 0$ for all $i \in I$. Define for each $i \in I$ an object $X_i = (X \oplus X, \phi_j)_{j \in I}$ by $\phi_j = \left[\begin{smallmatrix} 0 & 0 \\ \mathrm{id} & 0 \end{smallmatrix} \right]$ for $j = i$, and $\phi_j = 0$ for $j \neq i$. The object X_i fits into the following short exact sequence:

$$
\xi_i : \qquad 0 \longrightarrow \bar{X} \xrightarrow{\left[\begin{smallmatrix} \mathrm{id} \\ 0 \end{smallmatrix} \right]} X_i \xrightarrow{\left[\begin{smallmatrix} 0 & \mathrm{id} \end{smallmatrix} \right]} \bar{X} \longrightarrow 0.
$$

Lemma 4.4.20. *The map $I \to \mathrm{Ext}_{\mathcal{A}}^1(\bar{X}, \bar{X})$ given by $i \mapsto \xi_i$ is injective.*

Proof This is clear since $X_i \not\cong X_j$ for $i \neq j$. $\qquad \square$

Notes

Derived categories were introduced 1963 by Verdier in his thesis [199], following ideas of Grothendieck which he developed in the context of his duality theory outlined in [95]. The basic facts and properties of derived categories can be found in Verdier's work. The study of resolutions and extension has a much longer history. For pioneering work on Ext^1, see for instance Baer [20] or Eilenberg and Mac Lane [72]. The definition of Ext^n via long exact sequences is due to Yoneda [201], while the definition via resolutions is from Cartan and Eilenberg [46].

Abelian categories provided the original context for derived categories when they were introduced. We follow Neeman who generalised the definition to exact categories [147].

The construction of a derived category raises set-theoretic problems, but these can be safely ignored in most situations by reducing to a case where resolutions exist. For unbounded complexes, a first systematic study of resolutions was carried out by Spaltenstein [194]. The construction of K-injective resolutions via homotopy limits is due to Bökstedt and Neeman [37]. For Grothendieck categories the existence of K-injective resolutions can be deduced from resolutions of complexes of modules over a ring via the Gabriel–Popescu theorem [4]. There are various approaches towards defining derived functors. For our applications Deligne's definition seems to be most useful [62]. The discussion of homotopically minimal complexes follows [129]; see also Eilenberg's axiomatic treatment of minimal resolutions [71].

Grothendieck groups are among the basic invariants that are preserved when passing from an exact to its derived category; they were introduced for triangulated categories in the study of perfect complexes by Grothendieck [96].

For the derived category of an abelian category, it is a natural problem to identify the subcategory of complexes with cohomology in some fixed Serre subcategory. For instance, this arises in the study of derived categories of quasi-hereditary algebras [158]. Closely related is the concept of a homological epimorphism studied by Geigle and Lenzing [89].

Hereditary rings appear in Cartan–Eilenberg's book [46]. The term reflects the property that projectivity is inherited under passage to submodules. Hereditary abelian categories were introduced by Lenzing in his thesis [137]; for a comprehensive study see work of Reiten and Van den Bergh [168]. An important example of a Frobenius category is the category of maximal Cohen–Macaulay modules over a Gorenstein algebra. In this context the stable category and Tate cohomology have been studied extensively by Buchweitz [44].

5

Derived Categories of Representations

Contents

5.1	**Examples Related to the Projective Line**		**146**
	Tilting Hereditary Categories		147
	The Projective Line		149
	The Kronecker Quiver		155
	The Klein Four Group		159
5.2	**Derived Categories of Finitely Presented**		
	Modules		**164**
	The Trivial Extension Algebra		164
	Completing Perfect Complexes		167
Notes			**172**

In this chapter we focus on derived categories of representations of Artin algebras. We begin with three examples that are related to the projective line: coherent sheaves on the projective line, representations of the Kronecker quiver, and representations of the Klein four group. The connection between these examples is best explained via triangulated methods. Then we move on and give further descriptions of derived categories. For Artin algebras we provide a description of the derived category via modules over the trivial extension algebra. For a right coherent ring the bounded derived category of finitely presented modules is interpreted as a completion of the category of perfect complexes.

5.1 Examples Related to the Projective Line

We consider derived categories of finite dimensional algebras and discuss three examples in detail: coherent sheaves on the projective line, representations of the Kronecker quiver, and representations of the Klein four group. All three examples are closely related. The relation between the first two is via tilting. More precisely, we have a derived equivalence which is induced by a tilting object, and we explain this process in the general context of hereditary categories.

Tilting Hereditary Categories

'Tilting' can be described as a process of relating two abelian categories via a triangle equivalence between their derived categories. Here we consider the special case of an hereditary abelian category.

Let k be a commutative ring and fix a k-linear hereditary abelian category $\mathcal{A}$ such that $\mathrm{Hom}_{\mathcal{A}}(X, Y)$ and $\mathrm{Ext}^1_{\mathcal{A}}(X, Y)$ are finite length k-modules for all objects X, Y.

We fix a *tilting object* $T \in \mathcal{A}$, that is, $\mathrm{Ext}^1_{\mathcal{A}}(T, T) = 0$ and $\mathrm{Thick}(T) = \mathcal{A}$. Set $\Lambda := \mathrm{End}_{\mathcal{A}}(T)$ and consider the following full subcategories:

$$\mathcal{T} := \{X \in \mathcal{A} \mid \mathrm{Ext}^1_{\mathcal{A}}(T, X) = 0\}, \qquad \mathcal{F} := \{X \in \mathcal{A} \mid \mathrm{Hom}_{\mathcal{A}}(T, X) = 0\},$$

and

$$\mathcal{B} := \{X \in \mathbf{D}^b(\mathcal{A}) \mid \mathrm{Hom}_{\mathbf{D}(\mathcal{A})}(T, \Sigma^n X) = 0 \text{ for all } n \neq 0\}.$$

Given a pair $(\mathcal{T}, \mathcal{F})$ of subcategories of any exact category $\mathcal{A}$, one calls $(\mathcal{T}, \mathcal{F})$ a *torsion pair* for $\mathcal{A}$ if

$$\mathcal{T} = \{X \in \mathcal{A} \mid \mathrm{Hom}_{\mathcal{A}}(X, Y) = 0 \text{ for all } Y \in \mathcal{F}\},$$
$$\mathcal{F} = \{Y \in \mathcal{A} \mid \mathrm{Hom}_{\mathcal{A}}(X, Y) = 0 \text{ for all } X \in \mathcal{T}\},$$

and each object $X \in \mathcal{A}$ fits into an exact sequence $0 \to X' \to X \to X'' \to 0$ with $X' \in \mathcal{T}$ and $X'' \in \mathcal{F}$.

Lemma 5.1.1. *The objects in $\mathcal{T}$ are precisely the quotients of objects in* $\mathrm{add}\, T$, *and* $(\mathcal{T}, \mathcal{F})$ *is a torsion pair for* $\mathcal{A}$.

Proof First observe that $\mathcal{T} \cap \mathcal{F} = 0$, because $X \in \mathcal{T} \cap \mathcal{F}$ implies $\mathrm{Ext}^*_{\mathcal{A}}(A, X) = 0$ for all $A \in \mathrm{Thick}(T)$; see Example 4.4.11.

Now fix an object $X \in \mathcal{A}$ and choose generators $\phi_1, \ldots, \phi_r$ of $\mathrm{Hom}_{\mathcal{A}}(T, X)$. Let tX denote the image of $\phi \colon T^r \to X$. Clearly, $tX \in \mathcal{T}$ since $\mathrm{Ext}^2_{\mathcal{A}}(T, -) = 0$, and the inclusion $tX \to X$ induces an isomorphism

$$\mathrm{Hom}_{\mathcal{A}}(T, tX) \xrightarrow{\sim} \mathrm{Hom}_{\mathcal{A}}(T, X).$$

It follows that $X/tX \in \mathcal{F}$. We have $\mathrm{Hom}_{\mathcal{A}}(X, Y) = 0$ for $X \in \mathcal{T}$ and $Y \in \mathcal{F}$, since the image of any morphism $X \to Y$ lies in $\mathcal{T} \cap \mathcal{F} = 0$. Thus $tX = X$ if and only if $X \in \mathcal{T}$, and $tX = 0$ if and only if $X \in \mathcal{F}$. □

A tilting object provides a close relation between two abelian categories via a triangle equivalence between their derived categories. The following theorem makes this precise.

Theorem 5.1.2. *Let $\mathcal{A}$ be a k-linear hereditary abelian category with all Ext-groups of finite length over k. Suppose that $T \in \mathcal{A}$ is a tilting object. Then the following holds for*

$$\mathcal{B} := \{X \in \mathbf{D}^b(\mathcal{A}) \mid \operatorname{Hom}_{\mathbf{D}(\mathcal{A})}(T, \Sigma^n X) = 0 \text{ for all } n \neq 0\}.$$

(1) *The category $\mathcal{B}$ is abelian and the inclusion $\mathcal{B} \to \mathbf{D}^b(\mathcal{A})$ extends to a triangle equivalence $\mathbf{D}^b(\mathcal{B}) \xrightarrow{\sim} \mathbf{D}^b(\mathcal{A})$.*
(2) *The algebra $\Lambda = \operatorname{End}_{\mathcal{A}}(T)$ has finite global dimension and the functor $\operatorname{Hom}_{\mathbf{D}(\mathcal{A})}(T, -)$ induces an equivalence $\mathcal{B} \xrightarrow{\sim} \operatorname{mod} \Lambda$.*
(3) *$\mathcal{B} = \mathcal{T} \vee \Sigma \mathcal{F}$ and $(\Sigma \mathcal{F}, \mathcal{T})$ is a split torsion pair for $\mathcal{B}$.*

Proof   The functor $\operatorname{Hom}_{\mathcal{A}}(T, -)$ induces an equivalence $\operatorname{add} T \xrightarrow{\sim} \operatorname{proj} \Lambda$ and extends to a triangle equivalence $\mathbf{K}^b(\operatorname{add} T) \xrightarrow{\sim} \mathbf{K}^b(\operatorname{proj} \Lambda)$. We obtain the following commutative square

$$
\begin{array}{ccc}
\mathbf{K}^b(\operatorname{add} T) & \xrightarrow{\operatorname{Hom}_{\mathcal{A}}(T,-)} & \mathbf{K}^b(\operatorname{proj} \Lambda) \\
\downarrow & & \downarrow \\
\mathbf{D}^b(\mathcal{A}) & \longrightarrow & \mathbf{D}^b(\operatorname{proj} \Lambda)
\end{array}
$$

because the vertical functors are triangle equivalences. In fact, a dévissage argument (Lemma 3.1.8) shows that both vertical functors are fully faithful. Also, $\operatorname{Thick}(T) = \mathbf{D}^b(\mathcal{A})$ since T is tilting.

Next observe that Λ has finite global dimension. This either follows from Theorem 9.3.14, with bound $\operatorname{gl.dim} \Lambda \leq 2$, or one checks that Λ is a triangular ring. More precisely, let $T = \bigoplus_{i=1}^{r} T_i$ be a decomposition into indecomposable objects. Then each morphism $\phi \colon T_i \to T_j$ is either a monomorphism or an epimorphism since $\mathcal{A}$ is hereditary. This follows from the fact that the morphism ϕ fits into a split exact sequence $0 \to T_i \to \operatorname{Im} \phi \oplus E \to T_j \to 0$ by Remark 4.4.16. Thus $\operatorname{End}_{\mathcal{A}}(T_i)$ is a division ring for each i, and we may assume that the T_i are ordered in such a way that $\operatorname{Hom}_{\mathcal{A}}(T_i, T_j) = 0$ for all $i > j$. It follows that for each Λ-module the length of a minimal projective resolution is bounded by r. This yields a triangle equivalence

$$\mathbf{D}^b(\mathcal{A}) \xrightarrow{\sim} \mathbf{D}^b(\operatorname{proj} \Lambda) \xrightarrow{\sim} \mathbf{D}^b(\operatorname{mod} \Lambda)$$

which maps T to Λ; so it identifies $\mathcal{B}$ with $\operatorname{mod} \Lambda$.

It remains to show that $\mathcal{B} = \mathcal{T} \vee \Sigma \mathcal{F}$ and that $(\Sigma \mathcal{F}, \mathcal{T})$ is a split torsion pair for $\mathcal{B}$. First observe that $\mathcal{T} = \mathcal{B} \cap \mathcal{A}$ and $\Sigma\mathcal{F} = \mathcal{B} \cap \Sigma\mathcal{A}$. Choose $X \in \Sigma\mathcal{F}$ and $Y \in \mathcal{T}$. If $X = \Sigma X'$, then we have

$$\mathrm{Hom}_{\mathcal{B}}(X, Y) \cong \mathrm{Hom}_{\mathbf{D}(\mathcal{A})}(\Sigma X', Y) \cong \mathrm{Ext}_{\mathcal{A}}^{-1}(X', Y) = 0.$$

Thus it suffices to show that each indecomposable $X \in \mathcal{B}$ belongs either to $\mathcal{T}$ or to $\Sigma\mathcal{F}$; so we need to show that X lies in $\mathcal{A}$ or $\Sigma\mathcal{A}$. Suppose $X = \Sigma^n Y$ for some $Y \in \mathcal{A}$ and $n \neq 0, 1$. Then $\mathrm{Hom}_{\mathcal{A}}(T, Y) = 0$ and $\mathrm{Ext}_{\mathcal{A}}^1(T, Y) = 0$. Thus $Y = 0$ since $\mathrm{Thick}(T) = \mathcal{A}$; see Example 4.4.11. $\qquad\square$

Corollary 5.1.3. *The triangle equivalence* $\mathbf{D}^b(\mathcal{A}) \xrightarrow{\sim} \mathbf{D}^b(\mathrm{mod}\,\Lambda)$ *restricts to a pair of equivalences*

$$\mathcal{A} \supseteq \mathcal{T} \xrightarrow{\mathrm{Hom}_{\mathcal{A}}(T,-)} \mathrm{Hom}_{\mathbf{D}(\mathcal{A})}(T, \mathcal{T}) \subseteq \mathrm{mod}\,\Lambda,$$

$$\mathcal{A} \supseteq \mathcal{F} \xrightarrow{\mathrm{Ext}_{\mathcal{A}}^1(T,-)} \mathrm{Hom}_{\mathbf{D}(\mathcal{A})}(T, \Sigma\mathcal{F}) \subseteq \mathrm{mod}\,\Lambda.$$

Moreover, if $\mathcal{B}$ is hereditary, then $\mathcal{A} = \mathcal{F} \vee \mathcal{T}$.

Proof The first assertion is clear. When $\mathcal{B}$ is hereditary, we have

$$\mathcal{A} \subseteq \mathbf{D}^b(\mathcal{A}) = \bigvee_{n \in \mathbb{Z}} \Sigma^n \mathcal{B} = \left(\bigvee_{n \in \mathbb{Z}} \Sigma^n \mathcal{T} \right) \vee \left(\bigvee_{n \in \mathbb{Z}} \Sigma^n \mathcal{F} \right). \qquad\square$$

The following diagram illustrates the tilting process from $\mathcal{A}$ to $\mathcal{B}$.

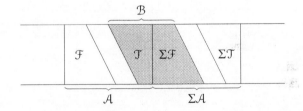

Later on we will discuss concrete examples of hereditary abelian categories with specific tilting objects; see Proposition 5.1.17 and Proposition 7.2.24.

The Projective Line

Let k be a field and $\mathbb{P}_k^1$ the projective line over k. We view $\mathbb{P}_k^1$ as a scheme and consider the category $\mathrm{coh}\,\mathbb{P}_k^1$ of coherent sheaves on $\mathbb{P}_k^1$. This is an example of an hereditary abelian category. So we describe this category and its derived category. This is meant as an illustration; so we do not give all details and refer to the literature when appropriate.

We begin with a description of the underlying set of points of $\mathbb{P}_k^1$. Let

$k[x_0, x_1]$ be the polynomial ring in two variables with the usual $\mathbb{Z}$-grading by total degree. Denote by Proj $k[x_0, x_1]$ the set of homogeneous prime ideals of $k[x_0, x_1]$ that are different from the unique maximal ideal consisting of positive degree elements. Note that $k[x_0, x_1]$ is a two-dimensional graded factorial domain. Thus homogeneous irreducible polynomials correspond to non-zero homogeneous prime ideals by taking a polynomial p to the ideal (p) generated by p, and $(p') = (p)$ if and only if $p' = \alpha p$ for some $\alpha \in k \setminus \{0\}$.

The elements of Proj $k[x_0, x_1]$ form the *points* of $\mathbb{P}^1_k$. A point $\mathfrak{p} \in \mathbb{P}^1_k$ is *closed* if $\mathfrak{p} \neq 0$, and $\mathfrak{p}$ is *generic* if $\mathfrak{p} = 0$. Using homogeneous coordinates, a *rational point* of $\mathbb{P}^1_k$ is a pair $[\lambda_0 : \lambda_1]$ of elements of k which are not both zero, subject to the relation $[\lambda_0 : \lambda_1] = [\alpha\lambda_0 : \alpha\lambda_1]$ for all $\alpha \in k$, $\alpha \neq 0$. We identify each rational point $[\lambda_0 : \lambda_1]$ with the prime ideal $(\lambda_1 x_0 - \lambda_0 x_1)$ of $k[x_0, x_1]$. If k is algebraically closed then all closed points are rational.

Using the identification $y = x_1/x_0$, we cover $\mathbb{P}^1_k$ by two copies $U' = \operatorname{Spec} k[y]$ and $U'' = \operatorname{Spec} k[y^{-1}]$ of the affine line, with $U' \cap U'' = \operatorname{Spec} k[y, y^{-1}]$. More precisely, the evaluation map $k[x_0, x_1] \to k[y]$ sending f to $f(1, y)$ induces an isomorphism $k[x_0, x_1]/(x_0 - 1) \xrightarrow{\sim} k[y]$ and yields a bijection

$$\operatorname{Proj} k[x_0, x_1] \setminus \{(x_0)\} \xrightarrow{\sim} \operatorname{Spec} k[y].$$

Analogously, the map $k[x_0, x_1] \to k[y^{-1}]$ sending f to $f(y^{-1}, 1)$ induces a bijection

$$\operatorname{Proj} k[x_0, x_1] \setminus \{(x_1)\} \xrightarrow{\sim} \operatorname{Spec} k[y^{-1}].$$

Based on the covering $\mathbb{P}^1_k = U' \cup U''$, the category coh $\mathbb{P}^1_k$ of coherent sheaves admits a description in terms of the following pullback of abelian categories

$$
\begin{array}{ccc}
\operatorname{coh} \mathbb{P}^1_k & \longrightarrow & \operatorname{coh} U' \\
\downarrow & & \downarrow \\
\operatorname{coh} U'' & \longrightarrow & \operatorname{coh} U' \cap U''
\end{array}
$$

where each functor is given by restricting a sheaf to the appropriate open subset; see [79, Proposition VI.2]. More concretely, this pullback diagram has, up to equivalence, the form

$$
\begin{array}{ccc}
\mathcal{A} & \longrightarrow & \operatorname{mod} k[y] \\
\downarrow & & \downarrow \\
\operatorname{mod} k[y^{-1}] & \longrightarrow & \operatorname{mod} k[y, y^{-1}]
\end{array}
$$

where the category $\mathcal{A}$ is defined as follows. The objects of $\mathcal{A}$ are given by triples (M', M'', μ), where M' is a finitely generated $k[y]$-module, M'' is a finitely generated $k[y^{-1}]$-module, and $\mu\colon M'_y \xrightarrow{\sim} M''_{y^{-1}}$ is an isomorphism of $k[y, y^{-1}]$-modules. Here, we use for any R-module M the notation M_x to denote the localisation with respect to an element $x \in R$. A morphism from (M', M'', μ) to (N', N'', ν) in $\mathcal{A}$ is a pair (ϕ', ϕ'') of morphisms, where $\phi'\colon M' \to N'$ is $k[y]$-linear and $\phi''\colon M'' \to N''$ is $k[y^{-1}]$-linear such that $\nu\phi'_y = \phi''_{y^{-1}}\mu$.

Given a sheaf $\mathscr{F}$ on $\mathbb{P}^1_k$, we denote for any open subset $U \subseteq \mathbb{P}^1_k$ by $\Gamma(U, \mathscr{F})$ the *sections* over U.

Lemma 5.1.4. *The assignment*

$$\mathscr{F} \longmapsto (\Gamma(U', \mathscr{F}), \Gamma(U'', \mathscr{F}), \mathrm{id}_{\Gamma(U'\cap U'', \mathscr{F})})$$

gives an equivalence $\operatorname{coh}\mathbb{P}^1_k \xrightarrow{\sim} \mathcal{A}$.

Proof The description of a sheaf $\mathscr{F}$ on $\mathbb{P}^1_k = U' \cup U''$ in terms of its restrictions $\mathscr{F}|_{U'}$, $\mathscr{F}|_{U''}$, and $\mathscr{F}|_{U'\cap U''}$ is standard; see [79, Proposition VI.2]. Thus it remains to observe that taking global sections identifies $\operatorname{coh} U' = \operatorname{mod} k[y]$, $\operatorname{coh} U'' = \operatorname{mod} k[y^{-1}]$, and $\operatorname{coh} U' \cap U'' = \operatorname{mod} k[y, y^{-1}]$. $\square$

From now on we identify the categories $\operatorname{coh}\mathbb{P}^1_k$ and $\mathcal{A}$ via the above equivalence.

Let $\operatorname{grmod} k[x_0, x_1]$ denote the category of finitely generated $\mathbb{Z}$-graded modules over $k[x_0, x_1]$ and write $\operatorname{grmod}_0 k[x_0, x_1]$ for the Serre subcategory consisting of all finite length modules. The property of a pullback yields an exact functor

$$F\colon \operatorname{grmod} k[x_0, x_1] \longrightarrow \operatorname{coh}\mathbb{P}^1_k \tag{5.1.5}$$

that fits into the following commutative diagram.

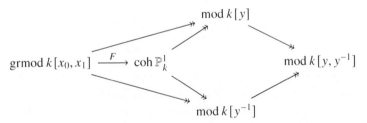

We give an explicit description of F; it takes a graded $k[x_0, x_1]$-module M to the triple

$$\widetilde{M} = ((M_{x_0})_0, (M_{x_1})_0, \sigma_M),$$

where the variable y acts on the degree zero part of M_{x_0} via the identification $y = x_1/x_0$, the variable y^{-1} acts on the degree zero part of M_{x_1} via the identification $y^{-1} = x_0/x_1$, and the isomorphism σ_M equals the obvious identification $[(M_{x_0})_0]_{x_1/x_0} = [(M_{x_1})_0]_{x_0/x_1}$. Note that F annihilates precisely the finite length modules.

The following result is due to Serre [188].

Proposition 5.1.6. *The functor* (5.1.5) *induces an equivalence*

$$\frac{\mathrm{grmod}\,k[x_0, x_1]}{\mathrm{grmod}_0\,k[x_0, x_1]} \xrightarrow{\sim} \mathrm{coh}\,\mathbb{P}^1_k. \qquad\qquad \square$$

For any $n \in \mathbb{Z}$ and $\mathscr{F} = (M', M'', \mu)$ in $\mathrm{coh}\,\mathbb{P}^1_k$, denote by $\mathscr{F}(n)$ the *twisted sheaf* $(M', M'', \mu^{(n)})$, where $\mu^{(n)}$ is the map μ followed by multiplication by y^{-n}. Given a graded $k[x_0, x_1]$-module M, the *twisted module* $M(n)$ is obtained by shifting the grading, that is, $M(n)_i = M_{i+n}$ for $i \in \mathbb{Z}$. Note that $\widetilde{M(n)} = \widetilde{M}(n)$.

The *structure sheaf* is the sheaf $\mathscr{O} = (k[y], k[y^{-1}], \mathrm{id}_{k[y,y^{-1}]})$; it is the image of the free $k[x_0, x_1]$-module of rank one under the functor (5.1.5). For any pair $m, n \in \mathbb{Z}$, we have a natural bijection

$$k[x_0, x_1]_{n-m} \xrightarrow{\sim} \mathrm{Hom}(\mathscr{O}(m), \mathscr{O}(n)). \qquad (5.1.7)$$

The map sends a homogeneous polynomial p of degree $n - m$ to the morphism (ϕ', ϕ''), where $\phi'\colon k[y] \to k[y]$ is multiplication by $p(1, y)$ and $\phi''\colon k[y^{-1}] \to k[y^{-1}]$ is multiplication by $p(y^{-1}, 1)$.

From the above formula one can deduce the following [24, Corollary 6]:

$$\mathrm{Ext}^1(\mathscr{O}(m), \mathscr{O}(n)) = 0 \quad \text{for all} \quad m \le n + 1. \qquad (5.1.8)$$

Each coherent sheaf $\mathscr{F}$ admits an essentially unique decomposition $\mathscr{F} = \bigoplus_{i=1}^r \mathscr{F}_i$ into indecomposable sheaves. The indecomposable sheaves come in two types:

(1) for each $n \in \mathbb{Z}$, the sheaf $\mathscr{O}(n)$, and
(2) for each closed point $\mathfrak{p} \in \mathbb{P}^1_k$ and $r \ge 1$, a sheaf $\mathscr{O}_{\mathfrak{p}^r}$.

Let $\mathfrak{p}$ be a closed point and choose a homogeneous irreducible polynomial p of degree d that generates $\mathfrak{p}$. The bijection (5.1.7) gives for each power p^r a monomorphism $\mathscr{O} \to \mathscr{O}(rd)$ whose cokernel we denote by $\mathscr{O}_{\mathfrak{p}^r}$. Thus there is an exact sequence

$$0 \longrightarrow \mathscr{O} \longrightarrow \mathscr{O}(rd) \longrightarrow \mathscr{O}_{\mathfrak{p}^r} \longrightarrow 0.$$

Note that for $r, s \ge 1$ the composite $\mathscr{O} \to \mathscr{O}(rd) \to \mathscr{O}(rd + sd)$ yields an exact sequence

$$0 \longrightarrow \mathscr{O}_{\mathfrak{p}^r} \longrightarrow \mathscr{O}_{\mathfrak{p}^{r+s}} \longrightarrow \mathscr{O}_{\mathfrak{p}^s} \longrightarrow 0.$$

Direct sums of sheaves of the form $\mathcal{O}(n)$ are called *vector bundles* or *locally free sheaves*; direct sums of sheaves of the form $\mathcal{O}_{\mathfrak{p}^r}$ are called *torsion sheaves*. Let $\mathrm{coh}_0\, \mathbb{P}^1_k$ denote the full subcategory of finite length objects of $\mathrm{coh}\, \mathbb{P}^1_k$.

Lemma 5.1.9. *For a torsion sheaf $\mathcal{O}_{\mathfrak{p}^r}$ and a locally free sheaf $\mathcal{O}(n)$ we have $\mathrm{Hom}(\mathcal{O}_{\mathfrak{p}^r}, \mathcal{O}(n)) = 0$. The full subcategory of torsion sheaves equals the category $\mathrm{coh}_0\, \mathbb{P}^1_k$ of finite length objects. The objects $\mathcal{O}_{\mathfrak{p}}$ ($\mathfrak{p}$ a closed point) form a representative set of simple objects.*

Proof The first assertion follows from (5.1.7) by applying $\mathrm{Hom}(-, \mathcal{O}(n))$ to the sequence defining $\mathcal{O}_{\mathfrak{p}^r}$.

For a closed point $\mathfrak{p}$ and $r \geq 1$, it is not difficult to show that

$$0 \subseteq \mathcal{O}_{\mathfrak{p}^1} \subseteq \mathcal{O}_{\mathfrak{p}^2} \subseteq \cdots \subseteq \mathcal{O}_{\mathfrak{p}^r}$$

is a composition series. $\qquad\square$

Given a sheaf $\mathscr{F}$ on $\mathbb{P}^1_k$ and a point $\mathfrak{p} \in \mathbb{P}^1_k$, the *stalk* of $\mathscr{F}$ at $\mathfrak{p}$ is the colimit

$$\mathscr{F}_{\mathfrak{p}} = \underset{\mathfrak{p} \in U}{\mathrm{colim}}\, \mathscr{F}(U)$$

where U runs through all open subsets of $\mathbb{P}^1_k$. The *support* of $\mathscr{F}$ is by definition

$$\mathrm{Supp}\, \mathscr{F} = \{\mathfrak{p} \in \mathbb{P}^1_k \mid \mathscr{F}_{\mathfrak{p}} \neq 0\}.$$

The functor (5.1.5) provides an alternative description of the support. In fact, for each graded $k[x_0, x_1]$-module M and $\mathfrak{p} \in \mathbb{P}^1_k$, the functor induces an isomorphism

$$(M_{\mathfrak{p}})_0 \xrightarrow{\sim} (\widetilde{M})_{\mathfrak{p}}.$$

Composing the natural homomorphism

$$k[x_0, x_1] \longrightarrow \mathrm{End}^*(M) = \bigoplus_{n \in \mathbb{Z}} \mathrm{Hom}(M, M(n))$$

with the induced homomorphism $\mathrm{End}^*(M) \to \mathrm{End}^*(\widetilde{M})$ yields for each $\mathscr{F}$ in $\mathrm{coh}\, \mathbb{P}^1_k$ a homomorphism

$$\chi_{\mathscr{F}} : k[x_0, x_1] \longrightarrow \mathrm{End}^*(\mathscr{F})$$

and

$$\mathrm{Supp}\, \mathscr{F} = \{\mathfrak{p} \in \mathbb{P}^1_k \mid \mathrm{Ker}\, \chi_{\mathscr{F}} \subseteq \mathfrak{p}\}.$$

It is not difficult to compute the support of each object.

Proposition 5.1.10. *Let* $\mathfrak{p} \in \mathbb{P}^1_k$ *be a closed point and* $n \in \mathbb{Z}$. *Then we have*

$$\operatorname{Supp} \mathcal{O}(n) = \mathbb{P}^1_k \qquad and \qquad \operatorname{Supp} \mathcal{O}_{\mathfrak{p}^n} = \{\mathfrak{p}\}. \qquad \square$$

Let us mention an immediate consequence. For each pair of closed points $\mathfrak{p} \neq \mathfrak{q}$ in $\mathbb{P}^1_k$ we have

$$\operatorname{Hom}(\mathcal{O}_{\mathfrak{p}^m}, \mathcal{O}_{\mathfrak{q}^n}) = 0 \quad \text{for all} \quad m, n \geq 1,$$

since the image of any morphism has support contained in $\{\mathfrak{p}\} \cap \{\mathfrak{q}\}$.

Proposition 5.1.11. *The assignment*

$$\operatorname{coh} \mathbb{P}^1_k \supseteq \mathcal{C} \longmapsto \operatorname{Supp} \mathcal{C} = \bigcup_{X \in \mathcal{C}} \operatorname{Supp} X$$

induces an inclusion preserving bijection between the Serre subcategories of $\operatorname{coh} \mathbb{P}^1_k$ *that are closed under twists, and the specialisation closed subsets of* $\mathbb{P}^1_k$.

Proof We combine the description of Serre subcategories of $\operatorname{grmod} k[x_0, x_1]$ from Proposition 2.4.21 with Proposition 5.1.6, keeping in mind that the minimal Serre subcategory $\operatorname{grmod}_0 k[x_0, x_1]$ corresponds to $\{\mathfrak{m}\} \subseteq \operatorname{Spec} k[x_0, x_1]$, where $\mathfrak{m}$ denotes the unique maximal ideal of positive degree elements. $\square$

We extend this to a description of all thick subcategories of $\operatorname{coh} \mathbb{P}^1_k$ and begin with the following observation.

Lemma 5.1.12. *Let* $\mathcal{C} \subseteq \operatorname{coh} \mathbb{P}^1_k$ *be a thick subcategory closed under twists. Then* $\mathcal{C}$ *is a Serre subcategory.*

Proof If $\mathcal{C}$ contains a non-zero vector bundle, then $\mathcal{C} = \operatorname{coh} \mathbb{P}^1_k$. Thus we assume $\mathcal{C} \subseteq \operatorname{coh}_0 \mathbb{P}^1_k$. A thick subcategory $\mathcal{C}$ of a length category is a Serre subcategory if for each object in $\mathcal{C}$ its *socle* (sum of simple subobjects) is in $\mathcal{C}$. Thus it suffices to show that $\mathcal{O}_{\mathfrak{p}^r} \in \mathcal{C}$ implies $\mathcal{O}_{\mathfrak{p}} \in \mathcal{C}$; see Lemma 5.1.9. But this is clear, because there is an endomorphism $\phi \colon \mathcal{O}_{\mathfrak{p}^r} \to \mathcal{O}_{\mathfrak{p}^r}$ with $\operatorname{Im} \phi = \mathcal{O}_{\mathfrak{p}}$. A thick subcategory of an hereditary category is closed under images of morphisms; see Remark 4.4.16. $\square$

Thus we can reformulate Proposition 5.1.11 as follows. The assignment $\mathcal{C} \mapsto \operatorname{Supp} \mathcal{C}$ induces a lattice isomorphism

$$\{\text{thick subcategories of } \operatorname{coh} \mathbb{P}^1_k \text{ closed under twists}\} \xrightarrow{\sim} \{\text{spc subsets of } \mathbb{P}^1_k\}$$

where 'spc' is an abbreviation for 'specialisation closed'.

For each $i \in \mathbb{Z}$ there is a thick subcategory

$$\operatorname{Thick}(\mathcal{O}(i)) = \operatorname{add} \mathcal{O}(i),$$

and these are the only proper non-trivial thick subcategories which are generated by vector bundles. Thus we have a lattice isomorphism

{thick subcategories of $\operatorname{coh} \mathbb{P}^1_k$ generated by vector bundles} $\xrightarrow{\sim} \mathbb{Z}$

where $\mathbb{Z}$ denotes the lattice given by the following Hasse diagram:

One can combine these classifications, where for a pair of lattices L', L'' with smallest elements $0', 0''$ and greatest elements $1', 1''$, we denote by $L' \sqcup L''$ the new lattice which is obtained from the disjoint union $L' \sqcup L''$ (viewed as a sum of posets) by identifying $0' = 0''$ and $1' = 1''$.

Proposition 5.1.13. *We have a lattice isomorphism*

$$\{\text{thick subcategories of } \operatorname{coh} \mathbb{P}^1_k\} \xrightarrow{\sim} \{\text{spc subsets of } \mathbb{P}^1_k\} \sqcup \mathbb{Z}.$$

Proof The verification that the evident map is indeed a lattice isomorphism as claimed is elementary: the line bundles $\mathcal{O}(i)$ are supported everywhere so are not contained in any proper thick subcategory closed under twists, and any line bundle and a torsion sheaf, or any pair of line bundles, generate the category. For any set $\mathcal{V}$ of points set

$$\operatorname{coh}_{\mathcal{V}} \mathbb{P}^1_k = \{\mathcal{F} \in \operatorname{coh} \mathbb{P}^1_k \mid \operatorname{Supp} \mathcal{F} \subseteq \mathcal{V}\}.$$

Thus for $i \neq j$ and $\mathcal{V}$ proper non-empty and specialisation closed in $\mathbb{P}^1_k$ we have

$$\operatorname{coh}_{\mathcal{V}} \mathbb{P}^1_k \vee \operatorname{Thick}(\mathcal{O}(i)) = \operatorname{coh} \mathbb{P}^1_k = \operatorname{Thick}(\mathcal{O}(i)) \vee \operatorname{Thick}(\mathcal{O}(j))$$

and

$$\operatorname{coh}_{\mathcal{V}} \mathbb{P}^1_k \wedge \operatorname{Thick}(\mathcal{O}(i)) = 0 = \operatorname{Thick}(\mathcal{O}(i)) \wedge \operatorname{Thick}(\mathcal{O}(j)). \qquad \square$$

Corollary 5.1.14. *We have a lattice isomorphism*

$$\{\text{thick subcategories of } \mathbf{D}^b(\operatorname{coh} \mathbb{P}^1_k)\} \xrightarrow{\sim} \{\text{spc subsets of } \mathbb{P}^1_k\} \sqcup \mathbb{Z}.$$

Proof Combine Proposition 4.4.17 and Proposition 5.1.13. $\qquad \square$

The Kronecker Quiver

We consider the following *Kronecker quiver*

$$\circ \rightrightarrows \circ$$

and fix a field k. A k-linear *representation* (V, W, ϕ, ψ) consists of a pair of vector spaces together with a pair of linear maps between them

$$V \underset{\psi}{\overset{\phi}{\rightrightarrows}} W.$$

The finite dimensional representations of the Kronecker quiver form an hereditary abelian category. We describe this category and its relation with the category coh $\mathbb{P}^1_k$ of coherent sheaves on the projective line. In particular, the indecomposable regular representations of the Kronecker quiver are parametrised by points of the projective line $\mathbb{P}^1_k$ over k. We recall briefly some definitions.

Let $k[x, y]$ be the polynomial ring in two variables with the usual $\mathbb{Z}$-grading by total degree. Denote by Proj $k[x, y]$ the set of homogeneous prime ideals of $k[x, y]$ that are different from the unique maximal ideal consisting of positive degree elements. The ring $k[x, y]$ is a two-dimensional graded factorial domain. Thus homogeneous irreducible polynomials correspond to non-zero homogeneous prime ideals by taking a polynomial p to the ideal generated by p. A *closed point* of $\mathbb{P}^1_k$ is by definition an element in Proj $k[x, y]$ that is different from the zero ideal.

We begin by listing the indecomposable representations. For each integer $n \geq 0$ let V_n denote the $(n+1)$-dimensional space of homogeneous polynomials of degree n in two variables x and y of degree one, and for $n < 0$ set $V_n = 0$. Thus

$$k[x, y] = \bigoplus_{n \geq 0} V_n.$$

For a vector space X let $X^* = \mathrm{Hom}_k(X, k)$ denote the dual space.

There are the indecomposable *preprojective* representations

$$P_n: \quad V_{n-1} \underset{y}{\overset{x}{\rightrightarrows}} V_n \qquad\qquad (n \geq 0)$$

and the indecomposable *postinjective* representations

$$I_n: \quad V_n^* \underset{y^*}{\overset{x^*}{\rightrightarrows}} V_{n-1}^* \qquad\qquad (n \geq 0).$$

Each $0 \neq f \in V_n$ gives rise to a *regular* representation

$$R_f: \quad V_{n-1} \underset{y}{\overset{x}{\rightrightarrows}} V_n / \langle f \rangle$$

where $\langle f \rangle$ is the k-linear subspace generated by f. Often we identify f with the ideal (f) generated by f and set set $R_{(f)} := R_f$.

This yields a complete list of the indecomposable Kronecker representations; see for example [18, Theorem 7.5].

Proposition 5.1.15. *The representations P_n, I_n ($n \geq 0$), and $R_{\mathfrak{p}^n}$ ($n \geq 1$ and $\mathfrak{p} \in \mathbb{P}_k^1$ a closed point) form, up to isomorphism, a complete list of finite dimensional indecomposable representations of the Kronecker quiver.* □

Let us compare the category of Kronecker representations with the category $\operatorname{coh} \mathbb{P}_k^1$ of coherent sheaves on the projective line. We consider the sheaf

$$\mathscr{T} = \mathscr{O} \oplus \mathscr{O}(1)$$

and observe that its endomorphism algebra $\Lambda = \operatorname{End}(\mathscr{T})$ identifies with the *Kronecker algebra* (path algebra of the Kronecker quiver), because of (5.1.7).

Lemma 5.1.16. *The sheaf $\mathscr{T}$ is a tilting object of $\operatorname{coh} \mathbb{P}_k^1$.*

Proof The formula (5.1.8) implies that $\operatorname{Ext}^1(\mathscr{T}, \mathscr{T}) = 0$. The formula (5.1.7) provides for each $n \in \mathbb{Z}$ a canonical monomorphism $\mathscr{O}(n) \to \mathscr{O}(n+1)^2$ with cokernel $\mathscr{O}(n+2)$. Thus $\operatorname{Thick}(\mathscr{T})$ contains $\mathscr{O}(n)$ for all $n \in \mathbb{Z}$, and then also each torsion sheaf $\mathscr{O}_{\mathfrak{p}^n}$ □

Next we apply Theorem 5.1.2 and the functor $\operatorname{Hom}(\mathscr{T}, -)$ induces a triangle equivalence

$$\mathbf{D}^b(\operatorname{coh} \mathbb{P}_k^1) \xrightarrow{\sim} \mathbf{D}^b(\operatorname{mod} \Lambda).$$

We illustrate this by some explicit calculations.

For each $n \geq 0$ we have

$$\operatorname{Hom}(\mathscr{T}, \mathscr{O}(n)) = P_n$$

since

$$P_n: \qquad \operatorname{Hom}(\mathscr{O}(1), \mathscr{O}(n)) \underset{y}{\overset{x}{\rightrightarrows}} \operatorname{Hom}(\mathscr{O}, \mathscr{O}(n)).$$

For each homogeneous polynomial $0 \neq f \in k[x, y]$ of degree n we define the sheaf $\mathscr{O}_{(f)}$ via the exact sequence

$$0 \longrightarrow \mathscr{O} \xrightarrow{f} \mathscr{O}(n) \longrightarrow \mathscr{O}_{(f)} \longrightarrow 0.$$

A product $fg \in k[x, y]$ of homogeneous polynomials yields an exact sequence

$$0 \longrightarrow \mathscr{O}_{(f)} \longrightarrow \mathscr{O}_{(fg)} \longrightarrow \mathscr{O}_{(g)} \longrightarrow 0$$

which is split exact when f and g are coprime. We have

$$\operatorname{Hom}(\mathscr{T}, \mathscr{O}_{(f)}) = R_f$$

since

$$R_f : \qquad \mathrm{Hom}(\mathscr{O}(1), \mathscr{O}_{(f)}) \underset{y}{\overset{x}{\rightrightarrows}} \mathrm{Hom}(\mathscr{O}, \mathscr{O}_{(f)}).$$

In fact, applying $\mathrm{Hom}(\mathscr{T}, -)$ to the defining sequence for $\mathscr{O}_{(f)}$ yields the following exact sequence of Kronecker representations.

$$
\begin{array}{ccccccccc}
0 & \longrightarrow & \mathrm{Hom}(\mathscr{O}(1), \mathscr{O}) & \overset{f}{\longrightarrow} & \mathrm{Hom}(\mathscr{O}(1), \mathscr{O}(n)) & \longrightarrow & \mathrm{Hom}(\mathscr{O}(1), \mathscr{O}_{(f)}) & \longrightarrow & 0 \\
& & x \big\downarrow\big\downarrow y & & x \big\downarrow\big\downarrow y & & x \big\downarrow\big\downarrow y & & \\
0 & \longrightarrow & \mathrm{Hom}(\mathscr{O}, \mathscr{O}) & \overset{f}{\longrightarrow} & \mathrm{Hom}(\mathscr{O}, \mathscr{O}(n)) & \longrightarrow & \mathrm{Hom}(\mathscr{O}, \mathscr{O}_{(f)}) & \longrightarrow & 0
\end{array}
$$

A representation is called *regular* if there are no indecomposable preprojective or postinjective direct summands. The regular representations form a thick subcategory of $\mathrm{mod}\, \Lambda$ which we denote by $\mathrm{reg}\, \Lambda$.

Proposition 5.1.17. *The tilting object $\mathscr{T} = \mathscr{O} \oplus \mathscr{O}(1)$ induces a triangle equivalence $\mathbf{D}^b(\mathrm{coh}\,\mathbb{P}^1_k) \overset{\sim}{\to} \mathbf{D}^b(\mathrm{mod}\,\Lambda)$ which restricts to equivalences*

$$\mathrm{coh}_0\,\mathbb{P}^1_k \xrightarrow{\ \mathrm{Hom}(\mathscr{T}, -)\ } \mathrm{reg}\,\Lambda,$$

$$\mathrm{add}\{\mathscr{O}(n) \mid n \geq 0\} \xrightarrow{\ \mathrm{Hom}(\mathscr{T}, -)\ } \mathrm{add}\{P_n \mid n \geq 0\},$$

$$\mathrm{add}\{\mathscr{O}(n) \mid n < 0\} \xrightarrow{\ \mathrm{Ext}^1(\mathscr{T}, -)\ } \mathrm{add}\{I_n \mid n \geq 0\}.$$

Proof   It follows from Lemma 5.1.2 that the tilting object $\mathscr{T}$ induces a torsion pair $(\mathcal{T}, \mathcal{F})$ for $\mathrm{coh}\,\mathbb{P}^1_k$, and we compute

$$\mathcal{T} = (\mathrm{coh}_0\,\mathbb{P}^1_k) \vee (\mathrm{add}\{\mathscr{O}(n) \mid n \geq 0\}) \qquad \text{and} \qquad \mathcal{F} = \mathrm{add}\{\mathscr{O}(n) \mid n < 0\}.$$

Now the assertion follows from Theorem 5.1.2 and Corollary 5.1.3, combined with the above computations. $\square$

The following is an immediate consequence of Corollary 5.1.14.

Corollary 5.1.18. *We have lattice isomorphisms*

$$\{\text{thick subcategories of } \mathbf{D}^b(\mathrm{mod}\,\Lambda)\} \overset{\sim}{\longrightarrow} \{\text{spc subsets of } \mathbb{P}^1_k\} \amalg \mathbb{Z}$$

and

$$\{\text{thick subcategories of } \mathrm{mod}\,\Lambda\} \overset{\sim}{\longrightarrow} \{\text{spc subsets of } \mathbb{P}^1_k\} \amalg \mathbb{Z}. \qquad \square$$

The Klein Four Group

We consider the group

$$G = \langle g_1, g_2 \rangle \cong \mathbb{Z}/2 \times \mathbb{Z}/2$$

and let k be a field of characteristic two. Let kG denote the group algebra of G over k, and set $x_1 := g_1 - 1$, $x_2 := g_2 - 1$ as elements of kG. Then $x_1^2 = x_2^2 = 0$, and we have

$$kG = k[x_1, x_2]/(x_1^2, x_2^2).$$

This is an exterior algebra on a two-dimensional space. The algebra is self-injective, so mod kG is a Frobenius category; see Example 3.3.4. It is not difficult to describe all finite dimensional kG-modules in this case. There is a notion of cohomological support for each kG-module, and using this we are able to classify all thick subcategories of the stable module category $\underline{\mathrm{mod}}\, kG = \mathrm{St}(\mathrm{mod}\, kG)$ and the bounded derived category $\mathbf{D}^b(\mathrm{mod}\, kG)$.

We describe kG-modules by diagrams in which the vertices represent basis elements as a k-vector space, and an edge

indicates that $ax_i = b$. If there is no edge labelled x_i in the downwards direction from a vertex then x_i sends the corresponding basis vector to zero. For example, the group algebra kG has the following diagram:

As a vector space, $kG = k1 \oplus kx_1 \oplus kx_2 \oplus ky$, where $y := x_1x_2 = x_2x_1$. We have $\mathrm{rad}\, kG = \mathrm{soc}^2 kG = kx_1 \oplus kx_2 \oplus ky$ and $\mathrm{rad}^2 kG = \mathrm{soc}\, kG = ky$.

Here are the diagrams for the syzygies of the trivial module:

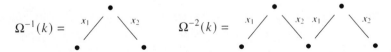

$$\Omega^1(k) = \quad \diagdown_{x_2} \diagup^{x_1} \qquad\qquad \Omega^2(k) = \quad \diagdown_{x_2} \diagup^{x_1} \diagdown_{x_2} \diagup^{x_1} \qquad\qquad \text{etc.}$$

Observe that in each diagram for $\Omega^n(k)$, the vertices of the bottom row correspond to a basis of $\operatorname{rad}\Omega^n(k) = \operatorname{soc}\Omega^n(k)$.

For each integer $n \geq 0$ we have an isomorphism

$$\underline{\operatorname{Hom}}_{kG}(k, \Omega^{-n}(k)) \xrightarrow{\;\sim\;} \operatorname{Ext}^n_{kG}(k,k) \qquad\qquad (5.1.19)$$

by Lemma 4.4.19, and so $\operatorname{rank}_k \operatorname{Ext}^n_{kG}(k,k) = n + 1$, since

$$\underline{\operatorname{Hom}}_{kG}(k, \Omega^{-n}(k)) \cong \operatorname{soc}\Omega^{-n}(k).$$

In fact, the full cohomology algebra is the $\mathbb{Z}$-graded algebra

$$H^*(G,k) = \operatorname{Ext}^*_{kG}(k,k) = k[\zeta_1, \zeta_2]$$

with $\deg(\zeta_1) = \deg(\zeta_2) = 1$. The ring $H^*(G,k)$ is a two-dimensional graded factorial domain. Thus homogeneous irreducible elements correspond to non-zero homogeneous prime ideals by taking an element p to the ideal generated by p. We write $\mathfrak{m} = H^+(G,k)$ for the unique maximal ideal consisting of positive degree elements. Let $\operatorname{Spec} H^*(G,k)$ denote the set of homogeneous prime ideals of $H^*(G,k)$.

Let $\mathfrak{p} \in \operatorname{Spec} H^*(G,k) \setminus \{0, \mathfrak{m}\}$ and choose a homogeneous irreducible element p of degree d that generates $\mathfrak{p}$. The bijection (5.1.19) gives for each power p^n a monomorphism $k \to \Omega^{-nd}(k)$ whose cokernel we denote by $L_{\mathfrak{p}^n}$. Thus there is an exact sequence

$$0 \longrightarrow k \longrightarrow \Omega^{-nd}(k) \longrightarrow L_{\mathfrak{p}^n} \longrightarrow 0.$$

Proposition 5.1.20. *The kG-modules kG, $\Omega^n(k)$ ($n \in \mathbb{Z}$), and $L_{\mathfrak{p}^n}$ ($n \geq 1$ and $\mathfrak{p} \in \operatorname{Spec} H^*(G,k) \setminus \{0, \mathfrak{m}\}$) form, up to isomorphism, a complete list of finite dimensional indecomposable kG-modules.*

The proof requires some preparations. We consider the category $\operatorname{Rep}(\Gamma, k)$ of k-linear representations of the Kronecker quiver $\Gamma\colon \circ \rightrightarrows \circ$. The *radical* $\operatorname{rad} X$ (intersection of all maximal subobjects) of a representation $X = (X' \overset{\phi_1}{\underset{\phi_2}{\rightrightarrows}} X'')$ identifies with $(0 \rightrightarrows \sum_i \operatorname{Im}\phi_i)$. We call X *separated* if $(\operatorname{rad} X)'' = X''$.

Lemma 5.1.21. *Each Kronecker representation X admits a decomposition $X = X_s \oplus X_t$ such that X_s is separated and X_t is a direct sum of copies of $(0 \rightrightarrows k)$.*

Proof Set $X_s = (X' \xrightarrow[\phi_2]{\phi_1} \sum_i \operatorname{Im} \phi_i)$ and $X_t = (0 \Longrightarrow X''/\sum_i \operatorname{Im} \phi_i)$.

$\square$

There is a similar decomposition for kG-modules. Call a kG-module X *stable* if X is annihilated by $x_1 x_2 = x_2 x_1$.

Lemma 5.1.22. *Each kG-module X admits a decomposition $X = X_s \oplus X_t$ such that X_s is stable and X_t is a direct sum of copies of kG.*

Proof Set $X_s = \{x \in X \mid xx_1 x_2 = 0\}$ and $X_t = X/X_s$. Each element $x \in X_t$ such that $xx_1 x_2 \neq 0$ yields a monomorphism $\phi \colon kG \to X_t$ such that $\phi(1) = x$. Clearly, ϕ splits since kG is injective. Thus X_t is a direct sum of copies of kG.

$\square$

We define a pair of functors

$$\operatorname{Rep}(\Gamma, k) \xrightleftharpoons[T]{S} \operatorname{Mod} \tfrac{k[x_1, x_2]}{(x_i x_j)} \subseteq \operatorname{Mod} \tfrac{k[x_1, x_2]}{(x_1^2, x_2^2)} \qquad (5.1.23)$$

as follows. For a representation $X = (X' \xrightarrow[\phi_2]{\phi_1} X'')$ set

$$S(X) = X' \oplus X'' \qquad \text{with} \qquad (x' + x'')x_i = \phi_i(x'),$$

and for a $k[x_1, x_2]/(x_i x_j)$-module Y set

$$T(Y) = (Y/\operatorname{rad} Y \xrightarrow[\psi_2]{\psi_1} \operatorname{rad} Y) \qquad \text{with} \qquad \psi_i(y + \operatorname{rad} Y) = yx_i.$$

For example, we use the description of the indecomposable Kronecker representations (Proposition 5.1.15) and compute for $n \geq 0$ and a homogeneous prime ideal $\mathfrak{p} \neq 0$

$$S(P_n) = \Omega^{-n}(k), \qquad S(I_n) = \Omega^n(k), \qquad S(R_{\mathfrak{p}^n}) = L_{\mathfrak{p}^n}.$$

Proof of Proposition 5.1.20 It is easily checked that $TS(X) \cong X$ when X is separated, and $ST(Y) \cong Y$ when Y is stable. Now the assertion follows from the classification of the indecomposable Kronecker representations; see Proposition 5.1.15.

$\square$

Given a finite dimensional kG-module X, consider the homomorphism

$$\chi_X \colon H^*(G, k) \longrightarrow \operatorname{Ext}^*_{kG}(X, X), \qquad \eta \mapsto X \otimes_k \eta.$$

Here, we use the tensor product $X \otimes_k Y$ for kG-modules X, Y where G acts diagonally, so $(x \otimes y)g = xg \otimes yg$ for $x \in X$, $y \in Y$, $g \in G$. The *support* of X is by definition the set

$$\operatorname{Supp} X = \{\mathfrak{p} \in \operatorname{Spec} H^*(G,k) \mid \operatorname{Ker} \chi_X \subseteq \mathfrak{p}\}.$$

Alternatively, $\operatorname{Supp} X$ can be computed as the support of the $H^*(G,k)$-module $\operatorname{Ext}^*_{kG}(k, X)$. For instance, this shows for each exact sequence $0 \to X \to Y \to Z \to 0$ that $\operatorname{Supp} Y \subseteq \operatorname{Supp} X \cup \operatorname{Supp} Z$.

It is not difficult to compute the support of each object.

Proposition 5.1.24. *Let* $\mathfrak{p} \in \operatorname{Spec} H^*(G,k) \setminus \{0, \mathfrak{m}\}$ *and* $n \in \mathbb{Z}$*. Then we have*

$$\operatorname{Supp} kG = \{\mathfrak{m}\}, \quad \operatorname{Supp} \Omega^n(k) = \operatorname{Spec} H^*(G,k), \quad \operatorname{Supp} L_{\mathfrak{p}^n} = \{\mathfrak{p}, \mathfrak{m}\}. \quad \square$$

The definition of the homomorphism χ_X extends without any changes to objects $X \in \mathbf{D}^b(\operatorname{mod} kG)$ if we set

$$\operatorname{Ext}^*_{kG}(X,Y) = \operatorname{Hom}^*_{\mathbf{D}(kG)}(X,Y) = \bigoplus_{n \in \mathbb{Z}} \operatorname{Hom}^n_{\mathbf{D}(kG)}(X, \Sigma^n Y).$$

For example, we have

$$X \in \mathbf{D}^b(\operatorname{proj} kG) \quad \Longleftrightarrow \quad \operatorname{Supp} X \subseteq \{\mathfrak{m}\}.$$

Now fix a pair of objects X, Y either in $\underline{\operatorname{mod}}\, kG$ or in $\mathbf{D}^b(\operatorname{mod} kG)$. Then we have

$$\operatorname{Thick}(X) \subseteq \operatorname{Thick}(Y) \quad \Longleftrightarrow \quad \operatorname{Supp} X \subseteq \operatorname{Supp} Y.$$

This is a consequence of the following classification result.

A subset $\mathcal{V}$ of $\operatorname{Spec} H^*(G,k)$ is *specialisation closed* if for any pair $\mathfrak{p} \subseteq \mathfrak{q}$ of prime ideals, $\mathfrak{p} \in \mathcal{V}$ implies $\mathfrak{q} \in \mathcal{V}$

Proposition 5.1.25. *The assignment*

$$\operatorname{mod} kG \supseteq \mathcal{C} \longmapsto \operatorname{Supp} \mathcal{C} = \bigcup_{X \in \mathcal{C}} \operatorname{Supp} X$$

induces an inclusion preserving bijection between the thick subcategories of $\operatorname{mod} kG$ *and the specialisation closed subsets of* $\operatorname{Spec} H^*(G,k)$*.*

Proof First observe that every thick subcategory $0 \neq \mathcal{C} \subseteq \operatorname{mod} kG$ contains $\operatorname{proj} kG$. Indeed, for each kG-module X we have that $X \otimes_k kG$ is isomorphic to $\operatorname{rank}_k X$ copies of kG. Thus $kG \in \operatorname{Thick}(k)$ implies that

$$X \otimes_k kG \in \operatorname{Thick}(X \otimes_k k) = \operatorname{Thick}(X).$$

In particular, $\operatorname{Supp} \mathcal{C}$ is specialisation closed since $\operatorname{Supp} kG = \{\mathfrak{m}\}$.

Let Λ denote the Kronecker algebra. We use the exact functor $S\colon \operatorname{mod} \Lambda \to$

mod kG from (5.1.23). So we can apply the classification of thick subcategories of mod Λ from Corollary 5.1.18. For example, we have for $X, Y \in \text{mod } \Lambda$

$$X \in \text{Thick}(Y) \quad \Longrightarrow \quad S(X) \in \text{Thick}(S(Y)).$$

Thus $L_{\mathfrak{p}^m} \in \text{Thick}(L_{\mathfrak{p}^n})$ for all $m, n \geq 1$, and this calculation implies

$$\mathcal{C} = \{X \in \text{mod } kG \mid \text{Supp } X \subseteq \text{Supp } \mathcal{C}\}$$

for every thick $\mathcal{C} \subseteq \text{mod } kG$. A similar calculation shows that every speciali-sation closed subset $\mathcal{V} \subseteq \text{Spec } H^*(G, k)$ is of the form $\text{Supp } \mathcal{C}$ for some thick subcategory $\mathcal{C}$. More precisely, $\mathcal{V} \neq \text{Spec } H^*(G, k)$ corresponds to a thick subcategory $\mathcal{C} \subseteq \text{reg } \Lambda$ (given by $\mathcal{V} \setminus \{\mathfrak{m}\} \subseteq \mathbb{P}^1_k$), and then $\text{Supp } S(\mathcal{C}) = \mathcal{V}$. $\quad\square$

We now obtain a classification of all thick subcategories for the triangulated categories $\underline{\text{mod}}\, kG$ and $\mathbf{D}^b(\text{mod } kG)$ since mod kG is a Frobenius category; see Example 4.4.13 and keeping in mind the triangle equivalence

$$\underline{\text{mod}}\, kG \xrightarrow{\ \sim\ } \frac{\mathbf{D}^b(\text{mod } kG)}{\mathbf{D}^b(\text{proj } kG)}$$

from Proposition 4.4.18.

Corollary 5.1.26. *The assignment*

$$\underline{\text{mod}}\, kG \supseteq \mathcal{C} \longmapsto \text{Supp } \mathcal{C} = \bigcup_{X \in \mathcal{C}} \text{Supp } X$$

induces an inclusion preserving bijection between the thick subcategories of $\underline{\text{mod}}\, kG$ *and the specialisation closed subsets of* $\text{Spec } H^*(G, k)$ *containing* $\mathfrak{m}$.

Proof Thick subcategories of mod kG containing proj kG correspond bijec-tively to thick subcategories of $\underline{\text{mod}}\, kG$; see Example 4.4.13. $\quad\square$

Corollary 5.1.27. *The assignment*

$$\mathbf{D}^b(\text{mod } kG) \supseteq \mathcal{C} \longmapsto \text{Supp } \mathcal{C} = \bigcup_{X \in \mathcal{C}} \text{Supp } X$$

induces an inclusion preserving bijection between the thick subcategories of $\mathbf{D}^b(\text{mod } kG)$ *and the specialisation closed subsets of* $\text{Spec } H^*(G, k)$.

Proof First observe that every thick subcategory $0 \neq \mathcal{C} \subseteq \mathbf{D}^b(\text{mod } kG)$ contains $\mathbf{D}^b(\text{proj } kG)$. To see this we copy the argument from the proof of Proposition 5.1.25. Thus for $X \in \mathcal{C}$ we have $X \otimes_k kG \in \mathbf{D}^b(\text{proj } kG)$, and

$$X \otimes_k kG \in \text{Thick}(X \otimes_k k) = \text{Thick}(X).$$

Moreover, the complex $X \otimes_k kG$ is split, so $H^n(X \otimes_k kG) \cong H^n(X) \otimes_k$

kG belongs to Thick(X); see Example 4.1.6. Thus $\mathcal{C}$ contains Thick$(kG) = \mathbf{D}^b(\operatorname{proj} kG)$.

To complete the proof one notes that thick subcategories of $\operatorname{mod} kG$ containing $\operatorname{proj} kG$ correspond bijectively to thick subcategories of $\mathbf{D}^b(\operatorname{mod} kG)$ containing $\mathbf{D}^b(\operatorname{proj} kG)$; see Example 4.4.13. □

5.2 Derived Categories of Finitely Presented Modules

For a right coherent ring we provide descriptions of the bounded derived category of finitely presented modules. When the ring is an Artin algebra, then we use the module category of the trivial extension algebra. Another approach describes the derived category as a completion of the category of perfect complexes.

The Trivial Extension Algebra

Let k be a commutative artinian ring and A an Artin k-algebra. We write $D = \operatorname{Hom}_k(-, E)$ for the Matlis duality over k, given by an injective k-module E, and consider the bimodule $_AD(A)_A$. This yields the *Nakayama functor*

$$\nu\colon \operatorname{Mod} A \longrightarrow \operatorname{Mod} A, \qquad X \mapsto X \otimes_A D(A).$$

Also, let

$$T(A) = A \ltimes D(A)$$

be the *trivial extension algebra*. Thus $T(A) = A \oplus D(A)$ with multiplication given by the formula

$$(x, y) \cdot (x', y') = (xx', xy' + yx').$$

The algebra $T(A)$ is $\mathbb{Z}$-graded with $T(A)^0 = A$ and $T(A)^1 = D(A)$. We consider $\mathbb{Z}$-graded $T(A)$-modules $X = \bigoplus_{n \in \mathbb{Z}} X^n$ with degree zero morphisms and denote this category by $\operatorname{GrMod} T(A)$. An object is given by a family $X = (X^n, x^n)_{n \in \mathbb{Z}}$ of A-modules X^n and A-linear maps $x^n\colon \nu X^n \to X^{n+1}$ satisfying $x^{n+1} \circ \nu x^n = 0$. A morphism $X \to Y$ is given by a family $(X^n \xrightarrow{\phi^n} Y^n)_{n \in \mathbb{Z}}$ of A-linear maps such that $y^n \circ \nu \phi^n = \phi^{n+1} \circ x^n$ for all n. We write

$$\cdots \longrightarrow X^{n-1} \xrightarrow{x^{n-1}} X^n \xrightarrow{x^n} X^{n+1} \longrightarrow \cdots$$

and keep in mind that the arrows represent the degree one morphism $\nu X \to X$.

Lemma 5.2.1. *The category* $\mathrm{GrMod}\, T(A)$ *is a Frobenius category. The projective and injective objects are the direct sums of objects of the form*

$$\cdots \longrightarrow 0 \longrightarrow P \xrightarrow{\ \mathrm{id}\ } \nu P \longrightarrow 0 \longrightarrow \cdots$$

where P is any projective A-module.

Proof   Adapt the proof of Lemma 4.1.1.   □

We consider the full subcategory $\mathrm{grmod}\, T(A)$ of finitely generated modules; it is also a Frobenius category. It follows from Proposition 4.4.18 that the composite

$$\mathrm{grmod}\, T(A) \longrightarrow \mathbf{D}^b(\mathrm{grmod}\, T(A)) \longrightarrow \frac{\mathbf{D}^b(\mathrm{grmod}\, T(A))}{\mathbf{D}^b(\mathrm{grproj}\, T(A))}$$

induces a triangle equivalence

$$f\colon \underline{\mathrm{grmod}\, T(A)} \xrightarrow{\ \sim\ } \frac{\mathbf{D}^b(\mathrm{grmod}\, T(A))}{\mathbf{D}^b(\mathrm{grproj}\, T(A))}.$$

Let f^- denote a quasi-inverse of f.

Let $p\colon T(A) \to A$ denote the canonical epimorphism; it induces via restriction of scalars an exact and fully faithful functor $p^*\colon \mathrm{mod}\, A \to \mathrm{grmod}\, T(A)$. The functor p^* identifies $\mathrm{mod}\, A$ with the full subcategory of modules X in $\mathrm{grmod}\, T(A)$ such that $X^i = 0$ for all $i \neq 0$. We use the left adjoint $p_! = -\otimes_{T(A)} A$ of p^*, which satisfies $p_! p^* \cong \mathrm{id}$ and preserves projectivity.

Lemma 5.2.2. *For X, Y in $\mathrm{mod}\, A$ and $n \in \mathbb{Z}$ we have natural isomorphisms*

$$\mathrm{Ext}^n_A(X,Y) \xrightarrow{\ p^*\ } \mathrm{Ext}^n_{T(A)}(X,Y) \xrightarrow{\ f^-\ } \underline{\mathrm{Hom}}_{T(A)}(X, \Sigma^n Y).$$

Proof   Choose a projective resolution

$$P\colon \quad \cdots \longrightarrow P_2 \longrightarrow P_1 \longrightarrow P_0 \longrightarrow X \longrightarrow 0$$

of X in $\mathrm{grmod}\, T(A)$. We may assume that $(P_i)^j = 0$ for all $i \geq 0 > j$. Applying $p_!$ yields a projective resolution in $\mathrm{mod}\, A$ that identifies with

$$\cdots \longrightarrow (P_2)^0 \longrightarrow (P_1)^0 \longrightarrow (P_0)^0 \longrightarrow X \longrightarrow 0.$$

Thus we have

$$\mathrm{Ext}^n_A(X,Y) = H^n \mathrm{Hom}_A(p_! P, Y) \xrightarrow{\ \sim\ } H^n \mathrm{Hom}_{T(A)}(P, p^* Y) = \mathrm{Ext}^n_{T(A)}(X,Y).$$

A direct calculation shows that $\underline{\mathrm{Hom}}_{T(A)}(X, \Sigma^n Y) = 0$ for all $n < 0$. Next observe that $\mathrm{Hom}_{T(A)}(X, P) = 0$ for every projective $T(A)$-module P. It is easily checked that we may reduce to the case $P = T(A)$. Then we have for $\phi \colon X \to T(A)$ that $\nu\phi^0 = \phi^1 \circ x^0 = 0$, since $X^i = 0$ for all $i \neq 0$. But $\nu \cong D\operatorname{Hom}_A(-, A)$, and therefore $\phi = 0$. This yields $\operatorname{Hom}_A(X, Y) \cong \underline{\operatorname{Hom}}_{T(A)}(X, Y)$, and then the second isomorphism follows from Lemma 4.4.19.

$\square$

Theorem 5.2.3. *The composite*

$$\mathbf{D}^b(\operatorname{mod} A) \xrightarrow{p^*} \mathbf{D}^b(\operatorname{grmod} T(A)) \longrightarrow \frac{\mathbf{D}^b(\operatorname{grmod} T(A))}{\mathbf{D}^b(\operatorname{grproj} T(A))} \xrightarrow{f^-} \underline{\operatorname{grmod}} T(A)$$

is fully faithful and exact. The functor makes the following square commutative

$$
\begin{array}{ccc}
\operatorname{mod} A & \xrightarrow{\;\;p^*\;\;} & \underline{\operatorname{grmod}} T(A) \\
\downarrow & & \downarrow \\
\mathbf{D}^b(\operatorname{mod} A) & \longrightarrow & \underline{\operatorname{grmod}} T(A)
\end{array}
$$

(up to a natural isomorphism) and is an equivalence if and only if A has finite global dimension.

Proof We denote by F the functor $\mathbf{D}^b(\operatorname{mod} A) \to \underline{\operatorname{grmod}} T(A)$. Combining the above lemma with Lemma 3.1.8 shows that F is fully faithful. From the construction it is clear that the square commutes.

Now suppose that A has finite global dimension. We claim that the objects of $\operatorname{mod} A$ generate $\underline{\operatorname{grmod}} T(A)$ as a triangulated category. Choose X in $\underline{\operatorname{grmod}} T(A)$. We use induction on

$$r := \min\{i \in \mathbb{Z} \mid X^i \neq 0\} \quad \text{and} \quad s := \max\{i \in \mathbb{Z} \mid X^i \neq 0\}.$$

We are done when $r = 0 = s$. If $r < 0$, then we use an induction on proj.dim X^r. Choose a projective cover $P^r \to X^r$ and extend this to a morphism $\phi \colon P \to X$ with P given by

$$\cdots \longrightarrow 0 \longrightarrow P^r \xrightarrow{\;\mathrm{id}\;} \nu P^r \longrightarrow 0 \longrightarrow \cdots .$$

For $Y = \operatorname{Ker} \phi$ we have that $Y^r = 0$ or proj.dim $Y^r <$ proj.dim X^r. There is an exact sequence $0 \to X' \to X \to X'' \to 0$ with $X' = \operatorname{Im} \phi = \Sigma Y$ and $(X'')^r = 0$. Thus X is generated by Y and X'', and we proceed until $r = 0$. A similar induction reduces to the case $s = 0$ when $s > 0$. Having shown that $\operatorname{mod} A$ generates, it follows that the essential image of F is $\underline{\operatorname{grmod}} T(A)$.

If the global dimension of A is infinite, then $\mathbf{D}^b(\operatorname{mod} A)$ admits no Serre functor (Theorem 6.4.13), while $\underline{\operatorname{grmod}} T(A)$ admits a Serre functor (Proposition 6.4.2). Thus there is no triangle equivalence $\mathbf{D}^b(\operatorname{mod} A) \xrightarrow{\sim} \underline{\operatorname{grmod}} T(A)$.

$\square$

Let us extract from the above proof a general result; it provides a conceptual explanation for the above functor $\mathbf{D}^b(\operatorname{mod} A) \to \underline{\operatorname{grmod} T(A)}$ to be fully faithful.

Proposition 5.2.4. *Let A be a Frobenius category and $\mathcal{C} \subseteq A$ a full exact subcategory such that*

(1) $\operatorname{Ext}^n_{\mathcal{C}}(X,Y) \xrightarrow{\sim} \operatorname{Ext}^n_A(X,Y)$ *for all $X, Y \in \mathcal{C}$ and $n \geq 0$, and*

(2) $\operatorname{Hom}_A(X,Y) = 0$ *for all $X \in \mathcal{C}$ and projective $Y \in A$.*

Then $\mathcal{C} \hookrightarrow A$ extends to a fully faithful exact functor $\mathbf{D}^b(\mathcal{C}) \to \operatorname{St} A$.

Proof From (1) it follows that $\mathcal{C} \hookrightarrow A$ induces a fully faithful and exact functor $\mathbf{D}^b(\mathcal{C}) \to \mathbf{D}^b(A)$, by Lemma 4.2.13.

Let $\mathcal{P} \subseteq A$ denote the full subcategory of projective (and injective) objects. Then (2) implies $\mathcal{C} \subseteq {}^\perp\mathcal{P}$ and therefore $\mathbf{D}^b(\mathcal{C}) \subseteq {}^\perp\mathbf{D}^b(\mathcal{P})$ in $\mathbf{D}^b(A)$. Thus the composite $\mathbf{D}^b(\mathcal{C}) \to \mathbf{D}^b(A) \twoheadrightarrow \mathbf{D}^b(A)/\mathbf{D}^b(\mathcal{P})$ is fully faithful, by Lemma 3.2.4. It remains to recall from Proposition 4.4.18 the triangle equivalence $\mathbf{D}^b(A)/\mathbf{D}^b(\mathcal{P}) \xrightarrow{\sim} \operatorname{St} A$. $\qquad\square$

Completing Perfect Complexes

Let Λ be a ring and consider the derived category $\mathbf{D}(\operatorname{Mod} \Lambda)$. A complex X is called *perfect* if it is quasi-isomorphic to a bounded complex of finitely generated projective Λ-modules. An equivalent condition is that X belongs to the thick subcategory generated by Λ, viewed as a complex concentrated in degree zero. We write $\mathbf{D}^{\operatorname{perf}}(\Lambda)$ for the full subcategory of perfect complexes. The inclusion $\operatorname{proj} \Lambda \to \operatorname{Mod} \Lambda$ induces a triangle equivalence

$$\mathbf{D}^b(\operatorname{proj} \Lambda) \xrightarrow{\sim} \mathbf{D}^{\operatorname{perf}}(\Lambda)$$

and it is convenient to view this as an identification.

The ring Λ is by definition *right coherent* if $\operatorname{mod} \Lambda$ is an abelian category. In this case we wish to understand how its derived category $\mathbf{D}^b(\operatorname{mod} \Lambda)$ is built from perfect complexes. We begin this analysis with a discussion of phantom morphisms.

Let $\mathcal{T}$ be a triangulated category and suppose that countable coproducts exist in $\mathcal{T}$. An object C in $\mathcal{T}$ is called *compact* if $\operatorname{Hom}(C, -)$ preserves all coproducts. A morphism $X \to Y$ is *phantom* if any composite $C \to X \to Y$ with C compact is zero. The phantom morphisms form an ideal and we write $\operatorname{Ph}(X,Y)$ for the subgroup of all phantoms in $\operatorname{Hom}(X,Y)$.

Lemma 5.2.5. *Let* $X = \mathrm{hocolim}_n X_n$ *be a homotopy colimit in* $\mathcal{T}$ *and let* $\phi\colon X \to Y$ *be a morphism. If* ϕ *factors through the canonical morphism* $X \to \coprod_{n \geq 0} \Sigma X_n$, *then* ϕ *is phantom. The converse holds when each* X_n *is a coproduct of compact objects.*

Proof Consider the defining triangle

$$\coprod_{n \geq 0} X_n \longrightarrow \coprod_{n \geq 0} X_n \xrightarrow{\ \alpha\ } X \xrightarrow{\ \beta\ } \coprod_{n \geq 0} \Sigma X_n$$

of $X = \mathrm{hocolim}_n X_n$. The proof of Lemma 3.4.3 shows that $\mathrm{Hom}(C, \alpha)$ is surjective for each compact $C \in \mathcal{T}$. Thus β is phantom. Conversely, if each X_n is a coproduct of compact objects and ϕ is phantom, then $\phi\alpha = 0$ and ϕ factors through β. $\qquad\square$

For any sequence $\cdots \to A_2 \xrightarrow{\phi_2} A_1 \xrightarrow{\phi_1} A_0$ of maps between abelian groups the limit and its first derived functor are given by the exact sequence

$$0 \longrightarrow \lim_n A_n \longrightarrow \prod_{n \geq 0} A_n \xrightarrow{\mathrm{id} - \phi} \prod_{n \geq 0} A_n \longrightarrow \lim_n^1 A_n \longrightarrow 0.$$

Note that $\lim_n^1 A_n = 0$ when $A_{n+1} \xrightarrow{\sim} A_n$ for $n \gg 0$.

Lemma 5.2.6. *Let* $X = \mathrm{hocolim}_n X_n$ *be a homotopy colimit in* $\mathcal{T}$ *such that each* X_n *is a coproduct of compact objects. Then we have for any* Y *in* $\mathcal{T}$ *a natural exact sequence*

$$0 \longrightarrow \mathrm{Ph}(X, Y) \longrightarrow \mathrm{Hom}(X, Y) \longrightarrow \lim_n \mathrm{Hom}(X_n, Y) \longrightarrow 0$$

and an isomorphism

$$\mathrm{Ph}(X, \Sigma Y) \cong \lim_n^1 \mathrm{Hom}(X_n, Y).$$

Proof Apply $\mathrm{Hom}(-, Y)$ to the exact triangle defining $\mathrm{hocolim}_n X_n$ and use the description of $\mathrm{Ph}(X, Y)$ in Lemma 5.2.5. $\qquad\square$

Let $\mathcal{C} \subseteq \mathcal{T}$ be a full additive subcategory consisting of compact objects and consider the restricted Yoneda functor

$$\mathcal{T} \longrightarrow \mathrm{Add}(\mathcal{C}^{\mathrm{op}}, \mathrm{Ab}), \quad X \mapsto h_X := \mathrm{Hom}(-, X)|_{\mathcal{C}}.$$

This functor induces for each pair of objects $X, Y \in \mathcal{T}$ a map

$$\mathrm{Hom}(X, Y) \longrightarrow \mathrm{Hom}(h_X, h_Y).$$

Clearly, this map is bijective when X is in $\mathcal{C}$, and it remains bijective when X is a coproduct of objects in $\mathcal{C}$.

Lemma 5.2.7. *Let* $X = \text{hocolim}_n X_n$ *be a homotopy colimit in* $\mathcal{T}$ *such that each* X_n *is a coproduct of objects in* $\mathcal{C}$. *Then we have for any* Y *in* $\mathcal{T}$ *a natural isomorphism*

$$\text{Hom}(X, Y)/\text{Ph}(X, Y) \xrightarrow{\sim} \text{Hom}(h_X, h_Y).$$

Proof We have

$$\begin{aligned}
\text{Hom}(X, Y)/\text{Ph}(X, Y) &\cong \lim_n \text{Hom}(X_n, Y) \\
&\cong \lim_n \text{Hom}(h_{X_n}, h_Y) \\
&\cong \text{Hom}(\text{colim}_n h_{X_n}, h_Y) \\
&\cong \text{Hom}(h_X, h_Y).
\end{aligned}$$

The first isomorphism follows from Lemma 5.2.6, the second follows from the observation about the restricted Yoneda functor, the third is clear, and the last follows from Lemma 3.4.3. $\qquad\square$

Let us apply the theory of phantoms to the study of $\mathbf{D}(\text{Mod}\,\Lambda)$. Recall that we have a triangle equivalence $\mathbf{K}_{\text{proj}}(\text{Mod}\,\Lambda) \xrightarrow{\sim} \mathbf{D}(\text{Mod}\,\Lambda)$ (Corollary 4.3.6). It is easily checked that each perfect complex is compact, because the compact objects form a thick subcategory containing Λ, viewed as a complex concentrated in degree zero.

Proposition 5.2.8. *Let* Λ *be a ring and set* $\mathcal{P} = \text{proj}\,\Lambda$. *Then the functor*

$$\mathbf{K}^{-,b}(\mathcal{P}) \longrightarrow \text{Add}(\mathbf{K}^b(\mathcal{P})^{\text{op}}, \text{Ab}), \quad X \mapsto h_X := \text{Hom}(-, X)|_{\mathbf{K}^b(\mathcal{P})},$$

is fully faithful.

Proof We view $\mathbf{K}^{-,b}(\mathcal{P})$ as a subcategory of $\mathbf{D}(\text{Mod}\,\Lambda)$, and the objects in $\mathbf{K}^b(\mathcal{P})$ are compact by the above remark. Let X, Y be objects in $\mathbf{K}^{-,b}(\mathcal{P})$ and write X as homotopy colimit of its truncations $X_n = \sigma_{\geq -n} X$ which lie in $\mathbf{K}^b(\mathcal{P})$; see Example 4.2.2. Let C_n denote the cone of $X_n \to X_{n+1}$. This complex is concentrated in degree $-n-1$; so $\text{Hom}(C_n, Y) = 0$ for $n \gg 0$. Thus $X_n \to X_{n+1}$ induces a bijection

$$\text{Hom}(X_{n+1}, Y) \xrightarrow{\sim} \text{Hom}(X_n, Y) \quad \text{for} \quad n \gg 0.$$

This implies

$$\text{Hom}(X, Y) \xrightarrow{\sim} \lim_n \text{Hom}(X_n, Y)$$

and therefore $\text{Ph}(X, Y) = 0$ by Lemma 5.2.6. From Lemma 5.2.7 we conclude that

$$\text{Hom}(X, Y) \xrightarrow{\sim} \text{Hom}(h_X, h_Y). \qquad\square$$

Let $\mathcal{D}$ be a triangulated category and $\mathcal{C} \subseteq \mathcal{D}$ an essentially small triangulated subcategory such that the functor

$$\mathcal{D} \longrightarrow \mathrm{Add}(\mathcal{C}^{\mathrm{op}}, \mathrm{Ab}), \quad X \mapsto h_X := \mathrm{Hom}(-, X)|_{\mathcal{C}}$$

is fully faithful.

For any $X \in \mathcal{D}$ let $\mathcal{C}/X$ denote the *slice category* consisting of pairs (C, ϕ) given by a morphism $\phi \colon C \to X$ with $C \in \mathcal{C}$. A morphism $(C, \phi) \to (C', \phi')$ is given by a morphism $\alpha \colon C \to C'$ in $\mathcal{C}$ such that $\phi'\alpha = \phi$. Then $\mathcal{C}/X$ is filtered.

Lemma 5.2.9. *Every object in $\mathcal{D}$ can be written canonically as a filtered colimit*

$$X = \operatorname*{colim}_{(C,\phi) \in \mathcal{C}/X} C$$

of the forgetful functor $\mathcal{C}/X \to \mathcal{D}$ that takes (C, ϕ) to C.

Given objects in $\mathcal{D}$ that are written as filtered colimits of objects (X_α) and (Y_β) in $\mathcal{C}$, then

$$\mathrm{Hom}(\operatorname{colim} X_\alpha, \operatorname{colim} Y_\beta) \cong \lim_\alpha \operatorname*{colim}_\beta \mathrm{Hom}(X_\alpha, Y_\beta).$$

Proof Fix an object $X \in \mathcal{D}$. The functor h_X is cohomological, so the morphisms $C \to X$ with $C \in \mathcal{C}$ form a filtered category. Thus h_X is the filtered colimit of representable functors h_C given by objects $(C, \phi) \in \mathcal{C}/X$; see Proposition 3.1.4. It follows that $X = \operatorname{colim}_{(C,\phi) \in \mathcal{C}/X} C$ since $\mathcal{D} \to \mathrm{Add}(\mathcal{C}^{\mathrm{op}}, \mathrm{Ab})$ is fully faithful.

For $X = \operatorname{colim}_\alpha X_\alpha$ and $Y = \operatorname{colim}_\beta Y_\beta$ we obtain

$$\begin{aligned}
\mathrm{Hom}(\operatorname{colim} X_\alpha, \operatorname{colim} Y_\beta) &\cong \mathrm{Hom}(\operatorname{colim} h_{X_\alpha}, \operatorname{colim} h_{Y_\beta}) \\
&\cong \lim_\alpha \mathrm{Hom}(h_{X_\alpha}, \operatorname{colim} h_{Y_\beta}) \\
&\cong \lim_\alpha \operatorname*{colim}_\beta \mathrm{Hom}(h_{X_\alpha}, h_{Y_\beta}) \\
&\cong \lim_\alpha \operatorname*{colim}_\beta \mathrm{Hom}(X_\alpha, Y_\beta). \qquad \square
\end{aligned}$$

Now suppose that Λ is right coherent. We study the derived category $\mathbf{D}^b(\mathrm{mod}\,\Lambda)$ using the following identifications (Corollary 4.2.9):

$$\begin{array}{ccc}
\mathbf{K}^b(\mathrm{proj}\,\Lambda) & \longrightarrow & \mathbf{K}^{-,b}(\mathrm{proj}\,\Lambda) \\
\downarrow\wr & & \downarrow\wr \\
\mathbf{D}^b(\mathrm{proj}\,\Lambda) & \longrightarrow & \mathbf{D}^b(\mathrm{mod}\,\Lambda)
\end{array}$$

The following result describes the derived category $\mathbf{D}^b(\mathrm{mod}\,\Lambda)$ as a completion of the category of perfect complexes.

Theorem 5.2.10. *Let Λ be a right coherent ring. Then each object in $\mathbf{D}^b(\mathrm{mod}\,\Lambda)$ can be written (canonically) as a filtered colimit of objects in $\mathbf{D}^b(\mathrm{proj}\,\Lambda)$. Given objects in $\mathbf{D}^b(\mathrm{mod}\,\Lambda)$ that are written as filtered colimits of objects (X_α) and (Y_β) in $\mathbf{D}^b(\mathrm{proj}\,\Lambda)$, then*

$$\mathrm{Hom}(\mathrm{colim}\,X_\alpha, \mathrm{colim}\,Y_\beta) \cong \lim_\alpha \mathrm{colim}_\beta \mathrm{Hom}(X_\alpha, Y_\beta).$$

Moreover, the inclusion $\mathrm{mod}\,\Lambda \to \mathrm{Mod}\,\Lambda$ induces a fully faithful triangle functor $\mathbf{D}^b(\mathrm{mod}\,\Lambda) \to \mathbf{D}(\mathrm{Mod}\,\Lambda)$ that identifies the objects in $\mathbf{D}^b(\mathrm{mod}\,\Lambda)$ with colimits of sequences $X_0 \to X_1 \to X_2 \to \cdots$ in $\mathbf{D}^b(\mathrm{proj}\,\Lambda)$ such that

(1) *for all $n \in \mathbb{Z}$ we have $H^n(X_i) \xrightarrow{\sim} H^n(X_{i+1})$ for $i \gg 0$, and*

(2) *for almost all $n \in \mathbb{Z}$ we have $H^n(X_i) = 0$ for $i \gg 0$.*

Proof   The first assertion follows from Proposition 5.2.8 and Lemma 5.2.9.

We know already that $\mathbf{D}^b(\mathrm{mod}\,\Lambda) \to \mathbf{D}(\mathrm{Mod}\,\Lambda)$ is fully faithful; see for instance Example 4.2.18. Now let X be a complex in $\mathbf{K}^{-,b}(\mathrm{proj}\,\Lambda)$ and write this as homotopy colimit of its truncations $\sigma_{\geq n} X$ which lie in $\mathbf{K}^b(\mathrm{proj}\,\Lambda)$; see Example 4.2.2. Using again Proposition 5.2.8 and combining it with Lemma 3.4.3, it follows that this homotopy colimit is actually a (filtered) colimit. Also, it is clear that the sequence $\sigma_{\geq 0} X \to \sigma_{\geq -1} X \to \cdots$ satisfies the conditions (1) and (2). On the other hand, if X equals the colimit of such a sequence, then $H^n X$ is finitely presented for all n and $H^n X = 0$ for $|n| \gg 0$, so X lies in $\mathbf{D}^b(\mathrm{mod}\,\Lambda)$. $\square$

The above theorem remains true when the assumption on the ring Λ to be coherent is removed, but we need to replace $\mathrm{mod}\,\Lambda$ by the category of pseudo-coherent Λ-modules.

A Λ-module X is *pseudo-coherent* if it admits a projective resolution

$$\cdots \longrightarrow P_1 \longrightarrow P_0 \longrightarrow X \longrightarrow 0$$

such that each P_i is finitely generated. We denote by $\mathrm{pcoh}\,\Lambda$ the full subcategory of pseudo-coherent Λ-modules; it is an extension closed subcategory of the category of all Λ-modules, thanks to the horseshoe lemma. In fact, it is the smallest full exact subcategory of $\mathrm{Mod}\,\Lambda$ containing $\mathrm{proj}\,\Lambda$ and having enough projective objects.

The category $\mathrm{pcoh}\,\Lambda$ is the appropriate generalisation of $\mathrm{mod}\,\Lambda$, and $\mathrm{pcoh}\,\Lambda$ equals $\mathrm{mod}\,\Lambda$ when Λ is right coherent. The following simple lemma provides the connection with the above theorem.

Lemma 5.2.11. *For a ring Λ the inclusion $\mathrm{proj}\,\Lambda \to \mathrm{pcoh}\,\Lambda$ induces a triangle equivalence*

$$\mathbf{K}^{-,b}(\mathrm{proj}\,\Lambda) \xrightarrow{\sim} \mathbf{D}^b(\mathrm{pcoh}\,\Lambda).$$

Proof This follows from Corollary 4.2.9 since pcoh Λ is an exact category with enough projective objects. □

Notes

The derived equivalence between coherent sheaves on $\mathbb{P}^1$ and representations of the Kronecker quiver is a special case of a theorem of Beilinson [25] for the category of coherent sheaves on the projective n-space. A systematic treatment of derived equivalences via torsion pairs was later developed by Happel, Reiten, and Smalø [105]. The Klein four group is an example of an elementary abelian p-group, and there is in fact an elaborated theory connecting representations of elementary abelian p-groups with sheaves on projective spaces [30].

The description of the derived category of an Artin algebra via the stable category of the trivial extension algebra is due to Happel [101]; see also [123]. The equivalent notion of a repetitive algebra was introduced by Hughes and Waschbüsch [114]. The description of the derived category as a completion of the category of perfect complexes is taken from [132]. This uses the notion of a 'phantom' from homotopy theory, and the observation that the first derived functor of the inverse limit functor describes the phantom maps goes back to Milnor [144].

PART TWO

ORTHOGONAL DECOMPOSITIONS

6

Gorenstein Algebras, Approximations, Serre Duality

Contents

	6.1	**Approximations**	**176**
		Cotorsion Pairs	176
		A Decomposition via Resolutions	177
	6.2	**Gorenstein Rings**	**179**
		Gorenstein Projective Modules	179
		Gorenstein Approximations	181
		The Stable Category	182
		Examples of Gorenstein Rings	183
	6.3	**Serre Duality**	**189**
		Serre Functors	189
		Auslander–Reiten Duality	190
		Gorenstein Projective and Injective Modules	192
		Serre Duality for the Stable Category	194
	6.4	**The Derived Nakayama Functor**	**195**
		The Nakayama Functor	195
		Serre Duality for Perfect Complexes	196
		Serre Duality for the Singularity Category	198
		Finite Global Dimension	200
	6.5	**Examples**	**203**
		Hereditary Categories	203
		A Gorenstein Algebra of Dimension One	204
	Notes		**205**

This chapter discusses the homological theory of modules over Gorenstein rings. A characteristic feature is the decomposition of the module category into two orthogonal subcategories: the Gorenstein projective (or maximal Cohen–Macaulay) modules and the modules of finite projective dimension. These subcategories are glued together via certain approximation sequences. The orthogonality refers to $\mathrm{Ext}^n(-,-)$ for $n > 0$ and this leads to the notion of a cotorsion pair. The stable category of Gorenstein projective modules admits a natural triangulated structure and is triangle equivalent to the singularity category, which is obtained from the derived category by forming the quotient modulo the subcategory of perfect complexes.

In the second part we focus on Artin algebras and study Serre functors for the stable category of Gorenstein projective modules and the category of perfect complexes.

6.1 Approximations

We establish the existence of approximations in exact categories. To formulate these results we use the concept of a cotorsion pair. Later on we will take up cotorsion pairs in the context of tilting.

Cotorsion Pairs

Let $\mathcal{A}$ be an exact category and $\mathcal{C} \subseteq \mathcal{A}$ a class of objects. The right and left *perpendicular categories* are the full subcategories

$$^{\perp}\mathcal{C} = \{X \in \mathcal{A} \mid \operatorname{Ext}^n(X, Y) = 0 \text{ for all } Y \in \mathcal{C},\, n > 0\}$$

and

$$\mathcal{C}^{\perp} = \{Y \in \mathcal{A} \mid \operatorname{Ext}^n(X, Y) = 0 \text{ for all } X \in \mathcal{C},\, n > 0\}.$$

Let $\mathcal{A}$ be an exact category and $\mathcal{X}, \mathcal{Y}$ full subcategories of $\mathcal{A}$. Then $(\mathcal{X}, \mathcal{Y})$ is a (hereditary and complete) *cotorsion pair* for $\mathcal{A}$ if

$$\mathcal{X}^{\perp} = \mathcal{Y} \qquad \text{and} \qquad \mathcal{X} = {}^{\perp}\mathcal{Y}$$

and every object $A \in \mathcal{A}$ fits into admissible exact sequences

$$0 \longrightarrow Y_A \longrightarrow X_A \longrightarrow A \longrightarrow 0 \qquad \text{and} \qquad 0 \longrightarrow A \longrightarrow Y^A \longrightarrow X^A \longrightarrow 0 \tag{6.1.1}$$

with $X_A, X^A \in \mathcal{X}$ and $Y_A, Y^A \in \mathcal{Y}$.

The sequences (6.1.1) are called *approximation sequences*, because every morphism $X \to A$ with $X \in \mathcal{X}$ factors through $X_A \to A$ and every morphism $A \to Y$ with $Y \in \mathcal{Y}$ factors through $A \to Y^A$. One may think of a cotorsion pair as a *decomposition* of the ambient category.

Remark 6.1.2. Let $(\mathcal{X}, \mathcal{Y})$ be a cotorsion pair for $\mathcal{A}$ and set $\mathcal{C} = \mathcal{X} \cap \mathcal{Y}$. We write $\mathcal{A}/\mathcal{C}$ for the additive quotient category which is obtained from $\mathcal{A}$ by annihilating all morphisms that factor through an object in $\mathcal{C}$.

(1) We have $X_A \in \mathcal{C}$ if $A \in \mathcal{Y}$, and $Y^A \in \mathcal{C}$ if $A \in \mathcal{X}$. In particular, any morphism from $\mathcal{X}$ to $\mathcal{Y}$ factors through an object in $\mathcal{C}$.

(2) The exact sequences in (6.1.1) are uniquely determined up to isomorphism in the quotient category $\mathcal{A}/\mathcal{C}$. In fact, the assignment $A \mapsto X_A$ gives a right adjoint of the inclusion $\mathcal{X}/\mathcal{C} \to \mathcal{A}/\mathcal{C}$, while the assignment $A \mapsto Y^A$ gives a left adjoint of the inclusion $\mathcal{Y}/\mathcal{C} \to \mathcal{A}/\mathcal{C}$.

A Decomposition via Resolutions

Let $\mathcal{A}$ be an exact category and $\mathcal{C} \subseteq \mathcal{A}$ a full additive subcategory. A *finite* $\mathcal{C}$-*resolution* of an object A in $\mathcal{A}$ is an admissible exact sequence (that is, an acyclic complex)

$$0 \longrightarrow X_r \longrightarrow \cdots \longrightarrow X_1 \longrightarrow X_0 \longrightarrow A \longrightarrow 0$$

such that $X_i \in \mathcal{C}$ for all i. We write $\mathrm{Res}(\mathcal{C})$ for the full subcategory of objects in $\mathcal{A}$ that admit a finite $\mathcal{C}$-resolution.

The following theorem establishes a decomposition for exact categories; it yields a procedure for constructing cotorsion pairs and is the basis for the existence of approximations.

Theorem 6.1.3. *Let $\mathcal{A}$ be an exact category and $\mathcal{C} \subseteq \mathcal{A}$ a full additive subcategory. Set $\mathcal{X} = {}^{\perp}\mathcal{C}$ and let $\mathcal{Y}$ be the closure under direct summands of $\mathrm{Res}(\mathcal{C})$. Suppose that $\mathcal{A} = \mathrm{Res}(\mathcal{X})$ and that $\mathcal{C}$ cogenerates $\mathcal{X}$, that is, every object $X \in \mathcal{X}$ fits into an admissible exact sequence $0 \to X \to Y \to Z \to 0$ with $Y \in \mathcal{C}$ and $Z \in \mathcal{X}$. Then $(\mathcal{X}, \mathcal{Y})$ is a cotorsion pair for $\mathcal{A}$.*

Proof Let $A \in \mathcal{A}$ and choose an admissible exact sequence

$$0 \longrightarrow X_r \longrightarrow \cdots \longrightarrow X_1 \longrightarrow X_0 \longrightarrow A \longrightarrow 0$$

with $X_i \in \mathcal{X}$ for all i. We need to construct the sequences (6.1.1) and use induction on r. The case $r = 0$ is clear. Now suppose $r > 0$ and let B denote the image of $X_1 \to X_0$ given by an admissible monomorphism $B \to X_0$. By the inductive hypothesis there is an exact sequence $0 \to B \to Y^B \to X^B \to 0$

with $X^B \in \mathcal{X}$ and $Y^B \in \mathcal{Y}$. We form the pushout diagram

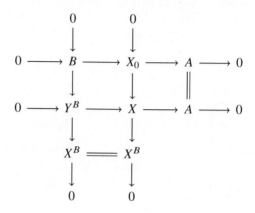

and obtain an exact sequence $0 \to Y^B \to X \to A \to 0$ with $X \in \mathcal{X}$ and $Y^B \in \mathcal{Y}$. This gives the first approximation sequence. Now take this sequence and complete the admissible monomorphism $Y^B \to X \to C$. This yields the following diagram

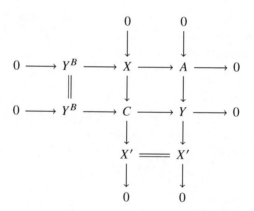

and the sequence $0 \to A \to Y \to X' \to 0$ has $X' \in \mathcal{X}$ and $Y \in \mathcal{Y}$. Thus we have constructed the second approximation sequence.

It remains to show that $\mathcal{X} = {}^\perp\mathcal{Y}$ and $\mathcal{X}^\perp = \mathcal{Y}$. The first equality is clear since

$$\mathcal{X} = {}^\perp\mathcal{C} = {}^\perp \operatorname{Res}(\mathcal{C}).$$

Also, the inclusion $\mathcal{X}^\perp \supseteq \mathcal{Y}$ is clear, since $\mathcal{X}^\perp \supseteq \mathcal{C}$. For the other inclusion, let $A \in \mathcal{X}^\perp$ and consider the sequence $0 \to A \to Y^A \to X^A \to 0$ which splits. Thus $A \in \mathcal{Y}$. □

6.2 Gorenstein Rings

Let Λ be a ring and suppose that Λ is two-sided noetherian. The ring Λ is called *Gorenstein* (or sometimes *Iwanaga-Gorenstein*) if the injective dimension of Λ is finite as a left and as a right module over itself. In that case one can show that both dimensions coincide. We denote this dimension by d and say Λ is Gorenstein *of dimension d*.

Gorenstein Projective Modules

We begin our discussion with the lemma that justifies the definition of the dimension of a Gorenstein ring. In fact, this numerical invariant admits other descriptions involving weak dimensions.

The *weak dimension* (or *flat dimension*) of a Λ-module X is by definition

$$\mathrm{w.dim}\, X = \inf\{n \geq 0 \mid \mathrm{Tor}_{n+1}^{\Lambda}(X, -) = 0\}.$$

Lemma 6.2.1. *Let Λ be a two-sided noetherian ring. If* $\mathrm{inj.dim}(\Lambda_\Lambda)$ *and* $\mathrm{inj.dim}(_\Lambda\Lambda)$ *are both finite, then they coincide.*

Proof   Given a finitely generated Λ-module X and an injective Λ^{op}-module I, we have a natural isomorphism

$$\mathrm{Hom}_\Lambda(\mathrm{Ext}_\Lambda^i(X, \Lambda), I) \cong \mathrm{Tor}_i^\Lambda(X, I) \qquad (i \geq 0).$$

Thus

$$\mathrm{inj.dim}(\Lambda_\Lambda) = \sup\{\mathrm{w.dim}(_\Lambda I) \mid {_\Lambda I} \text{ injective}\}.$$

Given a Λ-module X of finite weak dimension, one can test the vanishing of $\mathrm{Tor}_{n+1}^\Lambda(X, -)$ on injective modules, since any module embeds into an injective module. Thus

$$\sup\{\mathrm{w.dim}(X_\Lambda) \mid \mathrm{w.dim}(X_\Lambda) < \infty\} \leq \mathrm{inj.dim}(\Lambda_\Lambda)$$

and

$$\mathrm{inj.dim}(\Lambda_\Lambda) \leq \sup\{\mathrm{w.dim}(_\Lambda Y) \mid \mathrm{w.dim}(_\Lambda Y) < \infty\}.$$

This symmetry implies $\mathrm{inj.dim}(\Lambda_\Lambda) = \mathrm{inj.dim}(_\Lambda\Lambda)$.   □

Now suppose that the ring Λ is Gorenstein. A Λ-module X is called *Gorenstein projective* (or *maximal Cohen–Macaulay*) if $\mathrm{Ext}_\Lambda^i(X, \Lambda) = 0$ for all $i \neq 0$. We set

$$\mathrm{Gproj}\,\Lambda = \{X \in \mathrm{mod}\,\Lambda \mid X \text{ is Gorenstein projective}\}.$$

Fix a finitely presented Λ-module X and a projective resolution

$$\cdots \longrightarrow P_2 \xrightarrow{d_2} P_1 \xrightarrow{d_1} P_0 \longrightarrow X \longrightarrow 0$$

with all P_i finitely generated. We set $X^* = \mathrm{Hom}_\Lambda(X, \Lambda)$ and for $n \geq 1$ let $\Omega^n X = \mathrm{Im}\, d^n$ denote the nth *syzygy* of X.

Lemma 6.2.2. *Let Λ be a Gorenstein ring of dimension d and X a finitely presented Λ-module. Then the following holds.*

(1) *The module $\Omega^n X$ is Gorenstein projective for all $n \geq d$. In particular,*

$$\mathrm{proj.dim}\, X < \infty \quad \Longleftrightarrow \quad \mathrm{proj.dim}\, X \leq d.$$

(2) *If X is Gorenstein projective, then $\Omega^n X$ is Gorenstein projective for all $n \geq 1$.*

(3) *If X is Gorenstein projective, then the sequence*

$$0 \longrightarrow X^* \longrightarrow P_0^* \longrightarrow P_1^* \longrightarrow P_2^* \longrightarrow \cdots$$

is exact and X^* is Gorenstein projective. Moreover, $X \xrightarrow{\sim} X^{**}$.

(4) *The functor $\mathrm{Hom}_\Lambda(-, \Lambda)$ induces an exact duality*

$$(\mathrm{Gproj}\,\Lambda)^{\mathrm{op}} \xrightarrow{\sim} \mathrm{Gproj}(\Lambda^{\mathrm{op}}).$$

Proof We apply the dimension shift formula

$$\mathrm{Ext}_\Lambda^p(\Omega^q X, -) \cong \mathrm{Ext}_\Lambda^{p+q}(X, -) \qquad (p, q \geq 1).$$

Then (1) and (2) are clear. From this we obtain the exactness of

$$0 \longrightarrow X^* \longrightarrow P_0^* \longrightarrow P_1^* \longrightarrow \cdots$$

and therefore X^* is a syzygy of arbitrarily high order. Thus X^* is Gorenstein projective by (1). Applying $\mathrm{Hom}_\Lambda(-, \Lambda)$ to this coresolution of X^* gives a resolution of X^{**}, and we have $X \xrightarrow{\sim} X^{**}$ since $P_i \xrightarrow{\sim} P_i^{**}$ for all i. This completes (3) and the assertion in (4) is then a consequence. $\qquad\square$

We are now able to give another description of Gorenstein projective modules, which is usually taken as the definition.

Call a complex X in some additive category $\mathcal{A}$ *totally acyclic* if the complexes of abelian groups $\mathrm{Hom}_{\mathcal{A}}(A, X)$ and $\mathrm{Hom}_{\mathcal{A}}(A, X)$ are both acyclic for each object $A \in \mathcal{A}$.

Lemma 6.2.3. *Let Λ be a Gorenstein ring of dimension d. Then a finitely presented Λ-module X is Gorenstein projective if and only if*

$$X \cong \mathrm{Coker}(P_1 \to P_0)$$

for some totally acyclic complex P of finitely generated projective Λ-modules.

Proof Fix a complex of projective Λ-modules

$$P: \qquad \cdots \longrightarrow P_2 \longrightarrow P_1 \longrightarrow P_0 \longrightarrow P_{-1} \longrightarrow \cdots$$

and set

$$C_n = \mathrm{Coker}(P_{n+1} \to P_n) \qquad (n \in \mathbb{Z}).$$

We claim that P is totally acyclic when P is acyclic. This is clear since

$$H^n \mathrm{Hom}_\Lambda(P, \Lambda) \cong \mathrm{Ext}_\Lambda^n(C_0, \Lambda) \qquad (n > 0)$$

implies

$$H^n \mathrm{Hom}_\Lambda(P, \Lambda) \cong \mathrm{Ext}_\Lambda^{d+1}(C_{n-d-1}, \Lambda) = 0 \qquad (n \in \mathbb{Z}).$$

If X is Gorenstein projective, then we choose a projective resolution P of X and a projective resolution Q of X^*. Applying Lemma 6.2.2, we have $X^{**} \cong X$ and can splice together P and Q^* giving an acyclic complex

$$\cdots \longrightarrow P_2 \longrightarrow P_1 \longrightarrow P_0 \longrightarrow Q_0^* \longrightarrow Q_1^* \longrightarrow Q_2^* \longrightarrow \cdots$$

with cokernel of $P_1 \to P_0$ isomorphic to X. Conversely, if $X = \mathrm{Coker}(P_1 \to P_0)$ for some totally acyclic complex P in proj Λ, then

$$\mathrm{Ext}_\Lambda^n(X, \Lambda) = H^n \mathrm{Hom}_\Lambda(P, \Lambda) = 0 \qquad (n > 0)$$

and it follows that X is Gorenstein projective. $\square$

Gorenstein Approximations

For Gorenstein rings there is a good approximation theory. The category of finitely presented modules decomposes into two orthogonal subcategories which are glued together via approximation sequences.

Theorem 6.2.4. *Let Λ be a Gorenstein ring. Set $\mathfrak{X} = \mathrm{Gproj}\,\Lambda$ and write $\mathcal{Y}$ for the category of finitely presented Λ-modules of finite projective dimension. Then $(\mathfrak{X}, \mathcal{Y})$ is a cotorsion pair for $\mathrm{mod}\,\Lambda$ with $\mathfrak{X} \cap \mathcal{Y} = \mathrm{proj}\,\Lambda$.*

Proof We apply Theorem 6.1.3. Thus we set $\mathcal{A} = \mathrm{mod}\,\Lambda$ and $\mathcal{C} = \mathrm{proj}\,\Lambda$. This gives $\mathfrak{X} = {}^\perp\mathcal{C}$ and $\mathcal{Y} = \mathrm{Res}(\mathcal{C})$. The assumption on Λ implies that $\mathcal{A} = \mathrm{Res}(\mathfrak{X})$ and that $\mathcal{C}$ cogenerates $\mathfrak{X}$; this follows from Lemma 6.2.2. More precisely, if Λ is Gorenstein of dimension d, then any Λ-module X admits a resolution

$$0 \longrightarrow \Omega^d X \longrightarrow P_{d-1} \longrightarrow \cdots \longrightarrow P_1 \longrightarrow P_0 \longrightarrow X \longrightarrow 0$$

such that $P_0, \ldots, P_{d-1}$ are projective Λ-modules. Thus $X \in \mathrm{Res}(\mathfrak{X})$, since

$\Omega^d X$ is Gorenstein projective. If X is Gorenstein projective, choose an exact sequence $0 \to Y \to P \to X^* \to 0$ in mod Λ^{op} such that P is projective. This yields an exact sequence $0 \to X \to P^* \to Y^* \to 0$ in Gproj Λ, since $X \xrightarrow{\sim} X^{**}$.

It remains to show that $\mathcal{X} \cap \mathcal{Y} = \mathrm{proj}\,\Lambda$. One inclusion is obvious. Thus consider a module X that is Gorenstein projective and of finite projective dimension. Then an induction on the projective dimension of X shows that X is projective, keeping in mind that $\Omega^n X$ is Gorenstein projective for all $n \geq 1$. □

The Stable Category

For a noetherian ring Λ we consider the derived category $\mathbf{D}^b(\mathrm{mod}\,\Lambda)$ and obtain the *singularity category* (or *stabilised derived category*) by forming the triangulated quotient

$$\mathbf{D}_{\mathrm{sg}}(\Lambda) = \frac{\mathbf{D}^b(\mathrm{mod}\,\Lambda)}{\mathbf{D}^b(\mathrm{proj}\,\Lambda)}.$$

Also, we consider the triangulated category $\mathbf{K}_{\mathrm{ac}}(\mathrm{proj}\,\Lambda)$ of complexes of finitely generated projective Λ-modules that are acyclic.

An exact category $\mathcal{A}$ is a *Frobenius category* if $\mathcal{A}$ has enough projective objects and enough injective objects, and if projective and injective objects in $\mathcal{A}$ coincide. The *stable category* of $\mathcal{A}$ is obtained by annihilating all morphisms that factor through a projective object. The exact structure of $\mathcal{A}$ induces a triangulated structure for the stable category.

Theorem 6.2.5. *Let Λ be Gorenstein. Then the Gorenstein projective Λ-modules form a Frobenius category. Writing* $\underline{\mathrm{Gproj}\,\Lambda}$ *for its stable category, we have a triangle equivalence*

$$Z^0 \colon \mathbf{K}_{\mathrm{ac}}(\mathrm{proj}\,\Lambda) \xrightarrow{\sim} \underline{\mathrm{Gproj}\,\Lambda}.$$

Also, the composite

$$F \colon \mathrm{Gproj}\,\Lambda \rightarrowtail \mathbf{D}^b(\mathrm{mod}\,\Lambda) \twoheadrightarrow \mathbf{D}_{\mathrm{sg}}(\Lambda)$$

induces a triangle equivalence

$$\underline{\mathrm{Gproj}\,\Lambda} \xrightarrow{\sim} \mathbf{D}_{\mathrm{sg}}(\Lambda).$$

Proof It follows from Lemma 6.2.2 that Gproj Λ is a Frobenius category. The projective Λ-modules form the subcategory of objects that are projective and injective. Thus the first triangle equivalence follows from Proposition 4.4.18.

The functor F is exact: it takes an exact sequence $0 \to X \to Y \to Z \to 0$ in Gproj Λ to an exact triangle $F(X) \to F(Y) \to F(Z) \to F(X)[1]$. Also,

F annihilates all projective Λ-modules and yields therefore an exact functor $\bar{F}\colon \underline{\mathrm{Gproj}}\,\Lambda \to \mathbf{D}_{\mathrm{sg}}(\Lambda)$. The suspension in $\underline{\mathrm{Gproj}}\,\Lambda$ takes X to $\Omega^{-1}X$, and

$$F(\Omega^{-1}X) \cong F(X)[1].$$

We construct a quasi-inverse for $\bar{F}$ as follows.

Consider the category of complexes $\mathbf{K}(\mathrm{proj}\,\Lambda)$ of finitely generated projective Λ-modules up to homotopy. We identify the subcategories

$$\mathbf{K}^b(\mathrm{proj}\,\Lambda) \xrightarrow{\sim} \mathbf{D}^b(\mathrm{proj}\,\Lambda) \quad \text{and} \quad \mathbf{K}^{-,b}(\mathrm{proj}\,\Lambda) \xrightarrow{\sim} \mathbf{D}^b(\mathrm{mod}\,\Lambda).$$

For a complex X and $n \in \mathbb{Z}$ we use the following truncation:

$$
\begin{array}{ccccccccccc}
X & & \cdots \longrightarrow & X^{n-1} & \longrightarrow & X^n & \longrightarrow & X^{n+1} & \longrightarrow & X^{n+2} & \longrightarrow \cdots\\[4pt]
\Big\downarrow & & & \Big\downarrow{\scriptstyle\mathrm{id}} & & \Big\downarrow{\scriptstyle\mathrm{id}} & & \Big\downarrow & & \Big\downarrow &\\[4pt]
\sigma_{\le n}X & & \cdots \longrightarrow & X^{n-1} & \longrightarrow & X^n & \longrightarrow & 0 & \longrightarrow & 0 & \longrightarrow \cdots
\end{array}
$$

Now fix a complex X in $\mathbf{K}^{-,b}(\mathrm{proj}\,\Lambda)$ and choose $n \in \mathbb{Z}$ such that $H^i(X) = 0$ for all $i \le n + d$. Then $\mathrm{Coker}(X^{n-1} \to X^n)$ is Gorenstein projective, by Lemma 6.2.2. Note that the cone of $X \to \sigma_{\le n}X$ belongs to $\mathbf{K}^b(\mathrm{proj}\,\Lambda)$. Thus $X \cong \sigma_{\le n}X$ in $\mathbf{D}_{\mathrm{sg}}(\Lambda)$ and the assignment

$$X \longmapsto \Omega^n \,\mathrm{Coker}(X^{n-1} \to X^n)$$

yields a functor $G\colon \mathbf{D}_{\mathrm{sg}}(\Lambda) \to \underline{\mathrm{Gproj}}\,\Lambda$ which does not depend on n. It is not difficult to check that $G \circ \bar{F} \cong \mathrm{id}$ and $\bar{F} \circ G \cong \mathrm{id}$. $\qquad\square$

We observe that the stable category $\underline{\mathrm{Gproj}}\,\Lambda$ and the equivalent singularity category $\mathbf{D}_{\mathrm{sg}}(\Lambda)$ are idempotent complete when the Gorenstein projective modules form a Krull–Schmidt category. For instance, this holds when Λ is an Artin algebra. For an example when $\underline{\mathrm{Gproj}}\,\Lambda$ is not idempotent complete, see Lemma 6.2.12.

Examples of Gorenstein Rings

Gorenstein rings are ubiquitous and we provide several examples.

Commutative Rings

Let Λ be a commutative noetherian ring. In that context one uses a local property and calls Λ *Gorenstein* if for each prime ideal $\mathfrak{p}$ the localisation $\Lambda_{\mathfrak{p}}$ has finite injective dimension as a module over $\Lambda_{\mathfrak{p}}$. Clearly, this definition coincides with our original definition when Λ is local. Important examples are *hypersurface rings* and more generally *complete intersection rings*.

Non-commutative Rings

Let Λ be a (not necessarily commutative) two-sided noetherian ring. We begin with two extreme cases, where Gproj Λ is either all or nothing.

The ring Λ is Gorenstein of dimension zero if and only if it is a *quasi-Frobenius ring*. Then Λ is a two-sided artinian ring and projective and injective Λ-modules coincide. Clearly, in that case all Λ-modules are Gorenstein projective.

If Λ has finite global dimension, say d, then Λ is Gorenstein of dimension d. In that case only the projective Λ-modules are Gorenstein projective.

Let G be a finite group. Then the *integral group algebra* $\mathbb{Z}G$ is Gorenstein of dimension one. A $\mathbb{Z}G$-module is Gorenstein projective if and only if the underlying $\mathbb{Z}$-module is Gorenstein projective if and only if the underlying $\mathbb{Z}$-module is projective.

Further interesting examples of Gorenstein rings arise from the study of graded rings.

Artin Algebras

We fix a field k and discuss two specific constructions of Artin k-algebras that are Gorenstein.

For a Gorenstein ring Λ let Gor.dim Λ denote its dimension. Note that Gor.dim Λ equals the global dimension of Λ when both dimensions are finite.

Proposition 6.2.6. *Let Γ and Λ be finite dimensional k-algebras. If Γ and Λ are Gorenstein, then the tensor product $\Gamma \otimes_k \Lambda$ is Gorenstein and*

$$\text{Gor.dim } \Gamma \otimes_k \Lambda = \text{Gor.dim } \Gamma + \text{Gor.dim } \Lambda.$$

We need some preparation. Given chain complexes of k-modules X and Y, we consider the tensor product $X \otimes_k Y$ given by $(X \otimes_k Y)_n = \bigoplus_{i+j=n} X_i \otimes_k Y_j$ with differential $\partial(x \otimes y) = \partial x \otimes y + (-1)^{|x|} x \otimes \partial y$, where x and y are any homogeneous elements in X and Y respectively, and $|x|$ denotes the degree of x.

Lemma 6.2.7. *Let Γ and Λ be finite dimensional k-algebras. If P and Q are minimal projective resolutions of modules $_\Gamma M$ and N_Λ, then the tensor product $P \otimes_k Q$ is a minimal projective resolution of the $\Gamma \otimes_k \Lambda$-module $M \otimes_k N$.*

Proof The Künneth formula implies that $P \otimes_k Q$ is a projective resolution of $M \otimes_k N$; see [46, Theorem VI.3.1]. Next observe that $J(\Gamma) \otimes_k \Lambda + \Gamma \otimes_k J(\Lambda)$ is a nilpotent two-sided ideal in $\Gamma \otimes_k \Lambda$, and therefore it is contained in $J(\Gamma \otimes_k \Lambda)$.

For any modules $_\Gamma X$ and Y_Λ, we have $\mathrm{rad}(X) = J(\Gamma)X$ and $\mathrm{rad}(Y) = YJ(\Lambda)$. Thus

$$\mathrm{rad}(X) \otimes_k Y + X \otimes_k \mathrm{rad}(Y) \subseteq \mathrm{rad}(X \otimes_k Y)$$

as $\Gamma \otimes_k \Lambda$-module. The lemma now follows from the fact that a projective resolution is minimal if and only if the image of each differential lands in the radical of the next module (Lemma 2.1.21). □

Proof of Proposition 6.2.6 Set $c = \mathrm{Gor.dim}\,\Gamma$ and $d = \mathrm{Gor.dim}\,\Lambda$. We have

$$\mathrm{proj.dim}(_\Lambda D(\Lambda)) = \mathrm{inj.dim}(\Lambda_\Lambda) \quad \text{and} \quad \mathrm{proj.dim}(D(\Gamma)_\Gamma) = \mathrm{inj.dim}(_\Gamma\Gamma).$$

Thus the assumptions yield minimal projective resolutions

$$\cdots \longrightarrow 0 \longrightarrow P_d \longrightarrow \cdots \longrightarrow P_1 \longrightarrow P_0 \longrightarrow {}_\Lambda D(\Lambda) \longrightarrow 0$$

and

$$\cdots \longrightarrow 0 \longrightarrow Q_c \longrightarrow \cdots \longrightarrow Q_1 \longrightarrow Q_0 \longrightarrow D(\Gamma)_\Gamma \longrightarrow 0.$$

The tensor product $P \otimes_k Q$ yields a minimal projective resolution of $D(\Gamma \otimes_k \Lambda) \cong D(\Lambda) \otimes_k D(\Gamma)$ over $\Lambda \otimes_k \Gamma$ by the above lemma. This gives

$$\mathrm{inj.dim}(_\Gamma(\Gamma \otimes_k \Lambda)_\Lambda) = \mathrm{proj.dim}(_\Lambda D(\Gamma \otimes_k \Lambda)_\Gamma) = c + d.$$

The same computation for $(\Gamma \otimes_k \Lambda)^{\mathrm{op}}$ then shows that $\Gamma \otimes_k \Lambda$ has dimension $c + d$. □

Of particular interest is the case when Γ is the algebra $k[\varepsilon]$ of dual numbers. We have $k[\varepsilon] \otimes_k \Lambda \cong \Lambda[\varepsilon]$, and the $\Lambda[\varepsilon]$-modules identify with *differential modules* over Λ, that is, pairs (X, d) consisting of a Λ-module X with an endomorphism $d\colon X \to X$ satisfying $d^2 = 0$.

Gentle Algebras

Let k be a field. A k-algebra is called *gentle* if it is Morita equivalent to an algebra of the form kQ/I where $Q = (Q_0, Q_1, s, t)$ is a finite quiver and I is an ideal generated by paths of length two, subject to the following conditions.

(Ge1) For each $x \in Q_0$, there are at most two arrows starting at x.

(Ge2) For each $x \in Q_0$, there are at most two arrows ending at x.

(Ge3) For each $\alpha \in Q_1$, there is at most one arrow β such that $s(\beta) = t(\alpha)$ and $\beta\alpha \in I$, and there is at most one arrow β' such that $s(\beta') = t(\alpha)$ and $\beta'\alpha \notin I$.

(Ge4) For each $\beta \in Q_1$, there is at most one arrow α such that $t(\alpha) = s(\beta)$ and $\beta\alpha \in I$, and there is at most one arrow α' such that $t(\alpha') = s(\beta)$ and $\beta\alpha' \notin I$.

The following diagram shows the local shape of (Q, I) when kQ/I is gentle:

Now fix a pair (Q, I) such that the algebra $\Lambda = kQ/I$ satisfies the above conditions (Ge1)–(Ge4). A non-trivial path α in Q is a *primitive cycle* if $s(\alpha) = t(\alpha)$, $\alpha^r \notin I$ for all $r > 0$, and α is not a power of a cycle of smaller length. For $x \in Q_0$ let $c_x \in \Lambda$ denote the sum of all primitive cycles α with $t(\alpha) = x$. Note that there are at most two primitive cycles ending at x. If $\alpha \neq \beta$ are such cycles, then $\alpha\beta = 0 = \beta\alpha$ in Λ. Moreover, for any arrow $\alpha : x \to y$ we have $c_y\alpha = \alpha c_x$.

Let $k[c]$ denote the polynomial ring in one indeterminate c. Then the assignment $c \mapsto \sum_{x \in Q_0} c_x$ yields a $k[c]$-algebra structure for Λ.

Lemma 6.2.8. *The gentle algebra $\Lambda = kQ/I$ is a noetherian $k[c]$-algebra.*

Proof For each pair $x, y \in Q_0$ we consider the non-trivial paths in Q that generate $e_x\Lambda e_y$, where e_x and e_y denote the idempotents corresponding to x and y respectively. The conditions for a gentle algebra imply that all are of the form $\alpha^r\beta$ for some non-trivial paths α, β and some $r \geq 0$. If there are infinitely many such paths, then α is a primitive cycle, so $\alpha^r\beta = c_x^r\beta$ in Λ. Thus the $k[c]$-module $e_x\Lambda e_y$ is finitely generated. $\square$

Proposition 6.2.9. *A gentle algebra is Gorenstein.*

We fix a gentle algebra $\Lambda = kQ/I$. A possibly infinite path $\alpha_1\alpha_2\alpha_3 \cdots$ in Q is called *differential* if $\alpha_i\alpha_{i+1} \in I$ for all $i \geq 1$. Such a path is *maximal* if it has finite length, say n, and cannot be extended to a differential path of length $n + 1$. Note that there are only finitely many maximal differential paths in Q.

The proof of the proposition uses the following reduction argument.

Lemma 6.2.10. *Let Λ be a noetherian $k[c]$-algebra and Y a finitely generated Λ-module. Then $\mathrm{inj.dim}\, Y \leq d$ if $\mathrm{Ext}_\Lambda^n(S, Y) = 0$ for every simple Λ-module S and $n > d$.*

Proof Baer's criterion implies that it suffices to show $\mathrm{Ext}_\Lambda^n(X, Y) = 0$ for every finitely generated Λ-module X and $n > d$. Let X be finitely generated and set $t(X) = \{x \in X \mid xc^p = 0 \text{ for } p \gg 0\}$. Then $t(X)$ has finite length, and therefore it suffices to show that $\mathrm{Ext}_\Lambda^n(\bar{X}, Y) = 0$ for $\bar{X} = X/t(X)$ and every $n > d$. The exact sequence $0 \to \bar{X} \xrightarrow{c} \bar{X} \to \bar{X}/\bar{X}c \to 0$ induces a bijection

$\text{Ext}^n_\Lambda(\bar{X}, Y) \xrightarrow{c} \text{Ext}^n_\Lambda(\bar{X}, Y)$ for $n > d$. It follows that $\text{Ext}^n_\Lambda(\bar{X}, Y) = 0$ since Y is finitely generated.   □

Proof of Proposition 6.2.9   We wish to apply Lemma 6.2.10 and consider a simple Λ-module S. Then $\text{Hom}_\Lambda(P_x, S) \neq 0$ for some $x \in Q_0$, where P_x denotes the indecomposable projective Λ-module corresponding to $x \in Q_0$, with k-basis given by all paths in Q ending at x and not contained in I. There are two possible cases. Either $S\alpha = 0$ for every non-trivial path α, and then we write $S = S_x$. Otherwise, there is a primitive cycle α ending at x and an irreducible polynomial $f \in k[t, t^{-1}]$ such that S fits into an exact sequence

$$ 0 \longrightarrow P_x \xrightarrow{f(\alpha)} P_x \longrightarrow S \longrightarrow 0; $$

cf. [60, Theorem 1.2]. In this case we have proj.dim $S = 1$.

For $S = S_x$ we show that $\text{Ext}^n_\Lambda(S_x, \Lambda) \neq 0$ and $n > 0$ imply the existence of a maximal differential path of length n ending at x. There are at most two arrows ending at x

$$ u_1 \xrightarrow{\alpha_1} x \xleftarrow{\beta_1} v_1 $$

and then a projective resolution of S_x has the following form

$$ \cdots \longrightarrow P_{u_2} \oplus P_{v_2} \longrightarrow P_{u_1} \oplus P_{v_1} \longrightarrow P_x \longrightarrow S_x \longrightarrow 0. $$

The differentials are given by differential paths

$$ \cdots \longrightarrow u_3 \xrightarrow{\alpha_3} u_2 \xrightarrow{\alpha_2} u_1 \xrightarrow{\alpha_1} x \quad \text{and} \quad \cdots \longrightarrow v_3 \xrightarrow{\beta_3} v_2 \xrightarrow{\beta_2} v_1 \xrightarrow{\beta_1} x $$

which may be infinite. Let $y \in Q_0$ and $\text{Ext}^n_\Lambda(S_x, P_y) \neq 0$ with $n > 0$. A non-zero cocycle is given by a morphism $P_{u_n} \oplus P_{v_n} \to P_y$, and we may assume that the first component is non-zero. If $P_{u_n} \to P_y$ is invertible, then $\alpha_1 \cdots \alpha_n$ is maximal differential, because a composite $P_{u_{n+1}} \to P_{u_n} \to P_y$ would be zero. A radical morphism yields a path $u_n \xrightarrow{\gamma_1} \cdots \xrightarrow{\gamma_r} y$ with $\gamma_1 \neq \alpha_n$. This implies that $\alpha_1 \cdots \alpha_n$ is maximal differential, since $\alpha_n \alpha_{n+1} \in I$ and $\gamma_1 \alpha_{n+1} \in I$ is impossible. Here we use that the ideal I of kQ is generated by paths of length two, because a cocycle means that the composite $P_{u_{n+1}} \to P_{u_n} \to P_y$ is zero, and therefore necessarily $\gamma_1 \alpha_{n+1} \in I$ since $\gamma_r \cdots \gamma_1 \notin I$.

Now we apply Lemma 6.2.10 and conclude that inj.dim Λ equals the maximal length of a maximal differential path in Q. An exception is the case that this equals zero and there exists a primitive cycle; then inj.dim $\Lambda = 1$.   □

Corollary 6.2.11. *Let Λ be a gentle algebra. Then its dimension as a Gorenstein algebra equals the maximal length of a maximal differential path in its quiver.*

An exception is the case that this equals zero and there is a primitive cycle; then the dimension equals one.   □

A Complete Intersection

Let us consider the complete intersection ring

$$\Lambda = R_{(x,y)} \qquad \text{where} \qquad R = \mathbb{C}[x, y]/(x^2 - y^2(y + 1)).$$

The ring Λ is Gorenstein. Moreover, Λ is an integral domain with non-local integral closure. We use these facts to exhibit some phenomena of the stable category $\underline{\mathrm{mod}}\,\Lambda$.

Let Γ denote the integral closure of Λ. Since Γ is in the field of fractions of Λ, we know that Γ contains no proper direct summands. On the other hand the completion $\widehat{\Gamma}$ is the direct sum of two local rings $\widehat{\Gamma}_1$ and $\widehat{\Gamma}_2$, one for each maximal ideal. Now observe that for each $X \in \mathrm{mod}\,\Lambda$ the module $\mathrm{Ext}^1_\Lambda(\Gamma, X)$ is of finite length. This yields a decomposition of the functor $\mathrm{Ext}^1_\Lambda(\Gamma, -)$, since

$$\mathrm{Ext}^1_\Lambda(\Gamma, X) \cong \mathrm{Ext}^1_{\widehat{\Lambda}}(\widehat{\Gamma}, \widehat{X}) = \mathrm{Ext}^1_{\widehat{\Lambda}}(\widehat{\Gamma}_1, \widehat{X}) \oplus \mathrm{Ext}^1_{\widehat{\Lambda}}(\widehat{\Gamma}_2, \widehat{X}).$$

Next observe that $X \mapsto \mathrm{Ext}^1_\Lambda(X, -)$ provides a fully faithful functor

$$\underline{\mathrm{mod}}\,\Lambda \longrightarrow \mathrm{Fp}(\mathrm{mod}\,\Lambda, \mathrm{Ab})$$

into the category of finitely presented functors $\mathrm{mod}\,\Lambda \to \mathrm{Ab}$ by Lemma 2.1.26.

Lemma 6.2.12. *There is a proper idempotent in $\underline{\mathrm{End}}_\Lambda(\Gamma)$ reflecting the decomposition of $\mathrm{Ext}^1_\Lambda(\Gamma, -)$. This idempotent has no kernel in $\underline{\mathrm{mod}}\,\Lambda$. In particular, there is a direct summand of $\mathrm{Ext}^1_\Lambda(\Gamma, -)$ which is not isomorphic to $\mathrm{Ext}^1_\Lambda(X, -)$ for some Λ-module X.*

Proof   Suppose there is a decomposition $\Gamma = U \oplus V$ in $\underline{\mathrm{mod}}\,\Lambda$, and therefore $\Gamma \oplus P \cong U \oplus V \oplus Q$ for some projective Λ-modules P, Q by Lemma 2.1.27. The ring Λ is local, so all projective modules are free. Thus we may remove the indecomposable summands corresponding to P from the right-hand side, since these summands have local endomorphism rings. This yields a decomposition $\Gamma \cong U' \oplus V'$ in $\mathrm{mod}\,\Lambda$. Thus $U' = 0$ or $V' = 0$, and therefore $U = 0$ or $V = 0$ in $\underline{\mathrm{mod}}\,\Lambda$. The assertion about $\mathrm{Ext}^1_\Lambda(X, -)$ now follows, since any decomposition

$$\mathrm{Ext}^1_\Lambda(\Gamma, -) = \mathrm{Ext}^1_\Lambda(X, -) \oplus \mathrm{Ext}^1_\Lambda(Y, -)$$

is equivalent to a decomposition $\Gamma = X \oplus Y$ in $\underline{\mathrm{mod}}\,\Lambda$.   □

Finally, observe that the Λ-module Γ is Gorenstein projective. It follows that the stable category $\underline{\mathrm{Gproj}}\,\Lambda$ and the equivalent singularity category $\mathbf{D}_{\mathrm{sg}}(\Lambda)$ are not idempotent complete.

6.3 Serre Duality

A special feature of Gorenstein algebras is Serre duality. In fact, there are several results and we formulate them in the context of Artin algebras.

Let k be a commutative artinian ring and Λ an Artin k-algebra. We write $D = \mathrm{Hom}_k(-, E)$ for the Matlis duality over k, which is given by a minimal injective cogenerator E. Thus $X \xrightarrow{\sim} D^2 X$ for every k-module X of finite length.

The derived category $\mathbf{D}^b(\mathrm{mod}\,\Lambda)$ of a Gorenstein algebra Λ 'decomposes' into the category of *perfect complexes*

$$\mathbf{D}^{\mathrm{perf}}(\Lambda) = \mathbf{D}^b(\mathrm{proj}\,\Lambda)$$

and the singularity category

$$\underline{\mathrm{Gproj}}\,\Lambda \xrightarrow{\sim} \mathbf{D}_{\mathrm{sg}}(\Lambda).$$

This reflects the decomposition of the module category $\mathrm{mod}\,\Lambda$ from Theorem 6.2.4. We establish Serre duality for both categories. In fact, we derive Serre duality for the stable category of Gorenstein projective modules from Auslander–Reiten duality. The following section then discusses Serre duality for perfect complexes in terms of the derived Nakayama functor.

Serre Functors

Let $\mathcal{C}$ be a k-linear and Hom-finite additive category. Thus $\mathrm{Hom}(X, Y)$ is a k-module of finite length for all $X, Y \in \mathcal{C}$. A *Serre functor* is an equivalence $F: \mathcal{C} \to \mathcal{C}$ together with natural isomorphisms

$$\eta_{X,Y}: \mathrm{Hom}(X, Y) \xrightarrow{\sim} D\,\mathrm{Hom}(Y, FX)$$

for all objects X, Y in $\mathcal{C}$. Note that a Serre functor is determined by the natural isomorphisms $\eta_{X,Y}$ since FX represents the functor $D\,\mathrm{Hom}(X, -)$.

A Serre functor yields for each object X a morphism

$$\eta_X := \eta_{X,X}(\mathrm{id}_X): \mathrm{Hom}(X, FX) \longrightarrow E.$$

Lemma 6.3.1. *The morphisms* $(\eta_X)_{X \in \mathcal{C}}$ *have the following properties for all objects* X, Y *in* $\mathcal{C}$.

(1) *For all* $\phi: X \to Y$ *and* $\psi: Y \to FX$ *we have*

$$\eta_X(\psi\phi) = \eta_{X,Y}(\phi)(\psi) = \eta_Y(F(\phi)\psi).$$

(2) *The map* $F_{X,Y}$ *equals the composite*

$$\mathrm{Hom}(X, Y) \xrightarrow{\eta_{X,Y}} D\,\mathrm{Hom}(Y, FX) \xrightarrow{(D\eta_{Y,FX})^{-1}} \mathrm{Hom}(FX, FY).$$

(3) *The composite*

$$\operatorname{Hom}(Y, FX) \times \operatorname{Hom}(X, Y) \longrightarrow \operatorname{Hom}(X, FX) \xrightarrow{\eta_X} E$$

is a non-degenerate pairing.

Moreover, any bijection $X \mapsto FX$ on the isomorphism classes of objects in $\mathcal{C}$ together with a choice of k-linear maps $(\eta_X)_{X \in \mathcal{C}}$ satisfying (3) yield a Serre functor.

Proof   The calculations are straightforward. (1) uses the naturality of the $\eta_{X,Y}$. (2) follows from the identity in (1). The non-degeneracy in (3) follows from the fact that the $\eta_{X,Y}$ are isomorphisms. For the last assertion, observe that the $\eta_{X,Y}$ and $F_{X,Y}$ are obtained via the identities in (1) and (2). Also, (3) implies that the $\eta_{X,Y}$ are isomorphisms. In particular, F is fully faithful and therefore an equivalence. $\square$

The following remark collects some useful properties of Serre functors.

Remark 6.3.2.   Let $F \colon \mathcal{C} \to \mathcal{C}$ be a Serre functor with maps $(\eta_{X,Y})_{X,Y \in \mathcal{C}}$.

(1) For any Serre functor $F' \colon \mathcal{C} \to \mathcal{C}$, there is a canonical isomorphism $F \xrightarrow{\sim} F'$ which is compatible with the $\eta_{X,Y}$. This follows from Yoneda's lemma since for any object X we have the isomorphism

$$\operatorname{Hom}(-, FX) \xrightarrow{D\eta_{X,-}} D \operatorname{Hom}(X, -) \xrightarrow{(D\eta'_{X,-})^{-1}} \operatorname{Hom}(-, F'X).$$

(2) Let $\Sigma \colon \mathcal{C} \xrightarrow{\sim} \mathcal{C}$ be an autoequivalence. Then $\Sigma^- F\Sigma$ is a Serre functor, so $\Sigma^- F\Sigma \cong F$, and therefore $F\Sigma \cong \Sigma F$.

(3) When $\mathcal{C}$ is a triangulated category with suspension Σ, then F is exact. Thus there is a canonical isomorphism $F\Sigma \cong \Sigma F$ and F maps exact triangles to exact triangles.

(4) Given an abelian category $\mathcal{A}$ and a Serre functor $F \colon \mathbf{D}^b(\mathcal{A}) \to \mathbf{D}^b(\mathcal{A})$, we may choose $(\eta_X)_{X \in \mathcal{C}}$ such that $\eta_{\Sigma X}(\Sigma \phi) = \eta_X(\phi)$ for all $\phi \colon X \to FX$.

Auslander–Reiten Duality

Given Λ-modules X and Y, we set

$$\underline{\operatorname{Hom}}_\Lambda(X, Y) = \operatorname{Hom}_\Lambda(X, Y)/\{\phi \mid \phi \text{ factors through a projective module}\}$$

and

$$\overline{\operatorname{Hom}}_\Lambda(X, Y) = \operatorname{Hom}_\Lambda(X, Y)/\{\phi \mid \phi \text{ factors through an injective module}\}.$$

In this way we obtain the *projectively stable category* $\underline{\mathrm{mod}}\,\Lambda$ as additive quotient $(\mathrm{mod}\,\Lambda)/(\mathrm{proj}\,\Lambda)$. Analogously, the *injectively stable category* $\overline{\mathrm{mod}}\,\Lambda$ is defined.

For a finitely presented Λ-module X choose a projective presentation

$$P_1 \longrightarrow P_0 \longrightarrow X \longrightarrow 0$$

such that the P_i are finitely generated. The *transpose* $\mathrm{Tr}\,X$ is defined by the exactness of the following sequence of Λ^{op}-modules

$$P_0^* \longrightarrow P_1^* \longrightarrow \mathrm{Tr}\,X \longrightarrow 0$$

where $P^* = \mathrm{Hom}_\Lambda(P, \Lambda)$.

Lemma 6.3.3. *The transpose induces mutually inverse equivalences*

$$(\underline{\mathrm{mod}}\,\Lambda)^{\mathrm{op}} \xrightarrow{\ \sim\ } \underline{\mathrm{mod}}(\Lambda^{\mathrm{op}}) \qquad and \qquad \underline{\mathrm{mod}}(\Lambda^{\mathrm{op}}) \xrightarrow{\ \sim\ } (\underline{\mathrm{mod}}\,\Lambda)^{\mathrm{op}}.$$

Proof The transpose depends on the choice of a projective presentation and is therefore unique up to morphisms that factor through a projective module. For a finitely generated projective Λ-module P, we have a natural isomorphism $P \xrightarrow{\sim} P^{**}$. Thus $\mathrm{Tr}\,\mathrm{Tr}\,X \cong X$ in $\underline{\mathrm{mod}}\,\Lambda$. $\qquad\qquad\square$

From this it follows that the functors $D\,\mathrm{Tr}$ and $\mathrm{Tr}\,D$ induce mutually inverse equivalences:

$$\underline{\mathrm{mod}}\,\Lambda \underset{\mathrm{Tr}}{\overset{\mathrm{Tr}}{\rightleftarrows}} \underline{\mathrm{mod}}(\Lambda^{\mathrm{op}}) \underset{D}{\overset{D}{\rightleftarrows}} \overline{\mathrm{mod}}\,\Lambda.$$

Lemma 6.3.4. *We have a natural isomorphism*

$$\underline{\mathrm{Hom}}_\Lambda(X, -) \cong \mathrm{Tor}_1^\Lambda(-, \mathrm{Tr}\,X).$$

Proof A projective presentation $P_1 \to P_0 \to X \to 0$ induces for any Λ-module A the following commutative diagram with exact rows.

$$
\begin{array}{ccccccc}
A \otimes_\Lambda P_0^* & \longrightarrow & A \otimes_\Lambda P_1^* & \longrightarrow & A \otimes_\Lambda \mathrm{Tr}\,X & \to & 0 \\
\downarrow\wr & & \downarrow\wr & & & & \\
0 \to \mathrm{Hom}_\Lambda(X, A) \to & \mathrm{Hom}_\Lambda(P_0, A) & \to & \mathrm{Hom}_\Lambda(P_1, A) & & &
\end{array}
$$

Therefore an exact sequence

$$0 \longrightarrow A \xrightarrow{\ \phi\ } B \xrightarrow{\ \psi\ } C \longrightarrow 0$$

induces the following commutative diagram with exact rows and colums.

$$
\begin{array}{ccccccc}
0 & & 0 & & 0 & & \\
\downarrow & & \downarrow & & \downarrow & & \\
0 \to \operatorname{Hom}_\Lambda(X,A) \to & \operatorname{Hom}_\Lambda(P_0,A) \to & \operatorname{Hom}_\Lambda(P_1,A) \to & A \otimes_\Lambda \operatorname{Tr} X \to 0 \\
\downarrow & & \downarrow & & \downarrow & & \downarrow \\
0 \to \operatorname{Hom}_\Lambda(X,B) \to & \operatorname{Hom}_\Lambda(P_0,B) \to & \operatorname{Hom}_\Lambda(P_1,B) \to & B \otimes_\Lambda \operatorname{Tr} X \to 0 \\
\downarrow & & \downarrow & & \downarrow & & \downarrow \\
0 \to \operatorname{Hom}_\Lambda(X,C) \to & \operatorname{Hom}_\Lambda(P_0,C) \to & \operatorname{Hom}_\Lambda(P_1,C) \to & C \otimes_\Lambda \operatorname{Tr} X \to 0 \\
& & \downarrow & & \downarrow & & \downarrow \\
& & 0 & & 0 & & 0
\end{array}
$$

Now suppose that B is projective. Using the snake lemma we get

$$
\begin{aligned}
\underline{\operatorname{Hom}}_\Lambda(X,C) &\cong \operatorname{Coker}\operatorname{Hom}_\Lambda(X,\psi) \\
&\cong \operatorname{Ker}(\phi \otimes_\Lambda \operatorname{Tr} X) \\
&\cong \operatorname{Tor}_1^\Lambda(C,\operatorname{Tr} X). \qquad\qquad \square
\end{aligned}
$$

We obtain the following *Auslander–Reiten formulas*.

Proposition 6.3.5 (Auslander–Reiten). *For all $X \in \operatorname{mod}\Lambda$ there are natural isomorphisms*

$$
D\,\underline{\operatorname{Hom}}_\Lambda(X,-) \cong \operatorname{Ext}_\Lambda^1(-,D\operatorname{Tr} X)
$$

and

$$
D\operatorname{Ext}_\Lambda^1(X,-) \cong \overline{\operatorname{Hom}}_\Lambda(-,D\operatorname{Tr} X).
$$

Proof The first isomorphism follows from the above lemma together with Lemma 4.3.17. The second isomorphism follows from the first via Matlis duality. $\qquad\square$

Gorenstein Projective and Injective Modules

Let Λ be Gorenstein. We recall that a Λ-module X is *Gorenstein projective* if $\operatorname{Ext}_\Lambda^i(X,\Lambda) = 0$ for all $i \neq 0$. Dually, the module X is *Gorenstein injective* if $\operatorname{Ext}_\Lambda^i(D(\Lambda),X) = 0$ for all $i \neq 0$. We set

$$
\operatorname{Ginj}\Lambda = \{X \in \operatorname{mod}\Lambda \mid X \text{ is Gorenstein injective}\}.
$$

The duality D induces an equivalence

$$
(\operatorname{Gproj}\Lambda)^{\operatorname{op}} \xrightarrow{\sim} \operatorname{Ginj}(\Lambda^{\operatorname{op}}).
$$

Observe that a Λ-module has finite projective dimension if and only if it has finite injective dimension, because Λ is Gorenstein. Then Theorem 6.2.4 yields two cotorsion pairs

$$(\mathrm{Gproj}\,\Lambda, \mathcal{Y}) \qquad \text{and} \qquad (\mathcal{Y}, \mathrm{Ginj}\,\Lambda)$$

for mod Λ, where $\mathcal{Y}$ denotes the subcategory of modules having finite projective and finite injective dimension.

We consider the full subcategories

$$\underline{\mathrm{Gproj}\,\Lambda} \hookrightarrow \underline{\mathrm{mod}}\,\Lambda \qquad \text{and} \qquad \overline{\mathrm{Ginj}\,\Lambda} \hookrightarrow \overline{\mathrm{mod}}\,\Lambda.$$

The following lemma yields adjoints

$$\mathrm{GP} \colon \underline{\mathrm{mod}}\,\Lambda \to \underline{\mathrm{Gproj}}\,\Lambda \qquad \text{and} \qquad \mathrm{GI} \colon \overline{\mathrm{mod}}\,\Lambda \to \overline{\mathrm{Ginj}}\,\Lambda.$$

Lemma 6.3.6. *The inclusion* $\underline{\mathrm{Gproj}}\,\Lambda \to \underline{\mathrm{mod}}\,\Lambda$ *admits a right adjoint and the inclusion* $\overline{\mathrm{Ginj}}\,\Lambda \to \overline{\mathrm{mod}}\,\Lambda$ *admits a left adjoint.*

Proof The existence of the right adjoint follows from Theorem 6.2.4, using also Remark 6.1.2. Applying Matlis duality, we obtain the left adjoint. With the notation of the approximation sequence (6.1.1), we get $\mathrm{GP}(A) = X_A$ and $\mathrm{GI}(A) = Y^A$ for a Λ-module A. $\qquad\Box$

We collect further properties of Gorenstein projective and injective modules.

Lemma 6.3.7. *Let* $X \in \mathrm{mod}\,\Lambda$ *be Gorenstein projective and* $Y \in \mathrm{mod}\,\Lambda$ *be Gorenstein injective. Then the following holds.*

(1) $D\,\mathrm{Tr}\,X$ *is Gorenstein injective and* $\mathrm{Tr}\,DY$ *is Gorenstein projective.*
(2) $\mathrm{GP}(\mathrm{GI}(X)) \cong X$ *in* $\underline{\mathrm{mod}}\,\Lambda$ *and* $\mathrm{GI}(\mathrm{GP}(Y)) \cong Y$ *in* $\overline{\mathrm{mod}}\,\Lambda$.
(3) $\underline{\mathrm{Hom}}_\Lambda(X, Y) = \overline{\mathrm{Hom}}_\Lambda(X, Y)$.

Proof (1) If X is Gorenstein projective, then $\mathrm{Tr}\,X$ is a Gorenstein projective Λ^{op}-module by Lemma 6.2.2. Thus $D\,\mathrm{Tr}\,X$ is Gorenstein injective. The argument for Y is dual.

(2) Consider the approximation sequence $0 \to X \to \mathrm{GI}(X) \to Y \to 0$, where Y is of finite injective dimension and therefore of finite projective dimension. Let $P \to Y$ be a projective cover and form the following pullback.

$$
\begin{array}{ccccccccc}
0 & \longrightarrow & X & \longrightarrow & X \oplus P & \longrightarrow & P & \longrightarrow & 0 \\
  &   & \| &   & \downarrow &   & \downarrow &   & \\
0 & \longrightarrow & X & \longrightarrow & \mathrm{GI}(X) & \longrightarrow & Y & \longrightarrow & 0
\end{array}
$$

The morphism $X \oplus P \to \mathrm{GI}(X)$ is a Gorenstein projective approximation, since its kernel has finite projective dimension. Thus we have $GP(GI(X)) \cong X$ stably. The argument for Y is dual.

(3) Fix a morphism $\phi \colon X \to Y$ that factors through an injective module. Let $X \to P$ denote the approximation with P of finite projective dimension, which exists by Theorem 6.2.4. Then P is projective, by Remark 6.1.2, and ϕ factors through P, since any injective module has finite projective dimension. The dual argument shows that ϕ factors through an injective module when one assumes that it factors through a projective module.   □

Serre Duality for the Stable Category

Auslander–Reiten duality translates into Serre duality for the stable category of Gorenstein projective Λ-modules. Recall that $\operatorname{Gproj}\Lambda$ is a Frobenius category, and we denote by $\Omega^{-1} \colon \underline{\operatorname{Gproj}}\Lambda \xrightarrow{\sim} \underline{\operatorname{Gproj}}\Lambda$ the suspension of the stable category, which takes a module X to the cokernel of a monomorphism $X \to P$ into a projective module P.

Proposition 6.3.8. *Let Λ be Gorenstein. For Gorenstein projective Λ-modules X, Y there are natural isomorphisms*

$$\underline{\operatorname{Hom}}_\Lambda(\operatorname{Tr} D(GI\,\Omega Y), X) \cong D\,\underline{\operatorname{Hom}}_\Lambda(X, Y) \cong \underline{\operatorname{Hom}}_\Lambda(Y, \Omega^{-1}\,GP(D\,\operatorname{Tr} X)).$$

Proof   We have

$$
\begin{aligned}
D\,\underline{\operatorname{Hom}}_\Lambda(X, Y) &\cong D\operatorname{Ext}^1_\Lambda(X, \Omega Y)\\
&\cong \overline{\operatorname{Hom}}_\Lambda(\Omega Y, D\operatorname{Tr} X)\\
&\cong \overline{\operatorname{Hom}}_\Lambda(GI\,\Omega Y, D\operatorname{Tr} X)\\
&\cong \underline{\operatorname{Hom}}_\Lambda(\operatorname{Tr} D(GI\,\Omega Y), X).
\end{aligned}
$$

The first isomorphism is obtained by applying $\operatorname{Hom}_\Lambda(X, -)$ to an exact sequence

$$0 \longrightarrow \Omega Y \longrightarrow P \longrightarrow Y \longrightarrow 0$$

with P projective. The second isomorphism is Auslander–Reiten duality; see Proposition 6.3.5. The third isomorphism is induced by $\Omega Y \to GI(\Omega Y)$; see Lemma 6.3.7. The last isomorphism is obtained by applying $\operatorname{Tr} D$.

A similar sequence of arguments yields

$$
\begin{aligned}
D\,\underline{\mathrm{Hom}}_\Lambda(X,Y) &\cong D\,\mathrm{Ext}^1_\Lambda(X,\Omega Y) \\
&\cong \overline{\mathrm{Hom}}_\Lambda(\Omega Y, D\,\mathrm{Tr}\,X) \\
&= \underline{\mathrm{Hom}}_\Lambda(\Omega Y, D\,\mathrm{Tr}\,X) \\
&\cong \underline{\mathrm{Hom}}_\Lambda(\Omega Y, \mathrm{GP}(D\,\mathrm{Tr}\,X)) \\
&\cong \underline{\mathrm{Hom}}_\Lambda(Y, \Omega^{-1}\,\mathrm{GP}(D\,\mathrm{Tr}\,X)).
\end{aligned}
$$

$\square$

Corollary 6.3.9. *Let Λ be Gorenstein. The assignments*

$$
X \mapsto \Omega^{-1}\,\mathrm{GP}(D\,\mathrm{Tr}\,X) \quad and \quad Y \mapsto \mathrm{Tr}\,D(\mathrm{GI}\,\Omega Y)
$$

yield mutually inverse equivalences $\underline{\mathrm{Gproj}}\,\Lambda \xrightarrow{\sim} \underline{\mathrm{Gproj}}\,\Lambda$. *In particular, the composite* $\Omega^{-1} \circ \mathrm{GP} \circ D\,\mathrm{Tr}$ *is a Serre functor for* $\underline{\mathrm{Gproj}}\,\Lambda$.

Proof We have $\Omega^{-1} \circ \Omega \cong \mathrm{id} \cong \Omega \circ \Omega^{-1}$ since $\underline{\mathrm{Gproj}}\,\Lambda$ is a Frobenius category; see Theorem 6.2.5. Also, $D\,\mathrm{Tr} \circ \mathrm{Tr}\,D \cong \mathrm{id}$ and $\mathrm{Tr}\,D \circ D\,\mathrm{Tr} \cong \mathrm{id}$. Finally, $\mathrm{GP} \circ \mathrm{GI} \cong \mathrm{id}$ and $\mathrm{GI} \circ \mathrm{GP} \cong \mathrm{id}$ by Lemma 6.3.7. The isomorphism in Proposition 6.3.8 then shows that $\Omega^{-1} \circ \mathrm{GP} \circ D\,\mathrm{Tr}$ is a Serre functor. $\square$

6.4 The Derived Nakayama Functor

In this section we introduce the derived Nakayama functor and use this to establish Serre duality for the category of perfect complexes. We keep the setting from the previous section and consider an Artin algebra Λ over a commutative ring k. Also, we show that $\mathbf{D}^b(\mathrm{mod}\,\Lambda)$ admits a Serre functor if and only if Λ has finite global dimension.

The Nakayama Functor

The *Nakayama functor* $\mathrm{mod}\,\Lambda \to \mathrm{mod}\,\Lambda$ is given by

$$
\nu X := X \otimes_\Lambda D(\Lambda) \cong D\,\mathrm{Hom}_\Lambda(X,\Lambda).
$$

It restricts to an equivalence $\mathrm{proj}\,\Lambda \xrightarrow{\sim} \mathrm{inj}\,\Lambda$, where $\mathrm{inj}\,\Lambda$ denotes the category of finitely generated injective Λ-modules.

A projective presentation $P_1 \to P_0 \to X \to 0$ induces an exact sequence

$$
0 \longrightarrow D\,\mathrm{Tr}\,X \longrightarrow \nu P_1 \longrightarrow \nu P_0 \longrightarrow \nu X \longrightarrow 0 \tag{6.4.1}
$$

and therefore

$$
\Omega^{-2}(D\,\mathrm{Tr}\,X) \cong \nu X.
$$

Of particular interest is the following case.

Proposition 6.4.2. *Let Λ be self-injective. Then the functor*

$$\underline{\mathrm{mod}}\,\Lambda \longrightarrow \underline{\mathrm{mod}}\,\Lambda, \quad X \mapsto \Omega \nu X \cong \nu \Omega X,$$

is a Serre functor, and therefore

$$D\,\underline{\mathrm{Hom}}_\Lambda(X, -) \cong \underline{\mathrm{Hom}}_\Lambda(-, \Omega \nu X).$$

Proof This follows from Corollary 6.3.9 since $\underline{\mathrm{mod}}\,\Lambda = \underline{\mathrm{Gproj}}\,\Lambda$.   □

Suppose the algebra Λ is *symmetric* so that $D\Lambda \cong \Lambda$ as Λ-Λ-bimodules. Then we have $\nu = \mathrm{id}$. An example is the group algebra of a finite group, and in that case the above Serre duality is known as Tate duality.

Let us identify $\mathbf{K}^-(\mathrm{proj}\,\Lambda) \xrightarrow{\sim} \mathbf{D}^-(\mathrm{mod}\,\Lambda)$. Then the *derived Nakayama functor*

$$X \longmapsto X \otimes_\Lambda^L D(\Lambda)$$

is by definition the composite

$$\mathbf{D}^-(\mathrm{mod}\,\Lambda) \xrightarrow{\sim} \mathbf{K}^-(\mathrm{proj}\,\Lambda) \xrightarrow{-\otimes_\Lambda D(\Lambda)} \mathbf{K}^-(\mathrm{mod}\,\Lambda) \xrightarrow{\mathrm{can}} \mathbf{D}^-(\mathrm{mod}\,\Lambda).$$

Serre Duality for Perfect Complexes

We show that the category of perfect complexes

$$\mathbf{D}^{\mathrm{perf}}(\Lambda) = \mathbf{D}^b(\mathrm{proj}\,\Lambda)$$

admits a Serre functor if and only if the algebra Λ is Gorenstein. In fact, the derived Nakayama functor restricts to a Serre functor for $\mathbf{D}^{\mathrm{perf}}(\Lambda)$ when Λ is Gorenstein.

We recall the following standard isomorphisms and extend them to isomorphisms of complexes.

Lemma 6.4.3. *Let $(A_\Lambda, {}_\Gamma B_\Lambda, {}_\Gamma C)$ be modules and suppose that A_Λ is finitely generated projective. Then there are natural isomorphisms*

$$B \otimes_\Lambda \mathrm{Hom}_\Lambda(A, \Lambda) \xrightarrow{\sim} \mathrm{Hom}_\Lambda(A, B)$$

and

$$A \otimes_\Lambda \mathrm{Hom}_\Gamma(B, C) \xrightarrow{\sim} \mathrm{Hom}_\Gamma(\mathrm{Hom}_\Lambda(A, B), C).$$ □

A complex X is called *bounded* if $X^n = 0$ for almost all $n \in \mathbb{Z}$. A pair of complexes (X, Y) is *bounded* if for each $n \in \mathbb{Z}$ we have for almost all pairs of integers (p, q) with $-p + q = n$ that $X^p = 0$ or $Y^q = 0$.

Lemma 6.4.4. *Let* $(X_\Lambda, {}_\Gamma Y_\Lambda, {}_\Gamma Z)$ *be complexes of modules with each* X^n *finitely generated projective. Suppose that* (X, Y) *and* Z *are bounded. Then there are natural isomorphisms*

$$Y \otimes_\Lambda \operatorname{Hom}_\Lambda(X, \Lambda) \xrightarrow{\sim} \operatorname{Hom}_\Lambda(X, Y)$$

and

$$X \otimes_\Lambda \operatorname{Hom}_\Gamma(Y, Z) \xrightarrow{\sim} \operatorname{Hom}_\Gamma(\operatorname{Hom}_\Lambda(X, Y), Z).$$

Proof We may apply Lemma 6.4.3 degreewise, thanks to the boundedness assumptions. □

It is convenient to set $\mathbf{K}(\Lambda) = \mathbf{K}(\operatorname{Mod} \Lambda)$ and $\mathbf{D}(\Lambda) = \mathbf{D}(\operatorname{Mod} \Lambda)$.

Lemma 6.4.5. *Let* X, Y *be complexes of* Λ-*modules and suppose that* X *is perfect. Then we have natural isomorphisms*

$$D \operatorname{Hom}_{\mathbf{K}(\Lambda)}(X, Y) \cong \operatorname{Hom}_{\mathbf{K}(\Lambda)}(Y, X \otimes_\Lambda D(\Lambda))$$

and

$$D \operatorname{Hom}_{\mathbf{D}(\Lambda)}(X, Y) \cong \operatorname{Hom}_{\mathbf{D}(\Lambda)}(Y, X \otimes_\Lambda D(\Lambda)).$$

Proof Let X be a bounded complex of finitely generated projective modules. Then we have the following sequence of isomorphisms:

$$
\begin{aligned}
D \operatorname{Hom}_{\mathbf{K}(\Lambda)}(X, Y) &\cong \operatorname{Hom}_k(H^0 \operatorname{Hom}_\Lambda(X, Y), E) & \text{4.3.14} \\
&\cong H^0 \operatorname{Hom}_k(\operatorname{Hom}_\Lambda(X, Y), E) & \\
&\cong H^0 \operatorname{Hom}_k(Y \otimes_\Lambda \operatorname{Hom}_\Lambda(X, \Lambda), E) & \text{6.4.4} \\
&\cong H^0 \operatorname{Hom}_\Lambda(Y, \operatorname{Hom}_k(\operatorname{Hom}_\Lambda(X, \Lambda), E)) & \text{4.3.13} \\
&\cong H^0 \operatorname{Hom}_\Lambda(Y, X \otimes_\Lambda \operatorname{Hom}_k(\Lambda, E)) & \text{6.4.4} \\
&\cong \operatorname{Hom}_{\mathbf{K}(\Lambda)}(Y, X \otimes_\Lambda D(\Lambda)). & \text{4.3.14}
\end{aligned}
$$

The above isomorphism for a perfect complex X carries over to $\mathbf{D}(\Lambda)$ since

$$\operatorname{Hom}_{\mathbf{K}(\Lambda)}(X, -) \cong \operatorname{Hom}_{\mathbf{D}(\Lambda)}(X, -)$$

and

$$\operatorname{Hom}_{\mathbf{K}(\Lambda)}(-, X \otimes_\Lambda D(\Lambda)) \cong \operatorname{Hom}_{\mathbf{D}(\Lambda)}(-, X \otimes_\Lambda D(\Lambda)). \quad \square$$

Theorem 6.4.6. *An Artin algebra* Λ *is Gorenstein if and only if the category of perfect complexes* $\mathbf{D}^{\operatorname{perf}}(\Lambda)$ *admits a Serre functor. In this case, the Serre functor is given by the derived Nakayama functor.*

Proof   Set $\mathcal{P} = \text{Thick}(\text{proj}\,\Lambda)$ and $\mathcal{I} = \text{Thick}(\text{inj}\,\Lambda)$. We have equivalences

$$\mathbf{K}^b(\text{proj}\,\Lambda) \xrightarrow{\sim} \mathbf{D}^b(\text{proj}\,\Lambda) \xrightarrow{\sim} \mathbf{D}^b(\mathcal{P})$$

and analogously

$$\mathbf{K}^b(\text{inj}\,\Lambda) \xrightarrow{\sim} \mathbf{D}^b(\text{inj}\,\Lambda) \xrightarrow{\sim} \mathbf{D}^b(\mathcal{I}).$$

The Nakayama functor $- \otimes_\Lambda D(\Lambda)$ and its adjoint $\text{Hom}_\Lambda(D(\Lambda), -)$ induce mutually inverse equivalences:

$$\mathbf{D}^b(\mathcal{P}) \xleftarrow{\sim} \mathbf{K}^b(\text{proj}\,\Lambda) \underset{\text{Hom}_\Lambda(D(\Lambda),-)}{\overset{-\otimes_\Lambda D(\Lambda)}{\rightleftarrows}} \mathbf{K}^b(\text{inj}\,\Lambda) \xrightarrow{\sim} \mathbf{D}^b(\mathcal{I}).$$

If Λ is Gorenstein, then we have $\mathcal{P} = \mathcal{I}$. Thus the Nakayama functor gives an equivalence $\mathbf{D}^{\text{perf}}(\Lambda) \xrightarrow{\sim} \mathbf{D}^{\text{perf}}(\Lambda)$, and this is a Serre functor by the isomorphism from Lemma 6.4.5.

Now suppose that $F \colon \mathbf{D}^{\text{perf}}(\Lambda) \to \mathbf{D}^{\text{perf}}(\Lambda)$ is a Serre functor. Then we have isomorphisms

$$\text{Hom}_{\mathbf{D}(\Lambda)}(-, F\Lambda) \cong D\,\text{Hom}_{\mathbf{D}(\Lambda)}(\Lambda, -) \cong \text{Hom}_{\mathbf{D}(\Lambda)}(-, D(\Lambda))$$

of functors on $\mathbf{D}^{\text{perf}}(\Lambda)$. The first isomorphism is clear from the definition of a Serre functor, and the second by Lemma 6.4.5. The isomorphism of functors is induced by a morphism $F\Lambda \to D(\Lambda)$ and this is a quasi-isomorphism since

$$H^n(F\Lambda) \cong \text{Hom}_{\mathbf{D}(\Lambda)}(\Sigma^{-n}\Lambda, F\Lambda) \cong \text{Hom}_{\mathbf{D}(\Lambda)}(\Sigma^{-n}\Lambda, D(\Lambda)) \cong H^n(D(\Lambda)).$$

It follows that $D(\Lambda)_\Lambda$ has finite projective dimension. The functor $\text{Hom}_\Lambda(-, \Lambda)$ induces a triangle equivalence

$$\mathbf{D}^{\text{perf}}(\Lambda)^{\text{op}} \xrightarrow{\sim} \mathbf{D}^{\text{perf}}(\Lambda^{\text{op}}),$$

and it follows that $\mathbf{D}^{\text{perf}}(\Lambda^{\text{op}})$ admits a Serre functor. Thus $_\Lambda D(\Lambda)$ has finite projective dimension. Using Matlis duality, it follows that Λ_Λ and $_\Lambda\Lambda$ have finite injective dimension. We conclude that Λ is Gorenstein. $\square$

Serre Duality for the Singularity Category

We get back to the stable category of a Gorenstein algebra and provide another description of the Serre functor which uses the derived Nakayama functor.

Proposition 6.4.7. *Let Λ be Gorenstein. Then the derived Nakayama functor yields an equivalence*

$$\hat{\nu} \colon \mathbf{D}^b(\text{mod}\,\Lambda) \xrightarrow{\sim} \mathbf{D}^b(\text{mod}\,\Lambda)$$

satisfying $\hat{\nu}(\mathbf{D}^{\text{perf}}(\Lambda)) = \mathbf{D}^{\text{perf}}(\Lambda)$.

Proof The category $\mathbf{D}^b(\operatorname{mod}\Lambda)$ identifies with the full subcategory of objects $X \in \mathbf{K}^-(\operatorname{proj}\Lambda)$ such that the cohomology of X is concentrated in finitely many degrees. Suppose first that $H^n X = 0$ for all $n \neq 0$. Then we have

$$H^i(X \otimes_\Lambda D(\Lambda)) \cong \operatorname{Tor}_{-i}^\Lambda(H^0 X, D(\Lambda)).$$

Thus $X \otimes_\Lambda^L D(\Lambda)$ belongs to $\mathbf{D}^b(\operatorname{mod}\Lambda)$ since $D(\Lambda)$ has finite projective dimension. From the exactness of the derived Nakayama functor it follows that we get a functor $\mathbf{D}^b(\operatorname{mod}\Lambda) \to \mathbf{D}^b(\operatorname{mod}\Lambda)$, because the objects with cohomology concentrated in degree zero generate $\mathbf{D}^b(\operatorname{mod}\Lambda)$. A quasi-inverse is given by $\operatorname{RHom}_\Lambda(D(\Lambda), -)$, and that $\hat{\nu}$ identifies perfect complexes with perfect complexes follows from Theorem 6.4.6. $\square$

There is a converse of the above proposition for which we refer to Proposition 9.2.17.

Example 6.4.8. Let Λ be hereditary and X a Λ-module, viewed as a complex concentrated in degree zero. Then we have

$$X \otimes_\Lambda^L D(\Lambda) = \nu X \oplus (D \operatorname{Tr} X)[1].$$

This follows from the sequence (6.4.1).

The above proposition implies that the derived Nakayama functor induces an equivalence $\bar{\nu}\colon \mathbf{D}_{\mathrm{sg}}(\Lambda) \xrightarrow{\sim} \mathbf{D}_{\mathrm{sg}}(\Lambda)$ making the following diagram commutative.

$$
\begin{array}{ccccc}
\mathbf{D}^b(\operatorname{proj}\Lambda) & \rightarrowtail & \mathbf{D}^b(\operatorname{mod}\Lambda) & \twoheadrightarrow & \mathbf{D}_{\mathrm{sg}}(\Lambda) \\
\downarrow & & \downarrow{\scriptstyle \hat{\nu}} & & \downarrow{\scriptstyle \bar{\nu}} \\
\mathbf{D}^b(\operatorname{proj}\Lambda) & \rightarrowtail & \mathbf{D}^b(\operatorname{mod}\Lambda) & \twoheadrightarrow & \mathbf{D}_{\mathrm{sg}}(\Lambda)
\end{array}
$$

Moreover, the equivalence $\bar{\nu}$ makes the following square commutative

$$
\begin{array}{ccc}
\underline{\operatorname{Gproj}}\Lambda & \xrightarrow[p]{\sim} & \mathbf{D}_{\mathrm{sg}}(\Lambda) \\
{\scriptstyle \nu}\downarrow & & \downarrow{\scriptstyle \bar{\nu}} \\
\overline{\operatorname{Ginj}}\Lambda & \xrightarrow[q]{\sim} & \mathbf{D}_{\mathrm{sg}}(\Lambda)
\end{array}
$$

where the horizontal equivalence p is from Theorem 6.2.5 and q is its analogue for Gorenstein injectives.

Lemma 6.4.9. *The Gorenstein projective approximation* GP *induces a triangle equivalence* $\overline{\operatorname{Ginj}}\Lambda \xrightarrow{\sim} \underline{\operatorname{Gproj}}\Lambda$. *Moreover, we have*

$$q \cong p \circ \operatorname{GP} \qquad and \qquad \bar{\nu} \circ p \cong q \circ \nu \cong q \circ \Omega^{-2} \circ D \operatorname{Tr}.$$

Proof For any Λ-module X there is an exact sequence $0 \to X' \to \mathrm{GP}(X) \to X \to 0$ such that X' has finite projective dimension, by Theorem 6.2.4. Thus $q \cong p \circ \mathrm{GP}$. It follows that GP induces a triangle equivalence $\overline{\mathrm{Ginj}}\,\Lambda \xrightarrow{\sim} \underline{\mathrm{Gproj}}\,\Lambda$ since p and q are triangle equivalences. The isomorphism $\bar{\nu} \circ p \cong q \circ \nu$ is clear and the last one follows from (6.4.1). $\qquad\square$

In Corollary 6.3.9 we have already seen that the stable category of Gorenstein projectives admits a Serre functor and now we have an alternative description.

Corollary 6.4.10. *Let Λ be Gorenstein. Then*

$$\Sigma^{-1} \circ \bar{\nu} : \mathbf{D}_{\mathrm{sg}}(\Lambda) \xrightarrow{\sim} \mathbf{D}_{\mathrm{sg}}(\Lambda) \quad and \quad \Omega \circ \mathrm{GP} \circ \nu : \underline{\mathrm{Gproj}}\,\Lambda \xrightarrow{\sim} \underline{\mathrm{Gproj}}\,\Lambda$$

are Serre functors.

Proof We apply Proposition 6.3.8 and Lemma 6.4.9. Thus for Gorenstein projective Λ-modules X, Y we compute in $\mathbf{D}_{\mathrm{sg}}(\Lambda)$

$$
\begin{aligned}
D\,\mathrm{Hom}(pX, pY) &\cong \mathrm{Hom}(pY, p\Omega^{-1}\,\mathrm{GP}(D\,\mathrm{Tr}\,X)) \\
&\cong \mathrm{Hom}(pY, \Sigma^{-1}p\,\mathrm{GP}\,\Omega^{-2}(D\,\mathrm{Tr}\,X)) \\
&\cong \mathrm{Hom}(pY, \Sigma^{-1}q\Omega^{-2}(D\,\mathrm{Tr}\,X)) \\
&\cong \mathrm{Hom}(pY, \Sigma^{-1}\bar{\nu}pX).
\end{aligned}
$$

Thus $\Sigma^{-1} \circ \bar{\nu}$ is a Serre functor for $\mathbf{D}_{\mathrm{sg}}(\Lambda)$. We have $\bar{\nu} \circ p \cong p \circ \mathrm{GP} \circ \nu$, and it follows that $\Omega \circ \mathrm{GP} \circ \nu$ is a Serre functor for $\underline{\mathrm{Gproj}}\,\Lambda$. $\qquad\square$

Finite Global Dimension

We show that an Artin algebra Λ has finite global dimension if and only if $\mathbf{D}^b(\mathrm{mod}\,\Lambda)$ admits a Serre functor. This requires various computations in $\mathbf{K}(\Lambda) := \mathbf{K}(\mathrm{Mod}\,\Lambda)$. We begin with a statement about complexes of injective modules.

Lemma 6.4.11. *Let Λ be a right noetherian ring and X an object in $\mathbf{K}(\mathrm{Inj}\,\Lambda)$. For a Λ-module A an injective resolution $A \to iA$ induces an isomorphism*

$$\mathrm{Hom}_{\mathbf{K}(\Lambda)}(iA, X) \xrightarrow{\sim} \mathrm{Hom}_{\mathbf{K}(\Lambda)}(A, X).$$

Moreover, if $\mathrm{Hom}_{\mathbf{K}(\Lambda)}(A, \Sigma^n X) = 0$ for all $A \in \mathrm{mod}\,\Lambda$ and $n \in \mathbb{Z}$, then $X = 0$.

Proof For the first assertion see Lemma 4.2.6.

Now suppose $\mathrm{Hom}_{\mathbf{K}(\Lambda)}(A, \Sigma^n X) = 0$ for all $A \in \mathrm{mod}\,\Lambda$ and $n \in \mathbb{Z}$. Suppose first $H^n X \neq 0$ for some n. Choose $A \in \mathrm{mod}\,\Lambda$ and a morphism $A \to Z^n X$ inducing a non-zero morphism $A \to H^n X$. We obtain a morphism $A \to \Sigma^n X$ which induces a non-zero element in $\mathrm{Hom}_{\mathbf{K}(\Lambda)}(A, \Sigma^n X)$.

Now suppose $H^n X = 0$ for all n. We can choose n such that $Z^n X$ is non-injective. Applying Baer's criterion, there exists $A \in \text{mod}\,\Lambda$ such that $\text{Ext}_\Lambda^1(A, Z^n X)$ is non-zero. Observe that

$$\text{Hom}_{\mathbf{K}(\Lambda)}(A, \Sigma^{n+p} X) \cong \text{Ext}_\Lambda^p(A, Z^n X)$$

for all $p \geq 1$. Thus $\text{Hom}_{\mathbf{K}(\Lambda)}(A, \Sigma^{n+1} X) \neq 0$. $\qquad\qquad\square$

We consider the triangle equivalence

$$\mathbf{K}^{+,b}(\text{Inj}\,\Lambda) \xrightarrow{\sim} \mathbf{D}^b(\text{Mod}\,\Lambda) \xrightarrow{\sim} \mathbf{K}^{-,b}(\text{Proj}\,\Lambda), \qquad X \longmapsto \mathbf{p}X$$

from Corollary 4.2.9, which takes a complex to a homotopy projective resolution. This restricts to an equivalence

$$\mathcal{C} \xrightarrow{\sim} \mathbf{D}^b(\text{mod}\,\Lambda) \xrightarrow{\sim} \mathbf{K}^{-,b}(\text{proj}\,\Lambda)$$

where $\mathcal{C} \subseteq \mathbf{K}^{+,b}(\text{Inj}\,\Lambda)$ denotes the thick subcategory generated by all injective resolutions of finitely generated Λ-modules.

Lemma 6.4.12. *For* $X, Y \in \mathbf{K}(\text{Inj}\,\Lambda)$ *with* $X \in \mathcal{C}$ *we have a natural isomorphism*

$$D\,\text{Hom}_{\mathbf{K}(\Lambda)}(X, Y) \cong \text{Hom}_{\mathbf{K}(\Lambda)}(Y, \mathbf{p}X \otimes_\Lambda D(\Lambda)).$$

Proof We may assume that $Y^n = 0$ for $n \ll 0$ since Y can be written as a homotopy colimit of objects in $\mathbf{K}^+(\text{Inj}\,\Lambda)$ (using (4.2.3)). In fact, one checks that $D\,\text{Hom}_{\mathbf{K}(\Lambda)}(X, -)$ and $\text{Hom}_{\mathbf{K}(\Lambda)}(-, \mathbf{p}X \otimes_\Lambda D(\Lambda))$ preserve homotopy colimits. This is clear in the second case. In the first case we may reduce to the case that X is the injective resolution of a finitely generated module A, so there is a quasi-isomorphism $A \to X$ in $\mathbf{K}(\Lambda)$. Then

$$\text{Hom}_{\mathbf{K}(\Lambda)}(X, -) \cong \text{Hom}_{\mathbf{K}(\Lambda)}(A, -)$$

by Lemma 6.4.11, and this preserves coproducts since A is finitely generated.

Keeping in mind that Y is in $\mathbf{K}^+(\text{Inj}\,\Lambda)$, we have the following sequence of isomorphisms:

$$
\begin{aligned}
D\,\text{Hom}_{\mathbf{K}(\Lambda)}(X, Y) &\cong \text{Hom}_k(\text{Hom}_{\mathbf{K}(\Lambda)}(\mathbf{p}X, Y), E) && 4.2.5 \\
&\cong \text{Hom}_k(H^0\,\text{Hom}_\Lambda(\mathbf{p}X, Y), E) && 4.3.14 \\
&\cong H^0\,\text{Hom}_k(\text{Hom}_\Lambda(\mathbf{p}X, Y), E) && \\
&\cong H^0\,\text{Hom}_k(Y \otimes_\Lambda \text{Hom}_\Lambda(\mathbf{p}X, \Lambda), E) && 6.4.4 \\
&\cong H^0\,\text{Hom}_\Lambda(Y, \text{Hom}_k(\text{Hom}_\Lambda(\mathbf{p}X, \Lambda), E)) && 4.3.13 \\
&\cong H^0\,\text{Hom}_\Lambda(Y, \mathbf{p}X \otimes_\Lambda \text{Hom}_k(\Lambda, E)) && 6.4.4 \\
&\cong \text{Hom}_{\mathbf{K}(\Lambda)}(Y, \mathbf{p}X \otimes_\Lambda D(\Lambda)). && 4.3.14
\end{aligned}
$$

With our first observation the isomorphism follows for an arbitrary complex Y in $\mathbf{K}(\operatorname{Inj}\Lambda)$. $\square$

Theorem 6.4.13. *For an Artin algebra Λ the following are equivalent.*

(1) *The algebra Λ has finite global dimension.*
(2) *The canonical functor $\mathbf{D}^b(\operatorname{proj}\Lambda) \to \mathbf{D}^b(\operatorname{mod}\Lambda)$ is an equivalence.*
(3) *The category $\mathbf{D}^b(\operatorname{mod}\Lambda)$ admits a Serre functor.*
(4) *The algebra Λ is Gorenstein and each acyclic complex of finitely generated injective Λ-modules is contractible.*
(5) *The algebra Λ is Gorenstein and each acyclic complex of finitely generated projective Λ-modules is contractible.*

Proof (1) $\Leftrightarrow$ (2): The canonical functor $\mathbf{D}^b(\operatorname{proj}\Lambda) \to \mathbf{D}^b(\operatorname{mod}\Lambda)$ is fully faithful, and it is an equivalence if and only if every object in $\operatorname{mod}\Lambda$ has finite projective dimension. It remains to observe that the global dimension of Λ equals the maximum of the projective dimensions of the simple Λ-modules.

(1) $\Rightarrow$ (3): When Λ has finite global dimension, then we have $\mathbf{D}^{\operatorname{perf}}(\Lambda) \xrightarrow{\sim} \mathbf{D}^b(\operatorname{mod}\Lambda)$, and it follows from Theorem 6.4.6 that $\mathbf{D}^b(\operatorname{mod}\Lambda)$ admits a Serre functor.

(3) $\Rightarrow$ (4): Suppose that $F : \mathbf{D}^b(\operatorname{mod}\Lambda) \xrightarrow{\sim} \mathbf{D}^b(\operatorname{mod}\Lambda)$ is a Serre functor and set $S := \Lambda/\operatorname{rad}\Lambda$. Then

$$D\operatorname{Ext}^n_\Lambda(S,\Lambda) \cong D\operatorname{Hom}_{\mathbf{D}(\Lambda)}(\Sigma^{-n}S,\Lambda) \cong \operatorname{Hom}_{\mathbf{D}(\Lambda)}(\Lambda, \Sigma^{-n}F(S)) = H^{-n}F(S)$$

and therefore

$$\begin{aligned}
\operatorname{inj.dim}_\Lambda \Lambda &= \sup\{n \geq 0 \mid \operatorname{Ext}^n_\Lambda(S,\Lambda) \neq 0\} \\
&= \sup\{n \geq 0 \mid H^{-n}(FS) \neq 0\}
\end{aligned}$$

which is finite. It follows that Λ is Gorenstein, keeping in mind that $\mathbf{D}^b(\operatorname{mod}\Lambda^{\operatorname{op}})$ admits a Serre functor as well.

Let us identify $\mathcal{T} := \mathbf{K}^{+,b}(\operatorname{inj}\Lambda) = \mathbf{D}^b(\operatorname{mod}\Lambda)$. Then Lemma 6.4.12 implies for each $X \in \mathcal{T}$ a natural isomorphism

$$\operatorname{Hom}_{\mathbf{K}(\Lambda)}(-, F(X)) \cong D\operatorname{Hom}_{\mathbf{K}(\Lambda)}(X, -) \cong \operatorname{Hom}_{\mathbf{K}(\Lambda)}(-, \mathbf{p}X \otimes_\Lambda D(\Lambda))$$

of functors on $\mathcal{T}$. This is induced by a morphism $\phi : F(X) \to \mathbf{p}X \otimes_\Lambda D(\Lambda)$, which is an isomorphism by Lemma 6.4.11, since $\operatorname{Hom}_{\mathbf{K}(\Lambda)}(-, \operatorname{Cone}\phi)|_\mathcal{T} = 0$. Thus $\mathbf{p}X \otimes_\Lambda D(\Lambda)$ belongs to $\mathcal{T}$. Now choose a complex V in $\mathbf{K}(\operatorname{inj}\Lambda)$ which is acyclic. Then

$$D\operatorname{Hom}_{\mathbf{K}(\Lambda)}(X, V) \cong \operatorname{Hom}_{\mathbf{K}(\Lambda)}(V, \mathbf{p}X \otimes_\Lambda D(\Lambda)) = 0.$$

The first isomorphism is by Lemma 6.4.12 and the second by Lemma 4.2.5. Thus $V = 0$ by Lemma 6.4.11.

(4) $\Leftrightarrow$ (5): The Nakayama functor yields an equivalence $\mathbf{K}(\operatorname{proj}\Lambda) \xrightarrow{\sim} \mathbf{K}(\operatorname{inj}\Lambda)$ that identifies the acyclic complexes in both categories. This follows from Lemma 6.2.2.

(5) $\Rightarrow$ (1): Each Gorenstein projective module admits a complete resolution, so is of the form $Z^0 X$ for some acyclic complex X of projective Λ-modules, by Theorem 6.2.5. If X is contractible, then $Z^0 X$ is projective. Thus every Λ-module has finite projective dimension by Theorem 6.2.4. $\qquad\square$

6.5 Examples

We discuss two examples. The first one gives an application of Serre duality for hereditary abelian categories. The second example provides explicit computations for a Gorenstein algebra of dimension one.

Hereditary Categories

We give an application of Serre duality for hereditary abelian categories. Fix a commutative ring k and a k-linear hereditary abelian category $\mathcal{A}$ such that $\operatorname{Hom}_{\mathcal{A}}(X, Y)$ and $\operatorname{Ext}^1_{\mathcal{A}}(X, Y)$ are finite length k-modules for all objects X, Y. Let $D = \operatorname{Hom}_k(-E)$ denote Matlis duality given by an injective k-module E.

Proposition 6.5.1. *Suppose that $\mathcal{A}$ admits a tilting object and that $\mathcal{A}$ has no non-zero projective or injective objects. Then there is an equivalence $F \colon \mathcal{A} \xrightarrow{\sim} \mathcal{A}$ together with natural isomorphisms*

$$D \operatorname{Ext}^1_{\mathcal{A}}(X, Y) \xrightarrow{\sim} \operatorname{Hom}_{\mathcal{A}}(Y, FX) \qquad (X, Y \in \mathcal{A}). \qquad (6.5.2)$$

Proof Let $T \in \mathcal{A}$ be a tilting object and set $\Lambda = \operatorname{End}_{\mathcal{A}}(T)$. Then it follows from Theorem 5.1.2 that $\operatorname{RHom}_{\mathcal{A}}(T, -)$ induces a triangle equivalence $\mathbf{D}^b(\mathcal{A}) \xrightarrow{\sim} \mathbf{D}^b(\operatorname{mod}\Lambda)$. Moreover, the algebra Λ has finite global dimension. Next we apply Theorem 6.4.13. Thus there is a triangle equivalence $F \colon \mathbf{D}^b(\mathcal{A}) \xrightarrow{\sim} \mathbf{D}^b(\mathcal{A})$ together with natural isomorphisms

$$D \operatorname{Hom}_{\mathbf{D}(\mathcal{A})}(X, Y) \xrightarrow{\sim} \operatorname{Hom}_{\mathbf{D}(\mathcal{A})}(Y, FX) \qquad (X, Y \in \mathbf{D}^b(\mathcal{A})).$$

We identify $\mathcal{A}$ with the full subcategory of complexes in $\mathbf{D}^b(\mathcal{A})$ that are concentrated in degree zero and claim that $F(\mathcal{A}) \subseteq \Sigma(\mathcal{A})$. Fix an indecomposable object $X \in \mathcal{A}$. Then we find $Y \in \mathcal{A}$ such that $FX = \Sigma^n Y$ for some $n \in \mathbb{Z}$ since

$\mathcal{A}$ is hereditary (Proposition 4.4.15). We compute

$$D \operatorname{Hom}_{\mathbf{D}(\mathcal{A})}(X, \Sigma^n Y) \cong \operatorname{Hom}_{\mathbf{D}(\mathcal{A})}(\Sigma^n Y, \Sigma^n Y) \neq 0,$$

and therefore $n \in \{0, 1\}$. Suppose $n = 0$ and let $Z \in \mathcal{A}$. Then we have

$$D \operatorname{Ext}^1_{\mathcal{A}}(X, Z) = D \operatorname{Hom}_{\mathbf{D}(\mathcal{A})}(X, \Sigma Z) \cong \operatorname{Hom}_{\mathbf{D}(\mathcal{A})}(\Sigma Y, Y) = 0$$

which means that X is projective. This is a contradiction, and therefore $n = 1$. The dual result shows for a quasi-inverse F^- of F that $F^-(\mathcal{A}) \subseteq \Sigma^{-1}(\mathcal{A})$. Thus $F' = \Sigma^{-1} \circ F$ yields an equivalence $\mathcal{A} \xrightarrow{\sim} \mathcal{A}$ together with natural isomorphisms

$$D \operatorname{Ext}^1_{\mathcal{A}}(X, Y) \xrightarrow{\sim} \operatorname{Hom}_{\mathcal{A}}(Y, F'X) \qquad (X, Y \in \mathcal{A}). \qquad \square$$

Remark 6.5.3. Let $\mathcal{A}$ be an abelian category with an equivalence $F \colon \mathcal{A} \xrightarrow{\sim} \mathcal{A}$ and a natural isomorphism (6.5.2). Then $\mathcal{A}$ is hereditary and $\mathcal{A}$ has no non-zero projective or injective objects.

We obtain a version of Serre duality for the projective line over a field.

Example 6.5.4. The category $\operatorname{coh} \mathbb{P}^1_k$ of coherent sheaves on the projective line over a field k admits a tilting object (Lemma 5.1.16) and therefore an equivalence $\operatorname{coh} \mathbb{P}^1_k \xrightarrow{\sim} \operatorname{coh} \mathbb{P}^1_k$ providing a natural isomorphism

$$D \operatorname{Ext}^1(X, Y) \xrightarrow{\sim} \operatorname{Hom}(Y, X \otimes \Omega_{\mathbb{P}^1_k}) \qquad (X, Y \in \operatorname{coh} \mathbb{P}^1_k)$$

where $\Omega_{\mathbb{P}^1_k} \cong \mathcal{O}(-2)$ denotes the *sheaf of differential forms*.

A Gorenstein Algebra of Dimension One

We discuss a small example of an Artin algebra that is Gorenstein, but neither of finite global dimension nor self-injective.

Fix a field k and consider the finite dimensional algebra $\Lambda = k[\varepsilon] \otimes_k \left[\begin{smallmatrix} k & k \\ 0 & k \end{smallmatrix} \right]$, which is Gorenstein of dimension one and isomorphic to the path algebra of the quiver

$$\varepsilon_1 \,\circlearrowright\, 1 \xrightarrow{\ \alpha\ } 2 \,\circlearrowleft\, \varepsilon_2$$

modulo the relations

$$\varepsilon_1^2 = 0 = \varepsilon_2^2 \qquad \text{and} \qquad \alpha \varepsilon_1 = \varepsilon_2 \alpha.$$

There are two simple modules corresponding to the vertices 1 and 2. The indecomposable projective modules are given by idempotents e_1 and e_2 as follows:

$$e_1 \Lambda = k e_1 \oplus k \varepsilon_1 \quad \text{and} \quad e_2 \Lambda = k e_2 \oplus k \varepsilon_2 \oplus k \alpha \oplus k \varepsilon_2 \alpha.$$

Figure 6.1 Auslander–Reiten quiver of Λ

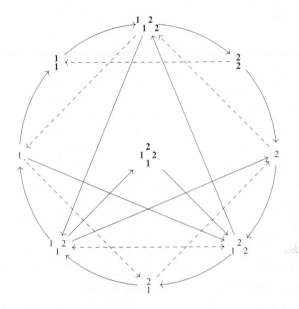

The algebra Λ is representation finite and has precisely nine indecomposable modules. The Auslander–Reiten quiver of Λ is shown in Figure 6.1. The vertices represent the indecomposables via their composition series. There is a solid arrow $X \to Y$ if there is an irreducible morphism, and a dashed arrow $X \dashrightarrow Y$ when $Y = D\,\mathrm{Tr}\,X$. The modules of finite projective and injective dimension have bold dimension vectors. The indecomposable Gorenstein projectives and Gorenstein injectives are given by

$$\mathrm{Gproj}\,\Lambda = \left\{ 1, \tfrac{1}{1}, \tfrac{2}{1}, {}^{1}{}_{1}2, {}^{1}_{1}{}^{2} \right\} \quad \text{and} \quad \mathrm{Ginj}\,\Lambda = \left\{ 2, \tfrac{2}{2}, \tfrac{2}{1}, {}_{1}{}^{2}2, {}^{1}_{1}{}^{2} \right\}.$$

A module belongs to all three classes if and only if it is projective and injective; there is a unique indecomposable with this property.

Notes

The ubiquity of Gorenstein rings was pointed out in a seminal article by Bass [22]. The terminology goes back to Grothendieck who studied duality phenomena for Gorenstein schemes, as explained in notes by Hartshorne [106].

For non-commutative rings it was Iwanaga who proposed to call a ring Gorenstein when it is of finite self-injective dimension on both sides [116]. The lemma that justifies the dimension of a Gorenstein ring is due to Zaks [202].

The decomposition theorem and the existence of approximations for modules over Gorenstein rings were established by Auslander and Buchweitz [13, 44]. The notion of a cotorsion pair provides a convenient language; it was introduced by Salce [179] in the context of abelian groups and we refer to work of Beligiannis and Reiten [28] for a comprehensive study in more general contexts.

For a Gorenstein ring the stable category of Gorenstein projective modules is discussed extensively in notes by Buchweitz [44]. In this work he introduced the singularity category (under the name 'stabilised derived category') that was later rediscovered by Orlov in the geometric context [152]. The example of a module M such that $\mathrm{Ext}^1(M, -)$ has a direct summand not given by a direct summand of M is due to Auslander [8]. This example shows that the singularity category need not be idempotent complete; see [153] for a geometric analysis of this phenomenon.

Serre duality for the derived category of a finite dimensional algebra of finite global dimension appears implicitly in Happel's work [101], while the notion of a Serre functor was introduced by Bondal and Kapranov [38] formalising Serre's duality for algebraic varieties [189]. In fact, Happel showed that the derived category of a finite dimensional algebra has Auslander–Reiten triangles if and only if the algebra has finite global dimension [103], while Reiten and Van den Bergh showed for any triangulated category that the existence of Auslander–Reiten triangles is equivalent to the existence of a Serre functor [168]. The Nakayama functor was introduced by Gabriel [80]; it is the categorical analogue of the Nakayama automorphism that permutes the isomorphism classes of simple modules over a self-injective algebra.

Auslander–Reiten duality based on the dual of transpose $D\,\mathrm{Tr}$ was initiated for Artin algebras in [15] by Auslander and Reiten and then extended to more general settings in [11]. For Gorenstein projective modules the Auslander–Reiten theory was developed in [17]. In the context of modular representations of finite groups a classical version of Serre duality is due to Tate [46].

Gentle algebras were introduced by Assem and Skowroński [6]. The fact that gentle algebras are Gorenstein is due to Geiß and Reiten [90]; the proof given here was suggested by Plamondon, with a modification by Briggs and Bennett-Tennenhaus. For the noetherianness of gentle algebras, see [60].

7

Tilting in Exact Categories

Contents

7.1	**Cotorsion Pairs**		**208**
	Thick Subcategories and Resolutions		208
	Self-Orthogonal Subcategories		208
	Cotorsion Pairs		211
	Resolving and Coresolving Subcategories		211
7.2	**Tilting in Exact Categories**		**215**
	Tilting Objects		215
	Tilting Modules		218
	Tilting Objects and Cotorsion Pairs		220
	Finite Global Dimension		221
	APR Tilting Modules		222
	Tilting Objects for Quivers of Type A_n		224
Notes			**226**

We discuss tilting objects in exact categories. An object T of an exact category $\mathcal{A}$ is a tilting object if it has no self-extensions and generates $\mathcal{A}$ as a thick subcategory. There is an analogous notion of a tilting object in a triangulated category and we see that T is tilting in $\mathcal{A}$ if and only if it is tilting when viewed as an object of the derived category $\mathbf{D}^b(\mathcal{A})$.

Any tilting object in $\mathcal{A}$ gives rise to a cotorsion pair for $\mathcal{A}$, and we characterise such cotorsion pairs. In fact, a cotorsion pair $(\mathcal{X}, \mathcal{Y})$ is determined either by $\mathcal{X}$ or by $\mathcal{Y}$. The subcategory $\mathcal{X}$ is resolving and contravariantly finite, while $\mathcal{Y}$ is coresolving and covariantly finite. This yields a correspondence between equivalence classes of tilting objects and appropriate subcategories.

7.1 Cotorsion Pairs

We introduce cotorsion pairs for exact categories and study their basic properties. A cotorsion pair is given by a pair of subcategories, and of particular interest are subcategories whose objects are defined via resolutions or coresolutions.

Thick Subcategories and Resolutions

Let $\mathcal{A}$ be an exact category. A full additive subcategory $\mathcal{C} \subseteq \mathcal{A}$ is *thick* if it is closed under direct summands and satisfies the following *two out of three property*: an admissible exact sequence $0 \to X \to Y \to Z \to 0$ lies in $\mathcal{C}$ if two of X, Y, Z are in $\mathcal{C}$. Given a class of objects $\mathcal{C} \subseteq \mathcal{A}$, we write Thick($\mathcal{C}$) for the smallest thick subcategory of $\mathcal{A}$ that contains $\mathcal{C}$.

Let $\mathcal{C} \subseteq \mathcal{A}$ be a full additive subcategory. A *finite $\mathcal{C}$-resolution* of an object A in $\mathcal{A}$ is an admissible exact sequence (that is, an acyclic complex)

$$0 \longrightarrow X_r \longrightarrow \cdots \longrightarrow X_1 \longrightarrow X_0 \longrightarrow A \longrightarrow 0$$

such that $X_i \in \mathcal{C}$ for all i. We write Res($\mathcal{C}$) for the full subcategory of objects in $\mathcal{A}$ that admit a finite $\mathcal{C}$-resolution. A *finite $\mathcal{C}$-coresolution* is defined dually, and we write Cores($\mathcal{C}$) for the full subcategory of objects in $\mathcal{A}$ that admit a finite $\mathcal{C}$-coresolution.

Self-Orthogonal Subcategories

Let $\mathcal{A}$ be an exact category. A full additive subcategory $\mathcal{C} \subseteq \mathcal{A}$ is *self-orthogonal* if it is closed under direct summands and $\mathrm{Ext}^n(X, Y) = 0$ for all X, Y in $\mathcal{C}$ and $n \neq 0$.

We wish to resolve objects in $\mathcal{A}$ via objects from a self-orthogonal subcategory $\mathcal{C}$. A basic tool is the derived category $\mathbf{D}(\mathcal{A})$. In fact, the inclusion $\mathcal{C} \to \mathcal{A}$ is exact and induces an exact functor $\mathbf{D}(\mathcal{C}) \to \mathbf{D}(\mathcal{A})$.

Lemma 7.1.1. *Let $\mathcal{A}$ be an exact category and $\mathcal{C} \subseteq \mathcal{A}$ a self-orthogonal subcategory. Then the canonical functor*

$$\mathbf{K}^b(\mathcal{C}) \xrightarrow{\sim} \mathbf{D}^b(\mathcal{C}) \longrightarrow \mathbf{D}(\mathcal{A})$$

is fully faithful and identifies $\mathbf{K}^b(\mathcal{C})$ with a thick subcategory of $\mathbf{D}(\mathcal{A})$.

Proof The functor $\mathbf{K}^b(\mathcal{C}) \to \mathbf{D}^b(\mathcal{C})$ is an equivalence since $\mathcal{C}$ is split exact, and $\mathbf{D}^b(\mathcal{C}) \to \mathbf{D}(\mathcal{A})$ is fully faithful by a dévissage argument (Lemma 3.1.8). Thus the composite identifies $\mathbf{K}^b(\mathcal{C})$ with a triangulated subcategory of $\mathbf{D}(\mathcal{A})$; it is thick because $\mathcal{C}$ is closed under direct summands. $\qquad\square$

Thick subcategories have been defined for exact and for triangulated categories. The next lemma shows that these two notions are compatible.

Let us write

$$\Phi \colon \mathcal{A} \longrightarrow \mathbf{D}(\mathcal{A})$$

for the inclusion that identifies $\mathcal{A}$ with the complexes concentrated in degree zero. We call a complex X in $\mathcal{A}$ *bounded* if $X^n = 0$ for $|n| \gg 0$.

Lemma 7.1.2. *Let $\mathcal{A}$ be an exact category and $\mathcal{C} \subseteq \mathcal{A}$ a self-orthogonal subcategory. For an object $A \in \mathcal{A}$ the following are equivalent.*

(1) $A \in \mathrm{Thick}(\mathcal{C})$.
(2) $\Phi(A) \in \mathrm{Thick}(\Phi(\mathcal{C}))$.
(3) *There is a bounded complex X in $\mathcal{C}$ that admits acyclic truncations*

$$\cdots \longrightarrow X^{-3} \longrightarrow X^{-2} \longrightarrow X^{-1} \longrightarrow \mathrm{Coker}\, d^{-2} \longrightarrow 0 \longrightarrow \cdots$$

and

$$\cdots \longrightarrow 0 \longrightarrow \mathrm{Ker}\, d^{0} \longrightarrow X^{0} \longrightarrow X^{1} \longrightarrow X^{2} \longrightarrow \cdots$$

which induce an admissible exact sequence

$$0 \longrightarrow \mathrm{Coker}\, d^{-2} \longrightarrow \mathrm{Ker}\, d^{0} \longrightarrow A \longrightarrow 0. \qquad (7.1.3)$$

(4) *There is a bounded complex X in $\mathcal{C}$ that admits acyclic truncations*

$$\cdots \longrightarrow 0 \longrightarrow \mathrm{Ker}\, d^{1} \longrightarrow X^{1} \longrightarrow X^{2} \longrightarrow X^{3} \longrightarrow \cdots$$

and

$$\cdots \longrightarrow X^{-2} \longrightarrow X^{-1} \longrightarrow X^{0} \longrightarrow \mathrm{Coker}\, d^{-1} \longrightarrow 0 \longrightarrow \cdots$$

which induce an admissible exact sequence

$$0 \longrightarrow A \longrightarrow \mathrm{Coker}\, d^{-1} \longrightarrow \mathrm{Ker}\, d^{1} \longrightarrow 0. \qquad (7.1.4)$$

Proof (1) $\Rightarrow$ (2): This is clear, since any admissible exact sequence in $\mathcal{A}$ induces an exact triangle in $\mathbf{D}(\mathcal{A})$.

(2) $\Rightarrow$ (3): An object X in $\mathrm{Thick}(\Phi(\mathcal{C}))$ is a bounded complex with $X^n \in \mathcal{C}$ for all $n \in \mathbb{Z}$, by Lemma 7.1.1. The truncations exist when X is quasi-isomorphic to $\Phi(A)$.

(3) $\Leftrightarrow$ (4): Reverse arrows and signs of the degrees.

(3) $\Rightarrow$ (1): Clear. $\qquad\qquad\qquad\qquad\qquad\qquad\qquad\qquad\qquad\qquad\qquad$ $\square$

We have the following immediate consequence.

Proposition 7.1.5. *For a self-orthogonal subcategory* $\mathcal{C} \subseteq \mathcal{A}$ *we have*

$$\Phi^{-1}(\mathrm{Thick}(\Phi(\mathcal{C}))) = \mathrm{Thick}(\mathcal{C}). \qquad \square$$

For a class of objects $\mathcal{C} \subseteq \mathcal{A}$ we set

$$^{\perp}\mathcal{C} = \{X \in \mathcal{A} \mid \mathrm{Ext}^n(X, Y) = 0 \text{ for all } Y \in \mathcal{C}, \, n > 0\}$$

and

$$\mathcal{C}^{\perp} = \{Y \in \mathcal{A} \mid \mathrm{Ext}^n(X, Y) = 0 \text{ for all } X \in \mathcal{C}, \, n > 0\}.$$

Lemma 7.1.6. *Let* $\mathcal{A}$ *be an exact category and* $\mathcal{C} \subseteq \mathcal{A}$ *a self-orthogonal subcategory. Then we have the following equalities:*

$$^{\perp}\mathcal{C} \cap \mathrm{Res}(\mathcal{C}) = \mathcal{C} = \mathrm{Cores}(\mathcal{C}) \cap \mathcal{C}^{\perp}$$

$$^{\perp}\mathcal{C} \cap \mathrm{Thick}(\mathcal{C}) = \mathrm{Cores}(\mathcal{C}) \qquad and \qquad \mathcal{C}^{\perp} \cap \mathrm{Thick}(\mathcal{C}) = \mathrm{Res}(\mathcal{C}).$$

Proof We show the first equality. Then the second follows by duality. The inclusion $^{\perp}\mathcal{C} \cap \mathrm{Res}(\mathcal{C}) \supseteq \mathcal{C}$ is clear. Thus we fix $A \in {}^{\perp}\mathcal{C} \cap \mathrm{Res}(\mathcal{C})$. An induction on the length n of a $\mathcal{C}$-resolution shows that A is in $\mathcal{C}$. The case $n = 0$ is clear. If $n > 0$, consider an exact sequence $\eta \colon 0 \to A' \to C \to A \to 0$ with $C \in \mathcal{C}$. Then $A' \in {}^{\perp}\mathcal{C} \cap \mathrm{Res}(\mathcal{C})$, and $A' \in \mathcal{C}$ by the inductive hypothesis. Thus the sequence η splits, and A is in $\mathcal{C}$ since $\mathcal{C}$ is closed under direct summands.

Next we verify the third equality. Then the last follows by duality. We have $^{\perp}\mathcal{C} \supseteq \mathrm{Cores}(\mathcal{C})$ since $^{\perp}\mathcal{C}$ contains $\mathcal{C}$ and is closed under kernels of admissible epimorphisms. The inclusion $\mathrm{Thick}(\mathcal{C}) \supseteq \mathrm{Cores}(\mathcal{C})$ is clear. Now fix A in $^{\perp}\mathcal{C} \cap \mathrm{Thick}(\mathcal{C})$. We apply Lemma 7.1.2 and choose a bounded complex X in $\mathcal{C}$ that is quasi-isomorphic to $\Phi(A)$. We have $\mathrm{Ker}\, d^1 \in {}^{\perp}\mathcal{C}$, and then the sequence (7.1.4) implies $\mathrm{Coker}\, d^{-1} \in {}^{\perp}\mathcal{C}$ since $^{\perp}\mathcal{C}$ is extension closed. From the first equality it follows that $\mathrm{Coker}\, d^{-1} \in \mathcal{C}$. Then

$$0 \longrightarrow A \longrightarrow \mathrm{Coker}\, d^{-1} \longrightarrow X^1 \longrightarrow X^2 \longrightarrow \cdots$$

yields a finite $\mathcal{C}$-coresolution of A. $\qquad \square$

The category $\mathrm{Proj}\,\mathcal{A}$ of projective objects in $\mathcal{A}$ is a particular example of a self-orthogonal subcategory.

Proposition 7.1.7. *Let* $\mathcal{A}$ *be an exact category and* $\mathcal{P} \subseteq \mathrm{Proj}\,\mathcal{A}$ *a full additive subcategory closed under direct summands. Then* $\mathrm{Thick}(\mathcal{P}) = \mathrm{Res}(\mathcal{P})$.

Proof The inclusion $\mathrm{Res}(\mathcal{P}) \subseteq \mathrm{Thick}(\mathcal{P})$ is clear. Thus we may assume that $\mathcal{A} = \mathrm{Thick}(\mathcal{P})$. Clearly, $\mathcal{P}^{\perp} = \mathcal{A}$. Then $\mathcal{A} = \mathrm{Res}(\mathcal{P})$ by Lemma 7.1.6. $\qquad \square$

Corollary 7.1.8. *Let Λ be a ring. Then a Λ-module X viewed as a complex concentrated in degree zero belongs to $\mathbf{D}^{\mathrm{perf}}(\Lambda)$ if and only if X admits a finite length projective resolution*

$$0 \longrightarrow P_n \longrightarrow \cdots \longrightarrow P_1 \longrightarrow P_0 \longrightarrow X \longrightarrow 0$$

such that each P_i is finitely generated.

Proof Combine Proposition 7.1.5 and Proposition 7.1.7. □

Cotorsion Pairs

Let $\mathcal{A}$ be an exact category and let $\mathfrak{X}, \mathcal{Y}$ be full subcategories of $\mathcal{A}$. Then $(\mathfrak{X}, \mathcal{Y})$ is a (hereditary and complete) *cotorsion pair* for $\mathcal{A}$ if

$$\mathfrak{X}^{\perp} = \mathcal{Y} \qquad \text{and} \qquad \mathfrak{X} = {}^{\perp}\mathcal{Y}$$

and every object $A \in \mathcal{A}$ fits into admissible exact sequences

$$0 \longrightarrow Y_A \longrightarrow X_A \longrightarrow A \longrightarrow 0 \qquad \text{and} \qquad 0 \longrightarrow A \longrightarrow Y^A \longrightarrow X^A \longrightarrow 0$$

with $X_A, X^A \in \mathfrak{X}$ and $Y_A, Y^A \in \mathcal{Y}$.

Remark 7.1.9. Let $(\mathfrak{X}, \mathcal{Y})$ be a cotorsion pair for $\mathcal{A}$ and set $\mathcal{C} = \mathfrak{X} \cap \mathcal{Y}$.

(1) We have $X_A \in \mathcal{C}$ if $A \in \mathcal{Y}$, and $Y^A \in \mathcal{C}$ if $A \in \mathfrak{X}$.

(2) The above exact sequences are uniquely determined up to isomorphism in the quotient category $\mathcal{A}/\mathcal{C}$ (that is obtained from $\mathcal{A}$ by annihilating all morphisms that factor through an object in $\mathcal{C}$). In fact, the assignment $A \mapsto X_A$ gives a right adjoint of the inclusion $\mathfrak{X}/\mathcal{C} \to \mathcal{A}/\mathcal{C}$, while the assignment $A \mapsto Y^A$ gives a left adjoint of the inclusion $\mathcal{Y}/\mathcal{C} \to \mathcal{A}/\mathcal{C}$.

Proposition 7.1.10. *Let $\mathcal{A}$ be an exact category and let $\mathcal{C} \subseteq \mathcal{A}$ be a self-orthogonal subcategory such that* $\mathrm{Thick}(\mathcal{C}) = \mathcal{A}$. *Then*

$$({}^{\perp}\mathcal{C}, \mathcal{C}^{\perp}) = (\mathrm{Cores}(\mathcal{C}), \mathrm{Res}(\mathcal{C}))$$

is a cotorsion pair for $\mathcal{A}$ and ${}^{\perp}\mathcal{C} \cap \mathcal{C}^{\perp} = \mathcal{C}$.

Proof Combine Lemma 7.1.2 and Lemma 7.1.6. □

Resolving and Coresolving Subcategories

Let $\mathcal{A}$ be an exact category and let $\mathfrak{X}, \mathcal{Y}$ be full subcategories of $\mathcal{A}$. The subcategory $\mathfrak{X}$ is *resolving* if $\mathfrak{X}$ is closed under extensions, direct summands, kernels of admissible epimorphisms, and for each object $A \in \mathcal{A}$ there is an admissible epimorphism $X \to A$ with $X \in \mathcal{X}$. Dually, the subcategory $\mathcal{Y}$ is *coresolving* if $\mathcal{Y}$ is resolving when viewed as a full subcategory of $\mathcal{A}^{\mathrm{op}}$.

Given an object $A \in \mathcal{A}$, a morphism $X \to A$ with $X \in \mathcal{X}$ is called a *right $\mathcal{X}$-approximation* of A if the induced map $\mathrm{Hom}(X', X) \to \mathrm{Hom}(X', A)$ is surjective for every object $X' \in \mathcal{X}$. The subcategory $\mathcal{X}$ is *contravariantly finite* if every object $A \in \mathcal{A}$ admits a right $\mathcal{X}$-approximation. Dually, a morphism $A \to Y$ with $Y \in \mathcal{Y}$ is called a *left $\mathcal{Y}$-approximation* of A if the induced map $\mathrm{Hom}(Y, Y') \to \mathrm{Hom}(A, Y')$ is surjective for every object $Y' \in \mathcal{Y}$. The subcategory $\mathcal{Y}$ is *covariantly finite* if every object $A \in \mathcal{A}$ admits a left $\mathcal{Y}$-approximation.

Example 7.1.11. A full subcategory $\mathcal{X} \subseteq \mathcal{A}$ is contravariantly finite if the inclusion admits a right adjoint $p \colon \mathcal{A} \to \mathcal{X}$. In that case the counit $p(A) \to A$ yields a right $\mathcal{X}$-approximation for each object $A \in \mathcal{A}$. Dually, $\mathcal{Y} \subseteq \mathcal{A}$ is covariantly finite if the inclusion admits a left adjoint $q \colon \mathcal{A} \to \mathcal{X}$, and then the unit $A \to q(A)$ yields a left $\mathcal{Y}$-approximation for $A \in \mathcal{A}$.

Lemma 7.1.12. *Let $(\mathcal{X}, \mathcal{Y})$ be a cotorsion pair for $\mathcal{A}$.*

(1) *The subcategory $\mathcal{X} \subseteq \mathcal{A}$ is resolving and contravariantly finite.*
(2) *The subcategory $\mathcal{Y} \subseteq \mathcal{A}$ is coresolving and covariantly finite.*

Proof Clear.   □

A morphism $\alpha \colon X \to Y$ is called *right minimal* if every endomorphism $\phi \colon X \to X$ with $\alpha\phi = \alpha$ is invertible. Dually, α is *left minimal* if every endomorphism $\psi \colon Y \to Y$ with $\psi\alpha = \alpha$ is invertible. Note that any morphism $\phi \colon X \to Y$ in a Krull–Schmidt category admits a decomposition $X = X' \oplus X''$ such that $\phi|_{X'}$ is right minimal and $\phi|_{X''} = 0$. There is an analogue for left minimal morphisms.

The following is known as *Wakamatsu's lemma*.

Lemma 7.1.13 (Wakamatsu). *Let $\mathcal{X}$ and $\mathcal{Y}$ be extension closed subcategories of $\mathcal{A}$ and $A \in \mathcal{A}$.*

(1) *Let $0 \to Y \to X \xrightarrow{\phi} A \to 0$ be an exact sequence in $\mathcal{A}$ such that ϕ is a right minimal $\mathcal{X}$-approximation. Then $\mathrm{Ext}^1(X', Y) = 0$ for all $X' \in \mathcal{X}$.*
(2) *Let $0 \to A \xrightarrow{\phi} Y \to X \to 0$ be an exact sequence in $\mathcal{A}$ such that ϕ is a left minimal $\mathcal{Y}$-approximation. Then $\mathrm{Ext}^1(X, Y') = 0$ for all $Y' \in \mathcal{Y}$.*

Proof We prove (1), and (2) is dual. An exact sequence $0 \to Y \to E \to X' \to 0$ gives rise to the following commutative diagram with exact rows and columns.

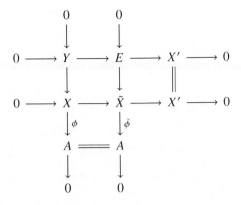

We have $\tilde{X} \in \mathcal{X}$ since $\mathcal{X}$ is extension closed, and $\tilde{\phi}$ factors through ϕ since ϕ is a right $\mathcal{X}$-approximation. Then the minimality of ϕ implies that $X \to \tilde{X}$ is a split monomorphism. The approximation ϕ induces the following exact sequence

$$0 \longrightarrow \mathrm{Hom}(X',Y) \longrightarrow \underset{\alpha}{\mathrm{Hom}(X',X)} \longrightarrow \mathrm{Hom}(X',A) \;\rightharpoondown$$
$$ \rightharpoondown\; \mathrm{Ext}^1(X',Y) \xrightarrow{\beta} \mathrm{Ext}^1(X',X) \longrightarrow \mathrm{Ext}^1(X',A) \longrightarrow \cdots$$

and we have shown that $\beta = 0$. On the other hand, $\alpha = 0$ since $\mathrm{Hom}(X',\phi)$ is surjective. Thus $\mathrm{Ext}^1(X',Y) = 0$. $\qquad\square$

Lemma 7.1.14. *Let $\mathcal{A}$ be an exact category and $\mathcal{Y} \subseteq \mathcal{A}$ a full additive sub-category. Then $({}^{\perp}\mathcal{Y}, \mathcal{Y})$ is a cotorsion pair for $\mathcal{A}$ if and only if the following holds.*

(1) *Each object $A \in \mathcal{A}$ fits into an admissible exact sequence*

$$0 \longrightarrow A \longrightarrow Y^A \longrightarrow X^A \longrightarrow 0$$

   with $X^A \in {}^{\perp}\mathcal{Y}$ and $Y^A \in \mathcal{Y}$.
(2) *For each object $A \in \mathcal{A}$ there is an admissible epimorphism $X \to A$ with $X \in {}^{\perp}\mathcal{Y}$.*
(3) *The subcategory $\mathcal{Y} \subseteq \mathcal{A}$ is closed under direct summands.*

Proof For an object $A \in \mathcal{A}$ we need to construct an admissible exact sequence $0 \to Y_A \to X_A \to A \to 0$. To this end choose an admissible epimorphism

$X \to A$ with $X \in {}^{\perp}\mathcal{Y}$ and form the following pushout diagram.

$$
\begin{array}{ccccccccc}
0 & \longrightarrow & U & \longrightarrow & X & \longrightarrow & A & \longrightarrow & 0 \\
& & \downarrow & & \downarrow & & \| & & \\
0 & \longrightarrow & Y^U & \longrightarrow & X' & \longrightarrow & A & \longrightarrow & 0
\end{array}
$$

Then the bottom row is the desired exact sequence. It is easily checked that $({}^{\perp}\mathcal{Y})^{\perp} = \mathcal{Y}$, since $\mathcal{Y}$ is closed under direct summands. $\qquad\square$

In order to apply the above lemma, we make the following observation. Let $\mathcal{A}$ be an exact category with enough projective objects and $\mathcal{X} \subseteq \mathcal{A}$ a resolving subcategory. Then a dimension shift argument shows for $Y \in \mathcal{A}$ that

$$
Y \in \mathcal{X}^{\perp} \quad \Longleftrightarrow \quad \mathrm{Ext}^1(X, Y) = 0 \text{ for all } X \in \mathcal{X}
$$

since $\mathrm{Ext}^{p+q}(X, Y) \cong \mathrm{Ext}^p(\Omega^q X, Y)$ for $p, q \geq 1$.

Corollary 7.1.15. *Let $\mathcal{A}$ be an exact category with enough projective objects and suppose that $\mathcal{A}$ is a Krull–Schmidt category. Then the assignment*

$$
\mathcal{Y} \longmapsto ({}^{\perp}\mathcal{Y}, \mathcal{Y})
$$

induces a bijection between the covariantly finite coresolving subcategories of $\mathcal{A}$ and the cotorsion pairs for $\mathcal{A}$.

Proof Any cotorsion pair yields a covariantly finite coresolving subcategory by Lemma 7.1.12. Conversely, if $\mathcal{Y} \subseteq \mathcal{A}$ is covariantly finite and coresolving, then the assumptions in Lemma 7.1.14 are satisfied, thanks to Lemma 7.1.13 and the fact that $\mathcal{A}$ is a Krull–Schmidt category. Thus $({}^{\perp}\mathcal{Y}, \mathcal{Y})$ is a cotorsion pair for $\mathcal{A}$. $\qquad\square$

Corollary 7.1.16. *Let $\mathcal{A}$ be an exact category with enough projective and enough injective objects. Suppose also that $\mathcal{A}$ is a Krull–Schmidt category. Then the assignments*

$$
\mathcal{X} \longmapsto \mathcal{X}^{\perp} \quad and \quad {}^{\perp}\mathcal{Y} \longleftarrow \mathcal{Y}
$$

induce mutually inverse bijections between the contravariantly finite resolving subcategories of $\mathcal{A}$ and the covariantly finite coresolving subcategories of $\mathcal{A}$.

Proof Apply Corollary 7.1.15 and the dual assertion. $\qquad\square$

Example 7.1.17. (1) Let Λ be an Artin algebra. Then $\mathrm{mod}\,\Lambda$ is an abelian Krull–Schmidt category with enough projective and enough injective objects.

(2) Let Λ be an Artin algebra and suppose that Λ is Gorenstein. Then the category of finitely generated Λ-modules of finite projective dimension is an

exact Krull–Schmidt category with enough projective and enough injective objects.

7.2 Tilting in Exact Categories

We introduce tilting objects in exact categories and discuss the connection with cotorsion pairs. Also, we show that each tilting object gives rise to a derived equivalence. The correspondence between tilting objects and cotorsion pairs is very explicit: a tilting object T corresponds to the pair $({}^{\perp}T, T^{\perp})$. The correspondence is of particular interest for modules over Artin algebras. Also, we characterise the subcategories which are of the form ${}^{\perp}T$ or $T^{\perp}$.

Tilting Objects

Before giving the definition of a tilting object, let us point out that there is a plethora of different definitions in the literature. Each definition depends on its context. There are definitions for module categories, abelian categories, triangulated categories etc. Also, a definition may require the existence of set-indexed coproducts.

Let $\mathcal{A}$ be an exact category. An object T is a *tilting object* if $\mathrm{Ext}^n(T, T) = 0$ for all $n \neq 0$ and $\mathrm{Thick}(T) = \mathcal{A}$.

For an object X in $\mathcal{A}$ we denote by add X the full subcategory consisting of the direct summands of finite direct sums of copies of X.

Proposition 7.2.1. *Let $\mathcal{A}$ be an exact category and $\mathcal{C} = $ add T for an object $T \in \mathcal{A}$. Then T is tilting if and only if $(\mathrm{Cores}(\mathcal{C}), \mathrm{Res}(\mathcal{C}))$ is a cotorsion pair for $\mathcal{A}$. In that case we have*

$$ {}^{\perp}T = \mathrm{Cores}(\mathcal{C}), \qquad T^{\perp} = \mathrm{Res}(\mathcal{C}), \qquad {}^{\perp}T \cap T^{\perp} = \mathcal{C}. $$

Proof Apply Proposition 7.1.10. □

The definition of a tilting object in an exact category is compatible with the definition of a tilting object in a triangulated category. Let $\mathcal{T}$ be a triangulated category with suspension $\Sigma \colon \mathcal{T} \xrightarrow{\sim} \mathcal{T}$. An object T is a *tilting object* if $\mathrm{Hom}(T, \Sigma^n T) = 0$ for all $n \neq 0$ and $\mathrm{Thick}(T) = \mathcal{T}$.

Proposition 7.2.2. *Let $\mathcal{A}$ be an exact category. An object T in $\mathcal{A}$ is a tilting object if and only if it is a tilting object of $\mathbf{D}^b(\mathcal{A})$ when viewed as a complex concentrated in degree zero.*

Proof Set $\mathcal{C} = $ add T and apply Proposition 7.1.5. □

A tilting object gives rise to a derived equivalence.

Theorem 7.2.3. *Let $\mathcal{A}$ be an exact and idempotent complete category. For an object T with $\Lambda = \mathrm{End}(T)$, the following are equivalent.*

(1) *The object T is a tilting object in $\mathcal{A}$.*

(2) *The functor $\mathrm{Hom}(T, -)$ induces a triangle equivalence $\mathbf{D}^b(\mathcal{A}) \xrightarrow{\sim} \mathbf{D}^{\mathrm{perf}}(\Lambda)$ that makes the following square commutative.*

$$
\begin{array}{ccc}
\mathbf{K}^b(\mathrm{add}\,T) & \xrightarrow{\ \mathrm{Hom}(T,-)\ } & \mathbf{K}^b(\mathrm{proj}\,\Lambda) \\
\downarrow{\wr} & & \downarrow{\wr} \\
\mathbf{D}^b(\mathcal{A}) & \xrightarrow{\quad\sim\quad} & \mathbf{D}^{\mathrm{perf}}(\Lambda)
\end{array}
$$

(3) *There is a triangle equivalence $\mathbf{D}^b(\mathcal{A}) \xrightarrow{\sim} \mathbf{D}^{\mathrm{perf}}(\Lambda)$ that maps T to Λ.*

Proof (1) $\Rightarrow$ (2): Suppose T is tilting. The functor $\mathrm{Hom}(T, -)$ induces an equivalence

$$
\mathrm{add}\,T \xrightarrow{\sim} \mathrm{proj}\,\Lambda
$$

while the vertical functors are fully faithful by Lemma 7.1.1. The functor on the right is surjective on objects by definition, and the functor on the left by Proposition 7.2.2.

(2) $\Rightarrow$ (3): Clear.

(3) $\Rightarrow$ (1): The object Λ is a tilting object in $\mathbf{D}^{\mathrm{perf}}(\Lambda)$. This property is preserved under a triangle equivalence, but also under the embedding $\mathcal{A} \to \mathbf{D}^b(\mathcal{A})$ by Proposition 7.2.2. $\square$

We will see another proof of the equivalence $\mathbf{D}^b(\mathcal{A}) \xrightarrow{\sim} \mathbf{D}^{\mathrm{perf}}(\Lambda)$ when we discuss tilting objects in $\mathbf{D}^b(\mathcal{A})$; see Proposition 9.1.20.

Next we consider Grothendieck groups and derive a consequence from the fact that a triangle equivalence preserves Grothendieck groups. Recall that $K_0(\mathcal{A})$ denotes the Grothendieck group of an exact category, and that $K_0(\Lambda) = K_0(\mathrm{proj}\,\Lambda)$ for any ring Λ.

Corollary 7.2.4. *Let $\mathcal{A}$ be an exact and idempotent complete category. Given a tilting object T with $\Lambda = \mathrm{End}(T)$, then $\mathrm{Hom}(T, -)$ induces an isomorphism*

$$
K_0(\mathcal{A}) \xrightarrow{\sim} K_0(\Lambda).
$$

Proof We have isomorphisms

$$
K_0(\mathcal{A}) \xrightarrow{\sim} K_0(\mathbf{D}^b(\mathcal{A})) \xrightarrow{\sim} K_0(\mathbf{D}^{\mathrm{perf}}(\Lambda)) \xrightarrow{\sim} K_0(\Lambda),
$$

where the first and the third follow from Lemma 4.1.17 and the middle one from Theorem 7.2.3. $\square$

Of particular interest are module categories. Let Λ be a right coherent ring and consider the abelian category $\operatorname{mod}\Lambda$ of finitely presented Λ-modules. If $T \in \operatorname{mod}\Lambda$ is a tilting object and $\Gamma = \operatorname{End}(T)$, then $\operatorname{Hom}(T, -)$ induces a triangle equivalence

$$\mathbf{D}^b(\operatorname{mod}\Lambda) \overset{\sim}{\longrightarrow} \mathbf{D}^{\operatorname{perf}}(\Gamma)$$

by the above theorem. For example, Λ_Λ is a tilting object in $\operatorname{mod}\Lambda$ if and only if every finitely presented Λ-module has finite projective dimension. This reflects the fact that the inclusion $\operatorname{proj}\Lambda \to \operatorname{mod}\Lambda$ induces a triangle equivalence $\mathbf{D}^{\operatorname{perf}}(\Lambda) \overset{\sim}{\to} \mathbf{D}^b(\operatorname{mod}\Lambda)$ if and only if every finitely presented Λ-module has finite projective dimension (Corollary 4.2.9).

The existence of a tilting object imposes some immediate constraints on objects and morphisms in $\mathcal{A}$ and $\mathbf{D}^b(\mathcal{A})$.

Lemma 7.2.5. *Let $\mathcal{A}$ be an exact category and suppose there is a tilting object in $\mathcal{A}$ or in $\mathbf{D}^b(\mathcal{A})$. Then for each pair of objects $X, Y \in \mathbf{D}^b(\mathcal{A})$ we have*

$$\operatorname{Hom}(X, \Sigma^n Y) = 0 \quad \text{for} \quad |n| \gg 0.$$

In particular, for all $X, Y \in \mathcal{A}$ we have $\operatorname{Ext}^n(X, Y) = 0$ for $n \gg 0$.

Proof Suppose there is a tilting object T in $\mathbf{D}^b(\mathcal{A})$. This includes the case that there is a tilting object in $\mathcal{A}$, by Proposition 7.2.2. It follows from the definition of a tilting object that T is homologically finite, so for all $Y \in \mathbf{D}^b(\mathcal{A})$ we have $\operatorname{Hom}(T, \Sigma^n Y) = 0$ for $|n| \gg 0$. The homologically finite objects form a thick subcategory (Example 3.1.7) and therefore all objects in $\mathbf{D}^b(\mathcal{A})$ are homologically finite. $\square$

We have further consequences when $\mathcal{A}$ is a length category.

Lemma 7.2.6. *Let $\mathcal{A}$ be a length category and $T \in \mathbf{D}^b(\mathcal{A})$ a tilting object. Then $\mathcal{A}$ has only finitely many isomorphism classes of simple objects and*

$$\operatorname{gl.dim}\mathcal{A} = \inf_{\substack{S, S' \\ \text{simple}}} \{i \in \mathbb{N} \mid \operatorname{Ext}^{i+1}(S, S') = 0\} < \infty.$$

Proof The length of $H = \bigoplus_n H^n T$ gives a bound for the number of isomorphism classes of simple objects in $\mathcal{A}$. More precisely, let $\mathcal{B} \subseteq \mathcal{A}$ denote the Serre subcategory generated by the composition factors of H. Then T belongs to the thick subcategory of objects $X \in \mathbf{D}^b(\mathcal{A})$ with $H^n X \in \mathcal{B}$ for all n. Thus $\mathcal{B} = \mathcal{A}$.

Having only finitely many simple objects in $\mathcal{A}$, the bound for $\operatorname{gl.dim}\mathcal{A}$ follows from the previous lemma. $\square$

Let us consider another class of exact categories.

Lemma 7.2.7. *Let $\mathcal{A}$ be a Frobenius category and $T \in \mathcal{A}$ a tilting object. Then* $\operatorname{Proj} \mathcal{A} \subseteq \operatorname{add} T$.

Proof Every projective (and injective) object belongs to $^{\perp}T \cap T^{\perp} = \operatorname{add} T$. □

Tilting Modules

We consider an exact category $\mathcal{A}$ and study its tilting objects. A useful assumption is that $\mathcal{A}$ contains a projective tilting object. For example, this holds for a ring Λ when $\mathcal{A}$ equals the category of Λ-modules X having a resolution

$$0 \longrightarrow P_r \longrightarrow \cdots \longrightarrow P_1 \longrightarrow P_0 \longrightarrow X \longrightarrow 0$$

such that each P_i is finitely generated projective. We set

$$\mathcal{P}(\Lambda) := \operatorname{Res}(\operatorname{proj} \Lambda)$$

and note that $\mathcal{P}(\Lambda) = \operatorname{Thick}(\Lambda)$ by Proposition 7.1.7. Let us give a criterion for when $\mathcal{P}(\Lambda)$ is trivial.

Lemma 7.2.8. *We have $\mathcal{P}(\Lambda) = \operatorname{proj} \Lambda$ if and only if $\operatorname{Hom}(X, \Lambda) \neq 0$ for every finitely presented $\Lambda^{\operatorname{op}}$-module $X \neq 0$.*

Proof Write $P^* = \operatorname{Hom}(P, \Lambda)$ for $P \in \operatorname{proj} \Lambda$. We have $\mathcal{P}(\Lambda) = \operatorname{proj} \Lambda$ if and only if every monomorphism $P \to Q$ in $\operatorname{proj} \Lambda$ splits. Such a monomorphism $P \to Q$ splits if and only if $Q^* \to P^*$ is an epimorphism. It remains to observe that $\operatorname{Hom}(X, \Lambda) = 0$ for $X = \operatorname{Coker}(Q^* \to P^*)$. □

We continue with an elementary characterisation of projective tilting objects; so all objects need to have finite projective dimension.

Lemma 7.2.9. *Let $\mathcal{A}$ be an exact category. Then a projective object P is a tilting object if and only if every object $A \in \mathcal{A}$ admits a finite resolution*

$$0 \longrightarrow P_r \longrightarrow \cdots \longrightarrow P_1 \longrightarrow P_0 \longrightarrow A \longrightarrow 0 \qquad (P_i \in \operatorname{add} P).$$

Proof If P is a tilting object, then $P^{\perp} = \operatorname{Res}(\operatorname{add} P)$, by Proposition 7.1.10. Now use that $P^{\perp} = \mathcal{A}$ since P is projective. The other direction is clear since $\operatorname{Res}(\operatorname{add} P) \subseteq \operatorname{Thick}(P)$. □

Proposition 7.2.10. *Let $\mathcal{A}$ be an exact category and $P \in \mathcal{A}$ a projective tilting object. Then an object $T \in \mathcal{A}$ is a tilting object if and only if*

(1) $\operatorname{Ext}^n(T, T) = 0$ *for all $n \neq 0$, and*

(2) *there is an exact sequence*

$$0 \longrightarrow P \longrightarrow T^0 \longrightarrow T^1 \longrightarrow \cdots \longrightarrow T^r \longrightarrow 0 \qquad (T^i \in \operatorname{add} T).$$

Proof If T is a tilting object, then $P \in {}^{\perp}T = \text{Cores}(\text{add}\,T)$, by Proposition 7.1.10. Conversely, if $P \in \text{Cores}(\text{add}\,T)$, then $\mathcal{A} = \text{Thick}(P) \subseteq \text{Thick}(T) \subseteq \mathcal{A}$. $\qquad\square$

When Λ is a ring, then a Λ-module T is called a *tilting module* (of finite projective dimension) if it is a tilting object of the exact category $\mathcal{P}(\Lambda)$. This means $\text{Ext}^n(T, T) = 0$ for all $n \neq 0$, and $\text{Thick}(T) = \text{Thick}(\Lambda)$. More concretely, it follows from the above proposition that a Λ-module T is a tilting module if and only if

(T1) there is an exact sequence
$$0 \longrightarrow P_r \longrightarrow \cdots \longrightarrow P_1 \longrightarrow P_0 \longrightarrow T \longrightarrow 0 \qquad (P_i \in \text{proj}\,\Lambda),$$

(T2) $\text{Ext}^n(T, T) = 0$ for all $n \neq 0$, and

(T3) there is an exact sequence
$$0 \longrightarrow \Lambda \longrightarrow T^0 \longrightarrow T^1 \longrightarrow \cdots \longrightarrow T^s \longrightarrow 0 \qquad (T^i \in \text{add}\,T).$$

Now let Λ be an Artin k-algebra, and write $D = \text{Hom}_k(-, E)$ for the Matlis duality given by an injective k-module E.

Example 7.2.11. The algebra Λ is Gorenstein if and only if $D(\Lambda)_\Lambda$ is a tilting module. In fact, the finite injective dimension of Λ_Λ corresponds to (T3), while the finite injective dimension of ${}_\Lambda\Lambda$ corresponds to (T1).

Example 7.2.12. Let Λ be an algebra such that every module of finite projective dimension is projective, so $\mathcal{P}(\Lambda) = \text{proj}\,\Lambda$. This holds if and only if $\text{Hom}(S, \Lambda) \neq 0$ for every simple Λ^{op}-module S, so for example when Λ is self-injective or local; see Lemma 7.2.8. Then a Λ-module T is tilting if and only if $\text{add}\,T = \text{proj}\,\Lambda$.

Any tilting object gives rise to a derived equivalence by Theorem 7.2.3, and the following result makes this more precise for modules over Artin algebras. For a generalisation involving tilting complexes, see Theorem 9.2.4.

Proposition 7.2.13. *Let Λ and Γ be Artin algebras of finite global dimension. Suppose that T_Λ is a tilting module and $\Gamma \cong \text{End}_\Lambda(T)$. Then we have an adjoint pair of triangle equivalences*

$$\mathbf{D}^b(\text{mod}\,\Lambda) \xrightleftharpoons[\text{RHom}_\Lambda(T,-)]{-\otimes^L_\Gamma T} \mathbf{D}^b(\text{mod}\,\Gamma).$$

Proof The pair of adjoint functors is taken from Proposition 4.3.15 and provides equivalences by Theorem 7.2.3, keeping in mind that $\mathbf{D}^{\text{perf}}(\Gamma) \xrightarrow{\sim} \mathbf{D}^b(\text{mod}\,\Gamma)$. $\qquad\square$

Remark 7.2.14. It suffices to assume that Λ has finite global dimension. Then for any tilting module T the algebra $\operatorname{End}_\Lambda(T)$ has finite global dimension, by Theorem 9.3.11.

Tilting Objects and Cotorsion Pairs

In Proposition 7.2.1 we have seen that each tilting object T yields a cotorsion pair $({}^\perp T, T^\perp)$. Now we wish to characterise the cotorsion pairs of an exact category that are induced by tilting objects. Let us keep the assumption that the category admits a projective tilting object.

Lemma 7.2.15. *Let $\mathcal{A}$ be an exact category and suppose $P \in \mathcal{A}$ is a projective tilting object. For a cotorsion pair $(\mathcal{X}, \mathcal{Y})$ the following are equivalent.*

(1) *There is an exact sequence*
$$0 \longrightarrow P \longrightarrow Y^0 \longrightarrow Y^1 \longrightarrow \cdots \longrightarrow Y^r \longrightarrow 0 \qquad (Y^i \in \mathcal{Y}).$$

(2) $\operatorname{gl.dim} \mathcal{X} < \infty$.

(3) *There exists a tilting object T in $\mathcal{A}$ such that $(\mathcal{X}, \mathcal{Y}) = ({}^\perp T, T^\perp)$.*

Proof (1) $\Rightarrow$ (2): Set $Z^i := \operatorname{Ker}(Y^i \to Y^{i+1})$. Then we have for any $X \in \mathcal{X}$
$$\operatorname{Ext}^1(X, Z^i) \cong \operatorname{Ext}^{i+1}(X, P)$$

and therefore $\operatorname{Ext}^i(X, P) = 0$ for all $i > r$. This implies $\operatorname{Ext}^i(X, -) = 0$ for all $i > r$ since every object in $\mathcal{A}$ has a finite projective resolution; see Lemma 7.2.9.

(2) $\Rightarrow$ (3): Suppose that $\operatorname{gl.dim} \mathcal{X} = r$. We apply successively Remark 7.1.9 and obtain an exact sequence
$$0 \longrightarrow P \longrightarrow T^0 \longrightarrow T^1 \longrightarrow \cdots \longrightarrow T^r \longrightarrow 0 \qquad (T^i \in \mathcal{X} \cap \mathcal{Y})$$

that terminates since $\operatorname{Ext}^{r+1}(X, P) = 0$. Thus $T = T^0 \oplus \cdots \oplus T^r$ is a tilting object by Proposition 7.2.10. We have
$${}^\perp T = {}^\perp(T^\perp) \subseteq {}^\perp \mathcal{Y} = \mathcal{X} \subseteq {}^\perp T$$

where the first equality holds by Proposition 7.2.1. Therefore ${}^\perp T = \mathcal{X}$. Analogously, $T^\perp = \mathcal{Y}$.

(3) $\Rightarrow$ (1): Apply Proposition 7.2.10. $\square$

We call tilting objects T and T' *equivalent* if $\operatorname{add} T = \operatorname{add} T'$.

Proposition 7.2.16. *Let $\mathcal{A}$ be an exact category and P a projective tilting object. Then the assignment $T \mapsto T^\perp = \operatorname{Res}(\operatorname{add} T)$ gives a bijection between the equivalence classes of tilting objects of $\mathcal{A}$ and full additive subcategories $\mathcal{Y} \subseteq \mathcal{A}$ satisfying the following.*

(1) *Each object $A \in \mathcal{A}$ fits into an admissible exact sequence*

$$0 \longrightarrow A \longrightarrow Y^A \longrightarrow X^A \longrightarrow 0$$

with $X^A \in {}^\perp \mathcal{Y}$ and $Y^A \in \mathcal{Y}$.

(2) *There is an exact sequence*

$$0 \longrightarrow P \longrightarrow Y^0 \longrightarrow Y^1 \longrightarrow \cdots \longrightarrow Y^r \longrightarrow 0 \qquad (Y^i \in \mathcal{Y}).$$

(3) *The subcategory $\mathcal{Y} \subseteq \mathcal{A}$ is closed under direct summands.*

Proof   We have a correspondence between tilting objects and cotorsion pairs by Proposition 7.2.1. Combining this with Lemma 7.2.15 the assertion follows, once we observe that a subcategory $\mathcal{Y} \subseteq \mathcal{A}$ satisfying (1)–(3) gives rise to a cotorsion pair $({}^\perp \mathcal{Y}, \mathcal{Y})$ by Lemma 7.1.14. $\square$

Let us consider the case $\mathcal{A} = \operatorname{mod} \Lambda$ when Λ is right coherent.

Proposition 7.2.17. *Let Λ be a right coherent ring of finite global dimension. Then the assignment $T \mapsto ({}^\perp T, T^\perp)$ gives a bijection between the equivalence classes of tilting objects of $\operatorname{mod} \Lambda$ and the cotorsion pairs for $\operatorname{mod} \Lambda$.*

Proof   The assertion follows from Proposition 7.2.1 and Lemma 7.2.15. The inverse maps sends $(\mathcal{X}, \mathcal{Y})$ to $\mathcal{X} \cap \mathcal{Y} = \operatorname{add} T$. $\square$

Finite Global Dimension

We consider an exact category $\mathcal{A}$ and make some additional assumptions:

(1) $\mathcal{A}$ is a Krull–Schmidt category, and
(2) $\mathcal{A}$ admits a projective tilting object.

Then the correspondence in Proposition 7.2.16 can be reformulated as follows.

Theorem 7.2.18. *The assignment $T \mapsto T^\perp = \operatorname{Res}(\operatorname{add} T)$ gives a bijection between the equivalence classes of tilting objects of $\mathcal{A}$ and full additive subcategories $\mathcal{Y} \subseteq \mathcal{A}$ that are covariantly finite and coresolving with $\operatorname{gl.dim} {}^\perp \mathcal{Y} < \infty$. The inverse map sends a subcategory $\mathcal{Y} \subseteq \mathcal{A}$ to an object T satisfying $\operatorname{add} T = {}^\perp \mathcal{Y} \cap \mathcal{Y}$.*

Clearly, the condition $\operatorname{gl.dim} {}^\perp \mathcal{Y} < \infty$ is obsolete when $\operatorname{gl.dim} \mathcal{A} < \infty$.

Proof   We apply the correspondence of Corollary 7.1.15 between cotorsion pairs and covariantly finite and coresolving subcategories. The cotorsion pairs corresponding to tilting objects are characterised in Lemma 7.2.15. Given a tilting cotorsion pair $(\mathcal{X}, \mathcal{Y})$, the tilting object T is determined by the equality $\operatorname{add} T = \mathcal{X} \cap \mathcal{Y}$; see Proposition 7.2.1. $\square$

For a ring Λ we consider the category $\mathcal{P}(\Lambda)$ of modules having a finite projective resolution via finitely generated projective modules. Then gl.dim $\mathcal{P}(\Lambda)$ is called the *finitistic dimension* of Λ; it is conjectured to be finite when Λ is an Artin algebra, and this has been established for many classes of algebras.

Corollary 7.2.19. *Let Λ be an Artin algebra and suppose that Λ is Gorenstein.*

(1) *The assignment $T \mapsto T^\perp = \operatorname{Res}(\operatorname{add} T)$ gives a bijection between the equivalence classes of tilting objects of $\mathcal{P}(\Lambda)$ and full additive subcategories $\mathcal{Y} \subseteq \mathcal{P}(\Lambda)$ that are covariantly finite and coresolving.*
(2) *The assignment $T \mapsto {}^\perp T = \operatorname{Cores}(\operatorname{add} T)$ gives a bijection between the equivalence classes of tilting objects of $\mathcal{P}(\Lambda)$ and full additive subcategories $\mathcal{X} \subseteq \mathcal{P}(\Lambda)$ that are contravariantly finite and resolving.*

Proof Let Λ be Gorenstein of dimension d. This means that the injective dimensions of Λ_Λ and ${}_\Lambda\Lambda$ equal d. Then the projective dimension of every injective Λ-module is bounded by d. Thus gl.dim $\mathcal{P}(\Lambda) = d < \infty$ (Lemma 6.2.2) and $\mathcal{P}(\Lambda)$ has enough injective objects. Then (1) follows from Theorem 7.2.18, and (2) follows from (1) with Corollary 7.1.16. $\square$

There are examples of rings such that the finitistic dimension is infinite. That means an exact category with a projective tilting object need not be of finite global dimension.

Example 7.2.20. Let k be a field and fix a partition $\mathbb{N} = \bigcup_i I_i$ into finite sets of unbounded cardinality. Consider the ring Λ which is obtained from localising the polynomial ring $A = k[x_0, x_1, x_2, \dots]$ at the complement of the union of the infinite set of prime ideals $\bigcup_i \mathfrak{p}_i$, where $\mathfrak{p}_i$ denotes the ideal generated by $\{x_n \mid n \in I_i\}$. Then this is an example of a commutative noetherian ring of infinite Krull dimension [146, Appendix, Example 1]. In fact, the height of $\mathfrak{p}_i$ in A is card I_i; so there is no bound for the height of a prime ideal in Λ. Moreover, there is no bound on the length of a regular sequence in Λ. It remains to note that for any commutative noetherian ring the supremum of the lengths of the regular sequences equals the finitistic dimension [12, Theorem 1.6].

APR Tilting Modules

Let Λ be an Artin algebra. We exhibit a particular class of tilting modules which have projective dimension one.

Proposition 7.2.21. *Let $e \in \Lambda$ be an idempotent such that no direct summand of $(1-e)\Lambda$ is isomorphic to $e\Lambda$. Suppose that $e\Lambda$ is a simple and non-injective Λ-module. Then $T = (1-e)\Lambda \oplus \operatorname{Tr} D(e\Lambda)$ is a tilting module.*

Proof Set $S = e\Lambda$ and denote by $P_1, \ldots, P_n$ a representative set of indecomposable projective Λ-modules which are not isomorphic to S. We need to check conditions (T1)–(T3) for T. The almost split sequence starting at S is of the form

$$0 \longrightarrow S \longrightarrow \bigoplus_i P_i^{d_i} \longrightarrow \operatorname{Tr} DS \longrightarrow 0 \qquad (7.2.22)$$

for some $d_i \geq 0$. This follows from the fact that for each indecomposable summand X of the middle term, the morphism $\phi \colon X \to \operatorname{Tr} DS$ yields a morphism $D\operatorname{Tr}\phi \colon D\operatorname{Tr}X \to S$, which is non-zero when $D\operatorname{Tr}X \neq 0$. Thus X is projective since S is simple projective. This sequence gives immediately (T1) and (T3). Condition (T2) is deduced from the Auslander-Reiten formula, so $D\operatorname{Ext}^1(T,T) \cong \overline{\operatorname{Hom}}(T, D\operatorname{Tr}T) = 0$, since $D\operatorname{Tr}T \cong S$ is simple projective. $\qquad\square$

Let us consider some specific algebras over a field k.

Example 7.2.23. Denote by Λ the k-algebra given by the following quiver with a commutativity relation:

$$
\begin{array}{ccc}
 & 2 & \\
{}^{\alpha}\nearrow & & \searrow {}^{\beta} \\
1 & & 4 \\
{}^{\gamma}\searrow & & \nearrow {}^{\delta} \\
 & 3 &
\end{array}
\qquad \text{with} \qquad \beta\alpha = \delta\gamma
$$

Let $P_i = e_i\Lambda$ denote the indecomposable projective module corresponding to the vertex i. Then P_1 is simple, so

$$T = \operatorname{Tr} DP_1 \oplus P_2 \oplus P_3 \oplus P_4$$

is a tilting module with $\Gamma = \operatorname{End}(T)$ isomorphic to the path algebra of the quiver of Dynkin type D_4:

$$
\begin{array}{cc}
2 & \\
 \searrow & \\
 & 1 \longrightarrow 4 \\
 \nearrow & \\
3 &
\end{array}
$$

We may also get back from Γ to Λ, because the Γ-module

$$T' = P_1 \oplus \operatorname{Tr} DP_2 \oplus \operatorname{Tr} DP_3 \oplus \operatorname{Tr} DP_4$$

is tilting and $\operatorname{End}(T') \cong \Lambda$.

Now let $\Lambda = kQ$ be the path algebra given by a finite quiver Q without oriented cycles. Suppose the vertex $i \in Q_0$ is a *source*, so no arrow ends in i, but at least one arrow starts at i. Denote by $Q(i)$ the quiver which is obtained from Q by reversing the orientation of each arrow starting at i.

Observe that the algebra Λ is hereditary. Thus the following illustrates the tilting process for hereditary abelian categories discussed in Theorem 5.1.2.

Proposition 7.2.24. *Let $i \in Q_0$ be a source of Q. Then the indecomposable projective module P_i is simple, and therefore*

$$T(i) = \operatorname{Tr} DP_i \oplus \left(\bigoplus_{i \neq j \in Q_0} P_j \right)$$

is a tilting module with $\operatorname{End}(T(i)) \cong kQ(i)$.

Proof The module $T(i)$ is tilting by Proposition 7.2.21. To compute its endomorphism algebra, we note that the almost split sequence (7.2.22) starting at P_i is of the form

$$0 \longrightarrow P_i \longrightarrow \bigoplus_{i \to j} P_j \longrightarrow \operatorname{Tr} DP_i \longrightarrow 0$$

where $i \to j$ runs through all arrows in Q starting at i. $\square$

We denote by $|Q|$ the *underlying diagram* of Q which is obtained by forgetting the orientation of each arrow.

Corollary 7.2.25. *Let Q and Q' be acyclic quivers such that $|Q| = |Q'|$. Then there is a triangle equivalence* $\mathbf{D}^b(\operatorname{mod} kQ) \overset{\sim}{\to} \mathbf{D}^b(\operatorname{mod} kQ')$.

Proof For a sequence $i_1, \ldots, i_n$ of vertices in Q one defines recursively

$$Q(i_1, \ldots, i_n) = Q(i_1, \ldots, i_{n-1})(i_n).$$

Because the quivers are acyclic, it is not difficult to construct from the assumption $|Q| = |Q'|$ a sequence $i_1, \ldots, i_n$ of vertices such that $Q' = Q(i_1, \ldots, i_n)$. Then we obtain a sequence of n tilting modules from Proposition 7.2.24. These yield triangle equivalences connecting $\mathbf{D}^b(\operatorname{mod} kQ)$ and $\mathbf{D}^b(\operatorname{mod} kQ')$, by applying Theorem 5.1.2 or Proposition 7.2.13, and keeping in mind that a path algebra is hereditary. $\square$

Tilting Objects for Quivers of Type A_n

We describe the lattice of tilting objects for the category of representations of a quiver of type A_n; it is isomorphic to the Tamari lattice of order n.

The Tamari Lattice

Fix an integer $n \geq 1$. The *Tamari lattice* of order n is a partially ordered set and is denoted by T_n. The elements consist of the meaningful bracketings of a string of $n + 1$ letters. The partial order is given by applying the rule $(xy)z \to x(yz)$ from left to right. For example, when $n = 3$, we have

$$((ab)c)d \geq (a(bc))d \geq a((bc)d) \geq a(b(cd)).$$

Here is the Hasse diagram of the lattice T_3:

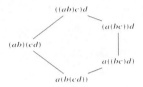

And here is the Hasse diagram of the lattice T_4:

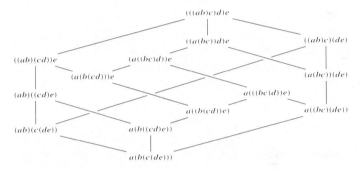

The cardinality of the Tamari lattice T_n equals the *Catalan number*

$$C_n = \frac{1}{n+1}\binom{2n}{n}.$$

Let $\mathcal{I}(n)$ denote the set of intervals $[i, j] = \{i, i+1, \ldots, j\}$ in $\mathbb{Z}$ with $0 \leq i < j \leq n$. For a pair of intervals I, J we set

$$I \perp J \quad :\Longleftrightarrow \quad I \subseteq J \quad \text{or} \quad J \subseteq I \quad \text{or} \quad I \cap J = \emptyset.$$

Let $\mathcal{T}(n)$ denote the set of all subsets $X \subseteq \mathcal{I}(n)$ of cardinality n such that $I \perp J$ for all $I, J \in X$.

Lemma 7.2.26. *Sending an interval $[i, j]$ to the bracketing*

$$x_0 \ldots (x_i \ldots x_j) \ldots x_n$$

of the string $x_0 \ldots x_n$ induces a bijection $\mathcal{T}(n) \xrightarrow{\sim} T_n$. $\qquad\square$

Representations of Type A_n

Fix an integer $n \geq 1$ and a field k. We consider the quiver of type A_n with linear orientation

$$1 \longrightarrow 2 \longrightarrow 3 \longrightarrow \cdots \longrightarrow n$$

and denote by Λ_n its path algebra over k. For each $j \in \{1, \ldots, n\}$ let P_j denote the indecomposable projective Λ_n-module having as a k-basis all paths ending in the vertex j, and for each interval $I = [i, j]$ in $\mathbb{Z}$ with $0 \leq i < j \leq n$ we set $M_I := P_j/\mathrm{rad}^{j-i} P_j$.

Lemma 7.2.27. *The following holds for the modules M_I.*

(1) *The set $\{M_I \mid I \in \mathcal{I}(n)\}$ is a complete set of isomorphism classes of indecomposable Λ_n-modules.*
(2) $\mathrm{Ext}^1(M_I, M_J) = 0 = \mathrm{Ext}^1(M_J, M_I)$ *if and only if $I \perp J$.*
(3) *There is an epimorphism $M_I \to M_J$ if and only if $J \subseteq I$ and $\sup J = \sup I$.*   □

A Λ_n-module T is a *basic tilting module* if T has precisely n pairwise non-isomorphic indecomposable direct summands and $\mathrm{Ext}^1(T, T) = 0$. Observe that the isomorphism classes of basic tilting modules correspond bijectively to the equivalence classes of tilting objects in $\mathrm{mod}\,\Lambda_n$, since $T \cong T'$ if and only if $\mathrm{add}\,T = \mathrm{add}\,T'$.

Write $T \geq T'$ if there is an epimorphism $T^r \to T'$ for some positive integer r. This induces a partial order on the isomorphism classes of basic tilting modules, and we have

$$T \geq T' \quad \Longleftrightarrow \quad T^\perp \supseteq T'^\perp$$

since an object M is in $T^\perp$ if and only if there is an epimorphism $T^r \to M$ for some positive integer r.

Proposition 7.2.28. *The assignment $X \mapsto \bigoplus_{I \in X} M_I$ induces a bijection between $\mathcal{I}(n)$ and the set of isomorphism classes of basic tilting modules over Λ_n. Composition with the bijection $T_n \xrightarrow{\sim} \mathcal{I}(n)$ yields a lattice isomorphism.*   □

Notes

Tilting theory has a rich history [5]. The notion of a tilting module over a finite dimensional algebra was introduced by Brenner and Butler [41], using the conditions of Proposition 7.2.10 and assuming projective dimension at most one. A generalisation of the Coxeter functors arising in the work of Bernšteĭn,

Gel'fand and Ponomarev [36] motivated the study of tilting modules; see also the contribution of Auslander, Platzeck and Reiten leading to the notion of APR tilting [14]. The original definition of a tilting module was later generalised in various directions.

According to Brenner and Butler, the term 'tilting' was chosen for the following reasons. Given a tilting object $T \in \mathcal{A}$, the functor $\mathrm{RHom}(T, -)$ swaps the components of the torsion pair $(\mathcal{T}, \mathcal{F})$ for $\mathcal{A}$ (Theorem 5.1.2). Inside the Grothendieck group $K_0(\mathcal{A}) \cong \mathbb{Z}^n$, the functor $\mathrm{RHom}(T, -)$ tilts the axes given by the standard basis vectors (Corollary 7.2.4). Moreover, the word 'tilting' inflicts well.

For representations of finite dimensional algebras, the link between tilting and derived categories was first established by Happel [101]. He proved that any tilting module induces a derived equivalence. A predecessor is a theorem of Beilinson that identifies a tilting object in the category of coherent sheaves on the projective n-space [25].

For modules over Artin algebras, the correspondence $T \mapsto T^{\perp}$ between tilting objects and covariantly finite and coresolving subcategories (Theorem 7.2.18) is due to Auslander and Reiten [16].

Gabriel noticed that the Catalan number C_n counts the tilting modules of the equioriented quiver of type A_n [81]. The connection with the Tamari lattice was pointed out in [43].

8

Polynomial Representations

Contents

8.1	**Quasi-hereditary Algebras**		**231**
	Standard Modules and Quasi-hereditary Algebras		231
	Approximations and Universal Extensions		236
	Canonical Cotorsion Pairs		238
	Characteristic Tilting Modules		240
8.2	**Symmetric Tensors**		**241**
	Partitions and Young Tableaux		241
	Finitely Generated Projective Modules		243
	Symmetric Tensors and Symmetric Powers		244
	The Category of Symmetric Tensors		244
	Exterior Powers		246
	The Algebra of Symmetric Tensors		246
	Polynomial Maps		249
8.3	**Polynomial Representations**		**250**
	Strict Polynomial Functors		251
	Tensor Products		252
	Decomposing Symmetric Tensors		254
	Representations of Schur Algebras		256
	Representations of Symmetric Groups		257
	Weight Space Decompositions		258
	Standard Morphisms		261
	Base Change		263
8.4	**Cauchy Decompositions**		**264**
	Standard Objects		264
	Straightening		264
	The Cauchy Decomposition of Symmetric Tensors		266
	The Cauchy Decomposition of Symmetric Powers		271
8.5	**Schur and Weyl Modules and Functors**		**272**
	Characteristic Zero		272
	Symmetric Tensors versus Exterior Powers		273
	A Square Root of the Nakayama Functor		275
	A Shuffling Morphism		279
	Schur and Weyl Modules		279
	Schur and Weyl Functors		280

	Reduced Presentations	281
	The Cauchy Decomposition of Exterior Powers	282
8.6	**Schur Algebras**	**284**
	Split Quasi-hereditary Algebras	285
	Finite Global Dimension	288
	Characteristic Tilting Modules	288
Notes		**291**

This chapter is devoted to polynomial representations of general linear groups, and we present this material in the context of quasi-hereditary algebras. These are associative algebras with an extra structure that provides a stratification of their module categories. The additional structure is encoded in a sequence of distinguished standard modules. There is also a distinguished class of tilting modules; these are modules that admit filtrations via standard modules and via costandard modules simultaneously. Important examples are Schur algebras which arise from the study of polynomial representations of general linear groups.

The first part of this chapter discusses representations of Artin algebras that are quasi-hereditary. The rest of the chapter is devoted to the study of polynomial representations, using the language of strict polynomial functors. In the following we provide a brief outline, explaining basic concepts and main results.

We wish to study the polynomial representations of the general linear group $GL_n(k)$ with coefficients in a commutative ring k. To this end fix a degree d and a free k-module $V = k^n$ of rank n. The symmetric group $\mathfrak{S}_d$ acts on $V^{\otimes d}$ via place permutation, and this action commutes with the natural action of $GL_n(k)$. We obtain a k-algebra homomorphism

$$k\,GL_n(k) \longrightarrow \mathrm{End}_{k\mathfrak{S}_d}(V^{\otimes d}) =: S_k(n,d)$$

into the corresponding *Schur algebra*, and modules over the Schur algebra identify via restriction of scalars with degree d polynomial representations of $GL_n(k)$. This idea goes back to Schur. Note that any polynomial representation decomposes into homogeneous parts of different degrees $d \geq 0$, and representations in different degrees do not interact.

We extend this approach by taking representations of $GL_n(k)$ for all $n \in \mathbb{N}$ simultaneously. This means we combine all Schur algebras (keeping d and k fixed) in one category, which we call the *category of symmetric tensors*. Objects are the finitely generated projective k-modules, and $\mathrm{End}(k^n) = S_k(n,d)$ for each n by definition. This category is denoted by $\Gamma^d \mathcal{P}_k$. A k-linear functor

$$F \colon \Gamma^d \mathcal{P}_k \longrightarrow \mathrm{Mod}\,k$$

is called *strict polynomial* of degree d and provides by construction for each n a degree d polynomial $\mathrm{GL}_n(k)$-representation by evaluation at k^n (via the action of $\mathrm{End}_{\Gamma^d \mathcal{P}_k}(k^n)$ on $F(k^n)$).

A natural choice of projective generators in the category of strict polynomial functors is parametrised by partitions $\lambda \vdash d$ and given by

$$\Gamma^\lambda : V \longmapsto \Gamma^{\lambda_1} V \otimes_k \Gamma^{\lambda_2} V \otimes_k \cdots \otimes_k \Gamma^{\lambda_n} V$$

where $\Gamma^{\lambda_i} V$ denotes the symmetric tensors of degree λ_i. The dominance order on the set of partitions provides standard objects $\Delta^\lambda \twoheadleftarrow \Gamma^\lambda$. The fact that each Schur algebra is quasi-hereditary amounts to the existence of a standard filtration of each projective Γ^μ. The associated graded object equals

$$\Gamma^\mu = \bigoplus_{\lambda \geq \mu} (\Delta^\lambda)^{K_{\lambda\mu}}$$

where $K_{\lambda\mu}$ denotes the Kostka number; this is also known as Cauchy decomposition. We prove this by reducing via base change to the characteristic zero case, using that this decomposition corresponds (via counting dimensions) to a classical identity for symmetric functions.

There is a duality which maps the standard projective Γ^λ to the standard injective object S^λ, given by the functor

$$S^\lambda : V \longmapsto S^{\lambda_1} V \otimes_k S^{\lambda_2} V \otimes_k \cdots \otimes_k S^{\lambda_n} V$$

where $S^{\lambda_i} V$ denotes the symmetric powers of degree λ_i. The costandard object $\nabla^\lambda \rightarrowtail S^\lambda$ identifies with the Schur functor, which is by definition the image of a canonical morphism $\Lambda^{\lambda'} \to S^\lambda$. Here, $\Lambda^{\lambda'}$ denotes the exterior power corresponding to the conjugate partition λ' and given by

$$\Lambda^{\lambda'} : V \longmapsto \Lambda^{\lambda'_1} V \otimes_k \Lambda^{\lambda'_2} V \otimes_k \cdots \otimes_k \Lambda^{\lambda'_n} V.$$

The following diagram combines the morphisms defining standard and costandard objects.

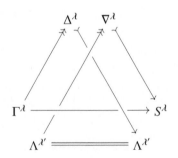

There is a functorial assignment $\Gamma^{\cdot\lambda} \mapsto \Lambda^{\cdot\lambda}$ which yields an equivalence

$$\Omega\colon \mathrm{add}\{\Gamma^{\cdot\lambda} \mid \lambda \vdash d\} \overset{\sim}{\longrightarrow} \mathrm{add}\{\Lambda^{\cdot\lambda} \mid \lambda \vdash d\}$$

and extends to a functor on all strict polynomial functors; this is induced by the twist of the action of $\mathfrak{S}_d$ on $V^{\otimes d}$ via the involution $\sigma \mapsto \mathrm{sgn}(\sigma)\sigma$. We have $\Omega(\Delta^\lambda) = \nabla^{\lambda'}$ and therefore the Cauchy decomposition of Γ^μ yields a filtration of Λ^μ with associated graded object

$$\Lambda^\mu = \bigoplus_{\lambda \geq \mu} (\nabla^{\lambda'})^{K_{\lambda\mu}}.$$

Because Λ^μ is self-dual, we also have a filtration with associated graded object

$$\Lambda^\mu = \bigoplus_{\lambda \geq \mu} (\Delta^{\lambda'})^{K_{\lambda\mu}}.$$

The fact that Λ^μ admits filtrations via standard objects and via costandard objects simultaneously implies that $\bigoplus_\mu \Lambda^\mu$ is a characteristic tilting object, which yields via evaluation at k^n characteristic tilting modules for each Schur algebra $S_k(n, d)$.

8.1 Quasi-hereditary Algebras

This section provides an introduction to representations of quasi-hereditary algebras. We show that the module category of a quasi-hereditary algebra admits a canonical cotorsion pair, which then determines a tilting object (the characteristic tilting module). The cotorsion pair requires the existence of approximation sequences, and we discuss a general method for constructing such approximations via universal extensions.

A characteristic feature of the representation theory of quasi-hereditary algebras is the choice of a partial order on the set of isomorphism classes of simple modules. A consequence is the fact that essentially all arguments are based on an induction.

Standard Modules and Quasi-hereditary Algebras

Let Λ be an Artin algebra. We introduce the notion of a quasi-hereditary algebra, which depends on the choice of a partial order on the set of isomorphism classes of simple Λ-modules. Let us label a representative set of simple modules $(S_i)_{i \in I}$ using the poset $(I, \leq)$. We denote by $(\Lambda, \leq)$ the pair consisting of the algebra and the partial order.

For each $i \in I$, choose a projective cover $P_i \to S_i$ and an injective envelope $S_i \to Q_i$. We define the *standard module* Δ_i to be the maximal quotient of P_i belonging to Filt$\{S_j \mid j \leq i\}$, and the *costandard module* ∇_i is the maximal submodule of Q_i belonging to Filt$\{S_j \mid j \leq i\}$. Then $(\Lambda, \leq)$ is a *quasi-hereditary algebra* if

(QH1) End$_\Lambda(\Delta_i)$ is a division ring for all $i \in I$,
(QH2) Ext$_\Lambda^1(\Delta_i, \Delta_j) \neq 0$ implies $i < j$, and
(QH3) $P_i \in$ Filt$\{\Delta_j \mid j \in I\}$ for all $i \in I$.

Let us fix a quasi-hereditary algebra $(\Lambda, \leq)$. Note that the standard modules $(\Delta_i)_{i \in I}$ determine the simple modules since $S_i \cong$ top Δ_i.

Fix a maximal element $i_0 \in I$ and set $I_0 = I \setminus \{i_0\}$. Let $\Gamma = $ End$_\Lambda(\Delta_{i_0})$ and for each $X \in$ mod Λ we consider the natural exact sequence

$$\text{Hom}_\Lambda(\Delta_{i_0}, X) \otimes_\Gamma \Delta_{i_0} \xrightarrow{\varepsilon_X} X \longrightarrow \bar{X} \longrightarrow 0.$$

In particular, the map $\Lambda \to \bar{\Lambda}$ is an algebra homomorphism.

Lemma 8.1.1. *The object Δ_{i_0} is projective and the counit ε_X is a monomorphism for X in* Filt$\{\Delta_i \mid i \in I\}$ *which induces a functorial exact sequence*

$$0 \longrightarrow \Delta_{i_0}^r \longrightarrow X \longrightarrow \bar{X} \longrightarrow 0$$

with $\bar{X}$ in Filt$\{\Delta_i \mid i \in I_0\}$. *Moreover, the assignment $X \mapsto \bar{X}$ provides an exact left adjoint of the inclusion*

$$\text{Filt}\{\Delta_i \mid i \in I_0\} \longrightarrow \text{Filt}\{\Delta_i \mid i \in I\}.$$

Proof There is an exact sequence $0 \to U \to P_{i_0} \to \Delta_j \to 0$ for some $j \in I$ with $U \in$ Filt$\{\Delta_i \mid i \in I\}$, since P_{i_0} is in Filt$\{\Delta_i \mid i \in I\}$. Clearly, $j = i_0$ since top $\Delta_j \cong$ top P_{i_0}. Then $U \in$ Filt$\{\Delta_i \mid i > i_0\}$, and therefore $U = 0$. Thus Δ_{i_0} is projective.

An induction on the length of a filtration of an object in Filt$\{\Delta_i \mid i \in I\}$ yields some $r \geq 0$ and an exact sequence $0 \to \Delta_{i_0}^r \to X \to X' \to 0$ with X' in Filt$\{\Delta_i \mid i \in I_0\}$. Then we have Hom$_\Lambda(\Delta_{i_0}, X) \otimes_\Gamma \Delta_{i_0} \cong \Delta_{i_0}^r$ and $\bar{X} \cong X'$. The exactness of the assignment $X \mapsto \bar{X}$ follows from the snake lemma since Hom$_\Lambda(\Delta_{i_0}, -) \otimes_\Gamma \Delta_{i_0}$ is exact. $\square$

The above lemma means that we have the following colocalisation sequence

$$\text{Filt}\{\Delta_i \mid i \in I_0\} \xleftarrow{\hspace{2cm}} \text{Filt}\{\Delta_i \mid i \in I\} \xleftarrow{\hspace{2cm}} \text{Filt}(\Delta_{i_0})$$

where all functors are exact.

Lemma 8.1.2. *Restriction via $\Lambda \twoheadrightarrow \bar{\Lambda}$ induces an equivalence*

$$\operatorname{mod} \bar{\Lambda} \xrightarrow{\sim} \{X \in \operatorname{mod} \Lambda \mid \operatorname{Hom}_\Lambda(\Delta_{i_0}, X) = 0\}$$

and $\bar{\Lambda}$ is a quasi-hereditary algebra with standard modules $(\Delta_i)_{i \in I_0}$. Moreover, we have for all $\bar{\Lambda}$-modules X, Y and $p \geq 0$ an isomorphism

$$\operatorname{Ext}_{\bar{\Lambda}}^p(X, Y) \xrightarrow{\sim} \operatorname{Ext}_\Lambda^p(X, Y).$$

Proof   The quasi-inverse is given by $- \otimes_\Lambda \bar{\Lambda}$. The $\bar{P}_i$ form a representative set of indecomposable projective $\bar{\Lambda}$-modules, and from the above lemma it is clear that $\bar{P}_i \in \operatorname{Filt}\{\Delta_i \mid i \in I_0\}$ for all $i \in I_0$. Thus $\bar{\Lambda}$ with standard modules $(\Delta_i)_{i \in I_0}$ is quasi-hereditary. The last assertion follows from Lemma 4.4.8. $\square$

We continue with a characterisation of quasi-hereditary algebras which has the advantage that it is transferable to other settings.

Proposition 8.1.3. *An Artin algebra Λ together with a partially ordered set of objects $(E_i)_{i \in I}$ in $\operatorname{mod} \Lambda$ is quasi-hereditary if and only if there are exact sequences*

$$0 \longrightarrow U_i \longrightarrow P_i \longrightarrow E_i \longrightarrow 0 \qquad (i \in I)$$

in $\operatorname{mod} \Lambda$ satisfying the following:

(1) $\operatorname{End}_\Lambda(E_i)$ *is a division ring for all i.*
(2) $\operatorname{Hom}_\Lambda(E_i, E_j) \neq 0$ *implies $i \leq j$.*
(3) U_i *belongs to $\operatorname{Filt}\{E_j \mid j > i\}$ for all i.*
(4) $\bigoplus_{i \in I} P_i$ *is a projective generator of $\operatorname{mod} \Lambda$.*

In this case $(\operatorname{top} E_i)_{i \in I}$ equals the partially ordered set of simple Λ-modules and $\Delta_i \cong E_i$ for all i.

Proof   We prove one direction; the other direction is similar. Let $(\Lambda, \leq)$ be quasi-hereditary. We take the canonical exact sequences

$$0 \longrightarrow U_i \longrightarrow P_i \longrightarrow \Delta_i \longrightarrow 0 \qquad (i \in I)$$

and need to show that $U_i \in \operatorname{Filt}\{\Delta_j \mid j > i\}$. The other conditions (1), (2), and (4) are clear. We use induction on $\operatorname{card} I$ and choose a maximal element $i_0 \in I$ as before. We have $U_{i_0} = 0$ by Lemma 8.1.1. For $i \neq i_0$ consider the following commutative diagram with exact rows and columns

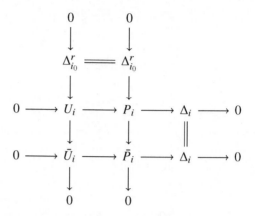

which is obtained from Lemma 8.1.1 by applying it to P_i. The algebra $\bar{\Lambda}$ is quasi-hereditary with standard modules $(\Delta_i)_{i \in I_0}$ by Lemma 8.1.2. Thus $\bar{U}_i \in$ Filt$\{\Delta_j \mid j > i, \ j \neq i_0\}$, and therefore $U_i \in$ Filt$\{\Delta_j \mid j > i\}$.   □

Next observe that there is a canonical bijection between the isomorphism classes of simple right and left Λ-modules; this takes for any primitive idempotent $e \in \Lambda$ the module $S = e(\Lambda/J)$ to $S' = (\Lambda/J)e$, where J denotes the Jacobson radical of Λ. This induces bijections $P_i \mapsto P_i'$ and $\Delta_i \mapsto \Delta_i'$ between the indecomposable projectives and the standard modules over Λ and Λ^{op}.

Lemma 8.1.4. *The pair* $(\Lambda, \leq)$ *is quasi-hereditary if and only if* $(\Lambda^{\mathrm{op}}, \leq)$ *is quasi-hereditary.*

Proof We assume that $(\Lambda, \leq)$ is quasi-hereditary and show by induction on the number of simple modules that $(\Lambda^{\mathrm{op}}, \leq)$ is quasi-hereditary, using Lemma 8.1.2. Thus the algebra $\bar{\Lambda}^{\mathrm{op}}$ is quasi-hereditary with standard modules $(\Delta_i')_{i \in I_0}$ by the induction hypothesis.

There is a primitive idempotent $e \in \Lambda$ such that $\Delta_{i_0} \cong e\Lambda$. We have

$$\mathrm{End}_\Lambda(\Delta_{i_0}) \cong \mathrm{End}_\Lambda(e\Lambda) \cong e\Lambda e = \mathrm{End}_\Lambda(\Lambda e)^{\mathrm{op}} \cong \mathrm{End}_\Lambda(\Delta_{i_0}')^{\mathrm{op}}.$$

Thus $\mathrm{End}_\Lambda(\Delta_{i_0}')$ is a division ring. The fact that Λ belongs to Filt$\{\Delta_i \mid i \in I\}$ implies that the counit ε_Λ is a monomorphism, by Lemma 8.1.1. An equivalent condition is that the multiplication map $\Lambda e \otimes_{e\Lambda e} e\Lambda \to \Lambda e\Lambda$ is bijective. But this is a symmetric condition. Thus each counit $\varepsilon_{P_i'}$ is a monomorphism. From the induction hypothesis we know that $\bar{P}_i'$ belongs to Filt$\{\Delta_i' \mid i \in I_0\}$ for all $i \in I_0$. Thus $P_i' \in$ Filt$\{\Delta_i' \mid i \in I\}$ for all i, and therefore $(\Lambda^{\mathrm{op}}, \leq)$ is quasi-hereditary.   □

Lemma 8.1.5. *The pair* $(\Lambda, \leq)$ *is quasi-hereditary if and only if we have the following.*

(1) $\mathrm{End}_\Lambda(\nabla_i)$ *is a division ring for all* $i \in I$,

(2) $\mathrm{Ext}^1_\Lambda(\nabla_i, \nabla_j) \neq 0$ *implies* $i > j$, *and*

(3) $Q_i \in \mathrm{Filt}\{\nabla_j \mid j \in I\}$ *for all* $i \in I$.

Proof We use Matlis duality, which identifies standard modules over Λ^{op} with costandard modules over Λ. Thus the assertion follows from Lemma 8.1.4. □

We are now able to establish finite global dimension; it is an important property of quasi-hereditary algebras.

Proposition 8.1.6. *A quasi-hereditary algebra* $(\Lambda, \leq)$ *with* n *isomorphism classes of simple modules has global dimension at most* $2(n - 1)$.

Proof We use induction on n. Fix a maximal element $i_0 \in I$ and identify the full subcategory of Λ-modules X such that $\mathrm{Hom}_\Lambda(\Delta_{i_0}, X) = 0$ with $\mathrm{mod}\,\bar\Lambda$; see Lemma 8.1.1. We claim that for each $\bar\Lambda$-module X we have

$$\mathrm{proj.dim}_\Lambda X \leq \mathrm{proj.dim}_{\bar\Lambda} X + 1.$$

This is clear when $\mathrm{proj.dim}_{\bar\Lambda} X = 0$, since every projective $\bar\Lambda$-module is a direct summand of an object of the form $\bar P$ for some projective Λ-module P and the counit ε_P is a monomorphism, by Lemma 8.1.1. So suppose $\mathrm{proj.dim}_{\bar\Lambda} X > 0$ and consider the following commutative diagram with exact rows and columns given by a projective cover $P \to X$ in $\mathrm{mod}\,\Lambda$.

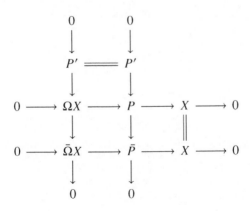

Note that $\bar P$ is a projective $\bar\Lambda$-module and that P' is a projective Λ-module.

Therefore by induction

$$\begin{aligned}
\text{proj.dim}_\Lambda X &= \text{proj.dim}_\Lambda \Omega X + 1 \\
&\le \text{proj.dim}_\Lambda \bar{\Omega} X + 1 \\
&\le \text{proj.dim}_{\bar\Lambda} \bar{\Omega} X + 2 \\
&= \text{proj.dim}_{\bar\Lambda} X + 1.
\end{aligned}$$

Now fix a Λ-module X. We use induction on n to show that $\text{proj.dim}_\Lambda X \le 2n - 2$. For $n = 1$ the assertion is clear. So assume $n > 1$. Consider the counit $\varepsilon_X \colon \text{Hom}_\Lambda(\Delta_{i_0}, X) \otimes_\Gamma \Delta_{i_0} \to X$. Then $\text{Ker}\,\varepsilon_X$ and $\bar{X} = \text{Coker}\,\varepsilon_X$ belong to $\text{mod}\,\bar\Lambda$. The algebra $\bar\Lambda$ is quasi-hereditary with $n - 1$ standard modules, by Lemma 8.1.2. Thus $\text{proj.dim}_{\bar\Lambda} A \le 2n - 4$ for every $A \in \text{mod}\,\bar\Lambda$, so $\text{proj.dim}_\Lambda A \le 2n - 3$ by our first observation. This implies for the image X' of ε_X that $\text{proj.dim}_\Lambda X' \le 2n - 2$. Thus $\text{proj.dim}_\Lambda X \le 2n - 2$.   □

Remark 8.1.7. Let $(\Lambda, \le)$ be a quasi-hereditary algebra and choose a bijection $\alpha \colon I \to \{1, \dots, n\}$ such that $i \le j$ implies $\alpha(i) \le \alpha(j)$ for all $i, j \in I$. Then Λ is quasi-hereditary with respect to the totally ordered sequence of simples $S_1, \dots, S_n$, where $S_{\alpha(i)} = S_i$ for $i \in I$. This follows from successive application of Lemma 8.1.2. Moreover, $\Delta_{\alpha(i)} = \Delta_i$ for all $i \in I$.

The above remark yields a chain of surjective algebra homomorphisms

$$\Lambda = \Lambda_n \twoheadrightarrow \Lambda_{n-1} \twoheadrightarrow \cdots \twoheadrightarrow \Lambda_1$$

such that for each i restriction via $\Lambda \twoheadrightarrow \Lambda_i$ identifies

$$\text{mod}\,\Lambda_i \xrightarrow{\sim} \{X \in \text{mod}\,\Lambda \mid \text{Hom}_\Lambda(\Delta_j, X) = 0 \text{ for } j > i\} = \text{Thick}(\Delta_1, \dots, \Delta_i)$$

and Λ_i is quasi-hereditary with standard modules $\Delta_1, \dots, \Delta_i$. Moreover, we obtain recollements

$$\mathbf{D}^b(\text{mod}\,\Lambda_{i-1}) \begin{smallmatrix} \twoheadleftarrow \\ \longrightarrow \\ \twoheadleftarrow \end{smallmatrix} \mathbf{D}^b(\text{mod}\,\Lambda_i) \begin{smallmatrix} \longleftarrow \\ \xrightarrow{\ \text{Hom}(\Delta_i, -)\ } \\ \longleftarrow \end{smallmatrix} \mathbf{D}^b(\text{mod}\,\Gamma_i)$$

where $\Gamma_i = \text{End}_\Lambda(\Delta_i)$; see Lemma 4.2.13.

Approximations and Universal Extensions

For exact categories we discuss a method of constructing approximations which uses universal extensions. Fix an exact category $\mathcal{A}$. For a pair of objects A, E in $\mathcal{A}$, a *universal extension* of A by E is an exact sequence

$$0 \longrightarrow E^r \longrightarrow B \longrightarrow A \longrightarrow 0$$

for which the connecting homomorphism $\mathrm{Hom}(E^r, E) \to \mathrm{Ext}^1(A, E)$ is surjective. If $\mathrm{Ext}^1(A, E)$ is finitely generated as a left $\mathrm{End}(E)$-module, say by $\eta_1, \ldots, \eta_r$, then the extension $\eta := (\eta_1, \ldots, \eta_r) \in \mathrm{Ext}^1(A, E^r)$ is universal. For, the corresponding connecting homomorphism is

$$\mathrm{Hom}(E^r, E) \to \mathrm{Ext}^1(A, E), \quad (\theta_1, \ldots, \theta_r) \mapsto \sum_i \theta_i \eta_i.$$

We now fix a sequence of objects $E_1, \ldots, E_n$ in $\mathcal{A}$ satisfying

(1) $\mathrm{Ext}^1(E_i, E_j) = 0$ for all $i \geq j$, and
(2) $\mathrm{Ext}^1(X, E_j)$ is finitely generated over $\mathrm{End}(E_j)$ for all $X \in \mathcal{A}$ and all j.

Proposition 8.1.8. *Suppose $A \in \mathcal{A}$ satisfies $\mathrm{Ext}^1(A, E_j) = 0$ for all $j < t$. Then there is an exact sequence $0 \to Y \to X \to A \to 0$ with $Y \in \mathrm{Filt}(E_t, \ldots, E_n)$ such that $\mathrm{Ext}^1(X, E_j) = 0$ for all j.*

Proof   We use descending induction on t. We first form a universal extension $0 \to E_t^r \to A' \to A \to 0$. Applying $\mathrm{Hom}(-, E_j)$ shows that $\mathrm{Ext}^1(A', E_j) = 0$ for all $j \leq t$, and from the induction hypothesis there is an exact sequence $0 \to Y' \to X \to A' \to 0$ with $Y' \in \mathrm{Filt}(E_{t+1}, \ldots, E_n)$ such that $\mathrm{Ext}^1(X, E_j) = 0$ for all j. Forming the pullback yields the following exact commutative diagram.

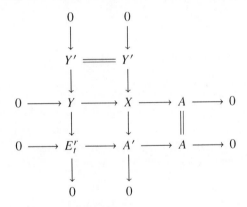

Then $Y \in \mathrm{Filt}(E_t, \ldots, E_n)$ as required.   □

Of particular interest is the case $t = 1$.

Corollary 8.1.9. *Given $A \in \mathcal{A}$ there is an exact sequence $0 \to Y \to X \to A \to 0$ with $Y \in \mathrm{Filt}(E_1, \ldots, E_n)$ such that $\mathrm{Ext}^1(X, E_j) = 0$ for all j.*   □

Example 8.1.10. Let $(\Lambda, \leq)$ be a quasi-hereditary algebra with standard modules $\Delta_1, \ldots, \Delta_n$. Then Proposition 8.1.8 yields exact sequences in $\mathrm{mod}\,\Lambda$

$$0 \longrightarrow U_i \longrightarrow P_i \longrightarrow \Delta_i \longrightarrow 0 \qquad (1 \leq i \leq n)$$

such that P_i is projective and U_i belongs to $\text{Filt}(\Delta_{i+1}, \ldots, \Delta_n)$ for all i; cf. Proposition 8.1.3.

Canonical Cotorsion Pairs

We fix a quasi-hereditary algebra $(\Lambda, \leq)$ and we may assume that the simple modules are totally ordered; see Remark 8.1.7. Let $\Delta_1, \ldots, \Delta_n$ denote the standard modules. To simplify the notation we set

$$\text{Filt}(\Delta) = \text{Filt}(\Delta_1, \ldots, \Delta_n) \qquad \text{and} \qquad \text{Filt}(\nabla) = \text{Filt}(\nabla_1, \ldots, \nabla_n).$$

The functors $\text{mod}\,\bar{\Lambda} \rightarrowtail \text{mod}\,\Lambda$ and $\text{Hom}_\Lambda(\Delta_n, -)\colon \text{mod}\,\Lambda \to \text{mod}\,\Gamma$ with $\Gamma = \text{End}_\Lambda(\Delta_n)$ induce the following recollement.

$$\text{mod}\,\bar{\Lambda} \underset{i^!}{\overset{i^*}{\underset{\longrightarrow}{\rightleftarrows}}} \xrightarrow{\ i_*=i_!\ } \text{mod}\,\Lambda \underset{j_*}{\overset{j_!}{\underset{\longrightarrow}{\rightleftarrows}}} \xrightarrow{\ j^!=j^*\ } \text{mod}\,\Gamma \qquad (8.1.11)$$

Note that $j_!(\Gamma) = \Delta_n$ and $j_*(\Gamma) = \nabla_n$.

Lemma 8.1.12. *For $X \in \text{Filt}(\Delta)$ and $Y \in \text{Filt}(\nabla)$ we have $\text{Ext}_\Lambda^p(X, Y) = 0$ for all $p > 0$. More precisely, for $1 \leq s, t \leq n$ and $p \geq 0$ we have*

$$\text{Ext}_\Lambda^p(\Delta_s, \nabla_t) \cong \begin{cases} \text{End}_\Lambda(\Delta_s) & \text{if } s = t \text{ and } p = 0, \\ 0 & \text{otherwise.} \end{cases}$$

Proof We use induction on n. For $s, t < n$ the assertion follows by induction, because $\Delta_s, \nabla_t \in \text{mod}\,\bar{\Lambda}$ and the inclusion $\text{mod}\,\bar{\Lambda} \to \text{mod}\,\Lambda$ preserves extension groups; see Lemma 8.1.2. If $s = n$ or $t = n$, then we use the fact that Δ_n is projective and ∇_n is injective. This gives the assertion for $p > 0$. For $p = 0$ we use the recollement (8.1.11). In fact, $\Delta_n = j_!(\Gamma)$ and $\nabla_n = j_*(\Gamma)$. Thus $\text{Hom}_\Lambda(\Delta_n, \nabla_n) \cong \Gamma$ by adjointness. □

Proposition 8.1.13. *Let $(\Lambda, \leq)$ be a quasi-hereditary algebra. For Λ-modules X, Y we have the following.*

(1) $X \in \text{Filt}(\Delta)$ *if and only if* $\text{Ext}_\Lambda^1(X, \nabla_t) = 0$ *for* $1 \leq t \leq n$.
(2) $Y \in \text{Filt}(\nabla)$ *if and only if* $\text{Ext}_\Lambda^1(\Delta_t, Y) = 0$ *for* $1 \leq t \leq n$.

Proof We prove (1). The proof of (2) is dual. One direction is clear by the above lemma. Thus assume that $\text{Ext}_\Lambda^1(X, \nabla_t) = 0$ for all t. We use induction on n and consider the recollement (8.1.11).

We claim that the counit $X' := j_!j^!(X) \to X$ is a monomorphism. To see this, fix an injective cogenerator Q of mod Λ. Note that Q belongs to Filt$(\nabla_1, \ldots, \nabla_n)$, by Lemma 8.1.5. Thus we have an exact sequence

$$0 \longrightarrow i_! i^!(Q) \longrightarrow Q \longrightarrow j_* j^*(Q) \longrightarrow 0$$

by the dual of Lemma 8.1.1 which induces the following commutative diagram with exact rows.

$$0 \to \mathrm{Hom}_\Lambda(X, i_! i^!(Q)) \to \mathrm{Hom}_\Lambda(X, Q) \to \mathrm{Hom}_\Lambda(X, j_* j^*(Q)) \to 0$$

$$\downarrow \qquad\qquad\qquad \downarrow \qquad\qquad\qquad \downarrow$$

$$0 \to \mathrm{Hom}_\Lambda(X', i_! i^!(Q)) \to \mathrm{Hom}_\Lambda(X', Q) \to \mathrm{Hom}_\Lambda(X', j_* j^*(Q)) \to 0$$

We have $\mathrm{Hom}_\Lambda(X', i_! i^!(Q)) = 0$, and the map

$$\mathrm{Hom}_\Lambda(X, j_* j^*(Q)) \longrightarrow \mathrm{Hom}_\Lambda(X', j_* j^*(Q))$$

is a bijection by adjointness. Thus the map

$$\mathrm{Hom}_\Lambda(X, Q) \longrightarrow \mathrm{Hom}_\Lambda(X', Q)$$

is surjective. It follows that the sequence

$$0 \longrightarrow j_! j^!(X) \longrightarrow X \longrightarrow i_* i^*(X) \longrightarrow 0$$

given by the unit and counit for X is exact.

The object $X'' := i_* i^*(X)$ belongs to mod $\bar{\Lambda}$ and satisfies $\mathrm{Ext}^1_\Lambda(X'', \nabla_t) = 0$ for all t. Thus X'' is in Filt$(\Delta_1, \ldots, \Delta_{n-1})$ by induction. It follows that X belongs to Filt$(\Delta_1, \ldots, \Delta_n)$. $\qquad\square$

Corollary 8.1.14. *For a Λ-module X the following are equivalent.*

(1) *X is a projective object of* Filt(∇).
(2) *X is an injective object of* Filt(Δ).
(3) *X belongs to* Filt$(\Delta) \cap$ Filt(∇).

Proof We apply Proposition 8.1.13. So any projective object of Filt(∇) belongs to Filt(Δ), and it is injective in Filt(Δ) because it belongs to Filt(∇). Conversely, if X is in Filt$(\Delta) \cap$ Filt(∇), then X is a projective object in Filt(∇) and an injective object in Filt(Δ). $\qquad\square$

The following is our main result because it provides a canonical tilting module for any quasi-hereditary algebra.

Theorem 8.1.15. *Let $(\Lambda, \le)$ be a quasi-hereditary algebra. Then the pair* (Filt(Δ), Filt(∇)) *is a cotorsion pair.*

Proof The equalities $\mathrm{Filt}(\Delta)^{\perp} = \mathrm{Filt}(\nabla)$ and $\mathrm{Filt}(\Delta) = {}^{\perp}\mathrm{Filt}(\nabla)$ follow from Proposition 8.1.13. It remains to construct approximation sequences for each Λ-module. For this we apply Corollary 8.1.9, using that the standard and costandard modules satisfy $\mathrm{Ext}^1_{\Lambda}(\Delta_i, \Delta_j) = 0$ and $\mathrm{Ext}^1_{\Lambda}(\nabla_j, \nabla_i) = 0$ for all $i \geq j$. $\quad\square$

Characteristic Tilting Modules

Fix a quasi-hereditary algebra $(\Lambda, \leq)$ with standard modules $\Delta_1, \ldots, \Delta_n$. Let us apply the correspondence between cotorsion pairs and tilting objects from Proposition 7.2.17. Then the cotorsion pair $(\mathrm{Filt}(\Delta), \mathrm{Filt}(\nabla))$ yields the *characteristic tilting module* T which is given by $\mathrm{add}\, T = \mathrm{Filt}(\Delta) \cap \mathrm{Filt}(\nabla)$.

Theorem 8.1.16. *A quasi-hereditary algebra $(\Lambda, \leq)$ determines, up to equivalence, a tilting module T via the equality* $\mathrm{add}\, T = \mathrm{Filt}(\Delta) \cap \mathrm{Filt}(\nabla)$.

Proof The cotorsion pair $(\mathrm{Filt}(\Delta), \mathrm{Filt}(\nabla))$ is given by Theorem 8.1.15, and then we use that Λ has finite global dimension, by Proposition 8.1.6. Thus Proposition 7.2.17 applies. $\quad\square$

A tilting module T has n pairwise non-isomorphic indecomposable direct summands, where n equals the number of simple Λ-modules. This follows from the fact that we have an isomorphism of Grothendieck groups $K_0(\mathrm{End}_{\Lambda}(T)) \xrightarrow{\sim} K_0(\Lambda)$; see Corollary 7.2.4. For an arbitrary tilting module T, there is no canonical bijection between the indecomposable summands of T and the simple Λ-modules. This is different for a characteristic tilting module.

Proposition 8.1.17. *Let $(\Lambda, \leq)$ be a quasi-hereditary algebra. Then there are exact sequences*

$$0 \longrightarrow U_i \longrightarrow T_i \longrightarrow \nabla_i \longrightarrow 0 \qquad (1 \leq i \leq n)$$

such that $U_i \in \mathrm{Filt}(\nabla_1, \ldots, \nabla_{i-1})$ and $T = \bigoplus_{i=1}^{n} T_i$ is a characteristic tilting module. Moreover, $i = \inf\{t \geq 1 \mid T_i \in \mathrm{Filt}(S_1, \ldots, S_t)\}$.

Proof We use that $\mathrm{Ext}^1_{\Lambda}(\nabla_i, \nabla_j) = 0$ for all $i \leq j$ and apply Proposition 8.1.8 to the pair $A = \nabla_i$ and $t = i - 1$ so that $\mathrm{Ext}^1_{\Lambda}(A, \nabla_j)$ for $j > t$. This yields a short exact sequence $0 \to U_i \to T_i \to \nabla_i \to 0$ with $U_i \in \mathrm{Filt}(\nabla_1, \ldots, \nabla_{i-1})$ and $\mathrm{Ext}^1_{\Lambda}(T, \nabla_j) = 0$ for all j. Thus $T_i \in \mathrm{Filt}(\Delta) \cap \mathrm{Filt}(\nabla)$ by Proposition 8.1.13. There is an admissible epimorphism $T_i \to \nabla_i$, and therefore $T = \bigoplus_i T_i$ is a projective generator for $\mathrm{Filt}(\nabla)$, so $\mathrm{add}\, T = \mathrm{Filt}(\Delta) \cap \mathrm{Filt}(\nabla)$ by Corollary 8.1.14. Finally, $T_i \in \mathrm{Filt}(S_1, \ldots, S_i)$ since U_i and ∇_i belong to $\mathrm{Filt}(S_1, \ldots, S_i)$, but $T_i \notin \mathrm{Filt}(S_1, \ldots, S_{i-1})$ since $\mathrm{soc}\, \nabla_i \cong S_i$. $\quad\square$

The explicit description of the characteristic tilting module T in Proposition 8.1.17 can be used to show that $\Lambda' = \mathrm{End}_\Lambda(T)$ is quasi-hereditary with standard modules $\Delta_i' = \mathrm{Hom}_\Lambda(T, \nabla_i)$, since $\mathrm{Hom}_\Lambda(T, -)$ yields an exact equivalence $\mathrm{Filt}(\nabla) \xrightarrow{\sim} \mathrm{Filt}(\Delta')$. The quasi-hereditary algebra Λ' is called the *Ringel dual* of Λ, and Λ'' is then Morita equivalent to Λ. In fact we have equivalences of exact categories

$$\mathrm{Filt}_{\Lambda''}(\Delta) \xrightarrow{\sim} \mathrm{Filt}_{\Lambda'}(\nabla) \xrightarrow{\sim} \mathrm{Filt}_{(\Lambda')^{\mathrm{op}}}(\Delta)^{\mathrm{op}} \xrightarrow{\sim} \mathrm{Filt}_{\Lambda^{\mathrm{op}}}(\nabla)^{\mathrm{op}} \xrightarrow{\sim} \mathrm{Filt}_\Lambda(\Delta).$$

Note that the first and third equivalences reverse the factors, while the second and fourth keep the order.

8.2 Symmetric Tensors

Important examples of quasi-hereditary algebras are Schur algebras. They arise from the study of polynomial representations of general linear groups. In fact, we use the language of strict polynomial functors. This requires a substantial discussion of symmetric tensors. So we begin with basic definitions and explain the connection between symmetric tensors and polynomial maps.

Throughout we keep fixed a commutative ring k.

Partitions and Young Tableaux

The following glossary collects basic definitions and facts that are used throughout.

Composition. Fix an integer $d \geq 0$. A *composition* of d into n parts is a sequence $\lambda = (\lambda_1, \lambda_2, \ldots, \lambda_n)$ of integers $\lambda_i \geq 0$ such that $\sum \lambda_i = d$. The set of such compositions is denoted by $\Lambda(n, d)$.

We say that two compositions λ and μ are *equivalent up to permutation* if there is a permutation $\sigma \in \mathfrak{S}_n$ such that $\mu_i = \lambda_{\sigma(i)}$ for all i.

Partition. Fix an integer $d \geq 0$. A *partition* of d is a sequence $\lambda = (\lambda_1, \lambda_2, \ldots)$ of integers $\lambda_i \geq 0$ satisfying $\lambda_1 \geq \lambda_2 \geq \ldots$ and $\sum \lambda_i = d$. In this case one writes $\lambda \vdash d$. The *conjugate partition* λ' is the partition where λ_i' equals the number of terms of λ that are greater than or equal to i.

Young diagram. Fix a partition λ of an integer d. The *Young diagram* corresponding to λ is given by d boxes which are arranged in rows and columns. Each integer $r \in \{1, \ldots, d\}$ can be written uniquely as a sum $r = \lambda_1 + \cdots + \lambda_{i-1} + j$ with $1 \leq j \leq \lambda_i$. The pair (i, j) describes the position (ith row and jth column) of the box corresponding to r.

Filling. A *filling* of a Young diagram is a map T which assigns to each pair (i, j) in the Young diagram corresponding to λ a positive integer $T(i, j)$. The fillings of a fixed Young diagram are ordered lexicographically via the lexicographic order on pairs of integers: $S \leq T$ if for every pair (i, j) we have $S(i, j) \leq T(i, j)$ whenever $S(e, f) = T(e, f)$ for all $(e, f) < (i, j)$.

Young tableau. A *Young tableau* is a filling T that is *weakly increasing along each row* $(T(i, j) \leq T(i, j + 1)$ for all $i, j)$ and *strictly increasing along each column* $(T(i, j) < T(i + 1, j)$ for all $i, j)$.

Content. Let λ be a partition and T a filling of the corresponding Young diagram. The *content* of T is by definition the sequence $\mu = (\mu_1, \mu_2, \ldots)$ such that μ_i equals the number of times the integer i occurs in T.

Dominance order. The set of partitions of an integer d is partially ordered via the following *dominance order*: $\mu \trianglelefteq \lambda$ if $\sum_{i=1}^{r} \mu_i \leq \sum_{i=1}^{r} \lambda_i$ for all integers $r \geq 1$.

Lemma 8.2.1. *Let λ and μ be partitions. Then there exists a Young tableau of shape λ with content equivalent up to permutation to μ if and only if $\mu \trianglelefteq \lambda$.*

Proof See [78, Section 2.2]. □

Lexicographic order. The set of partitions of an integer d is totally ordered via the following *lexicographic order*: $\mu \leq \lambda$ if for every integer $r \geq 1$ we have $\mu_r \leq \lambda_r$ whenever $\mu_i = \lambda_i$ for all $i < r$. For a partition λ let λ^+ denote its immediate successor and set $(d)^+ = +\infty$. Analogously, λ^- denotes the immediate predecessor of λ and $(1, \ldots, 1)^- = -\infty$.

Lemma 8.2.2. *Let λ and μ be partitions. Then $\mu \trianglelefteq \lambda$ implies $\mu \leq \lambda$.* □

Kostka number. Let λ and μ be partitions. The *Kostka number* $K_{\lambda\mu}$ denotes the number of Young tableaux of shape λ and content μ. Note that $K_{\lambda\mu} \neq 0$ if and only if $\mu \trianglelefteq \lambda$, and $K_{\lambda\mu} = 1$ for $\mu = \lambda$.

Symmetric function. For a partition λ let s_λ denote the *Schur function* and h_λ the *complete symmetric function*. Evaluating at 1^n for an integer $n > 0$ gives

$$s_\lambda(1^n) = \text{ number of Young tableaux of shape } \lambda \text{ with entries in } \{1, \ldots, n\},$$

and the following result is a consequence of the Cauchy identity.

Proposition 8.2.3. *Let μ be a partition of an integer d. Then*

$$h_\mu = \sum_\lambda K_{\lambda\mu} s_\lambda$$

where λ runs through all partitions of d.

Proof See (5.15) in Part I of [140] or Corollary 7.12.5 in [195]. □

Finitely Generated Projective Modules

Let $\mathcal{P}_k$ denote the category of finitely generated projective k-modules. Given V, W in $\mathcal{P}_k$, we write $V \otimes W$ for their tensor product over k and $\mathrm{Hom}(V, W)$ for the k-module of k-linear maps $V \to W$. This provides two bifunctors

$$- \otimes - : \mathcal{P}_k \times \mathcal{P}_k \longrightarrow \mathcal{P}_k$$
$$\mathrm{Hom}(-, -) : (\mathcal{P}_k)^{\mathrm{op}} \times \mathcal{P}_k \longrightarrow \mathcal{P}_k.$$

For any k-module V the assignment

$$V \longmapsto V^\vee = \mathrm{Hom}(V, k)$$

yields a duality $\mathrm{Mod}\, k \to \mathrm{Mod}\, k$. In particular, we have natural isomorphisms

$$\mathrm{Hom}(V, W^\vee) \cong \mathrm{Hom}(V \otimes W, k) \cong \mathrm{Hom}(W, V^\vee). \qquad (8.2.4)$$

For functors $F, G : \mathcal{P}_k \to \mathrm{Mod}\, k$ we define the dual $F^\circ : \mathcal{P}_k \to \mathrm{Mod}\, k$ by

$$F^\circ(V) = F(V^\vee)^\vee \qquad (V \in \mathcal{P}_k)$$

and write $\mathrm{Hom}(F, G)$ for the group of natural transformations $F \to G$. Then (8.2.4) yields a natural isomorphism

$$\mathrm{Hom}(F, G^\circ) \cong \mathrm{Hom}(G, F^\circ). \qquad (8.2.5)$$

For V, V', W, W' in $\mathcal{P}_k$ there is a canonical isomorphism

$$\mathrm{Hom}(V, W) \otimes \mathrm{Hom}(V', W') \cong \mathrm{Hom}(V \otimes V', W \otimes W'),$$

which is natural in all variables.

Symmetric Tensors and Symmetric Powers

Let G be a group acting on a k-module V. For each $g \in G$ the assignment $v \mapsto vg - v$ yields a map $V \to V$. One defines the *invariants* V^G as the kernel of the natural morphism $V \to \prod_{g \in G} V$ and the *coinvariants* V_G as the cokernel of the natural morphism $\coprod_{g \in G} V \to V$.

Now fix an integer $d > 0$ and denote by $\mathfrak{S}_d$ the *symmetric group* permuting d elements. For each k-module V, the group $\mathfrak{S}_d$ acts on $V^{\otimes d}$ by permuting the factors of the tensor product:

$$(v_1 \otimes \cdots \otimes v_d)\sigma = v_{\sigma(1)} \otimes \cdots \otimes v_{\sigma(d)}.$$

Denote by $\Gamma^d V$ the submodule $(V^{\otimes d})^{\mathfrak{S}_d}$ of $V^{\otimes d}$ consisting of the elements which are invariant under the action of $\mathfrak{S}_d$; this is called the module of *symmetric tensors* of degree d.[1] The largest quotient $(V^{\otimes d})_{\mathfrak{S}_d}$ of $V^{\otimes d}$ on which $\mathfrak{S}_d$ acts trivially is denoted by $S^d V$ and this module of coinvariants is called the *symmetric power* of degree d. For the image of $x_1 \otimes \cdots \otimes x_d$ under the canonical map $V^{\otimes d} \to S^d V$ we write $x_1 \cdots x_d$. Set $\Gamma^0 V = k$ and $S^0 V = k$.

Now assume $V \in \mathcal{P}_k$. From the definition, it follows that

$$\Gamma^d(V^{\vee}) \cong (S^d V)^{\vee}.$$

Note that $S^d V$ is a free k-module provided that V is free. If $(v_i)_{i \in I}$ is a basis of V, then a basis of $S^d V$ with respect to a total ordering $\leq$ on I is given by the products $v_{i_1} \cdots v_{i_d}$ with $i_1 \leq \ldots \leq i_d$. Thus $\Gamma^d V$ and $S^d V$ belong to $\mathcal{P}_k$ for all $V \in \mathcal{P}_k$, and we obtain functors $\Gamma^d, S^d : \mathcal{P}_k \to \mathcal{P}_k$ satisfying $S^d \cong (\Gamma^d)^{\circ}$.

The Category of Symmetric Tensors

We consider for $d \geq 0$ the *category of symmetric tensors* $\Gamma^d \mathcal{P}_k$ which is defined as follows. The objects are the finitely generated projective k-modules and for two objects V, W set

$$\operatorname{Hom}_{\Gamma^d \mathcal{P}_k}(V, W) = \Gamma^d \operatorname{Hom}(V, W).$$

This identifies with $\operatorname{Hom}(V^{\otimes d}, W^{\otimes d})^{\mathfrak{S}_d}$, since

$$\operatorname{Hom}(V, W)^{\otimes d} \cong (W \otimes V^{\vee})^{\otimes d} \cong W^{\otimes d} \otimes (V^{\otimes d})^{\vee} \cong \operatorname{Hom}(V^{\otimes d}, W^{\otimes d})$$

with $\mathfrak{S}_d$ acting on $\operatorname{Hom}(V^{\otimes d}, W^{\otimes d})$ via $(f\sigma)(v) = f(v\sigma^{-1})\sigma$ for $f : V^{\otimes d} \to W^{\otimes d}$ and $\sigma \in \mathfrak{S}_d$. Using this identification one defines the composition of

---

[1] The notation $\Gamma^d V$ is common practice but also misleading, because originally it referred to the module of divided powers, which actually is isomorphic to the module of symmetric tensors when V is a free k-module; see [39, IV.5, Exercise 8].

morphisms in $\Gamma^d \mathcal{P}_k$. The assignment $V \mapsto V^\vee$ induces a duality

$$(\Gamma^d \mathcal{P}_k)^{\mathrm{op}} \xrightarrow{\sim} \Gamma^d \mathcal{P}_k.$$

We denote by $k\mathfrak{S}_d$ the group algebra of $\mathfrak{S}_d$. For each $V \in \mathcal{P}_k$ we view $V^{\otimes d}$ as a module over $k\mathfrak{S}_d$ via the action of $\mathfrak{S}_d$.

Lemma 8.2.6. *The assignment $V \mapsto V^{\otimes d}$ induces a natural isomorphism*

$$\mathrm{Hom}_{\Gamma^d \mathcal{P}_k}(V, W) \xrightarrow{\sim} \mathrm{Hom}_{k\mathfrak{S}_d}(V^{\otimes d}, W^{\otimes d})$$

and therefore an equivalence

$$\Gamma^d \mathcal{P}_k \xrightarrow{\sim} \{V^{\otimes d} \mid V \in \mathcal{P}_k\} \subseteq \mathrm{Mod}\, k\mathfrak{S}_d.$$

Proof For $f \in \mathrm{Hom}(V^{\otimes d}, W^{\otimes d})$ we have $f\sigma = f$ for all $\sigma \in \mathfrak{S}_d$ if and only if $f(v\sigma) = (fv)\sigma$ for all $\sigma \in \mathfrak{S}_d$ and $v \in V^{\otimes d}$. $\qquad\square$

Let $n \geq 0$ be an integer and set $V = k^n$. Then the k-algebra

$$S_k(n, d) := \mathrm{End}_{\Gamma^d \mathcal{P}_k}(V)$$

is called a *Schur algebra*. Note that the isomorphism $\mathrm{End}(V)^{\otimes d} \xrightarrow{\sim} \mathrm{End}(V^{\otimes d})$ induces an isomorphism of k-algebras

$$\mathrm{End}_{\Gamma^d \mathcal{P}_k}(V) \xrightarrow{\sim} \mathrm{End}_{k\mathfrak{S}_d}(V^{\otimes d}).$$

Now consider $V^{\otimes d}$ as an $S_k(n, d)$-$k\mathfrak{S}_d$-bimodule. Then each element in $k\mathfrak{S}_d$ induces an endomorphism $V^{\otimes d} \to V^{\otimes d}$ which is given by right multiplication.

Proposition 8.2.7. *For $n \geq d$ the canonical homomorphism*

$$(k\mathfrak{S}_d)^{\mathrm{op}} \longrightarrow \mathrm{End}_{S_k(n,d)}(V^{\otimes d})$$

is an isomorphism of k-algebras.

Proof Fix a k-basis $v_1, \ldots, v_n$ of V and set $v = v_1 \otimes \cdots \otimes v_d$. The elements $v\sigma$ with $\sigma \in \mathfrak{S}_d$ form a k-basis of a $k\mathfrak{S}_d$-submodule of $V^{\otimes d}$, which is in fact a direct summand isomorphic to $k\mathfrak{S}_d$. Thus $k\mathfrak{S}_d \cong \mathrm{Ker}\,\varepsilon$ for some idempotent $\varepsilon \in \mathrm{End}_{k\mathfrak{S}_d}(V^{\otimes d})$, and the assertion follows from Lemma 8.2.8 below. $\qquad\square$

Lemma 8.2.8. *Let Λ be a ring and M a Λ-module with $\Gamma = \mathrm{End}_\Lambda(M)$. If for some integer $r \geq 1$ there is an idempotent $\varepsilon \in \mathrm{End}_\Lambda(M^r)$ with $\mathrm{Ker}\,\varepsilon \cong \Lambda$, then the canonical homomorphism $\Lambda^{\mathrm{op}} \to \mathrm{End}_\Gamma(M)$ is an isomorphism.*

Proof We consider for each Λ-module X the natural morphism

$$\phi_X \colon X \longrightarrow H(X) := \mathrm{Hom}_\Gamma(\mathrm{Hom}_\Lambda(X, M), M)$$

which is given by evaluation. The idempotent $\varepsilon \colon M^r \to M^r$ induces an exact sequence

$$0 \longrightarrow \Lambda \longrightarrow M^r \overset{\varepsilon}{\longrightarrow} M^r$$

of Λ-modules, and applying H gives an exact sequence

$$0 \longrightarrow H(\Lambda) \longrightarrow H(M^r) \longrightarrow H(M^r).$$

Then ϕ_{M^r} is an isomorphism, and therefore ϕ_Λ is an isomorphism, which identifies with the canonical homomorphism $\Lambda^{\mathrm{op}} \to \mathrm{End}_\Gamma(M)$. $\qquad\square$

Exterior Powers

For a k-module V let $T^*V = \bigoplus_{d \geq 0} V^{\otimes d}$ denote the *tensor algebra*. From this one obtains the *exterior algebra* $\Lambda^* V = \bigoplus_{d \geq 0} \Lambda^d V$ by taking the quotient with respect to the ideal generated by the elements $v \otimes v$, $v \in V$. The canonical map $V^{\otimes d} \to \Lambda^d V$ takes $v_1 \otimes \cdots \otimes v_d$ to $v_1 \wedge \cdots \wedge v_d$, and $\Lambda^d V$ is called the *exterior power* of degree d.

For each $d \geq 0$, the k-module $\Lambda^d V$ is free provided that V is free. Thus $\Lambda^d V$ belongs to $\mathcal{P}_k$ for all $V \in \mathcal{P}_k$, and this gives a functor $\mathcal{P}_k \to \mathcal{P}_k$. There is a natural isomorphism

$$\Lambda^d(V^\vee) \cong (\Lambda^d V)^\vee$$

induced by $(f_1 \wedge \cdots \wedge f_d)(v_1 \wedge \cdots \wedge v_d) = \det(f_i(v_j))$. Thus $(\Lambda^d)^\circ \cong \Lambda^d$.

The Algebra of Symmetric Tensors

Let V be a k-module. We set $\Gamma^* V = \bigoplus_{d \geq 0} \Gamma^d V$. For integers $d, e \geq 0$ the inclusion $\mathfrak{S}_d \times \mathfrak{S}_e \subseteq \mathfrak{S}_{d+e}$ induces the *multiplication map*

$$\Gamma^d V \otimes \Gamma^e V \longrightarrow \Gamma^{d+e} V \tag{8.2.9}$$

which sends $x \otimes y \in \Gamma^d V \otimes \Gamma^e V$ to

$$xy = \sum_{g \in \mathfrak{S}_d \times \mathfrak{S}_e \backslash \mathfrak{S}_{d+e}} (x \otimes y)g$$

where $(x \otimes y)g = (x \otimes y)\sigma$ for a coset $g = (\mathfrak{S}_d \times \mathfrak{S}_e)\sigma$. This multiplication is also known as the *shuffle product*; it is associative and gives $\Gamma^* V$ the structure of a commutative k-algebra.

Fix a set I and a total ordering $\leq$ on I. We write $\mathbb{N}^{(I)}$ for the set of sequences $\lambda = (\lambda_i)_{i \in I}$ of non-negative integers with $\lambda_i = 0$ for almost all i and set

$|\lambda| = \sum_{i \in I} \lambda_i$. Now fix $d \geq 0$ and let $I^d = \{\mathbf{i} = (i_1, \ldots, i_d) \mid i_l \in I\}$. The group $\mathfrak{S}_d$ acts on I^d via

$$(i_1, \ldots, i_d)\sigma = (i_{\sigma(1)}, \ldots, i_{\sigma(d)})$$

and let $I^d/\mathfrak{S}_d$ denote the set of orbits. Consider the map $I^d \to \mathbb{N}^{(I)}$ sending $\mathbf{i}$ to $\mathbf{i}^*$ given by $\mathbf{i}^*(i) = \text{card}\{l \mid i_l = i\}$ for $i \in I$. Then $\mathbf{i}_1^* = \mathbf{i}_2^*$ if and only if $\mathbf{i}_2 = \mathbf{i}_1\sigma$ for some $\sigma \in \mathfrak{S}_d$. Thus the assignment $\mathbf{i} \mapsto \mathbf{i}^*$ induces a bijection

$$I^d/\mathfrak{S}_d \xrightarrow{\sim} \{\lambda \in \mathbb{N}^{(I)} \mid |\lambda| = d\}.$$

Let $\lambda \in \mathbb{N}^{(I)}$ with $|\lambda| = d$ and write

$$\{1, \ldots, d\} = \bigcup_{i \in I} X_{\lambda,i} \quad \text{with} \quad X_{\lambda,i} = \left\{ \sum_{j < i} \lambda_j + 1, \ldots, \sum_{j \leq i} \lambda_j \right\}.$$

We denote by $\mathfrak{S}_\lambda \cong \prod_{i \in I} \mathfrak{S}_{\lambda_i}$ the *Young subgroup* of $\mathfrak{S}_d$ consisting of all $\sigma \in \mathfrak{S}_d$ such that $\sigma(X_{\lambda,i}) \subseteq X_{\lambda,i}$ for all $i \in I$. Let $\mathfrak{S}^\lambda$ denote the subset of all $\sigma \in \mathfrak{S}_d$ such that $\sigma|_{X_{\lambda,i}}$ is increasing for all $i \in I$. Then there are bijections

$$\mathfrak{S}^\lambda \times \mathfrak{S}_\lambda \xrightarrow{\sim} \mathfrak{S}_d, \qquad (\sigma, \tau) \mapsto \sigma\tau,$$

and

$$\mathfrak{S}_\lambda \times \mathfrak{S}^\lambda \xrightarrow{\sim} \mathfrak{S}_d, \qquad (\sigma, \tau) \mapsto \sigma\tau^{-1}.$$

Now suppose that V is a free k-module with basis $(v_i)_{i \in I}$, and for $\lambda \in \mathbb{N}^{(I)}$ with $|\lambda| = d$ set

$$v_\lambda = \prod_{i \in I} v_i^{\otimes \lambda_i}.$$

Given elements $x_i \in \Gamma^{\lambda_i} V$, we have in $\Gamma^* V$

$$\prod_{i \in I} x_i = \sum_{\mathfrak{S}_\lambda \sigma \in \mathfrak{S}_\lambda \backslash \mathfrak{S}_d} \left(\bigotimes_{i \in I} x_i \right)\sigma = \sum_{\sigma^{-1} \in \mathfrak{S}^\lambda} \left(\bigotimes_{i \in I} x_i \right)\sigma, \qquad (8.2.10)$$

where the first equality is by the definition of the multiplication in $\Gamma^* V$ and the second equality follows from the bijection $\mathfrak{S}^\lambda \xrightarrow{\sim} \mathfrak{S}_\lambda \backslash \mathfrak{S}_d$ given by $\sigma \mapsto \mathfrak{S}_\lambda \sigma^{-1}$.

Lemma 8.2.11. *The elements v_λ with $\lambda \in \mathbb{N}^{(I)}$ and $|\lambda| = d$ form a k-basis of $\Gamma^d V$.*

Proof The elements

$$v_\mathbf{i} = v_{i_1} \otimes v_{i_2} \otimes \cdots \otimes v_{i_d} \quad \text{for} \quad \mathbf{i} \in I^d$$

form a k-basis of $V^{\otimes d}$, and we claim that the elements

$$v(\omega) = \sum_{\mathbf{i} \in \omega} v_{i_1} \otimes v_{i_2} \otimes \cdots \otimes v_{i_d} \quad \text{for} \quad \omega \in I^d/\mathfrak{S}_d$$

form a basis of $\Gamma^d V = (V^{\otimes d})^{\mathfrak{S}_d}$. The elements $v(\omega)$ are invariant under $\mathfrak{S}_d$ and therefore belong to $\Gamma^d V$. They are linearly independent, since any equality $\sum_\omega \alpha_\omega v(\omega) = 0$ in $\Gamma^d V$ also holds in $V^{\otimes d}$, so $\alpha_\omega = 0$ for all ω. Finally, if $\sum_{\mathbf{i}} \alpha_{\mathbf{i}} v_{\mathbf{i}}$ is any element in $\Gamma^d V$, then $\alpha_{\mathbf{i}} = \alpha_{\mathbf{i}\sigma}$ for all $\sigma \in \mathfrak{S}_d$. Thus the $v(\omega)$ generate $\Gamma^d V$.

Now observe that $\mathbf{i} \mapsto \mathbf{i}^*$ induces a bijection $I^d/\mathfrak{S}_d \xrightarrow{\sim} \{\lambda \in \mathbb{N}^{(I)} \mid |\lambda| = d\}$. If $\omega = \mathbf{i}\mathfrak{S}_d$ corresponds to $\mathbf{i}^* = \lambda$, then the identity (8.2.10) yields

$$v_\lambda = \sum_{\mathbf{i}^* = \lambda} v_{i_1} \otimes v_{i_2} \otimes \cdots \otimes v_{i_d} = v(\omega).$$

Thus the elements v_λ form a k-basis of $\Gamma^d V$. $\qquad\square$

Next we consider the comultiplication for $\Gamma^* V$, assuming that V is a flat k-module. We need the following lemma.

Lemma 8.2.12. *For integers $d, e \geq 0$ the inclusion $\mathfrak{S}_d \times \mathfrak{S}_e \subseteq \mathfrak{S}_{d+e}$ induces an isomorphism*

$$(V^{\otimes d})^{\mathfrak{S}_d} \otimes (V^{\otimes e})^{\mathfrak{S}_e} \cong (V^{\otimes d+e})^{\mathfrak{S}_d \times \mathfrak{S}_e}.$$

Proof First observe that for any finite group G acting on k-modules X and Y we have an isomorphism

$$X^G \otimes Y \cong (X \otimes Y)^G$$

provided that Y is flat and satisfying $Y^G = Y$. Also, for any filtered colimit $\mathrm{colim}_i X_i$ of k-modules with a G-action we have

$$\mathrm{colim}_i(X_i^G) \cong (\mathrm{colim}_i X_i)^G$$

since X^G is by definition the kernel of the natural morphism $X \to \prod_{g \in G} X$. It follows from Lemma 8.2.11 that $(V^{\otimes d})^{\mathfrak{S}_d}$ is flat and we obtain

$$\begin{aligned} (V^{\otimes d})^{\mathfrak{S}_d} \otimes (V^{\otimes e})^{\mathfrak{S}_e} &\cong \left((V^{\otimes d})^{\mathfrak{S}_d} \otimes V^{\otimes e}\right)^{\mathfrak{S}_e} \\ &\cong \left((V^{\otimes d} \otimes V^{\otimes e})^{\mathfrak{S}_d}\right)^{\mathfrak{S}_e} \\ &\cong (V^{\otimes d+e})^{\mathfrak{S}_d \times \mathfrak{S}_e}. \end{aligned}$$

$\qquad\square$

Applying the above lemma, we obtain for integers $d, e \geq 0$ the *diagonal* or *comultiplication map*

$$\Gamma^{d+e} V \longrightarrow \Gamma^d V \otimes \Gamma^e V \qquad (8.2.13)$$

as composite

$$(V^{\otimes d+e})^{\mathfrak{S}_{d+e}} \subseteq (V^{\otimes d+e})^{\mathfrak{S}_d \times \mathfrak{S}_e} \cong (V^{\otimes d})^{\mathfrak{S}_d} \otimes (V^{\otimes e})^{\mathfrak{S}_e}.$$

When V is a free k-module of finite rank, fix a basis $\{v_1, \ldots, v_n\}$ and let $\{v_1^{\vee}, \ldots, v_n^{\vee}\}$ denote the dual basis of $V^{\vee}$. Then the elements

$$v_{\lambda} = v_1^{\otimes \lambda_1} v_2^{\otimes \lambda_2} \cdots v_n^{\otimes \lambda_n} \quad \text{for} \quad \lambda \in \Lambda(n, d)$$

form a k-basis of $\Gamma^d V$. If $\{v_{\lambda}^{\vee}\}_{\lambda \in \Lambda(n,d)}$ denotes the dual basis of $(\Gamma^d V)^{\vee}$, then the canonical isomorphism $(\Gamma^d V)^{\vee} \xrightarrow{\sim} S^d(V^{\vee})$ maps each $v_{\lambda}^{\vee}$ to $\prod_{i=1}^{n}(v_i^{\vee})^{\lambda_i}$.

The diagonal map $\Gamma^{d+e} V \to \Gamma^d V \otimes \Gamma^e V$ is given on basis elements by

$$v_{\lambda} \longmapsto \sum_{\substack{\mu \in \Lambda(n,d) \\ \lambda-\mu \in \Lambda(n,e)}} v_{\mu} \otimes v_{\lambda-\mu} \tag{8.2.14}$$

and the multiplication map $\Gamma^d V \otimes \Gamma^e V \to \Gamma^{d+e} V$ is given by

$$v_{\mu} \otimes v_{\nu} \longmapsto c v_{\mu+\nu} \quad \text{with} \quad c = |\mathfrak{S}_{\mu+\nu}/\mathfrak{S}_{\mu} \times \mathfrak{S}_{\nu}|. \tag{8.2.15}$$

Polynomial Maps

There is a close relation between symmetric tensors and polynomial maps which can be explained by the following lemma.

For a k-module M and $x \in M$ set $\gamma_d(x) = x^{\otimes d}$. For a set I, and elements $\nu \in \mathbb{N}^{(I)}$ and $(\alpha_i) \in k^{(I)}$, we write $\alpha^{\nu} = \prod_{i \in I} \alpha_i^{\nu_i}$.

Lemma 8.2.16. *Let M, N be k-modules such that M is free. Then the following are equivalent for a map $f : M \to N$.*

(1) *There exists a basis $(x_i)_{i \in I}$ of M and a family $(y_{\nu})_{\nu \in \mathbb{N}^{(I)}, |\nu|=d}$ of elements in N such that for all $(\alpha_i) \in k^{(I)}$*

$$f\left(\sum_{i \in I} \alpha_i x_i\right) = \sum_{\nu \in \mathbb{N}^{(I)}, |\nu|=d} \alpha^{\nu} y_{\nu}.$$

(2) *There exists a k-linear map $h : \Gamma^d M \to N$ such that $f(x) = h(\gamma_d(x))$ for all $x \in M$.*

Proof The proof uses the multiplication in $\Gamma^* M$. Given elements $x_1, \ldots, x_n$ in M, an induction on n shows that

$$\gamma_d(x_1 + \cdots + x_n) = \sum_{d_1 + \cdots + d_n = d} \gamma_{d_1}(x_1) \cdots \gamma_{d_n}(x_n).$$

From this we obtain for elements $(x_i)_{i \in I}$ in M and $(\alpha_i) \in k^{(I)}$

$$\gamma_d\left(\sum_{i \in I} \alpha_i x_i\right) = \sum_{\nu \in \mathbb{N}^{(I)}, |\nu|=d} \alpha^\nu x_\nu.$$

$(1) \Rightarrow (2)$: Let (x_i) and (y_ν) be elements satisfying the condition in (1). We note that $(x_\nu)_{|\nu|=d}$ is a basis of $\Gamma^d M$ and let $h \colon \Gamma^d M \to N$ be the homomorphism defined by $h(x_\nu) = y_\nu$. Then for $x = \sum_{i \in I} \alpha_i x_i$ in M we have

$$f(x) = f\left(\sum_{i \in I} \alpha_i x_i\right) = \sum_{\nu \in \mathbb{N}^{(I)}, |\nu|=d} \alpha^\nu y_\nu = h\left(\sum_{\nu \in \mathbb{N}^{(I)}, |\nu|=d} \alpha^\nu x_\nu\right) = h(\gamma_d(x)).$$

$(2) \Rightarrow (1)$: Let h be a map satisfying the condition in (2). When $(x_i)_{i \in I}$ is a basis of M, then

$$f\left(\sum_{i \in I} \alpha_i x_i\right) = h\left(\sum_{\nu \in \mathbb{N}^{(I)}, |\nu|=d} \alpha^\nu x_\nu\right) = \sum_{\nu \in \mathbb{N}^{(I)}, |\nu|=d} \alpha^\nu h(x_\nu). \qquad \square$$

A map $f \colon M \to N$ satisfying the equivalent conditions of the above lemma is called *homogeneous polynomial of degree d*. If M has a finite basis $\{x_1, \ldots, x_n\}$, then f is polynomial if and only if there is a polynomial F in n indeterminates with coefficients in N such that

$$f\left(\sum_i \alpha_i x_i\right) = F(\alpha) \qquad \text{for} \qquad \alpha = (\alpha_1, \ldots, \alpha_n) \in k^n.$$

This property does not depend on the basis chosen for M and justifies the term 'polynomial map'.

Remark 8.2.17. The lemma says that composition with $\gamma_d \colon M \to \Gamma^d M$ induces a surjection

$$\{h \colon \Gamma^d M \to N \mid h \text{ k-linear}\} \longrightarrow \{f \colon M \to N \mid f \text{ polynomial of deg } d\},$$

and we note that it is a bijection when k is an infinite field (see [39, IV.5, Proposition 16]).

8.3 Polynomial Representations

In this section we study polynomial representations of the *general linear group* $GL_n(k)$ given by the invertible $n \times n$ matrices over k, simultaneously for all integers $n > 0$. We use the language of strict polynomial functors, because a functor $F \colon \mathcal{P}_k \to \mathcal{P}_k$ induces maps $\operatorname{End}(k^n) \to \operatorname{End}(F(k^n))$ that yield representations $\mathrm{GL}_n(k) \to \mathrm{GL}(F(k''))$ for all $n > 0$. These representations are polynomial if the induced maps

$$\mathrm{Hom}(V, W) \longrightarrow \mathrm{Hom}(FV, FW)$$

are polynomial for all free modules $V, W \in \mathcal{P}_k$.

Polynomial representations of general linear groups can be identified with modules over Schur algebras. Moreover, there is a canonical functor connecting polynomial representations of degree d with linear representations of the symmetric group $\mathfrak{S}_d$.

Throughout we keep fixed a commutative ring k.

Strict Polynomial Functors

Fix an integer $d \geq 0$. Let $\gamma_d \colon \mathcal{P}_k \to \Gamma^d \mathcal{P}_k$ denote the functor which is the identity on objects and sends a morphism f to $f^{\otimes d}$.

A *strict polynomial functor* $\mathcal{P}_k \to \mathrm{Mod}\, k$ of degree d is by definition a k-linear functor $\Gamma^d \mathcal{P}_k \to \mathrm{Mod}\, k$. We denote by $\mathrm{Pol}^d \mathcal{P}_k$ the category of degree d strict polynomial functors $\mathcal{P}_k \to \mathrm{Mod}\, k$, and write $\mathrm{pol}^d \mathcal{P}_k$ for the full subcategory of strict polynomial functors $\mathcal{P}_k \to \mathcal{P}_k$.

Our aim is to develop a structure theory of the small category $\mathrm{pol}^d \mathcal{P}_k$, but for some constructions the full category $\mathrm{Pol}^d \mathcal{P}_k$ is needed.

It is often convenient to identify a k-linear functor $\Gamma^d \mathcal{P}_k \to \mathrm{Mod}\, k$ with the composite $F \colon \mathcal{P}_k \xrightarrow{\gamma_d} \Gamma^d \mathcal{P}_k \to \mathrm{Mod}\, k$. The fact that F is strict polynomial of degree d means that the induced map

$$\mathrm{Hom}(V, W) \longrightarrow \mathrm{Hom}(FV, FW)$$

is polynomial of degree d for all free modules $V, W \in \mathcal{P}_k$; see Lemma 8.2.16. The following diagram illustrates this correspondence.

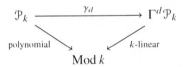

The strict polynomial functors form an exact category, where a sequence $0 \to F' \to F \to F'' \to 0$ is exact when $0 \to F'V \to FV \to F''V \to 0$ is an exact sequence for all $V \in \mathcal{P}_k$.

The functor

$$\Gamma^d \mathcal{P}_k \longrightarrow \mathrm{pol}^d \mathcal{P}_k, \qquad V \mapsto \Gamma^{d,V} := \Gamma^d \mathrm{Hom}(V, -)$$

is contravariant and fully faithful by definition. We identify $\Gamma^{d,k} = \Gamma^d$.

For $X \in \mathrm{Pol}^d \mathcal{P}_k$ there is the Yoneda isomorphism

$$\mathrm{Hom}(\Gamma^{d,V}, X) \xrightarrow{\sim} X(V) \qquad (8.3.1)$$

and it follows that $\Gamma^{d,V}$ is a projective object in $\mathrm{Pol}^d \mathcal{P}_k$.

Some of the constructions for strict polynomial functors require $V \in \mathcal{P}_k$ to be free. This is not a serious obstruction since an arbitrary $V \in \mathcal{P}_k$ is the colimit of a sequence $k^n \xrightarrow{\varepsilon} k^n \xrightarrow{\varepsilon} \cdots$ for some idempotent morphism ε. Then one sets $F(V) = \mathrm{colim}\, F(k^n)$ for any functor F given on free k-modules.

Using the Yoneda isomorphism it follows that each object X in $\mathrm{Pol}^d \mathcal{P}_k$ can be written canonically as a colimit of representable functors

$$\operatorname*{colim}_{\Gamma^{d,V} \to X} \Gamma^{d,V} \xrightarrow{\sim} X \qquad (8.3.2)$$

where the colimit is taken over the category of morphisms $\Gamma^{d,V} \to X$ and V runs through the objects of $\Gamma^d \mathcal{P}_k$ (Lemma 11.1.8).

Tensor Products

For integers $d, e \geq 0$ there is a *tensor product*

$$- \otimes -\colon \mathrm{Pol}^d \mathcal{P}_k \times \mathrm{Pol}^e \mathcal{P}_k \longrightarrow \mathrm{Pol}^{d+e} \mathcal{P}_k.$$

Let $X \in \mathrm{Pol}^d \mathcal{P}_k$ and $Y \in \mathrm{Pol}^e \mathcal{P}_k$. The functor $X \otimes Y$ acts on objects via

$$(X \otimes Y)(V) = X(V) \otimes Y(V) \qquad (V \in \mathcal{P}_k)$$

and on morphisms via the diagonal map (8.2.13)

$$\Gamma^{d+e}\, \mathrm{Hom}(V, W) \longrightarrow \Gamma^d\, \mathrm{Hom}(V, W) \otimes \Gamma^e\, \mathrm{Hom}(V, W)$$

by composing with

$$\Gamma^d\, \mathrm{Hom}(V, W) \otimes \Gamma^e\, \mathrm{Hom}(V, W)$$
$$\longrightarrow \mathrm{Hom}(X(V), X(W)) \otimes \mathrm{Hom}(Y(V), Y(W))$$
$$\longrightarrow \mathrm{Hom}(X(V) \otimes Y(V), X(W) \otimes Y(W)).$$

Main examples of strict polynomial functors of degree d are the following:

$$V \longmapsto \Gamma^d V \qquad V \longmapsto T^d V \qquad V \longmapsto S^d V \qquad V \longmapsto \Lambda^d V.$$

We have seen this already for Γ^d. Given any $\lambda \in \Lambda(n, d)$ it follows that

$$\Gamma^\lambda := \Gamma^{\lambda_1} \otimes \cdots \otimes \Gamma^{\lambda_n}$$

is strict polynomial of degree d. In particular, $T^d = \Gamma^{(1,\dots,1)}$ is strict polynomial. For S^d and Λ^d this property can be deduced from the lemma below, since any cokernel of a morphism between strict polynomial functors is again strict polynomial.

Lemma 8.3.3. *Given $d \geq 0$ and $V \in \mathcal{P}_k$, there are exact sequences*

$$\bigoplus_{i=1}^{d-1} V^{\otimes i-1} \otimes \Gamma^2 V \otimes V^{\otimes d-i-1} \xrightarrow{1 \otimes \Delta \otimes 1} V^{\otimes d} \longrightarrow \Lambda^d V \longrightarrow 0$$

$$\bigoplus_{i=1}^{d-1} V^{\otimes i-1} \otimes \Lambda^2 V \otimes V^{\otimes d-i-1} \xrightarrow{1 \otimes \Delta \otimes 1} V^{\otimes d} \longrightarrow S^d V \longrightarrow 0$$

where $\Delta\colon \Gamma^2 V \to V \otimes V$ is the inclusion and $\Delta\colon \Lambda^2 V \to V \otimes V$ is given by $\Delta(v \wedge w) = v \otimes w - w \otimes v$.

Proof This is clear from the definitions of $\Lambda^d V$ and $S^d V$ respectively. □

Also, the duality $F \mapsto F^\circ$ maps strict polynomial functors to strict polynomial functors. In particular, we have for $X, Y \in \text{pol}^d \mathcal{P}_k$ a natural isomorphism

$$(X \otimes Y)^\circ \cong X^\circ \otimes Y^\circ. \tag{8.3.4}$$

Now fix a free k-module V with basis $\{v_1, \ldots, v_r\}$ and a partition $\lambda = (\lambda_1, \ldots, \lambda_n)$. Each filling T with entries in $\{1, \ldots, r\}$ yields an element

$$v_T \in \Gamma^{\lambda_1} V \otimes \cdots \otimes \Gamma^{\lambda_n} V$$

by replacing each i in a box by v_i. Here is an example of a Young tableau

$$\lambda = (5, 3, 3, 2)$$

1	2	2	3	3
2	3	5
4	4	6
5	6

and here is the corresponding element

$$v_T = (v_1(v_2 \otimes v_2)(v_3 \otimes v_3)) \otimes (v_2 v_3 v_5) \otimes ((v_4 \otimes v_4)v_6) \otimes (v_5 v_6).$$

More precisely, let $T(i, j)$ denote the entry of the box (i, j) and define $\alpha^i \in \Lambda(r, \lambda_i)$ by setting $\alpha^i_j = \text{card}\{t \mid T(i, t) = j\}$. Then we set

$$v_T = v_{\alpha^1} \otimes \cdots \otimes v_{\alpha^n}.$$

Lemma 8.3.5. *The elements v_T form a k-basis of $\Gamma^\lambda V$ as T runs through all fillings that are weakly increasing along each row.*

Proof Observe that α^i only depends on the entries of the ith row of T and not on their order. Thus the fillings T that are weakly increasing along each row correspond bijectively to sequences of compositions $\alpha^i \in \Lambda(r, \lambda_i)$ ($i = 1, \ldots, n$). The elements v_{α^i} form a basis of $\Gamma^{\lambda_i} V$ by Lemma 8.2.11. Taking their tensor products then yields a basis of $\Gamma^\lambda V$. □

We keep a free k-module V with basis $\{v_1, \ldots, v_r\}$ and a partition $\mu = (\mu_1, \ldots, \mu_n)$. Each filling T with entries in $\{1, \ldots, r\}$ yields an element

$$v^T \in \Lambda^{\mu_1} V \otimes \cdots \otimes \Lambda^{\mu_n} V$$

as follows. For a subset $J = \{j_1, \ldots, j_s\} \subseteq \{1, \ldots, r\}$ set $v_J = v_{j_1} \wedge \cdots \wedge v_{j_s}$ and

$$v^T = v_{J_1} \otimes \cdots \otimes v_{J_n}$$

where $J_i = \{T(i, j) \mid 1 \le j \le \mu_i\}$.

Lemma 8.3.6. *The elements v^T form a k-basis of $\Lambda^\mu V$ as T runs through all fillings that are strictly increasing along each row.*

Proof The elements v_J form a basis of $\Lambda^s V$, where J runs through all s-element subsets of $\{1, \ldots, r\}$. Taking their tensor products then yields a basis of $\Lambda^\mu V$. □

Decomposing Symmetric Tensors

We study decompositions of symmetric tensors and employ the structure of a graded algebra which one obtains by combining all degrees.

A *graded functor* $X = (X^0, X^1, X^2, \ldots)$ is given by a sequence of functors $X^i \colon \mathcal{P}_k \to \mathcal{P}_k$. The tensor product $X \otimes Y$ of graded functors X, Y is defined in degree d by

$$(X \otimes Y)^d = \bigoplus_{i+j=d} X^i \otimes Y^j.$$

For a composition $\lambda \in \Lambda(n, d)$ we set

$$X^\lambda := X^{\lambda_1} \otimes \cdots \otimes X^{\lambda_n}.$$

Given $V \in \mathcal{P}_k$, we write $S^* V = \bigoplus_{d \ge 0} S^d V$ for the *symmetric algebra*. For objects $V, W \in \mathcal{P}_k$, there is an isomorphism of k-algebras

$$S^* V \otimes S^* W \xrightarrow{\sim} S^*(V \oplus W)$$

which takes an element $(x_1 \cdots x_i) \otimes (y_1 \cdots y_j)$ in degree $i + j$ to $x_1 \cdots x_i y_1 \cdots y_j$.

Now consider the composite

$$\Gamma^* V \otimes \Gamma^* W \longrightarrow \Gamma^*(V \oplus W) \otimes \Gamma^*(V \oplus W) \longrightarrow \Gamma^*(V \oplus W)$$

where the first map is given by the inclusions $V \to V \oplus W$ and $W \to V \oplus W$ and the second is the multiplication map (8.2.9). This yields an isomorphism of k-algebras

$$\Gamma^* V \otimes \Gamma^* W \xrightarrow{\sim} \Gamma^*(V \oplus W) \tag{8.3.7}$$

because it maps basis elements to basis elements. Note that this isomorphism is dual to the one for symmetric algebras.

From (8.3.7) one obtains for each integer $n \geq 0$ in degree d an isomorphism

$$\Gamma^{d,k^n} \cong \bigoplus_{i=0}^{d} (\Gamma^{d-i,k^{n-1}} \otimes \Gamma^{i,k})$$

and using induction plus the identification $\Gamma^{i,k} = \Gamma^i$ a canonical isomorphism

$$\Gamma^{d,k^n} \cong \bigoplus_{\lambda \in \Lambda(n,d)} \Gamma^\lambda. \tag{8.3.8}$$

We have already seen that the functors of the form $\Gamma^{d,V}$ with $V \in \mathcal{P}_k$ yield a set of projective generators for $\mathrm{pol}^d \, \mathcal{P}_k$, thanks to the Yoneda isomorphism (8.3.1). Thus the decomposition of symmetric tensors implies that the projective objects in $\mathrm{pol}^d \, \mathcal{P}_k$ are precisely the direct summands of finite direct sums of functors Γ^λ, where $\lambda = (\lambda_1, \dots, \lambda_n)$ is any sequence of integers $\lambda_i \geq 0$ satisfying $\sum \lambda_i = d$ and n is any positive integer. Note that $\Gamma^\lambda \cong \Gamma^\mu$ if the compositions λ and μ yield the same partition of d after reordering. In particular, given $\lambda \in \Lambda(n,d)$, we have $\Gamma^\lambda \cong \Gamma^\mu$ for some $\mu \in \Lambda(d,d)$.

Lemma 8.3.9. *Each* $X \in \mathrm{pol}^d \, \mathcal{P}_k$ *admits a projective resolution*

$$\cdots \longrightarrow \Gamma^{d,V_2} \longrightarrow \Gamma^{d,V_1} \longrightarrow \Gamma^{d,V_0} \longrightarrow X \longrightarrow 0 \qquad (V_i \in \mathcal{P}_k).$$

Proof It suffices to construct an epimorphism $\Gamma^{d,V} \to X$ for some $V \in \mathcal{P}_k$ because the kernel is again in $\mathrm{pol}^d \, \mathcal{P}_k$ and we can iterate this.

For each partition λ of d the k-module $\mathrm{Hom}(\Gamma^\lambda, X)$ is finitely generated, say by n_λ elements. Taking their sum yields a morphism $(\Gamma^\lambda)^{n_\lambda} \to X$ and then

$$\pi \colon \bigoplus_\lambda (\Gamma^\lambda)^{n_\lambda} \longrightarrow X$$

is an epimorphism, where λ runs through all partitions of d. This follows from the Yoneda isomorphism (8.3.1), since each morphism $\Gamma^{d,V} \to X$ factors through π by construction. Now choose $V = k^n$ sufficiently big such that for each partition λ there are at least n_λ compositions in $\Lambda(n,d)$ that are equivalent to λ up to a permutation. This yields an epimorphism $\Gamma^{d,V} \to X$ because of the decomposition (8.3.8). $\qquad\qquad\square$

There is an analogue of the decomposition (8.3.8) for the exterior powers. For objects $V, W \in \mathcal{P}_k$ there is an isomorphism

$$\Lambda^* V \otimes \Lambda^* W \xrightarrow{\sim} \Lambda^*(V \oplus W)$$

given by multiplication. This yields a canonical decomposition

$$\bigoplus_{\lambda \in \Lambda(n,d)} \Lambda^\lambda \cong \Lambda^{d,k^n} := \Lambda^d \operatorname{Hom}(k^n, -). \tag{8.3.10}$$

Representations of Schur Algebras

Strict polynomial functors and modules over Schur algebras are closely related, since for $X \in \operatorname{Pol}^d \mathcal{P}_k$ and an integer $n \geq 1$ the Schur algebra $S_k(n,d)$ acts on $X(k^n)$. The action is from the left since we consider covariant functors $\mathcal{P}_k \to \operatorname{Mod} k$.

Proposition 8.3.11. *Evaluation at k^n gives a functor*

$$\operatorname{Pol}^d \mathcal{P}_k \longrightarrow \operatorname{Mod} S_k(n,d)^{\mathrm{op}}, \qquad X \mapsto X(k^n)$$

which has a fully faithful left adjoint

$$\Gamma^{d,k^n} \otimes -: \operatorname{Mod} S_k(n,d)^{\mathrm{op}} \longrightarrow \operatorname{Pol}^d \mathcal{P}_k.$$

This left adjoint identifies $\operatorname{Mod} S_k(n,d)^{\mathrm{op}}$ with the full subcategory of functors X that admit a projective presentation

$$P_1 \longrightarrow P_0 \longrightarrow X \longrightarrow 0 \quad \text{with} \quad P_i \in \operatorname{Add}\{\Gamma^\lambda \mid \lambda \in \Lambda(n,d)\}.$$

Thus evaluation at k^n is an equivalence if and only if $n \geq d$.

Proof Set $P = \Gamma^{d,k^n}$. Then evaluation at k^n identifies with $\operatorname{Hom}(P,-)$ by the Yoneda isomorphism (8.3.1). This functor admits a left adjoint $P \otimes -$ which is right exact, preserves all coproducts, and takes $S_k(n,d)$ to P. Thus it takes an $S_k(n,d)$-module M to the functor

$$V \longmapsto \Gamma^d \operatorname{Hom}(k^n, V) \otimes_{S_k(n,d)} M \qquad (V \in \Gamma^d \mathcal{P}_k).$$

The functor $P \otimes -$ is fully faithful since the counit

$$\operatorname{Hom}(P, P \otimes M) \longrightarrow M$$

is an isomorphism (cf. Example 2.2.24). We observe that

$$\operatorname{Add} P = \operatorname{Add}\{\Gamma^\lambda \mid \lambda \in \Lambda(n,d)\}$$

because of the decomposition (8.3.8).

Now let $n \geq d$. Then each projective object Γ^λ occurs in $\operatorname{Add} P$, and therefore $\Gamma^{d,V} \in \operatorname{Add} P$ for all $V \in \mathcal{P}_k$. It follows that each $X \in \operatorname{Pol}^d \mathcal{P}_k$ admits a projective presentation $P_1 \to P_0 \to X \to 0$ with $P_i \in \operatorname{Add} P$, and therefore evaluation at k^n is an equivalence. For $n < d$ one checks that $\Lambda^d(k^n) = 0$. $\qquad\square$

Representations of Symmetric Groups

Schur–Weyl duality yields a relation between representations of the general linear groups and representations of the symmetric groups. In our context this takes the following form. Let $\omega = (1, \ldots, 1)$ be a sequence of length d. Then Γ^ω is the functor taking V to $V^{\otimes d}$.

Lemma 8.3.12. *Permuting the tensors via*

$$(v_1 \otimes \cdots \otimes v_d) \longmapsto (v_{\sigma(1)} \otimes \cdots \otimes v_{\sigma(d)}) \qquad (\sigma \in \mathfrak{S}_d)$$

induces an isomorphism of k-algebras $(k\mathfrak{S}_d)^{\mathrm{op}} \xrightarrow{\sim} \mathrm{End}(\Gamma^\omega)$.

Proof Let $V = k^n$ for some $n \geq d$. Evaluation at V yields a homomorphism

$$(k\mathfrak{S}_d)^{\mathrm{op}} \longrightarrow \mathrm{End}(\Gamma^\omega) \longrightarrow \mathrm{End}_{S_k(n,d)}(V^{\otimes d})$$

which is an isomorphism by Proposition 8.2.7. Then the assertion follows since the second map is an isomorphism by Proposition 8.3.11. $\qquad \square$

This observation gives rise to a functor into the category of left $k\mathfrak{S}_d$-modules

$$\mathrm{Hom}(\Gamma^\omega, -)\colon \mathrm{Pol}^d \, \mathcal{P}_k \longrightarrow \mathrm{Mod}(k\mathfrak{S}_d)^{\mathrm{op}}$$

which is also called the *Schur functor* (and not to be confused with the Schur functors parametrised by partitions). The functor admits a fully faithful left adjoint

$$\Gamma^\omega \otimes -\colon \mathrm{Mod}(k\mathfrak{S}_d)^{\mathrm{op}} \longrightarrow \mathrm{Pol}^d \, \mathcal{P}_k$$

which can be described as follows. The functor is right exact, preserves all coproducts, and takes $k\mathfrak{S}_d$ to Γ^ω. Thus it takes $M \in \mathrm{Mod}\, k\mathfrak{S}_d$ to the functor

$$V \longmapsto V^{\otimes d} \otimes_{k\mathfrak{S}_d} M \qquad (V \in \Gamma^d \mathcal{P}_k)$$

where $\mathfrak{S}_d$ acts on $V^{\otimes d}$ via

$$(v_1 \otimes \cdots \otimes v_d)\sigma = v_{\sigma(1)} \otimes \cdots \otimes v_{\sigma(d)}.$$

Lemma 8.3.13. *The functor $\Gamma^\omega \otimes -\colon \mathrm{Mod}(k\mathfrak{S}_d)^{\mathrm{op}} \to \mathrm{Pol}^d \, \mathcal{P}_k$ is fully faithful.*

Proof For every $k\mathfrak{S}_d$-module M the counit

$$\mathrm{Hom}(\Gamma^\omega, \Gamma^\omega \otimes M) \longrightarrow M$$

is an isomorphism, and therefore $\Gamma^\omega \otimes -$ is fully faithful. $\qquad \square$

The functor $F = \mathrm{Hom}(\Gamma^\omega, -)$ also admits a fully faithful right adjoint

$$G\colon \mathrm{Mod}(k\mathfrak{S}_d)^{\mathrm{op}} \longrightarrow \mathrm{Pol}^d \, \mathcal{P}_k,$$

which is given by the formula (8.3.15) below.

Proposition 8.3.14. *The adjoint pair* (F, G) *restricts to a pair of mutually quasi-inverse equivalences*

$$\{\Gamma^{d,V} \mid V \in \mathcal{P}_k\} \;\; \underset{\longleftarrow}{\overset{\longrightarrow}{}} \;\; \{V^{\otimes d} \mid V \in \mathcal{P}_k\}$$

identifying $\Gamma^{d,V}$ *and* $(V^\vee)^{\otimes d}$ *for any* $V \in \mathcal{P}_k$.

Proof   We consider the following pair of equivalences

$$\operatorname{Mod} k\mathfrak{S}_d \supseteq \{V^{\otimes d} \mid V \in \mathcal{P}_k\} \;\xleftarrow{\;\sim\;}\; \Gamma^d \mathcal{P}_k \;\xrightarrow{\;\sim\;}\; \{\Gamma^{d,V} \mid V \in \mathcal{P}_k\} \subseteq \operatorname{Pol}^d \mathcal{P}_k$$

$$k\mathfrak{S}_d \;\longleftarrow\!\!\!\longmapsto\; P^\omega \;\longmapsto\!\!\!\longrightarrow\; \Gamma^\omega$$

given by Yoneda's lemma and Lemma 8.2.6. The distinguished object P^ω of the idempotent completion of $\Gamma^d \mathcal{P}_k$ corresponds to Γ^ω and $k\mathfrak{S}_d$ respectively. Now we compute for $V \in \mathcal{P}_k$

$$
\begin{aligned}
\operatorname{Hom}(\Gamma^\omega, \Gamma^{d,V}) &\cong \operatorname{Hom}_{\Gamma^d \mathcal{P}_k}(V, P^\omega) \\
&\cong \operatorname{Hom}_{k\mathfrak{S}_d}(V^{\otimes d}, k\mathfrak{S}_d) \\
&\cong \operatorname{Hom}_k(V^{\otimes d}, k) \\
&\cong (V^\vee)^{\otimes d}.
\end{aligned}
$$

Next we compute $G(M)$ for a left $k\mathfrak{S}_d$-module M and have for $V \in \mathcal{P}_k$

$$(GM)(V) \cong \operatorname{Hom}(\Gamma^{d,V}, GM) \cong \operatorname{Hom}_{k\mathfrak{S}_d}((V^\vee)^{\otimes d}, M). \qquad (8.3.15)$$

Specialising $M = W^{\otimes d}$ we get $G(W^{\otimes d}) \cong \Gamma^{d,W^\vee}$ since

$$\operatorname{Hom}_{k\mathfrak{S}_d}((V^\vee)^{\otimes d}, W^{\otimes d}) \cong \operatorname{Hom}_{\Gamma^d \mathcal{P}_k}(V^\vee, W) \cong \operatorname{Hom}_{\Gamma^d \mathcal{P}_k}(W^\vee, V).$$

For another formal argument computing the right adjoint G, see Lemma 8.5.8. $\square$

It follows from the proposition that Schur's functor $\operatorname{Hom}(\Gamma^\omega, -)$ extends the equivalence from Lemma 8.2.6, making the following square commutative.

$$
\begin{array}{ccc}
\Gamma^d \mathcal{P}_k & \xrightarrow{\;\sim\;} & \{V^{\otimes d} \mid V \in \mathcal{P}_k\} \\
\downarrow & & \downarrow \\
\operatorname{Pol}^d \mathcal{P}_k & \xrightarrow{\operatorname{Hom}(\Gamma^\omega, -)} & \operatorname{Mod}(k\mathfrak{S}_d)^{\operatorname{op}}
\end{array}
$$

Weight Space Decompositions

Symmetric tensors admit canonical decompositions that are indexed by sequences of non-negative integers. These integer sequences are called weights, and there are induced decompositions for any strict polynomial functor.

Fix a free k-module V with ordered basis $\{v_1, \ldots, v_n\}$. For each $X \in \mathrm{pol}^d \mathcal{P}_k$ we describe a decomposition of $X(V)$ into weight spaces.

For $i \in \{1, \ldots, n\}$ let $e_i \colon V \to V$ denote the endomorphism that is given by $e_i(v_j) = \delta_{ij} v_j$. Then $\mathrm{id}_V = e_1 + \cdots + e_n$ and the e_i generate a k-subalgebra $E \subseteq \mathrm{End}(V)$ that is isomorphic to $k \times \cdots \times k$. Now set

$$I(n, d) := \{\mathbf{i} = (i_1, \ldots, i_d) \mid 1 \leq i_l \leq n\}.$$

Then the elements $v_{\mathbf{i}} = v_{i_1} \otimes \cdots \otimes v_{i_d}$ with $\mathbf{i} \in I(n, d)$ form a basis of $V^{\otimes d}$, and we consider the endomorphism $\varepsilon_{\mathbf{i}} \colon V^{\otimes d} \to V^{\otimes d}$ given by $\varepsilon_{\mathbf{i}}(v_{\mathbf{j}}) = \delta_{\mathbf{ij}} v_{\mathbf{j}}$.

Lemma 8.3.16. *We have an inclusion of k-algebras*

$$\Gamma^d E \subseteq \Gamma^d \, \mathrm{End}(V) = \mathrm{End}_{\Gamma^d \mathcal{P}_k}(V) = S_k(n, d).$$

The elements e_λ with $\lambda \in \Lambda(n, d)$ form a basis of $\Gamma^d E$ consisting of pairwise orthogonal idempotents such that $\mathrm{id} = \sum_\lambda e_\lambda$.

Proof It follows from Lemma 8.2.11 that the elements e_λ with $\lambda \in \Lambda(n, d)$ form a basis of $\Gamma^d E$.

For $\mathbf{i} \in I(n, d)$ the isomorphism $\mathrm{End}(V)^{\otimes d} \xrightarrow{\sim} \mathrm{End}(V^{\otimes d})$ identifies the element $e_{\mathbf{i}} = e_{i_1} \otimes \cdots \otimes e_{i_d}$ with $\varepsilon_{\mathbf{i}}$, and the $\varepsilon_{\mathbf{i}}$ are pairwise orthogonal idempotents such that $\mathrm{id}_{V^{\otimes d}} = \sum_{\mathbf{i}} \varepsilon_{\mathbf{i}}$. The proof of Lemma 8.2.11 shows that e_λ identifies with $\sum_{\mathbf{i}^* = \lambda} \varepsilon_{\mathbf{i}}$. Thus

$$\mathrm{id} = \sum_{\mathbf{i}} \varepsilon_{\mathbf{i}} = \sum_\lambda \sum_{\mathbf{i}^* = \lambda} \varepsilon_{\mathbf{i}} = \sum_\lambda e_\lambda$$

and it is clear that the elements e_λ are pairwise orthogonal idempotents. $\square$

Let us consider the decomposition $\mathrm{End}(V) = \bigoplus_{i=1}^n V_i$ with $V_i = \mathrm{End}(V)e_i$. The isomorphism (8.3.7) given by multiplication yields the isomorphism

$$\bigoplus_{\lambda \in \Lambda(n, d)} \Gamma^{\lambda_1} V_1 \otimes \cdots \otimes \Gamma^{\lambda_n} V_n \xrightarrow{\sim} \Gamma^d \, \mathrm{End}(V)$$

which sends $e_1^{\otimes \lambda_1} \otimes \cdots \otimes e_n^{\otimes \lambda_n}$ to e_λ. Now identify $V \xrightarrow{\sim} V_i$ for each i via

$$V \xrightarrow{\sim} V \otimes v_i^\vee \subseteq V \otimes V^\vee \cong \mathrm{End}(V).$$

Then it follows that the canonical isomorphism (8.3.8)

$$\bigoplus_{\lambda \in \Lambda(n, d)} \Gamma^\lambda \xrightarrow{\sim} \Gamma^{d, V}$$

evaluated at V yields an isomorphism

$$\Gamma^\lambda(V) \xrightarrow{\sim} \Gamma^{d, V}(V) e_\lambda \tag{8.3.17}$$

that maps $v_1^{\otimes \lambda_1} \otimes \cdots \otimes v_n^{\otimes \lambda_n}$ to e_λ, since $V \xrightarrow{\sim} V_i$ identifies v_i with e_i.

Now fix $X \in \mathrm{pol}^d \mathcal{P}_k$ and view $X(V)$ as a left $S_k(n,d)$-module. Then Lemma 8.3.16 yields the *weight space decomposition*

$$X(V) = \bigoplus_{\lambda \in \Lambda(n,d)} X(V)_\lambda \qquad \text{with} \qquad X(V)_\lambda := e_\lambda X(V).$$

Lemma 8.3.18. *There is a canonical isomorphism*

$$\mathrm{Hom}(\Gamma^\lambda, X) \xrightarrow{\sim} X(V)_\lambda$$

that sends ϕ to $\phi_V(v_1^{\otimes \lambda_1} \otimes \cdots \otimes v_n^{\otimes \lambda_n})$.

Proof We restrict the Yoneda isomorphism $\mathrm{Hom}(\Gamma^{d,V}, X) \xrightarrow{\sim} X(V)$ by multiplying with e_λ. So the isomorphism can be written as the composite of

$$\mathrm{Hom}(\Gamma^\lambda, X) \xrightarrow{\sim} \mathrm{Hom}_{S_k(n,d)}(\Gamma^\lambda(V), X(V)), \qquad \phi \mapsto \phi_V$$

and

$$\mathrm{Hom}_{S_k(n,d)}(\Gamma^\lambda(V), X(V)) \xrightarrow{\sim} X(V)_\lambda, \qquad \psi \mapsto \psi(v_1^{\otimes \lambda_1} \otimes \cdots \otimes v_n^{\otimes \lambda_n}).$$

Here we use that $v_1^{\otimes \lambda_1} \otimes \cdots \otimes v_n^{\otimes \lambda_n}$ generates $\Gamma^\lambda(V)$ as an $S_k(n,d)$-module, which follows from the isomorphism (8.3.17). $\qquad \square$

Remark 8.3.19. The weight space decomposition depends on a choice, but the weight spaces are unique up to an isomorphism, which reflects the choice. Let V' be a free k-module with ordered basis $\{v'_1, \ldots, v'_n\}$. Then the isomorphism $V \xrightarrow{\sim} V'$ sending each v_i to v'_i induces an isomorphism $e_\lambda X(V) \xrightarrow{\sim} e'_\lambda X(V')$.

The following says that the duality on $\mathrm{pol}^d \mathcal{P}_k$ preserves weight spaces.

Lemma 8.3.20. *Let V be a k-module with ordered basis $\{v_1, \ldots, v_n\}$. For $\lambda \in \Lambda(n,d)$ and $X \in \mathrm{pol}^d \mathcal{P}_k$ there is an isomorphism of weight spaces $X(V)_\lambda^\vee \xrightarrow{\sim} X^\circ(V)_\lambda$.*

Proof Consider the isomorphism $V \xrightarrow{\sim} V^\vee$ that maps each v_i to $v_i^\vee$. This yields an isomorphism

$$X(V)_\lambda = e_\lambda X(V) \xrightarrow{\sim} e_\lambda^\vee X(V^\vee);$$

see Remark 8.3.19. The identity $X(V^\vee)^\vee = X^\circ(V)$ then identifies the direct summand $(e_\lambda^\vee X(V^\vee))^\vee$ with $e_\lambda X^\circ(V)$. $\qquad \square$

Standard Morphisms

We compute the weight spaces for Γ^λ. Let $\lambda = (\lambda_1, \lambda_2, \ldots)$ and $\mu = (\mu_1, \mu_2, \ldots)$ be sequences of non-negative integers satisfying $\sum \lambda_i = d = \sum \mu_j$. Given a matrix $A = (a_{ij})_{i,j \geq 1}$ of non-negative integers with $\lambda_i = \sum_j a_{ij}$ and $\mu_j = \sum_i a_{ij}$ for all i, j, there is a *standard morphism*

$$\gamma_A: \ \Gamma^\mu = \bigotimes_j \Gamma^{\mu_j} \to \bigotimes_j \left(\bigotimes_i \Gamma^{a_{ij}} \right) = \bigotimes_i \left(\bigotimes_j \Gamma^{a_{ij}} \right) \to \bigotimes_i \Gamma^{\lambda_i} = \Gamma^\lambda$$

where the first morphism is the tensor product of the diagonal maps $\Gamma^{\mu_j} \to \bigotimes_i \Gamma^{a_{ij}}$ and the second morphism is the tensor product of the multiplication maps $\bigotimes_j \Gamma^{a_{ij}} \to \Gamma^{\lambda_i}$, as given by (8.2.9).

Lemma 8.3.21. *Let $\lambda = (\lambda_1, \lambda_2, \ldots)$ and $\mu = (\mu_1, \mu_2, \ldots)$ be sequences of non-negative integers with $\sum \lambda_i = d = \sum \mu_j$. Then the standard morphisms γ_A form a k-basis of* $\mathrm{Hom}(\Gamma^\mu, \Gamma^\lambda)$.[2]

Proof We may assume that $\lambda, \mu \in \Lambda(n, d)$ and apply Lemma 8.3.18. Fix a free k-module V with basis $\{v_1, \ldots, v_n\}$. Then we have an isomorphism

$$\mathrm{Hom}(\Gamma^\mu, \Gamma^\lambda) \xrightarrow{\sim} (\Gamma^\lambda V)_\mu.$$

A standard morphism γ_A evaluated at V takes the element $v_1^{\otimes \mu_1} \otimes \cdots \otimes v_n^{\otimes \mu_n}$ to $v_A = v_{\alpha^1} \otimes \cdots \otimes v_{\alpha^n}$ with $\alpha^i \in \Lambda(n, \lambda_i)$ and $\alpha^i_j = a_{ij}$. Now the assertion follows from the fact that the elements v_A form a basis of $\Gamma^\lambda V$ as μ runs through $\Lambda(n, d)$; see Lemma 8.3.5 and cf. Lemma 8.3.22 below. $\square$

For example, let $\lambda = (5, 3, 3, 2)$ and $\mu = (1, 3, 3, 2, 2, 2)$. For

$$A = \begin{bmatrix} 1 & 2 & 2 & 0 & 0 & 0 \\ 0 & 1 & 1 & 0 & 1 & 0 \\ 0 & 0 & 0 & 2 & 0 & 1 \\ 0 & 0 & 0 & 0 & 1 & 1 \end{bmatrix}$$

the morphism γ_A evaluated at $V = k^6$ takes $v_1^{\otimes \mu_1} \otimes \cdots \otimes v_6^{\otimes \mu_6}$ to the element

$$(v_1(v_2 \otimes v_2)(v_3 \otimes v_3)) \otimes (v_2 v_3 v_5) \otimes ((v_4 \otimes v_4)v_6) \otimes (v_5 v_6).$$

Lemma 8.3.22. *Let λ be a partition and set $V = k^n$. For a filling T of the corresponding Young diagram with entries in $\{1, \ldots, n\}$, the element v_T belongs to $(\Gamma^\lambda V)_\mu$ where μ equals the content of T.*

If $\mu = \lambda$ and T is a Young tableau, then it is the unique tableau such that all boxes of the ith row have entry i.

---

[2] This yields a basis of the Schur algebra $S_k(n, d) \cong \bigoplus_{\lambda, \mu \in \Lambda(n,d)} \mathrm{Hom}(\Gamma^\mu, \Gamma^\lambda)$.

Proof The filling T yields integers $a_{ij} = \text{card}\{t \mid T(i,t) = j\}$ for $i, j \geq 1$ and a standard morphism $\gamma_A\colon \Gamma^\mu \to \Gamma^\lambda$ for $A = (a_{ij})$. Evaluated at V this morphism sends $v_1^{\otimes \mu_1} \otimes \cdots \otimes v_n^{\otimes \mu_n}$ to v_T. $\qquad\square$

We have the following analogue for weight spaces of exterior powers.

Lemma 8.3.23. *Let μ be a partition and set $V = k^n$. For a filling T of the corresponding Young diagram with entries in $\{1, \ldots, n\}$, the element v^T belongs to $(\Lambda^\mu V)_\lambda$ where λ equals the content of T. In particular, $(\Lambda^\mu V)_\lambda \neq 0$ for a partition λ implies $\lambda \trianglelefteq \mu'$, where μ' denotes the conjugate partition.*

If $\lambda = \mu'$, then T is the unique filling such that all boxes of the ith column have entry i.

Proof The filling T yields integers $a_{ij} = \text{card}\{t \mid T(i,t) = j\}$ for $i, j \geq 1$ and a canonical morphism

$$\Gamma^\lambda = \bigotimes_j \Gamma^{\lambda_j} \to \bigotimes_j \left(\bigotimes_i \Gamma^{a_{ij}} \right) = \bigotimes_i \left(\bigotimes_j \Gamma^{a_{ij}} \right) \to \bigotimes_i \Lambda^{\mu_i} = \Lambda^\mu$$

where the first morphism is the tensor product of the diagonal maps

$$\Gamma^{\lambda_j} \to \bigotimes_i \Gamma^{a_{ij}} \rightarrowtail \bigotimes_i \Gamma^{a_{ij}}$$

and the second morphism is the tensor product of the multiplication maps

$$\bigotimes_j \Gamma^{a_{ij}} \twoheadrightarrow \bigotimes_j \Lambda^{a_{ij}} \to \Lambda^{\mu_i}.$$

Evaluated at V this morphism sends $v_1^{\otimes \lambda_1} \otimes \cdots \otimes v_n^{\otimes \lambda_n}$ to v^T. From the description of the basis of $\Lambda^\mu V$ given in Lemma 8.3.6 it follows that the content of each filling is bounded (with respect to the dominance order) by the content of the unique filling such that all boxes of the ith column have entry i. $\qquad\square$

The above lemma amounts to a description of a k-basis of $\text{Hom}(\Gamma^\lambda, \Lambda^\mu)$ for any pair of partitions λ, μ, and this yields an immediate consequence.

Lemma 8.3.24. *Let λ and μ be partitions of an integer d. Then*

$$\text{Hom}(\Gamma^\lambda, \Lambda^\mu) \neq 0 \quad \Longrightarrow \quad \lambda \trianglelefteq \mu'.$$

Proof This follows from Lemma 8.3.18 and the above lemma, since the elements v^T form a basis of $\Lambda^\mu V$ by Lemma 8.3.6. $\qquad\square$

The following example provides another description of the isomorphism $(k\mathfrak{S}_d)^{\text{op}} \xrightarrow{\sim} \text{End}(\Gamma^\omega)$ from Lemma 8.3.12.

Example 8.3.25. For $\sigma \in \mathfrak{S}_d$ we consider the $d \times d$ matrix $A_\sigma = (a_{ij})$ given by $a_{ij} = \delta_{\sigma(i)j}$ (so $a_{ij} = 1$ if $\sigma(i) = j$, and $a_{ij} = 0$ otherwise). Then for $\omega = (1, \ldots, 1)$ the set $\{\gamma_{A_\sigma} \mid \sigma \in \mathfrak{S}_d\}$ is a k-basis of $\mathrm{Hom}(\Gamma^\omega, \Gamma^\omega)$. Note that $\gamma_{A_\tau}\gamma_{A_\sigma} = \gamma_{A_{\sigma\tau}}$. Thus $(k\mathfrak{S}_d)^{\mathrm{op}} \xrightarrow{\sim} \mathrm{End}(\Gamma^\omega)$ as k-algebras.

Base Change

Let $k \to \ell$ be a homomorphism of commutative rings. Then we have for each integer $d \geq 0$ and $V, W \in \mathcal{P}_k$ a natural isomorphism

$$\Gamma^d \mathrm{Hom}(V, W) \otimes_k \ell \xrightarrow{\sim} \Gamma^d \mathrm{Hom}(V \otimes_k \ell, W \otimes_k \ell).$$

Thus the functor $- \otimes_k \ell$ induces a functor $\Gamma^d \mathcal{P}_k \to \Gamma^d \mathcal{P}_\ell$ which we denote again by $- \otimes_k \ell$. The functor $\Gamma^d \mathcal{P}_k \to \Gamma^d \mathcal{P}_\ell$ extends to a right exact functor $\mathrm{pol}^d \mathcal{P}_k \to \mathrm{pol}^d \mathcal{P}_\ell$ by sending $\Gamma^{d,V}$ to $\Gamma^{d,V\otimes_k\ell}$, keeping in mind that each object in $\mathrm{pol}^d \mathcal{P}_k$ admits a projective presentation via the representable functors $\Gamma^{d,V}$. Again, we denote this functor by $- \otimes_k \ell$.

Let $X_k \in \mathrm{pol}^d \mathcal{P}_k$ be an object or a morphism that is defined for every commutative ring k. We say that X_k is *stable under base change* when the following equivalent conditions are satisfied:

(1) $X_k(V) \otimes_k \ell \cong X_\ell(V \otimes_k \ell)$ for all $V \in \mathcal{P}_k$ and $k \to \ell$,
(2) $X_k \otimes_k \ell \cong X_\ell$ for all $k \to \ell$,
(3) $X_\mathbb{Z} \otimes_\mathbb{Z} k \cong X_k$ for all k.

For example, the symmetric tensors Γ^λ are stable under base change for all $\lambda \in \Lambda(n, d)$.

Base change allows the reduction of proofs to the case that k is a field of characteristic zero. The following lemma gives a useful argument.

Lemma 8.3.26. *Let $\phi_k \colon X_k \to Y_k$ be an epimorphism in $\mathrm{Pol}^d \mathcal{P}_k$ that is stable under base change. Suppose that $X_\mathbb{Z}(V)$ is torsion free for all $V \in \mathcal{P}_\mathbb{Z}$ and that $\phi_\mathbb{Q}$ is an isomorphism. Then ϕ_k is an isomorphism for all k.*

Proof We have that $\mathrm{Ker}\,\phi_\mathbb{Z}$ evaluated at $V \in \mathcal{P}_\mathbb{Z}$ is torsion since

$$(\mathrm{Ker}\,\phi_\mathbb{Z}) \otimes_\mathbb{Z} \mathbb{Q} \cong \mathrm{Ker}(\phi_\mathbb{Z} \otimes_\mathbb{Z} \mathbb{Q}) \cong \mathrm{Ker}\,\phi_\mathbb{Q} = 0.$$

On the other hand, $X_\mathbb{Z}(V)$ is torsion free. Thus $\mathrm{Ker}\,\phi_\mathbb{Z} = 0$. It remains to observe that $\phi_k \cong \phi_\mathbb{Z} \otimes_\mathbb{Z} k$ for all k. $\qquad\square$

8.4 Cauchy Decompositions

In this section we introduce the standard objects in the category of strict polynomial functors and describe a standard basis in terms of Young tableaux.
Standard objects are indexed by partitions, and we consider on the set of partitions the lexicographic order. Closely related is the Cauchy decomposition
of symmetric tensors, because the factors of this decomposition are given by
standard objects. This decomposition is an analogue of the Cauchy identity for
symmetric functions.

Throughout we keep fixed a commutative ring k.

Standard Objects

Let λ be a partition of an integer d. The *standard object* corresponding to λ is
defined via the following presentation

$$\bigoplus_{\substack{\gamma_A : \Gamma^\mu \to \Gamma^\lambda \\ \mu > \lambda}} \Gamma^\mu \xrightarrow{\ \alpha\ } \Gamma^\lambda \longrightarrow \Delta^\lambda \longrightarrow 0 \tag{8.4.1}$$

which is given by all standard morphisms $\Gamma^\mu \to \Gamma^\lambda$ with μ a partition satisfying
$\mu > \lambda$ (lexicographic order). Note that Δ^λ is stable under base change, since
the standard morphisms $\Gamma^\mu \to \Gamma^\lambda$ are stable under base change.

Straightening

We need a more explicit description of the standard objects. This is based on a
technique which is known as *straightening*.

Fix a k-module V with basis $\{v_1, \dots, v_n\}$ and a partition $\lambda = (\lambda_1, \dots, \lambda_p)$.

Lemma 8.4.2. *Let T be a filling of shape λ that is weakly increasing along
each row but not a Young tableau. Then there exist fillings $T_l < T$ of shape λ
that are weakly increasing along each row and integers c_l such that*

$$v_T + \sum_l c_l v_{T_l} \in \mathrm{Im}(\alpha_V).$$

Proof   We consider two consecutive rows

$$T(i) = (x_1, \dots, x_{\lambda_i})$$
$$T(i+1) = (y_1, \dots, y_{\lambda_{i+1}})$$

of T such that the entries along columns do not strictly increase. Suppose
that $y_j > x_j$ for $1 \le j \le r$ but $y_{r+1} \le x_{r+1}$. Let s be maximal such that

$y_{r+1} = y_{r+2} = \cdots = y_s$. Now consider the composition $\mu = (\mu_1, \ldots, \mu_{p+1})$ which is obtained from λ by replacing λ_i and λ_{i+1} by the three entries $\mu_i = r$, $\mu_{i+1} = \lambda_i - r + s$, and $\mu_{i+2} = \lambda_{i+1} - s$. The filling T is replaced by the filling $\bar{T}$ of shape μ with the following rows:

$$\bar{T}(i) = (x_1, \ldots, x_r)$$
$$\bar{T}(i+1) = (y_1, \ldots, y_s, x_{r+1}, \ldots, x_{\lambda_i})$$
$$\bar{T}(i+2) = (y_{s+1}, \ldots, y_{\lambda_{i+1}}).$$

Consider the standard morphism $\phi \colon \Gamma^\mu \to \Gamma^\lambda$ of the form

$$(\Gamma^{\mu_1} \otimes \cdots \otimes \Gamma^{\mu_{i-1}}) \otimes (\Gamma^{\mu_i} \otimes \Gamma^{\mu_{i+1}} \otimes \Gamma^{\mu_{i+2}}) \otimes (\Gamma^{\mu_{i+3}} \otimes \cdots \otimes \Gamma^{\mu_{p+1}})$$

$$\xrightarrow{1 \otimes \phi' \otimes 1} (\Gamma^{\lambda_1} \otimes \cdots \otimes \Gamma^{\lambda_{i-1}}) \otimes (\Gamma^{\lambda_i} \otimes \Gamma^{\lambda_{i+1}}) \otimes (\Gamma^{\lambda_{i+2}} \otimes \cdots \otimes \Gamma^{\lambda_p})$$

with ϕ' given by $\begin{bmatrix} r & \lambda_i - r & 0 \\ 0 & s & \lambda_{i+1} - s \end{bmatrix}$.

Let $v_{\bar{T}} = v_{\alpha^1} \otimes \cdots \otimes v_{\alpha^{p+1}}$ and write $\sum_l v_{\beta'_l} \otimes v_{\beta''_l}$ for the image of $v_{\alpha^{i+1}}$ under the diagonal map

$$\Gamma^{\mu_{i+1}} = \Gamma^{\lambda_i - r + s} V \longrightarrow \Gamma^{\lambda_i - r} V \otimes \Gamma^s V;$$

see (8.2.14). Then

$$\phi_V(v_{\bar{T}}) = \sum_l v_{\alpha^1} \otimes \cdots \otimes v_{\alpha^{i-1}} \otimes v_{\alpha^i} v_{\beta'_l} \otimes v_{\beta''_l} v_{\alpha^{i+2}} \otimes v_{\alpha^{i+3}} \otimes \cdots \otimes v_{\alpha^{p+1}}.$$

Each summand corresponds to a basis element $v_{T_l} \in \Gamma^\lambda V$ that is given by a filling T_l of shape λ, and multiplicities arise when v_{α^i} and $v_{\beta'_l}$ or $v_{\beta''_l}$ and $v_{\alpha^{i+2}}$ have common factors; see (8.2.15). There is precisely one choice of l such that $v_{\beta'_l}$ corresponds to $(x_{r+1}, \ldots, x_{\lambda_i})$. In that case $T_l = T$ and there are no repetitions, because v_{α^i} and $v_{\beta'_l}$ have no common factors (since $x_r < x_{r+1}$), and $v_{\beta''_l}$ and $v_{\alpha^{i+2}}$ have no common factors (since $y_s < y_{s+1}$). For any other choice of l, the filling T_l is obtained from T by replacing entries from $x_{r+1}, \ldots, x_{\lambda_i}$ in the ith row by entries from $y_1, \ldots, y_s$ in the $i+1$st row, and the latter ones are smaller or equal by construction. Thus $T_l < T$.

It remains to observe that $\phi(v_{\bar{T}}) \in \mathrm{Im}(\alpha_V)$ since $\tilde{\mu} > \lambda$ for the partition $\tilde{\mu}$ which is equivalent to μ up to permutation. $\qquad\square$

The lemma has the following immediate consequence, since the elements of the form v_T form a basis of $\Gamma^\lambda V$ by Lemma 8.3.5. For an improvement, see Corollary 8.4.14.

For any set X we write $\mathrm{span}_k X$ for the set of k-linear combinations of elements from X.

Proposition 8.4.3. *We have*

$$\Gamma^\lambda V = \operatorname{Im}(\alpha_V) + \operatorname{span}_k \{v_T \mid T \text{ is a Young tableau}\}. \qquad \square$$

The description of the standard objects via Young tableaux yields a criterion for Hom vanishing.

Proposition 8.4.4. *Let λ and μ be partitions of an integer d. Then*

$$\mu > \lambda \implies \mu \ntrianglelefteq \lambda \implies \operatorname{Hom}(\Delta^\mu, \Delta^\lambda) = \operatorname{Hom}(\Gamma^\mu, \Delta^\lambda) = 0.$$

Proof Consider the weight space decomposition

$$\Gamma^\lambda V = \bigoplus_{\mu \in \Lambda(n,d)} (\Gamma^\lambda V)_\mu.$$

An element $v_T \in \Gamma^\lambda V$ given by a filling T belongs to $(\Gamma^\lambda V)_\mu$, where μ denotes the content of T; see Lemma 8.3.22. If $\mu \ntrianglelefteq \lambda$, then Lemma 8.2.1 and Proposition 8.4.3 imply $(\Delta^\lambda V)_\mu = 0$. Thus $\operatorname{Hom}(\Gamma^\mu, \Delta^\lambda) = 0$ by Lemma 8.3.18, and clearly then $\operatorname{Hom}(\Delta^\mu, \Delta^\lambda) = 0$.

The other implication follows from Lemma 8.2.2. $\qquad \square$

The Cauchy Decomposition of Symmetric Tensors

We establish a filtration of the symmetric tensors and identify the factors of this filtration in terms of the standard objects. There is an analogue for symmetric powers which amounts to a filtration involving costandard objects.

Fix $V, W \in \mathcal{P}_k$. For every integer $r \geq 0$ there is a unique map

$$\psi^r : \Gamma^r V \otimes \Gamma^r W \longrightarrow \Gamma^r(V \otimes W)$$

making the following square commutative.

$$
\begin{array}{ccc}
\Gamma^r V \otimes \Gamma^r W & \xrightarrow{\;\psi^r\;} & \Gamma^r(V \otimes W) \\
\Big\downarrow & & \Big\downarrow \\
V^{\otimes r} \otimes W^{\otimes r} & \xrightarrow{\;\sim\;} & (V \otimes W)^{\otimes r}
\end{array}
$$

Extend this map for a partition $\lambda = (\lambda_1, \ldots, \lambda_n)$ of an integer d to a map

$$\psi^\lambda : \Gamma^\lambda V \otimes \Gamma^\lambda W \longrightarrow \Gamma^d(V \otimes W)$$

which is given as a composite

$$\Gamma^\lambda V \otimes \Gamma^\lambda W \xrightarrow{\;\sim\;} (\Gamma^{\lambda_1} V \otimes \Gamma^{\lambda_1} W) \otimes \cdots \otimes (\Gamma^{\lambda_n} V \otimes \Gamma^{\lambda_n} W)$$

$$\xrightarrow{\;\psi^{\lambda_1} \otimes \cdots \otimes \psi^{\lambda_n}\;} \Gamma^{\lambda_1}(V \otimes W) \otimes \cdots \otimes \Gamma^{\lambda_n}(V \otimes W) \longrightarrow \Gamma^d(V \otimes W)$$

with the last map given by multiplication. We call ψ^λ the *comparison morphism*. The *Cauchy filtration* for symmetric tensors is by definition the chain

$$0 = F_{+\infty} \subseteq F_{(d)} \subseteq F_{(d-1,1)} \subseteq \cdots \subseteq F_{(2,1,\ldots,1)} \subseteq F_{(1,\ldots,1)} \subseteq \Gamma^d(V \otimes W)$$
$$(8.4.5)$$

where $F_\lambda = \sum_{\mu \geq \lambda} \operatorname{Im} \psi^\mu$.

The following result describes the factors of the Cauchy filtration.

Theorem 8.4.6. *Let* $V, W \in \mathcal{P}_k$. *Then* $F_{(1,\ldots,1)} = \Gamma^d(V \otimes W)$ *and for every partition* λ *of an integer* d *the morphism* $\psi^\lambda \colon \Gamma^\lambda V \otimes \Gamma^\lambda W \to F_\lambda$ *induces an isomorphism*

$$\Delta^\lambda V \otimes \Delta^\lambda W \xrightarrow{\sim} F_\lambda / F_{\lambda^+}$$

which is functorial in V *and* W. *Therefore the associated graded object of the filtration* (F_λ) *is*

$$\bigoplus_\lambda (\Delta^\lambda V \otimes \Delta^\lambda W) = \Gamma^d(V \otimes W).$$

The proof will be postponed until the proof of Theorem 8.4.11. The first step is to show that the comparison morphism ψ^λ induces a morphism $\Delta^\lambda V \otimes \Delta^\lambda W \to F_\lambda / F_{\lambda^+}$.

Lemma 8.4.7. *For a standard morphism* $\gamma_A \colon \Gamma^\mu \to \Gamma^\lambda$ *the following square commutes.*

$$
\begin{array}{ccc}
\Gamma^\mu V \otimes \Gamma^\lambda W & \xrightarrow{\gamma_A V \otimes \mathrm{id}} & \Gamma^\lambda V \otimes \Gamma^\lambda W \\
\downarrow{\scriptstyle \mathrm{id} \otimes \gamma_{A^{\mathrm{tr}}} W} & & \downarrow{\scriptstyle \psi^\lambda} \\
\Gamma^\mu V \otimes \Gamma^\mu W & \xrightarrow{\psi^\mu} & \Gamma^d(V \otimes W)
\end{array}
$$

Proof Straightforward calculation. $\qquad\qquad\qquad\qquad\qquad\qquad\qquad\square$

We use the presentation (8.4.1) of Δ^λ and write $p_V \colon \Gamma^\lambda V \to \Delta^\lambda V$ for the canonical morphism. Also, we use the following fact. For any pair of exact sequences of k-modules $X_i \xrightarrow{\alpha_i} Y_i \xrightarrow{\beta_i} Z_i \to 0$ with $i = 1, 2$, we have

$$\operatorname{Im}(\alpha_1 \otimes \mathrm{id}_{Y_2} + \mathrm{id}_{Y_1} \otimes \alpha_2) = \operatorname{Ker}(\beta_1 \otimes \beta_2).$$

Lemma 8.4.8. *There is a morphism* $\bar{\psi}^\lambda$ *making the following square commutative.*

$$
\begin{array}{ccc}
\Gamma^\lambda V \otimes \Gamma^\lambda W & \xrightarrow{p_V \otimes p_W} & \Delta^\lambda V \otimes \Delta^\lambda W \\
\downarrow{\scriptstyle \psi^\lambda} & & \downarrow{\scriptstyle \bar{\psi}^\lambda} \\
F_\lambda & \xrightarrow{q} & F_\lambda / F_{\lambda^+}
\end{array}
$$

Moreover, $\bar{\psi}^\lambda$ *is an epimorphism.*

Proof We have $\text{Ker}(p_V \otimes p_W) = \text{Im}(f + g)$ for

$$f: \bigoplus_{\substack{\gamma_A : \Gamma^\mu \to \Gamma^\lambda \\ \mu > \lambda}} \Gamma^\mu V \otimes \Gamma^\lambda W \xrightarrow{\gamma_A V \otimes \text{id}} \Gamma^\lambda V \otimes \Gamma^\lambda W$$

and

$$g: \bigoplus_{\substack{\gamma_A : \Gamma^\mu \to \Gamma^\lambda \\ \mu > \lambda}} \Gamma^\lambda V \otimes \Gamma^\mu W \xrightarrow{\text{id} \otimes \gamma_A W} \Gamma^\lambda V \otimes \Gamma^\lambda W.$$

Then it follows from Lemma 8.4.7 that ψ^λ maps the kernel of $p_V \otimes p_W$ into F_{λ^+}. This yields $\bar{\psi}^\lambda$.

It is immediate from the definitions of F_λ and F_{λ^+} that the composite $q \circ \psi^\lambda$ is surjective and, by the commutativity of the diagram, so is $\bar{\psi}^\lambda$. $\qquad\square$

Next we show that the Cauchy filtration exhausts all of $\Gamma^d(V \otimes W)$.

Lemma 8.4.9. *We have* $F_{(1,\dots,1)} = \Gamma^d(V \otimes W)$.

Proof We may assume that V and W are free. Let $\{v_1, \dots, v_s\}$ be a basis of V and $\{w_1, \dots, v_t\}$ be a basis of W. Then the elements $x_{ij} = v_i \otimes w_j$ yield a basis of $V \otimes W$, and the elements $x_\lambda = \prod_{i,j} x_{ij}^{\otimes \lambda_{ij}}$ with $\sum \lambda_{ij} = d$ form a basis of $\Gamma^d(V \otimes W)$. We fix λ and by reordering the basis of $V \otimes W$ we may assume that λ is a partition of d. Then it is easily checked that x_λ is the image of $\left(\bigotimes_{i,j} v_i^{\otimes \lambda_{ij}} \right) \otimes \left(\bigotimes_{i,j} w_j^{\otimes \lambda_{ij}} \right)$ under the map ψ^λ. $\qquad\square$

The Cauchy filtration (8.4.5) induces filtrations for finitely generated projective objects in $\text{pol}^d \mathcal{P}_k$. More precisely, replacing in the filtration (8.4.5) the object V by $V^\vee$ and using its functoriality in W gives the filtration

$$0 = X_{+\infty} \subseteq X_{(d)} \subseteq X_{(d-1,1)} \subseteq \cdots \subseteq X_{(2,1,\dots,1)} \subseteq X_{(1,\dots,1)} \subseteq \Gamma^{d,V}.$$

Note that the comparison morphism $\bar{\psi}^\lambda$ induces an epimorphism

$$\Delta^\lambda(V^\vee) \otimes \Delta^\lambda \longrightarrow X_\lambda / X_{\lambda^+}. \tag{8.4.10}$$

The filtration of $\Gamma^{d,V}$ induces a filtration for each direct summand of $\Gamma^{d,V}$. This follows from the functoriality of the filtration (8.4.5) in V via the canonical isomorphism

$$\text{End}_{\Gamma^d \mathcal{P}_k}(V)^{\text{op}} \xrightarrow{\sim} \text{End}(\Gamma^{d,V}).$$

Theorem 8.4.11. *Let μ be a partition of an integer d. There is a filtration*

$$0 = Y_{+\infty} \subseteq Y_{(d)} \subseteq Y_{(d-1,1)} \subseteq \cdots \subseteq Y_{\mu^+} \subseteq Y_\mu = \Gamma^\mu$$

such that for each partition $\lambda \geq \mu$

$$Y_\lambda / Y_{\lambda^+} \cong (\Delta^\lambda)^{K_{\lambda\mu}}.$$

Proof Let $\mu \in \Lambda(n, d)$. The functor Γ^μ is a direct summand of $\Gamma^{d,k''}$ thanks to the decomposition (8.3.8). Then the functoriality of the filtration (8.4.5) in V yields the filtration of Γ^μ by passing for each partition λ from $X_\lambda \subseteq \Gamma^{d,k''}$ to the direct summand $Y_\lambda \subseteq \Gamma^\mu$ corresponding to μ. The epimorphism (8.4.10) restricts for each partition λ to an epimorphism

$$\Delta^\lambda(k'')_\mu \otimes \Delta^\lambda \longrightarrow Y_\lambda / Y_{\lambda^+}$$

where $\Delta^\lambda(k'')_\mu$ is the weight space corresponding to μ. This induces an exact sequence

$$\Delta^\lambda(k'')_\mu \otimes \Delta^\lambda \longrightarrow \Gamma^\mu / Y_{\lambda^+} \longrightarrow \Gamma^\mu / Y_\lambda \longrightarrow 0 \qquad (8.4.12)$$

which is stable under base change by construction. Note that $Y_\lambda / Y_{\lambda^+}$ is stable under flat base change, since it is the kernel of a morphism that is stable under base change.

For $\lambda < \mu$ we have $\Delta^\lambda(k'')_\mu = 0$ by Proposition 8.4.4, since $\Delta^\lambda(k'')_\mu \cong \mathrm{Hom}(\Gamma^\mu, \Delta^\lambda)$ by Lemma 8.3.18. Thus $Y_\mu = \Gamma^\mu$ by Lemma 8.4.9.

For $V = k''$ we set

$$D^\lambda V = \mathrm{span}_k \{v_T \mid T \text{ Young tableau}\} \subseteq \Gamma^\lambda V$$

and recall from Proposition 8.4.3 that the canonical map $D^\lambda V \to \Delta^\lambda V$ is an epimorphism. This gives a pair of epimorphisms

$$D^\lambda(k'')_\mu \otimes D^\lambda(k'') \longrightarrow \Delta^\lambda(k'')_\mu \otimes \Delta^\lambda(k'') \longrightarrow Y_\lambda / Y_{\lambda^+}(k'') \qquad (8.4.13)$$

with

$$\mathrm{rank}_k \, D^\lambda(k'') = s_\lambda(1^n) \qquad \text{and} \qquad \mathrm{rank}_k \, D^\lambda(k'')_\mu = K_{\lambda\mu}.$$

When k is a field, we have

$$h_\mu(1^n) = \mathrm{rank}_k \, \Gamma^\mu(k'') = \sum_\lambda \mathrm{rank}_k \, Y_\lambda / Y_{\lambda^+}(k'') \leq \sum_\lambda K_{\lambda\mu} s_\lambda(1^n).$$

The first equality is clear, since $h_\lambda = \prod_i h_{\lambda_i}$ and $h_{\lambda_i}(1^n)$ equals the number of monomials of degree λ_i in n variables. The second equality is obtained by taking the sum of all the factors in the filtration of $\Gamma^\mu(k'')$, and the inequality follows from the epimorphism (8.4.13). The identity for symmetric functions in Proposition 8.2.3 then implies equality. Thus

$$\mathrm{rank}_k \, Y_\lambda / Y_{\lambda^+}(k'') = \mathrm{rank}_k \, D^\lambda(k'') \cdot \mathrm{rank}_k \, D^\lambda(k'')_\mu$$

for all λ. Using flat base change via $\mathbb{Z} \to \mathbb{Q}$, it follows that the epimorphism

(8.4.13) is an isomorphism for $k = \mathbb{Z}$, since its kernel is torsion free. Thus we obtain isomorphisms

$$k^{s_\lambda(1^n)} \cong D^\lambda(k^n) \xrightarrow{\sim} \Delta^\lambda(k^n) \qquad \text{and} \qquad (\Delta^\lambda)^{K_{\lambda\mu}} \xrightarrow{\sim} Y_\lambda/Y_{\lambda^+}$$

for $k = \mathbb{Z}$, because we can specialise $\mu = \lambda$ and have $D^\lambda(k^n)_\lambda \cong k$. Also, we may assume that $n \geq d$ so that evaluation at k^n yields the equivalence (8.3.11). It follows that the sequence

$$0 \longrightarrow (\Delta^\lambda)^{K_{\lambda\mu}} \longrightarrow \Gamma^\mu/Y_{\lambda^+} \longrightarrow \Gamma^\mu/Y_\lambda \longrightarrow 0$$

is exact. Evaluating at k^n and starting with $\lambda = \mu$, an induction on λ shows that the sequence splits. Thus Y_λ/Y_{λ^+} is stable under arbitrary base change, and we obtain the isomorphism $(\Delta^\lambda)^{K_{\lambda\mu}} \cong Y_\lambda/Y_{\lambda^+}$ for all k.   □

A consequence of Theorem 8.4.11 is Theorem 8.4.6.

Proof of Theorem 8.4.6   First observe that the decomposition (8.3.8) yields for $V = k^n$ a decomposition

$$\bigoplus_{\mu \in \Lambda(n,d)} \Gamma^\mu(W) \cong \Gamma^{d,k^n}(W) \cong \Gamma^d(V \otimes W).$$

The identity $F_{(1,\dots,1)} = \Gamma^d(V \otimes W)$ has already been shown in Lemma 8.4.9. The comparison morphism (8.4.10) is an isomorphism by Theorem 8.4.11, and therefore the comparison morphism $\bar{\psi}^\lambda$ is an isomorphism.   □

Another immediate consequence of the proof of Theorem 8.4.11 is the following *standard basis theorem* that improves Proposition 8.4.3.

Corollary 8.4.14. *Let λ be a partition of d and let V be a free k-module of rank n. Then the canonical map $\Gamma^\lambda V \to \Delta^\lambda V$ sends the elements v_T with T a Young tableau on λ with entries in $\{1, \dots, n\}$ to a k-basis of $\Delta^\lambda V$. In particular, $(\Delta^\lambda V)_\mu$ has rank $K_{\lambda\mu}$ for a partition $\mu \in \Lambda(n,d)$.*   □

Corollary 8.4.15. *We have $\mathrm{End}(\Delta^\lambda) \cong k$.*

Proof   For $\lambda \in \Lambda(n,d)$ and $V = k^n$ we have an embedding

$$\mathrm{End}(\Delta^\lambda) \subseteq \mathrm{Hom}(\Gamma^\lambda, \Delta^\lambda) \cong (\Delta^\lambda V)_\lambda \cong k$$

by Lemma 8.3.18, and using that $K_{\lambda\lambda} = 1$.   □

The Cauchy Decomposition of Symmetric Powers

For $V \in \mathcal{P}_k$ and $d \geq 0$, we consider the functor

$$S^{d,V} : \mathcal{P}_k \longrightarrow \mathcal{P}_k, \quad W \mapsto S^d(V \otimes W)$$

and have a natural isomorphism $(\Gamma^{d,V})^\circ \cong S^{d,V}$ since

$$(\Gamma^d \operatorname{Hom}(V, W^\vee))^\vee \cong (\Gamma^d (V \otimes W)^\vee)^\vee \cong S^d(V \otimes W).$$

If follows for $\lambda \in \Lambda(n,d)$ and $S^\lambda = S^{\lambda_1} \otimes \cdots \otimes S^{\lambda_n}$ that $(\Gamma^\lambda)^\circ \cong S^\lambda$, using the isomorphism (8.3.4). Thus

$$\operatorname{Hom}(\Gamma^\lambda, \Gamma^\mu) \cong \operatorname{Hom}(S^\mu, S^\lambda),$$

and we denote by $\gamma'_A : S^\mu \to S^\lambda$ the morphism corresponding to the standard morphism $\gamma_A : \Gamma^\lambda \to \Gamma^\mu$.

For a partition λ of an integer d, the *costandard object* is defined via the following copresentation

$$0 \longrightarrow \nabla^\lambda \longrightarrow S^\lambda \longrightarrow \bigoplus_{\substack{\gamma'_A : S^\lambda \to S^\mu \\ \mu > \lambda}} S^\mu$$

which is given by all standard morphisms $S^\lambda \to S^\mu$ with μ a partition satisfying $\mu > \lambda$. Applying the duality to the presentation (8.4.1) of Δ^λ yields a canonical isomorphism

$$(\Delta^\lambda)^\circ \cong \nabla^\lambda.$$

Now fix $V, W \in \mathcal{P}_k$. We have the Cauchy filtration (8.4.5) of $\Gamma^d(V \otimes W)$ and apply the duality. This yields the *Cauchy filtration* of the symmetric powers

$$S^d(V \otimes W) = G_{(1,\dots,1)} \twoheadrightarrow G_{(2,1,\dots,1)} \twoheadrightarrow \cdots \twoheadrightarrow G_{(d-1,1)} \twoheadrightarrow G_{(d)} \twoheadrightarrow G_{+\infty} = 0$$

with a canonical exact sequence

$$0 \longrightarrow \nabla^\lambda V \otimes \nabla^\lambda W \longrightarrow G_\lambda \longrightarrow G_{\lambda^+} \longrightarrow 0$$

for each partition λ. Therefore the associated graded object of the filtration (G_λ) is

$$\bigoplus_\lambda (\nabla^\lambda V \otimes \nabla^\lambda W) = S^d(V \otimes W).$$

Analogously, we obtain from the filtration of Γ^μ the filtration

$$S^\mu = Z_\mu \twoheadrightarrow Z_{\mu^+} \twoheadrightarrow \cdots \twoheadrightarrow Z_{(d-1,1)} \twoheadrightarrow Z_{(d)} \twoheadrightarrow Z_{+\infty} = 0$$

with a canonical exact sequence

$$0 \longrightarrow (\nabla^\lambda)^{K_{\lambda\mu}} \longrightarrow Z_\lambda \longrightarrow Z_{\lambda^+} \longrightarrow 0$$

for each partition $\lambda \geq \mu$.

8.5 Schur and Weyl Modules and Functors

We introduce Schur and Weyl modules because they provide a useful description of the standard and costandard objects. We proceed in several steps and begin with some preparations. The characteristic zero case is important because we reduce to this via base change. Also, we need to connect symmetric tensors and exterior powers via a canonical equivalence.

Throughout we keep fixed a commutative ring k.

Characteristic Zero

Let k be a field of characteristic zero and $d \geq 0$ an integer. Then the group algebra $k\mathfrak{S}_d$ is semisimple by Maschke's theorem. This has the following consequence.

Proposition 8.5.1. *The functor* $\mathrm{Hom}(\Gamma^\omega, -)\colon \mathrm{Pol}^d \mathcal{P}_k \to \mathrm{Mod}(k\mathfrak{S}_d)^{\mathrm{op}}$ *is an equivalence. Therefore the category* $\mathrm{Pol}^d \mathcal{P}_k$ *is semisimple, and the standard objects* Δ^λ *(λ a partition of d) form a complete set of simple objects.*

Proof The algebra $k\mathfrak{S}_d$ is semisimple. Thus each $k\mathfrak{S}_d$-module of the form $V^{\otimes d}$, given by $V \in \mathcal{P}_k$, is in $\mathrm{add}(k\mathfrak{S}_d)$. Then the functor $\mathrm{Hom}(\Gamma^\omega, -)$ induces the following commutative square; see Proposition 8.3.14.

$$
\begin{array}{ccc}
\mathrm{add}(\Gamma^\omega) & \xrightarrow{\quad\sim\quad} & \mathrm{add}(k\mathfrak{S}_d) \\
\uparrow & & \uparrow \\
\mathrm{add}\{\Gamma^{d,V} \mid V \in \mathcal{P}_k\} & \xrightarrow[\mathrm{Hom}(\Gamma^\omega,-)]{\quad\sim\quad} & \mathrm{add}\{V^{\otimes d} \mid V \in \mathcal{P}_k\}
\end{array}
$$

It follows that all functors in this diagram are equivalences, and therefore $\mathrm{Hom}(\Gamma^\omega, -)$ induces an equivalence

$$\mathrm{add}\{\Gamma^{d,V} \mid V \in \mathcal{P}_k\} \xrightarrow{\sim} \mathrm{add}(k\mathfrak{S}_d)$$

between the categories of finitely generated projective objects. This yields an equivalence $\mathrm{Pol}^d \mathcal{P}_k \xrightarrow{\sim} \mathrm{Mod}(k\mathfrak{S}_d)^{\mathrm{op}}$.

Each object Δ^λ has a local endomorphism ring by Corollary 8.4.15 and is therefore simple. Also, the objects Δ^λ are pairwise non-isomorphic for different partitions, by Proposition 8.4.4. We have $\Gamma^\mu \in \mathrm{Filt}\{\Delta^\lambda \mid \lambda$ partition of $d\}$ for each partition μ by Theorem 8.4.11, and therefore each simple object is of the form Δ^λ for some partition λ. $\qquad\square$

Remark 8.5.2. The above proposition reflects the well-known fact that the irreducible representations of the symmetric group correspond via their characters to conjugacy classes which are parametrised by partitions.

Symmetric Tensors versus Exterior Powers

Fix an integer $d \geq 0$. For $V \in \mathcal{P}_k$ we consider the following canonical maps:

$$\nabla : \quad V^{\otimes d} \longrightarrow \Gamma^d V, \qquad x_1 \otimes \cdots \otimes x_d \mapsto \sum_\sigma x_{\sigma(1)} \otimes \cdots \otimes x_{\sigma(d)}$$

$$\hat{\nabla} : \quad V^{\otimes d} \longrightarrow \Lambda^d V, \qquad x_1 \otimes \cdots \otimes x_d \mapsto x_1 \wedge \cdots \wedge x_d$$

$$\Delta : \quad \Gamma^d V \longrightarrow V^{\otimes d}, \qquad\qquad \text{canonical inclusion}$$

$$\hat{\Delta} : \quad \Lambda^d V \longrightarrow V^{\otimes d}, \qquad x_1 \wedge \cdots \wedge x_d \mapsto \sum_\sigma \operatorname{sgn}(\sigma) x_{\sigma(1)} \otimes \cdots \otimes x_{\sigma(d)}.$$

These maps induce morphisms between the corresponding strict polynomial functors. For $n \geq 1$ and $\lambda \in \Lambda(n, d)$ we write

$$\nabla : T^d \xrightarrow{\;\nabla \otimes \cdots \otimes \nabla\;} \Gamma^{\lambda_1} \otimes \cdots \otimes \Gamma^{\lambda_n} = \Gamma^\lambda$$

and

$$\Delta : \Gamma^\lambda = \Gamma^{\lambda_1} \otimes \cdots \otimes \Gamma^{\lambda_n} \xrightarrow{\;\Delta \otimes \cdots \otimes \Delta\;} T^d$$

for the n-fold tensor product of the above morphisms. Analogously, $\hat{\nabla}$ and $\hat{\Delta}$ are defined for Λ^λ.

Permuting the tensors of T^d induces an isomorphism $(k\mathfrak{S}_d)^{\mathrm{op}} \xrightarrow{\sim} \operatorname{End}(T^d)$ by Lemma 8.3.12, and $\sigma \mapsto \operatorname{sgn}(\sigma)\sigma$ induces an involution

$$\omega : k\mathfrak{S}_d \xrightarrow{\sim} k\mathfrak{S}_d.$$

Proposition 8.5.3. *For $\lambda, \mu \in \Lambda(n, d)$ there is a canonical isomorphism*

$$\operatorname{Hom}(\Gamma^\lambda, \Gamma^\mu) \xrightarrow{\sim} \operatorname{Hom}(\Lambda^\lambda, \Lambda^\mu)$$

that makes the following diagram commutative.

$$
\begin{array}{ccc}
\operatorname{Hom}(\Gamma^\lambda, \Gamma^\mu) & \xrightarrow{\;\sim\;} & \operatorname{Hom}(\Lambda^\lambda, \Lambda^\mu) \\
\downarrow{\scriptstyle (\nabla, \Delta)} & & \downarrow{\scriptstyle (\hat{\nabla}, \hat{\Delta})} \\
\operatorname{Hom}(T^d, T^d) & \xrightarrow{\;\omega\;} & \operatorname{Hom}(T^d, T^d)
\end{array}
$$

Proof First observe that the labeled maps of the diagram are stable under base change. This follows for (∇, Δ) from Lemma 8.3.21, and for $(\hat{\nabla}, \hat{\Delta})$ one may use Lemma 8.3.23. Thus it suffices to prove the assertion for $k = \mathbb{Z}$ and then it follows for arbitrary k by base change.

We may assume $n \geq d$ and use the identification of $\mathrm{Pol}^d \, \mathcal{P}_k$ with the category of modules over the Schur algebra $A := S_k(n,d)$ via evaluation at $E := k^n$ (Proposition 8.3.11). Thus we need to show that there is a commutative diagram of the following form

$$
\begin{array}{ccc}
\mathrm{Hom}_A(\Gamma^\lambda E, \Gamma^\mu E) & \xrightarrow{\quad\sim\quad} & \mathrm{Hom}_A(\Lambda^\lambda E, \Lambda^\mu E) \\
\downarrow{\scriptstyle(\nabla,\Delta)} & & \downarrow{\scriptstyle(\hat{\nabla},\hat{\Delta})} \\
\mathrm{Hom}_A(E^{\otimes d}, E^{\otimes d}) & \xrightarrow{\quad\omega\quad} & \mathrm{Hom}_A(E^{\otimes d}, E^{\otimes d})
\end{array}
$$

and we proceed in several steps.

We consider the case $k = \mathbb{Z}$ and claim that the vertical maps are $\mathbb{Z}$-split monomorphisms. This is clear for $(\hat{\nabla}, \hat{\Delta})$ since $\hat{\Delta} \colon \Lambda^\mu E \to E^{\otimes d}$ is a $\mathbb{Z}$-split monomorphism and $\hat{\nabla} \colon E^{\otimes d} \to \Lambda^\lambda E$ is a $\mathbb{Z}$-split epimorphism. Also $\Delta \colon \Gamma^\mu E \to E^{\otimes d}$ is a $\mathbb{Z}$-split monomorphism. It remains to oberserve that the map $\nabla \colon E^{\otimes d} \to \Gamma^\lambda E$ induces a $\mathbb{Z}$-split monomorphism

$$(\nabla, \mathrm{id}) \colon \mathrm{Hom}_A(\Gamma^\lambda E, E^{\otimes d}) \longrightarrow \mathrm{Hom}_A(E^{\otimes d}, E^{\otimes d}).$$

To see this we use the identification $(k\mathfrak{S}_d)^{\mathrm{op}} \xrightarrow{\sim} \mathrm{End}_A(E^{\otimes d})$ from Proposition 8.2.7 and then a $\mathbb{Z}$-basis of $\mathrm{Hom}_A(\Gamma^\lambda E, E^{\otimes d})$ is given by maps $\Gamma^\lambda E \xrightarrow{\Delta} E^{\otimes d} \xrightarrow{\sigma} E^{\otimes d}$ where σ runs through a representative set of right cosets of the Young subgroup $\mathfrak{S}_\lambda \subseteq \mathfrak{S}_d$. The map (∇, id) sends $\sigma \circ \Delta$ to $\sum_{\tau \in \mathfrak{S}_\lambda} \tau\sigma$, which is just the sum of all permutations in the right coset of $\mathfrak{S}_\lambda$ represented by σ. Thus (∇, id) is a $\mathbb{Z}$-split monomorphism.

Now we need to show that

$$\mathrm{Im}(\omega \circ (\nabla, \Delta)) \subseteq \mathrm{Im}(\hat{\nabla}, \hat{\Delta}) \qquad \text{and} \qquad \mathrm{Im}(\omega \circ (\hat{\nabla}, \hat{\Delta})) \subseteq \mathrm{Im}(\nabla, \Delta).$$

View $E^{\otimes d}$ as an $S_k(n,d)$-$k\mathfrak{S}_d$-bimodule. We consider the *symmetriser* e_λ and the *antisymmetriser* $\hat{e}_\lambda$ in the group algebra $k\mathfrak{S}_d$ by setting

$$e_\lambda = \sum_{\sigma \in \mathfrak{S}_\lambda} \sigma \qquad \text{and} \qquad \hat{e}_\lambda = \sum_{\sigma \in \mathfrak{S}_\lambda} \mathrm{sgn}(\sigma)\sigma,$$

where $\mathfrak{S}_\lambda$ denotes the Young subgroup of $\mathfrak{S}_d$ corresponding to λ. Note that the isomorphism $\mathrm{End}_A(E^{\otimes d}) \xrightarrow{\sim} (k\mathfrak{S}_d)^{\mathrm{op}}$ from Proposition 8.2.7 identifies $e_\lambda = \Delta \circ \nabla$ and $\hat{e}_\lambda = \hat{\Delta} \circ \hat{\nabla}$. Now let $k = \mathbb{Q}$. Then we can identify $\Gamma^\lambda E = (E^{\otimes d})e_\lambda$ and $\Lambda^\lambda E = (E^{\otimes d})\hat{e}_\lambda$. Thus the isomorphism $\mathrm{End}_A(E^{\otimes d}) \xrightarrow{\sim} (k\mathfrak{S}_d)^{\mathrm{op}}$ yields

$$\mathrm{Hom}_A(\Gamma^\lambda E, \Gamma^\mu E) \xrightarrow{\sim} e_\lambda(k\mathfrak{S}_d)e_\mu$$

and

$$\mathrm{Hom}_A(\Lambda^\lambda E, \Lambda^\mu E) \xrightarrow{\sim} \hat{e}_\lambda(k\mathfrak{S}_d)\hat{e}_\mu.$$

It is clear that ω identifies $e_\lambda(k\mathfrak{S}_d)e_\mu$ with $\hat{e}_\lambda(k\mathfrak{S}_d)\hat{e}_\mu$. The case $k = \mathbb{Z}$ now follows since

$$\mathrm{Hom}_A(\Gamma^\lambda E, \Gamma^\mu E) = (\mathrm{Hom}_A(\Gamma^\lambda E, \Gamma^\mu E) \otimes_\mathbb{Z} \mathbb{Q}) \cap \mathrm{Hom}_A(E^{\otimes d}, E^{\otimes d})$$

and

$$\mathrm{Hom}_A(\Lambda^\lambda E, \Lambda^\mu E) = (\mathrm{Hom}_A(\Lambda^\lambda E, \Lambda^\mu E) \otimes_\mathbb{Z} \mathbb{Q}) \cap \mathrm{Hom}_A(E^{\otimes d}, E^{\otimes d})$$

by our previous discussion of the maps (∇, Δ) and $(\hat{\nabla}, \hat{\Delta})$. This yields the claim for all k via base change. $\square$

Corollary 8.5.4. *The assignment* $\Gamma^\lambda \mapsto \Lambda^\lambda$ *extends to an equivalence*

$$\Omega \colon \mathrm{add}\{\Gamma^\lambda \mid \lambda \text{ partition of } d\} \xrightarrow{\ \sim\ } \mathrm{add}\{\Lambda^\lambda \mid \lambda \text{ partition of } d\}$$

which maps $\Gamma^{d,V}$ *to* $\Lambda^{d,V}$ *for each* $V \in \mathcal{P}_k$.

Proof The first part follows from Proposition 8.5.3 and the second part then follows from a computation using the decompositions (8.3.8) and (8.3.10). Note that the action on morphisms is determined by the map

$$\mathrm{End}(\Gamma^{(1,\ldots,1)}) \cong (k\mathfrak{S}_d)^{\mathrm{op}} \xrightarrow{\ \omega\ } (k\mathfrak{S}_d)^{\mathrm{op}} \cong \mathrm{End}(\Lambda^{(1,\ldots,1)}). \qquad \square$$

We write k_{sgn} for the one-dimensional *sign representation* of $\mathfrak{S}_d$. Thus

$$\omega^*(M) = M \otimes_k k_{\mathrm{sgn}}$$

for every $k\mathfrak{S}_d$-module M. The following commutative diagram shows the interaction of the functor Ω with the Schur functor $\mathrm{Hom}(\Gamma^{(1,\ldots,1)}, -)$ into the representations of the symmetric group $\mathfrak{S}_d$ (cf. Proposition 8.3.14).

$$
\begin{array}{ccc}
\mathrm{add}\{\Gamma^{d,V} \mid V \in \mathcal{P}_k\} & \xrightarrow{\ \ \Omega\ \ } & \mathrm{add}\{\Lambda^{d,V} \mid V \in \mathcal{P}_k\} \\
\Big\downarrow{\scriptstyle \mathrm{Hom}(\Gamma^{(1,\ldots,1)},-)} & & \Big\downarrow{\scriptstyle \mathrm{Hom}(\Gamma^{(1,\ldots,1)},-)} \\
\mathrm{add}\{V^{\otimes d} \mid V \in \mathcal{P}_k\} & \xrightarrow{\ \ \omega^*\ \ } & \mathrm{add}\{\omega^*(V^{\otimes d}) \mid V \in \mathcal{P}_k\}
\end{array}
$$

A Square Root of the Nakayama Functor

The assignment $\Gamma^\lambda \mapsto \Lambda^\lambda$ from Corollary 8.5.4 can be extended to a functor which is right exact and preserves all coproducts. This yields an adjoint pair of functors

$$\mathrm{Pol}^d\,\mathcal{P}_k \underset{\Omega^-}{\overset{\Omega}{\rightleftarrows}} \mathrm{Pol}^d\,\mathcal{P}_k.$$

Each object in $\mathrm{Pol}^d\,\mathcal{P}_k$ can be written canonically as a colimit of representable functors via (8.3.2). Thus for an object $X \in \mathrm{Pol}^d\,\mathcal{P}_k$ the assignment

$$\Gamma^{d,V} \longmapsto \Lambda^{d,V} = \Lambda^d\,\mathrm{Hom}(V,-) \qquad (V \in \mathcal{P}_k)$$

extends to a colimit preserving functor $\Omega\colon \mathrm{Pol}^d\,\mathcal{P}_k \to \mathrm{Pol}^d\,\mathcal{P}_k$ by setting

$$\Omega(X) := \operatorname*{colim}_{\Gamma^{d,V}\to X} \Lambda^{d,V} \qquad (X \in \mathrm{Pol}^d\,\mathcal{P}_k)$$

where the colimit is taken over the category of morphisms $\Gamma^{d,V} \to X$ and V runs through the objects of $\mathcal{P}_k$.

The functor Ω admits a right adjoint $\Omega^-\colon \mathrm{Pol}^d\,\mathcal{P}_k \to \mathrm{Pol}^d\,\mathcal{P}_k$. For an object $Y \in \mathrm{Pol}^d\,\mathcal{P}_k$ this is given by

$$\Omega^-(Y)(V) = \mathrm{Hom}(\Gamma^{d,V}, \Omega^-(Y)) = \mathrm{Hom}(\Lambda^{d,V}, Y) \qquad (V \in \mathcal{P}_k).$$

We record some properties of the functors Ω and Ω^-.

Lemma 8.5.5. *For $X, Y \in \mathrm{Pol}^d\,\mathcal{P}_k$ we have a natural isomorphism*

$$\mathrm{Hom}(X, \Omega(Y)^\circ) \cong \mathrm{Hom}(Y, \Omega^-(X^\circ)).$$

Proof   This is clear since (Ω, Ω^-) is an adjoint pair, using the duality (8.2.5).
□

Next we collect some elementary facts about the duality $X \mapsto X^\circ$. For $V \in \mathcal{P}_k$ and a partition λ of d we have the following natural isomorphisms:

$$(\Gamma^{d,V})^\circ \cong S^{d,V} \quad (\Lambda^{d,V})^\circ \cong \Lambda^{d,V^\vee} \quad (\Gamma^\lambda)^\circ \cong S^\lambda \quad (\Lambda^\lambda)^\circ \cong \Lambda^\lambda. \quad (8.5.6)$$

Lemma 8.5.7. *For $V, W \in \mathcal{P}_k$ we have natural isomorphisms*

$$\Omega(\Lambda^{d,V}) \cong S^{d,V^\vee} \qquad \text{and} \qquad \Omega^-(S^{d,W}) \cong \Lambda^{d,W^\vee}.$$

For partitions λ, μ of d we have natural isomorphisms

$$\Omega(\Lambda^\lambda) \cong S^\lambda \qquad \text{and} \qquad \Omega^-(S^\mu) \cong \Lambda^\mu.$$

Proof   We combine Lemma 8.5.5 with the identities in (8.5.6). This yields

$$\begin{aligned}
\mathrm{Hom}(\Gamma^{d,V}, \Omega^-(S^{d,W})) &\cong \mathrm{Hom}(\Gamma^{d,W}, \Omega(\Gamma^{d,V})^\circ) \\
&\cong \mathrm{Hom}(\Gamma^{d,W}, (\Lambda^{d,V})^\circ) \\
&\cong \Lambda^d\,\mathrm{Hom}(V^\vee, W) \\
&\cong \Lambda^d\,\mathrm{Hom}(W^\vee, V) \\
&\cong \mathrm{Hom}(\Gamma^{d,V}, \Lambda^{d,W^\vee}).
\end{aligned}$$

Thus $\Omega^-(S^{d,W}) \cong \Lambda^{d,W^\vee}$. Similarly, we compute

$$\mathrm{Hom}(\Gamma^{d,W}, \Omega(\Lambda^{d,V})^\circ) \cong \mathrm{Hom}(\Lambda^{d,V}, \Omega^-(S^{d,W}))$$
$$\cong \mathrm{Hom}(\Lambda^{d,V}, \Lambda^{d,W^\vee})$$
$$\cong \mathrm{Hom}(\Lambda^{d,W}, \Lambda^{d,V^\vee})$$
$$\cong \mathrm{Hom}(\Gamma^{d,W}, \Gamma^{d,V^\vee}).$$

The second isomorphism uses the first computation, and the third isomorphism uses the duality (8.2.5). Thus $\Omega(\Lambda^{d,V}) \cong S^{d,V^\vee}$.

For the second assertion we compute

$$\mathrm{Hom}(\Gamma^{d,V}, \Omega(\Lambda^\lambda)^\circ) \cong \mathrm{Hom}(\Lambda^\lambda, \Omega^-(S^{d,V}))$$
$$\cong \mathrm{Hom}(\Lambda^\lambda, \Lambda^{d,V^\vee})$$
$$\cong \mathrm{Hom}(\Lambda^{d,V}, \Lambda^\lambda)$$
$$\cong \mathrm{Hom}(\Gamma^{d,V}, \Gamma^\lambda).$$

Thus $\Omega(\Lambda^\lambda) \cong S^\lambda$, and the computation for $\Omega^-(S^\mu)$ is similar. $\square$

Recall that the *Nakayama functor*

$$\nu \colon \mathrm{Pol}^d \mathcal{P}_k \longrightarrow \mathrm{Pol}^d \mathcal{P}_k$$

identifies projectives and injectives. More precisely, it is determined by

$$\nu(\Gamma^{d,V}) = (\Gamma^{d,V^\vee})^\circ = S^{d,V^\vee} \qquad (V \in \mathcal{P}_k)$$

and the fact that it preserves colimits, so

$$\nu(X) = \operatorname*{colim}_{\Gamma^{d,V} \to X} S^{d,V^\vee} \qquad (X \in \mathrm{Pol}^d \mathcal{P}_k)$$

where the colimit is taken over the category of morphisms $\Gamma^{d,V} \to X$ and V runs through the objects of $\mathcal{P}_k$.

We need the following basic fact about adjoint functors.

Lemma 8.5.8. *Let (F, G) be a pair of adjoint functors $\mathcal{C} \rightleftarrows \mathcal{D}$ such that F restricts to an equivalence $\mathcal{C}_0 \xrightarrow{\sim} \mathcal{D}_0$ for a pair of full subcategories $\mathcal{C}_0 \subseteq \mathcal{C}$ and $\mathcal{D}_0 \subseteq \mathcal{D}$. If $\mathcal{C}_0$ is generating in the sense that any morphism ϕ in $\mathcal{C}$ is invertible provided that $\mathrm{Hom}_{\mathcal{C}}(C, \phi)$ is bijective for all $C \in \mathcal{C}_0$, then G restricts to a quasi-inverse $\mathcal{D}_0 \xrightarrow{\sim} \mathcal{C}_0$.*

Proof Let $X, X' \in \mathcal{C}_0$. The composite

$$\mathrm{Hom}_{\mathcal{C}}(X', X) \xrightarrow{F} \mathrm{Hom}_{\mathcal{D}}(F(X'), F(X)) \xrightarrow{\sim} \mathrm{Hom}_{\mathcal{C}}(X', GF(X))$$

is given by composition with the unit $\eta_X \colon X \to GF(X)$. Thus η_X is invertible. It follows for every $Y \in \mathcal{D}_0$ that $G(Y)$ belongs to $\mathcal{C}_0$, up to an isomorphism. $\square$

Theorem 8.5.9. *The functor Ω induces equivalences*

$$\mathrm{add}\{\Gamma^\lambda \mid \lambda \ \textit{partition of} \ d\} \xrightarrow{\ \sim\ } \mathrm{add}\{\Lambda^\lambda \mid \lambda \ \textit{partition of} \ d\}$$
$$\mathrm{add}\{\Lambda^\lambda \mid \lambda \ \textit{partition of} \ d\} \xrightarrow{\ \sim\ } \mathrm{add}\{S^\lambda \mid \lambda \ \textit{partition of} \ d\}$$

with quasi-inverses induced by Ω^-. In particular, we have a natural isomorphism

$$\Omega \circ \Omega \cong \nu.$$

Proof   The first equivalence is from Corollary 8.5.4. Then Lemma 8.5.8 shows that the right adjoint provides a quasi-inverse. The second equivalence follows from Lemma 8.5.7. $\square$

Remark 8.5.10.   (1) The functor $\Omega \colon \mathrm{Pol}^d \, \mathcal{P}_k \to \mathrm{Pol}^d \, \mathcal{P}_k$ is compatible with base change. For $X \in \mathrm{Pol}^d \, \mathcal{P}_k$ and a homomorphism $k \to \ell$ of commutative rings we have a natural isomorphism

$$\Omega_k(X) \otimes_k \ell \cong \Omega_\ell(X \otimes_k \ell).$$

For $X = \Gamma^{d,V}$ this is clear, and it follows for an arbitrary $X \in \mathrm{Pol}^d \, \mathcal{P}_k$ because the tensor functors preserve colimits.

(2) When k is a Dedekind domain, then the pair (Ω, Ω^-) restricts to functors $\mathrm{pol}^d \, \mathcal{P}_k \rightleftarrows \mathrm{pol}^d \, \mathcal{P}_k$. For Ω^- this follows from the fact that each $X \in \mathrm{pol}^d \, \mathcal{P}_k$ admits a copresentation $0 \to X \to S^{d,V_0} \to S^{d,V_1}$ so that $\Omega^-(X)$ is a subfunctor of $\Lambda^{d,V_0^\vee}$. On the other hand, $\Omega(X) \cong \Omega^-(X^\circ)^\circ$. This holds for $X = \Gamma^{d,V}$ and follows for an arbitrary $X \in \mathrm{pol}^d \, \mathcal{P}_k$ because there is a presentation $\Gamma^{d,V_1} \to \Gamma^{d,V_0} \to X \to 0$.

The following diagram shows the equivalences given by Ω and $(-)^\circ$.

$$
\begin{array}{ccc}
\mathrm{add}\{\Gamma^\lambda \mid \lambda \vdash d\} & \underset{\nu}{\overset{(-)^\circ}{\rightrightarrows}} & \mathrm{add}\{S^\lambda \mid \lambda \vdash d\} \\[4pt]
& \searrow^{\Omega} \qquad \nearrow^{\Omega} & \\[4pt]
& \mathrm{add}\{\Lambda^\lambda \mid \lambda \vdash d\} & \\
& \underset{(-)^\circ}{\circlearrowleft} &
\end{array}
$$

We have $\Omega \circ \Omega \cong \nu$ and $\Omega \circ (-)^\circ \circ \Omega \cong (-)^\circ$.

A Shuffling Morphism

Fix a partition λ of an integer d and denote by λ' its conjugate partition. Each integer $r \in \{1, \ldots, d\}$ can be written uniquely as a sum $r = \lambda_1 + \cdots + \lambda_{i-1} + j$ with $1 \leq j \leq \lambda_i$. The pair (i, j) describes the position (ith row and jth column) of r in the Young diagram corresponding to λ. The partition λ determines a permutation $\tau_\lambda \in \mathfrak{S}_d$ by $\tau_\lambda(r) = \lambda'_1 + \cdots + \lambda'_{j-1} + i$, where $1 \leq i \leq \lambda_j$. Note that $\tau_{\lambda'} = \tau_\lambda^{-1}$. Here is an example:

$$\lambda = (3, 2) \quad \begin{array}{|c|c|c|} \hline 1 & 2 & 3 \\ \hline 4 & 5 \\ \cline{1-2} \end{array} \qquad \lambda' = (2, 2, 1) \quad \begin{array}{|c|c|} \hline 1 & 2 \\ \hline 3 & 4 \\ \hline 5 \\ \cline{1-1} \end{array} \qquad \tau_\lambda = \begin{pmatrix} 1 & 2 & 3 & 4 & 5 \\ 1 & 3 & 5 & 2 & 4 \end{pmatrix}.$$

For any k-module V the permutation τ_λ induces an automorphism

$$t_\lambda : V^{\otimes d} \xrightarrow{\sim} V^{\otimes d}, \qquad v_1 \otimes \cdots \otimes v_d \mapsto v_{\tau_\lambda(1)} \otimes \cdots \otimes v_{\tau_\lambda(d)}$$

by exchanging rows and columns of the Young diagram corresponding to λ. For the above example this yields the following:

$$V^{\otimes 5} = V^{\otimes 2} \otimes V^{\otimes 2} \otimes V^{\otimes 1} \quad \longrightarrow \quad \begin{array}{|c|c|c|} \hline 1 & 3 & 5 \\ \hline 2 & 4 \\ \cline{1-2} \end{array} \xrightarrow{\tau_\lambda} \begin{array}{|c|c|} \hline 1 & 4 \\ \hline 2 & 5 \\ \hline 3 \\ \cline{1-1} \end{array} \quad \longrightarrow \quad V^{\otimes 3} \otimes V^{\otimes 2} = V^{\otimes 5}$$

$$(v_1 \otimes v_2) \otimes (v_3 \otimes v_4) \otimes v_5 \quad \longmapsto \quad (v_1 \otimes v_3 \otimes v_5) \otimes (v_2 \otimes v_4).$$

Schur and Weyl Modules

Fix a partition λ of an integer d so that $\lambda_1 + \cdots + \lambda_n = d = \lambda'_1 + \cdots + \lambda'_n$. We set

$$\Lambda^\lambda = \Lambda^{\lambda_1} \otimes \cdots \otimes \Lambda^{\lambda_n} \qquad \text{and} \qquad S^\lambda = S^{\lambda_1} \otimes \cdots \otimes S^{\lambda_n}.$$

For $V \in \mathcal{P}_k$ one defines the *Schur module* $\mathrm{Sch}^\lambda V$ as the image of the map

$$\Lambda^{\lambda'_1} V \otimes \cdots \otimes \Lambda^{\lambda'_n} V \xrightarrow{\Delta \otimes \cdots \otimes \Delta} V^{\otimes d} \xrightarrow{t_\lambda} V^{\otimes d} \xrightarrow{\nabla \otimes \cdots \otimes \nabla} S^{\lambda_1} V \otimes \cdots \otimes S^{\lambda_n} V.$$

Here, for an integer r, we denote by $\Delta : \Lambda^r V \to V^{\otimes r}$ the map given by

$$\Delta(v_1 \wedge \cdots \wedge v_r) = \sum_{\sigma \in \mathfrak{S}_r} \mathrm{sgn}(\sigma) v_{\sigma(1)} \otimes \cdots \otimes v_{\sigma(r)},$$

$\nabla : V^{\otimes r} \to S^r V$ is the canonical projection, and $t_\lambda : V^{\otimes d} \to V^{\otimes d}$ is the shuffling morphism given by

$$t_\lambda(v_1 \otimes \cdots \otimes v_d) = v_{\tau_\lambda(1)} \otimes \cdots \otimes v_{\tau_\lambda(d)}.$$

The *Weyl module* $\mathrm{Weyl}^\lambda V$ is by definition the image of the analogous map

$$\Gamma^{\lambda_1} V \otimes \cdots \otimes \Gamma^{\lambda_n} V \xrightarrow{\Delta \otimes \cdots \otimes \Delta} V^{\otimes d} \xrightarrow{t_{\lambda'}} V^{\otimes d} \xrightarrow{\nabla \otimes \cdots \otimes \nabla} \Lambda^{\lambda'_1} V \otimes \cdots \otimes \Lambda^{\lambda'_n} V,$$

where $\Delta \colon \Gamma^r V \to V^{\otimes r}$ is the inclusion and $\nabla \colon V^{\otimes r} \to \Lambda^r V$ is the canonical projection. Note that both maps are related via the duality as follows:

$$(\Gamma^\lambda V \to \Lambda^{\lambda'} V)^\vee \cong (\Lambda^{\lambda'}(V^\vee) \to S^\lambda(V^\vee)).$$

Remark 8.5.11. In Lemma 8.3.23 we have already seen that there is a distinguished morphism $\Gamma^\lambda \to \Lambda^{\lambda'}$. Evaluated at $V \in \mathcal{P}_k$ this equals the morphism which defines the Weyl module $\mathrm{Weyl}^\lambda V$.

Schur and Weyl Functors

The definition of Schur and Weyl modules gives rise to the *Schur functor* $V \mapsto \mathrm{Sch}^\lambda(V)$ and the *Weyl functor* $V \mapsto \mathrm{Weyl}^\lambda(V)$ in $\mathrm{Pol}^d \mathcal{P}_k$ for each partition λ of d. For example, we have

$$\mathrm{Sch}^{(1,\dots,1)} = \Lambda^d, \quad \mathrm{Sch}^{(d)} = S^d, \quad \mathrm{Weyl}^{(1,\dots,1)} = \Lambda^d, \quad \mathrm{Weyl}^{(d)} = \Gamma^d.$$

Proposition 8.5.12. *Let λ be a partition. Then the canonical morphisms $\Gamma^\lambda \twoheadrightarrow \mathrm{Weyl}^\lambda$ and $\mathrm{Sch}^\lambda \rightarrowtail S^\lambda$ induce isomorphisms*

$$\Delta^\lambda \xrightarrow{\sim} \mathrm{Weyl}^\lambda \qquad and \qquad \mathrm{Sch}^\lambda \xrightarrow{\sim} \nabla^\lambda.$$

Proof First observe that for any morphism $\Gamma^\mu \to \Gamma^\lambda$ with $\mu > \lambda$ the composite $\Gamma^\mu \to \Gamma^\lambda \twoheadrightarrow \mathrm{Weyl}^\lambda$ is zero. This follows from Lemma 8.3.24 since $\mathrm{Weyl}^\lambda \subseteq \Lambda^{\lambda'}$. This observation yields an epimorphism $\Delta^\lambda \to \mathrm{Weyl}^\lambda$. When k is a field of characteristic zero, then this is an isomorphism since Δ^λ is simple; see Proposition 8.5.1. For $k = \mathbb{Z}$ it follows that the kernel is torsion. But Δ^λ is torsion free. Thus $\Delta^\lambda \to \mathrm{Weyl}^\lambda$ is an isomorphism when $k = \mathbb{Z}$. Given any free module $V = \mathbb{Z}^n$ we have already seen in Corollary 8.4.14 a $\mathbb{Z}$-basis of $\Delta^\lambda V$. The canonical map $\Gamma^\lambda V \to \Lambda^{\lambda'} V$ takes its elements to basis elements of $\Lambda^{\lambda'} V$; this follows from the discussion of weight spaces of exterior powers in Lemma 8.3.23. Thus the inclusion $\mathrm{Weyl}^\lambda V \to \Lambda^{\lambda'} V$ is a $\mathbb{Z}$-split monomorphism, and therefore the canonical morphism $\Delta^\lambda \to \mathrm{Weyl}^\lambda$ is stable under base change. From the case $k = \mathbb{Z}$ the isomorphism $\Delta^\lambda \xrightarrow{\sim} \mathrm{Weyl}^\lambda$ follows.

The assertion about $\mathrm{Sch}^\lambda \to \nabla^\lambda$ follows from the first part using the duality, since

$$\mathrm{Sch}^\lambda \cong (\mathrm{Weyl}^\lambda)^\circ \cong (\Delta^\lambda)^\circ \cong \nabla^\lambda. \qquad \square$$

From now on we identify $\Delta^\lambda = \mathrm{Weyl}^\lambda$ and $\nabla^\lambda = \mathrm{Sch}^\lambda$ for each partition λ.

Next we establish a presentation of $\nabla^{\lambda'}$ which is an analogue of the presentation (8.4.1) of Δ^λ. This requires the following definition. A *standard morphism* $\Lambda^\mu \to \Lambda^\lambda$ is a morphism corresponding to a standard morphism $\gamma_A \colon \Gamma^\mu \to \Gamma^\lambda$ under the bijection from Proposition 8.5.3.

Proposition 8.5.13. *Let λ be a partition. The functor $\nabla^{\lambda'}$ admits a presentation*

$$\bigoplus_{\substack{\Lambda^\mu \to \Lambda^\lambda \\ \mu > \lambda}} \Lambda^\mu \longrightarrow \Lambda^\lambda \longrightarrow \nabla^{\lambda'} \longrightarrow 0 \qquad (8.5.14)$$

which is given by all standard morphisms $\Lambda^\mu \to \Lambda^\lambda$ with partition $\mu > \lambda$. In particular, we have

$$\Omega(\Delta^\lambda) \cong \nabla^{\lambda'}.$$

Proof First observe that each costandard object ∇^λ is stable under base change. This follows from the presentation (8.4.1) of the standard object Δ^λ in which the first map $\bigoplus_{\mu > \lambda} \Gamma^\mu \to \Gamma^\lambda$ evaluated at any free module $V = k^n$ is the composite of a k-split epimorphism followed by a k-split monomorphism; see Corollary 8.4.14. Thus the dual of this presentation yields a copresentation of ∇^λ which is stable under base change.

The right exact functor Ω identifies the morphism $\Gamma^\lambda \to \Lambda^{\lambda'}$ defining the Weyl functor Δ^λ with the morphism $\Lambda^\lambda \to S^{\lambda'}$ defining the Schur functor $\nabla^{\lambda'}$; see Theorem 8.5.9. On the other hand, applying Ω to (8.4.1) yields a presentation

$$\bigoplus_{\substack{\Lambda^\mu \to \Lambda^\lambda \\ \mu > \lambda}} \Lambda^\mu \longrightarrow \Lambda^\lambda \longrightarrow \Omega(\Delta^\lambda) \longrightarrow 0,$$

and we obtain an epimorphism $\phi \colon \Omega(\Delta^\lambda) \to \nabla^{\lambda'}$ which is stable under base change. This morphism is an isomorphism for $k = \mathbb{Q}$ since in this case Ω is an exact functor; see Proposition 8.5.1. Now observe that the values of Ω are torsion free for $k = \mathbb{Z}$ by Remark 8.5.10. Thus base change yields that ϕ is an isomorphism for all k; see Lemma 8.3.26. $\qquad\square$

The following diagram summarises our discussion. The commutative triangle in the midle describes the standard object Δ^λ as the image of the distinguished morphism $\Gamma^\lambda \to \Lambda^{\lambda'}$. The triangle on the left is its image under the functor Ω, while the triangle on the right is its image under the duality $(-)^\circ$.

Reduced Presentations

The presentations describing the standard and costandard objects can be reduced, using a particular sort of standard morphisms. For a triple (r, s, t) of non-negative integers we consider the standard morphisms

$$\gamma(r,s,t)\colon \Gamma^{r+s} \otimes \Gamma^t \xrightarrow{\ \Delta \otimes 1\ } \Gamma^r \otimes \Gamma^s \otimes \Gamma^t \xrightarrow{\ 1 \otimes \nabla\ } \Gamma^r \otimes \Gamma^{s+t}$$

and

$$\lambda(r,s,t)\colon \Lambda^{r+s} \otimes \Lambda^t \xrightarrow{\ \Delta \otimes 1\ } \Lambda^r \otimes \Lambda^s \otimes \Lambda^t \xrightarrow{\ 1 \otimes \nabla\ } \Lambda^r \otimes \Lambda^{s+t},$$

where ∇ and Δ denote multiplication and comultiplication for $\Gamma^* V$ and $\Lambda^* V$, respectively. These standard morphisms serve as building blocks for the presentations of standard and costandard objects, generalising the presentations of $\Lambda^d \cong \Delta^{(1,\dots,1)}$ and $S^d \cong \nabla^{(d)}$ from Lemma 8.3.3.

Proposition 8.5.15. *Let $\lambda = (\lambda_1, \dots, \lambda_n)$ be a partition.*

(1) *The standard object Δ^λ admits a presentation*

$$\bigoplus_{i=1}^{n-1} \bigoplus_{r=1}^{\lambda_{i+1}} \Gamma^\mu \longrightarrow \Gamma^\lambda \longrightarrow \Delta^\lambda \longrightarrow 0$$

where

$$\Gamma^\mu = (\Gamma^{\lambda_1} \otimes \cdots \otimes \Gamma^{\lambda_{i-1}}) \otimes (\Gamma^{\lambda_i + r} \otimes \Gamma^{\lambda_{i+1} - r}) \otimes (\Gamma^{\lambda_{i+2}} \otimes \cdots \otimes \Gamma^{\lambda_n}) \longrightarrow \Gamma^\lambda$$

is the standard morphism given by $1 \otimes \gamma(\lambda_i, r, \lambda_{i+1} - r) \otimes 1$.

(2) *The costandard object $\nabla^{\lambda'}$ admits a presentation*

$$\bigoplus_{i=1}^{n-1} \bigoplus_{r=1}^{\lambda_{i+1}} \Lambda^\mu \longrightarrow \Lambda^\lambda \longrightarrow \nabla^{\lambda'} \longrightarrow 0$$

where

$$\Lambda^\mu = (\Lambda^{\lambda_1} \otimes \cdots \otimes \Lambda^{\lambda_{i-1}}) \otimes (\Lambda^{\lambda_i + r} \otimes \Lambda^{\lambda_{i+1} - r}) \otimes (\Lambda^{\lambda_{i+2}} \otimes \cdots \otimes \Lambda^{\lambda_n}) \longrightarrow \Lambda^\lambda$$

is the standard morphism given by $1 \otimes \lambda(\lambda_i, r, \lambda_{i+1} - r) \otimes 1$.

Proof The proof for Δ^λ follows the straightening argument of Lemma 8.4.2, because only a particular class of standard morphisms is used in the proof of that lemma. For $\nabla^{\lambda'}$ the assertion then follows by applying the functor Ω; see Proposition 8.5.13. We leave details to the interested reader since the reduced presentations will not be used in the rest of this chapter. $\square$

The Cauchy Decomposition of Exterior Powers

Fix $V, W \in \mathcal{P}_k$. For every integer $r \geq 0$ there is a unique map

$$\Lambda^r(V \otimes W) \longrightarrow S^r V \otimes \Lambda^r W$$

making the following square commutative.

$$
\begin{array}{ccc}
(V \otimes W)^{\otimes r} & \xrightarrow{\ \sim\ } & V^{\otimes r} \otimes W^{\otimes r} \\
\downarrow & & \downarrow \\
\Lambda^r(V \otimes W) & \longrightarrow & S^r V \otimes \Lambda^r W
\end{array}
$$

The map sends $(v_1 \otimes w_1) \wedge \cdots \wedge (v_r \otimes w_r)$ to $(v_1 \cdots v_r) \otimes (w_1 \wedge \cdots \wedge w_r)$. Dualising this diagram yields a unique map

$$
\psi^r : \Gamma^r V \otimes \Lambda^r W \longrightarrow \Lambda^r(V \otimes W)
$$

making the following square commutative.

$$
\begin{array}{ccc}
\Gamma^r V \otimes \Lambda^r W & \xrightarrow{\ \psi^r\ } & \Lambda^r(V \otimes W) \\
\uparrow & & \uparrow \\
V^{\otimes r} \otimes W^{\otimes r} & \xrightarrow{\ \sim\ } & (V \otimes W)^{\otimes r}
\end{array}
$$

Extend this map for a partition $\lambda = (\lambda_1, \ldots, \lambda_n)$ of an integer d to a map

$$
\psi^\lambda : \Gamma^\lambda V \otimes \Lambda^\lambda W \longrightarrow \Lambda^d(V \otimes W)
$$

which is given as the composite

$$
\Gamma^\lambda V \otimes \Lambda^\lambda W \xrightarrow{\ \sim\ } (\Gamma^{\lambda_1} V \otimes \Lambda^{\lambda_1} W) \otimes \cdots \otimes (\Gamma^{\lambda_n} V \otimes \Lambda^{\lambda_n} W)
$$

$$
\xrightarrow{\ \psi^{\lambda_1} \otimes \cdots \otimes \psi^{\lambda_n}\ } \Lambda^{\lambda_1}(V \otimes W) \otimes \cdots \otimes \Lambda^{\lambda_n}(V \otimes W) \longrightarrow \Lambda^d(V \otimes W)
$$

with the last map given by multiplication.

The *Cauchy filtration* for exterior powers is by definition the chain

$$
0 = F_{+\infty} \subseteq F_{(d)} \subseteq F_{(d-1,1)} \subseteq \cdots \subseteq F_{(2,1,\ldots,1)} \subseteq F_{(1,\ldots,1)} \subseteq \Lambda^d(V \otimes W) \tag{8.5.16}
$$

where $F_\lambda = \sum_{\mu \geq \lambda} \operatorname{Im} \psi^\mu$.

The following result describes the factors of this Cauchy filtration; it is an analogue of Theorem 8.4.6.

Theorem 8.5.17. *Let* $V, W \in \mathcal{P}_k$. *Then* $F_{(1,\ldots,1)} = \Lambda^d(V \otimes W)$ *and for every partition* λ *of an integer* d *the morphism* $\psi^\lambda : \Gamma^\lambda V \otimes \Lambda^\lambda W \to F_\lambda$ *induces an isomorphism*

$$
\Delta^\lambda V \otimes \nabla^{\lambda'} W \xrightarrow{\ \sim\ } F_\lambda / F_{\lambda^+}
$$

which is functorial in V *and* W. *Therefore the associated graded object of the filtration* (F_λ) *is*

$$
\bigoplus_\lambda (\Delta^\lambda V \otimes \nabla^{\lambda'} W) = \Lambda^d(V \otimes W).
$$

Proof Adapt the proof of Theorem 8.4.6, using the presentation (8.4.1) of Δ^λ and the presentation (8.5.14) of $\nabla^{\lambda'}$.   □

For $V \in \mathcal{P}_k$ set $\Lambda^{d,V} = \Lambda^d \operatorname{Hom}(V, -)$. Replacing in the filtration (8.5.16) the object V by $V^\vee$ and using its functoriality in W gives a filtration

$$0 = X_{+\infty} \subseteq X_{(d)} \subseteq X_{(d-1,1)} \subseteq \cdots \subseteq X_{(2,1,\ldots,1)} \subseteq X_{(1,\ldots,1)} = \Lambda^{d,V}$$

with an isomorphism

$$\Delta^\lambda(V^\vee) \otimes \nabla^{\lambda'} \xrightarrow{\;\sim\;} X_\lambda / X_{\lambda^+}$$

for each partition λ. The filtration of $\Lambda^{d,V}$ induces a filtration for each direct summand of $\Lambda^{d,V}$. This follows from the functoriality of the filtration (8.5.16) in V.

Corollary 8.5.18. *Let μ be a partition of an integer d. There are filtrations*

$$0 = Y_{+\infty} \rightarrowtail Y_{(d)} \rightarrowtail Y_{(d-1,1)} \rightarrowtail \cdots \rightarrowtail Y_{\mu^+} \rightarrowtail Y_\mu = \Lambda^\mu$$
$$\Lambda^\mu = Z_\mu \twoheadrightarrow Z_{\mu^+} \twoheadrightarrow \cdots \twoheadrightarrow Z_{(d-1,1)} \twoheadrightarrow Z_{(d)} \twoheadrightarrow Z_{+\infty} = 0$$

with canonical exact sequences

$$0 \longrightarrow Y_{\lambda^+} \longrightarrow Y_\lambda \longrightarrow (\nabla^{\lambda'})^{K_{\lambda\mu}} \longrightarrow 0$$
$$0 \longrightarrow (\Delta^{\lambda'})^{K_{\lambda\mu}} \longrightarrow Z_\lambda \longrightarrow Z_{\lambda^+} \longrightarrow 0$$

for each partition $\lambda \geq \mu$.

Proof For the first filtration adapt the proof of Theorem 8.4.11. Applying the duality yields the second filtration, since $(\Lambda^\mu)^\circ \cong \Lambda^\mu$ and $(\nabla^\lambda)^\circ \cong \Delta^\lambda$.   □

Remark 8.5.19. Let μ be a partition of d. Then

$$\Lambda^{\mu'} \in \operatorname{Filt}\{\nabla^\lambda \mid \lambda \trianglelefteq \mu\} \cap \operatorname{Filt}\{\Delta^\lambda \mid \lambda \trianglelefteq \mu\}$$

since

$$K_{\lambda'\mu'} \neq 0 \implies \mu' \trianglelefteq \lambda' \implies \lambda \trianglelefteq \mu.$$

8.6 Schur Algebras

In this section we consider Schur algebras and establish their quasi-hereditary structure, using the Cauchy filtration for symmetric tensors. Then we identify the characteristic tilting modules, using the Cauchy filtration for exterior powers.

Throughout we keep fixed a commutative ring k.

Split Quasi-hereditary Algebras

Given a k-algebra A, we write $\text{rep}(A, k)$ for the category of A-modules that are finitely generated projective when restricted to k. This category is k-linear and carries a natural exact structure. A sequence in $\text{rep}(A, k)$ is exact if its underlying sequence of k-modules is split exact.

For the Schur algebra $S_k(n, d)$ given by parameters d, n let $\text{rep}\, S_k(n, d)$ denote the category of left $S_k(n, d)$-modules that are finitely generated projective over k.

A k-algebra A together with a partially ordered set of objects $(\Delta_i)_{i \in I}$ in $\text{rep}(A, k)$ is called *split quasi-hereditary* if A is finitely generated projective over k, and there are exact sequences

$$0 \longrightarrow U_i \longrightarrow P_i \longrightarrow \Delta_i \longrightarrow 0 \qquad (i \in I) \qquad (8.6.1)$$

in $\text{rep}(A, k)$ satisfying the following.

(SQ1) $\text{End}_A(\Delta_i) \cong k$ for all i.
(SQ2) $\text{Hom}_A(\Delta_i, \Delta_j) \neq 0$ implies $i \leq j$.
(SQ3) U_i belongs to $\text{Filt}\{\Delta_j \mid j > i\}$ for all i.
(SQ4) $\bigoplus_{i \in I} P_i$ is a projective generator of $\text{rep}(A, k)$.

The Δ_i are called *standard modules* and we set $\text{Filt}_A(\Delta) = \text{Filt}\{\Delta_i \mid i \in I\}$.

It follows immediately from the definition that

$$\text{Ext}_A^1(\Delta_i, \Delta_j) \neq 0 \quad \Longrightarrow \quad i < j.$$

The definition is consistent with that for a quasi-hereditary Artin algebra, thanks to the characterisation given in Proposition 8.1.3. In order to see this, let A be an Artin k-algebra over a commutative local ring k, together with a partial order $\leq$ on the set of isomorphism classes of simple A-modules. Suppose that each standard module Δ_i is finitely generated projective over k and satisfying $\text{End}_A(\Delta_i) \cong k$. Then $(A, \leq)$ is quasi-hereditary if and only if A together with the sequence of standard modules is split quasi-hereditary.

Next we show that any split quasi-hereditary algebra A gives rise to families of split quasi-hereditary algebras A_J and A^J, where J runs through all coideals of the poset I that parametrises the standard modules.

Let us fix a split quasi-hereditary algebra A given by exact sequences (8.6.1). Let $J \subseteq I$ and set $\bar{J} = I \setminus J$. Then J is a *coideal* (so $i \leq j$ and $i \in J$ imply $j \in J$) if and only if $\bar{J}$ is an *ideal* (so $i \leq j$ and $j \in \bar{J}$ imply $i \in \bar{J}$).

Now suppose that $J \subseteq I$ is a coideal. Set $P_J = \bigoplus_{i \in J} P_i$ and $A_J = \text{End}_A(P_J)$. For each A-module X we consider the natural exact sequence

$$\text{Hom}_A(P_J, X) \otimes_{A_J} P_J \xrightarrow{\varepsilon_X} X \longrightarrow X^{\bar{J}} \longrightarrow 0$$

and set $X_J = \mathrm{Hom}_A(P_J, X) \otimes_{A_J} P_J$. In particular, the map $A \to A^{\bar{J}}$ is an algebra homomorphism.

Lemma 8.6.2. *Let $X \in \mathrm{Filt}\{\Delta_i \mid i \in I\}$. Then the counit ε_X is a monomorphism and we have $X_J \in \mathrm{Filt}\{\Delta_i \mid i \in J\}$ and $X^{\bar{J}} \in \mathrm{Filt}\{\Delta_i \mid i \in \bar{J}\}$. The assignments $X \mapsto X_J$ and $X \mapsto X^{\bar{J}}$ yield a colocalisation sequence*

$$\mathrm{Filt}\{\Delta_i \mid i \in \bar{J}\} \overset{\longleftarrow}{\underset{\longrightarrow}{\quad\quad}} \mathrm{Filt}\{\Delta_i \mid i \in I\} \overset{\longleftarrow}{\underset{\twoheadrightarrow}{\quad\quad}} \mathrm{Filt}\{\Delta_i \mid i \in J\}$$

where all functors are exact.

Proof The functor $- \otimes_{A_J} P_J$ identifies $\mathrm{Mod}\, A_J$ with the full subcategory of A-modules X such that the counit ε_X is invertible. For each $i \in J$ we have a presentation $P_J^n \to P_J \to \Delta_i \to 0$ for some $n \geq 0$, and therefore the sequence (8.6.1) lies in the image of $- \otimes_{A_J} P_J$ when $i \in J$. On the other hand, the functor $\mathrm{Hom}_A(P_J, -)$ annihilates Δ_i for all $i \in \bar{J}$. Thus ε_{Δ_i} is a monomorphism for all $i \in I$. Now an induction on the length of a filtration of X in $\mathrm{Filt}\{\Delta_i \mid i \in I\}$ shows that ε_X is a monomorphism, with $X_J \in \mathrm{Filt}\{\Delta_i \mid i \in J\}$ and $X^{\bar{J}} \in \mathrm{Filt}\{\Delta_i \mid i \in \bar{J}\}$. Then each morphism $X' \to X$ with $X' \in \mathrm{Filt}\{\Delta_i \mid i \in J\}$ factors uniquely through ε_X. Thus the assignment $X \mapsto X_J$ provides a right adjoint to the inclusion of $\mathrm{Filt}\{\Delta_i \mid i \in J\}$. In particular, the assignment is left exact; it is right exact by construction. The analogous properties of $X \mapsto X^{\bar{J}}$ easily follow. $\square$

Recall that the functor $- \otimes_{A_J} P_J$ induces an equivalence

$$\mathrm{Mod}\, A_J \overset{\sim}{\longrightarrow} \{X \in \mathrm{Mod}\, A \mid \varepsilon_x \text{ invertible}\}$$

while restriction along the homomorphism $A \twoheadrightarrow A^{\bar{J}}$ induces an equivalence

$$\mathrm{Mod}\, A^{\bar{J}} \overset{\sim}{\longrightarrow} \{X \in \mathrm{Mod}\, A \mid \varepsilon_X = 0\}$$

(cf. Example 2.2.23). The following result says that both functors induce quasi-hereditary structures.

Proposition 8.6.3. *Let A be a split quasi-hereditary k-algebra with a partially ordered set of standard modules $(\Delta_i)_{i \in I}$. If $J \subseteq I$ is a coideal, then A_J and $A^{\bar{J}}$ carry canonical split quasi-hereditary structures given by exact equivalences*

$$\mathrm{Filt}_{A_J}(\Delta) \overset{\sim}{\to} \mathrm{Filt}\{\Delta_i \mid i \in J\} \quad\quad and \quad\quad \mathrm{Filt}_{A^{\bar{J}}}(\Delta) \overset{\sim}{\to} \mathrm{Filt}\{\Delta_i \mid i \in \bar{J}\}$$

which map standard modules to standard modules.

Proof Each sequence (8.6.1) lies in the image of $- \otimes_{A_J} P_J$ when $i \in J$. Thus A_J is split quasi-hereditary with partially ordered set of standard modules $(\mathrm{Hom}_A(P_J, \Delta_i))_{i \in J}$.

For the algebra $A^{\bar{J}}$ observe that the functor $X \mapsto X^{\bar{J}}$ maps each sequence (8.6.1) with $i \in \bar{J}$ to a sequence

$$0 \longrightarrow (U_i)^{\bar{J}} \longrightarrow (P_i)^{\bar{J}} \longrightarrow \Delta_i \longrightarrow 0$$

in $\mathrm{rep}(A^{\bar{J}}, k)$ where $(U_i)^{\bar{J}}$ is in $\mathrm{Filt}\{\Delta_j \mid j \in \bar{J}, \ j > i\}$ and $(P_i)^{\bar{J}}$ is projective. Thus $A^{\bar{J}}$ is split quasi-hereditary with partially ordered set of standard modules $(\Delta_i)_{i \in \bar{J}}$.   □

Remark 8.6.4. There is an analogue of Proposition 8.6.3 for quasi-hereditary Artin algebras with the same proof.

Example 8.6.5. For $i \in I$ we have a pair of canonical algebra homomorphisms $A \twoheadrightarrow A_i \twoheadrightarrow \bar{A}_i$ given by ideals of I as follows:

$$A_i := A^{\{j \in I \mid j \not> i\}} \qquad \text{and} \qquad \bar{A}_i := A^{\{j \in I \mid j \not\geq i\}}.$$

We are now ready to establish that Schur algebras are quasi-hereditary.

Theorem 8.6.6. *For all parameters d, n the Schur algebra $S_k(n, d)$ is split quasi-hereditary via the collection of canonical sequences*

$$0 \longrightarrow U^\lambda(k^n) \longrightarrow \Gamma^\lambda(k^n) \longrightarrow \Delta^\lambda(k^n) \longrightarrow 0$$

where $\lambda = (\lambda_1, \dots, \lambda_n)$ runs through the set of partitions of d (partially ordered by dominance).

Proof We verify all properties in the category $\mathrm{Pol}^d \mathcal{P}_k$, keeping in mind that $\mathrm{rep}\, S_k(n, d)$ identifies with a full subcategory of $\mathrm{Pol}^d \mathcal{P}_k$ via evaluation at k^n; see Proposition 8.3.11.

(SQ1) follows from Corollary 8.4.15.
(SQ2) follows from Proposition 8.4.4.
(SQ3) follows from Theorem 8.4.11.
(SQ4) follows from the canonical decomposition (8.3.8) which gives

$$S_k(n, d) = \bigoplus_{\lambda \in \Lambda(n,d)} \Gamma^\lambda(k^n). \qquad\qquad □$$

Remark 8.6.7. For any integer $n \geq 1$ the partitions of d belonging to $\Lambda(n, d)$ form a coideal $\Lambda^+(n, d) \subseteq \Lambda^+(d)$ of the set of all partitions of d (partially ordered by dominance), with equality if and only if $n \geq d$. If A denotes a 'full' Schur algebra A satisfying $\mathrm{pol}^d \mathcal{P}_k \xrightarrow{\sim} \mathrm{rep}(A, k)$, then $S_k(n, d)$ identifies (up to Morita equivalence) with the algebra $A_{\Lambda^+(n,d)}$. This follows from Proposition 8.3.11 and its proof, since $\Gamma^{d,k^n} = \bigoplus_{\lambda \in \Lambda(n,d)} \Gamma^\lambda$ identifies with a projective generator of $\mathrm{rep}\, S_k(n, d)$.

Finite Global Dimension

The Schur algebra $S_k(n, d)$ has finite global dimension provided that k is a field. This follows from Proposition 8.1.6 since the algebra $S_k(n, d)$ is quasi-hereditary. Note that there is a bound depending only on d since the number of isomorphism classes of simple modules equals the number of partitions of d. We extend this result as follows.

Proposition 8.6.8. *Suppose that k is a commutative noetherian ring. Then for all parameters d, n the exact category* rep $S_k(n, d)$ *has finite global dimension.*

The proof is based on the following lemma. For a prime ideal $\mathfrak{p} \subseteq k$, let $k(\mathfrak{p})$ denote the residue field $k_\mathfrak{p}/\mathfrak{p}_\mathfrak{p}$.

Lemma 8.6.9. *Let A be a noetherian k-algebra and M a finitely generated A-module. Suppose that A and M are k-projective. Then M is projective over A if and only if $M \otimes_k k(\mathfrak{p})$ is projective over $A \otimes_k k(\mathfrak{p})$ for all prime ideals $\mathfrak{p} \subseteq k$.*

Proof One direction is clear. So suppose that $M \otimes_k k(\mathfrak{p})$ is projective over $A \otimes_k k(\mathfrak{p})$ for all $\mathfrak{p}$. It suffices to prove the assertion when k is local with maximal ideal $\mathfrak{m}$, and we may assume that k is complete since k is noetherian. Thus A is semi-perfect and a projective cover $P \to M \otimes_k k(\mathfrak{m})$ lifts to a projective cover $P \to M$, which is an isomorphism since $P \otimes_k k(\mathfrak{m}) \to M \otimes_k k(\mathfrak{m})$ is one. It follows that M is projective over A. $\square$

Proof of Proposition 8.6.8 Fix a module $X \in$ rep $S_k(n, d)$. We choose a projective resolution

$$\cdots \longrightarrow P_2 \longrightarrow P_1 \longrightarrow P_0 \longrightarrow X \longrightarrow 0$$

and set $\Omega_r(X) = \mathrm{Coker}(P_{r+1} \to P_r)$ for $r \geq 0$. Base change yields for each ring homomorphism $k \to \ell$ an isomorphism

$$S_k(n, d) \otimes_k \ell \xrightarrow{\sim} S_\ell(n, d),$$

and $\Omega_r(X \otimes_k \ell) = \Omega_r(X) \otimes_k \ell$ for all $r \geq 0$. Now the assertion follows from Lemma 8.6.9 since the global dimension of $S_{k(\mathfrak{p})}(n, d)$ is bounded in terms of d for all prime ideals $\mathfrak{p} \subseteq k$, by Proposition 8.1.6. $\square$

Characteristic Tilting Modules

Let A be a k-algebra and suppose it is split quasi-hereditary with partially ordered set of standard modules $(\Delta_i)_{i \in I}$. Set $\mathcal{A} = \mathrm{rep}(A, k)$.

We show that there is canonically defined a set of *costandard modules*

$(\nabla_i)_{i \in I}$ in rep(A, k). This corresponds to the fact that the opposite algebra A^{op} is canonically a split quasi-hereditary with standard modules $(\Delta_i')_{i \in I}$ given by $\Delta_i' = \text{Hom}_k(\nabla_i, k)$.

For $i \in I$ we consider the full subcategories

$$\mathcal{A}_i = \{X \in \mathcal{A} \mid \text{Hom}_{\mathcal{A}}(\Delta_j, X) = 0 \text{ for } j > i\} \xrightarrow{\sim} \text{rep}(A_i, k)$$

and

$$\bar{\mathcal{A}}_i = \{X \in \mathcal{A} \mid \text{Hom}_{\mathcal{A}}(\Delta_j, X) = 0 \text{ for } j \geq i\} \xrightarrow{\sim} \text{rep}(\bar{A}_i, k);$$

see Example 8.6.5. Then there is an idempotent $e_i \in A_i$ and the module $e_i A_i$ identifies via $A \twoheadrightarrow A_i$ with Δ_i. Note that $\text{End}_A(\Delta_i) \cong e_i A_i e_i \cong k$. The functor $\text{Hom}_A(\Delta_i, -) \colon \mathcal{A}_i \to \text{proj}\,\text{End}_A(\Delta_i) \xrightarrow{\sim} \mathcal{P}_k$ induces a recollement of exact categories

$$\bar{\mathcal{A}}_i \begin{array}{c} \overleftarrow{} \\ \xrightarrow{} \\ \overleftarrow{} \end{array} \mathcal{A}_i \begin{array}{c} \overleftarrow{E_i} \\ \xrightarrow{} \\ \overleftarrow{F_i} \end{array} \mathcal{P}_k \qquad (8.6.10)$$

with $E_i \cong - \otimes_{e_i A_i e_i} e_i A_i$ and $F_i \cong \text{Hom}_{e_i A_i e_i}(A_i e_i, -)$ (cf. Example 2.2.23). Then we have $E_i(k) \cong \Delta_i$ and set $\nabla_i = F_i(k)$.

To simplify notation we set

$$\text{Filt}(\Delta) = \text{Filt}\{\Delta_i \mid i \in I\} \qquad \text{and} \qquad \text{Filt}(\nabla) = \text{Filt}\{\nabla_i \mid i \in I\}.$$

Proposition 8.6.11. *For an object X in $\mathcal{A}$ the following are equivalent.*

(1) *X is a projective object of* Filt(∇).
(2) *X is an injective object of* Filt(Δ).
(3) *X belongs to* Filt$(\Delta) \cap$ Filt(∇).

Proof Adapt the proof of Corollary 8.1.14. □

A module satisfying the equivalent conditions of the following theorem is called a *characteristic tilting module*.

Theorem 8.6.12. *Let A be a split quasi-hereditary k-algebra. For an A-module T the following are equivalent.*

(1) *T is a projective generator of* Filt(∇).
(2) *T is an injective cogenerator of* Filt(Δ).
(3) *$T \in$ Filt$(\Delta) \cap$ Filt$(\nabla) \subseteq$ add T.*

An object satisfying these equivalent conditions is a tilting object of rep(A, k), *provided that* rep(A, k) *has finite global dimension.*

Proof First observe that $\text{Ext}^1_A(\nabla_i, \nabla_j) = 0$ for all $i \le j$ since the ∇_i identify with standard modules over A^{op}. The argument from the proof of Proposition 8.1.17 shows that $\text{Filt}(\nabla)$ admits a projective generator. Then the equivalence of (1) and (3) follows from Proposition 8.6.11. The equivalence of (2) and (3) is dual.

Now set $\mathcal{A} = \text{rep}(A, k)$ and let T be an injective cogenerator of $\text{Filt}(\Delta)$. Then

$$\text{Thick}(T) = \text{Thick}(\text{Filt}(\Delta)) = \mathcal{A}$$

since $\text{Filt}(\Delta)$ contains a projective generator of $\mathcal{A}$ and the global dimension of $\mathcal{A}$ is finite. Thus T is a tilting object of $\mathcal{A}$. □

Corollary 8.6.13. *Let A be a split quasi-hereditary k-algebra with partially ordered set of standard modules $(\Delta_i)_{i \in I}$. If $J \subseteq I$ is a coideal, then the canonical functor $\text{Filt}_A(\Delta) \to \text{Filt}_{A_J}(\Delta)$ maps a characteristic tilting module over A to a characteristic tilting module over A_J.*

Proof The functor is right adjoint to an exact functor by Lemma 8.6.2. Thus it maps an injective cogenerator to an injective cogenerator. □

We wish to identify the characteristic tilting modules over Schur algebras. Our first step is to identify the costandard modules. Set $\mathcal{A} = \text{pol}^d \mathcal{P}_k$ and write $D \colon \mathcal{A} \to \mathcal{A}$ for the duality sending X to X°.

Lemma 8.6.14. *For a partition λ of d we have $D(\mathcal{A}_\lambda) = \mathcal{A}_\lambda$.*

Proof We have

$$\mathcal{A}_\lambda = \{X \in \mathcal{A} \mid \text{Hom}(\Delta^\mu, X) = 0 \text{ for } \mu > \lambda\}$$
$$= \{X \in \mathcal{A} \mid \text{Hom}(\Gamma^\mu, X) = 0 \text{ for } \mu > \lambda\}.$$

Now observe that $\text{Hom}(\Gamma^\mu, X)$ identifies with a weight space $X(V)_\mu$ when $\mu \in \Lambda(n, d)$ and $V = k^n$; see Lemma 8.3.18. Then $X(V)^\vee_\mu \cong X^\circ(V)_\mu$ by Lemma 8.3.20, and the assertion follows. □

Lemma 8.6.15. *In $\text{pol}^d \mathcal{P}_k$ the objects $\nabla^\lambda = (\Delta^\lambda)^\circ$ (λ partition of d) are the costandard objects corresponding to the sequence of standard objects Δ^λ.*

Proof Fix a partition λ and consider the corresponding recollement (8.6.10)

$$\mathcal{A}_{\lambda^-} \;\rightleftarrows\; \mathcal{A}_\lambda \;\xrightarrow[\;F_\lambda\;]{\;E_\lambda\;}\; \mathcal{P}_k$$

with $E_\lambda(k) = \Delta^\lambda$. We claim that $F_\lambda(k) = (\Delta^\lambda)^\circ$. This follows from the previous lemma. Because $D(\mathcal{A}_\mu) = \mathcal{A}_\mu$ for all μ, the duality D maps the above diagram to another recollement

$$
\mathcal{A}_{\lambda^-} \rightleftarrows \mathcal{A}_\lambda \xrightarrow[D\circ E_\lambda\circ(-)^\vee]{D\circ F_\lambda\circ(-)^\vee} \mathcal{P}_k
$$

where the functors from left to right do not change. From this we conclude that $F_\lambda(k) = (D \circ E_\lambda)(k) = (\Delta^\lambda)^\circ$. □

Theorem 8.6.16. *Let k be a commutative noetherian ring. Then for all parameters d, n the module $T = \bigoplus_{\mu\in\Lambda(n,d)} \Lambda^{\mu'}(k^n)$ is a characteristic tilting module over $S_k(n,d)$. For $n \geq d$ the algebras $\mathrm{End}(T)$ and $S_k(n,d)^{\mathrm{op}}$ are Morita equivalent.*

Proof As before, all properties are verified in the category $\mathrm{Pol}^d \, \mathcal{P}_k$, keeping in mind that $\mathrm{rep}\, S_k(n,d)$ identifies with a full subcategory of $\mathrm{Pol}^d \, \mathcal{P}_k$ via evaluation at k^n; see Proposition 8.3.11.

We apply Theorem 8.6.12, using the fact that $\mathrm{rep}\, S_k(n,d)$ has finite global dimension by Proposition 8.6.8. For each partition μ we have $\Lambda^\mu \in \mathrm{Filt}(\Delta) \cap \mathrm{Filt}(\nabla)$ by Corollary 8.5.18. The same result yields exact sequences $0 \to U^\mu \to \Lambda^\mu \to \nabla^\mu \to 0$ in $\mathrm{Filt}(\nabla)$. It follows from Proposition 8.6.11 that $\bigoplus_\mu \Lambda^{\mu'}$ is a projective generator of $\mathrm{Filt}(\nabla)$ and therefore a characteristic tilting object for $\mathrm{pol}^d \, \mathcal{P}_k$. It remains to apply Corollary 8.6.13 in combination with Remark 8.6.7. Taking only summands of the tilting object corresponding to the coideal of all partitions in $\Lambda(n,d)$, we set $\tilde{T} = \bigoplus_{\mu\in\Lambda(n,d)} \Lambda^{\mu'}$. In fact, $\Lambda^{\mu'}(k^n) = 0$ for $\mu \notin \Lambda(n,d)$ by Lemma 8.3.23, since we have a weight space decomposition

$$
\Lambda^{\mu'}(k^n) = \bigoplus_{\lambda\in\Lambda(n,d)} \Lambda^{\mu'}(k^n)_\lambda.
$$

It follows that $T = \tilde{T}(k^n)$ is a characteristic tilting module over $S_k(n,d)$.

For $n \geq d$ we have isomorphisms

$$
S_k(n,d)^{\mathrm{op}} \xrightarrow{\sim} \mathrm{End}(\Gamma^{d,k^n}) \xrightarrow{\sim} \mathrm{End}(\Lambda^{d,k^n})
$$

and

$$
\mathrm{End}(\tilde{T}) \xrightarrow{\sim} \mathrm{End}(T),
$$

using Proposition 8.5.3 and Proposition 8.3.11. Moreover, these algebras are Morita equivalent since all partitions of d arise in $\Lambda(n,d)$ and keeping in mind the decomposition (8.3.10). □

Notes

Quasi-hereditary algebras were introduced by Scott [187] and then studied in joint work with Cline and Parshall [52, 158]; see also the work of Dlab and Ringel [64, 65]. There is a close connection between highest weight categories and quasi-hereditary algebras [52]. A motivating example arises in the study of representations of complex semisimple Lie algebras. In this context the *BGG category $\mathcal{O}$* was introduced in the early 1970s by Bernšteĭn, Gel'fand, and Gel'fand [34]; see [115] for an introduction. This category decomposes into blocks. Each block is equivalent to the module category of a finite dimensional algebra and naturally is a highest weight category.

The notion of a recollement is due to Beilinson, Bernšteĭn and Deligne [26]. The use of recollements of abelian and triangulated categories in the context of highest weight categories was first explained in [158].

Characteristic tilting modules for quasi-hereditary algebras were introduced by Ringel [173], using the correspondence between tilting objects and covariantly finite and coresolving subcategories due to Auslander and Reiten [16]. Another essential ingredient is the construction of approximation sequences via iterated universal extensions, which has been used in various contexts. In commutative algebra the argument is known as 'Serre's trick', which he used in the study of projective modules [190].

The polynomial representations of the complex general linear group $GL_n(\mathbb{C})$ were determined by Schur in his thesis [184], and these ideas were extended to infinite fields of arbitrary characteristic by Green [92]. A basic tool is the Schur functor into the category of representations of the symmetric group. Another important insight of Schur is the fact that polynomial representations of general linear groups can be identified with modules over Schur algebras [92, 185].

Strict polynomial functors were introduced by Friedlander and Suslin in [76] as a tool for studying functor cohomology. The original definition in terms of polynomial maps uses an infinite field as a basis while the definition in terms of symmetric tensors is more flexible. The precise connection between symmetric tensors and polynomial maps goes back to work of Roby [176]. A tensor product for strict polynomial functors of some fixed degree has been studied in [130].

A predecessor of the Cauchy decomposition for polynomial representations appears under the name *Cauchy identity* in the theory of symmetric functions as the expansion

$$\prod_{i,j}(1 - x_i y_j)^{-1} = \sum_{\lambda} s_\lambda(x)s_\lambda(y),$$

where the sum runs over all partitions and s_λ denotes the Schur polynomial corresponding to λ. This is attributed to Cauchy. Although he does not state the formula explicitly, it is easily deduced from Cauchy's work [48] which amounts to the computation of a double version of a Vandermonde determinant.

Symmetric functions identify with characters of polynomial representations of general linear groups, and this provides an alternative way to deduce the Cauchy identity.

The characteristic-free Cauchy decomposition of symmetric powers is due to Akin, Buchsbaum and Weyman [3]. This amounts to a direct sum decomposition

$$S(V \otimes W) = \bigoplus_\lambda \mathrm{Sch}^\lambda(V) \otimes \mathrm{Sch}^\lambda(W)$$

into irreducible $GL(V) \times GL(W)$-modules when k is a field of characteristic zero. Here, $\mathrm{Sch}^\lambda(V)$ denotes the Schur module corresponding to the partition λ, and the assignment $V \mapsto \mathrm{Sch}^\lambda(V)$ yields the corresponding Schur functor. The analogue of this decomposition for symmetric tensors involves as factors the Weyl modules. These were first defined by Carter and Lusztig [47]. In particular, their work contains the result that any Weyl module over k admits a k-basis. Similar decompositions were obtained independently by de Concini, Eisenbud and Procesi [61], and also by Doubilet, Rota and Stein [69].

The fact that Schur algebras are quasi-hereditary follows from work of Donkin [66, 67]. He considers an arbitrary commutative ring k and defines a k-algebra $S_k(\pi)$ for each finite saturated set π of dominant weights of a semisimple complex finite dimensional Lie algebra $\mathfrak{g}$, which is free of finite rank over k. These algebras are all quasi-hereditary and for some particular choices of π and $\mathfrak{g}$ one obtains the Schur algebra. The filtration of $S^d(V \otimes W)$ amounts in this context to a *good filtration* of the injective $S_k(\pi)$-modules. For some further proofs of the fact that Schur algebras are quasi-hereditary, see Parshall [157] and Green [93]. The characteristic tilting modules over Schur algebras were identified by Donkin [68].

Finite global dimension is a well-known fact for quasi-hereditary algebras [65]. For Schur algebras Akin and Buchsbaum [2] as well as Donkin [67] gave independent proofs. Precise values are given in a beautiful paper of Totaro [198].

PART THREE

DERIVED EQUIVALENCES

9

Derived Equivalences

Contents

9.1	**Differential Graded Algebras**		**298**
	Differential Graded Algebras and Modules		298
	Perfect dg Modules		300
	Filtering dg Modules		302
	Compact Objects and Dévissage		304
	Compactly Generated Triangulated Categories		305
	Quasi-isomorphisms of dg Algebras		306
	Tilting Objects		307
	The Stable Homotopy Category Is Not Algebraic		308
9.2	**Derived Equivalences**		**309**
	Compact Objects		309
	Homologically Finite Objects		309
	A Morita Theorem for Derived Categories		310
	Homologically Perfect Objects		312
	Coherent Rings		313
	Gorenstein Algebras		316
	Serre Duality		316
9.3	**Finite Global Dimension**		**318**
	T-Structures		319
	Tilting for Unbounded Derived Categories		323
	Locally Noetherian Grothendieck Categories		325
	Tilting for Bounded Derived Categories		326
Notes			**328**

This chapter is devoted to a Morita theory for derived categories. We discuss the question when derived categories are equivalent as triangulated categories. It is convenient to work with differential graded algebras and their derived categories; the notion of a tilting object also plays a crucial role. For the derived category of a module category we need to look at various subcategories, asking in each case for an intrinsic description. The final section is devoted to proving that tilting preserves finite global dimension.

9.1 Differential Graded Algebras

We introduce differential graded algebras and study their derived categories. These are algebraic triangulated categories, and basically all algebraic triangulated categories arise in this way. Then we discuss tilting objects in algebraic triangulated categories and show that any tilting object T induces a triangle equivalence that identifies the triangulated category with the derived category of the endomorphism ring of T. This yields a Morita theorem for derived categories.

The construction of the derived category of a differential graded algebra is analogous to the construction of the derived category of an abelian category. This means that terminology, notation, and arguments are very similar. So we keep the exposition short and refer to the previous chapter on derived categories for more details.

Differential Graded Algebras and Modules

A *differential graded algebra* or *dg algebra* is a $\mathbb{Z}$-graded associative algebra

$$A = \bigoplus_{n \in \mathbb{Z}} A^n$$

over some fixed commutative ring k, together with a *differential* $d \colon A \to A$, that is, a homogeneous k-linear map of degree $+1$ satisfying $d^2 = 0$ and the *Leibniz rule*

$$d(xy) = d(x)y + (-1)^n x d(y) \quad \text{for all} \quad x \in A^n \quad \text{and} \quad y \in A.$$

A *dg A-module* is a $\mathbb{Z}$-graded (right) A-module X, together with a *differential* $d \colon X \to X$, that is, a homogeneous k-linear map of degree one satisfying $d^2 = 0$ and the *Leibniz rule*

$$d(xy) = d(x)y + (-1)^n x d(y) \quad \text{for all} \quad x \in X^n \quad \text{and} \quad y \in A.$$

A morphism of dg A-modules is an A-linear map which is homogeneous of degree zero and commutes with the differential. We denote by $\mathbf{C}(A)$ the category of dg A-modules.

Example 9.1.1. (1) An associative algebra Λ can be viewed as a dg algebra A if one defines $A^0 = \Lambda$ and $A^n = 0$ otherwise. In this case $\mathbf{C}(A) = \mathbf{C}(\mathrm{Mod}\,\Lambda)$.

(2) Let X, Y be complexes in some additive category $\mathcal{C}$. Define a new complex $\mathcal{H}om_{\mathcal{C}}(X, Y)$ as follows. The nth component is

$$\prod_{p \in \mathbb{Z}} \mathrm{Hom}_{\mathcal{C}}(X^p, Y^{p+n})$$

and the differential is given by

$$d''(\phi^p) = d_Y \circ \phi^p - (-1)^n \phi^{p+1} \circ d_X.$$

Note that

$$H'' \mathcal{H}om_\mathcal{C}(X, Y) \cong \mathrm{Hom}_{\mathbf{K}(\mathcal{C})}(X, \Sigma'' Y) \tag{9.1.2}$$

because $\mathrm{Ker}\, d''$ identifies with $\mathrm{Hom}_{\mathbf{C}(\mathcal{C})}(X, \Sigma'' Y)$ and $\mathrm{Im}\, d''^{-1}$ identifies with the ideal of null-homotopic maps $X \to \Sigma'' Y$. The composition of graded maps yields a dg algebra structure for

$$\mathcal{E}nd_\mathcal{C}(X) = \mathcal{H}om_\mathcal{C}(X, X)$$

and $\mathcal{H}om_\mathcal{C}(X, Y)$ is a dg module over $\mathcal{E}nd_\mathcal{C}(X)$.

A morphism $\phi \colon X \to Y$ of dg A-modules is *null-homotopic* if there is a morphism $\rho \colon X \to Y$ of graded A-modules which is homogeneous of degree -1 such that $\phi = d_Y \circ \rho + \rho \circ d_X$. The null-homotopic morphisms form an ideal and the *homotopy category* $\mathbf{K}(A)$ is the quotient of $\mathbf{C}(A)$ with respect to this ideal. The homotopy category carries a triangulated structure which is defined as before for the homotopy category $\mathbf{K}(\mathcal{A})$ of an additive category $\mathcal{A}$; see Lemma 4.1.1.

A morphism $X \to Y$ of dg A-modules is a *quasi-isomorphism* if it induces isomorphisms $H'' X \to H'' Y$ for all $n \in \mathbb{Z}$. A dg A-module X is *acyclic* if $H'' X = 0$ for all $n \in \mathbb{Z}$. As in the case of an abelian category, a morphism is a quasi-isomorphism if and only if its cone is acyclic; see Lemma 4.1.4.

Let Qis denote the class of all quasi-isomorphisms in $\mathbf{C}(A)$ and denote by $\mathbf{Ac}(A)$ the full subcategory of all acyclic dg A-modules. Then $(\mathbf{C}(A), \mathbf{Ac}(A))$ is a Frobenius pair; see Lemma 4.1.9. The *derived category* of the dg algebra A is by definition the localisation

$$\mathbf{D}(A) = \mathbf{C}(A)[\mathrm{Qis}^{-1}].$$

This identifies with the derived category of the Frobenius pair $(\mathbf{C}(A), \mathbf{Ac}(A))$ and the canonical functor $\mathbf{C}(A) \to \mathbf{D}(A)$ induces a triangle equivalence

$$\mathbf{K}(A)/\mathbf{Ac}(A) \xrightarrow{\sim} \mathbf{D}(A).$$

Lemma 9.1.3. *Let X be a dg A-module and $n \in \mathbb{Z}$. Then we have natural isomorphisms*

$$H'' X \cong \mathrm{Hom}_{\mathbf{K}(A)}(A, \Sigma'' X) \cong \mathrm{Hom}_{\mathbf{D}(A)}(A, \Sigma'' X).$$

Proof We identify X'' with the A-linear maps $A \to \Sigma'' X$ that are homogeneous of degree zero. Then the morphisms $A \to \Sigma'' X$ that commute with the differential identify with $Z^n X = \operatorname{Ker} d|_{X^n}$, and taking morphisms up to null-homotopic maps gives $H^n X \cong \operatorname{Hom}_{\mathbf{K}(A)}(A, \Sigma^n X)$. Thus $\operatorname{Hom}_{\mathbf{K}(A)}(A, X) = 0$ when X is acyclic, and the second isomorphism follows from Lemma 3.2.4. $\square$

Remark 9.1.4. A *differential graded category* or *dg category* $\mathcal{A}$ is by definition a dg algebra with several objects. More precisely, $\mathcal{A}$ is a $\mathbb{Z}$-graded k-linear category, that is,

$$\operatorname{Hom}_{\mathcal{A}}(A, B) = \bigoplus_{n \in \mathbb{Z}} \operatorname{Hom}_{\mathcal{A}}(A, B)^n$$

is a $\mathbb{Z}$-graded k-module for all A, B in $\mathcal{A}$, and the composition maps

$$\operatorname{Hom}_{\mathcal{A}}(B, C) \times \operatorname{Hom}_{\mathcal{A}}(A, B) \longrightarrow \operatorname{Hom}_{\mathcal{A}}(A, C)$$

are bilinear and homogeneous of degree zero. In addition, there are *differentials* $d \colon \operatorname{Hom}_{\mathcal{A}}(A, B) \to \operatorname{Hom}_{\mathcal{A}}(A, B)$ for all A, B in $\mathcal{A}$, that is, homogeneous k-linear maps of degree one satisfying $d^2 = 0$ and the Leibniz rule.

A dg algebra can be viewed as a dg category with one object. Conversely, a dg category $\mathcal{A}$ with one object A gives a dg algebra $\operatorname{Hom}_{\mathcal{A}}(A, A)$.

A *dg $\mathcal{A}$-module* is the analogue of a dg module over a dg algebra, that is, a graded functor $X \colon \mathcal{A}^{\mathrm{op}} \to \operatorname{GrMod} k$ into the category of $\mathbb{Z}$-graded k-modules, together with a *differential* $d \colon X \to X$ given by homogeneous k-linear maps of degree one satisfying $d^2 = 0$ and the Leibniz rule.

For an essentially small dg category $\mathcal{A}$, the category of dg $\mathcal{A}$-modules gives rise to the derived category $\mathbf{D}(\mathcal{A})$ which is obtained by formally inverting the class of quasi-isomorphisms. We omit details and refer instead to [121].

Perfect dg Modules

Let A be a dg algebra. A dg A-module is called *perfect* if it belongs to the thick subcategory of $\mathbf{D}(A)$ that is generated by the dg module A. We denote by $\mathbf{D}^{\mathrm{perf}}(A)$ the full subcategory of $\mathbf{D}(A)$ whose objects are the perfect dg A-modules.

We wish to characterise the triangulated categories that are triangle equivalent to $\mathbf{D}^{\mathrm{perf}}(A)$ for some dg algebra A. A triangulated category is called *algebraic* if it is triangle equivalent to the stable category of a Frobenius category.

Proposition 9.1.5. *For a triangulated category $\mathcal{T}$ the following are equivalent.*

(1) *$\mathcal{T}$ is algebraic, idempotent complete, and $\mathcal{T} = \operatorname{Thick}(X)$ for some $X \in \mathcal{T}$.*

(2) *There is a triangle equivalence $\mathcal{T} \xrightarrow{\sim} \mathbf{D}^{\mathrm{perf}}(A)$ for some dg algebra A.*

In that case there is an equivalence that takes X to A.

Proof It follows from the construction of $\mathbf{D}(A)$ that $\mathbf{D}^{\mathrm{perf}}(A)$ is algebraic. Clearly, $\mathbf{K}(A)$ is algebraic. The canonical functor $\mathbf{K}(A) \to \mathbf{D}(A)$ is fully faithful when restricted to $\mathrm{Thick}(A)$, by Lemma 3.1.8 and Lemma 9.1.3. So it remains to observe that any triangulated subcategory of an algebraic triangulated category is again algebraic.

The other direction follows from the lemma below.   □

Let $\mathcal{A}$ be a Frobenius category and denote by $\mathcal{P}$ the full subcategory of projective (and injective) objects. We write $\mathbf{K}_{\mathrm{ac}}(\mathcal{P})$ for the full subcategory of $\mathbf{K}(\mathcal{P})$ consisting of the complexes in $\mathcal{P}$ that are acyclic when viewed in $\mathcal{A}$. Taking a complex X to $Z^0 X$ induces a triangle equivalence $\mathbf{K}_{\mathrm{ac}}(\mathcal{P}) \overset{\sim}{\to} \mathrm{St}\,\mathcal{A}$ (Proposition 4.4.18). The quasi-inverse takes an object X of $\mathcal{A}$ to a *complete resolution* which we denote by $\bar{X}$.

An additive functor $F\colon \mathcal{C} \to \mathcal{D}$ is an equivalence *up to direct summands* if F is fully faithful and every object in $\mathcal{D}$ is isomorphic to a direct summand of an object in the image of F.

Lemma 9.1.6. *Let $\mathcal{T}$ be an algebraic triangulated category and fix an object X. Identifying $\mathcal{T} = \mathrm{St}\,\mathcal{A}$ for some Frobenius category $\mathcal{A}$, we set $A = \mathcal{E}nd_{\mathcal{A}}(\bar{X})$. Then the functor $\mathcal{H}om_{\mathcal{A}}(\bar{X}, -)\colon \mathbf{K}(\mathcal{A}) \to \mathbf{D}(A)$ induces up to direct summands a triangle equivalence*

$$\mathrm{Thick}(X) \overset{\sim}{\longrightarrow} \mathbf{D}^{\mathrm{perf}}(A).$$

Proof Set $F = \mathcal{H}om_{\mathcal{A}}(\bar{X}, -)$. The functor takes $\bar{X}$ to A. Combining the isomorphism (9.1.2) and Lemma 9.1.3 gives for all $n \in \mathbb{Z}$

$$\mathrm{Hom}_{\mathbf{K}(\mathcal{A})}(\bar{X}, \Sigma^n \bar{X}) \cong H^n \mathcal{H}om_{\mathcal{A}}(\bar{X}, \bar{X}) \cong H^n A \cong \mathrm{Hom}_{\mathbf{D}(A)}(A, \Sigma^n A).$$

Thus F is fully faithful on $\{\Sigma^n \bar{X} \mid n \in \mathbb{Z}\}$. Because F is exact, it follows by dévissage that F is fully faithful on $\mathrm{Thick}(\bar{X})$; see Lemma 3.1.8.   □

Example 9.1.7. Let Λ be a ring. Then the inclusion $\mathrm{proj}\,\Lambda \to \mathrm{Mod}\,\Lambda$ induces a triangle equivalence $\mathbf{D}^b(\mathrm{proj}\,\Lambda) \overset{\sim}{\to} \mathbf{D}^{\mathrm{perf}}(\Lambda)$. If Λ is right coherent and of finite global dimension, then $\mathbf{D}^b(\mathrm{mod}\,\Lambda) \overset{\sim}{\to} \mathbf{D}^{\mathrm{perf}}(\Lambda)$.

Remark 9.1.8. There is an analogue of Proposition 9.1.5 for any essentially small triangulated category $\mathcal{T}$, without the assumption that $\mathcal{T} = \mathrm{Thick}(X)$ for a single object X. Then $\mathcal{T} \overset{\sim}{\to} \mathbf{D}^{\mathrm{perf}}(\mathcal{A})$ for some dg category $\mathcal{A}$; see Remark 9.1.4.

Filtering dg Modules

Let A be a dg algebra. We describe a method of 'filtering' a dg A-module such that each subquotient is a direct sum of suspensions of A.

Lemma 9.1.9. *Let X be a dg A-module. Then there exists a sequence of morphisms $\mu_n \colon X_n \to X_{n+1}$ and $\phi_n \colon X_n \to X$ ($n \geq 0$) in $\mathbf{K}(A)$ such that*

(1) *$X_0 = 0$ and the cone of μ_n is a coproduct of objects $\Sigma^i A$ with $|i| \leq n + 1$,*

(2) *$\phi_n = \phi_{n+1}\mu_n$, and*

(3) *the ϕ_n induce a quasi-isomorphism $\mathrm{hocolim}_n X_n \to X$.*

Proof   We construct inductively exact triangles

$$U_n \longrightarrow X_n \xrightarrow{\phi_n} X \longrightarrow \Sigma U_n.$$

Set $X_0 = 0$ and $\phi_0 = 0$. Each $x \in H^i(U_n)$ corresponds to a morphism $\Sigma^{-i} A \to U_n$ by Lemma 9.1.3. Composing with $U_n \to X_n$ yields for $n \geq 0$ exact triangles making the following diagram commutative.

$$
\begin{array}{ccccccc}
\coprod_{\substack{|i| \leq n \\ x \in H^i(U_n)}} \Sigma^{-i} A & \longrightarrow & U_n & \longrightarrow & U_{n+1} & \longrightarrow & \Sigma\bigl(\coprod_{\substack{|i| \leq n \\ x \in H^i(U_n)}} \Sigma^{-i} A\bigr) \\
\Big\| & & \Big\downarrow & & \Big\downarrow & & \Big\| \\
\coprod_{\substack{|i| \leq n \\ x \in H^i(U_n)}} \Sigma^{-i} A & \longrightarrow & X_n & \xrightarrow{\ \mu_n\ } & X_{n+1} & \longrightarrow & \Sigma\bigl(\coprod_{\substack{|i| \leq n \\ x \in H^i(U_n)}} \Sigma^{-i} A\bigr) \\
 & & \Big\downarrow{\scriptstyle \phi_n} & & \Big\downarrow{\scriptstyle \phi_{n+1}} & & \\
 & & X & =\!=\!= & X & & \\
 & & \Big\downarrow & & \Big\downarrow & & \\
 & & \Sigma U_n & \longrightarrow & \Sigma U_{n+1} & &
\end{array}
$$

Thus we obtain a morphism $\phi \colon \mathrm{hocolim}_n X_n \to X$. The construction of the ϕ_n yields for $|i| < n$ a commutative diagram

$$
\begin{array}{ccccccc}
H^i U_n & \longrightarrow & H^i X_n & \longrightarrow & H^i X & \longrightarrow & H^{i+1} U_n \\
\Big\downarrow{\scriptstyle 0} & & \Big\downarrow & & \Big\| & & \Big\downarrow{\scriptstyle 0} \\
H^i U_{n+1} & \longrightarrow & H^i X_{n+1} & \longrightarrow & H^i X & \longrightarrow & H^{i+1} U_{n+1}
\end{array}
$$

with exact rows, and therefore an isomorphism $\mathrm{colim}_n H^i(X_n) \xrightarrow{\sim} H^i X$. On the other hand, the defining exact triangle of $\mathrm{hocolim}_n X_n$ induces for each i an isomorphism

$$\mathop{\mathrm{colim}}_{n} H^i(X_n) \xrightarrow{\sim} H^i(\mathop{\mathrm{hocolim}}_{n} X_n)$$

by Lemma 3.4.3. It follows that ϕ induces an isomorphism

$$H^i(\operatorname*{hocolim}_n X_n) \xrightarrow{\sim} H^i X. \qquad \square$$

Let us write $\mathbf{K}_{\mathrm{proj}}(A)$ for the full subcategory of $\mathbf{K}(A)$ consisting of all K-projective objects, where a dg module P is *K-projective* if $\operatorname{Hom}_{\mathbf{K}(A)}(P, X) = 0$ for each acyclic module X. A *K-projective resolution* of a dg module X is a quasi-isomorphism $P \to X$ such that P is K-projective. For the formal properties of K-projective resolutions, see the dual of Proposition 4.3.1.

We set $\mathbf{p}(X) = \operatorname{hocolim}_n X_n$ for the object constructed in the above lemma and observe that the morphism $\mathbf{p}(X) \to X$ is a K-projective resolution, since $\mathbf{K}_{\mathrm{proj}}(A)$ is a thick subcategory containing A and closed under arbitrary coproducts. The following proposition provides an intrinsic description of this resolution and a convenient description of the derived category $\mathbf{D}(A)$.

Proposition 9.1.10. *Every dg A-module X admits a K-projective resolution* $\mathbf{p}(X) \to X$, *and the assignment $X \mapsto \mathbf{p}(X)$ induces a right adjoint for the inclusion $\mathbf{K}_{\mathrm{proj}}(A) \to \mathbf{K}(A)$. The K-projective resolution equals the counit and can be extended to a functorial exact triangle*

$$\mathbf{p}(X) \longrightarrow X \longrightarrow \mathbf{a}(X) \longrightarrow \Sigma\mathbf{p}(X)$$

such that $\mathbf{a}(X)$ lies in $\mathbf{Ac}(A)$. This yields the following colocalisation sequence of exact functors

$$\mathbf{K}_{\mathrm{proj}}(A) \underset{\mathbf{p}}{\overset{\longrightarrow}{\longleftarrow}} \mathbf{K}(A) \underset{\longleftarrow}{\overset{\mathbf{a}}{\longrightarrow}} \mathbf{Ac}(A)$$

and therefore the canonical functor $\mathbf{K}(A) \to \mathbf{D}(A)$ restricts to a triangle equivalence $\mathbf{K}_{\mathrm{proj}}(A) \xrightarrow{\sim} \mathbf{D}(A)$.

Proof We complete $\mathbf{p}(X) \to X$ to an exact triangle

$$\Sigma^{-1}\mathbf{a}(X) \longrightarrow \mathbf{p}(X) \longrightarrow X \longrightarrow \mathbf{a}(X).$$

Clearly, $\mathbf{a}(X)$ is acyclic, and therefore $\operatorname{Hom}_{\mathbf{K}(A)}(X', \mathbf{a}(X)) = 0$ for all $X' \in \mathbf{K}_{\mathrm{proj}}(A)$. Thus $\mathbf{p}(X) \to X$ induces for all $X' \in \mathbf{K}_{\mathrm{proj}}(A)$ a bijection

$$\operatorname{Hom}_{\mathbf{K}(A)}(X', \mathbf{p}(X)) \xrightarrow{\sim} \operatorname{Hom}_{\mathbf{K}(A)}(X', X).$$

This means that $X \mapsto \mathbf{p}(X)$ provides a right adjoint. Also, we have $\mathbf{K}_{\mathrm{proj}}(A) = {}^{\perp}\mathbf{Ac}(A)$, and then the rest of the assertion follows from Proposition 3.2.8. $\square$

Corollary 9.1.11. *The morphisms between two objects in $\mathbf{D}(A)$ form a set.* $\square$

Corollary 9.1.12. *Let $\mathcal{C} \subseteq \mathbf{D}(A)$ be a triangulated subcategory of $\mathbf{D}(A)$ that is closed under all coproducts and contains A. Then $\mathcal{C} = \mathbf{D}(A)$. If Λ is a ring, then every object in $\mathbf{D}(\Lambda)$ is a homotopy colimit of objects from $\mathbf{K}^b(\operatorname{Proj}\Lambda)$.* $\square$

Compact Objects and Dévissage

An object X in a triangulated category $\mathcal{T}$ is called *compact* if for any morphism $\phi\colon X \to \coprod_{i \in I} Y_i$ in $\mathcal{T}$ there is a finite set $J \subseteq I$ such that ϕ factors through $\coprod_{i \in J} Y_i$. An equivalent condition is that the canonical map

$$\coprod_{i \in I} \mathrm{Hom}_{\mathcal{T}}(X, Y_i) \longrightarrow \mathrm{Hom}_{\mathcal{T}}\left(X, \coprod_{i \in I} Y_i\right)$$

is bijective for all coproducts $\coprod_{i \in I} Y_i$ in $\mathcal{T}$.

Proposition 9.1.13. *Let A be a dg algebra and X a dg A-module. Then X is a compact object in $\mathbf{D}(A)$ if and only if X is perfect.*

Proof We have $\mathbf{D}^{\mathrm{perf}}(A) = \mathrm{Thick}(A)$ by definition. The module A is compact since $\mathrm{Hom}_{\mathbf{D}(A)}(A, X) \cong H^0 X$ for all $X \in \mathbf{D}(A)$. Thus every perfect dg module is compact, since the compact objects form a thick subcategory. Now suppose that $X \in \mathbf{D}(A)$ is compact. Then we combine the description of $X \cong \mathrm{hocolim}\, X_n$ as a homotopy colimit in Lemma 9.1.9 with Proposition 3.4.13. More precisely, id_X factors through $X_n \to X$ for some n by Lemma 3.4.3, and therefore $X \in \mathrm{Thick}(A)$ since X_n is an extension of coproducts of suspensions of A. $\qquad\square$

Let $F\colon \mathcal{T} \to \mathcal{U}$ be an exact functor between triangulated categories and suppose that F preserves coproducts. Let $\mathcal{T}_0 \subseteq \mathcal{T}$ be a full triangulated subcategory that consists of compact objects and denote by $\mathcal{T}_1 \subseteq \mathcal{T}$ the smallest full triangulated subcategory that contains $\mathcal{T}_0$ and is closed under coproducts.

For objects X, Y in $\mathcal{T}$ we set

$$\mathrm{Hom}_{\mathcal{T}}^*(X, Y) = \bigoplus_{n \in \mathbb{Z}} \mathrm{Hom}_{\mathcal{T}}(X, \Sigma^n Y).$$

Lemma 9.1.14. *Suppose the restriction $F|_{\mathcal{T}_0}$ is fully faithful and $F(\mathcal{T}_0)$ consists of compact objects. Then $F|_{\mathcal{T}_1}$ is fully faithful.*

Proof Fix $X \in \mathcal{T}_0$. Then

$$\mathcal{T}_X = \{Y \in \mathcal{T} \mid \mathrm{Hom}_{\mathcal{T}}^*(X, Y) \xrightarrow{\sim} \mathrm{Hom}_{\mathcal{U}}^*(FX, FY)\}$$

is a triangulated subcategory that contains $\mathcal{T}_0$ and is closed under coproducts; thus it contains $\mathcal{T}_1$. Now fix $Y \in \mathcal{T}_1$. Then

$$\mathcal{T}^Y = \{X \in \mathcal{T} \mid \mathrm{Hom}_{\mathcal{T}}^*(X, Y) \xrightarrow{\sim} \mathrm{Hom}_{\mathcal{U}}^*(FX, FY)\}$$

contains $\mathcal{T}_0$ by our first observation. Also, $\mathcal{T}^Y$ is a triangulated subcategory and is closed under coproducts; thus it contains $\mathcal{T}_1$. $\qquad\square$

Compactly Generated Triangulated Categories

Let $\mathcal{T}$ be a triangulated category that admits arbitrary coproducts. Then $\mathcal{T}$ is *compactly generated* if $\mathcal{T} = \mathrm{Loc}(\mathfrak{X})$ for a set $\mathfrak{X}$ of compact objects.

We characterise the algebraic triangulated categories that are generated by a compact object. This is an analogue of Proposition 9.1.5.

Proposition 9.1.15. *For a triangulated category $\mathcal{T}$ which admits arbitrary coproducts the following are equivalent.*

(1) *$\mathcal{T}$ is algebraic and $\mathcal{T} = \mathrm{Loc}(X)$ for some compact object $X \in \mathcal{T}$.*

(2) *There is a triangle equivalence $\mathcal{T} \xrightarrow{\sim} \mathbf{D}(A)$ for some dg algebra A.*

In that case there is an equivalence that takes X to A.

Proof (1) $\Rightarrow$ (2): Fix a Frobenius category $\mathcal{A}$ and suppose $\mathcal{T} = \mathrm{St}\,\mathcal{A}$. In fact, we identify each $X \in \mathcal{T}$ with a complete resolution $\bar{X}$ in $\mathbf{K}(\mathcal{A})$ (Proposition 4.4.18). Now fix a compact object $X \in \mathcal{T}$ and set $A = \mathcal{E}nd_{\mathcal{A}}(\bar{X})$. We claim that the functor $F = \mathcal{H}om_{\mathcal{A}}(\bar{X}, -)\colon \mathbf{K}(\mathcal{A}) \to \mathbf{D}(A)$ induces a triangle equivalence

$$\mathrm{Loc}(X) \xrightarrow{\sim} \mathbf{D}(A).$$

The functor takes $\bar{X}$ to A. Combining the isomorphism (9.1.2) and Lemma 9.1.3 gives for all $n \in \mathbb{Z}$

$$\mathrm{Hom}_{\mathbf{K}(\mathcal{A})}(\bar{X}, \Sigma^n \bar{X}) \cong H^n \mathcal{H}om_{\mathcal{A}}(\bar{X}, \bar{X}) \cong H^n A \cong \mathrm{Hom}_{\mathbf{D}(A)}(A, \Sigma^n A).$$

Thus F is fully faithful on $\{\Sigma^n \bar{X} \mid n \in \mathbb{Z}\}$. The functor F is exact and preserves all coproducts since X is compact. It follows by dévissage that F is fully faithful on $\mathrm{Loc}(\bar{X})$; see Lemma 3.1.8 and Lemma 9.1.14. The functor F is essentially surjective since $\mathbf{D}(A) = \mathrm{Loc}(A)$ by Corollary 9.1.12.

(2) $\Rightarrow$ (1): Let A be a dg algebra. Clearly, $\mathbf{K}(A)$ is algebraic. The canonical functor $\mathbf{K}(A) \to \mathbf{D}(A)$ restricts to an equivalence $\mathrm{Loc}(A) \xrightarrow{\sim} \mathbf{D}(A)$ by Corollary 9.1.12. Thus $\mathbf{D}(A)$ is algebraic, since any triangulated subcategory of an algebraic triangulated category is again algebraic. It remains to observe that $A \in \mathbf{D}(A)$ is a compact object, since $\mathrm{Hom}_{\mathbf{D}(A)}(A, X) \cong H^0 X$ for all $X \in \mathbf{D}(A)$. □

Remark 9.1.16. There is an analogue of Proposition 9.1.15 where the assumption $\mathcal{T} = \mathrm{Loc}(X)$ for a single object X is replaced by $\mathcal{T} = \mathrm{Loc}(\mathfrak{X})$ for a set $\mathfrak{X}$ of compact objects. Then $\mathcal{T} \xrightarrow{\sim} \mathbf{D}(\mathcal{A})$ for some dg category $\mathcal{A}$; see Remark 9.1.4.

Quasi-isomorphisms of dg Algebras

Let $\phi\colon A \to B$ be a morphism of dg algebras. Then ϕ induces via restriction of scalars a functor $\phi^*\colon \mathbf{C}(B) \to \mathbf{C}(A)$ because every dg B-module becomes a dg A-module via ϕ. This functor admits a left adjoint:

$$\mathbf{C}(A) \underset{\mathrm{Hom}_B(B,-)}{\overset{-\otimes_A B}{\rightleftarrows}} \mathbf{C}(B).$$

For $X \in \mathbf{C}(A)$, the (usual ungraded) tensor product $X \otimes_A B$ admits a grading; the degree n part is the quotient of $\bigoplus_{p+q=n} X^p \otimes_{A^0} B^q$ modulo the submodule generated by $x \otimes ay - xa \otimes y$ where $x \in X^p$, $y \in B^q$, and $a \in A^{n-p-q}$. The differential is given by

$$d(x \otimes y) = d(x) \otimes y + (-1)^i x \otimes d(y)$$

for $x \in X^i$ and $y \in B$. Clearly, ϕ^* preserves quasi-isomorphisms and induces therefore a functor

$$\phi^* = \mathrm{RHom}_B(B, -)\colon \mathbf{D}(B) \longrightarrow \mathbf{D}(A).$$

The left adjoint is the composite

$$\phi_! = - \otimes_A^L B \colon \mathbf{D}(A) \overset{\sim}{\longrightarrow} \mathbf{K}_{\mathrm{proj}}(A) \overset{-\otimes_A B}{\longrightarrow} \mathbf{K}(B) \overset{\mathrm{can}}{\longrightarrow} \mathbf{D}(B).$$

A dg algebra morphism $A \to B$ is by definition a *quasi-isomorphism* if it induces an isomorphism $H^n A \overset{\sim}{\to} H^n B$ for all n.

Lemma 9.1.17. *For a morphism of dg algebras $\phi\colon A \to B$ the following are equivalent.*

(1) *ϕ is a quasi-isomorphism.*
(2) *$\phi_!$ induces a triangle equivalence $\mathbf{D}^{\mathrm{perf}}(A) \overset{\sim}{\to} \mathbf{D}^{\mathrm{perf}}(B)$.*
(3) *$\phi_!$ induces a triangle equivalence $\mathbf{D}(A) \overset{\sim}{\to} \mathbf{D}(B)$.*

Proof (1) ⇔ (2): The functor $\phi_!$ induces for all n a map

$$H^n(A) \cong \mathrm{Hom}_{\mathbf{D}(A)}(A, \Sigma^n A) \longrightarrow \mathrm{Hom}_{\mathbf{D}(B)}(B, \Sigma^n B) \cong H^n B \qquad (9.1.18)$$

where the isomorphisms at both ends are from Lemma 9.1.3. This is an isomorphism if and only if ϕ is a quasi-isomorphism. In that case $\phi_!$ identifies $\mathbf{D}^{\mathrm{perf}}(A) = \mathrm{Thick}(A)$ with $\mathbf{D}^{\mathrm{perf}}(B) = \mathrm{Thick}(B)$, by Lemma 3.1.8.

(1) ⇒ (3): Consider for each $X \in \mathbf{D}(A)$ the unit $\eta_X\colon X \to \phi^*\phi_!(X)$. The fact that ϕ is a quasi-isomorphism means that η_A is an isomorphism. The functors ϕ^* and $\phi_!$ are exact and preserve coproducts. Thus the full subcategory

$$\{X \in \mathbf{D}(A) \mid \Sigma^n X \overset{\sim}{\to} \phi^*\phi_!(\Sigma^n X) \text{ for all } n \in \mathbb{Z}\}$$

is a triangulated category that is closed under coproducts and contains A. It follows from Corollary 9.1.12 that $\mathrm{id}_{\mathbf{D}(A)} \xrightarrow{\sim} \phi^* \phi_!$. Therefore $\phi_!$ is fully faithful. The essential image of $\phi_!$ contains $B = \phi_!(A)$ and is a triangulated subcategory of $\mathbf{D}(B)$ that is closed under coproducts. Thus $\phi_!$ is essentially surjective on objects, again by Corollary 9.1.12, and therefore a triangle equivalence.

(3) $\Rightarrow$ (1): Use the map (9.1.18). $\qquad\qquad\qquad\qquad\qquad\qquad\square$

Example 9.1.19. Let A be a dg algebra. We denote by $H^0 A$ the dg algebra B with $B^0 = H^0 A$ and $B^n = 0$ for $n \neq 0$. Also, let $\tau_{\leq 0} A$ denote the dg subalgebra $B \subseteq A$ obtained via truncation, so $B^n = A^n$ for $n < 0$, $B^0 = Z^0 A$, and $B^n = 0$ for $n > 0$. We have canonical morphisms

$$A \hookleftarrow \tau_{\leq 0} A \twoheadrightarrow H^0 A$$

and these are quasi-isomorphisms when $H^* A = H^0 A$.

Tilting Objects

Let $\mathcal{T}$ be a triangulated category. We consider two settings for an object $T \in \mathcal{T}$ to be a tilting object, depending on whether the category $\mathcal{T}$ is essentially small or not.

Suppose first that $\mathcal{T}$ is essentially small. Then $T \in \mathcal{T}$ is called a *tilting object* if $\mathrm{Hom}_{\mathcal{T}}(T, \Sigma^n T) = 0$ for all $n \neq 0$ and $\mathrm{Thick}(T) = \mathcal{T}$.

Proposition 9.1.20. *Let $\mathcal{T}$ be an essentially small algebraic triangulated category. Then an object T in $\mathcal{T}$ is a tilting object if and only if there exists a ring A and a triangle equivalence up to direct summands*

$$\mathcal{T} \xrightarrow{\sim} \mathbf{D}^{\mathrm{perf}}(A)$$

that sends T to A.

Proof Suppose that T is a tilting object and set $A = \mathrm{End}_{\mathcal{T}}(T)$. Lemma 9.1.6 yields a triangle equivalence $\mathcal{T} \xrightarrow{\sim} \mathbf{D}^{\mathrm{perf}}(E)$ for some dg algebra $E = \mathcal{E}nd_A(\bar{T})$, and we have $H^* E \cong A$. Now apply Lemma 9.1.17 with Example 9.1.19 to get the equivalence $\mathcal{T} \xrightarrow{\sim} \mathbf{D}^{\mathrm{perf}}(A)$.

For the other implication observe that A is a tilting object of $\mathbf{D}^{\mathrm{perf}}(A)$. This property is preserved under a triangle equivalence. $\qquad\qquad\qquad\qquad\square$

Remark 9.1.21. Let $\mathcal{T}$ be an essentially small triangulated category. Then $T \in \mathcal{T}$ is a tilting object if and only if T is a tilting object in $\mathcal{T}^{\mathrm{op}}$. For $A = \mathrm{End}_{\mathcal{T}}(T)$ the functor $\mathrm{Hom}_A(-, A)$ induces a triangle equivalence

$$\mathbf{D}^{\mathrm{perf}}(A)^{\mathrm{op}} \xrightarrow{\sim} \mathbf{D}^{\mathrm{perf}}(A^{\mathrm{op}}).$$

Now suppose that $\mathcal{T}$ admits arbitrary coproducts. Then $T \in \mathcal{T}$ is called a *tilting object* if T is a compact object, $\mathrm{Hom}_{\mathcal{T}}(T, \Sigma^n T) = 0$ for all $n \neq 0$, and $\mathrm{Loc}(T) = \mathcal{T}$.

Proposition 9.1.22. *Let $\mathcal{T}$ be an algebraic triangulated category which admits arbitrary coproducts. Then an object T in $\mathcal{T}$ is a tilting object if and only if there exists a ring A and a triangle equivalence*

$$\mathcal{T} \xrightarrow{\sim} \mathbf{D}(A)$$

that sends T to A.

Proof   Suppose that T is a tilting object and set $A = \mathrm{End}_{\mathcal{T}}(T)$. Proposition 9.1.15 yields a triangle equivalence $\mathcal{T} \xrightarrow{\sim} \mathbf{D}(E)$ for some dg algebra $E = \mathcal{E}nd_A(\tilde{T})$, and we have $H^* E \cong A$. Now apply Lemma 9.1.17 with Example 9.1.19 to get the equivalence $\mathcal{T} \xrightarrow{\sim} \mathbf{D}(A)$.

For the other implication observe that A is a tilting object of $\mathbf{D}(A)$. This property is preserved under a triangle equivalence.   □

Remark 9.1.23.   The two notions of a tilting object are related as follows. If $\mathcal{T}$ admits arbitrary coproducts and $T \in \mathcal{T}$ is tilting, then the full subcategory $\mathcal{T}^c$ of compact objects in $\mathcal{T}$ is essentially small and $T \in \mathcal{T}^c$ is tilting. On the other hand, if $T \in \mathcal{T}^c$ is tilting, then T is a tilting object in $\mathcal{T}$.

The Stable Homotopy Category Is Not Algebraic

There are triangulated categories which are not algebraic. For instance, the *Spanier–Whitehead category* of finite CW-complexes SW_f is not algebraic. Given any endomorphism $\phi \colon X \to X$ in a triangulated category $\mathcal{T}$, denote by X/ϕ its cone. If $\mathcal{T}$ is algebraic, then we can identify X with a complex and ϕ induces an endomorphism of the mapping cone X/ϕ which is null-homotopic. Thus for $\phi = 2 \cdot \mathrm{id}_X$ we have $2 \cdot \mathrm{id}_{X/\phi} = 0$ in $\mathcal{T}$. On the other hand, let S denote the sphere spectrum. Then it is well known (and can be shown using Steenrod operations, cf. [186, Proposition 4]) that the identity map of the mod 2 Moore spectrum $M(2) = S/(2 \cdot \mathrm{id}_S)$ has order different from 2. In fact, its order is 4.

Proposition 9.1.24. *There is no faithful exact functor $\mathrm{SW}_f \to \mathcal{T}$ into an algebraic triangulated category $\mathcal{T}$.*

Proof   Let $F \colon \mathrm{SW}_f \to \mathcal{T}$ be an exact functor to an algebraic triangulated category $\mathcal{T}$ and let $X = F(S)$. Then we have $F(M(2)) = X/(2 \cdot \mathrm{id}_X)$ and therefore $F(2 \cdot \mathrm{id}_{M(2)}) = 2 \cdot \mathrm{id}_{X/(2 \cdot \mathrm{id}_X)} = 0$. On the other hand, $2 \cdot \mathrm{id}_{M(2)} \neq 0$. Thus F is not faithful.   □

9.2 Derived Equivalences

In this section we study the derived category of a module category and identify various important subcategories. Then we establish a Morita theorem by giving for any pair of rings a criterion for when these subcategories are triangle equivalent.

Compact Objects

Recall that an object X in a triangulated category $\mathcal{T}$ is *compact* if the functor $\text{Hom}_{\mathcal{T}}(X, -)$ preserves coproducts. We describe the compact objects for various subcategories of the derived category $\mathbf{D}(\Lambda)$ of a ring Λ.

Lemma 9.2.1. *Let Λ be a ring and let $\mathcal{T}$ be one of the categories $\mathbf{D}(\Lambda)$, $\mathbf{K}^-(\text{Proj}\,\Lambda)$, or $\mathbf{K}^b(\text{Proj}\,\Lambda)$. Then an object in $\mathcal{T}$ is compact if and only if it is isomorphic to an object in $\mathbf{K}^b(\text{proj}\,\Lambda)$.*

Proof One direction is clear, since Λ is compact when viewed as a complex concentrated in degree zero, and therefore any object in $\mathbf{K}^b(\text{proj}\,\Lambda) = \text{Thick}(\Lambda)$ is compact.

Now fix a compact object X in $\mathbf{D}(\Lambda)$ and write X as a homotopy colimit of a sequence $X_0 \to X_1 \to X_2 \to \cdots$ as in Lemma 9.1.9. Then compactness implies that the identity id_X factors through the canonical morphism $X_n \to X$ for some n; see Lemma 3.4.3. Thus X belongs up to isomorphism to $\mathbf{K}^b(\text{Proj}\,\Lambda)$.

A compact object X in $\mathbf{K}^-(\text{Proj}\,\Lambda)$ can be written as a homotopy colimit of its truncations

$$\sigma_{\geq 0}X \longrightarrow \sigma_{\geq -1}X \longrightarrow \sigma_{\geq -2}X \longrightarrow \cdots$$

via the exact triangle (4.2.3). As before, compactness implies that the identity id_X factors through the truncation $\sigma_{\geq n}X \to X$ for some n, and therefore X belongs up to isomorphism to $\mathbf{K}^b(\text{Proj}\,\Lambda)$.

Now fix a compact object X in $\mathbf{K}^b(\text{Proj}\,\Lambda)$. We consider the class $\mathcal{C}$ consisting of complexes $C \in \mathbf{K}^b(\text{Proj}\,\Lambda)$ such that for some $n \in \mathbb{Z}$ we have $C^n \in \text{proj}\,\Lambda$ and $C^i = 0$ for all $i \neq n$. Then we can apply Proposition 3.4.13. The object X is an extension of coproducts of objects in $\mathcal{C}$, and it follows that X belongs to $\text{Thick}(\mathcal{C}) = \mathbf{K}^b(\text{proj}\,\Lambda)$. $\qquad\square$

Homologically Finite Objects

Let $\mathcal{T}$ be a triangulated category. An object X in $\mathcal{T}$ is *homologically finite* if for every object Y in $\mathcal{T}$ we have $\text{Hom}_{\mathcal{T}}(X, \Sigma^n Y) = 0$ for almost all $n \in \mathbb{Z}$. Note that the homologically finite objects form a thick subcategory of $\mathcal{T}$.

Example 9.2.2. Let $\mathcal{A}$ be an exact category. If $X \in \mathcal{A}$ has finite projective dimension, then X is homologically finite when viewed as an object in $\mathbf{D}^b(\mathcal{A})$. If all objects in $\mathcal{A}$ have finite projective dimension, then all objects in $\mathbf{D}^b(\mathcal{A})$ are homologically finite.

The following describes the homologically finite objects in $\mathbf{D}^b(\mathrm{Mod}\, A)$.

Lemma 9.2.3. *Let Λ be a ring. For X in $\mathbf{K}^{-,b}(\mathrm{Proj}\,\Lambda)$ the following are equivalent.*

(1) *X is isomorphic to an object in $\mathbf{K}^b(\mathrm{Proj}\,\Lambda)$.*
(2) *The truncation $X \to \sigma_{\leq n}X$ is null-homotopic for $n \ll 0$.*
(3) *X is homologically finite.*

Proof (1) $\Rightarrow$ (2): Clear.

(2) $\Rightarrow$ (1): For $n \in \mathbb{Z}$ we have an exact triangle $\sigma_{>n}X \to X \to \sigma_{\leq n}X \to \Sigma(\sigma_{>n}X)$. If $X \to \sigma_{\leq n}X$ is null-homotopic, then id_X factors through $\sigma_{>n}X \to X$ and therefore X is a direct summand of $\sigma_{>n}X$ which belongs to $\mathbf{K}^b(\mathrm{Proj}\,\Lambda)$.

(1) $\Rightarrow$ (3): Let $\mathcal{P}$ denote the subcategory consisting of all complexes concentrated in degree zero. Clearly, every object in $\mathcal{P}$ is homologically finite, and therefore every object in $\mathbf{K}^b(\mathrm{Proj}\,\Lambda) = \mathrm{Thick}(\mathcal{P})$ is homologically finite.

(3) $\Rightarrow$ (2): Suppose the truncation $X \to \sigma_{\leq n}X$ is not null-homotopic for $n \ll 0$. Set $\bar{X} = \coprod_{n \leq 0} \Sigma^{-n}\sigma_{\leq n}X$ and note that it belongs to $\mathbf{K}^{-,b}(\mathrm{Proj}\,\Lambda)$. Then $\mathrm{Hom}_{\mathcal{T}}(X, \Sigma^n \bar{X}) \neq 0$ for $n \ll 0$. Thus X is not homologically finite. $\square$

A Morita Theorem for Derived Categories

For a ring Λ we consider the derived category $\mathbf{D}(\Lambda) = \mathbf{D}(\mathrm{Mod}\,\Lambda)$ and identify the following subcategories:

$$
\begin{array}{ccccccc}
\mathbf{K}^b(\mathrm{proj}\,\Lambda) & \longhookrightarrow & \mathbf{K}^{-,b}(\mathrm{Proj}\,\Lambda) & \longhookrightarrow & \mathbf{K}^-(\mathrm{Proj}\,\Lambda) & \longhookrightarrow & \mathbf{K}_{\mathrm{proj}}(\Lambda) \\
\downarrow{\wr} & & \downarrow{\wr} & & \downarrow{\wr} & & \downarrow{\wr} \\
\mathbf{D}^{\mathrm{perf}}(\Lambda) & \longhookrightarrow & \mathbf{D}^b(\mathrm{Mod}\,\Lambda) & \longhookrightarrow & \mathbf{D}^-(\mathrm{Mod}\,\Lambda) & \longhookrightarrow & \mathbf{D}(\mathrm{Mod}\,\Lambda)
\end{array}
$$

Given a pair of rings, the following theorem describes the equivalences for each of these subcategories.

Theorem 9.2.4. *For rings Λ and Γ the following are equivalent.*

(1) *There is a triangle equivalence $\mathbf{D}(\mathrm{Mod}\,\Lambda) \xrightarrow{\sim} \mathbf{D}(\mathrm{Mod}\,\Gamma)$.*
(2) *There is a triangle equivalence $\mathbf{D}^-(\mathrm{Mod}\,\Lambda) \xrightarrow{\sim} \mathbf{D}^-(\mathrm{Mod}\,\Gamma)$.*
(3) *There is a triangle equivalence $\mathbf{D}^b(\mathrm{Mod}\,\Lambda) \xrightarrow{\sim} \mathbf{D}^b(\mathrm{Mod}\,\Gamma)$.*
(4) *There is a triangle equivalence $\mathbf{D}^{\mathrm{perf}}(\Lambda) \xrightarrow{\sim} \mathbf{D}^{\mathrm{perf}}(\Gamma)$.*

(5) *There is a tilting object $T \in \mathbf{D}^{\mathrm{perf}}(\Lambda)$ with $\mathrm{End}_{\mathbf{D}(\Lambda)}(T) \cong \Gamma$.*

The rings Λ and Γ are called *derived equivalent* if the equivalent conditions of the above theorem are satisfied.

The proof of the theorem shows that any of the above triangle equivalences induces an equivalence when restricted to one of the subcategories.

The equivalence (4) $\Leftrightarrow$ (5) is a direct consequence of Proposition 9.1.20. The other implications require some further analysis, involving compact objects and a dévissage argument. We begin with the following lemma.

Lemma 9.2.5. *Let T be a bounded complex of finitely generated projective Λ-modules and set $E = \mathcal{E}nd_\Lambda(T)$. Then $\mathcal{H}om_\Lambda(T, -)$ induces an exact functor $\mathbf{D}(\Lambda) \to \mathbf{D}(E)$ that preserves coproducts.*

Proof The functor $F = \mathcal{H}om_\Lambda(T, -)$ maps complexes of Λ-modules to dg modules over E, and the formula (9.1.2) shows that F preserves quasi-isomorphisms. Thus we obtain an exact functor $\mathbf{D}(\Lambda) \to \mathbf{D}(E)$. From the definition of F it follows that F preserves coproducts, because $\mathrm{Hom}_\Lambda(T^i, -)$ preserves coproducts for all $i \in \mathbb{Z}$. $\qquad\square$

Let T be a tilting object in $\mathbf{D}^{\mathrm{perf}}(\Lambda)$ with $\mathrm{End}(T) \cong \Gamma$. Set $E = \mathcal{E}nd_\Lambda(T)$ and note that $H^* E \cong \Gamma$. Thus we have a quasi-isomorphism

$$\phi: \Gamma \twoheadleftarrow \tau_{\leq 0} E \hookrightarrow E$$

and obtain an exact functor

$$F_T: \mathbf{D}(\Lambda) \xrightarrow{\;\mathcal{H}om_\Lambda(T,-)\;} \mathbf{D}(E) \xrightarrow{\quad\phi^*\quad} \mathbf{D}(\Gamma)$$

where ϕ^* is given by restriction along $\tau_{\leq 0} E \hookrightarrow E$ composed with the left adjoint of restriction along $\tau_{\leq 0} E \twoheadrightarrow \Gamma$.

Lemma 9.2.6. *The functor F_T is a triangle equivalence and restricts to an equivalence $\mathbf{D}^{\mathrm{perf}}(\Lambda) \xrightarrow{\sim} \mathbf{D}^{\mathrm{perf}}(\Gamma)$.*

Proof The functor preserves coproducts by Lemma 9.2.5. Also, ϕ^* is a triangle equivalence by Lemma 9.1.17, which identifies $\mathbf{D}^{\mathrm{perf}}(E)$ with $\mathbf{D}^{\mathrm{perf}}(\Gamma)$. The functor $\mathcal{H}om_\Lambda(T, -)$ maps T to E and for all $n \in \mathbb{Z}$ induces an isomorphism

$$\mathrm{Hom}_{\mathbf{D}(\Lambda)}(T, \Sigma^n T) \cong H^n \mathcal{H}om_\Lambda(T, T) \cong H^n E \cong \mathrm{Hom}_{\mathbf{D}(E)}(E, \Sigma^n E).$$

This follows by combining the isomorphisms in (9.1.2) and Lemma 9.1.3. Thus we get an equivalence $\mathbf{D}^{\mathrm{perf}}(\Lambda) \xrightarrow{\sim} \mathbf{D}^{\mathrm{perf}}(E)$, thanks to Lemma 3.1.8. It remains to apply Lemma 9.1.14 and it follows that F_T is a triangle equivalence. Here, we use the description of the compact objects from Lemma 9.2.1 and that every object in $\mathbf{D}(\Lambda)$ is built from compact objects by Corollary 9.1.12. $\qquad\square$

Proof of Theorem 9.2.4   Each of (1)–(3) implies (4). We identify $\mathbf{D}(\Lambda) = \mathbf{K}_{\mathrm{proj}}(\Lambda)$, using Proposition 9.1.10, and observe that the full subcategory of compact objects identifies with $\mathbf{D}^{\mathrm{perf}}(\Lambda)$ by Lemma 9.2.1. Clearly, an equivalence $\mathbf{D}(\Lambda) \xrightarrow{\sim} \mathbf{D}(\Gamma)$ restricts to an equivalence between the full subcategories of compact objects. The argument for $\mathbf{D}^-(\mathrm{Mod}\,\Lambda)$ is analogous. Now identify $\mathbf{D}^b(\mathrm{Mod}\,\Lambda) = \mathbf{K}^{-,b}(\mathrm{Proj}\,\Lambda)$ and observe that the full subcategory of compact and homologically finite objects identifies with $\mathbf{D}^{\mathrm{perf}}(\Lambda)$ by Lemma 9.2.1 and Lemma 9.2.3. Thus an equivalence $\mathbf{D}^b(\mathrm{Mod}\,\Lambda) \xrightarrow{\sim} \mathbf{D}^b(\mathrm{Mod}\,\Gamma)$ restricts to an equivalence $\mathbf{D}^{\mathrm{perf}}(\Lambda) \xrightarrow{\sim} \mathbf{D}^{\mathrm{perf}}(\Gamma)$.

(4) $\Rightarrow$ (5): A triangle equivalence $\mathbf{D}^{\mathrm{perf}}(\Gamma) \xrightarrow{\sim} \mathbf{D}^{\mathrm{perf}}(\Lambda)$ sends Γ to a tilting object T in $\mathbf{D}^{\mathrm{perf}}(\Lambda)$ with $\mathrm{End}(T) \cong \Gamma$.

(5) implies each of (1)–(4). This follows from Lemma 9.2.6 by taking the functor F_T that is given by a tilting object T. It remains to observe that for any object X in $\mathbf{D}(\Lambda)$ we have

$$H^n X = 0 \text{ for } n \gg 0 \iff \mathrm{Hom}_{\mathbf{D}(\Lambda)}(\Lambda, \Sigma^n X) = 0 \text{ for } n \gg 0$$
$$\iff \mathrm{Hom}_{\mathbf{D}(\Lambda)}(T, \Sigma^n X) = 0 \text{ for } n \gg 0$$
$$\iff \mathrm{Hom}_{\mathbf{D}(\Gamma)}(\Gamma, \Sigma^n F_T X) = 0 \text{ for } n \gg 0$$
$$\iff H^n(F_T X) = 0 \text{ for } n \gg 0.$$

Thus F_T induces a triangle equivalence $\mathbf{D}^-(\mathrm{Mod}\,\Lambda) \xrightarrow{\sim} \mathbf{D}^-(\mathrm{Mod}\,\Gamma)$. Also, we have $H^n X = 0$ for $|n| \gg 0$ if and only if $H^n(F_T X) = 0$ for $|n| \gg 0$. Thus F_T induces a triangle equivalence $\mathbf{D}^b(\mathrm{Mod}\,\Lambda) \xrightarrow{\sim} \mathbf{D}^b(\mathrm{Mod}\,\Gamma)$. $\square$

Remark 9.2.7. Two rings Λ and Γ are derived equivalent if and only if Λ^{op} and Γ^{op} are derived equivalent; see Remark 9.1.21.

Homologically Perfect Objects

Let Λ be a ring. We call a complex $X \in \mathbf{D}(\Lambda)$ *homologically perfect* if X can be written as homotopy colimit of a sequence $X_0 \to X_1 \to X_2 \to \cdots$ in $\mathbf{D}^{\mathrm{perf}}(\Lambda)$ such that

(HP1) $H^n X = 0$ for almost all $n \in \mathbb{Z}$, and
(HP2) for every $n \in \mathbb{Z}$ we have $H^n(X_i) \xrightarrow{\sim} H^n(X_{i+1})$ for $i \gg 0$.

Note that we can reformulate this as follows. For every $P \in \mathbf{D}^{\mathrm{perf}}(\Lambda)$ we have

(HP1') $\mathrm{Hom}(P, \Sigma^n X) = 0$ for almost all $n \in \mathbb{Z}$, and
(HP2') $\mathrm{Hom}(P, X_i) \xrightarrow{\sim} \mathrm{Hom}(P, X_{i+1})$ for $i \gg 0$.

It is not difficult to check that the homologically perfect objects form a thick subcategory of $\mathbf{D}(\Lambda)$.

Lemma 9.2.8. *A triangle equivalence* $\mathbf{D}(\Lambda) \xrightarrow{\sim} \mathbf{D}(\Gamma)$ *restricts to an equivalence between the full subcategories of homologically perfect objects.*

Proof Fix a triangle equivalence $F \colon \mathbf{D}(\Lambda) \xrightarrow{\sim} \mathbf{D}(\Gamma)$. Then F restricts to an equivalence $\mathbf{D}^{\mathrm{perf}}(\Lambda) \xrightarrow{\sim} \mathbf{D}^{\mathrm{perf}}(\Gamma)$ by Lemma 9.2.1, and F preserves homotopy colimits. From this the assertion follows, using the reformulation of the definition of a homologically perfect object via the conditions (HP1′) and (HP2′). □

Coherent Rings

Let Λ be a *right coherent* ring. Then by definition the category of finitely presented Λ-modules is abelian. We study the derived category $\mathbf{D}^{b}(\mathrm{mod}\,\Lambda)$ using the following identifications:

$$
\begin{array}{ccc}
\mathbf{K}^{b}(\mathrm{proj}\,\Lambda) & \lhook\joinrel\longrightarrow & \mathbf{K}^{-,b}(\mathrm{proj}\,\Lambda) \\
\downarrow{\scriptstyle\wr} & & \downarrow{\scriptstyle\wr} \\
\mathbf{D}^{\mathrm{perf}}(\Lambda) & \lhook\joinrel\longrightarrow & \mathbf{D}^{b}(\mathrm{mod}\,\Lambda)
\end{array}
$$

Note that $\mathbf{D}^{\mathrm{perf}}(\Lambda) = \mathbf{D}^{b}(\mathrm{mod}\,\Lambda)$ if and only if every finitely presented Λ-module has finite projective dimension.

We wish to analyse how the categories $\mathbf{D}^{\mathrm{perf}}(\Lambda)$ and $\mathbf{D}^{b}(\mathrm{mod}\,\Lambda)$ determine each other. We begin with the following intrinsic description of the objects from $\mathbf{D}^{b}(\mathrm{mod}\,\Lambda)$.

Lemma 9.2.9. *Let Λ be a right coherent ring. Then X in $\mathbf{D}(\Lambda)$ is homologically perfect if and only if X belongs to $\mathbf{D}^{b}(\mathrm{mod}\,\Lambda)$.*

Proof Let X be a complex in $\mathbf{K}^{-,b}(\mathrm{proj}\,\Lambda) = \mathbf{D}^{b}(\mathrm{mod}\,\Lambda)$ and write this as the homotopy colimit of its truncations $\sigma_{\geq n} X$ which lie in $\mathbf{K}^{b}(\mathrm{proj}\,\Lambda)$. It is clear that X is homologically perfect. On the other hand, if X is homologically perfect, then $H^{n} X$ is finitely presented for all n and $H^{n} X = 0$ for $|n| \gg 0$, so X lies in $\mathbf{D}^{b}(\mathrm{mod}\,\Lambda)$. □

We are now able to show that $\mathbf{D}^{b}(\mathrm{mod}\,\Lambda)$ is a derived invariant.

Proposition 9.2.10. *Let Λ and Γ be right coherent rings. Then a triangle equivalence* $\mathbf{D}(\Lambda) \xrightarrow{\sim} \mathbf{D}(\Gamma)$ *restricts to an equivalence* $\mathbf{D}^{b}(\mathrm{mod}\,\Lambda) \xrightarrow{\sim} \mathbf{D}^{b}(\mathrm{mod}\,\Gamma)$.

Proof Combine Lemma 9.2.8 and Lemma 9.2.9. □

The converse of this proposition requires an intrinsic description of the objects from $\mathbf{D}^{\mathrm{perf}}(\Lambda)$ inside $\mathbf{D}^b(\mathrm{mod}\,\Lambda)$. We begin with some notation. Set $\mathfrak{T} := \mathbf{K}^{-,b}(\mathrm{proj}\,\Lambda)$. For $n \in \mathbb{Z}$ set

$$\mathfrak{T}^{>n} := \{X \in \mathfrak{T} \mid H^i X = 0 \text{ for all } i \le n\}$$

and

$$\mathfrak{T}^{\le n} := \{X \in \mathfrak{T} \mid H^i X = 0 \text{ for all } i > n\}.$$

For an object $U \in \mathfrak{T}$ and $n \in \mathbb{Z}$ set

$$\mathfrak{T}_U^{>n} := \{X \in \mathfrak{T} \mid \mathrm{Hom}(U, \Sigma^i X) = 0 \text{ for all } i \le n\}$$

and

$$\mathfrak{T}_U^{\le n} := \{X \in \mathfrak{T} \mid \mathrm{Hom}(X, \mathfrak{T}_U^{>n}) = 0\}.$$

Note that $\mathrm{Hom}(X, Y) = 0$ for all $X \in \mathfrak{T}^{\le n}$ and $Y \in \mathfrak{T}^{>n}$. Also, we have $\mathfrak{T}^{>n} = \mathfrak{T}_\Lambda^{>n}$ since $H^i X = \mathrm{Hom}(\Lambda, \Sigma^i X)$ for all $X \in \mathfrak{T}$ and $i \in \mathbb{Z}$.

Lemma 9.2.11. *If $U \in \mathfrak{T}^{\le n}$, then $\mathfrak{T}^{>n} \subseteq \mathfrak{T}_U^{>0}$.*

Proof Clearly, $U \in \mathfrak{T}^{\le n}$ implies $\Sigma^i U \in \mathfrak{T}^{\le n}$ for all $i \ge 0$. This implies $\mathrm{Hom}(U, \Sigma^i X) = 0$ for all $i \le 0$ and $X \in \mathfrak{T}^{>n}$. Therefore $\mathfrak{T}^{>n} \subseteq \mathfrak{T}_U^{>0}$. $\square$

For objects $U, V \in \mathfrak{T}$ write $U \le V$ if there are $p, q \in \mathbb{Z}$ such that $\mathfrak{T}_U^{>p} \subseteq \mathfrak{T}_V^{>q}$. Call $U \in \mathfrak{T}$ *initial* if $U \le V$ for all $V \in \mathfrak{T}$. The preceding lemma shows that Λ is initial. Set

$$\mathfrak{T}_U := \{X \in \mathfrak{T} \mid \mathrm{Hom}(X, \mathfrak{T}_U^{\le n}) = 0 \text{ for all } n \ll 0\}.$$

Lemma 9.2.12. *If $U \in \mathfrak{T}$ is initial, then $\mathbf{K}^b(\mathrm{proj}\,\Lambda) = \mathfrak{T}_U$.*

Proof When U and V are initial, then $\mathfrak{T}_U = \mathfrak{T}_V$. Thus we may assume $U = \Lambda$. Then $\mathfrak{T}_U^{\le n} = \mathfrak{T}^{\le n}$. The inclusion $\mathfrak{T}_U^{\le n} \supseteq \mathfrak{T}^{\le n}$ is automatic; the other one uses that Λ is right coherent, because then $\mathrm{Hom}(X, \mathfrak{T}^{>n}) = 0$ implies $X \in \mathfrak{T}^{\le n}$. Clearly, $\mathfrak{T}_U$ is a thick subcategory that contains Λ. Thus $\mathfrak{T}_U$ contains $\mathbf{K}^b(\mathrm{proj}\,\Lambda)$. Now fix an object $X \in \mathfrak{T}$ that is not in $\mathbf{K}^b(\mathrm{proj}\,\Lambda)$. Then the canonical morphism $X \to \sigma_{\le n} X$ is not null-homotopic for $n \ll 0$, by Lemma 9.2.3. Note that $\sigma_{\le n} X$ lies in $\mathfrak{T}^{\le n}$. Thus $X \notin \mathfrak{T}_U$. $\square$

Proposition 9.2.13. *Let Λ and Γ be right coherent rings. Then a triangle equivalence $\mathbf{D}^b(\mathrm{mod}\,\Lambda) \xrightarrow{\sim} \mathbf{D}^b(\mathrm{mod}\,\Gamma)$ restricts to an equivalence $\mathbf{D}^{\mathrm{perf}}(\Lambda) \xrightarrow{\sim} \mathbf{D}^{\mathrm{perf}}(\Gamma)$.*

Proof We fix a triangle equivalence $F : \mathbf{D}^b(\mathrm{mod}\,\Lambda) \xrightarrow{\sim} \mathbf{D}^b(\mathrm{mod}\,\Gamma)$ and identify $\mathbf{K}^{-,b}(\mathrm{proj}\,\Lambda) = \mathbf{D}^b(\mathrm{mod}\,\Lambda)$. Clearly, F maps initial objects to initial objects. If $U \in \mathbf{D}^b(\mathrm{mod}\,\Lambda)$ is initial, then Lemma 9.2.12 implies

$$\mathbf{D}^{\mathrm{perf}}(\Lambda) = \mathfrak{T}_U \xrightarrow{\sim} F(\mathfrak{T}_U) = \mathfrak{T}_{FU} = \mathbf{D}^{\mathrm{perf}}(\Gamma). \qquad \square$$

The above Lemma 9.2.12 provides an intrinsic characterisation of perfect complexes in $\mathbf{D}^b(\mathrm{mod}\,\Lambda)$. Another more practical criterion says that a Λ-module X viewed as a complex concentrated in degree zero belongs to $\mathbf{D}^{\mathrm{perf}}(\Lambda)$ if and only if X admits a projective resolution

$$0 \longrightarrow P_n \longrightarrow \cdots \longrightarrow P_1 \longrightarrow P_0 \longrightarrow X \longrightarrow 0$$

such that each P_i is finitely generated (Corollary 7.1.8).

Proposition 9.2.14. *Let Λ be a right coherent ring and consider for a complex $X \in \mathbf{D}^b(\mathrm{mod}\,\Lambda)$ the following conditions.*

(1) *$H^n X$ has finite projective dimension for all $n \in \mathbb{Z}$.*
(2) *X is perfect.*
(3) *X is a homologically finite object.*

Then (1) $\Rightarrow$ (2) $\Rightarrow$ (3). Moreover, (2) $\Leftarrow$ (3) when Λ is semiperfect.

Proof (1) $\Rightarrow$ (2): A complex X belongs to the thick subcategory generated by the cohomology objects $H^n X$ (Lemma 4.2.1). Now let $\mathcal{P}$ denote the thick subcategory of modules $M \in \mathrm{mod}\,\Lambda$ such that proj.dim $M < \infty$. Then the inclusion $\mathrm{mod}\,\Lambda \to \mathbf{D}^b(\mathrm{mod}\,\Lambda)$ maps $\mathcal{P} = \mathrm{Thick}(\Lambda)$ into $\mathbf{D}^{\mathrm{perf}}(\Lambda)$. Thus $H^n X \in \mathcal{P}$ for all $n \in \mathbb{Z}$ implies that X is perfect.

(2) $\Rightarrow$ (3): This is clear, since the homologically finite objects form a thick subcategory containing Λ.

(3) $\Rightarrow$ (2): Suppose Λ is semiperfect and set $S = \Lambda/J(\Lambda)$, where $J(\Lambda)$ denotes the Jacobson radical. We use that $M \in \mathrm{mod}\,\Lambda$ is projective if and only if every morphism $M \to S$ factors through a projective module.

Let us identify $\mathbf{K}^{-,b}(\mathrm{proj}\,\Lambda) = \mathbf{D}^b(\mathrm{mod}\,\Lambda)$, and suppose X is an object not isomorphic to an object from $\mathbf{K}^b(\mathrm{proj}\,\Lambda)$. For $n \in \mathbb{Z}$ let Y^n denote the cokernel of $X^{n-1} \to X^n$, and observe that Y^n is not projective for infinitely many $n \in \mathbb{Z}$. Choose a projective resolution $P \to S$ of S. Then we have $\mathrm{Hom}_{\mathbf{K}(\Lambda)}(X, \Sigma^n P) \neq 0$ whenever Y^n is not projective. Thus X is not homologically finite. $\square$

Remark 9.2.15. The above implication (3) $\Rightarrow$ (2) also holds when Λ is commutative noetherian; then one reduces to the local case and uses the quasi-compactness of the prime ideal spectrum Spec Λ, cf. [23, Lemma 4.5].

Example 9.2.16. Let k be a field and V an n-dimensional space with dual $V^\vee$. Consider the categories of finitely generated $\mathbb{Z}$-graded modules grmod $S(V)$ and grmod $\Lambda(V^\vee)$ over the symmetric algebra and the exterior algebra, respectively. Koszul duality (10.3.1) provides a triangle equivalence

$$\mathbf{D}^b(\mathrm{grmod}\,S(V)) \xrightarrow{\sim} \mathbf{D}^b(\mathrm{grmod}\,\Lambda(V^\vee))$$

and therefore all objects are homologically finite, since $S(V)$ has finite global dimension; see Example 9.2.2. On the other hand, $\Lambda(V^\vee)$ has infinite global dimension when $n > 0$, so not all homologically finite objects are perfect in $\mathbf{D}^b(\text{grmod } \Lambda(V))$.

Gorenstein Algebras

Let Λ be an Artin k-algebra with Matlis duality $D = \text{Hom}_k(-, E)$ given by an injective k-module E. The *derived Nakayama functor* is the left derived functor

$$\mathbf{D}(\text{Mod } \Lambda) \xrightarrow{\;-\otimes_\Lambda^L D(\Lambda)\;} \mathbf{D}(\text{Mod } \Lambda)$$

of the Nakayama functor $\nu = - \otimes_\Lambda D(\Lambda)\colon \text{Mod } \Lambda \to \text{Mod } \Lambda$.

Proposition 9.2.17. *For an Artin algebra Λ the following are equivalent.*

(1) *The algebra Λ is Gorenstein.*
(2) *The module $D(\Lambda)$ is a tilting object in $\mathbf{D}(\text{Mod } \Lambda)$.*
(3) *The derived Nakayama functor is a triangle equivalence.*

Proof Set $T = D(\Lambda)$ and view it as a complex concentrated in degree zero.

(1) $\Leftrightarrow$ (2): The algebra Λ is Gorenstein by definition if Λ_Λ and $_\Lambda\Lambda$ have finite injective dimension. An equivalent condition is that Λ_Λ has finite injective dimension and $D(\Lambda)_\Lambda$ has finite projective dimension. It follows that Λ is Gorenstein if and only if T belongs to $\mathbf{D}^{\text{perf}}(\Lambda)$ and is a tilting object, so $\text{Thick}(T) = \mathbf{D}^{\text{perf}}(\Lambda)$.

(2) $\Rightarrow$ (3): If T is a tilting object, then the right adjoint $\text{RHom}_\Lambda(T, -)$ of the derived Nakayama functor is a triangle equivalence, by Theorem 9.2.4 and its proof.

(3) $\Rightarrow$ (2): A triangle equivalence maps tilting objects to tilting objects. So it suffices to observe that $T = \Lambda \otimes_\Lambda^L D(\Lambda)$. $\qquad\qquad\square$

Serre Duality

Let us get back to Serre duality, but working in the unbounded derived category. In fact, we consider two cases, namely the derived category $\mathbf{D}(\text{Mod } \Lambda)$ for an algebra Λ and the derived category $\mathbf{D}(\text{Qcoh } \mathbb{X})$ for a scheme $\mathbb{X}$. In both cases we describe a dualising complex and find an interesting parallel.

Let us fix a commutative ring k and a k-algebra Λ. We write $D = \text{Hom}_k(-, E)$ for Matlis duality given by an injective k-module E.

We begin with a simple observation. Recall from Corollary 9.1.12 that the triangulated category $\mathbf{D}(\text{Mod } \Lambda)$ is compactly generated, and let $X \in \mathbf{D}(\text{Mod } \Lambda)$

be a perfect complex. Then the functor $D \operatorname{Hom}_{\mathbf{D}(\Lambda)}(X, -) : \mathbf{D}(\operatorname{Mod}\Lambda)^{\mathrm{op}} \to \mathrm{Ab}$ is representable by Theorem 3.4.16 since X is compact. Thus we have a complex X' such that

$$D \operatorname{Hom}_{\mathbf{D}(\Lambda)}(X, -) \cong \operatorname{Hom}_{\mathbf{D}(\Lambda)}(-, X').$$

Our aim is to give an explicit description of the representing object X'.

Fix complexes X and Y in $\mathbf{D}(\Lambda) = \mathbf{D}(\operatorname{Mod}\Lambda)$ and suppose that X is perfect. Note that we have isomorphisms

$$\operatorname{RHom}_\Lambda(X, Y) \cong Y \otimes_\Lambda^L X^* \quad \text{and} \quad X \cong X^{**}$$

where $X^* = \operatorname{RHom}_\Lambda(X, \Lambda)$. Using the adjointness of $\otimes^L$ and RHom, and viewing the injective k-module E as a complex concentrated in degree zero, we obtain the following isomorphism

$$\begin{aligned}
D \operatorname{Hom}_{\mathbf{D}(\Lambda)}(X, Y) &\cong \operatorname{Hom}_k(H^0(\operatorname{RHom}_\Lambda(X, Y)), E) \\
&\cong \operatorname{Hom}_{\mathbf{D}(k)}(\operatorname{RHom}_\Lambda(X, Y), E) \\
&\cong \operatorname{Hom}_{\mathbf{D}(k)}(Y \otimes_\Lambda^L X^*, E) \\
&\cong \operatorname{Hom}_{\mathbf{D}(\Lambda)}(Y, \operatorname{RHom}_k(X^*, E)) \\
&\cong \operatorname{Hom}_{\mathbf{D}(\Lambda)}(Y, \operatorname{RHom}_k(\Lambda \otimes_\Lambda^L X^*, E)) \\
&\cong \operatorname{Hom}_{\mathbf{D}(\Lambda)}(Y, \operatorname{RHom}_\Lambda(X^*, \operatorname{RHom}_k(\Lambda, E))) \\
&\cong \operatorname{Hom}_{\mathbf{D}(\Lambda)}(Y, X \otimes_\Lambda^L \operatorname{Hom}_k(\Lambda, E))
\end{aligned}$$

which is natural in X and Y. Thus the derived Nakayama functor

$$X \longmapsto X \otimes_\Lambda^L \operatorname{Hom}_k(\Lambda, E)$$

provides an analogue of Serre duality where the *dualising complex* is the Λ^{op}-module $D\Lambda = \operatorname{Hom}_k(\Lambda, E)$ concentrated in degree zero.

Now let k be a field and $\mathbb{X}$ a *projective scheme* over k, which is given by a projective morphism

$$f : \mathbb{X} \longrightarrow \mathbb{Y} = \operatorname{Spec} k.$$

We consider the derived category $\mathbf{D}(\operatorname{Qcoh}\mathbb{X})$ of the category of quasi-coherent sheaves on $\mathbb{X}$. This is a compactly generated triangulated category, and the subcategory of compact objects is equivalent to the bounded derived category $\mathbf{D}^b(\operatorname{vect}\mathbb{X})$ of vector bundles on $\mathbb{X}$; see [148, Example 1.10]. Note that the inclusion $\operatorname{vect}\mathbb{X} \to \operatorname{coh}\mathbb{X}$ into the category of coherent sheaves induces a triangle equivalence

$$\mathbf{D}^b(\operatorname{vect}\mathbb{X}) \xrightarrow{\sim} \mathbf{D}^b(\operatorname{coh}\mathbb{X})$$

provided that $\mathbb{X}$ is smooth.

The derived direct image functor

$$\mathbf{R}f_* \colon \mathbf{D}(\mathrm{Qcoh}\,\mathbb{X}) \longrightarrow \mathbf{D}(\mathrm{Qcoh}\,\mathbb{Y})$$

preserves coproducts (cf. [148, Lemma 1.4]) and then Brown representability (Theorem 3.4.16) provides a right adjoint

$$f^! \colon \mathbf{D}(\mathrm{Qcoh}\,\mathbb{Y}) \longrightarrow \mathbf{D}(\mathrm{Qcoh}\,\mathbb{X}).$$

This yields Serre duality for $\mathbb{X}$, as we now explain.

Fix complexes X and Y in $\mathbf{D}(\mathbb{X}) = \mathbf{D}(\mathrm{Qcoh}\,\mathbb{X})$ and suppose that X is perfect, so isomorphic to an object in $\mathbf{D}^b(\mathrm{vect}\,\mathbb{X})$. Note that we have isomorphisms

$$\mathrm{RHom}_{\mathbb{X}}(X, Y) \cong Y \otimes^L_{\mathscr{O}_{\mathbb{X}}} X^* \quad \text{and} \quad X \cong X^{**}$$

where $X^* = \mathrm{RHom}_{\mathbb{X}}(X, \mathscr{O}_{\mathbb{X}})$. Using the adjointness of $f^!$ and $\mathbf{R}f_*$, we obtain the following isomorphism

$$
\begin{aligned}
D\,\mathrm{Hom}_{\mathbf{D}(\mathbb{X})}(X, Y) &\cong \mathrm{Hom}_{\mathbf{D}(\mathbb{Y})}(\mathbf{R}f_*\,\mathrm{RHom}_{\mathbb{X}}(X, Y), k) \\
&\cong \mathrm{Hom}_{\mathbf{D}(\mathbb{X})}(\mathrm{RHom}_{\mathbb{X}}(X, Y), f^! k) \\
&\cong \mathrm{Hom}_{\mathbf{D}(\mathbb{X})}(Y \otimes^L_{\mathscr{O}_{\mathbb{X}}} X^*, f^! k) \\
&\cong \mathrm{Hom}_{\mathbf{D}(\mathbb{X})}(Y, \mathrm{RHom}_{\mathbb{X}}(X^*, f^! k)) \\
&\cong \mathrm{Hom}_{\mathbf{D}(\mathbb{X})}(Y, X \otimes^L_{\mathscr{O}_{\mathbb{X}}} f^! k)
\end{aligned}
$$

which is natural in X and Y.

There is a notion for a scheme to be *Gorenstein* (cf. [106, p. 144]), which means that the *dualising complex* $f^! k$ is isomorphic to an invertible sheaf and therefore induces an equivalence

$$- \otimes^L_{\mathscr{O}_{\mathbb{X}}} f^! k \colon \mathbf{D}(\mathrm{Qcoh}\,\mathbb{X}) \xrightarrow{\sim} \mathbf{D}(\mathrm{Qcoh}\,\mathbb{X}).$$

For instance, if $\mathbb{X}$ is smooth of dimension n, then this equivalence takes a familiar form since $f^! k = \bigwedge^n \Omega_{\mathbb{X}/k}[n]$, where $\Omega_{\mathbb{X}/k}$ denotes the *sheaf of differential forms*.

9.3 Finite Global Dimension

In this section we show that tilting preserves finite global dimension. So given a tilting object of the derived category of an abelian category of finite global dimension, we give a bound for the global dimension of its endomorphism ring. The proof involves the use of t-structures. In fact, the derived category of an abelian category admits a canonical t-structure. A tilting object induces a derived equivalence; so we need to compare different t-structures.

We consider two settings. First we study a tilting object in the unbounded derived category of a Grothendieck category. Then we consider a tilting object in the bounded derived category of an essentially small category. Both settings are related. In fact, in the second case we deduce the result about the global dimension of the endomorphism ring from the first case.

T-Structures

Let $\mathcal{T}$ be a triangulated category with suspension $\Sigma \colon \mathcal{T} \xrightarrow{\sim} \mathcal{T}$. A pair $(\mathcal{U}, \mathcal{V})$ of full additive subcategories is called a *t-structure* provided the following holds.

(TS1) $\Sigma \mathcal{U} \subseteq \mathcal{U}$ and $\Sigma^{-1}\mathcal{V} \subseteq \mathcal{V}$.

(TS2) $\mathrm{Hom}(X, Y) = 0$ for all $X \in \mathcal{U}$ and $Y \in \mathcal{V}$.

(TS3) For each $X \in \mathcal{T}$ there exists an exact triangle $X' \to X \to X'' \to \Sigma X'$ such that $X' \in \mathcal{U}$ and $X'' \in \mathcal{V}$.

The following characterisation of a t-structure shows that this concept only involves the suspension and not the choice of exact triangles.

Lemma 9.3.1. *A pair $(\mathcal{U}, \mathcal{V})$ of full additive subcategories of $\mathcal{T}$ is a t-structure if and only if the following holds.*

(1) *$\Sigma \mathcal{U} \subseteq \mathcal{U}$ and $\Sigma^{-1}\mathcal{V} \subseteq \mathcal{V}$.*

(2) *$\mathrm{Hom}(X, Y) = 0$ for all $Y \in \mathcal{V}$ if and only if $X \in \mathcal{U}$, and $\mathrm{Hom}(X, Y) = 0$ for all $X \in \mathcal{U}$ if and only if $Y \in \mathcal{V}$.*

(3) *The inclusion $\mathcal{U} \hookrightarrow \mathcal{T}$ admits a right adjoint and $\mathcal{V} \hookrightarrow \mathcal{T}$ admits a left adjoint.*

Proof Suppose the pair $(\mathcal{U}, \mathcal{V})$ is a t-structure. Then the assignment $X \mapsto X'$ given by the triangle $X' \to X \to X'' \to \Sigma X'$ yields a right adjoint of the inclusion $\mathcal{U} \to \mathcal{T}$, and analogously the assignment $X \mapsto X''$ yields a left adjoint of the inclusion $\mathcal{V} \to \mathcal{T}$. If $X \in \mathcal{T}$ satisfies $\mathrm{Hom}(X, Y) = 0$ for all $Y \in \mathcal{V}$, then $X' \xrightarrow{\sim} X$ and therefore $X \in \mathcal{U}$. Analogously, $\mathrm{Hom}(X, Y) = 0$ for all $X \in \mathcal{U}$ implies $Y \in \mathcal{V}$.

Now suppose the pair $(\mathcal{U}, \mathcal{V})$ satisfies (1)–(3). Let $X \mapsto X_{\mathcal{U}}$ denote the right adjoint of the inclusion $\mathcal{U} \to \mathcal{T}$, and let $X \mapsto X^{\mathcal{V}}$ denote the left adjoint of the inclusion $\mathcal{V} \to \mathcal{T}$. We claim that the counit $X_{\mathcal{U}} \to X$ and the unit $X \to X^{\mathcal{V}}$ fit into an exact triangle $X_{\mathcal{U}} \to X \to X^{\mathcal{V}} \to \Sigma X_{\mathcal{U}}$. To see this complete the counit to an exact triangle $X_{\mathcal{U}} \to X \to Y \to \Sigma X_{\mathcal{U}}$. It is easily checked that $Y \in \mathcal{V}$. Thus the property of the counit implies that $X \to Y$ factors through $X \to X^{\mathcal{V}}$. Also $X \to X^{\mathcal{V}}$ factors through $X \to Y$ since the composite $X_{\mathcal{U}} \to X \to X^{\mathcal{V}}$ is zero. The composite $X^{\mathcal{V}} \to Y \to X^{\mathcal{V}}$ equals the identity, and we obtain a decomposition $Y = X^{\mathcal{V}} \oplus Y'$. The induced morphism $Y' \to \Sigma X_{\mathcal{U}}$ is then a split monomorphism. Thus $Y' \in \mathcal{U} \cap \mathcal{V}$, and therefore $Y' = 0$. This yields the claim and it follows that $(\mathcal{U}, \mathcal{V})$ is a t-structure. □

We consider the following example. Let $\mathcal{A}$ be an abelian category and $\mathcal{T} = \mathbf{D}(\mathcal{A})$ its derived category. For $n \in \mathbb{Z}$ set

$$\mathcal{T}^{\leq n} := \{X \in \mathcal{T} \mid H^i X = 0 \text{ for all } i > n\},$$

and

$$\mathcal{T}^{>n} := \{X \in \mathcal{T} \mid H^i X = 0 \text{ for all } i \leq n\}.$$

Then we have $\mathcal{T}^{\leq n} = \Sigma^{-n}\mathcal{T}^{\leq 0}$ and $\mathcal{T}^{>n} = \Sigma^{-n}\mathcal{T}^{>0}$ for all $n \in \mathbb{Z}$. For each $X \in \mathcal{T}$ the truncations in degree n provide an exact triangle

$$\tau_{\leq n} X \longrightarrow X \longrightarrow \tau_{>n} X \longrightarrow \Sigma(\tau_{\leq n} X)$$

with $\tau_{\leq n} X \in \mathcal{T}^{\leq n}$ and $\tau_{>n} X \in \mathcal{T}^{>n}$.

Lemma 9.3.2. *The pair $(\mathcal{T}^{\leq 0}, \mathcal{T}^{>0})$ is a t-structure on $\mathbf{D}(\mathcal{A})$; it restricts to a t-structure on $\mathbf{D}^b(\mathcal{A})$.* □

The t-structure $(\mathcal{T}^{\leq 0}, \mathcal{T}^{>0})$ on $\mathbf{D}(\mathcal{A})$ and its restriction to $\mathbf{D}^b(\mathcal{A})$ are called *canonical t-structures*.

Lemma 9.3.3. *Let $(\mathcal{D}^{\leq 0}, \mathcal{D}^{>0})$ denote the canonical t-structure on $\mathbf{D}^b(\mathcal{A})$. Then the global dimension of $\mathcal{A}$ is bounded by d if and only if $\mathrm{Hom}(X, Y) = 0$ for all $X \in \mathcal{D}^{\geq 0}$ and $Y \in \mathcal{D}^{<-d}$.*

Proof For objects $A, A' \in \mathcal{A}$ and $i \in \mathbb{Z}$ we have $\mathrm{Ext}^i(A, A') \cong \mathrm{Hom}(A, \Sigma^i A')$. Thus the global dimension of $\mathcal{A}$ is bounded by d if and only if for all objects $X, Y \in \mathbf{D}^b(\mathcal{A})$ with cohomology concentrated in a single degree we have $\mathrm{Hom}(X, Y) = 0$ when $X \in \mathcal{D}^{\geq 0}$ and $Y \in \mathcal{D}^{<-d}$. The assertion of the lemma follows since for $X \in \mathcal{D}^{\geq 0}$ and $Y \in \mathcal{D}^{<-d}$, the truncations induce finite filtrations

$$X = \tau_{\geq 0} X \twoheadrightarrow \tau_{\geq 1} X \twoheadrightarrow \tau_{\geq 2} X \twoheadrightarrow \cdots$$

and

$$\cdots \rightarrowtail \tau_{<-d-2} Y \rightarrowtail \tau_{<-d-1} Y \rightarrowtail \tau_{<-d} Y = Y$$

such that each subquotient has its cohomology concentrated in a single degree i, with $i \geq 0$ for the subquotients of X and $i < -d$ for the subquotients of Y. □

We wish to extend this lemma from $\mathbf{D}^b(\mathcal{A})$ to $\mathbf{D}(\mathcal{A})$. To this end fix a Grothendieck category $\mathcal{A}$, and let us recall some basic facts about derived limits and colimits in $\mathbf{D}(\mathcal{A})$.

Let I denote a small category. The category $\mathbf{C}(\mathcal{A})$ of chain complexes may be viewed as a subcategory of the functor category $\mathrm{Fun}(\mathbb{Z}, \mathcal{A})$, where $\mathbb{Z}$ denotes the category of integers (with a single morphism $i \to j$ if and only if $i \le j$). We have a canonical equivalence

$$\mathrm{Fun}(I, \mathrm{Fun}(\mathbb{Z}, \mathcal{A})) \cong \mathrm{Fun}(I \times \mathbb{Z}, \mathcal{A}) \cong \mathrm{Fun}(\mathbb{Z}, \mathrm{Fun}(I, \mathcal{A}))$$

which restricts to an equivalence

$$\mathrm{Fun}(I, \mathbf{C}(\mathcal{A})) \cong \mathbf{C}(\mathrm{Fun}(I, \mathcal{A})).$$

Thus we consider the derived category

$$\mathbf{D}(\mathcal{A}^I) := \mathbf{D}(\mathrm{Fun}(I, \mathcal{A}))$$

for the study of (co)limits of chain complexes in $\mathcal{A}$. Next we derive the functors colim: $\mathrm{Fun}(I, \mathcal{A}) \to \mathcal{A}$ and lim: $\mathrm{Fun}(I, \mathcal{A}) \to \mathcal{A}$. When I is filtered, then colim is exact. Thus only lim needs to be derived, and we obtain functors

$$\mathrm{colim} \colon \mathbf{D}(\mathcal{A}^I) \longrightarrow \mathbf{D}(\mathcal{A}) \qquad \text{and} \qquad \mathrm{Rlim} \colon \mathbf{D}(\mathcal{A}^I) \longrightarrow \mathbf{D}(\mathcal{A}).$$

For $X \in \mathbf{D}(\mathcal{A}^I)$ we compute $\mathrm{Rlim}\, X = \lim\, iX$ using a K-injective (homotopy injective) resolution $X \to iX$ in $\mathbf{K}(\mathcal{A}^I)$ (cf. Theorem 4.3.9 and Proposition 4.3.11).

Lemma 9.3.4. *For each complex $X \in \mathbf{D}(\mathcal{A})$ its truncations induce exact triangles*

$$\Sigma^{-1} X \longrightarrow \coprod_{p \ge 0} \tau_{\le p} X \longrightarrow \coprod_{p \ge 0} \tau_{\le p} X \longrightarrow X$$

and

$$\mathrm{Rlim}_{q \le 0} \tau_{\ge q} X \longrightarrow \prod_{q \le 0} \tau_{\ge q} X \longrightarrow \prod_{q \le 0} \tau_{\ge q} X \longrightarrow \Sigma\Big(\mathrm{Rlim}_{q \le 0} \tau_{\ge q} X \Big).$$

Moreover, we have $X \xrightarrow{\sim} \mathrm{Rlim}\, \tau_{\ge q} X$ when the injective dimension of each $H^n X$ admits a global bound not depending on n and $H^n X = 0$ for $n \gg 0$.

Note that the products in the second triangle are computed in the derived category, not the category of complexes (cf. Example 4.3.12).

Proof For the first triangle we observe that the colimit of the $\tau_{\le p} X$ in the category of complexes can be computed degreewise. This gives an exact sequence

$$0 \longrightarrow \coprod_{p \ge 0} \tau_{\le p} X \longrightarrow \coprod_{p \ge 0} \tau_{\le p} X \longrightarrow X \longrightarrow 0$$

of complexes and therefore an exact triangle in $\mathbf{D}(\mathcal{A})$, as in the assertion of the lemma.

For the second triangle we need to construct a K-injective resolution of $(\tau_{\geq q}X)$ in the category of complexes of inverse systems. For each $q < 0$, choose an injective resolution $H^q X \to J_q$. Then choose a K-injective resolution $\tau_{\geq 0} X \to I_0$ and, for $q < 0$, recursively define morphisms $\varepsilon_q \colon I_{q+1} \to \Sigma^{q+1} J_q$ such that we have morphisms of triangles in $\mathbf{D}(\mathcal{A})$

$$
\begin{array}{ccccccc}
\Sigma^q H^q X & \longrightarrow & \tau_{\geq q} X & \longrightarrow & \tau_{\geq q+1} X & \longrightarrow & \Sigma^{q+1} H^q X \\
\downarrow & & \downarrow & & \downarrow & & \downarrow \\
\Sigma^q J_q & \longrightarrow & I_q & \longrightarrow & I_{q+1} & \xrightarrow{\ \varepsilon_q\ } & \Sigma^{q+1} J_q
\end{array}
$$

where the vertical morphisms are quasi-isomorphisms and ΣI_q is the cone over a lift to a morphism of complexes of ε_q. The system (I_q) is then quasi-isomorphic to $(\tau_{\geq q}X)$ and K-injective in the homotopy category of complexes of inverse systems. Thus, it may be used to compute the right derived limit of $(\tau_{\geq q}X)$. We obtain a degreewise split exact sequence of complexes

$$
0 \longrightarrow \lim I_q \longrightarrow \prod_{q \leq 0} I_q \longrightarrow \prod_{q \leq 0} I_q \longrightarrow 0
$$

and therefore an exact triangle in $\mathbf{D}(\mathcal{A})$, as in the assertion of the lemma, with

$$
\mathrm{Rlim}\, \tau_{\geq q} X \cong \mathrm{Rlim}\, I_q \cong \lim I_q.
$$

Now suppose that the injective dimension of $H^q X$ admits a global bound, say d, and we may assume that $H^q X = 0$ for all $q > 0$. To show the isomorphism $X \xrightarrow{\sim} \mathrm{Rlim}\, \tau_{\geq q} X$ we modify the above construction of a K-injective resolution of $(\tau_{\geq q}X)$ as follows. For each $q \leq 0$, choose an injective resolution $H^q X \to J_q$, where the components of J_q vanish in all degrees strictly greater than d. We put $I_0 = J_0$ and, for $q < 0$, recursively define morphisms $\varepsilon_q \colon I_{q+1} \to \Sigma^{q+1} J_q$ as before. Again, the system (I_q) yields the right derived limit of $(\tau_{\geq q}X)$. Since the J_q are uniformly right bounded, the system (I_q) becomes stationary in each degree. This yields in $\mathbf{D}(\mathcal{A})$ the required isomorphism

$$
X \cong \lim I_q \cong \mathrm{Rlim}(\tau_{\geq q}X). \qquad \square
$$

We record an immediate consequence.

Proposition 9.3.5. *Let $\mathcal{A}$ be a Grothendieck category and suppose the global dimension of $\mathcal{A}$ is finite. Then the canonical functor $\mathbf{K}(\mathrm{Inj}\,\mathcal{A}) \to \mathbf{D}(\mathcal{A})$ is a triangle equivalence.*

Proof For any complex X in $\mathcal{A}$ the morphism $X \to \mathrm{Rlim}\, \tau_{\geq q} X$ is a K-injective resolution, and we may assume that $\mathrm{Rlim}\, \tau_{\geq q} X$ is a complex of injective objects. Clearly, any acyclic complex of injectives is contractible. Thus $\mathbf{K}(\mathrm{Inj}\,\mathcal{A}) \to \mathbf{D}(\mathcal{A})$ is essentially surjective and fully faithful. $\qquad\square$

The following is the analogue of Lemma 9.3.3.

Lemma 9.3.6. *Let* $(\mathcal{D}^{\leq 0}, \mathcal{D}^{>0})$ *denote the canonical t-structure on* $\mathbf{D}(\mathcal{A})$ *and suppose the global dimension of* $\mathcal{A}$ *is bounded by* d. *Then for* $X \in \mathcal{D}^{\geq 0}$ *and* $Y \in \mathcal{D}^{<-d-2}$ *we have* $\mathrm{Hom}(X, Y) = 0$.

Proof We apply Lemma 9.3.4. Thus X fits into an exact triangle given by the truncations $\tau_{\leq p} X$, and it suffices to show that $\mathrm{Hom}(\tau_{\leq p} X, Y)$ and $\mathrm{Hom}(\Sigma \tau_{\leq p} X, Y)$ vanish for all p. On the other hand, Y fits into an exact triangle given by the truncations $\tau_{\geq q} Y$, and therefore it suffices to show that $\mathrm{Hom}(\tau_{\leq p} X, \tau_{\geq q} Y)$, $\mathrm{Hom}(\Sigma \tau_{\leq p} X, \tau_{\geq q} Y)$, and $\mathrm{Hom}(\Sigma \tau_{\leq p} X, \Sigma^{-1} \tau_{\geq q} Y)$ vanish for all p and q. This holds by Lemma 9.3.3 since both arguments belong to $\mathbf{D}^b(\mathcal{A})$. $\qquad\square$

Tilting for Unbounded Derived Categories

Let $\mathcal{A}$ be a Grothendieck category and $\mathbf{D}(\mathcal{A})$ its unbounded derived category. Recall that the category $\mathbf{D}(\mathcal{A})$ has arbitrary (set-indexed) coproducts given by coproducts in the category of complexes. Notice that the right derived product functor yields arbitrary products in $\mathbf{D}(\mathcal{A})$. In particular, the product of a family of left bounded complexes with injective components is also their product in $\mathbf{D}(\mathcal{A})$.

Lemma 9.3.7. *If* C *is a compact object of* $\mathbf{D}(\mathcal{A})$, *then the cohomology* $H^p C$ *vanishes for all but finitely many integers* p.

Proof For each $p \in \mathbb{Z}$, choose a monomorphism $i_p \colon H^p C \to I_p$ into an injective object. Using the identification

$$\mathrm{Hom}_{\mathbf{D}(\mathcal{A})}(C, \Sigma^{-p} I) = \mathrm{Hom}_{\mathcal{A}}(H^p C, I)$$

valid for each injective I of $\mathcal{A}$, the i_p yield a morphism i from C to the product (in the category of complexes and in the derived category) of the $\Sigma^{-p} I_p$. Clearly, in the category of complexes (and hence in the derived category), this product is canonically isomorphic to the corresponding coproduct. So we obtain a morphism from C to the coproduct of the $\Sigma^{-p} I_p$ which in cohomology induces the i_p. By the compactness of C, this morphism factors through a finite subcoproduct of the $\Sigma^{-p} I_p$ so that all but finitely many of the i_p have to vanish. Since they are monomorphisms, the same holds for the $H^p C$. $\qquad\square$

Now let T be a tilting object of $\mathbf{D}(\mathcal{A})$. Thus T is compact, the group $\mathrm{Hom}(T, \Sigma^p T)$ vanishes for all $p \neq 0$, and $\mathbf{D}(\mathcal{A})$ equals its localising subcategory generated by T.

Let Λ be the endomorphism ring of T. Then Λ is quasi-isomorphic to the derived endomorphism algebra $\mathrm{RHom}(T, T)$ and so the functor $\mathrm{RHom}(T, -)$ yields a triangle equivalence

$$\mathbf{D}(\mathcal{A}) \xrightarrow{\sim} \mathbf{D}(\mathrm{Mod}\,\Lambda)$$

by Proposition 9.1.22. We use it to identify $\mathbf{D}(\mathcal{A})$ with $\mathbf{D}(\mathrm{Mod}\,\Lambda)$. The canonical t-structure on $\mathbf{D}(\mathcal{A})$ is denoted by $(\mathcal{D}^{\leq 0}, \mathcal{D}^{> 0})$, while the canonical t-structure on $\mathbf{D}(\mathrm{Mod}\,\Lambda)$ is denoted by $(\mathcal{D}(\Lambda)^{\leq 0}, \mathcal{D}(\Lambda)^{> 0})$.

Lemma 9.3.8. *Suppose that $\mathcal{A}$ and $\mathrm{Mod}\,\Lambda$ have finite global dimension. Then the functor $\mathrm{RHom}(T, -)$ restricts to an equivalence $\mathbf{D}^b(\mathcal{A}) \xrightarrow{\sim} \mathbf{D}^b(\mathrm{Mod}\,\Lambda)$.*

Proof Given objects $X, Y \in \mathbf{D}^b(\mathcal{A})$ we have $\mathrm{Hom}(X, \Sigma^i Y) = 0$ for almost all i since $\mathcal{A}$ has finite global dimension. This is easily shown by induction on the number of integers n such that $H^n(X \oplus Y) \neq 0$. It follows that $\mathrm{RHom}(T, -)$ restricts to a functor $F \colon \mathbf{D}^b(\mathcal{A}) \to \mathbf{D}^b(\mathrm{Mod}\,\Lambda)$, since

$$H^i \mathrm{RHom}(T, X) \cong \mathrm{Hom}(T, \Sigma^i X)$$

and $T \in \mathbf{D}^b(\mathcal{A})$ by Lemma 9.3.7. On the other hand, $\mathbf{D}^b(\mathrm{Mod}\,\Lambda)$ equals the thick subcategory of $\mathbf{D}(\mathrm{Mod}\,\Lambda)$ which is generated by the category $\mathrm{Proj}\,\Lambda$ of projective Λ-modules, viewed as complexes concentrated in degree zero, since Λ has finite global dimension. It follows that F is essentially surjective since F identifies the closure of T under arbitrary coproducts and direct summands with $\mathrm{Proj}\,\Lambda$. $\square$

From now on suppose that the global dimension of $\mathcal{A}$ is bounded by d, and fix $t \geq 0$ such that $H^p T = 0$ for all $p \notin [-t, 0]$, cf. Lemma 9.3.7.

Lemma 9.3.9. *We have $\mathcal{D}(\Lambda)^{\leq 0} \subseteq \mathcal{D}^{\leq 0}$.*

Proof For $X \in \mathcal{D}^{> 0}$ and $i \leq 0$ we have $\mathrm{Hom}(T, \Sigma^i X) = 0$ since $T \in \mathcal{D}^{\leq 0}$. It follows that $X \in \mathcal{D}(\Lambda)^{> 0}$, since $\mathbf{D}(\mathcal{A}) \xrightarrow{\sim} \mathbf{D}(\mathrm{Mod}\,\Lambda)$ identifies T with Λ and $H^i X \cong \mathrm{Hom}(\Lambda, \Sigma^i X)$ in $\mathbf{D}(\mathrm{Mod}\,\Lambda)$. Thus $\mathcal{D}(\Lambda)^{\leq 0} \subseteq \mathcal{D}^{\leq 0}$. $\square$

Lemma 9.3.10. *We have $\mathcal{D}(\Lambda)^{\geq 0} \subseteq \mathcal{D}^{\geq -d-t-2}$.*

Proof Let $X \in \mathcal{D}^{\leq 0}$. Then $H^i T = 0$ for all $i \notin [-t, 0]$ implies $\mathrm{Hom}(T, \Sigma^i X) = 0$ for all $i > d + t + 2$ by Lemma 9.3.6. It follows that $\mathcal{D}^{\leq 0} \subseteq \mathcal{D}(\Lambda)^{\leq d+t+2}$, and therefore $\mathcal{D}(\Lambda)^{\geq 0} \subseteq \mathcal{D}^{\geq -d-t-2}$. $\square$

Theorem 9.3.11. *Let $\mathcal{A}$ be a Grothendieck category and $T \in \mathbf{D}(\mathcal{A})$ a tilting object. Then* $\mathrm{RHom}(T, -)$ *induces a triangle equivalence* $\mathbf{D}(\mathcal{A}) \xrightarrow{\sim} \mathbf{D}(\mathrm{Mod}\,\Lambda)$ *for* $\Lambda = \mathrm{End}(T)$, *and the global dimension of* Λ *is at most $2d + t$, where d is the global dimension of $\mathcal{A}$ and t is the smallest integer such that $H^i T = 0$ for all i outside an interval of length t.*

Proof Let $X, Y \in \mathrm{Mod}\,\Lambda$ and $i > 2d + t + 4$. Then

$$X \in \mathcal{D}(\Lambda)^{\geq 0} \subseteq \mathcal{D}^{\geq -d-t-2} \quad \text{and} \quad \Sigma^i Y \in \mathcal{D}(\Lambda)^{<-2d-t-4} \subseteq \mathcal{D}^{<-2d-t-4}$$

by Lemma 9.3.9 and Lemma 9.3.10. It follows from Lemma 9.3.6 that

$$\mathrm{Ext}^i(X, Y) = \mathrm{Hom}(X, \Sigma^i Y) = 0.$$

Thus the global dimension of Λ is bounded by $2d + t + 4$. In order to improve this bound, observe that $\mathrm{RHom}(T, -)$ restricts to an equivalence $\mathbf{D}^b(\mathcal{A}) \xrightarrow{\sim} \mathbf{D}^b(\mathrm{Mod}\,\Lambda)$ by Lemma 9.3.8. Then we compare t-structures on $\mathbf{D}^b(\mathcal{A})$ and use Lemma 9.3.3 instead of Lemma 9.3.6. It follows that the global dimension of Λ is bounded by $2d + t$. $\qquad\square$

Locally Noetherian Grothendieck Categories

Let $\mathcal{A}$ be a Grothendieck category and suppose $\mathcal{A}$ is *locally noetherian*, that is, $\mathcal{A}$ has a generating set of noetherian objects. Let us denote by $\mathrm{noeth}\,\mathcal{A}$ the full subcategory of noetherian objects. The full subcategory of injective objects $\mathrm{Inj}\,\mathcal{A}$ is closed under coproducts (Theorem 11.2.12) and therefore $\mathbf{K}(\mathrm{Inj}\,\mathcal{A})$ has arbitrary coproducts.

Proposition 9.3.12. *The triangulated category* $\mathbf{K}(\mathrm{Inj}\,\mathcal{A})$ *is compactly generated (so equals the localising subcategory generated by all compact objects) and the inclusion* $\mathrm{noeth}\,\mathcal{A} \to \mathcal{A}$ *induces a fully faithful functor* $\mathbf{D}^b(\mathrm{noeth}\,\mathcal{A}) \to \mathbf{K}(\mathrm{Inj}\,\mathcal{A})$ *that identifies* $\mathbf{D}^b(\mathrm{noeth}\,\mathcal{A})$ *with the full subcategory of compact objects.*

Proof For an object $A \in \mathcal{A}$ let $A \to iA$ denote an injective resolution. Then

$$\mathrm{Hom}_{\mathbf{K}(\mathcal{A})}(iA, X) \xrightarrow{\sim} \mathrm{Hom}_{\mathbf{K}(\mathcal{A})}(A, X)$$

for all $X \in \mathbf{K}(\mathrm{Inj}\,\mathcal{A})$ (Lemma 4.2.6). If A is noetherian, then $\mathrm{Hom}_{\mathcal{A}}(A, -)$ preserves coproducts and it follows that iA is compact in $\mathbf{K}(\mathrm{Inj}\,\mathcal{A})$. There are exact and fully faithful functors

$$\mathbf{D}^b(\mathrm{noeth}\,\mathcal{A}) \to \mathbf{D}^b(\mathcal{A}) \xrightarrow{\sim} \mathbf{K}^{+,b}(\mathrm{Inj}\,\mathcal{A}) \to \mathbf{K}(\mathrm{Inj}\,\mathcal{A})$$

(Proposition 4.2.19 and Corollary 4.2.9), which map $\mathbf{D}^b(\text{noeth}\,\mathcal{A})$ into the subcategory of compact objects by our previous observation, since

$$\mathbf{D}^b(\text{noeth}\,\mathcal{A}) \xrightarrow{\sim} \text{Thick}(\{iA \mid A \in \text{noeth}\,\mathcal{A}\}).$$

Given an object $X \in \mathbf{K}(\text{Inj}\,\mathcal{A})$, we have that $\text{Hom}_{\mathbf{K}(\mathcal{A})}(A, \Sigma^n X) = 0$ for all $A \in \text{noeth}\,\mathcal{A}$ and $n \in \mathbb{Z}$ implies $X = 0$ (Lemma 6.4.11). This implies that $\mathbf{D}^b(\text{noeth}\,\mathcal{A})$ generates $\mathbf{K}(\text{Inj}\,\mathcal{A})$ and that $\mathbf{D}^b(\text{noeth}\,\mathcal{A})$ identifies with the full subcategory of compact objects; see Lemma 9.3.13 below.   □

Lemma 9.3.13. *Let $\mathcal{T}$ be a triangulated category with arbitrary coproducts and let $\mathcal{C} \subseteq \mathcal{T}$ be a set of compact objects. Suppose $\text{Hom}(C, \Sigma^n X) = 0$ for all $C \in \mathcal{C}$ and $n \in \mathbb{Z}$ implies $X = 0$. Then $\mathcal{T} = \text{Loc}(\mathcal{C})$ and every compact object in $\mathcal{T}$ belongs to $\text{Thick}(\mathcal{C})$.*

Proof Set $\mathcal{T}' = \text{Loc}(\mathcal{C})$ and fix $X \in \mathcal{T}$. Then the functor $\text{Hom}(-, X)|_{\mathcal{T}'}$ is representable, say by $X' \in \mathcal{T}'$; this follows from Brown's representability theorem (Theorem 3.4.16). We obtain a morphism $X' \to X$ by Yoneda's lemma and complete this to an exact triangle $X' \to X \to X'' \to \Sigma X'$. Then $\text{Hom}(C, \Sigma^n X'') = 0$ for all $C \in \mathcal{C}$ and $n \in \mathbb{Z}$, and therefore $X' \xrightarrow{\sim} X$. If X is compact, then the construction of the representing object implies that $X \in \text{Thick}(\mathcal{C})$ (Proposition 3.4.15).   □

Tilting for Bounded Derived Categories

Let $\mathcal{A}$ be an essentially small abelian category and let $T \in \mathbf{D}^b(\mathcal{A})$ be a tilting object; recall this means $\text{Hom}(T, \Sigma^i T) = 0$ for all $i \neq 0$ and $\mathbf{D}^b(\mathcal{A})$ equals the thick subcategory generated by T. Then we know from Theorem 7.2.3 that for $\Lambda = \text{End}(T)$ the composite $\text{proj}\,\Lambda \xrightarrow{\sim} \text{add}\,T \hookrightarrow \mathbf{D}^b(\mathcal{A})$ induces a triangle equivalence

$$\mathbf{D}^b(\text{proj}\,\Lambda) \xrightarrow{\sim} \mathbf{K}^b(\text{add}\,T) \xrightarrow{\sim} \mathbf{D}^b(\mathcal{A}).$$

Theorem 9.3.14. *Let $\mathcal{A}$ be an essentially small abelian category, and suppose every object in $\mathcal{A}$ is noetherian. Let $T \in \mathbf{D}^b(\mathcal{A})$ be tilting. Then the global dimension of $\Lambda = \text{End}(T)$ is at most $2d + t$, where d is the global dimension of $\mathcal{A}$ and t is the smallest integer such that $H^i T = 0$ for all i outside an interval of length t. Moreover, $\text{RHom}(T, -)$ induces a triangle equivalence $\mathbf{D}^b(\mathcal{A}) \xrightarrow{\sim} \mathbf{D}^b(\text{mod}\,\Lambda)$ when Λ is right coherent.*

Proof Let $\bar{\mathcal{A}} := \text{Lex}(\mathcal{A}^{\text{op}}, \text{Ab})$ denote the category of left exact functors $\mathcal{A}^{\text{op}} \to \text{Ab}$. Then $\bar{\mathcal{A}}$ is a Grothendieck category and the Yoneda embedding $\mathcal{A} \to \bar{\mathcal{A}}$ which sends $X \in \mathcal{A}$ to $\text{Hom}(-, X)$ is fully faithful and exact (Corollary 11.1.19). In fact, $\bar{\mathcal{A}}$ is locally noetherian since $\mathcal{A}$ is noetherian, and the inclusion $\mathcal{A} \to \bar{\mathcal{A}}$ identifies $\mathcal{A}$ with the full subcategory of noetherian objects in $\bar{\mathcal{A}}$ (Proposition 11.2.5). It follows from Baer's criterion that an object I of $\bar{\mathcal{A}}$ is injective if and only if $\mathrm{Ext}^1(-, I)$ vanishes on all noetherian objects (Lemma 11.2.10). This implies that the global dimension of $\bar{\mathcal{A}}$ equals that of $\mathcal{A}$. Thus we have an equivalence $\mathbf{K}(\mathrm{Inj}\,\mathcal{A}) \xrightarrow{\sim} \mathbf{D}(\mathcal{A})$ by Proposition 9.3.5, and we can apply Proposition 9.3.12. The functor $\mathbf{D}^b(\mathcal{A}) \to \mathbf{D}(\bar{\mathcal{A}})$ identifies a tilting object T of $\mathbf{D}^b(\mathcal{A})$ with a tilting object of $\mathbf{D}(\bar{\mathcal{A}})$. Let $\Lambda = \mathrm{End}(T)$. Then Theorem 9.3.11 provides the bound for the global dimension of Λ. When Λ is right coherent, then the triangle equivalence $\mathbf{D}(\bar{\mathcal{A}}) \xrightarrow{\sim} \mathbf{D}(\mathrm{Mod}\,\Lambda)$ restricts to an equivalence

$$\mathbf{D}^b(\mathcal{A}) \xrightarrow{\sim} \mathbf{D}^b(\mathrm{proj}\,\Lambda) \xrightarrow{\sim} \mathbf{D}^b(\mathrm{mod}\,\Lambda)$$

on the full subcategory of compact objects □

We end our discussion of tilting objects with some remarks. Let us fix an essentially small abelian category $\mathcal{A}$ with a tilting object $T \in \mathbf{D}^b(\mathcal{A})$, and set $\Lambda = \mathrm{End}(T)$.

Remark 9.3.15. The assumption on $\mathcal{A}$ to be noetherian is needed for the global dimension of Λ to be finite. Consider for $\mathcal{A}$ the category of vector spaces of dimension at most $\aleph_\omega$. Then a vector space of dimension $\aleph_\omega$ is a tilting object and its endomorphism ring has infinite global dimension [154].

Recall that $\mathrm{pcoh}\,\Lambda$ denotes the full subcategory of pseudo-coherent Λ-modules; it is a full exact subcategory of the category of all Λ-modules. The category $\mathrm{pcoh}\,\Lambda$ is the appropriate generalisation of $\mathrm{mod}\,\Lambda$, and $\mathrm{pcoh}\,\Lambda$ equals $\mathrm{mod}\,\Lambda$ when Λ is right coherent.

Remark 9.3.16. Suppose that $\mathcal{A}$ is noetherian and of finite global dimension. Then $\mathrm{RHom}(T, -)$ induces a triangle equivalence $\mathbf{D}^b(\mathcal{A}) \xrightarrow{\sim} \mathbf{D}^b(\mathrm{pcoh}\,\Lambda)$ (Lemma 5.2.11).

For each pair of objects $X, X' \in \mathcal{A}$ we have $\mathrm{Ext}^i(X, X') = 0$ for $i \gg 0$, since each object in $\mathcal{A}$ is finitely built from T. This provides some restriction on the global dimension of $\mathcal{A}$.

Remark 9.3.17. Let $\mathcal{A}$ be a length category. Then

$$\mathrm{gl.dim}\,\mathcal{A} = \inf_{\substack{S,S' \\ \mathrm{simple}}} \{i \in \mathbb{N} \mid \mathrm{Ext}^{i+1}(S, S') = 0\} < \infty$$

since the length of $\bigoplus_n H^n T$ gives a bound for the number of isoclasses of simple objects (Lemma 7.2.6).

Remark 9.3.18. The global dimension of $\mathcal{A}$ need not be finite when $\mathbf{D}^b(\mathcal{A})$ admits a tilting object. Let Λ be a right noetherian ring and set $\mathcal{A} = \mathrm{mod}\,\Lambda$.

Then $\Lambda \in \mathbf{D}^b(\mathcal{A})$ is tilting if and only if each object in $\mathcal{A}$ has finite projective dimension. In this case the global dimension of $\mathcal{A}$ equals the finitistic dimension of Λ, which may be infinite even when Λ is commutative (Example 7.2.20).

Notes

For representations of algebras, the first link between tilting and derived categories was established by Happel [101]. The Morita theorem for derived categories is due to Rickard [170]. The use of differential graded algebras was then explained in Keller's work [121], which inspired much of our exposition. For instance, the description of algebraic triangulated categories via derived categories of differential graded algebras is taken from [121]. Differential graded algebras were introduced by Cartan in order to study the cohomology of Eilenberg–MacLane spaces [45]. The argument for the stable homotopy category of spectra to be not algebraic was suggested by Dwyer. The intrinsic description of perfect complexes over coherent rings via initial objects (Proposition 9.2.13) is due to Neeman.

Derived equivalences appear already in Grothendieck's work on duality, as explained in notes by Hartshorne [106]. Grothendieck duality extends Serre duality for coherent sheaves on algebraic varieties [189]; a modern version based on Brown representability is due to Neeman [148].

A derived equivalence between two algebras preserves various homological invariants, for instance finiteness of global dimension [83, 102]. The more general result for tilting from abelian to module categories is taken from work with Keller [122]. The proof involves t-structures; these formalise truncations of complexes and were introduced by Beilinson, Bernšteĭn and Deligne [26]. The study of the homotopy category of complexes of injective objects for locally noetherian categories was initiated in [129].

10

Examples of Derived Equivalences

Contents

10.1	Coherent Sheaves on Projective Space		329
10.2	Koszul Duality		330
10.3	The BGG Correspondence		332
10.4	Koszul Duality for the Beilinson Algebra		333
10.5	Weighted Projective Lines		334
10.6	Gentle Algebras		336

A derived equivalence provides a tool for transferring homological information between two abelian categories that are not necessarily equivalent. We discuss several important examples of such derived equivalences. This amounts to identifying tilting objects. We do not give proofs but include references to the literature for further details.

10.1 Coherent Sheaves on Projective Space

Let k be a field and fix an integer $n \geq 0$. A theorem of Beilinson [25] provides a tilting object for the category $\operatorname{coh} \mathbb{P}^n$ of coherent sheaves on the projective n-space over k.

More precisely, let $A = k[x_0, \ldots, x_n]$ be the polynomial ring on $n + 1$ variables, graded by degree, so that $\mathbb{P}^n = \operatorname{Proj} A$ (the projective variety or scheme given by the set of homogeneous prime ideals). Let $\operatorname{grmod} A$ be the category of finitely generated $\mathbb{Z}$-graded modules. Then a theorem of Serre [188] provides an equivalence of abelian categories

$$\frac{\operatorname{grmod} A}{\operatorname{grmod}_0 A} \xrightarrow{\sim} \operatorname{coh} \mathbb{P}^n$$

where $\mathrm{grmod}_0 A$ denotes the Serre subcategory of finite length objects.

Let $\mathcal{O}$ denote the structure sheaf on $\mathbb{P}^n$, so the image of the graded projective module $A \in \mathrm{grmod}\, A$. The object

$$T = \mathcal{O} \oplus \mathcal{O}(1) \oplus \cdots \oplus \mathcal{O}(n)$$

is a tilting object of $\mathrm{coh}\,\mathbb{P}^n$ and we have

$$\mathrm{Hom}(\mathcal{O}(i), \mathcal{O}(j)) \cong A_{j-i} \qquad (0 \le i, j \le n),$$

where $A_p \subseteq A$ denotes the subgroup of homogeneous degree p elements. Thus the endomorphism ring of T is isomorphic to the *Beilinson algebra* B_n given by the path algebra of the following quiver

$$0 \xrightarrow[\;\;x_n\;\;]{\overset{x_0}{\cdots}} 1 \xrightarrow[\;\;x_n\;\;]{\overset{x_0}{\cdots}} 2 \xrightarrow[\;\;x_n\;\;]{\overset{x_0}{\cdots}} \cdots \xrightarrow[\;\;x_n\;\;]{\overset{x_0}{\cdots}} n$$

modulo all relations of the form $x_i x_j - x_j x_i$. It follows from Theorem 7.2.3 that the functor $\mathrm{Hom}(T, -)$ induces a triangle equivalence

$$\mathbf{D}^b(\mathrm{coh}\,\mathbb{P}^n) \xrightarrow{\sim} \mathbf{D}^b(\mathrm{mod}\, B_n),$$

and Proposition 9.1.22 extends this to a triangle equivalence

$$\mathbf{D}(\mathrm{Qcoh}\,\mathbb{P}^n) \xrightarrow{\sim} \mathbf{D}(\mathrm{Mod}\, B_n).$$

10.2 Koszul Duality

Let k be a field and let $A = \bigoplus_{i \ge 0} A_i$ be a graded k-algebra. Suppose that $A_0 = k$ and that each A_i is finite dimensional over k. We consider the category $\mathrm{GrMod}\, A$ of graded A-modules with morphisms of degree zero, and $\mathrm{grmod}\, A$ denotes the full subcategory of finitely generated modules.

We collect the basic facts about Koszul algebras and Koszul duality [27]. The algebra A is *Koszul* if there is a projective resolution $\cdots \to P_1 \to P_0 \to k \to 0$ of the trivial A-module k such that each P_i is generated in degree i. Let us write

$$E(A) = \bigoplus_{n \ge 0} \mathrm{Ext}_A^n(k, k)$$

for the Ext-algebra of the trivial A-module k.

The algebra A is *quadratic* if $A = T(V)/\langle R \rangle$, where

$$T(V) = \bigoplus_{n \ge 0} V^{\otimes n}$$

denotes the *tensor algebra* of a finite dimensional k-space V and $\langle R \rangle$ denotes an ideal generated by a subspace $R \subseteq V \otimes V$. The *quadratic dual* of A is

$$A^! = T(V^\vee)/\langle R^\perp \rangle$$

where $V^\vee$ denotes the dual space of V and

$$R^\perp = \{\phi \in (V \otimes V)^\vee \mid \phi(R) = 0\} \subseteq V^\vee \otimes V^\vee.$$

Clearly, we have $(A^!)^! \cong A$ as graded k-algebras. Also, a Koszul algebra A is quadratic. Indeed, the maximal ideal $\bigoplus_{i>0} A_i$ is generated in degree one since the first step of a projective resolution $\cdots \to P_1 \to P_0 \to k \to 0$ is generated in degree one, and the second step gives the relations, which then must be quadratic as P_2 is generated in degree two.

Consider for a quadratic algebra $A = T(V)/\langle R \rangle$ the *Koszul complex*

$$K(A) \quad \cdots \longrightarrow (A^!_n)^\vee \otimes_k A \xrightarrow{d_n} (A^!_{n-1})^\vee \otimes_k A \longrightarrow \cdots \longrightarrow A \longrightarrow 0 \longrightarrow \cdots$$

with differential d_n taking $\phi \in \mathrm{Hom}_k(A^!_n, A) = (A^!_n)^\vee \otimes_k A$ to

$$A^!_{n-1} = A^!_{n-1} \otimes_k k \to A^!_{n-1} \otimes_k V^\vee \otimes_k V \xrightarrow{\mu \otimes V} A^!_n \otimes_k V \xrightarrow{\phi \otimes V} A \otimes_k V \xrightarrow{\mu} A$$

where μ denotes the multiplication. When the algebra A is Koszul, then $K(A)$ provides a projective resolution of the trivial A-module k.

Given chain complexes $X \in \mathbf{K}(\mathrm{GrMod}\, A)$ and $Y \in \mathbf{K}(\mathrm{GrMod}\, A^!)$ the assignments

$$X \longmapsto \mathrm{Hom}_A(K(A), X) \qquad \text{and} \qquad Y \longmapsto Y \otimes_{A^!} K(A)$$

induce an adjoint pair of functors $\mathbf{K}(\mathrm{GrMod}\, A) \leftrightarrows \mathbf{K}(\mathrm{GrMod}\, A^!)$. We write $\mathbf{D}^\downarrow(A)$ for the full subcategory of objects $X \in \mathbf{D}(\mathrm{GrMod}\, A)$ such that for some integer n (depending on X)

$$X^j_i \neq 0 \implies j \geq -n \text{ and } i + j \leq n$$

(with $\bigoplus_{i \in \mathbb{Z}} X^j_i$ the A-module in cohomological degree j). Analogously, $\mathbf{D}^\uparrow(A^!)$ denotes the full subcategory of objects $Y \in \mathbf{D}(\mathrm{GrMod}\, A^!)$ such that for some integer n (depending on Y)

$$Y^j_i \neq 0 \implies j \leq n \text{ and } i + j \geq -n.$$

If the algebra A is Koszul, then $\mathrm{Hom}_A(K(A), -)$ induces an isomorphism of k-algebras

$$A^! \cong H^* \mathrm{End}_A(K(A)) \cong E(A).$$

Moreover, $\text{Hom}_A(K(A), -)$ yields the derived functor

$$\text{RHom}_A(k, -) \colon \mathbf{D}(\text{GrMod}\,A) \longrightarrow \mathbf{D}(\text{GrMod}\,A^!)$$

which restricts to triangle equivalences

$$\mathbf{D}^{\downarrow}(A) \xrightarrow{\sim} \mathbf{D}^{\uparrow}(A^!) \qquad \text{and} \qquad \mathbf{D}^b(\text{grmod}\,A) \xrightarrow{\sim} \mathbf{D}^b(\text{grmod}\,A^!)$$

that identify $A_0 = k$ with $A^!$.

10.3 The BGG Correspondence

Let k be a field. Fix an integer $n \geq 0$ and an $(n+1)$-dimensional space V over k. We consider the *symmetric algebra*

$$S(V) = T(V)/\langle x \otimes y - y \otimes x \mid x, y \in V \rangle$$

and the *exterior algebra*

$$\Lambda(V) = T(V)/\langle x \otimes x \mid x \in V \rangle.$$

These algebras are graded via the canonical grading of the tensor algebra $T(V)$.

Now let $A = S(V)$ and let $A^! = \Lambda(V^{\vee})$ be its Koszul dual. Both algebras are $\mathbb{Z}$-graded and we consider the categories of finitely generated $\mathbb{Z}$-graded modules grmod A and grmod $A^!$. The above Koszul duality [27, 35] provides a triangle equivalence

$$\mathbf{D}^b(\text{grmod}\,A) \xrightarrow{\sim} \mathbf{D}^b(\text{grmod}\,A^!) \qquad\qquad (10.3.1)$$

which identifies $A_0 = k$ with $A^!$. In particular $\text{Ext}^*_A(A_0, A_0) \cong A^!$.

Observe that the inclusion $\text{grmod}_0\,A \to \text{grmod}\,A$ of the category of finite length modules induces a fully faithful functor

$$\mathbf{D}^b(\text{grmod}_0\,A) \longrightarrow \mathbf{D}^b(\text{grmod}\,A);$$

see Example 4.2.20. A theorem of Serre [188] provides an equivalence

$$\frac{\text{grmod}\,A}{\text{grmod}_0\,A} \xrightarrow{\sim} \text{coh}\,\mathbb{P}^n.$$

Thus $\mathbf{D}^b(\text{grmod}_0\,A)$ identifies with the kernel of the functor $\mathbf{D}^b(\text{grmod}\,A) \to \mathbf{D}^b(\text{coh}\,\mathbb{P}^n)$ and this yields a triangle equivalence

$$\frac{\mathbf{D}^b(\text{grmod}\,A)}{\mathbf{D}^b(\text{grmod}_0\,A)} \xrightarrow{\sim} \mathbf{D}^b(\text{coh}\,\mathbb{P}^n);$$

see Lemma 4.4.1. On the other hand, $\operatorname{grmod} A^!$ is a Frobenius category. It follows from Proposition 4.4.18 that the inclusion

$$\mathbf{D}^b(\operatorname{grproj} A^!) \longrightarrow \mathbf{D}^b(\operatorname{grmod} A^!)$$

induces a triangle equivalence

$$\frac{\mathbf{D}^b(\operatorname{grmod} A^!)}{\mathbf{D}^b(\operatorname{grproj} A^!)} \xrightarrow{\ \sim\ } \underline{\operatorname{grmod} A^!}.$$

The equivalence (10.3.1) identifies $\mathbf{D}^b(\operatorname{grmod}_0 A)$ with $\mathbf{D}^b(\operatorname{grproj} A^!)$ because Koszul duality identifies simples with indecomposable projectives. This yields the following triangle equivalence

$$\mathbf{D}^b(\operatorname{coh} \mathbb{P}^n) \xrightarrow{\ \sim\ } \underline{\operatorname{grmod} A^!}$$

which is due to Bernšteĭn, Gel'fand and Gel'fand [35].

10.4 Koszul Duality for the Beilinson Algebra

Let k be a field. Fix an integer $n \geq 0$ and an $(n + 1)$-dimensional space V over k. We consider again the symmetric algebra and the exterior algebra

$$S(V) = \bigoplus_{i \geq 0} S^i \quad \text{and} \quad \Lambda(V^\vee) = \bigoplus_{i \geq 0} \Lambda^i.$$

The Beilinson algebra can be written as an algebra of $(n + 1) \times (n + 1)$-matrices

$$A := \begin{bmatrix} S^0 & & & & \\ S^1 & S^0 & & & \\ S^2 & S^1 & S^0 & & \\ \vdots & \vdots & & \ddots & \\ S^n & S^{n-1} & S^{n-2} & \cdots & S^0 \end{bmatrix}$$

and we obtain a grading $A = A_0 \oplus A_1 \oplus \cdots \oplus A_n$ by setting $A_i = J^i/J^{i+1}$, where $J = J(A)$ denotes the Jacobson radical of A. The diagonal matrices provide the degree zero component

$$A_0 = S_0 \oplus S_1 \oplus \cdots \oplus S_n,$$

where S_i equals the simple A-module given by projecting onto the $(i+1)$-entry. The Ext-algebra is given by the exterior powers

$$B := \operatorname{Ext}^*_A(A_0, A_0) \cong \begin{bmatrix} \Lambda^0 & \Lambda^1 & \Lambda^2 \cdots \cdots \Lambda^n \\ & \Lambda^0 & \Lambda^1 \cdots \Lambda^{n-1} \\ & & \Lambda^0 \cdots \Lambda^{n-2} \\ & & & \ddots & \vdots \\ & & & & \Lambda^0 \end{bmatrix}$$

and this computation shows that

$$T = \Sigma^n S_0 \oplus \Sigma^{n-1} S_1 \oplus \cdots \oplus \Sigma^0 S_n$$

is a tilting object in $\mathbf{D}(\operatorname{Mod} A)$ with $\operatorname{End}_{\mathbf{D}(A)}(T) = B$. Thus Theorem 9.2.4 yields a triangle equivalence

$$\operatorname{RHom}_A(T, -) : \mathbf{D}(\operatorname{Mod} A) \xrightarrow{\sim} \mathbf{D}(\operatorname{Mod} B).$$

10.5 Weighted Projective Lines

Let k be an algebraically closed field, let $\mathbb{P}^1$ denote the projective line over k, let $\lambda = (\lambda_1, \ldots, \lambda_n)$ be a (possibly empty) collection of distinct closed points of $\mathbb{P}^1$, and let $\mathbf{p} = (p_1, \ldots, p_n)$ be a *weight sequence*, that is, a sequence of positive integers. The triple $\mathbb{X} = (\mathbb{P}^1, \lambda, \mathbf{p})$ is called a *weighted projective line*. Geigle and Lenzing [88] have associated to each weighted projective line a category $\operatorname{coh} \mathbb{X}$ of coherent sheaves on $\mathbb{X}$, which is the quotient category of the category of finitely generated $G(\mathbf{p})$-graded $S(\mathbf{p}, \lambda)$-modules, modulo the Serre subcategory of finite length modules. Here $G(\mathbf{p})$ is the rank one additive group

$$G(\mathbf{p}) = \langle \vec{x}_1, \ldots, \vec{x}_n, \vec{c} \mid p_1 \vec{x}_1 = \cdots = p_n \vec{x}_n = \vec{c} \rangle,$$

and

$$S(\mathbf{p}, \lambda) = k[u, v, x_1, \ldots, x_n] / (x_i^{p_i} + \lambda_{i1} u - \lambda_{i0} v),$$

with grading $\deg u = \deg v = \vec{c}$ and $\deg x_i = \vec{x}_i$, where $\lambda_i = [\lambda_{i0} : \lambda_{i1}]$ in $\mathbb{P}^1$. Note that $\operatorname{coh} \mathbb{X}$ is an hereditary abelian category with finite dimensional Hom and Ext spaces.

Consider the class of connected hereditary abelian categories (k-linear, with finite dimensional Hom and Ext spaces) which admit a tilting object. A theorem of Happel [104] shows that each derived equivalence class contains the module category of an hereditary algebra or the category of coherent sheaves on a weighted projective line.

Let us describe a specific tilting object for the category $\operatorname{coh} \mathbb{X}$. The free module $S(\mathbf{p}, \lambda)$ yields a structure sheaf $\mathcal{O}$, and shifting the grading gives twists $E(\vec{x})$ for any sheaf E and $\vec{x} \in G(\mathbf{p})$. Then

$$T = \bigoplus_{0 \le \vec{x} \le \vec{c}} \mathcal{O}(\vec{x})$$

is a tilting object for $\text{coh}\,\mathbb{X}$ and its endomorphism algebra is isomorphic to the *canonical algebra* $C(\mathbf{p}, \lambda)$ in the sense of Ringel [172], which is the finite dimensional associative algebra given by the quiver

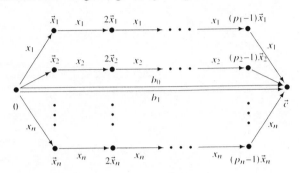

modulo the relations[1]

$$x_i^{p_i} = \lambda_{i0} b_1 - \lambda_{i1} b_0 \quad (i = 1, \dots, n).$$

This yields a triangle equivalence

$$\mathbf{D}^b(\text{coh}\,\mathbb{X}) \xrightarrow{\sim} \mathbf{D}^b(\text{mod}\,S(\mathbf{p}, \lambda)).$$

Every object in $\text{coh}\,\mathbb{X}$ is the direct sum of a torsion free sheaf and a finite length sheaf. A torsion free sheaf has a finite filtration by line bundles, that is, sheaves of the form $\mathcal{O}(\vec{x})$. The finite length sheaves are easily described as follows. There are simple sheaves S_x ($x \in \mathbb{P}^1 \setminus \lambda$) and S_{ij} ($1 \le i \le n$, $1 \le j \le p_i$) satisfying for any $r \in \mathbb{Z}$ that $\text{Hom}(\mathcal{O}(r\vec{c}), S_{ij}) \ne 0$ if and only if $j = 1$, and the only extensions between them are

$$\text{Ext}^1(S_x, S_x) = k, \quad \text{Ext}^1(S_{ij}, S_{ij'}) = k \quad (j' \equiv j - 1 \,(\text{mod } p_i)).$$

For each simple sheaf S and $p > 0$ there is a unique sheaf $S^{[p]}$ of length p and with top S, which is a uniserial object, so it has a unique composition series. These are all the finite length indecomposable sheaves.

The category $\text{coh}\,\mathbb{X}$ admits the following tilting object

$$\mathcal{O} \oplus \mathcal{O}(\vec{c}) \oplus (S_1^{[1]} \oplus \cdots \oplus S_1^{[p_1-1]}) \oplus \cdots \oplus (S_n^{[1]} \oplus \cdots \oplus S_n^{[p_n-1]})$$

---

[1] Note that the relations do not generate an admissible ideal of the path algebra, except when the collection λ is empty. In that case $C(\mathbf{p}, \lambda)$ equals the Kronecker algebra.

where $S_i = S_{i1}$ as above. The endomorphism algebra is the *squid algebra* $\mathrm{Sq}(\mathbf{p}, \lambda)$, in the sense of Brenner and Butler [41], which is the finite dimensional associative algebra given by the quiver

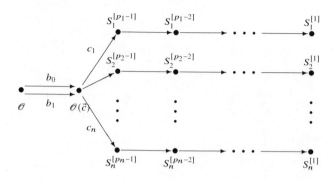

modulo the relations

$$c_i(\lambda_{i0}b_1 - \lambda_{i1}b_0) = 0 \quad (i = 1, \ldots, n).$$

This yields a triangle equivalence

$$\mathbf{D}^b(\mathrm{coh}\,\mathbb{X}) \xrightarrow{\sim} \mathbf{D}^b(\mathrm{mod}\,\mathrm{Sq}(\mathbf{p}, \lambda)).$$

10.6 Gentle Algebras

Let k be a field and $Q = (Q_0, Q_1, s, t)$ a finite quiver. We denote by Q^{op} the *opposite quiver* which is obtained by reversing the arrows, so

$$Q_0^{\mathrm{op}} = Q_0 \qquad \text{and} \qquad Q_1^{\mathrm{op}} = \{\alpha^- \mid \alpha \in Q_1\}$$

with $s(\alpha^-) = t(\alpha)$ and $t(\alpha^-) = s(\alpha)$. Let Q_2 denote the set of paths of length 2 and fix a partition $Q_2 = P_+ \sqcup P_-$. It is not difficult to check that the algebra $kQ/\langle P_-\rangle$ (not necessarily finite dimensional) is gentle if and only if $kQ^{\mathrm{op}}/\langle P_-^{\mathrm{op}}\rangle$ is gentle, where $P_-^{\mathrm{op}} = \{\alpha^- \mid \alpha \in P_+\}$.

Now suppose that the algebra $A = kQ/\langle P_-\rangle$ is gentle. Then A is graded by path length and we have a decomposition $A_0 = \bigoplus_{i \in Q_0} S_i$ into simple A-modules. Each arrow $\alpha: i \to j$ corresponds to an extension $\xi_\alpha: 0 \to S_i \to E_\alpha \to S_j \to 0$. A theorem of Green and Zacharia [91] shows that the algebra A is Koszul, and the algebra $kQ^{\mathrm{op}}/\langle P_-^{\mathrm{op}}\rangle$ identifies with the Koszul dual $A^! \cong \mathrm{Ext}_A^*(A_0, A_0)$ via the assignment $\alpha^- \mapsto \xi_\alpha$. For example, consider the quiver

$$x \, \circlearrowright \, \circ \, \circlearrowleft \, y$$

with $P_+ = \{xy, yx\}$ and $P_- = \{x^2, y^2\}$. Then the algebra $A = k\langle x, y \rangle / (x^2, y^2)$ is gentle with Koszul dual $A^! \cong k\langle x, y \rangle / (xy, yx)$.

PART FOUR

PURITY

11

Locally Finitely Presented Categories

Contents

11.1	**Locally Finitely Presented Categories**	**342**
	Filtered Colimits	342
	Locally Finitely Presented Categories	343
	Cofinal Subcategories	344
	Categories of Additive Functors	345
	Linear Representations	347
	Categories of Left Exact Functors	349
	Categories of Exact Functors	353
	Change of Categories	354
11.2	**Grothendieck Categories**	**356**
	Finitely Generated and Finitely Presented Objects	357
	Locally Noetherian Categories	359
	Locally Finite Categories	360
	Injective Objects	362
11.3	**Gröbner Categories**	**367**
	Hilbert's Basis Theorem	367
	Noetherian Posets	368
	Functor Categories	369
	Noetherian Functors	369
	Gröbner Categories	372
	Base Change	372
	Categories of Finite Sets	373
	FI-Modules	375
	Generic Representations	375
Notes		**376**

We study additive categories that are locally finitely presented. This means that every object is the filtered colimit of finitely presented objects. The categorical notion of being finitely presented means for an object X that the functor $\mathrm{Hom}(X, -)$ preserves filtered colimits. Of particular interest is the case of an abelian category. Every locally finitely presented abelian category is a Grothendieck category; so it is a category with injective envelopes and we can study its injective objects.

The theory of locally finitely presented categories applies in particular to locally noetherian Grothendieck categories, that is, Grothendieck categories having a generating set of noetherian objects. Then finitely presented and noetherian objects coincide. Also, in that case every injective object decomposes into a direct sum of indecomposable objects. We include a discussion of Gröbner categories and provide criteria for when a functor category is locally noetherian; this can be thought of as a generalisation of Hilbert's basis theorem.

11.1 Locally Finitely Presented Categories

We introduce the concept of a locally finitely presented additive category. Any locally finitely presented category $\mathcal{A}$ is completely determined by its subcategory $\mathrm{fp}\,\mathcal{A}$ of finitely presented objects, because $\mathcal{A}$ identifies with the category of left exact functors $(\mathrm{fp}\,\mathcal{A})^{\mathrm{op}} \to \mathrm{Ab}$.

Filtered Colimits

A category $\mathfrak{J}$ is called *filtered* if

(Fil1) the category is non-empty,
(Fil2) given objects i, i' there is an object j with morphisms $i \to j \leftarrow i'$, and
(Fil3) given morphisms $\alpha, \alpha' \colon i \to j$ there is a morphism $\beta \colon j \to k$ such that $\beta\alpha = \beta\alpha'$.

For a functor $F \colon \mathfrak{J} \to \mathcal{C}$, we denote by $\operatorname{colim} F$ or $\operatorname{colim}_{i \in \mathfrak{J}} F(i)$ its colimit, provided it exists in $\mathcal{C}$. The term *filtered colimit* is used for the colimit of a functor $F \colon \mathfrak{J} \to \mathcal{C}$ such that the category $\mathfrak{J}$ is filtered.

Example 11.1.1. (1) A partially ordered set $(I, \leq)$ can be viewed as a category. The objects are the elements of I and there is a unique morphism $i \to j$ whenever $i \leq j$. This category is filtered if and only if $(I, \leq)$ is non-empty and directed. A colimit $\operatorname{colim}_{i \in \mathfrak{J}} F(i)$ is called a *directed colimit* if $\mathfrak{J}$ is given by a directed set.

(2) The coproduct of a family of objects $(X_i)_{i \in I}$ can be written as

$$\coprod_{i \in I} X_i = \operatorname*{colim}_{J \in \mathfrak{J}} \left(\coprod_{i \in J} X_i \right)$$

where $\mathfrak{J}$ denotes the filtered category of finite subsets $J \subseteq I$.

(3) Let $\mathcal{A}$ be an additive category and $\mathcal{C} \subseteq \mathcal{A}$ a full additive subcategory that

is essentially small. For any $X \in \mathcal{A}$ let $\mathcal{C}/X$ denote the *slice category* consisting of pairs (C, ϕ) given by a morphism $\phi \colon C \to X$ with $C \in \mathcal{C}$. A morphism $(C, \phi) \to (C', \phi')$ is given by a morphism $\alpha \colon C \to C'$ in $\mathcal{C}$ such that $\phi'\alpha = \phi$. Then $\mathcal{C}/X$ is filtered, provided that each morphism in $\mathcal{C}$ admits a cokernel in $\mathcal{A}$ that lies in $\mathcal{C}$. In fact, having weak cokernels is sufficient.

Locally Finitely Presented Categories

Let $\mathcal{A}$ be an additive category and suppose that $\mathcal{A}$ is cocomplete. Thus each functor $F \colon \mathcal{J} \to \mathcal{A}$ from an essentially small category $\mathcal{J}$ admits a colimit. Let us recall the construction of the colimit because it is very explicit. For a morphism $\alpha \colon i \to j$ in $\mathcal{J}$ we set $s(\alpha) = i$ and $t(\alpha) = j$. For $j \in \mathcal{J}$ we write $\iota_j \colon F(j) \to \coprod_{i \in \mathcal{J}} F(i)$ for the canonical inclusion and set $\phi_\alpha = \iota_{s(\alpha)} - \iota_{t(\alpha)} \circ F(\alpha)$. Then colim F is computed as the cokernel of $\phi = (\phi_\alpha)_{\alpha \in \mathcal{J}}$ and fits into an exact sequence

$$\coprod_{\alpha \in \mathcal{J}} F(s(\alpha)) \xrightarrow{\phi} \coprod_{i \in \mathcal{J}} F(i) \longrightarrow \operatorname{colim} F \longrightarrow 0.$$

Often we write $F_i = F(i)$ for $i \in \mathcal{J}$ and then $\operatorname{colim}_i F_i = \operatorname{colim} F$. A consequence of this construction is the fact that an additive category is cocomplete if and only if it has coproducts and every morphism admits a cokernel.

An object $X \in \mathcal{A}$ is *finitely presented* if the functor $\operatorname{Hom}_{\mathcal{A}}(X, -)$ preserves filtered colimits. This means that for every filtered colimit $\operatorname{colim}_i Y_i$ in $\mathcal{A}$ the canonical map

$$\operatorname*{colim}_i \operatorname{Hom}_{\mathcal{A}}(X, Y_i) \longrightarrow \operatorname{Hom}_{\mathcal{A}}(X, \operatorname*{colim}_i Y_i)$$

is bijective. Let $\operatorname{fp}\mathcal{A}$ denote the full subcategory of finitely presented objects. We record the following elementary facts.

Lemma 11.1.2. *The subcategory* $\operatorname{fp}\mathcal{A}$ *is closed under finite coproducts, direct summands, and cokernels. If $X \in \operatorname{fp}\mathcal{A}$ is written as a filtered colimit $X = \operatorname{colim} X_i$, then for some index i_0 the canonical morphism $X_{i_0} \to X$ is a split epimorphism.* □

The category $\mathcal{A}$ is called *locally finitely presented* if $\operatorname{fp}\mathcal{A}$ is essentially small and every object in $\mathcal{A}$ is a filtered colimit of finitely presented objects. In that case any object $X \in \mathcal{A}$ can be written canonically as a filtered colimit

$$X = \operatorname*{colim}_{(C, \phi) \in \operatorname{fp}\mathcal{A}/X} C$$

of the forgetful functor $\operatorname{fp}\mathcal{A}/X \to \mathcal{A}$ that takes (C, ϕ) to C, as we will see in Corollary 11.1.16.

From now on the term 'locally finitely presented' for a category $\mathcal{A}$ includes the properties that $\mathcal{A}$ is additive and cocomplete.

Remark 11.1.3. Let $\mathcal{A}^2$ denote the category of morphisms in $\mathcal{A}$. If $\mathcal{A}$ is locally finitely presented, then $\mathcal{A}^2$ is locally finitely presented and $(\mathrm{fp}\,\mathcal{A})^2 \xrightarrow{\sim} \mathrm{fp}(\mathcal{A}^2)$. This means that each morphism in $\mathcal{A}$ can be written canonically as a filtered colimit of morphisms in $\mathrm{fp}\,\mathcal{A}$.

Example 11.1.4. (1) Let Λ be any ring. Then the category of Λ-modules is locally finitely presented. The finitely presented objects are precisely the modules M that admit a presentation $\Lambda^r \to \Lambda^s \to M \to 0$ for some integers $r, s \geq 0$.

(2) Let $\mathbb{X}$ be a *scheme* and suppose it is quasi-compact and quasi-separated. Then the category $\mathrm{Qcoh}\,\mathbb{X}$ of quasi-coherent $\mathcal{O}_{\mathbb{X}}$-modules is locally finitely presented. The finitely presented objects are precisely the finitely presented $\mathcal{O}_{\mathbb{X}}$-modules [97, I.6.9.12]. When $\mathbb{X}$ is noetherian, then the category of finitely presented objects identifies with the category $\mathrm{coh}\,\mathbb{X}$ of coherent sheaves.

Cofinal Subcategories

For the computation of filtered colimits it is often useful to vary the index category. We consider an essentially small filtered category $\mathcal{I}$ and a fully faithful functor $\phi\colon \mathcal{J} \to \mathcal{I}$. Then ϕ is called *cofinal* if it satisfies the equivalent conditions of the following lemma. When ϕ is an inclusion we call $\mathcal{J}$ a *cofinal subcategory* of $\mathcal{I}$.

Lemma 11.1.5. *Let $\mathcal{I}$ be an essentially small filtered category. For a fully faithful functor $\phi\colon \mathcal{J} \to \mathcal{I}$ the following are equivalent.*

(1) *For every object $i \in \mathcal{I}$ there exists $j \in \mathcal{J}$ and a morphism $i \to \phi(j)$.*

(2) *Every functor $F\colon \mathcal{I}^{\mathrm{op}} \to \mathrm{Set}$ induces an isomorphism $\lim F \xrightarrow{\sim} \lim(F \circ \phi)$.*

(3) *For every category $\mathcal{C}$ which admits filtered colimits, every functor $F\colon \mathcal{I} \to \mathcal{C}$ induces an isomorphism $\mathrm{colim}(F \circ \phi) \xrightarrow{\sim} \mathrm{colim}\,F$.*

Moreover, in this case the category $\mathcal{J}$ is filtered.

Proof (1) $\Rightarrow$ (2): Limits in the category of sets can be calculated explicitly. Thus the condition (1) implies that $\lim F \to \lim(F \circ \phi)$ is injective. Combined with the fact that $\mathcal{I}$ is filtered, the map is also bijective.

(2) $\Rightarrow$ (3): We have for each $X \in \mathcal{C}$ a canonical bijection

$$\mathrm{Hom}(\mathrm{colim}_i F(i), X) \xrightarrow{\sim} \lim_i \mathrm{Hom}(F(i), X).$$

Thus we can use the functor $F_X\colon \mathcal{I}^{\mathrm{op}} \to \mathrm{Set}$ given by $i \mapsto \mathrm{Hom}(F(i), X)$. Then the isomorphism $\lim F_X \xrightarrow{\sim} \lim(F_X \circ \phi)$ for all X implies that $\operatorname{colim}(F \circ \phi) \xrightarrow{\sim}$ $\operatorname{colim} F$.

(3) $\Rightarrow$ (1): Consider the Yoneda functor $F : \mathcal{I} \to \operatorname{Fun}(\mathcal{I}^{\mathrm{op}}, \mathrm{Set})$. Colimits in $\operatorname{Fun}(\mathcal{I}^{\mathrm{op}}, \mathrm{Set})$ are computed pointwise. Thus we have for each $x \in \mathcal{I}$ a bijection

$$\operatorname*{colim}_{j \in \mathcal{J}} \operatorname{Hom}(x, \phi(j)) \xrightarrow{\sim} \operatorname*{colim}_{i \in \mathcal{I}} \operatorname{Hom}(x, i).$$

Choosing $x = i$, we find $j \in \mathcal{J}$ and a morphism $i \to \phi(j)$.

Using condition (1), the fact that $\mathcal{I}$ is filtered implies that $\mathcal{J}$ is filtered. $\square$

Let $\mathcal{A}$ be a locally finitely presented category. For a full additive subcategory $\mathcal{C} \subseteq \operatorname{fp} \mathcal{A}$ let $\vec{\mathcal{C}}$ denote the full subcategory of $\mathcal{A}$ consisting of the filtered colimits of objects in $\mathcal{C}$.

Lemma 11.1.6. *An object $X \in \mathcal{A}$ belongs to $\vec{\mathcal{C}}$ if and only if every morphism $C \to X$ with $C \in \operatorname{fp} \mathcal{A}$ factors through an object in $\mathcal{C}$.*

Proof Let $X = \operatorname{colim} X_i$ be written as a filtered colimit of objects in $\mathcal{C}$. Then every morphism $C \to X$ with $C \in \operatorname{fp} \mathcal{A}$ factors through $X_i \to X$ for some i. Conversely, let $X = \operatorname{colim}_{(C,\phi) \in \operatorname{fp} \mathcal{A}/X} C$ and suppose that each $\phi : C \to X$ factors through an object in $\mathcal{C}$. This means that the inclusion $\mathcal{C}/X \to (\operatorname{fp} \mathcal{A})/X$ is cofinal, so $\operatorname{colim}_{(C,\phi) \in \mathcal{C}/X} C \xrightarrow{\sim} X$ by Lemma 11.1.5. Thus $X \in \vec{\mathcal{C}}$. $\square$

Example 11.1.7. Let Λ be a ring and set $\mathcal{C} = \operatorname{proj} \Lambda$. Then $\vec{\mathcal{C}}$ equals the category of flat Λ-modules.

Categories of Additive Functors

Let $\mathcal{C}$ be an essentially small additive category and let $\operatorname{Add}(\mathcal{C}^{\mathrm{op}}, \mathrm{Ab})$ denote the category of additive functors $\mathcal{C}^{\mathrm{op}} \to \mathrm{Ab}$. This functor category inherits (co)kernels and (co)products from Ab, because these are computed 'pointwise'. In particular, $\operatorname{Add}(\mathcal{C}^{\mathrm{op}}, \mathrm{Ab})$ is an abelian category. Also, filtered colimits of exact sequences are exact.

For an additive functor $F : \mathcal{C}^{\mathrm{op}} \to \mathrm{Ab}$ let $\mathcal{C}/F$ denote the category consisting of pairs (C, f) with $C \in \mathcal{C}$ and $f \in F(C)$. A morphism $(C, f) \to (C', f')$ is given by a morphism $\alpha : C \to C'$ in $\mathcal{C}$ such that $F(\alpha)(f') = f$.

Lemma 11.1.8. *An additive functor $F : \mathcal{C}^{\mathrm{op}} \to \mathrm{Ab}$ equals the colimit of the functor*

$$\Phi_{\mathcal{C}} : \mathcal{C}/F \longrightarrow \operatorname{Add}(\mathcal{C}^{\mathrm{op}}, \mathrm{Ab}), \quad (C, f) \mapsto \operatorname{Hom}_{\mathcal{C}}(-, C).$$

Proof Each pair (C, f) in $\mathcal{C}/F$ yields a morphism $\mathrm{Hom}_{\mathcal{C}}(-, C) \to F$ and these induce a morphism

$$\operatorname*{colim}_{(C,f)\in\mathcal{C}/F} \mathrm{Hom}_{\mathcal{C}}(-, C) \longrightarrow F.$$

We obtain an inverse by giving for each $X \in \mathcal{C}$ a morphism

$$F(X) \longrightarrow \operatorname*{colim}_{(C,f)\in\mathcal{C}/F} \mathrm{Hom}_{\mathcal{C}}(X, C)$$

as follows. An element $x \in F(X)$ is sent to the image of id_X under the canonical map

$$\mathrm{Hom}_{\mathcal{C}}(X, X) \longrightarrow \operatorname*{colim}_{(C,f)\in\mathcal{C}/F} \mathrm{Hom}_{\mathcal{C}}(X, C)$$

corresponding to (X, x) in $\mathcal{C}/F$. □

We write $\mathrm{Fp}(\mathcal{C}^{\mathrm{op}}, \mathrm{Ab})$ for the category of functors $F \colon \mathcal{C}^{\mathrm{op}} \to \mathrm{Ab}$ that admit a presentation

$$\mathrm{Hom}_{\mathcal{C}}(-, C) \longrightarrow \mathrm{Hom}_{\mathcal{C}}(-, D) \longrightarrow F \longrightarrow 0.$$

It follows from Yoneda's lemma that each representable functor is a finitely presented object in $\mathrm{Add}(\mathcal{C}^{\mathrm{op}}, \mathrm{Ab})$. Thus a cokernel of a morphism between representable functors is a finitely presented object.

We obtain another presentation of an additive functor $F \colon \mathcal{C}^{\mathrm{op}} \to \mathrm{Ab}$ using the slice category $\mathrm{Fp}(\mathcal{C}^{\mathrm{op}}, \mathrm{Ab})/F$ which is filtered; see Example 11.1.1.

Proposition 11.1.9. *An additive functor* $F \colon \mathcal{C}^{\mathrm{op}} \to \mathrm{Ab}$ *equals the filtered colimit of the forgetful functor*

$$\Psi \colon \mathrm{Fp}(\mathcal{C}^{\mathrm{op}}, \mathrm{Ab})/F \longrightarrow \mathrm{Add}(\mathcal{C}^{\mathrm{op}}, \mathrm{Ab}).$$

Therefore the additive category $\mathrm{Add}(\mathcal{C}^{\mathrm{op}}, \mathrm{Ab})$ *is locally finitely presented and*

$$\mathrm{fp}\,\mathrm{Add}(\mathcal{C}^{\mathrm{op}}, \mathrm{Ab}) = \mathrm{Fp}(\mathcal{C}^{\mathrm{op}}, \mathrm{Ab}).$$

Proof We consider the Yoneda functor $h \colon \mathcal{C} \to \mathcal{D} := \mathrm{Fp}(\mathcal{C}^{\mathrm{op}}, \mathrm{Ab})$ and set $\bar{F} = \mathrm{Hom}(-, F)|_{\mathcal{D}}$. Then $\bar{F} = \operatorname{colim} \Phi_{\mathcal{D}}$ by Lemma 11.1.8. We have $\Psi = h^* \circ \Phi_{\mathcal{D}}$ and therefore

$$\operatorname{colim} \Psi = h^*(\operatorname{colim} \Phi_{\mathcal{D}}) = h^*(\bar{F}) = F.$$

The second assertion is an immediate consequence of the first. □

Let us add another useful presentation of an additive functor as a colimit which uses a directed set.

Proposition 11.1.10. *Every additive functor* $\mathcal{C}^{\mathrm{op}} \to \mathrm{Ab}$ *is a directed colimit of functors in* $\mathrm{Fp}(\mathcal{C}^{\mathrm{op}}, \mathrm{Ab})$.

Proof An additive functor $F\colon \mathcal{C}^{\mathrm{op}} \to \mathrm{Ab}$ admits a presentation

$$\coprod_{i \in I} \mathrm{Hom}_{\mathcal{C}}(-, C_i) \longrightarrow \coprod_{j \in J} \mathrm{Hom}_{\mathcal{C}}(-, D_j) \longrightarrow F \longrightarrow 0$$

because $\mathcal{C}$ is essentially small. Let U denote the set of pairs $u = (I', J')$ consisting of finite subsets $I' \subseteq I$ and $J' \subseteq J$ making the following square commutative

$$
\begin{array}{ccc}
\coprod_{i \in I'} \mathrm{Hom}_{\mathcal{C}}(-, C_i) & \longrightarrow & \coprod_{j \in J'} \mathrm{Hom}_{\mathcal{C}}(-, D_j) \\
\downarrow & & \downarrow \\
\coprod_{i \in I} \mathrm{Hom}_{\mathcal{C}}(-, C_i) & \longrightarrow & \coprod_{j \in J} \mathrm{Hom}_{\mathcal{C}}(-, D_j)
\end{array}
$$

and denote by $F_u \to F$ the induced morphism between the cokernels of the horizontal morphisms. The set U is partially ordered by inclusion, and we have $\sup(u_1, u_2) \in U$ for $u_1, u_2 \in U$. Thus U is directed and it is easily checked that $\mathrm{colim}_{u \in U} F_u \xrightarrow{\sim} F$. $\square$

Linear Representations

A category $\mathcal{C}$ is *preadditive* if each morphism set $\mathrm{Hom}_{\mathcal{C}}(X, Y)$ is an abelian group, and the composition maps

$$\mathrm{Hom}_{\mathcal{C}}(Y, Z) \times \mathrm{Hom}_{\mathcal{C}}(X, Y) \longrightarrow \mathrm{Hom}_{\mathcal{C}}(X, Z)$$

are biadditive. An additive category carries an intrinsic structure of a preadditive category, but in general this is an additional structure. It is often convenient to consider functor categories $\mathrm{Add}(\mathcal{C}^{\mathrm{op}}, \mathrm{Ab})$ when $\mathcal{C}$ is preadditive, and the above results generalise with same proofs.

The *centre* $Z(\mathcal{C})$ of a preadditive category $\mathcal{C}$ is the ring of all natural transformations $\mathrm{id}_{\mathcal{C}} \to \mathrm{id}_{\mathcal{C}}$ of the identity functor on $\mathcal{C}$. For a commutative ring k the structure of a k-*linear category* on $\mathcal{C}$ is given by a ring homomorphism $k \to Z(\mathcal{C})$.

Let $\mathcal{C}$ be a k-linear category $\mathcal{C}$. Then for any additive functor $F\colon \mathcal{C} \to \mathrm{Ab}$ there is a canonical k-module structure on FX for each $X \in \mathcal{C}$ via the homomorphism $k \to \mathrm{End}_{\mathcal{C}}(X) \to \mathrm{End}_{\mathbb{Z}}(FX)$. Thus we may view F as a k-*linear functor* $\mathcal{C} \to \mathrm{Mod}\, k$.

Example 11.1.11. A ring Λ may be viewed as a preadditive category with a single object, and then $Z(\Lambda)$ identifies with the usual centre given by all elements $x \in \Lambda$ satisfying $xy = yx$ for all $y \in \Lambda$. Moreover, $Z(\Lambda) \xrightarrow{\sim} Z(\mathrm{Mod}\,\Lambda)$.

Let $\mathcal{C}$ be an essentially small category and k a commutative ring. The forgetful functor $\operatorname{Mod} k \to \operatorname{Set}$ admits a left adjoint which sends a set S to a free k-module $k[S]$ with basis S. Thus there is a natural bijection

$$\operatorname{Hom}_k(k[S], X) \xrightarrow{\sim} \operatorname{Hom}_{\operatorname{Set}}(S, X)$$

for any k-module X. The *k-linearisation* $k\mathcal{C}$ of $\mathcal{C}$ is the k-linear category obtained by setting $\operatorname{Ob} k\mathcal{C} = \operatorname{Ob} \mathcal{C}$ and

$$\operatorname{Hom}_{k\mathcal{C}}(X, Y) = k[\operatorname{Hom}_{\mathcal{C}}(X, Y)]$$

for each pair of objects X, Y.

Consider the category $\operatorname{Fun}(\mathcal{C}, \operatorname{Mod} k)$ of all functors $\mathcal{C} \to \operatorname{Mod} k$. We think of a functor $\mathcal{C} \to \operatorname{Mod} k$ as a *k-linear representation* of $\mathcal{C}$.

Lemma 11.1.12. *Restriction via the inclusion $i \colon \mathcal{C} \to k\mathcal{C}$ gives an equivalence*

$$\operatorname{Add}(k\mathcal{C}, \operatorname{Ab}) \xrightarrow{\sim} \operatorname{Fun}(\mathcal{C}, \operatorname{Mod} k).$$

Proof The quasi-inverse functor $\operatorname{Fun}(\mathcal{C}, \operatorname{Mod} k) \to \operatorname{Add}(k\mathcal{C}, \operatorname{Ab})$ is obtained by applying the left adjoint of the forgetful functor $\operatorname{Mod} k \to \operatorname{Set}$. Thus any functor $F \colon \mathcal{C} \to \operatorname{Mod} k$ extends uniquely to a k-linear functor $F' \colon k\mathcal{C} \to \operatorname{Mod} k$ such that $F' \circ i = F$. □

Example 11.1.13. (1) Let Λ be a ring. Then evaluation at Λ yields an equivalence

$$\operatorname{Add}((\operatorname{proj} \Lambda)^{\operatorname{op}}, \operatorname{Ab}) \xrightarrow{\sim} \operatorname{Mod} \Lambda.$$

Taking a Λ-module X to $\operatorname{Hom}(-, X)|_{\operatorname{proj} \Lambda}$ gives a quasi-inverse.

(2) Let Q be a quiver, k a commutative ring, and $\operatorname{Rep}(Q, k)$ the category of *k-linear representations* of Q. The *path category* is the k-linearisation kQ of the category of paths in Q. Then restriction to Q yields an equivalence

$$\operatorname{Add}(kQ, \operatorname{Ab}) \xrightarrow{\sim} \operatorname{Rep}(Q, k).$$

(3) Let G be a group, k a commutative ring, and $\operatorname{Rep}(G, k)$ the category of *k-linear representations* of G. We view the group as a category with a single object, and then its k-linearisation identifies with the *group algebra* kG. This yields an equivalence

$$\operatorname{Mod}(kG^{\operatorname{op}}) \xrightarrow{\sim} \operatorname{Rep}(G, k).$$

Categories of Left Exact Functors

Let $\mathcal{C}$ be an essentially small additive category and suppose that $\mathcal{C}$ has cokernels. We consider the functor category $\mathrm{Add}(\mathcal{C}^{\mathrm{op}}, \mathrm{Ab})$ and denote by $\mathrm{Lex}(\mathcal{C}^{\mathrm{op}}, \mathrm{Ab})$ the full subcategory of additive functors $F \colon \mathcal{C}^{\mathrm{op}} \to \mathrm{Ab}$ that are *left exact*, so taking an exact sequence $X \to Y \to Z \to 0$ in $\mathcal{C}$ to an exact sequence $0 \to FZ \to FY \to FX$.[1] This category has filtered colimits, kernels, and products, because left exact functors are closed under these operations. Note that every representable functor is a finitely presented object in $\mathrm{Lex}(\mathcal{C}^{\mathrm{op}}, \mathrm{Ab})$.

Lemma 11.1.14. *Let $F \colon \mathcal{C}^{\mathrm{op}} \to \mathrm{Ab}$ be an additive functor. Then the category $\mathcal{C}/F$ is filtered if and only if F is left exact.*

Proof When $\mathcal{C}/F$ is filtered then F is a filtered colimit of left exact functors since each representable functor is left exact; see Lemma 11.1.8. Thus F is left exact.

Now suppose that F is left exact. We need to show that $\mathcal{C}/F$ is filtered. Given pairs (C, f) and (C', f'), we have canonical morphisms

$$(C, f) \to (C \oplus C', f + f') \leftarrow (C', f')$$

since F is additive. Given morphisms $\alpha_1, \alpha_2 \colon (C, f) \to (C', f')$, we obtain a morphism $\beta \colon (C', f') \to (C'', f'')$ by taking $C'' = \mathrm{Coker}(\alpha_1 - \alpha_2)$. Because F is left exact, there is $f'' \in F(C'')$ which is sent to $f' \in F(C')$ since $F(\alpha_1 - \alpha_2)(f') = 0$. Thus $\beta\alpha_1 = \beta\alpha_2$. $\qquad\square$

The following correspondence provides a useful description of locally finitely presented categories.

Theorem 11.1.15. *We have a correspondence between locally finitely presented categories and essentially small additive categories with cokernels.*

(1) *Let $\mathcal{C}$ be an essentially small additive category that admits cokernels and set $\mathcal{A} = \mathrm{Lex}(\mathcal{C}^{\mathrm{op}}, \mathrm{Ab})$. Then $\mathcal{A}$ is locally finitely presented with $\mathcal{C} \xrightarrow{\sim} \mathrm{fp}\,\mathcal{A}$.*

(2) *Let $\mathcal{A}$ be a locally finitely presented category and set $\mathcal{C} = \mathrm{fp}\,\mathcal{A}$. Then*

$$\mathcal{A} \longrightarrow \mathrm{Lex}(\mathcal{C}^{\mathrm{op}}, \mathrm{Ab}), \quad X \longmapsto h_X := \mathrm{Hom}_{\mathcal{A}}(-, X)|_{\mathcal{C}}$$

is an equivalence.

Proof (1) We consider the category $\mathcal{A} = \mathrm{Lex}(\mathcal{C}^{\mathrm{op}}, \mathrm{Ab})$. Clearly, each representable functor is a finitely presented object in $\mathcal{A}$, by Yoneda's lemma. Then it follows from Lemma 11.1.8 and Lemma 11.1.14 that every object in $\mathcal{A}$ is

---

[1] When $\mathcal{C}$ is an exact category with cokernels, there are two notions of a left exact functor $\mathcal{C}^{\mathrm{op}} \to \mathrm{Ab}$. In general, these are different.

a filtered colimit of finitely presented objects. Any finitely presented object is isomorphic to a representable functor by Lemma 11.1.2. Thus $\mathcal{C} \xrightarrow{\sim} \mathrm{fp}\,\mathcal{A}$. A simple calculation shows that the Yoneda embedding $\mathcal{C} \to \mathcal{A}$ is right exact, so it takes cokernels to cokernels.

Any morphism $\phi \colon X \to Y$ in $\mathcal{A}$ can be written as a filtered colimit $\phi = \mathrm{colim}\,\phi_i$ of morphisms in $\mathrm{fp}\,\mathcal{A}$. Then $\mathrm{Coker}\,\phi = \mathrm{colim}\,\mathrm{Coker}\,\phi_i$. Thus $\mathcal{A}$ has cokernels and is therefore cocomplete, since $\mathcal{A}$ has coproducts.

(2) We show that the assignment $X \mapsto h_X$ is fully faithful and essentially surjective. Let X, Y be objects in $\mathcal{A}$ and $X = \mathrm{colim}\,X_i$ written as a filtered colimit of objects in $\mathrm{fp}\,\mathcal{A}$. Then

$$\mathrm{Hom}(\operatorname*{colim}_i X_i, Y) \cong \lim_i \mathrm{Hom}(X_i, Y)$$

$$\cong \lim_i \mathrm{Hom}(h_{X_i}, h_Y)$$

$$\cong \mathrm{Hom}(\operatorname*{colim}_i h_{X_i}, h_Y)$$

$$\cong \mathrm{Hom}(h_X, h_Y),$$

where we use Yoneda's lemma and the fact that $X \mapsto h_X$ preserves filtered colimits. Any object $F \in \mathrm{Lex}(\mathcal{C}^{\mathrm{op}}, \mathrm{Ab})$ can be written as a filtered colimit

$$F = \operatorname*{colim}_{(C, f) \in \mathcal{C}/F} h_C$$

by Lemma 11.1.8. Thus for $X = \mathrm{colim}_{(C, f) \in \mathcal{C}/F}\,C$ in $\mathcal{A}$ we have $h_X \cong F$.

We conclude that a quasi-inverse $\mathrm{Lex}(\mathcal{C}^{\mathrm{op}}, \mathrm{Ab}) \to \mathcal{A}$ sends F to $\mathrm{colim}\,\tilde{F}$ where $\tilde{F} \colon \mathcal{C}/F \to \mathcal{A}$ is the functor that sends (C, f) to C. □

Let us collect some consequences.

Corollary 11.1.16. *An object X in a locally finitely presented category can be written canonically as a filtered colimit*

$$X = \operatorname*{colim}_{(C, \phi) \in \mathrm{fp}\,\mathcal{A}/X} C \tag{11.1.17}$$

of the forgetful functor $\mathrm{fp}\,\mathcal{A}/X \to \mathcal{A}$ that takes (C, ϕ) to C. □

Corollary 11.1.18. *A locally finitely presented category is complete.*

Proof A limit of left exact functors is again left exact. □

Corollary 11.1.19. *Let $\mathcal{A}$ be a locally finitely presented category.*

(1) *If $\mathcal{A}$ is abelian, then filtered colimits in $\mathcal{A}$ are exact, and therefore $\mathcal{A}$ is a Grothendieck category. In particular, $\mathcal{A}$ has injective envelopes.*

(2) *If $\mathrm{fp}\,\mathcal{A}$ is abelian, then $\mathcal{A}$ is abelian and the inclusion $\mathrm{fp}\,\mathcal{A} \to \mathcal{A}$ is exact.*

Proof Set $\mathcal{C} = \mathrm{fp}\,\mathcal{A}$. We can identify $\mathcal{A}$ with $\mathrm{Lex}(\mathcal{C}^{\mathrm{op}}, \mathrm{Ab})$ and can compute filtered colimits in $\mathrm{Add}(\mathcal{C}^{\mathrm{op}}, \mathrm{Ab})$, where they are exact, keeping in mind that filtered colimits of left exact functors are left exact.

Now suppose that $\mathcal{C}$ is abelian. Given a morphism $\phi = \mathrm{colim}\,\phi_i$, written as a filtered colimit of morphisms in $\mathrm{fp}\,\mathcal{A}$, we have $\mathrm{Ker}\,\phi = \mathrm{colim}\,\mathrm{Ker}\,\phi_i$, since kernels are computed in $\mathrm{Add}(\mathcal{C}^{\mathrm{op}}, \mathrm{Ab})$ and filtered colimits in $\mathrm{Add}(\mathcal{C}^{\mathrm{op}}, \mathrm{Ab})$ are exact. Thus $\mathcal{A}$ has kernels. The Yoneda embedding $\mathcal{C} \to \mathcal{A}$ is left exact since the embedding $\mathcal{C} \to \mathrm{Add}(\mathcal{C}^{\mathrm{op}}, \mathrm{Ab})$ is left exact. On the other hand, $\mathcal{C}$ is closed under cokernels. Thus the inclusion $\mathcal{C} \to \mathcal{A}$ is exact. □

Remark 11.1.20. Let $\mathcal{A}$ be locally finitely presented and $\mathrm{fp}\,\mathcal{A}$ abelian. Then every exact sequence $\eta: 0 \to X \xrightarrow{\alpha} Y \xrightarrow{\beta} Z \to 0$ in $\mathcal{A}$ can be written as a filtered colimit of exact sequences in $\mathrm{fp}\,\mathcal{A}$. To see this, write $\alpha = \mathrm{colim}\,\alpha_i$ with $\alpha_i: X_i \to Y_i$ in $\mathrm{fp}\,\mathcal{A}$ for all i. Let $\beta_i: Y_i \to Z_i$ denote the cokernel of each α_i, and let $\alpha_i': X_i' \to Y_i$ denote the kernel of β_i. Then η is the filtered colimit of the exact sequences $0 \to X_i' \xrightarrow{\alpha_i'} Y_i \xrightarrow{\beta_i} Z_i \to 0$.

Lemma 11.1.21. *The inclusion* $\mathrm{Lex}(\mathcal{C}^{\mathrm{op}}, \mathrm{Ab}) \hookrightarrow \mathrm{Add}(\mathcal{C}^{\mathrm{op}}, \mathrm{Ab})$ *admits a left adjoint.*

Proof The adjoint maps a finitely presented functor $\mathrm{Coker}\,\mathrm{Hom}_{\mathcal{C}}(-, \phi)$ (given by a morphism ϕ in $\mathcal{C}$) to $\mathrm{Hom}_{\mathcal{C}}(-, \mathrm{Coker}\,\phi)$; see Example 1.1.4. This extends to

$$\operatorname*{colim}_{i \in \mathcal{J}} \mathrm{Coker}\,\mathrm{Hom}_{\mathcal{C}}(-, \phi_i) \longmapsto \operatorname*{colim}_{i \in \mathcal{J}} \mathrm{Hom}_{\mathcal{C}}(-, \mathrm{Coker}\,\phi_i).$$

Alternatively, take $F \in \mathrm{Add}(\mathcal{C}^{\mathrm{op}}, \mathrm{Ab})$ to

$$\operatorname*{colim}_{(C,f) \in \mathcal{C}/F} \mathrm{Hom}_{\mathcal{C}}(-, C)$$

in $\mathrm{Lex}(\mathcal{C}^{\mathrm{op}}, \mathrm{Ab})$; see Lemma 11.1.8. □

Corollary 11.1.22. *In a locally finitely presented category every object can be written as a directed colimit of finitely presented objects.*

Proof Any locally finitely presented category is equivalent to one of the form $\mathrm{Lex}(\mathcal{C}^{\mathrm{op}}, \mathrm{Ab})$. Write $F \in \mathrm{Lex}(\mathcal{C}^{\mathrm{op}}, \mathrm{Ab})$ as a directed colimit $F = \mathrm{colim}\,F_i$ of objects $F_i \in \mathrm{Fp}(\mathcal{C}^{\mathrm{op}}, \mathrm{Ab})$; see Proposition 11.1.10. Let $Q: \mathrm{Add}(\mathcal{C}^{\mathrm{op}}, \mathrm{Ab}) \to \mathrm{Lex}(\mathcal{C}^{\mathrm{op}}, \mathrm{Ab})$ denote the left adjoint of the inclusion; see Lemma 11.1.21. Then $F = Q(F) = \mathrm{colim}\,Q(F_i)$ is a directed colimit of finitely presented objects. □

Example 11.1.23. Let $\mathcal{A}$ be a locally finitely presented category and $\mathcal{C} \subseteq \mathrm{fp}\,\mathcal{A}$ a full additive subcategory. Suppose the category $\mathcal{C}$ admits cokernels (not necessarily the same as in $\mathcal{A}$). Then $\tilde{\mathcal{C}}$ is locally finitely presented with $\mathcal{C} \xrightarrow{\sim} \mathrm{fp}\,\tilde{\mathcal{C}}$.

Proof Clearly, $\vec{\mathcal{C}}$ is a category with filtered colimits and $\mathcal{C} \subseteq \mathrm{fp}\,\vec{\mathcal{C}}$. On the other hand, when $X \in \mathrm{fp}\,\vec{\mathcal{C}}$ is written as a filtered colimit $X = \mathrm{colim}\,X_i$ of objects in $\mathcal{C}$, then id_X factors through some X_i, so X is a direct summand of an object in $\mathcal{C}$. Thus $\mathcal{C} \xrightarrow{\sim} \mathrm{fp}\,\vec{\mathcal{C}}$, and $X \mapsto \mathrm{Hom}(-, X)|_\mathcal{C}$ yields an equivalence $\vec{\mathcal{C}} \xrightarrow{\sim} \mathrm{Lex}(\mathcal{C}^{\mathrm{op}}, \mathrm{Ab})$. $\qquad\qquad\square$

Recall that a full subcategory $\mathcal{C} \subseteq \mathcal{D}$ of some category $\mathcal{D}$ is *covariantly finite* if for every object $X \in \mathcal{D}$ there is a morphism $X \to C^X$ (called a *left $\mathcal{C}$-approximation*) such that $C^X \in \mathcal{C}$ and every morphism $X \to C$ with $C \in \mathcal{C}$ factors through $X \to C^X$. For example, $\mathcal{C}$ is covariantly finite if the inclusion $\mathcal{C} \to \mathcal{D}$ admits a left adjoint. Then a left approximation $X \to C^X$ is given by the unit of the adjunction..

Example 11.1.24. Let $\mathcal{A}$ be a locally finitely presented category and $\mathcal{C} \subseteq \mathrm{fp}\,\mathcal{A}$ a full additive subcategory. Then $\vec{\mathcal{C}}$ is closed under products in $\mathcal{A}$ if and only if $\mathcal{C}$ is covariantly finite in $\mathrm{fp}\,\mathcal{A}$.

Proof We apply the criterion of Lemma 11.1.6. Suppose first that $\mathcal{C}$ is covariantly finite in $\mathrm{fp}\,\mathcal{A}$. If $X := \prod_{i \in I} X_i$ is a product of objects in $\vec{\mathcal{C}}$, then every morphism $F \to X$ with $F \in \mathrm{fp}\,\mathcal{A}$ factors through a product $\prod_{i \in I} C_i$ of objects in $\mathcal{C}$, and this factors through $F \to C^F$. Thus $X \in \vec{\mathcal{C}}$. Conversely, suppose that $\vec{\mathcal{C}}$ is closed under products. Fix $X \in \mathrm{fp}\,\mathcal{A}$ and consider the product $X_\mathcal{C} := \prod_{X \to C} C$ where $X \to C$ runs through all morphisms with $C \in \mathcal{C}$. This product belongs to $\vec{\mathcal{C}}$, and therefore the canonical morphism $X \to X_\mathcal{C}$ factors through an object in $\mathcal{C}$ via a morphism $X \to C^X$. Clearly, this is a left $\mathcal{C}$-approximation. $\qquad\qquad\square$

Example 11.1.25. Let $\mathcal{A}$ be a locally finitely presented category and suppose that $\mathcal{A}$ is abelian. If $(\mathcal{T}, \mathcal{F})$ is a torsion pair for $\mathrm{fp}\,\mathcal{A}$, then $(\vec{\mathcal{T}}, \vec{\mathcal{F}})$ is a torsion pair for $\mathcal{A}$.

Proof Each object $X \in \mathrm{fp}\,\mathcal{A}$ fits into an exact sequence $0 \to X' \to X \to X'' \to 0$ with $X' \in \mathcal{T}$ and $X'' \in \mathcal{F}$. If $X = \mathrm{colim}\,X_i$ is written as a filtered colimit of finitely presented objects, then $0 \to \mathrm{colim}_i X_i' \to X \to \mathrm{colim}_i X_i'' \to 0$ is the desired exact sequence in $\mathcal{A}$, using that filtered colimits in $\mathcal{A}$ are exact. The formula

$$\mathrm{Hom}(\mathrm{colim}_i X_i, \mathrm{colim}_j Y_j) \cong \lim_i \mathrm{colim}_j \mathrm{Hom}(X_i, Y_j)$$

then shows that $\mathrm{Hom}(X, Y) = 0$ for $X \in \vec{\mathcal{T}}$ and $Y \in \vec{\mathcal{F}}$. $\qquad\qquad\square$

Categories of Exact Functors

Let $\mathcal{C}$ be an essentially small additive category and consider $\mathrm{Add}(\mathcal{C}^{\mathrm{op}}, \mathrm{Ab})$. Suppose that $\mathcal{C}$ is abelian. Then we denote by $\mathrm{Ex}(\mathcal{C}^{\mathrm{op}}, \mathrm{Ab})$ the full subcategory of additive functors $F\colon \mathcal{C}^{\mathrm{op}} \to \mathrm{Ab}$ that are exact. This category has filtered colimits and products, because exact functors are closed under these operations.

The following lemma identifies the exact functors in the category of left exact functors $\mathcal{C}^{\mathrm{op}} \to \mathrm{Ab}$.

Lemma 11.1.26. *Let $\mathcal{C}$ be an essentially small abelian category and consider $\mathcal{A} = \mathrm{Lex}(\mathcal{C}^{\mathrm{op}}, \mathrm{Ab})$. Then $X \in \mathcal{A}$ is exact if and only if $\mathrm{Ext}^1_{\mathcal{A}}(C, X) = 0$ for all $C \in \mathrm{fp}\,\mathcal{A}$.*

Proof Using the identification $\mathcal{C} \xrightarrow{\sim} \mathrm{fp}\,\mathcal{A}$, the functor X is exact if and only if for every exact sequence $\eta\colon 0 \to A \to B \to C \to 0$ in $\mathrm{fp}\,\mathcal{A}$ the induced sequence

$$\mathrm{Hom}_{\mathcal{A}}(\eta, X)\colon 0 \to \mathrm{Hom}_{\mathcal{A}}(C, X) \to \mathrm{Hom}_{\mathcal{A}}(B, X) \to \mathrm{Hom}_{\mathcal{A}}(A, X) \to 0$$

is exact.

Now suppose $\mathrm{Ext}^1_{\mathcal{A}}(C, X) = 0$. This implies the exactness of $\mathrm{Hom}_{\mathcal{A}}(\eta, X)$ for any exact $\eta\colon 0 \to A \to B \to C \to 0$ in $\mathrm{fp}\,\mathcal{A}$. Conversely, let $\mu\colon 0 \to X \to Y \to C \to 0$ be exact in $\mathcal{A}$ and write $Y = \mathrm{colim}\, Y_i$ as a filtered colimit of finitely presented objects. This yields an exact sequence $\mu_j\colon 0 \to X_j \to Y_j \to C \to 0$ in $\mathrm{fp}\,\mathcal{A}$ for some j. Now exactness of $\mathrm{Hom}_{\mathcal{A}}(\mu_j, X)$ implies that μ splits. $\square$

The next proposition provides an explicit construction that turns every left exact functor into an exact functor.

Proposition 11.1.27. *Let $\mathcal{C}$ be an essentially small abelian category. Then $\mathrm{Ex}(\mathcal{C}^{\mathrm{op}}, \mathrm{Ab})$ is a covariantly finite subcategory of $\mathrm{Lex}(\mathcal{C}^{\mathrm{op}}, \mathrm{Ab})$.*

Proof Let $F\colon \mathcal{C}^{\mathrm{op}} \to \mathrm{Ab}$ be a left exact functor and choose a representative set of monomorphisms $\alpha\colon A \to B$ in $\mathcal{C}$. We construct inductively a sequence

$$F = F_0 \longrightarrow F_1 \longrightarrow F_2 \longrightarrow \cdots$$

such that $\mathrm{colim}\, F_n$ is exact and $F \to \mathrm{colim}\, F_n$ is the left approximation of F. Set

$$\Gamma_n = \bigsqcup_{\alpha\colon A \to B} F_n A \setminus \mathrm{Im}\, F_n \alpha.$$

Then Yoneda's lemma yields a morphism $\bigsqcup_{i \in \Gamma_n} \mathrm{Hom}(-, A_i) \to F_n$ and we can form the pushout

$$
\begin{array}{ccc}
\coprod_{i \in \Gamma_n} \mathrm{Hom}(-, A_i) & \xrightarrow{\ \coprod_i(-, \alpha_i)\ } & \coprod_{i \in \Gamma_n} \mathrm{Hom}(-, B_i) \\
\downarrow & & \downarrow \\
F_n & \longrightarrow & F_{n+1}
\end{array}
$$

in $\mathrm{Lex}(\mathcal{C}^{\mathrm{op}}, \mathrm{Ab})$. It is clear from the construction that $F' = \mathrm{colim}\, F_n$ is exact, because for any monomorphism $\alpha \colon A \to B$ each element in $F'A = \mathrm{colim}\, F_n A$ lies in the image of $F_n A \to F'A$ for some n, and therefore also in the image of $F_{n+1} B \to F'B \to F'A$. Now let $F \to G$ be a morphism such that G is exact. Then in each step $F_n \to G$ factors through $F_n \to F_{n+1}$. Thus $F \to G$ factors through $F \to F'$. $\qquad \square$

Corollary 11.1.28. *Let $\mathcal{C}$ be an essentially small abelian category. Then $\mathrm{Ex}(\mathcal{C}^{\mathrm{op}}, \mathrm{Ab})$ is a covariantly finite subcategory of $\mathrm{Add}(\mathcal{C}^{\mathrm{op}}, \mathrm{Ab})$.*

Proof Observe that $\mathrm{Lex}(\mathcal{C}^{\mathrm{op}}, \mathrm{Ab}) \subseteq \mathrm{Add}(\mathcal{C}^{\mathrm{op}}, \mathrm{Ab})$ is covariantly finite, since the inclusion admits a left adjoint by Lemma 11.1.21. Then for each F in $\mathrm{Add}(\mathcal{C}^{\mathrm{op}}, \mathrm{Ab})$ the unit $F \to F^{\mathrm{Lex}}$ yields a left approximation. This approximation one composes with a left approximation $F^{\mathrm{Lex}} \to F^{\mathrm{Ex}}$ from the preceding proposition. $\qquad \square$

Change of Categories

Let $f \colon \mathcal{C} \to \mathcal{D}$ be an additive functor between essentially small additive categories. Then

$$
f^* \colon \mathrm{Add}(\mathcal{D}^{\mathrm{op}}, \mathrm{Ab}) \longrightarrow \mathrm{Add}(\mathcal{C}^{\mathrm{op}}, \mathrm{Ab}), \quad X \mapsto X \circ f
$$

admits a left adjoint $f_!$ that is defined by

$$
f_!(X) = \operatorname*{colim}_{(C, x) \in \mathcal{C}/X} \mathrm{Hom}_{\mathcal{D}}(-, f(C))
$$

for $X \in \mathrm{Add}(\mathcal{C}^{\mathrm{op}}, \mathrm{Ab})$. In particular, for $C \in \mathcal{C}$ one has

$$
f_!(\mathrm{Hom}_{\mathcal{C}}(-, C)) = \mathrm{Hom}_{\mathcal{D}}(-, f(C)).
$$

When $\mathcal{C}$ and $\mathcal{D}$ admit cokernels and $f \colon \mathcal{C} \to \mathcal{D}$ is right exact, this yields an adjoint pair

$$
\mathrm{Lex}(\mathcal{C}^{\mathrm{op}}, \mathrm{Ab}) \underset{f^*}{\overset{f_!}{\rightleftarrows}} \mathrm{Lex}(\mathcal{D}^{\mathrm{op}}, \mathrm{Ab}).
$$

We collect some basic properties of f^* and $f_!$.

Lemma 11.1.29. *Let* $f \colon \mathcal{C} \to \mathcal{D}$ *be an additive functor that inverts universally a class of morphisms in* $\mathcal{C}$. *Then* f^* *is fully faithful.*

Proof The assertion follows from the definition of the quotient functor $\mathcal{C} \to \mathcal{C}[S^{-1}]$ with respect to a class of morphisms S in $\mathcal{C}$; see also Lemma 1.1.1. $\square$

We can be more specific when $f \colon \mathcal{C} \to \mathcal{D}$ is exact.

Lemma 11.1.30. *Let* $f \colon \mathcal{C} \to \mathcal{D}$ *be an exact functor between abelian categories. Then* $f_! \colon \mathrm{Lex}(\mathcal{C}^{\mathrm{op}}, \mathrm{Ab}) \to \mathrm{Lex}(\mathcal{D}^{\mathrm{op}}, \mathrm{Ab})$ *is exact. Moreover,* f^* *is fully faithful if and only if* f *induces an equivalence* $\mathcal{C}/(\mathrm{Ker}\, f) \xrightarrow{\sim} \mathcal{D}$.

Proof We embed $\mathcal{C}$ into $\mathcal{A} = \mathrm{Lex}(\mathcal{C}^{\mathrm{op}}, \mathrm{Ab})$ via the Yoneda functor. Any exact sequence in $\mathcal{A}$ can be written as a filtered colimit of exact sequences in $\mathcal{C}$; see Remark 11.1.20. Now use that $f_!$ preserves filtered colimits and that filtered colimits in $\mathrm{Lex}(\mathcal{D}^{\mathrm{op}}, \mathrm{Ab})$ are exact.

We have already seen in Lemma 11.1.29 that f^* is fully faithful when f induces an equivalence $\mathcal{C}/(\mathrm{Ker}\, f) \xrightarrow{\sim} \mathcal{D}$. For the converse we apply Proposition 2.2.11. Thus $f_!$ induces an equivalence $\mathcal{A}/(\mathrm{Ker}\, f_!) \xrightarrow{\sim} \mathrm{Lex}(\mathcal{D}^{\mathrm{op}}, \mathrm{Ab})$. One checks that the subcategory $\mathcal{C} \subseteq \mathcal{A}$ is right cofinal with respect to the morphisms that are inverted by $f_!$, using that each object in $\mathcal{A}$ is a filtered colimit of objects in $\mathcal{C}$. Then it follows from Lemma 1.2.5 that $f_!$ restricts to an equivalence $\mathcal{C}/(\mathrm{Ker}\, f) \xrightarrow{\sim} \mathcal{D}$. $\square$

Proposition 11.1.31. *Let* $\mathcal{C}$ *be an essentially small abelian category and* $\mathcal{D} = \mathcal{C}/\mathcal{B}$ *the quotient with respect to a Serre subcategory* $\mathcal{B} \subseteq \mathcal{C}$. *Then the diagram*

$$\mathcal{B} \xrightarrow{\;\;i\;\;} \mathcal{C} \xrightarrow{\;\;p\;\;} \mathcal{D}$$

induces a localisation sequence of abelian categories

$$\mathrm{Lex}(\mathcal{B}^{\mathrm{op}}, \mathrm{Ab}) \underset{i^*}{\overset{i_!}{\rightleftarrows}} \mathrm{Lex}(\mathcal{C}^{\mathrm{op}}, \mathrm{Ab}) \underset{p^*}{\overset{p_!}{\rightleftarrows}} \mathrm{Lex}(\mathcal{D}^{\mathrm{op}}, \mathrm{Ab}).$$

In particular, the functors $i_!$ *and* $p_!$ *are exact and induce equivalences*

$$\mathrm{Lex}(\mathcal{B}^{\mathrm{op}}, \mathrm{Ab}) \xrightarrow{\;\sim\;} \mathrm{Ker}\, p_!$$

and

$$(\mathrm{Lex}(\mathcal{C}^{\mathrm{op}}, \mathrm{Ab}))/(\mathrm{Ker}\, p_!) \xrightarrow{\;\sim\;} \mathrm{Lex}(\mathcal{D}^{\mathrm{op}}, \mathrm{Ab}).$$

Proof We use the fact that every functor in $\mathrm{Lex}(\mathcal{C}^{\mathrm{op}}, \mathrm{Ab})$ is a filtered colimit of representable functors, by Lemma 11.1.8 and Lemma 11.1.14. This is combined with the fact that all functors $i_!, i^*, p_!, p^*$ preserve colimits.

The functors $i_!$ and $p_!$ are exact by Lemma 11.1.30. The functor p^* is fully faithful by Lemma 11.1.29. On the other hand, id $\cong i^* i_!$ since $i^* i_!$ equals the identity on all representable functors. Thus $i_!$ is fully faithful (Proposition 1.1.3).

It remains to show that $\operatorname{Im} i_! = \operatorname{Ker} p_!$. Then the rest follows from the localisation theory of abelian categories (Proposition 2.2.11).

We have $\operatorname{Im} i_! \subseteq \operatorname{Ker} p_!$ since $pi = 0$. For the other inclusion fix an object X in $\operatorname{Lex}(\mathcal{C}^{\mathrm{op}}, \mathrm{Ab})$ and consider the exact sequence $0 \to X' \to X \to p^* p_!(X)$. We claim that $X' \in \operatorname{Im} i_!$. Then $p_! X = 0$ implies $X \in \operatorname{Im} i_!$. It suffices to show this when $X = h_C$ is representable, given by $C \in \mathcal{C}$. For this we show that every morphism $h_{C_0} \to X'$ with $C_0 \in \mathcal{C}$ factors through h_B for some $B \in \mathcal{B}$; then Lemma 11.1.6 implies that X' is a filtered colimit of representable functors in the image of $i_!$. Now observe that a morphism $h_{C_0} \to h_C$ given by $\phi \colon C_0 \to C$ in $\mathcal{C}$ factors through X' if and only if $p\phi = 0$, by the adjointness of $p_!$ and p^*. This happens if and only if $\operatorname{Im} \phi \in \mathcal{B}$. Thus $h_{C_0} \to X'$ factors through h_B for some $B \in \mathcal{B}$. $\qquad\square$

Remark 11.1.32. The injective objects in $\operatorname{Lex}(\mathcal{D}^{\mathrm{op}}, \mathrm{Ab})$ identify via p^* with the injective objects in $\operatorname{Lex}(\mathcal{C}^{\mathrm{op}}, \mathrm{Ab})$ that vanish on $\mathcal{B}$ when viewed as functors on $\mathcal{C}$ (Corollary 2.2.15).

Corollary 11.1.33. *Let $\mathcal{A}$ be a locally finitely presented Grothendieck category such that* $\operatorname{fp} \mathcal{A}$ *is abelian. If* $\mathcal{S} \subseteq \operatorname{fp} \mathcal{A}$ *is a Serre subcategory, then $\vec{\mathcal{S}}$ is a localising subcategory of $\mathcal{A}$ satisfying $\vec{\mathcal{S}} \cap \operatorname{fp} \mathcal{A} = \mathcal{C}$. Moreover, the canonical functor $\mathcal{A} \twoheadrightarrow \mathcal{A}/\vec{\mathcal{S}}$ restricts to an equivalence $\mathcal{S}^\perp \xrightarrow{\sim} \mathcal{A}/\vec{\mathcal{S}}$.*

Proof This follows from Proposition 11.1.31, using the equivalence $\mathcal{A} \xrightarrow{\sim} \operatorname{Lex}((\operatorname{fp} \mathcal{A})^{\mathrm{op}}, \mathrm{Ab})$ which identifies the subcategory $\vec{\mathcal{S}}$ with $\operatorname{Lex}(\mathcal{S}^{\mathrm{op}}, \mathrm{Ab})$. $\qquad\square$

11.2 Grothendieck Categories

In this section we study a hierarchy of finiteness conditions for Grothendieck categories. This involves the notion of a generating set of objects.

Given an additive category $\mathcal{A}$, a set of objects $\mathcal{C}$ is *generating* if for any non-zero morphism $\phi \colon X \to Y$ in $\mathcal{A}$ there is $\alpha \colon C \to X$ with $C \in \mathcal{C}$ such that $\phi\alpha \neq 0$. If $\mathcal{A}$ has coproducts, then $\mathcal{C}$ is generating if and only if for every object $X \in \mathcal{A}$ there is an epimorphism $\coprod_{i \in I} C_i \to X$ such that $C_i \in \mathcal{C}$ for all i.

Now fix a Grothendieck category $\mathcal{A}$. We have the following hierarchy of finiteness conditions for an object $X \in \mathcal{A}$:

X of finite length $\implies$ X noetherian

$\qquad\qquad\qquad\qquad \implies$ X finitely generated $\impliedby$ X finitely presented.

The Grothendieck category $\mathcal{A}$ is called

- *locally finitely generated*, if $\mathcal{A}$ has a generating set of finitely generated objects,
- *locally finitely presented*, if $\mathcal{A}$ has a generating set of finitely presented objects,[2]
- *locally noetherian*, if $\mathcal{A}$ has a generating set of noetherian objects,
- *locally finite*, if $\mathcal{A}$ has a generating set of finite length objects.

Suppose that $\mathcal{A}$ has a set $\mathcal{C}$ of generating objects such that for every pair of objects $C, C' \in \mathcal{C}$ and every subobject $D \subseteq C$ the direct sum $C \oplus C'$ and the quotient C/D are isomorphic to objects in $\mathcal{C}$. Then every object $X \in \mathcal{A}$ can be written as the directed union $X = \sum_i X_i$ of subobjects $X_i \subseteq X$ such that $X_i \in \mathcal{C}$ for all i.

Finitely Generated and Finitely Presented Objects

Let $\mathcal{A}$ be an abelian category, and suppose that filtered colimits in $\mathcal{A}$ are exact. An object X is *finitely generated* whenever $X = \sum_{i \in I} X_i$ for a directed family of subobjects $X_i \subseteq X$ implies $X = X_{i_0}$ for some $i_0 \in I$. We record the following elementary fact.

Lemma 11.2.1. *For an exact sequence* $0 \to X' \to X \to X'' \to 0$ *we have*

$$X', X'' \ finitely \ generated \ \Longrightarrow \ X \ finiteley \ generated$$
$$\Longrightarrow \ X'' \ finitely \ generated. \qquad \square$$

We wish to compare finitely generated and finitely presented objects. Observe that 'finitely generated' is a local property, depending only on the lattice of subobjects. The property of an object to be finitely presented is different; it depends on the ambient category.

We have the following characterisation. In particular, we see that every finitely presented object is finitely generated.

Lemma 11.2.2. *For an object X the following are equivalent.*

(1) *X is finitely generated.*
(2) *The canonical map* $\operatorname{colim}_i \operatorname{Hom}(X, Y_i) \to \operatorname{Hom}(X, \operatorname{colim}_i Y_i)$ *is injective for every filtered colimit* $\operatorname{colim}_i Y_i$.

---

[2] This terminology is consistent: a Grothendieck category $\mathcal{A}$ has a generating set of finitely presented objects if and only if fp $\mathcal{A}$ is essentially small and every object in $\mathcal{A}$ is a filtered colimit of finitely presented objects.

(3) *The canonical map $\sum_i \operatorname{Hom}(X, Y_i) \to \operatorname{Hom}(X, \sum_i Y_i)$ is bijective for every directed family of subobjects $Y_i \subseteq Y$.*

Proof (1) $\Rightarrow$ (2): A morphism $\phi \in \operatorname{colim}_i \operatorname{Hom}(X, Y_i)$ is given by a morphism $X \to Y_j$ for some index j. For all $j \to i$ consider the composite with $Y_j \to Y_i$ which yields an exact sequence $0 \to X_i \to X \to Y_i$. The colimit $0 \to \operatorname{colim}_i X_i \to X \to \operatorname{colim}_i Y_i$ is exact, and if $X \to \operatorname{colim}_i Y_i$ is zero, then $X = \sum_i X_i$. Thus $X = X_{i_0}$ for some index i_0, and therefore $\phi = 0$.

(2) $\Rightarrow$ (3): Consider the following commutative diagram with exact rows.

$$
\begin{array}{ccccc}
0 \longrightarrow & \operatorname{colim}_i \operatorname{Hom}(X, Y_i) & \longrightarrow & \operatorname{Hom}(X, Y) & \longrightarrow & \operatorname{colim}_i \operatorname{Hom}(X, Y/Y_i) \\
& \downarrow{\alpha} & & \downarrow{\text{id}} & & \downarrow{\gamma} \\
0 \longrightarrow & \operatorname{Hom}(X, \operatorname{colim}_i Y_i) & \longrightarrow & \operatorname{Hom}(X, Y) & \longrightarrow & \operatorname{Hom}(X, \operatorname{colim}_i Y/Y_i)
\end{array}
$$

Then α and γ are injective, and therefore α is bijective.

(3) $\Rightarrow$ (1): Let $X = \sum_i X_i$. Then the identity $X \to \sum_i X_i$ factors through $X_{i_0} \to \sum_i X_i$ for some index i_0. Thus $X = X_{i_0}$. □

Let $\mathcal{A}$ be a Grothendieck category with a generating set of finitely generated objects. This means that each object is a directed union of its finitely generated subobjects. Also, if $\phi \colon X \to Y$ is an epimorphism such that Y is finitely generated, then there exists a finitely generated subobject $X' \subseteq X$ such that $\phi|_{X'} \colon X' \to Y$ is an epimorphism.

Lemma 11.2.3. *Let $\mathcal{A}$ be a Grothendieck category with a generating set of finitely generated objects. For an object $X \in \mathcal{A}$ the following are equivalent.*

(1) *X is finitely presented.*

(2) *X is finitely generated and every epimorphism $X' \to X$ from a finitely generated object X' has a finitely generated kernel.*

Proof (1) $\Rightarrow$ (2): We have already seen that X is finitely generated. Now fix an epimorphism $\phi \colon X' \to X$ and write $\operatorname{Ker} \phi = \sum_i X_i$ as a directed union of finitely generated subobjects $X_i \subseteq X'$. Then $\operatorname{colim}_i X'/X_i \cong X$, so the identity id_X factors through $X'/X_{i_0} \to X$ for some index i_0. Thus the sequence $0 \to \operatorname{Ker} \phi/X_{i_0} \to X'/X_{i_0} \to X \to 0$ is split exact. It follows that $\operatorname{Ker} \phi/X_{i_0}$ is finitely generated, if X' is finitely generated. Thus $\operatorname{Ker} \phi$ is finitely generated.

(2) $\Rightarrow$ (1): In view of Lemma 11.2.2, it suffices to show that the canonical map $\operatorname{colim}_i \operatorname{Hom}(X, Y_i) \to \operatorname{Hom}(X, \operatorname{colim}_i Y_i)$ is surjective for every filtered colimit $\operatorname{colim}_i Y_i$. Given a morphism $\phi \colon X \to \operatorname{colim}_i Y_i$, we consider the following

pullback.

$$
\begin{array}{ccc}
P & \longrightarrow & X \\
\downarrow & & \downarrow \phi \\
\coprod_i Y_i & \xrightarrow{\text{can}} & \operatorname{colim}_i Y_i
\end{array}
$$

We find a finitely generated subobject $P' \subseteq P$ and an index i_0 such that the pullback restricts to a commutative square

$$
\begin{array}{ccc}
P' & \xrightarrow{\ \pi\ } & X \\
\downarrow & & \downarrow \phi \\
Y_{i_0} & \longrightarrow & \operatorname{colim}_i Y_i
\end{array}
$$

and π is an epimorphism. Since $\operatorname{Ker} \pi$ is finitely generated, there is an index i_1 such that the composite $\operatorname{Ker} \pi \to P' \to Y_{i_0} \to Y_{i_1}$ is zero, by Lemma 11.2.2. It follows that ϕ factors through $Y_{i_1} \to \operatorname{colim}_i Y_i$, and this yields an element in $\operatorname{colim}_i \operatorname{Hom}(X, Y_i)$ which is mapped to ϕ. $\qquad\square$

Locally Noetherian Categories

Let $\mathcal{A}$ be an abelian category. An object in $\mathcal{A}$ is *noetherian* if it satisfies the ascending chain condition on subobjects. We record the following elementary facts.

Lemma 11.2.4. *For an exact sequence* $0 \to X' \to X \to X'' \to 0$ *we have*

$$X', X'' \text{ noetherian} \iff X \text{ noetherian}.$$

If filtered colimits are exact, then an object is noetherian if and only if every subobject is finitely generated. $\qquad\square$

A Grothendieck category is called *locally noetherian* if there exists a generating set of noetherian objects. Locally noetherian categories form an important class of locally finitely presented categories.

Proposition 11.2.5. *For a Grothendieck category* $\mathcal{A}$ *the following are equivalent.*

(1) *The category* $\mathcal{A}$ *is locally noetherian.*
(2) *The category* $\mathcal{A}$ *is locally finitely presented and for each* $X \in \mathcal{A}$ *we have*

$$X \text{ finitely presented} \iff X \text{ finitely generated} \iff X \text{ noetherian}.$$

(3) *The category* $\mathcal{A}$ *is locally finitely presented and* $\operatorname{fp} \mathcal{A}$ *is an abelian category consisting of noetherian objects.*

Proof (1) $\Rightarrow$ (2): Suppose that $\mathcal{A}$ is locally noetherian. Then finitely generated objects and noetherian objects coincide. In particular, finitely generated objects are closed under subobjects. The characterisation of finitely presented objects in Lemma 11.2.3 then implies that finitely generated objects and finitely presented objects coincide. In particular, $\mathcal{A}$ is a locally finitely presented category since every object is a directed union of its finitely generated subobjects, so a filtered colimit of finitely presented objects.

(2) $\Rightarrow$ (3): Clear.

(3) $\Rightarrow$ (1): If $\mathcal{A}$ is locally finitely presented, then the finitely presented objects generate $\mathcal{A}$. If an object $X \in \mathrm{fp}\,\mathcal{A}$ satisfies the ascending chain condition on subobjects in $\mathrm{fp}\,\mathcal{A}$, then each subobject $U \subseteq X$ in $\mathcal{A}$ is finitely presented since $U = \bigcup_{X' \subseteq U} X'$ where X' runs through all $X' \subseteq X$ in $\mathrm{fp}\,\mathcal{A}$. Thus X is noetherian in $\mathcal{A}$. □

Corollary 11.2.6. *The assignments* $\mathcal{A} \mapsto \mathrm{fp}\,\mathcal{A}$ *and* $\mathcal{C} \mapsto \mathrm{Lex}(\mathcal{C}^{\mathrm{op}}, \mathrm{Ab})$ *induce, up to equivalence, a bijective correspondence between locally noetherian Grothendieck categories and essentially small abelian categories such that every object is noetherian.*

Proof This correspondence is obtained by restricting the correspondence from Theorem 11.1.15 and Corollary 11.1.19 between locally finitely presented Grothendieck categories and essentially small additive categories. Then apply Proposition 11.2.5 to identify the locally noetherian categories. □

Locally Finite Categories

An object X of an abelian category has *finite length* if it has a finite composition series

$$0 = X_0 \subseteq X_1 \subseteq \cdots \subseteq X_n = X,$$

that is, each subquotient X_i/X_{i-1} is simple. Note that X has finite length if and only if X satisfies both chain conditions on subobjects.

A Grothendieck category $\mathcal{A}$ is called *locally finite* if there exists a generating set of finite length objects. When $\mathcal{A}$ is a locally finite category, then every noetherian object has finite length, since any object is the directed union of finite length subobjects. Thus for every object $X \in \mathcal{A}$ we have

$$X \text{ finitely presented} \iff X \text{ noetherian} \iff X \text{ of finite length.}$$

Let us discuss some further finiteness properties of locally finite categories. To this end fix an object X of an abelian category. The composition length of X is denoted by $\ell(X)$. The height $\mathrm{ht}(X)$ is the smallest $n \geq 0$ such that

$\mathrm{soc}^n(X) = X$. When $\ell(X) < \infty$, then $\mathrm{ht}(X) \le \ell(X)$, and $\mathrm{ht}(X)$ equals the smallest $n \ge 0$ such that $\mathrm{rad}^n(X) = 0$.

Lemma 11.2.7. *Let X, E be objects of an abelian category and suppose that E is injective. The assignment*

$$X \supseteq U \longmapsto H(U) := \mathrm{Hom}(X/U, E) \subseteq \mathrm{Hom}(X, E)$$

gives a lattice anti-homomorphism into the lattice of $\mathrm{End}(E)$-submodules of $\mathrm{Hom}(X, E)$. Every finitely generated $\mathrm{End}(E)$-submodule is in its image, and the homomorphism is injective when E is a cogenerator.

Proof Given subobjects U, V of X, the Noether isomorphisms imply that

$$H(U \cap V) = H(U) + H(V) \qquad \text{and} \qquad H(U + V) = H(U) \cap H(V).$$

Clearly, $U \ne V$ implies $H(U) \ne H(V)$ when E is a cogenerator. Now let $\phi \colon X \to E$ be a morphism and set $U = \mathrm{Ker}\,\phi$. Then a morphism $X \to E$ factors through $X \twoheadrightarrow X/U$ if and only if it factors through ϕ. Thus $\mathrm{End}(E)\phi = H(U)$. It follows that every cyclic $\mathrm{End}(E)$-submodule is in the image of H, and therefore so is every finitely generated submodule by the first part of the proof. $\qquad\square$

Proposition 11.2.8. *Let $\mathcal{A}$ be a locally finite Grothendieck category and J the Jacobson radical of the endomorphism ring of an injective object E. Then $\bigcap_{n \ge 0} J^n = 0$. Moreover, for $n \ge 0$ we have*

(1) $\mathrm{ht}(C) \le n$ *for all* $C \in \mathrm{fp}\,\mathcal{A}$ *implies* $J^n = 0$, *and*
(2) $J^n = 0$ *implies* $\mathrm{ht}(C) \le n$ *for all* $C \in \mathrm{fp}\,\mathcal{A}$ *when* E *cogenerates* $\mathcal{A}$.

Proof Let $C \in \mathrm{fp}\,\mathcal{A}$. A radical morphism $E \to E$ annihilates all simple objects in $\mathcal{A}$, and therefore

$$J^n \,\mathrm{Hom}(C, E) \subseteq \mathrm{Hom}(C/\mathrm{soc}^n C, E)$$

by induction on n. This implies $\bigcap_{n \ge 0} J^n = 0$ and part (1).

To show (2), assume that E is a cogenerator. An induction on $\ell(C)$ gives

$$\ell_{\mathrm{End}(E)}(\mathrm{Hom}(C, E)) = \ell(C).$$

Thus every submodule of $\mathrm{Hom}(C, E)$ is finitely generated. Then Lemma 11.2.7 implies

$$\mathrm{rad}^n \,\mathrm{Hom}(C, E) = \mathrm{Hom}(C/\mathrm{soc}^n C, E)$$

for all $n \ge 0$. Observe that $JM = \mathrm{rad}\,M$ for every $\mathrm{End}(E)$-module M, since $\mathrm{End}(E)/J$ is a product of division rings by Theorem 11.2.12 below. Thus $J^n = 0$ implies $\mathrm{soc}^n C = C$. $\qquad\square$

Remark 11.2.9. For $C \in \mathrm{fp}\,\mathcal{A}$ we have

$$\ell_{\mathrm{End}(E)}(\mathrm{Hom}(C, E)) \le \ell(C) \qquad \text{and} \qquad \mathrm{ht}_{\mathrm{End}(E)}(\mathrm{Hom}(C, E)) \le \mathrm{ht}(C)$$

with equalities when E is a cogenerator.

Injective Objects

In a locally noetherian Grothendieck category we have a very satisfactory decomposition theory for injective objects.

We need some preparations and begin with a version of *Baer's criterion*.

Lemma 11.2.10 (Baer). *Let $\mathcal{A}$ be a Grothendieck category and let $\mathcal{C}$ be a class of objects that is generating and closed under quotients. If $X \in \mathcal{A}$ satisfies $\mathrm{Ext}^1(C, X) = 0$ for all $C \in \mathcal{C}$, then X is injective.*

Proof Choose an injective envelope $\alpha \colon X \to E(X)$. If $\mathrm{Coker}\,\alpha \ne 0$, then there exists a subobject $0 \ne C \subseteq \mathrm{Coker}\,\alpha$ with $C \in \mathcal{C}$. The pullback of $0 \to X \to E(X) \to \mathrm{Coker}\,\alpha \to 0$ along the inclusion $C \to \mathrm{Coker}\,\alpha$ is a split exact sequence. Thus α factors through a monomorphism $X \oplus C \to E(X)$, contradicting the property of an injective envelope. It follows that α is an isomorphism and X is injective. $\qquad\square$

We continue with a technical lemma which is crucial for the decomposition of injective objects into indecomposables; it is known as *Chase's lemma*.

For a sequence of morphisms $\gamma = (C_n \to C_{n+1})_{n \in \mathbb{N}}$ we denote by $\gamma_n \colon C_0 \to C_n$ the composite of the first n morphisms. Recall that an object X is *compact* if for any morphism $\phi \colon X \to \coprod_{i \in I} Y_i$ there is a finite set $J \subseteq I$ such that ϕ factors through $\coprod_{i \in J} Y_i$.

Lemma 11.2.11 (Chase). *Let $(X_n)_{n \in \mathbb{N}}$ and $(Y_i)_{i \in I}$ be families of objects in an additive category and let*

$$\phi \colon \prod_{n \in \mathbb{N}} X_n \longrightarrow \coprod_{i \in I} Y_i$$

be a morphism. If $\gamma = (C_n \to C_{n+1})_{n \in \mathbb{N}}$ is a sequence of morphisms and $C = C_0$ is compact, then there exists $m \in \mathbb{N}$ such that for almost all $j \in I$ each composite

$$C \xrightarrow{\gamma_m} C_m \xrightarrow{\theta} \prod_{n \in \mathbb{N}} X_n \xrightarrow{\phi} \coprod_{i \in I} Y_i \twoheadrightarrow Y_j$$

with $\theta_n = 0$ for $n < m$ factors through $\gamma_n \colon C \to C_n$ for all $n \in \mathbb{N}$.

It is convenient to introduce further notation. For a morphism $\gamma \colon C \to D$ and an object X we denote by X_γ the image of the map

$$\mathrm{Hom}(D, X) \xrightarrow{-\circ\gamma} \mathrm{Hom}(C, X).$$

Then a sequence of morphisms $\gamma = (C_n \to C_{n+1})_{n\in\mathbb{N}}$ yields a descending chain

$$\cdots \subseteq X_{\gamma_2} \subseteq X_{\gamma_1} \subseteq X_{\gamma_0} = \mathrm{Hom}(C_0, X).$$

We can now rephrase the statement of the lemma as follows. Set $X = \prod_{n\in\mathbb{N}} X_n$, $Y = \coprod_{i\in I} Y_i$, and write

$$\phi_i \colon \mathrm{Hom}(C, X) \xrightarrow{\phi\circ-} \mathrm{Hom}(C, Y) \longrightarrow \mathrm{Hom}(C, Y_i) \qquad (i \in I).$$

There exists $m \in \mathbb{N}$ such that for almost all $i \in I$ we have

$$\phi_i\!\left(\Big(\prod_{n\geq m} X_n\Big)_{\gamma_m}\right) \subseteq \bigcap_{n\geq 0}(Y_i)_{\gamma_n}.$$

Proof Assume the conclusion to be false. We construct inductively sequences of elements $n_j \in \mathbb{N}$, $i_j \in I$, and $\theta_j \in \mathrm{Hom}(C, X)$ with $j \in \mathbb{N}$ and satisfying

(1) $n_{j+1} > n_j$,
(2) $\theta_j \in (\prod_{n\geq n_j} X_n)_{\gamma_{n_j}}$,
(3) $\phi_{i_j}(\theta_j) \notin (Y_{i_j})_{\gamma_{n_{j+1}}}$,
(4) $\phi_{i_j}(\theta_k) = 0$ for $k < j$.

We proceed as follows. Set $n_0 = 0$. Then there exists $i_0 \in I$ such that

$$\phi_{i_0}(X_{\gamma_0}) \nsubseteq \bigcap_{n\geq 0}(Y_{i_0})_{\gamma_n},$$

and hence we may select $\theta_0 \in X_{\gamma_0}$ and $n_1 > 0$ such that $\phi_{i_0}(\theta_0) \notin (Y_{i_0})_{\gamma_{n_1}}$. Thus conditions (1)–(4) are satisfied for $j = 0$.

Proceeding by induction on j, assume that elements $n_{k+1} \in \mathbb{N}$, $i_k \in I$ and $\theta_k \in \mathrm{Hom}(C, X)$ have been constructed for $k < j$ such that conditions (1)–(4) are satisfied. Using that C is compact, there exists a finite subset $I' \subseteq I$ such that for $i \in I \setminus I'$ we have $\phi_i(\theta_k) = 0$ for $k < j$. We may then select $i_j \in I \setminus I'$ such that

$$\phi_{i_j}\!\left(\Big(\prod_{n\geq n_j} X_n\Big)_{\gamma_{n_j}}\right) \nsubseteq \bigcap_{n\geq 0}(Y_{i_j})_{\gamma_n},$$

because otherwise the lemma would be true. Thus there exists an element $\theta_j \in (\prod_{n\geq n_j} X_n)_{\gamma_{n_j}}$ and $n_{j+1} > n_j$ such that $\phi_{i_j}(\theta_j) \notin (Y_{i_j})_{\gamma_{n_{j+1}}}$. It is then clear that the elements $n_{k+1} \in \mathbb{N}$, $i_k \in I$, and $\theta_k \in \mathrm{Hom}(C, X)$ for $k \leq j$ satisfy the conditions (1)–(4).

Now let $\theta = \sum_{j\in\mathbb{N}} \theta_j \in \mathrm{Hom}(C, X)$, which is well defined since the sum for each component $C \to X_n$ is finite. For each $j \in \mathbb{N}$ we have $\phi_{i_j}(\theta) = \phi_{i_j}(\theta_j) + \phi_{i_j}(\sum_{k>j} \theta_k) \neq 0$, since the second summand lies in $(Y_{i_j})_{\gamma_{n_{j+1}}}$, whereas the first does not. On the other hand, the morphism $\phi\theta$ factors through a finite sum $\coprod_{i \in J} Y_i$ for some $J \subseteq I$, since C is compact. This contradiction finishes the proof. $\qquad\qquad\qquad\qquad\qquad\qquad\qquad\qquad\qquad\qquad\qquad\qquad\quad\square$

We have the following characterisation of local noetherianness.

Theorem 11.2.12. *For a locally finitely generated Grothendieck category $\mathcal{A}$ the following are equivalent.*

(1) *The category $\mathcal{A}$ is locally noetherian.*
(2) *The subcategory of injective objects in $\mathcal{A}$ is closed under filtered colimits.*
(3) *The subcategory of injective objects in $\mathcal{A}$ is closed under coproducts.*
(4) *Every injective object decomposes into a coproduct of indecomposable objects with local endomorphism rings.*
(5) *There is an object E such that every object in $\mathcal{A}$ is a subobject of a coproduct of copies of E.*

Proof (1) $\Rightarrow$ (2): If $\mathcal{A}$ is locally noetherian, then $\mathrm{fp}\,\mathcal{A}$ is closed under quotients. Thus the equivalence $\mathcal{A} \xrightarrow{\sim} \mathrm{Lex}((\mathrm{fp}\,\mathcal{A})^{\mathrm{op}}, \mathrm{Ab})$ identifies the injective objects with the exact functors $(\mathrm{fp}\,\mathcal{A})^{\mathrm{op}} \to \mathrm{Ab}$, by Lemma 11.1.26 and Lemma 11.2.10. It remains to note that a filtered colimit of exact functors is exact.

(2) $\Rightarrow$ (3): Clear.

(3) $\Rightarrow$ (1): Fix an injective cogenerator E and let $C_1 \subseteq C_2 \subseteq C_3 \subseteq \cdots$ be an ascending chain of subobjects of a finitely generated object C. Choose morphisms $C/C_i \to E$ for all i such that the restriction to C_{i+1}/C_i is non-zero provided that $C_{i+1}/C_i \neq 0$, and consider for $j \leq n$ the composite $\phi_{nj}: C_n \to C \to C/C_j \to E$. For each n these yield a morphism $\phi_n: C_n \to \coprod_{i=1}^{n} E$ and we obtain a morphism $\phi: \sum_{n \geq 1} C_n \to \coprod_{i \geq 1} E$, since the ϕ_n are compatible. The morphism ϕ extends to a morphism $C \to \coprod_{i \geq 1} E$, since we assume $\coprod_{i \geq 1} E$ to be injective, and this factors through a finite sum $\coprod_{i=1}^{m} E$ for some m, since C is finitely generated. Thus $C_n = C_m$ for $n \geq m$, so C is noetherian.

(2) $\Rightarrow$ (4): Let $X \neq 0$ be injective and fix a finitely generated subobject $0 \neq C \subseteq X$. Using Zorn's lemma and the fact that injectives are closed under filtered colimits, there exists a maximal injective subobject $X' \subseteq X$ not containing C. Then $X = X' \oplus X''$, and we claim that X'' is indecomposable. For, if $X'' = U \oplus V$, then $(X'+U) \cap (X'+V) = X'$ implies that one of the objects $X'+U$ and $X'+V$ does not contain C. Thus $U = 0$ or $V = 0$ by the maximality of X'.

Using again Zorn's lemma, there exists a maximal family of indecomposable injective subobjects $(X_i)_{i \in I}$ of X such that the sum $X' = \sum_{i \in I} X_i$ is direct. This yields a decomposition $X = X' \oplus X''$, and $X'' = 0$ by the previous observation.

It remains to note that every indecomposable injective object has a local endomorphism ring (Lemma 2.5.7).

$(4) \Rightarrow (5)$: Let E be the coproduct of indecomposable injective objects, taking one representative from each isomorphism class. Note that there is only a set of such representatives, since each indecomposable injective is the injective envelope of a quotient G/U when G is a generator of $\mathcal{A}$. Then every object in $\mathcal{A}$ is a subobject of a coproduct of copies of E, since every object embeds into an injective object (Corollary 2.5.4).

$(5) \Rightarrow (1)$: Let $C \in \mathcal{A}$ be a finitely generated object. We wish to show that C is noetherian. To this end fix a chain of finitely generated subobjects $0 = B_0 \subseteq B_1 \subseteq B_2 \subseteq \cdots$ and set $C_n = C/B_n$. This yields a sequence of epimorphisms $\gamma = (C_n \twoheadrightarrow C_{n+1})_{n \in \mathbb{N}}$. For $X \in \mathcal{A}$ we set $X_{\bar{\gamma}_n} = \mathrm{Hom}(B_{n+1}/B_n, X)$ and obtain an exact sequence

$$0 \longrightarrow X_{\gamma_{n+1}} \longrightarrow X_{\gamma_n} \longrightarrow X_{\bar{\gamma}_n} \longrightarrow 0$$

provided that X is injective or a coproduct of injective objects.

Now consider a cogenerator E such that each object of $\mathcal{A}$ embeds into a coproduct of copies of E. We may assume that E is injective by replacing E with its injective envelope. Let $\kappa = \max(\aleph_0, \mathrm{card}\,\mathrm{Hom}(C, E))$ and choose a monomorphism

$$\phi \colon \prod_{n \in \mathbb{N}} E^\kappa \longrightarrow \coprod_{i \in I} E.$$

For each $m \in \mathbb{N}$ we apply $\mathrm{Hom}(C_m, -)$ and obtain a monomorphism

$$\phi_m \colon \prod_{n \in \mathbb{N}} (E_{\gamma_m})^\kappa \longrightarrow \coprod_{i \in I} E_{\gamma_m}$$

since $X \mapsto X_{\gamma_m}$ preserves products and coproducts. Then it follows from Lemma 11.2.11 that for some $m \in \mathbb{N}$ the map ϕ_m restricts to an embedding

$$\prod_{n \geq m} (E_{\gamma_m})^\kappa \longrightarrow \left(\coprod_{i \in J} E_{\gamma_\infty} \right) \amalg \left(\coprod_{\text{finite}} E_{\gamma_m} \right)$$

for some cofinite subset $J \subseteq I$, where $E_{\gamma_\infty} = \bigcap_{n \geq 0} E_{\gamma_n}$. Comparing this with ϕ_{m+1} and passing to the quotient yields a commutative diagram with exact rows

$$
\begin{array}{ccccccccc}
0 & \longrightarrow & \prod\limits_{n \geq m} (E_{\gamma_{m+1}})^\kappa & \longrightarrow & \prod\limits_{n \geq m} (E_{\gamma_m})^\kappa & \longrightarrow & \prod\limits_{n \geq m} (E_{\bar{\gamma}_m})^\kappa & \to & 0 \\
& & \downarrow & & \downarrow & & \downarrow & & \\
0 \to & (\coprod\limits_{i \in J} E_{\gamma_\infty}) \amalg (\coprod\limits_{\text{finite}} E_{\gamma_{m+1}}) & \to & (\coprod\limits_{i \in J} E_{\gamma_\infty}) \amalg (\coprod\limits_{\text{finite}} E_{\gamma_m}) & \longrightarrow & \coprod\limits_{\text{finite}} E_{\bar{\gamma}_m} & \longrightarrow & 0
\end{array}
$$

where we use the fact that E is injective. The vertical map on the right is a monomorphism because it is a restriction of $\text{Hom}(B_{m+1}/B_m, \phi)$. From the choice of κ it follows that $E_{\bar{\gamma}_m} = 0$, cf. Lemma 11.2.13 below. Thus $C_m = C_{m+1}$ since E cogenerates $\mathcal{A}$. We conclude that C is noetherian. □

Lemma 11.2.13. *Let A be an abelian group with $\alpha = \text{card } A$ and let $\kappa \geq \max(\aleph_0, \alpha)$. If there is a monomorphism $A^\kappa \to A^n$ for some $n \in \mathbb{N}$, then $A = 0$.*

Proof Suppose $A \neq 0$. Then we have

$$\text{card}(A^\kappa) = \alpha^\kappa \geq 2^\kappa > \kappa = \kappa^n \geq \alpha^n = \text{card}(A^n).$$

This contradicts the fact that there is an injective map $A^\kappa \to A^n$. □

Remark 11.2.14. The Krull–Remak–Schmidt–Azumaya theorem implies that a decomposition into indecomposable objects with local endomorphism rings is essentially unique (Theorem 2.5.8).

We formulate some consequences of Theorem 11.2.12 and its proof.

Corollary 11.2.15. *Let $\mathcal{C}$ be an essentially small abelian category. Then all exact functors in $\text{Lex}(\mathcal{C}^{\text{op}}, \text{Ab})$ are injective if and only if all objects in $\mathcal{C}$ are noetherian.* □

A variation of the above theorem will be needed later.

Proposition 11.2.16. *Let $\mathcal{A}$ be a locally finitely presented Grothendieck category such that $\text{fp } \mathcal{A}$ is abelian. Suppose that $X \in \mathcal{A}$ is an object satisfying $\text{Ext}^1(C, X) = 0$ for all $C \in \text{fp } \mathcal{A}$ and that $\text{Hom}(C, X) = 0$ implies $C = 0$ for all $C \in \text{fp } \mathcal{A}$. Then the following are equivalent.*

(1) *The category $\mathcal{A}$ is locally noetherian*
(2) *The canonical monomorphism $X^{(\mathbb{N})} \to X^{\mathbb{N}}$ splits.*
(3) *There exists a decomposition $X^{\mathbb{N}} = \coprod_{i \in I} X_i$ such that $\text{End}(X_i)$ is local for all $i \in I$.*
(4) *There exists an object Y such that every product of copies of X is a subobject of a coproduct of copies of Y.*

Moreover, in this case the object X is injective.

Proof (1) $\Rightarrow$ (2) & (3) & (4): If $\mathcal{A}$ is locally noetherian, then $\text{fp } \mathcal{A}$ is closed under quotients. It follows from Lemma 11.2.10 that X is injective. Now apply Theorem 11.2.12.

(2) $\Rightarrow$ (1): Choose a splitting $\phi: X^{\mathbb{N}} \to X^{(\mathbb{N})}$. Let $C = C_0$ be a finitely presented object and let $\gamma = (C_i \twoheadrightarrow C_{i+1})_{i \in \mathbb{N}}$ be a chain of epimorphisms. We wish to show that C is noetherian and apply Lemma 11.2.11 to ϕ as above.

Thus $X_{\gamma_m} \subseteq \bigcap_{n \geq 0} X_{\gamma_n}$ for some $m \in \mathbb{N}$, and $C_m = C_{m+1} = \cdots$ follows, since X cogenerates $\mathrm{fp}\,\mathcal{A}$. We conclude that every object in $\mathrm{fp}\,\mathcal{A}$ is noetherian.

(3) $\Rightarrow$ (1): First observe that any indecomposable direct summand Y of X occurs, up to isomorphism, an infinite number of times in the family $(X_i)_{i \in I}$, by the Krull–Remak–Schmidt–Azumaya theorem (Theorem 2.5.8). Now let $\gamma = (C_i \twoheadrightarrow C_{i+1})_{i \in \mathbb{N}}$ be a chain of epimorphisms in $\mathrm{fp}\,\mathcal{A}$. Then it follows from Lemma 11.2.11 that there exists $m \in \mathbb{N}$ such that $Y_{\gamma_m} \subseteq \bigcap_{n \in \mathbb{N}} Y_{\gamma_n}$ for all indecomposable direct summands Y of X. Therefore $X_{\gamma_m} \subseteq \bigcap_{n \in \mathbb{N}} X_{\gamma_n}$, and $C_m = C_{m+1} = \cdots$ follows, since X cogenerates $\mathrm{fp}\,\mathcal{A}$. Thus every object in $\mathrm{fp}\,\mathcal{A}$ is noetherian.

(4) $\Rightarrow$ (1): Adapt the proof of Theorem 11.2.12, keeping in mind that X cogenerates $\mathrm{fp}\,\mathcal{A}$. $\qquad\square$

11.3 Gröbner Categories

Given an essentially small category $\mathcal{C}$, we study the problem when for any locally noetherian Grothendieck category $\mathcal{A}$ the functor category $\mathrm{Fun}(\mathcal{C}, \mathcal{A})$ is again locally noetherian. This problem is motivated by Hilbert's basis theorem and leads to the notion of a Gröbner category.

Hilbert's Basis Theorem

Let A be a (not necessarily commutative) ring and denote by $A[t]$ the polynomial ring in one variable. We can identify modules over $A[t]$ with pairs (X, ϕ) given by an A-module X and a morphism $\phi \colon X \to X$ that sends $x \in X$ to xt.

We view the set of non-negative integers as a category $\bar{\mathbb{N}}$ with a single object $*$, morphisms given by $\mathrm{Hom}(*, *) = \mathbb{N}$, and composition given by addition. Then there is an obvious equivalence

$$\mathrm{Fun}(\bar{\mathbb{N}}, \mathrm{Mod}\,A) \xrightarrow{\sim} \mathrm{Mod}\,A[t]$$

which sends a functor $F \colon \bar{\mathbb{N}} \to \mathrm{Mod}\,A$ to $F(*)$.

Now consider the partially ordered set of non-negative integers as a category $\vec{\mathbb{N}}$ with set of objects $\mathbb{N}$ and a single morphism $m \to n$ if and only if $m \leq n$. We view $A[t] = \bigoplus_{n \geq 0} A[t]_n$ as an $\mathbb{N}$-graded ring where $A[t]_n$ denotes the set of homogeneous polynomials of degree n. If we denote by $\mathrm{GrMod}\,A[t]$ the category of $\mathbb{N}$-graded $A[t]$-modules (with degree zero morphisms), then there is an obvious equivalence

$$\mathrm{Fun}(\vec{\mathbb{N}}, \mathrm{Mod}\,A) \xrightarrow{\sim} \mathrm{GrMod}\,A[t]$$

which sends a functor $F\colon \vec{\mathbb{N}} \to \operatorname{Mod} A$ to $\bigoplus_{n \geq 0} F(n)$.

The following is a reformulation of *Hilbert's basis theorem*.

Theorem 11.3.1 (Hilbert). *Let A be a right noetherian ring. Then the polynomial ring $A[t]$ is right noetherian; it is also right noetherian as a graded ring. Therefore $\operatorname{Mod} A[t]$ and $\operatorname{GrMod} A[t]$ are both locally noetherian Grothendieck categories.* □

Noetherian Posets

Let $\mathcal{C}$ be a poset. A subset $\mathcal{D} \subseteq \mathcal{C}$ is an *ideal* if the conditions $x \leq y$ in $\mathcal{C}$ and $y \in \mathcal{D}$ imply $x \in \mathcal{D}$. The ideals in $\mathcal{C}$ are partially ordered by inclusion.

A poset $\mathcal{C}$ is *noetherian* if every ascending chain of elements in $\mathcal{C}$ stabilises, and $\mathcal{C}$ is *strongly noetherian* if every ascending chain of ideals in $\mathcal{C}$ stabilises.

For a poset $\mathcal{C}$ and $x \in \mathcal{C}$, set $\mathcal{C}(x) = \{t \in \mathcal{C} \mid t \leq x\}$. The assignment $x \mapsto \mathcal{C}(x)$ yields an embedding of $\mathcal{C}$ into the poset of ideals in $\mathcal{C}$.

Lemma 11.3.2. *For a poset $\mathcal{C}$ the following are equivalent.*

(1) *The poset $\mathcal{C}$ is strongly noetherian.*

(2) *For every infinite sequence $(x_i)_{i \in \mathbb{N}}$ of elements in $\mathcal{C}$ there exists $i \in \mathbb{N}$ such that $x_j \leq x_i$ for infinitely many $j \in \mathbb{N}$.*

(3) *For every infinite sequence $(x_i)_{i \in \mathbb{N}}$ of elements in $\mathcal{C}$ there is a map $\alpha \colon \mathbb{N} \to \mathbb{N}$ such that $i < j$ implies $\alpha(i) < \alpha(j)$ and $x_{\alpha(j)} \leq x_{\alpha(i)}$.*

(4) *For every infinite sequence $(x_i)_{i \in \mathbb{N}}$ of elements in $\mathcal{C}$ there are $i < j$ in $\mathbb{N}$ such that $x_j \leq x_i$.*

Proof (1) $\Rightarrow$ (2): Suppose that $\mathcal{C}$ is strongly noetherian and let $(x_i)_{i \in \mathbb{N}}$ be elements in $\mathcal{C}$. For $n \in \mathbb{N}$ set $\mathcal{C}_n = \bigcup_{i \leq n} \mathcal{C}(x_i)$. The chain $(\mathcal{C}_n)_{n \in \mathbb{N}}$ stabilises, say $\mathcal{C}_n = \mathcal{C}_N$ for all $n \geq N$. Thus there exists $i \leq N$ such that $x_j \leq x_i$ for infinitely many $j \in \mathbb{N}$.

(2) $\Rightarrow$ (3): Define $\alpha \colon \mathbb{N} \to \mathbb{N}$ recursively by taking for $\alpha(0)$ the smallest $i \in \mathbb{N}$ such that $x_j \leq x_i$ for infinitely many $j \in \mathbb{N}$. For $n > 0$ set

$$\alpha(n) = \min\{i > \alpha(n-1) \mid x_j \leq x_i \leq x_{\alpha(n-1)} \text{ for infinitely many } j \in \mathbb{N}\}.$$

(3) $\Rightarrow$ (4): Clear.

(4) $\Rightarrow$ (1): Suppose there is a properly ascending chain $(\mathcal{C}_n)_{n \in \mathbb{N}}$ of ideals in $\mathcal{C}$. Choose $x_n \in \mathcal{C}_{n+1} \setminus \mathcal{C}_n$ for each $n \in \mathbb{N}$. There are $i < j$ in $\mathbb{N}$ such that $x_j \leq x_i$. This implies $x_j \in \mathcal{C}_{i+1} \subseteq \mathcal{C}_j$ which is a contradiction. □

Functor Categories

Let $\mathcal{C}$ be an essentially small category. We simplify the notation by setting

$$\mathcal{C}(x, y) := \mathrm{Hom}_{\mathcal{C}}(x, y) \qquad \text{for objects } x, y \in \mathcal{C}.$$

For a Grothendieck category $\mathcal{A}$ we denote by $\mathrm{Fun}(\mathcal{C}^{\mathrm{op}}, \mathcal{A})$ the category of functors $\mathcal{C}^{\mathrm{op}} \to \mathcal{A}$. The morphisms between two functors are the natural transformations. Note that $\mathrm{Fun}(\mathcal{C}^{\mathrm{op}}, \mathcal{A})$ is a Grothendieck category.

Given an object $x \in \mathcal{C}$, the evaluation functor

$$\mathrm{Fun}(\mathcal{C}^{\mathrm{op}}, \mathcal{A}) \longrightarrow \mathcal{A}, \qquad F \mapsto F(x)$$

admits a left adjoint

$$\mathcal{A} \longrightarrow \mathrm{Fun}(\mathcal{C}^{\mathrm{op}}, \mathcal{A}), \qquad M \mapsto M[\mathcal{C}(-, x)]$$

where for any set X we denote by $M[X]$ a coproduct of copies of M indexed by the elements of X. Thus we have for objects $M \in \mathcal{A}$ and $F \in \mathrm{Fun}(\mathcal{C}^{\mathrm{op}}, \mathcal{A})$ a natural isomorphism

$$\mathrm{Hom}(M[\mathcal{C}(-, x)], F) \cong \mathrm{Hom}(M, F(x)). \tag{11.3.3}$$

Lemma 11.3.4. *Let $(M_i)_{i \in I}$ be a set of generators of $\mathcal{A}$. Then the functors $M_i[\mathcal{C}(-, x)]$ with $i \in I$ and $x \in \mathcal{C}$ generate $\mathrm{Fun}(\mathcal{C}^{\mathrm{op}}, \mathcal{A})$.*

Proof Use the adjointness isomorphism (11.3.3). □

Recall that a Grothendieck category $\mathcal{A}$ is *locally noetherian* if $\mathcal{A}$ has a generating set of noetherian objects. In that case an object $M \in \mathcal{A}$ is noetherian if and only if M is *finitely presented*, that is, the representable functor $\mathrm{Hom}(M, -)$ preserves filtered colimits; see Proposition 11.2.5

Lemma 11.3.5. *Let $\mathcal{A}$ be locally noetherian. Then $\mathrm{Fun}(\mathcal{C}^{\mathrm{op}}, \mathcal{A})$ is locally noetherian if and only if $M[\mathcal{C}(-, x)]$ is noetherian for every noetherian $M \in \mathcal{A}$ and $x \in \mathcal{C}$.*

Proof First observe that $M[\mathcal{C}(-, x)]$ is finitely presented if M is finitely presented. This follows from the isomorphism (11.3.3) since evaluation at $x \in \mathcal{C}$ preserves colimits. Now the assertion of the lemma is an immediate consequence of Lemma 11.3.4. □

Noetherian Functors

Let $\mathcal{C}$ be a small category and fix an object $x \in \mathcal{C}$. Set

$$\mathcal{C}(x) := \coprod_{t \in \mathcal{C}} \mathcal{C}(t, x).$$

Given $f, g \in \mathcal{C}(x)$, let $\langle f \rangle$ denote the set of morphisms in $\mathcal{C}(x)$ that factor through f, and set $f \leq_x g$ if $\langle f \rangle \subseteq \langle g \rangle$. We identify f and g when $\langle f \rangle = \langle g \rangle$. This yields a poset which we denote by $\bar{\mathcal{C}}(x)$.

A functor is *noetherian* if every ascending chain of subfunctors stabilises.

Lemma 11.3.6. *The functor* $\mathcal{C}(-, x) \colon \mathcal{C}^{\mathrm{op}} \to \mathrm{Set}$ *is noetherian if and only if the poset* $\bar{\mathcal{C}}(x)$ *is strongly noetherian.*

Proof Sending $F \subseteq \mathcal{C}(-, x)$ to $\bigcup_{t \in \mathcal{C}} F(t)$ induces an inclusion preserving bijection between the subfunctors of $\mathcal{C}(-, x)$ and the ideals in $\bar{\mathcal{C}}(x)$. □

For a poset $\mathcal{T}$ let $\mathrm{Set} \wr \mathcal{T}$ denote the category consisting of pairs (X, ξ) given by a set X and a map $\xi \colon X \to \mathcal{T}$. A morphism $(X, \xi) \to (X', \xi')$ is a map $f \colon X \to X'$ such that $\xi(a) \leq \xi' f(a)$ for all $a \in X$.

A functor $\mathcal{C}^{\mathrm{op}} \to \mathrm{Set} \wr \mathcal{T}$ is given by a pair (F, ϕ) consisting of a functor $F \colon \mathcal{C}^{\mathrm{op}} \to \mathrm{Set}$ and a map $\phi \colon \bigsqcup_{t \in \mathcal{C}} F(t) \to \mathcal{T}$ such that $\phi(a) \leq \phi(F(f)(a))$ for every $a \in F(t)$ and $f \colon t' \to t$ in $\mathcal{C}$.

Lemma 11.3.7. *Let* $\mathcal{T}$ *be a noetherian poset. If the functor* $\mathcal{C}(-, x) \colon \mathcal{C}^{\mathrm{op}} \to \mathrm{Set}$ *is noetherian, then every functor* $\mathcal{C}^{\mathrm{op}} \to \mathrm{Set} \wr \mathcal{T}$ *whose composite with the canonical functor* $\mathrm{Set} \wr \mathcal{T} \to \mathrm{Set}$ *equals* $\mathcal{C}(-, x)$ *is also noetherian.*

Proof Fix a functor $(F, \phi) \colon \mathcal{C}^{\mathrm{op}} \to \mathrm{Set} \wr \mathcal{T}$, and let $(F_n, \phi_n)_{n \in \mathbb{N}}$ be a strictly ascending chain of subfunctors of (F, ϕ). The chain $(F_n)_{n \in \mathbb{N}}$ stabilises since $\mathcal{C}(-, x)$ is noetherian. Thus we may assume that $F_n = F$ for all $n \in \mathbb{N}$, and we find $f_n \in \bigsqcup_{t \in \mathcal{C}} F(t)$ such that $\phi_n(f_n) < \phi_{n+1}(f_n)$. The poset $\bar{\mathcal{C}}(x)$ is strongly noetherian by Lemma 11.3.6. It follows from Lemma 11.3.2 that there is a map $\alpha \colon \mathbb{N} \to \mathbb{N}$ such that $i < j$ implies $\alpha(i) < \alpha(j)$ and $f_{\alpha(j)} \leq_x f_{\alpha(i)}$. Thus

$$\phi_{\alpha(n)}(f_{\alpha(n)}) < \phi_{\alpha(n)+1}(f_{\alpha(n)}) \leq \phi_{\alpha(n+1)}(f_{\alpha(n)}) \leq \phi_{\alpha(n+1)}(f_{\alpha(n+1)}).$$

This yields a strictly ascending chain in $\mathcal{T}$, contradicting the assumption on $\mathcal{T}$ to be noetherian. □

A partial order $\leq$ on $\mathcal{C}(x)$ is *admissible* if the following holds.

(Ad1) The order $\leq$ restricted to $\mathcal{C}(t, x)$ is total and noetherian for every $t \in \mathcal{C}$.
(Ad2) For $f, f' \in \mathcal{C}(t, x)$ and $e \in \mathcal{C}(s, t)$, the condition $f < f'$ implies $fe < f'e$.

Assume there is given an admissible partial order $\leq$ on $\mathcal{C}(x)$ and an object M in a Grothendieck category $\mathcal{A}$. Let $\mathrm{Sub}(M)$ denote the poset of subobjects of M and consider the functor

$$\mathcal{C}(-, x) \wr M \colon \mathcal{C}^{\mathrm{op}} \longrightarrow \mathrm{Set} \wr \mathrm{Sub}(M), \quad t \mapsto \left(\mathcal{C}(t, x), (M)_{f \in \mathcal{C}(t, x)} \right).$$

For a subfunctor $F \subseteq M[\mathcal{C}(-,x)]$ define a subfunctor $\tilde{F} \subseteq \mathcal{C}(-,x) \wr M$ as follows:

$$\tilde{F}: \mathcal{C}^{\mathrm{op}} \longrightarrow \mathrm{Set} \wr \mathrm{Sub}(M), \quad t \mapsto \left(\mathcal{C}(t,x), \left(\pi_f(M[\mathcal{C}(t,x)_f] \cap F(t)) \right)_{f \in \mathcal{C}(t,x)} \right)$$

where $\mathcal{C}(t,x)_f = \{ g \in \mathcal{C}(t,x) \mid f \leq g \}$ and $\pi_f: M[\mathcal{C}(t,x)_f] \to M$ is the projection onto the factor corresponding to f. For a morphism $e: t' \to t$ in $\mathcal{C}$, the morphism $\tilde{F}(e)$ is induced by precomposition with e. Note that

$$\pi_f(M[\mathcal{C}(t,x)_f] \cap F(t)) \subseteq \pi_{fe}(M[\mathcal{C}(t',x)_{fe}] \cap F(t'))$$

since $\leq$ is compatible with the composition in $\mathcal{C}$.

Lemma 11.3.8. *Suppose there is an admissible partial order on $\mathcal{C}(x)$. Then the assignment which sends a subfunctor $F \subseteq M[\mathcal{C}(-,x)]$ to $\tilde{F}$ preserves proper inclusions. Therefore $M[\mathcal{C}(-,x)]$ is noetherian provided that $\mathcal{C}(-,x) \wr M$ is noetherian.*

Proof Let $F \subseteq G \subseteq M[\mathcal{C}(-,x)]$. Then $\tilde{F} \subseteq \tilde{G}$. Now suppose that $F \neq G$. Thus there exists $t \in \mathcal{C}$ such that $F(t) \neq G(t)$. We have $\mathcal{C}(t,x) = \bigcup_{f \in \mathcal{C}(t,x)} \mathcal{C}(t,x)_f$, and this union is directed since $\leq$ is total. Thus

$$F(t) = \sum_{f \in \mathcal{C}(t,x)} \left(M[\mathcal{C}(t,x)_f] \cap F(t) \right)$$

since filtered colimits in $\mathcal{A}$ are exact. This yields f such that

$$M[\mathcal{C}(t,x)_f] \cap F(t) \neq M[\mathcal{C}(t,x)_f] \cap G(t).$$

Choose $f \in \mathcal{C}(t,x)$ maximal with respect to this property, using that $\leq$ is noetherian. Now observe that the projection π_f induces an exact sequence

$$0 \longrightarrow \sum_{f < g} \left(M[\mathcal{C}(t,x)_g] \cap F(t) \right) \longrightarrow F(t) \longrightarrow \pi_f(M[\mathcal{C}(t,x)_f] \cap F(t)) \longrightarrow 0$$

since the kernel of π_f equals the directed union $\sum_{f < g} M[\mathcal{C}(t,x)_g]$. For the directedness one uses again that $\leq$ is total. Thus

$$\pi_f(M[\mathcal{C}(t,x)_f] \cap F(t)) \neq \pi_f(M[\mathcal{C}(t,x)_f] \cap G(t))$$

and therefore $\tilde{F} \neq \tilde{G}$. $\qquad\square$

Proposition 11.3.9. *Let $x \in \mathcal{C}$. Suppose that $\mathcal{C}(-,x)$ is noetherian and that $\mathcal{C}(x)$ has an admissible partial order. If $M \in \mathcal{A}$ is noetherian, then $M[\mathcal{C}(-,x)]$ is noetherian.*

Proof Combine Lemma 11.3.7 and Lemma 11.3.8. $\qquad\square$

Gröbner Categories

A small category $\mathcal{C}$ is a *Gröbner category* if the following holds.

(Gr1) The functor $\mathcal{C}(-, x)$ is noetherian for every $x \in \mathcal{C}$.
(Gr2) There is an admissible partial order on $\mathcal{C}(x)$ for every $x \in \mathcal{C}$.

Theorem 11.3.10. *Let $\mathcal{C}$ be a Gröbner category and $\mathcal{A}$ a Grothendieck category. If $\mathcal{A}$ is locally noetherian, then $\mathrm{Fun}(\mathcal{C}^{\mathrm{op}}, \mathcal{A})$ is locally noetherian.*

Proof   Combine Lemma 11.3.4 and Proposition 11.3.9.   □

Example 11.3.11. A strongly noetherian poset (viewed as a category) is a Gröbner category.

Example 11.3.12. Consider the additive monoid $\bar{\mathbb{N}}$ of non-negative integers, viewed as a category with a single object, and the poset $\vec{\mathbb{N}}$ of non-negative integers, again viewed as a category. Then $\bar{\mathbb{N}}^{\mathrm{op}}$ and $\vec{\mathbb{N}}^{\mathrm{op}}$ are Gröbner categories. Let $\mathcal{A}$ be the module category of a right noetherian ring A. Then $\mathrm{Fun}(\bar{\mathbb{N}}, \mathcal{A})$ and $\mathrm{Fun}(\vec{\mathbb{N}}, \mathcal{A})$ identify with categories of modules over the polynomial ring in one variable over A (ungraded and graded). Thus Theorem 11.3.10 generalises Hilbert's basis theorem (Theorem 11.3.1).

Base Change

Given functors $F, G \colon \mathcal{C}^{\mathrm{op}} \to \mathrm{Set}$, we write $F \rightsquigarrow G$ if there is a finite chain

$$F = F_0 \twoheadrightarrow F_1 \hookleftarrow F_2 \twoheadrightarrow \cdots \twoheadrightarrow F_{n-1} \hookleftarrow F_n = G$$

of epimorphisms and monomorphisms of functors $\mathcal{C}^{\mathrm{op}} \to \mathrm{Set}$.

A functor $\phi \colon \mathcal{C} \to \mathcal{D}$ is *contravariantly finite* if the following holds.

(Con1) Every object $y \in \mathcal{D}$ is isomorphic to $\phi(x)$ for some $x \in \mathcal{C}$.
(Con2) For every object $y \in \mathcal{D}$ there are objects $x_1, \ldots, x_n$ in $\mathcal{C}$ such that

$$\bigsqcup_{i=1}^{n} \mathcal{C}(-, x_i) \rightsquigarrow \mathcal{D}(\phi-, y).$$

The functor ϕ is *covariantly finite* if $\phi^{\mathrm{op}} \colon \mathcal{C}^{\mathrm{op}} \to \mathcal{D}^{\mathrm{op}}$ is contravariantly finite.

Note that a composite of contravariantly finite functors is contravariantly finite.

Lemma 11.3.13. *Let $f \colon \mathcal{C} \to \mathcal{D}$ be a contravariantly finite functor and $\mathcal{A}$ a Grothendieck category. Fix $M \in \mathcal{A}$ and suppose that $M[\mathcal{C}(-, x)]$ is noetherian for all $x \in \mathcal{C}$. Then $M[\mathcal{D}(-, y)]$ is noetherian for all $y \in \mathcal{D}$.*

Proof   A finite chain

$$\coprod_{i=1}^{n} \mathcal{C}(-, x_i) = F_0 \twoheadrightarrow F_1 \hookleftarrow F_2 \twoheadrightarrow \cdots \twoheadrightarrow F_{n-1} \hookleftarrow F_n = \mathcal{D}(\phi-, y)$$

of epimorphisms and monomorphisms induces a chain

$$\coprod_{i=1}^{n} M[\mathcal{C}(-, x_i)] = \bar{F}_0 \twoheadrightarrow \bar{F}_1 \hookleftarrow \bar{F}_2 \twoheadrightarrow \cdots \twoheadrightarrow \bar{F}_{n-1} \hookleftarrow \bar{F}_n = M[\mathcal{D}(\phi-, y)]$$

of epimorphisms and monomorphisms in $\mathrm{Fun}(\mathcal{C}^{\mathrm{op}}, \mathcal{A})$. Thus $M[\mathcal{D}(\phi-, y)]$ is noetherian. It follows that $M[\mathcal{D}(-, y)]$ is noetherian, since precomposition with ϕ yields a faithful and exact functor $\mathrm{Fun}(\mathcal{D}^{\mathrm{op}}, \mathcal{A}) \to \mathrm{Fun}(\mathcal{C}^{\mathrm{op}}, \mathcal{A})$.   □

Proposition 11.3.14. *Let $f \colon \mathcal{C} \to \mathcal{D}$ be a contravariantly finite functor and $\mathcal{A}$ a locally noetherian Grothendieck category. If the category $\mathrm{Fun}(\mathcal{C}^{\mathrm{op}}, \mathcal{A})$ is locally noetherian, then $\mathrm{Fun}(\mathcal{D}^{\mathrm{op}}, \mathcal{A})$ is locally noetherian.*

Proof   Combine Lemma 11.3.5 and Lemma 11.3.13.   □

Categories of Finite Sets

Let Γ denote the category of finite sets; a skeleton is given by the sets $\mathbf{n} = \{1, 2, \ldots, n\}$. The subcategory of finite sets with surjective morphisms is denoted by Γ_{sur}. A surjection $f \colon \mathbf{m} \to \mathbf{n}$ is *ordered* if $i < j$ implies $\min f^{-1}(i) < \min f^{-1}(j)$. We write Γ_{os} for the subcategory of finite sets whose morphisms are ordered surjections. Given a surjection $f \colon \mathbf{m} \to \mathbf{n}$, let $f^! \colon \mathbf{n} \to \mathbf{m}$ denote the map given by $f^!(i) = \min f^{-1}(i)$. Note that $f f^! = \mathrm{id}$, and $g f = f^! g^!$ provided that f and g are ordered surjections.

Lemma 11.3.15. *The inclusions $\Gamma_{\mathrm{os}} \to \Gamma_{\mathrm{sur}}$ and $\Gamma_{\mathrm{sur}} \to \Gamma$ are both contravariantly finite.*

Proof   For each integer $n \geq 0$ there is an isomorphism

$$\Gamma_{\mathrm{os}}(-, \mathbf{n}) \times \mathfrak{S}_n \xrightarrow{\sim} \Gamma_{\mathrm{sur}}(-, \mathbf{n})$$

which sends a pair (f, σ) to σf. The inverse sends a surjective map $g \colon \mathbf{m} \to \mathbf{n}$ to $(\tau^{-1} g, \tau)$ where $\tau \in \mathfrak{S}_n$ is the unique permutation such that $g^! \tau$ is increasing.

For each integer $n \geq 0$ there is an isomorphism

$$\coprod_{\mathbf{m} \hookrightarrow \mathbf{n}} \Gamma_{\mathrm{sur}}(-, \mathbf{m}) \xrightarrow{\sim} \Gamma(-, \mathbf{n})$$

which is induced by the injective maps $\mathbf{m} \to \mathbf{n}$.   □

Fix an integer $n \geq 0$. Given $f, g \in \Gamma(\mathbf{n})$ we set $f \leq g$ if there exists an ordered surjection h such that $f = gh$.

Lemma 11.3.16. *The poset* $(\Gamma(\mathbf{n}), \leq)$ *is strongly noetherian.*

Proof We fix some notation for each $f \in \Gamma(\mathbf{m}, \mathbf{n})$. Set $\lambda(f) = m$. If f is not injective, set

$$\mu(f) = m - \max\{i \in \mathbf{m} \mid \text{there exists } j < i \text{ such that } f(i) = f(j)\}$$

and $\pi(f) = f(m - \mu(f))$. Define $\tilde{f} \in \Gamma(\mathbf{m} - \mathbf{1}, \mathbf{n})$ by setting $\tilde{f}(i) = f(i)$ for $i < m - \mu(f)$ and $\tilde{f}(i) = f(i+1)$ otherwise.

Note that $f \leq \tilde{f}$. Moreover, $\mu(f) = \mu(g)$, $\pi(f) = \pi(g)$, and $\tilde{f} \leq \tilde{g}$ imply $f \leq g$.

Suppose that $(\Gamma(\mathbf{n}), \leq)$ is not strongly noetherian. Then there exists an infinite sequence $(f_r)_{r \in \mathbb{N}}$ in $\Gamma(\mathbf{n})$ such that $i < j$ implies $f_j \not\leq f_i$; see Lemma 11.3.2. Call such a sequence *bad*. Choose the sequence *minimal* in the sense that $\lambda(f_i)$ is minimal for all bad sequences $(g_r)_{r \in \mathbb{N}}$ with $g_j = f_j$ for all $j < i$. There is an infinite subsequence $(f_{\alpha(r)})_{r \in \mathbb{N}}$ (given by some increasing map $\alpha \colon \mathbb{N} \to \mathbb{N}$) such that μ and π agree on all $f_{\alpha(r)}$, since the values of μ and π are bounded by n. Now consider the sequence $f_0, f_1, \ldots, f_{\alpha(0)-1}, \tilde{f}_{\alpha(0)}, \tilde{f}_{\alpha(1)}, \ldots$ and denote this by $(g_r)_{r \in \mathbb{N}}$. This sequence is not bad, since $(f_r)_{r \in \mathbb{N}}$ is minimal. Thus there are $i < j$ in $\mathbb{N}$ with $g_j \leq g_i$. Clearly, $j < \alpha(0)$ is impossible. If $i < \alpha(0)$, then

$$f_{\alpha(j-\alpha(0))} \leq \tilde{f}_{\alpha(j-\alpha(0))} = g_j \leq g_i = f_i,$$

which is a contradiction, since $i < \alpha(0) \leq \alpha(j - \alpha(0))$. If $i \geq \alpha(0)$, then $f_{\alpha(j-\alpha(0))} \leq f_{\alpha(i-\alpha(0))}$; this is a contradiction again. Thus $(\Gamma(\mathbf{n}), \leq)$ is strongly noetherian. $\square$

Proposition 11.3.17. *The category* Γ_{os} *is a Gröbner category.*

Proof Fix an integer $n \geq 0$. The poset $\bar{\Gamma}_{os}(\mathbf{n})$ is strongly noetherian by Lemma 11.3.16, and it follows from Lemma 11.3.6 that the functor $\Gamma_{os}(-, \mathbf{n})$ is noetherian.

The admissible partial order on $\Gamma_{os}(\mathbf{n})$ is given by the lexicographic order. Thus for $f, g \in \Gamma_{os}(\mathbf{m}, \mathbf{n})$, we have $f < g$ if there exists $j \in \mathbf{m}$ with $f(j) < g(j)$ and $f(i) = g(i)$ for all $i < j$. $\square$

Theorem 11.3.18. *Let* $\mathcal{A}$ *be a locally noetherian Grothendieck category. Then the category* $\mathrm{Fun}(\Gamma^{\mathrm{op}}, \mathcal{A})$ *is locally noetherian.*

Proof The category Γ_{os} is a Gröbner category by Proposition 11.3.17. It follows from Theorem 11.3.10 that $\mathrm{Fun}((\Gamma_{os})^{\mathrm{op}}, \mathcal{A})$ is locally noetherian. The inclusion $\Gamma_{os} \rightarrow \Gamma$ is contravariantly finite by Lemma 11.3.15. Thus $\text{Fun}(\Gamma^{op}, \mathcal{A})$ is locally noetherian by Proposition 11.3.14.   □

FI-Modules

Let Γ_{inj} denote the category whose objects are finite sets and whose morphisms are injective maps. When $\mathcal{A}$ is the category of modules over a ring, then a functor $\Gamma_{inj} \rightarrow \mathcal{A}$ is called an *FI-module* (F = finite sets, I = injective maps).

Theorem 11.3.19. *Let $\mathcal{A}$ be a locally noetherian Grothendieck category. Then the category $\text{Fun}(\Gamma_{inj}, \mathcal{A})$ is locally noetherian.*

Proof Consider the functor $\phi: \Gamma_{os} \rightarrow (\Gamma_{inj})^{op}$ which is the identity on objects and takes a map $f: \mathbf{m} \rightarrow \mathbf{n}$ to $f^!: \mathbf{n} \rightarrow \mathbf{m}$ given by $f^!(i) = \min f^{-1}(i)$. This functor is contravariantly finite, since for each integer $n \geq 0$ the morphism

$$\Gamma_{os}(-, \mathbf{n}) \times \mathfrak{S}_n \longrightarrow \Gamma_{inj}(\mathbf{n}, \phi-)$$

which sends a pair (f, σ) to $f^! \sigma$ is an epimorphism.

It follows from Proposition 11.3.14 that the category $\text{Fun}(\Gamma_{inj}, \mathcal{A})$ is locally noetherian, since $\text{Fun}((\Gamma_{os})^{op}, \mathcal{A})$ is locally noetherian by Proposition 11.3.17 and Theorem 11.3.10.   □

Generic Representations

Let A be a ring. We denote by $\mathcal{F}(A)$ the category of finitely generated free A-modules. Note that $\mathcal{F}(A)^{op} \xrightarrow{\sim} \mathcal{F}(A^{op})$. Now fix the module category $\mathcal{A} = \text{Mod}\, k$ of a commutative ring k. Then a functor $F: \mathcal{F}(A) \rightarrow \mathcal{A}$ yields a family $F(A^n)$ of k-linear representations of $\text{GL}_n(A)$ for $n \geq 0$ via evaluation; so one calls F a *generic representation* of A. In fact, F is equivalent to a compatible family of k-linear representations of $M_n(A)$, where $M_n(A)$ denotes the semigroup of all $n \times n$ matrices over A.

Suppose that A is *finite*, that is, the underlying set has finite cardinality. Then the functor $\Gamma \rightarrow \mathcal{F}(A)$ sending X to $A[X]$ is a left adjoint of the forgetful functor $\mathcal{F}(A) \rightarrow \Gamma$.

Lemma 11.3.20. *Let A be finite. Then the functor $\Gamma \rightarrow \mathcal{F}(A)$ is contravariantly finite.*

Proof The assertion follows from the adjointness isomorphism

$$\mathcal{F}(A)(A[X], P) \cong \Gamma(X, P).$$ □

Theorem 11.3.21. *Let A be a finite ring and $\mathcal{A}$ a locally noetherian Grothendieck category. Then the category* $\mathrm{Fun}(\mathcal{F}(A),\mathcal{A})$ *is locally noetherian.*

Proof We reduce the assertion about $\mathcal{F}(A)$ to the category of finite sets, using Lemma 11.3.20 and Proposition 11.3.14. Then one applies Theorem 11.3.18.

□

There is the following immediate consequence, also known as the *artinian conjecture*, because it amounts to the fact that the standard injective objects are artinian.

Corollary 11.3.22. *For a finite field $\mathbb{F}$ the category of generic representations* $\mathrm{Fun}(\mathrm{mod}\,\mathbb{F}, \mathrm{Mod}\,\mathbb{F})$ *is locally noetherian.*   □

Notes

Locally finitely presented categories were introduced by Gabriel and Ulmer [84]. For the special case of abelian categories and the properties of injective objects, see Gabriel's thesis [79]. In particular, that work contains the idea of using categories of left exact functors. The decomposition theory of injective objects in locally noetherian categories goes back to results for modules by Matlis [143] and Papp [155]; see also the exposition of Roos [177, 178].

Chase's lemma appears as an argument in [49] and is formulated explicitly in [50].

In a seminal paper Mitchell pointed out the parallel between modules and additive functors, introducing the term *ring with several objects* for a preadditive category [145].

The concept of a Gröbner category and the corresponding generalisation of Hilbert's basis theorem [112] is due to Richter [169] and was rediscovered by Sam and Snowden [180]. In particular, [180] contains a proof of the artinian conjecture. Lannes and Schwartz formulated this conjecture and were motivated by their study of unstable modules over the Steeenrod algebra [109]. The fact that FI-modules over a noetherian ring form a locally noetherian category is due to Church, Ellenberg, Farb and Nagpal [51]. Our exposition follows notes of Djament [63] which are motivated by applications to generic representation theory; see also the expository articles by Kuhn, Powell and Schwartz in [133].

12

Purity

Contents

12.1	**Purity**		**378**
	From Left Exact to Exact Functors		378
	Pure-Exactness and Pure-Injectives		380
	The Spectrum of Indecomposable Injectives		381
	Compactness		383
	The Spectrum of Indecomposable Pure-Injectives		384
12.2	**Definable Subcategories**		**384**
	Definable Subcategories		385
	Closure Properties of Definable Subcategories		387
	Change of Categories		390
12.3	**Indecomposable Pure-Injective Objects**		**392**
	Subgroups of Finite Definition		392
	Σ-Pure-Injectivity		393
	Product-Complete Objects		395
	Prüfer Objects		396
	Compactness		397
	Left Almost Split Morphisms		398
	Fp-Injective Objects		399
12.4	**Pure-Injective Modules**		**400**
	The Free Abelian Category		400
	A Criterion for Pure-Injectivity		402
	Duality		403
	Pure-Semisimplicity		404
	Modules of Finite Projective Dimension		405
	The Ziegler Spectrum of an Artin Algebra		406
	The Zariski Spectrum		407
	Injective Cohomology Representations		408
	Notes		**412**

We study the notion of purity for additive categories that are locally finitely presented. A typical example is the category of modules over a ring. We are mostly interested in pure-injective objects; they enjoy decomposition properties that are analogous to those of injective objects in Grothendieck categories.

A basic idea is to assign to a locally finitely presented category $\mathcal{A}$ an essentially small abelian category $\mathrm{Ab}(\mathcal{A})$ such that the objects in $\mathcal{A}$ identify with exact functors $\mathrm{Ab}(\mathcal{A})^{\mathrm{op}} \to \mathrm{Ab}$. For instance, when $\mathcal{A} = \mathrm{Mod}\, \Lambda$ is the category of modules over a ring Λ, then $\mathrm{Ab}(\mathcal{A})$ equals the free abelian category $\mathrm{Ab}(\Lambda)$ over Λ.

Viewing objects of $\mathcal{A}$ as exact functors leads naturally to the notion of a definable subcategory of $\mathcal{A}$ if we consider all exact functors which vanish on a specific Serre subcategory of $\mathrm{Ab}(\mathcal{A})$. In particular, we see that any such definable subcategory is determined by its indecomposable pure-injective objects.

12.1 Purity

In this section we introduce for locally finitely presented categories the notion of purity. This is based on the concept of a pure-exact sequence, and there are several ways to define this. For example, a sequence is pure-exact if it is a filtered colimit of split exact sequences. We can embed any locally finitely presented category $\mathcal{A}$ into a Grothendieck category $\mathbf{P}(\mathcal{A})$ such that pure-exactness identifies with the usual notion of exactness in abelian categories. We call this the purity category of $\mathcal{A}$. From this embedding we deduce that every object admits a pure-injective envelope.

From Left Exact to Exact Functors

Let $\mathcal{A}$ be a locally finitely presented category and set $\mathcal{C} = \mathrm{fp}\,\mathcal{A}$. We introduce the embedding $\mathcal{A} \hookrightarrow \mathbf{P}(\mathcal{A})$ into a Grothendieck category, which is our main tool.

A functor $F\colon \mathcal{C} \to \mathrm{Ab}$ is *finitely presented* if it admits a presentation

$$\mathrm{Hom}_{\mathcal{C}}(D,-) \longrightarrow \mathrm{Hom}_{\mathcal{C}}(C,-) \longrightarrow F \longrightarrow 0. \qquad (12.1.1)$$

We denote by $\mathrm{Fp}(\mathcal{C}, \mathrm{Ab})$ the category of finitely presented functors and observe that $\mathrm{Fp}(\mathcal{C}, \mathrm{Ab})$ is abelian since $\mathcal{C}$ admits cokernels.

A functor F in $\mathrm{Fp}(\mathcal{C}, \mathrm{Ab})$ induces the functor

$$\bar{F}\colon \mathcal{A} \longrightarrow \mathrm{Ab}, \quad X \mapsto \operatorname*{colim}_{(C,\phi)\in\mathcal{C}/X} F(C)$$

using the presentation (11.1.17) of X as a filtered colimit of finitely presented objects. A presentation (12.1.1) of F then yields the presentation

$$\mathrm{Hom}_{\mathcal{A}}(D,-) \longrightarrow \mathrm{Hom}_{\mathcal{A}}(C,-) \longrightarrow \bar{F} \longrightarrow 0. \qquad (12.1.2)$$

Remark 12.1.3. A functor of the form $\bar{F}\colon \mathcal{A} \to \mathrm{Ab}$ with $F \in \mathrm{Fp}(\mathcal{C}, \mathrm{Ab})$ preserves filtered colimits and products. This is clear when $\bar{F} = \mathrm{Hom}_{\mathcal{A}}(C, -)$ for some $C \in \mathrm{fp}\,\mathcal{A}$, and the general case follows from the presentation (12.1.2); for a converse see Corollary 12.2.11.

For $X \in \mathcal{A}$ we consider the *evaluation*

$$\bar{X}\colon \mathrm{Fp}(\mathcal{C}, \mathrm{Ab}) \longrightarrow \mathrm{Ab}, \quad F \mapsto \bar{F}(X).$$

Clearly, the functor $\bar{X}$ is exact when $X \in \mathcal{C}$, and $\bar{X} = \mathrm{colim}\,\bar{X}_i$ is exact when $X = \mathrm{colim}\,X_i$, since taking filtered colimits is exact. This yields the functor

$$\mathrm{ev}\colon \mathcal{A} \longrightarrow \mathbf{P}(\mathcal{A}) := \mathrm{Lex}(\mathrm{Fp}(\mathcal{C}, \mathrm{Ab}), \mathrm{Ab}), \quad X \mapsto \bar{X}.$$

The category $\mathbf{P}(\mathcal{A})$ is by definition the *purity category* of $\mathcal{A}$. It is a locally finitely presented Grothendieck category and the finitely presented objects form an abelian category that is equivalent to $\mathrm{Fp}(\mathcal{C}, \mathrm{Ab})^{\mathrm{op}}$.

Let us collect some basic properties of this evaluation functor. We write $\mathrm{Ex}(\mathrm{Fp}(\mathcal{C}, \mathrm{Ab}), \mathrm{Ab})$ for the category of exact functors $\mathrm{Fp}(\mathcal{C}, \mathrm{Ab}) \to \mathrm{Ab}$.

Lemma 12.1.4. *The functor* $\mathrm{ev}\colon \mathcal{A} \to \mathbf{P}(\mathcal{A})$ *is fully faithful and induces an equivalence*

$$\mathcal{A} \xrightarrow{\sim} \mathrm{Ex}(\mathrm{Fp}(\mathcal{C}, \mathrm{Ab}), \mathrm{Ab}).$$

Moreover, the functor preserves filtered colimits, products, and cokernels.

Proof First observe that $\bar{X}$ is an exact functor for any $X \in \mathcal{A}$, since evaluation is exact. When we identify $\mathcal{A} = \mathrm{Lex}(\mathcal{C}^{\mathrm{op}}, \mathrm{Ab})$, then the quasi-inverse functor

$$\mathrm{Ex}(\mathrm{Fp}(\mathcal{C}, \mathrm{Ab}), \mathrm{Ab}) \longrightarrow \mathcal{A}$$

sends F to $F \circ h$, where

$$h\colon \mathcal{C}^{\mathrm{op}} \longrightarrow \mathrm{Fp}(\mathcal{C}, \mathrm{Ab}), \quad C \mapsto \mathrm{Hom}_{\mathcal{C}}(C, -)$$

denotes the Yoneda functor.

For $F \in \mathrm{Fp}(\mathcal{C}, \mathrm{Ab})$ the corresponding functor $\bar{F}\colon \mathcal{A} \to \mathrm{Ab}$ preserves filtered colimits and products. Thus ev preserves filtered colimits and products, since in $\mathbf{P}(\mathcal{A})$ these are computed pointwise. It remains to consider cokernels. For $C \in \mathcal{C}$ we have $\bar{C} = \mathrm{Hom}(\mathrm{Hom}_{\mathcal{C}}(C, -), -)$. Thus the restriction $\mathrm{ev}\,|_{\mathcal{C}}$ preserves cokernels. It follows that ev preserves cokernels, since any cokernel sequence in $\mathcal{A}$ can be written as a filtered colimit of cokernel sequences in $\mathcal{C}$. $\square$

Remark 12.1.5. The category $\mathcal{A}$ viewed as a subcategory of $\mathbf{P}(\mathcal{A})$ is covariantly finite; this follows from Proposition 11.1.27.

Pure-Exactness and Pure-Injectives

A sequence of morphisms $0 \to X \to Y \to Z \to 0$ in $\mathcal{A}$ is called *pure-exact* if the induced sequence

$$0 \longrightarrow \operatorname{Hom}_{\mathcal{A}}(C, X) \longrightarrow \operatorname{Hom}_{\mathcal{A}}(C, Y) \longrightarrow \operatorname{Hom}_{\mathcal{A}}(C, Z) \longrightarrow 0$$

of abelian groups is exact for all finitely presented $C \in \mathcal{A}$. In that case the morphism $X \to Y$ is called a *pure monomorphism*. An object $Q \in \mathcal{A}$ is *pure-injective* if every pure monomorphism $X \to Y$ induces a surjective map $\operatorname{Hom}_{\mathcal{A}}(Y, Q) \to \operatorname{Hom}_{\mathcal{A}}(X, Q)$.

Lemma 12.1.6. *For a sequence of morphisms $\eta \colon 0 \to X \to Y \to Z \to 0$ in $\mathcal{A}$ the following are equivalent.*

(1) *The sequence η is pure-exact.*
(2) *The sequence η is a filtered colimit of split exact sequences.*
(3) *The sequence $\bar{\eta} \colon 0 \to \bar{X} \to \bar{Y} \to \bar{Z} \to 0$ is exact in $\mathbf{P}(\mathcal{A})$.*

Proof (1) $\Rightarrow$ (2): Write $Z = \operatorname{colim} Z_i$ as a filtered colimit of finitely presented objects. Composing η with $Z_i \to Z$ yields a split exact sequence $\eta_i \colon 0 \to X \to Y_i \to Z_i \to 0$, and $\eta = \operatorname{colim} \eta_i$.

(2) $\Rightarrow$ (3): The assignment $X \mapsto \bar{X}$ preserves filtered colimits, and in $\mathbf{P}(\mathcal{A})$ a filtered colimit of exact sequences is exact.

(3) $\Rightarrow$ (1): For $C \in \operatorname{fp} \mathcal{A}$ the sequence

$$0 \longrightarrow \operatorname{Hom}_{\mathbf{P}(\mathcal{A})}(\bar{C}, \bar{X}) \longrightarrow \operatorname{Hom}_{\mathbf{P}(\mathcal{A})}(\bar{C}, \bar{Y}) \longrightarrow \operatorname{Hom}_{\mathbf{P}(\mathcal{A})}(\bar{C}, \bar{Z}) \longrightarrow 0$$

is exact by Lemma 11.1.26. Thus η is pure-exact. □

Lemma 12.1.7. *A morphism $X \to Y$ in $\mathcal{A}$ is a pure monomorphism if and only if $\bar{X} \to \bar{Y}$ is a monomorphism in $\mathbf{P}(\mathcal{A})$.*

Proof Complete the morphism $\alpha \colon X \to Y$ to an exact sequence $X \xrightarrow{\alpha} Y \to Z \to 0$ in $\mathcal{A}$. If α is a pure monomorphism, then $\bar{\alpha}$ is a monomorphism by Lemma 12.1.6. Conversely, if $\bar{\alpha}$ is a monomorphism, then the sequence $0 \to \bar{X} \xrightarrow{\bar{\alpha}} \bar{Y} \to \bar{Z} \to 0$ in $\mathbf{P}(\mathcal{A})$ is exact, since ev is right exact by Lemma 12.1.4. Thus α is a pure monomorphism by Lemma 12.1.6. □

Lemma 12.1.8. *The functor* ev$\colon \mathcal{A} \to \mathbf{P}(\mathcal{A})$ *identifies the pure-injective objects in $\mathcal{A}$ with the injective objects in $\mathbf{P}(\mathcal{A})$.*

Proof An injective object in $\mathbf{P}(\mathcal{A})$ is of the form $\bar{X}$ for some $X \in \mathcal{A}$ by Lemma 11.1.26 and Lemma 12.1.4. Clearly, X is pure-injective, since ev sends any pure monomorphism in $\mathcal{A}$ to a monomorphism in $\mathbf{P}(\mathcal{A})$ by Lemma 12.1.7.

Now suppose that $X \in \mathcal{A}$ is pure-injective and choose an injective envelope

$\bar{\alpha}\colon \bar{X} \to \bar{Y}$ in $\mathbf{P}(\mathcal{A})$. Then α is a pure monomorphism by Lemma 12.1.7, and therefore a split monomorphism. It follows that $\bar{X}$ is an injective object. $\quad\square$

A pure monomorphism $\phi\colon X \to Y$ in $\mathcal{A}$ is called a *pure-injective envelope* of X, if Y is pure-injective and if every endomorphism $\alpha\colon Y \to Y$ satisfying $\phi = \alpha\phi$ is invertible.

Theorem 12.1.9. *Every object $X \in \mathcal{A}$ admits a pure-injective envelope. Moreover, a morphism $X \to Y$ is a pure-injective envelope if and only if the induced morphism $\bar{X} \to \bar{Y}$ is an injective envelope in $\mathbf{P}(\mathcal{A})$.*

Proof Choose a morphism $\phi\colon X \to Y$ such that $\bar{\phi}\colon \bar{X} \to \bar{Y}$ is an injective envelope in $\mathbf{P}(\mathcal{A})$ (Corollary 2.5.4). Then ϕ is a pure monomorphism by Lemma 12.1.7 and Y is pure-injective by Lemma 12.1.8. The additional minimality property for every endomorphism $Y \to Y$ follows from the corresponding characterisation of injective envelopes (Lemma 2.1.19). Clearly, a pure-injective envelope is essentially unique, and this yields the second part of the assertion. $\quad\square$

The pure-exact sequences provide an exact structure on $\mathcal{A}$. We give an application which is a variation of Example 11.1.25.

Example 12.1.10. Let $(\mathcal{T}, \mathcal{F})$ be a split torsion pair for fp$\mathcal{A}$. Then $(\vec{\mathcal{T}}, \vec{\mathcal{F}})$ is a torsion pair for $\mathcal{A}$ and each object $X \in \mathcal{A}$ fits into a pure-exact sequence $0 \to X' \to X \to X'' \to 0$ with $X' \in \vec{\mathcal{T}}$ and $X'' \in \vec{\mathcal{F}}$.

The Spectrum of Indecomposable Injectives

Let $\mathcal{A}$ be a Grothendieck category. We denote by Sp $\mathcal{A}$ a representative set of the isomorphism classes of indecomposable injective objects in $\mathcal{A}$ (the *spectrum* of indecomposable injectives). Note that Sp $\mathcal{A}$ is a set, because $\mathcal{A}$ has a set of generators and each object in Sp $\mathcal{A}$ is the injective envelope of X/U for some generating object X and some subobject $U \subsetneq X$.

Lemma 12.1.11. *Let $\mathcal{A}$ be a locally finitely presented Grothendieck category. Then the objects in Sp $\mathcal{A}$ form a set of cogenerators for $\mathcal{A}$.*

Proof Let $X \in \mathcal{A}$ be a non-zero object. Thus we find $C \in$ fp$\mathcal{A}$ and a non-zero monomorphism $C/U \to X$ for some subobject $U \subsetneq C$. Using Zorn's lemma, we choose a maximal subobject $V \subseteq C$ containing U and an injective envelope $C/V \to Q$. This yields a non-zero morphism $X \to Q$. Clearly, Q is indecomposable. $\quad\square$

Our next goal is the definition of a topology on the spectrum of $\mathcal{A}$. We fix a locally finitely presented Grothendieck category $\mathcal{A}$ such that $\mathrm{fp}\,\mathcal{A}$ is abelian. For classes $\mathcal{C} \subseteq \mathrm{fp}\,\mathcal{A}$ and $\mathcal{U} \subseteq \mathrm{Sp}\,\mathcal{A}$ set

$$\mathcal{C}^{\perp} = \{X \in \mathrm{Sp}\,\mathcal{A} \mid \mathrm{Hom}_A(C, X) = 0 \text{ for all } C \in \mathcal{C}\} \subseteq \mathrm{Sp}\,\mathcal{A}$$

$$^{\perp}\mathcal{U} = \{C \in \mathrm{fp}\,\mathcal{A} \mid \mathrm{Hom}_A(C, X) = 0 \text{ for all } X \in \mathcal{U}\} \subseteq \mathrm{fp}\,\mathcal{A}.$$

Lemma 12.1.12. *The assignment* $\mathcal{U} \mapsto \overline{\mathcal{U}} := (^{\perp}\mathcal{U})^{\perp}$ *defines a closure operator on* $\mathrm{Sp}\,\mathcal{A}$. *Thus the subsets* $\mathcal{U} \subseteq \mathrm{Sp}\,\mathcal{A}$ *satisfying* $\mathcal{U} = \overline{\mathcal{U}}$ *form the closed subsets of a topology on* $\mathrm{Sp}\,\mathcal{A}$.

Proof Following Kuratowski's axiomatisation of a topological space we need to verify that

(1) $\overline{\varnothing} = \varnothing$,

(2) $\mathcal{U} \subseteq \overline{\mathcal{U}}$ for every subset $\mathcal{U}$,

(3) $\overline{\mathcal{U}} = \overline{\overline{\mathcal{U}}}$ for every subset $\mathcal{U}$,

(4) $\overline{\mathcal{U}_1 \cup \mathcal{U}_2} = \overline{\mathcal{U}_1} \cup \overline{\mathcal{U}_2}$ for every pair of subsets $\mathcal{U}_1$ and $\mathcal{U}_2$.

The conditions (1)–(3) are easily checked; so it remains to show (4). From $^{\perp}(\mathcal{U}_1 \cup \mathcal{U}_2) \subseteq {}^{\perp}\mathcal{U}_1 \cap {}^{\perp}\mathcal{U}_2$ it follows that $\overline{\mathcal{U}_1} \cup \overline{\mathcal{U}_2} \subseteq \overline{\mathcal{U}_1 \cup \mathcal{U}_2}$. Now choose $X \notin \overline{\mathcal{U}_1} \cup \overline{\mathcal{U}_2}$, and we claim this implies $X \notin \overline{\mathcal{U}_1 \cup \mathcal{U}_2}$. Choose non-zero morphisms $\phi_i : C_i \to X$ with $C_i \in {}^{\perp}\mathcal{U}_i$. We have $\mathrm{Im}\,\phi_1 \cap \mathrm{Im}\,\phi_2 \neq 0$ since X is indecomposable. Choosing a finitely generated subobject $0 \neq U \subseteq \mathrm{Im}\,\phi_1 \cap \mathrm{Im}\,\phi_2$, there are finitely generated subobjects $U_i \subseteq C_i$ such that $\phi_i(U_i) = U$. We obtain the following commutative diagram with exact rows.

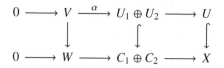

The morphisms $\alpha_i : V \to U_i$ are epimorphisms. Thus there are finitely generated subobjects $V_i \subseteq V$ such that $\alpha_i(V_i) = U_i$. Now set $C = (U_1 \oplus U_2)/\alpha(V_1 + V_2)$. We have $C \in \mathrm{fp}\,\mathcal{A}$ since $\mathrm{fp}\,\mathcal{A}$ is abelian, and one checks that $\mathrm{Hom}_A(C, X) \neq 0$. On the other hand, $C \in {}^{\perp}(\mathcal{U}_1 \cup \mathcal{U}_2)$ since C is a quotient of each U_i. Therefore $X \notin \overline{\mathcal{U}_1 \cup \mathcal{U}_2}$ and the proof is complete. □

Proposition 12.1.13. *Let* $\mathcal{A}$ *be a locally finitely presented Grothendieck category and suppose that* $\mathrm{fp}\,\mathcal{A}$ *is abelian. Then the assignments* $\mathcal{C} \mapsto \mathcal{C}^{\perp}$ *and* $\mathcal{U} \mapsto {}^{\perp}\mathcal{U}$ *provide mutually inverse and inclusion reversing bijections between the Serre subcategories of* $\mathrm{fp}\,\mathcal{A}$ *and the closed subsets of* $\mathrm{Sp}\,\mathcal{A}$.

Proof Clearly, both maps are well defined. Let $\mathcal{U} \subseteq \mathrm{Sp}\,\mathcal{A}$ be closed. Then $(^{\perp}\mathcal{U})^{\perp} = \mathcal{U}$ by definition. Now let $\mathcal{C} \subseteq \mathrm{fp}\,\mathcal{A}$ be a Serre subcategory. The inclusion $\mathcal{C} \subseteq \,^{\perp}(\mathcal{C}^{\perp})$ is clear. For the other inclusion we apply Corollary 11.1.33. Thus $\vec{\mathcal{C}}$ is a localising subcategory satisfying $\vec{\mathcal{C}} \cap \mathrm{fp}\,\mathcal{A} = \mathcal{C}$. Furthermore, $\mathcal{C}^{\perp} = \vec{\mathcal{C}}^{\perp}$ and $\mathcal{C}^{\perp}$ identifies with $\mathrm{Sp}(\mathcal{A}/\vec{\mathcal{C}})$. The category $\mathcal{A}/\vec{\mathcal{C}}$ is locally finitely presented. Thus $\mathcal{C} = \,^{\perp}(\mathcal{C}^{\perp})$ by Lemma 12.1.11, since $\vec{\mathcal{C}} = \,^{\perp}(\vec{\mathcal{C}}^{\perp})$. □

We discuss briefly an alternative closure operation. Let $\mathcal{A}$ be a Grothendieck category and fix $\mathcal{U} \subseteq \mathrm{Sp}\,\mathcal{A}$. We denote by $\widehat{\mathcal{U}}$ the set of objects $X \in \mathrm{Sp}\,\mathcal{A}$ such that $X \subseteq \prod_{i \in I} Y_i$ for some set of objects $Y_i \in \mathcal{U}$. Now consider the localising subcategory $\mathcal{A}_{\mathcal{U}} = \{X \in \mathcal{A} \mid \mathrm{Hom}_{\mathcal{A}}(X, \mathcal{U}) = 0\}$. Then we have

$$\widehat{\mathcal{U}} = \{X \in \mathrm{Sp}\,\mathcal{A} \mid \mathrm{Hom}_{\mathcal{A}}(\mathcal{A}_{\mathcal{U}}, X) = 0\}$$

by Corollary 2.2.18.

Lemma 12.1.14. *Let $\mathcal{A}$ be a locally noetherian Grothendieck category and* $\mathcal{U} \subseteq \mathrm{Sp}\,\mathcal{A}$. *Then* $\widehat{\mathcal{U}} = \overline{\mathcal{U}}$.

Proof The inclusion $\widehat{\mathcal{U}} \subseteq \overline{\mathcal{U}}$ is automatic since $^{\perp}\mathcal{U} \subseteq \mathcal{A}_{\mathcal{U}}$. On the other hand, $\mathcal{A}_{\mathcal{U}}$ is generated by $^{\perp}\mathcal{U}$ since $\mathcal{A}$ is locally noetherian. Thus we have equality. □

Compactness

Let $\mathcal{C}$ be an abelian category and $\mathcal{X}$ a class of objects in $\mathcal{C}$. We write $S\langle\mathcal{X}\rangle$ for the smallest Serre subcategory containing $\mathcal{X}$.

Lemma 12.1.15. *If an object $X \in \mathcal{C}$ belongs to $S\langle\mathcal{X}\rangle$, then $X \in S\langle\mathcal{X}_0\rangle$ for some finite set of objects $\mathcal{X}_0 \subseteq \mathcal{X}$.*

Proof The objects in $S\langle\mathcal{X}\rangle$ are obtained by closing the objects in $\mathcal{X}$ under forming subobjects, quotients, and extensions. For each $X \in S\langle\mathcal{X}\rangle$, finitely many such operations suffice. □

Let us call $\mathcal{C}$ *finitely generated* if $\mathcal{C} = S\langle X\rangle$ for some object $X \in \mathcal{C}$. An equivalent condition is the following. For any family $(\mathcal{C}_i)_{i \in I}$ of Serre subcategories $\bigvee_{i \in I} \mathcal{C}_i = \mathcal{C}$ implies $\bigvee_{i \in J} \mathcal{C}_i = \mathcal{C}$ for some finite subset $J \subseteq I$.

Recall that a topological space T is *quasi-compact* if for any family $(U_i)_{i \in I}$ of open subsets $\bigcup_{i \in I} U_i = T$ implies $\bigcup_{i \in J} U_i = T$ for some finite subset $J \subseteq I$.

Lemma 12.1.16. *Let $\mathcal{A}$ be a locally finitely presented Grothendieck category and suppose that $\mathrm{fp}\,\mathcal{A}$ is abelian. For a closed subset $\mathcal{V} \subseteq \mathrm{Sp}\,\mathcal{A}$ and $\mathcal{U} = \mathrm{Sp}\,\mathcal{A} \setminus \mathcal{V}$, we have*

(1) $\mathcal{U}$ is quasi-compact if and only if $^{\perp}\mathcal{V}$ is finitely generated, and

(2) $\mathcal{V}$ *is quasi-compact if and only if* $(\mathrm{fp}\,\mathcal{A})/(^{\perp}\mathcal{V})$ *is finitely generated.*

Proof   This is an immediate consequence of the correspondence in Proposition 12.1.13, since for any family $(\mathcal{C}_i)_{i\in I}$ of Serre subcategories of $\mathrm{fp}\,\mathcal{A}$

$$\left(\bigvee_{i\in I} \mathcal{C}_i\right)^{\perp} = \bigcap_{i\in I}(\mathcal{C}_i^{\perp}). \qquad\qquad \square$$

The Spectrum of Indecomposable Pure-Injectives

Let $\mathcal{A}$ be a locally finitely presented category and denote by $\mathrm{Ind}\,\mathcal{A}$ a representative set of the isomorphism classes of indecomposable pure-injective objects in $\mathcal{A}$ (the *spectrum* of indecomposable pure-injectives or *Ziegler spectrum*).

Lemma 12.1.17. *Let $\mathcal{A}$ be a locally finitely presented category. The functor* $\mathrm{ev}\colon \mathcal{A} \to \mathbf{P}(\mathcal{A})$ *induces a bijection* $\mathrm{Ind}\,\mathcal{A} \xrightarrow{\sim} \mathrm{Sp}\,\mathbf{P}(\mathcal{A})$. *Therefore every object admits a pure monomorphism into a pure-injective object which is a product of indecomposable objects.*

Proof   The first assertion is clear from Lemma 12.1.8. Let $X \in \mathcal{A}$ and consider the canonical morphism

$$X \longrightarrow \prod_{\substack{Q\in\mathrm{Ind}\,\mathcal{A} \\ \phi\in\mathrm{Hom}_{\mathcal{A}}(X,Q)}} Q.$$

It follows from Lemma 12.1.7 and Lemma 12.1.11 that this is a pure monomorphism. $\square$

We use the identification $\mathrm{Ind}\,\mathcal{A} \xrightarrow{\sim} \mathrm{Sp}\,\mathbf{P}(\mathcal{A})$ and obtain a topology on $\mathrm{Ind}\,\mathcal{A}$. For classes $\mathcal{C} \subseteq \mathrm{fp}\,\mathbf{P}(\mathcal{A})$ and $\mathcal{U} \subseteq \mathrm{Ind}\,\mathcal{A}$ we set

$$\mathcal{C}^{\perp} = \{X \in \mathrm{Ind}\,\mathcal{A} \mid \mathrm{Hom}_{\mathbf{P}(\mathcal{A})}(C, \bar{X}) = 0 \text{ for all } C \in \mathcal{C}\} \subseteq \mathrm{Ind}\,\mathcal{A}$$
$$^{\perp}\mathcal{U} = \{C \in \mathrm{fp}\,\mathbf{P}(\mathcal{A}) \mid \mathrm{Hom}_{\mathbf{P}(\mathcal{A})}(C, \bar{X}) = 0 \text{ for all } X \in \mathcal{U}\} \subseteq \mathrm{fp}\,\mathbf{P}(\mathcal{A}).$$

Lemma 12.1.18. *The assignment* $\mathcal{U} \mapsto \overline{\mathcal{U}} := (^{\perp}\mathcal{U})^{\perp}$ *defines a closure operator on* $\mathrm{Ind}\,\mathcal{A}$. *Thus the subsets* $\mathcal{U} \subseteq \mathrm{Ind}\,\mathcal{A}$ *satisfying* $\mathcal{U} = \overline{\mathcal{U}}$ *form the closed subsets of a topology on* $\mathrm{Ind}\,\mathcal{A}$.

Proof   Apply Lemma 12.1.12. $\square$

12.2 Definable Subcategories

Let $\mathcal{A}$ be a locally finitely presented category. We set

$$\mathcal{C} := \mathrm{fp}\,\mathcal{A} \qquad \text{and} \qquad \mathrm{Ab}(\mathcal{A}) := \mathrm{Fp}(\mathcal{C}, \mathrm{Ab})^{\mathrm{op}}.$$

Note that $\mathrm{Ab}(\mathcal{A}) \xrightarrow{\sim} \mathrm{fp}\,\mathbf{P}(\mathcal{A})$ via $F \mapsto \mathrm{Hom}(-, F)$; see Theorem 11.1.15. The category $\mathrm{Ab}(\mathcal{A})$ is abelian and $\mathcal{A}$ identifies with the category of exact functors $\mathrm{Ab}(\mathcal{A})^{\mathrm{op}} \to \mathrm{Ab}$ via the assignment $X \mapsto \bar{X}$; see Lemma 12.1.4. The kernel of any exact functor $\bar{X}$ is a Serre subcategory of $\mathrm{Ab}(\mathcal{A})$ and therefore a natural invariant of X.

In this section we study the class of definable subcategories of $\mathcal{A}$. The terminology is justified by the fact that any definable subcategory is given by a family of morphisms in $\mathcal{C}$. Observe that a morphism ϕ in $\mathcal{C}$ yields a functor $F = \mathrm{Coker}\,\mathrm{Hom}_{\mathcal{C}}(\phi, -)$ in $\mathrm{Fp}(\mathcal{C}, \mathrm{Ab})$, and for $X \in \mathcal{A}$ we have

$$\bar{F}(X) = 0 \quad \Longleftrightarrow \quad \mathrm{Hom}_{\mathcal{A}}(\phi, X) \text{ is surjective} \quad \Longleftrightarrow \quad \bar{X}(F) = 0.$$

Definable Subcategories

A full subcategory $\mathcal{B} \subseteq \mathcal{A}$ is called *definable* if it is of the form

$$\mathcal{B} = \{X \in \mathcal{A} \mid \mathrm{Hom}_{\mathcal{A}}(\phi_i, X) \text{ is surjective for all } i \in I\}$$

for a family of morphisms $(\phi_i)_{i \in I}$ in $\mathrm{fp}\,\mathcal{A}$; thus it is 'defined' by the ϕ_i. Similarly, a subset $\mathcal{U} \subseteq \mathrm{Ind}\,\mathcal{A}$ is *Ziegler closed* if there is a family $(\phi_i)_{i \in I}$ of morphisms in $\mathrm{fp}\,\mathcal{A}$ such that

$$\mathcal{U} = \{X \in \mathrm{Ind}\,\mathcal{A} \mid \mathrm{Hom}_{\mathcal{A}}(\phi_i, X) \text{ is surjective for all } i \in I\}.$$

Let us consider the pairing

$$\mathrm{Fp}(\mathcal{C}, \mathrm{Ab}) \times \mathcal{A} \longrightarrow \mathrm{Ab}, \quad (F, X) \mapsto \bar{F}(X) = \bar{X}(F).$$

For classes $\mathcal{F} \subseteq \mathrm{Fp}(\mathcal{C}, \mathrm{Ab})$ and $\mathcal{X} \subseteq \mathcal{A}$ we set

$$\mathcal{F}^{\perp} = \{X \in \mathcal{A} \mid \bar{F}(X) = 0 \text{ for all } F \in \mathcal{F}\} \subseteq \mathcal{A}$$
$$^{\perp}\mathcal{X} = \{F \in \mathrm{Fp}(\mathcal{C}, \mathrm{Ab}) \mid \bar{F}(X) = 0 \text{ for all } X \in \mathcal{X}\} \subseteq \mathrm{Fp}(\mathcal{C}, \mathrm{Ab}).$$

The pairing admits another interpretation. To this end identify $\mathrm{Fp}(\mathcal{C}, \mathrm{Ab})$ with the full subcategory of finitely presented objects in the purity category $\mathbf{P}(\mathcal{A})$ via the Yoneda embedding $F \mapsto \mathrm{Hom}(F, -)$. Then we have for all $X \in \mathcal{A}$

$$\bar{F}(X) \cong \mathrm{Hom}_{\mathbf{P}(\mathcal{A})}(F, \bar{X}).$$

Lemma 12.2.1. *The following holds.*

(1) $\mathcal{F}^{\perp}$ *is a definable subcategory of* $\mathcal{A}$.
(2) $^{\perp}\mathcal{X}$ *is a Serre subcategory of* $\mathrm{Fp}(\mathcal{C}, \mathrm{Ab})$.

Proof (1) For $F = \operatorname{Coker}\operatorname{Hom}_{\mathcal{C}}(\phi, -)$ in $\mathcal{F}$, we have $\bar{F}(X) = 0$ if and only if $\operatorname{Hom}_{\mathcal{A}}(\phi, X)$ is surjective. Thus $\mathcal{F}^{\perp}$ is a definable subcategory for any choice of $\mathcal{F}$.

(2) The assignment $F \mapsto \bar{F}(X)$ is exact for fixed $X \in \mathcal{A}$. Thus $^{\perp}\mathcal{X}$ is a Serre subcategory for any choice of $\mathcal{X}$. $\qquad\square$

We obtain for $\mathcal{X} \subseteq \mathcal{A}$ an abelian category by forming the quotient

$$\operatorname{Ab}(\mathcal{A}) \twoheadrightarrow \operatorname{Ab}(\mathcal{X}) := (\operatorname{Fp}(\mathcal{C}, \operatorname{Ab})/^{\perp}\mathcal{X})^{\operatorname{op}}.$$

Note that any inclusion $\mathcal{X}' \subseteq \mathcal{X}$ induces an exact functor $\operatorname{Ab}(\mathcal{X}) \twoheadrightarrow \operatorname{Ab}(\mathcal{X}')$.

There is the following fundamental correspondence for definable subcategories.

Theorem 12.2.2. *Let $\mathcal{A}$ be a locally finitely presented category.*

(1) *The assignments $\mathcal{F} \mapsto \mathcal{F}^{\perp}$ and $\mathcal{X} \mapsto {}^{\perp}\mathcal{X}$ provide mutually inverse and inclusion reversing bijections between the Serre subcategories of $\operatorname{Ab}(\mathcal{A})$ and the definable subcategories of $\mathcal{A}$.*

(2) *The assignment*

$$\mathcal{A} \supseteq \mathcal{B} \longmapsto \mathcal{B} \cap \operatorname{Ind}\mathcal{A} \subseteq \operatorname{Ind}\mathcal{A}$$

provides an inclusion preserving bijection between the definable subcategories of $\mathcal{A}$ and the Ziegler closed subsets of $\operatorname{Ind}\mathcal{A}$.

The first part of Theorem 12.2.2 has an immediate consequence.

Corollary 12.2.3. *For a definable subcategory $\mathcal{B} \subseteq \mathcal{A}$ the assignment $X \mapsto \bar{X}$ (Lemma 12.1.4) induces the following commutative square*

$$\begin{array}{ccc}
\mathcal{B} & \xrightarrow{\ \sim\ } & \operatorname{Ex}(\operatorname{Ab}(\mathcal{B})^{\operatorname{op}}, \operatorname{Ab}) \\
\Big\downarrow & & \Big\downarrow \\
\mathcal{A} & \xrightarrow{\ \sim\ } & \operatorname{Ex}(\operatorname{Ab}(\mathcal{A})^{\operatorname{op}}, \operatorname{Ab})
\end{array}$$

where the inclusion on the right is induced by composing with the canonical functor $\operatorname{Ab}(\mathcal{A}) \twoheadrightarrow \operatorname{Ab}(\mathcal{B})$.

Proof Let $\mathcal{S} \subseteq \operatorname{Fp}(\mathcal{C}, \operatorname{Ab})$ be a Serre subcategory. Then the exact functors $\operatorname{Fp}(\mathcal{C}, \operatorname{Ab}) \to \operatorname{Ab}$ that vanish on $\mathcal{S}$ identify with the exact functors $\frac{\operatorname{Fp}(\mathcal{C}, \operatorname{Ab})}{\mathcal{S}} \to \operatorname{Ab}$; see Proposition 11.1.31. $\qquad\square$

Proof of Theorem 12.2.2 (1) It is convenient to work in the purity category $\mathbf{P}(\mathcal{A})$ and we identify $\operatorname{Fp}(\mathcal{C}, \operatorname{Ab})$ with the full subcategory of finitely presented objects in $\mathbf{P}(\mathcal{A})$ via the Yoneda embedding $F \mapsto \mathrm{Hom}(F, -)$. Then we have for all $X \in \mathcal{A}$

$$\bar{F}(X) \cong \mathrm{Hom}_{\mathbf{P}(\mathcal{A})}(F, \bar{X}).$$

We use the bijection $\mathrm{Ind}\,\mathcal{A} \xrightarrow{\sim} \mathrm{Sp}\,\mathbf{P}(\mathcal{A})$ from Lemma 12.1.17 and combine this with the bijection from Proposition 12.1.13. Thus the assignments $\mathcal{F} \mapsto \mathcal{F}^\perp \cap \mathrm{Ind}\,\mathcal{A}$ and $\mathcal{U} \mapsto {}^\perp\mathcal{U}$ provide mutually inverse and inclusion reversing bijections between the Serre subcategories of $\mathrm{Fp}(\mathcal{C}, \mathrm{Ab})$ and the Ziegler closed subsets of $\mathrm{Ind}\,\mathcal{A}$.

Fix a Serre subcategory $\mathcal{S}$ of $\mathrm{Fp}(\mathcal{C}, \mathrm{Ab})$. Then the above bijections imply $\mathcal{S} = {}^\perp(\mathcal{S}^\perp)$.

Now fix a definable subcategory $\mathcal{B} = \mathcal{F}^\perp$ of $\mathcal{A}$, which is given by some $\mathcal{F} \subseteq \mathrm{Fp}(\mathcal{C}, \mathrm{Ab})$. Let $\mathcal{S} \subseteq \mathrm{Fp}(\mathcal{C}, \mathrm{Ab})$ denote the smallest Serre subcategory containing $\mathcal{F}$. Clearly, $\mathcal{B} = \mathcal{S}^\perp$. Thus we have

$$({}^\perp\mathcal{B})^\perp = ({}^\perp(\mathcal{S}^\perp))^\perp = \mathcal{S}^\perp = \mathcal{B},$$

where one uses the equality $\mathcal{S} = {}^\perp(\mathcal{S}^\perp)$ from the first part of the proof.

(2) The assertion claims that a definable subcategory $\mathcal{B}$ is determined by $\mathcal{B} \cap \mathrm{Ind}\,\mathcal{A}$. This follows from (1). In fact, $\mathcal{B}$ identifies with $\mathrm{Ex}(\mathrm{Ab}(\mathcal{B})^{\mathrm{op}}, \mathrm{Ab})$ as in Corollary 12.2.3, and it remains to observe that $\mathcal{B} \cap \mathrm{Ind}\,\mathcal{A}$ identifies with the indecomposable injective objects in $\mathrm{Lex}(\mathrm{Ab}(\mathcal{B})^{\mathrm{op}}, \mathrm{Ab})$, which form a set of cogenerators; see Lemma 12.1.11. $\qquad\square$

Closure Properties of Definable Subcategories

Definable subcategories are characterised by some natural closure properties. The proof of this requires some preparations.

Lemma 12.2.4. *A filtered colimit* $\mathrm{colim}_{i \in \mathfrak{I}} X_i$ *and a family of monomorphisms* $(X_i \to Y_i)_{i \in \mathfrak{I}}$ *in a Grothendieck category induce a monomorphism*

$$\mathrm{colim}_{i \in \mathfrak{I}} X_i \longrightarrow \mathrm{colim}_{i \in \mathfrak{I}}\left(\prod_{i \to j} Y_j\right).$$

Proof For each $i \in \mathfrak{I}$ we have a canonical monomorphisms $X_i \to \prod_{i \to j} Y_j$, where $i \to j$ runs through all morphisms in $\mathfrak{I}$ starting at i and each component is given by the composite $X_i \to X_j \to Y_j$. A morphism $i \to i'$ in $\mathfrak{I}$ yields a commuting square.

$$\begin{array}{ccc} X_i & \longrightarrow & \prod_{i \to j} Y_j \\ \downarrow & & \downarrow \\ X_{i'} & \longrightarrow & \prod_{i' \to j} Y_{j'} \end{array}$$

Taking colimits yields the desired morphism, which is a monomorphism since filtered colimits preserve monomorphisms. □

Theorem 12.2.5. *A full subcategory of a locally finitely presented category is definable if and only if it is closed under taking products, filtered colimits, and pure subobjects.*

Proof Let $\mathcal{A}$ be a locally finitely presented category. As before, it is convenient to work in $\mathbf{P}(\mathcal{A})$ and we identify $\mathrm{Fp}(\mathcal{C}, \mathrm{Ab})$ with the full subcategory of finitely presented objects in $\mathbf{P}(\mathcal{A})$.

One direction is clear, since for any $F \in \mathrm{Fp}(\mathcal{C}, \mathrm{Ab})$, the functor $\bar{F} \colon \mathcal{A} \to \mathrm{Ab}$ preserves filtered colimits, products, and sends pure monomorphisms to monomorphisms. This follows easily from the presentation (12.1.2).

Now suppose that $\mathcal{B} \subseteq \mathcal{A}$ is closed under taking products, filtered colimits, and pure subobjects. Set

$$\mathcal{F} = \{X \in \mathbf{P}(\mathcal{A}) \mid X \subseteq \bar{Y} \text{ for some } Y \in \mathcal{B}\}$$

and

$$\mathcal{T} = \{X \in \mathbf{P}(\mathcal{A}) \mid \mathrm{Hom}_{\mathbf{P}(\mathcal{A})}(X, \bar{Y}) = 0 \text{ for all } Y \in \mathcal{B}\}.$$

We claim that this gives a torsion pair $(\mathcal{T}, \mathcal{F})$ for $\mathbf{P}(\mathcal{A})$. First observe that $\mathcal{F}$ is closed under filtered colimits, by Lemma 12.2.4. The inclusion $\mathcal{F} \hookrightarrow \mathbf{P}(\mathcal{A})$ has a left adjoint $f \colon \mathbf{P}(\mathcal{A}) \to \mathcal{F}$ which is constructed as follows. For $X \in \mathbf{P}(\mathcal{A})$ let $(Y_i)_{i \in I}$ be the set of quotient objects of X which are in $\mathcal{F}$. Define $f(X)$ to be the image and $t(X)$ the kernel of the canonical morphism $X \to \prod_{i \in I} Y_i$. Next observe that $\mathcal{S} = \mathcal{T} \cap \mathrm{fp}\,\mathbf{P}(\mathcal{A})$ is a Serre subcategory of $\mathrm{fp}\,\mathbf{P}(\mathcal{A})$. We write $\mathcal{T}' = \vec{\mathcal{S}}$ for the full subcategory consisting of the filtered colimits $\mathrm{colim}\, X_i$ with $X_i \in \mathcal{S}$ for all i. We claim that $\mathcal{T}' = \mathcal{T}$.

For each $X \in \mathrm{fp}\,\mathbf{P}(\mathcal{A})$ we show that $t(X) \in \mathcal{T}'$. To this end write $t(X) = \mathrm{colim}\, U_i$ as filtered colimit of its finitely generated subobjects. We need to show that $U_i \in \mathcal{S}$ for all i. Suppose that $U = U_i \notin \mathcal{S}$. Then there is a non-zero morphism $\phi \colon U \to \bar{Y}$ for some $Y \in \mathcal{B}$, and ϕ extends to a morphism $\psi \colon X \to \bar{Y}$ since $\bar{Y}$ is an exact functor; see Lemma 12.1.4. But the adjointness property of f implies that ψ factors through $X \to f(X)$. Therefore $\phi(U) = 0$, a contradiction to our assumption. Thus $t(X) \in \mathcal{T}'$. Now let $X = \mathrm{colim}\, X_i$ be an arbitrary object in $\mathbf{P}(\mathcal{A})$, written as a filtered colimit of objects in $\mathrm{fp}\,\mathbf{P}(\mathcal{A})$. We obtain an exact sequence

$$0 \longrightarrow \mathrm{colim}\, t(X_i) \longrightarrow \mathrm{colim}\, X_i \longrightarrow \mathrm{colim}\, f(X_i) \longrightarrow 0$$

with $\mathrm{colim}\, t(X_i) \in \mathcal{T}'$ and $\mathrm{colim}\, f(X_i) \in \mathcal{F}$, since both $\mathcal{T}'$ and $\mathcal{F}$ are closed under filtered colimits. We conclude that $\mathcal{T}' = \mathcal{T}$ and $(\mathcal{T}, \mathcal{F})$ is a torsion pair. Thus for $X \in \mathcal{A}$ we have $X \in \mathcal{B}$ if and only if $\bar{X} \in \mathcal{F}$ if and only if $\mathrm{Hom}_{\mathbf{P}(\mathcal{A})}(\mathcal{S}, \bar{X}) = 0$ if and only if $X \in \mathcal{S}^{\perp}$. It follows that $\mathcal{B}$ is definable.   □

Example 12.2.6. Let $\mathcal{A}$ be a locally finitely presented category and $\mathcal{C} \subseteq \mathrm{fp}\,\mathcal{A}$ a full additive subcategory. Then $\vec{\mathcal{C}}$ is a definable subcategory of $\mathcal{A}$ if and only if $\mathcal{C}$ is covariantly finite in $\mathrm{fp}\,\mathcal{A}$.

Proof For any $\mathcal{C}$, it is easily checked that $\vec{\mathcal{C}}$ is closed under filtered colimits and pure subobjects. Thus it remains to check when $\vec{\mathcal{C}}$ is closed under products; see Example 11.1.24.   □

Let us mention another important property of definable subcategories.

Proposition 12.2.7. *A definable subcategory of a locally finitely presented category is covariantly finite.*

Proof Let $\mathcal{B} \subseteq \mathcal{A}$ be a definable subcategory. We use the identification $\mathcal{B} \xrightarrow{\sim} \mathrm{Ex}(\mathrm{Ab}(\mathcal{B})^{\mathrm{op}}, \mathrm{Ab})$ from Corollary 12.2.3 and view this as a subcategory of $\mathcal{A}$ via the canonical functor $p\colon \mathrm{Ab}(\mathcal{A}) \twoheadrightarrow \mathrm{Ab}(\mathcal{B})$. Now observe that

$$\mathrm{Lex}(\mathrm{Ab}(\mathcal{B})^{\mathrm{op}}, \mathrm{Ab}) \subseteq \mathrm{Lex}(\mathrm{Ab}(\mathcal{A})^{\mathrm{op}}, \mathrm{Ab})$$

is covariantly finite, since the restriction p^* admits a left adjoint; see Proposition 11.1.31. On the other hand,

$$\mathrm{Ex}(\mathrm{Ab}(\mathcal{B})^{\mathrm{op}}, \mathrm{Ab}) \subseteq \mathrm{Lex}(\mathrm{Ab}(\mathcal{B})^{\mathrm{op}}, \mathrm{Ab})$$

is covariantly finite by Proposition 11.1.27.

Let X be an object in $\mathcal{A}$, viewed as an exact functor $\mathrm{Ab}(\mathcal{A})^{\mathrm{op}} \to \mathrm{Ab}$. Compose the approximations $X \to X^{\mathrm{Lex}\,\mathcal{B}}$ and $X^{\mathrm{Lex}\,\mathcal{B}} \to X^{\mathrm{Ex}\,\mathcal{B}}$, which are obtained from the above inclusions. This gives a left $\mathcal{B}$-approximation of X.   □

We add one more closure property of definable subcategories.

Proposition 12.2.8. *A definable subcategory of a locally finitely presented category is closed under taking pure-injective envelopes.*

Proof Let $\mathcal{B} \subseteq \mathcal{A}$ be a definable subcategory. As before, we use the identification $\mathcal{B} \xrightarrow{\sim} \mathrm{Ex}(\mathrm{Ab}(\mathcal{B})^{\mathrm{op}}, \mathrm{Ab})$ from Corollary 12.2.3 and view this as a subcategory of $\mathcal{A}$ via the canonical functor $\mathrm{Ab}(\mathcal{A}) \twoheadrightarrow \mathrm{Ab}(\mathcal{B})$. Also, we use that pure-injectives in $\mathcal{A}$ identify with injectives in $\mathrm{Lex}(\mathrm{Ab}(\mathcal{A})^{\mathrm{op}}, \mathrm{Ab})$, by Lemma 12.1.8. Then the assertion follows from the fact that

$$\mathrm{Lex}(\mathrm{Ab}(\mathcal{B})^{\mathrm{op}}, \mathrm{Ab}) \subseteq \mathrm{Lex}(\mathrm{Ab}(\mathcal{A})^{\mathrm{op}}, \mathrm{Ab})$$

is closed under taking injective envelopes; see Corollary 2.2.15 and Proposition 11.1.31.   □

Change of Categories

We study functors between locally finitely presented categories. For any locally finitely presented category $\mathcal{A}$ we use the canonical identification $\mathcal{A} = \mathrm{Ex}(\mathrm{Ab}(\mathcal{A})^{\mathrm{op}}, \mathrm{Ab})$; see Lemma 12.1.4. Now let $\mathcal{A}$ and $\mathcal{B}$ be locally finitely presented categories. Then an exact functor $f \colon \mathrm{Ab}(\mathcal{B}) \to \mathrm{Ab}(\mathcal{A})$ induces a functor $f^* \colon \mathcal{A} \to \mathcal{B}$ by sending $X \in \mathcal{A}$ to $X \circ f$.

Theorem 12.2.9. *For a functor $F \colon \mathcal{A} \to \mathcal{B}$ between locally finitely presented categories the following are equivalent.*

(1) *F preserves filtered colimits and products.*
(2) *$F \cong f^*$ for some exact functor $f \colon \mathrm{Ab}(\mathcal{B}) \to \mathrm{Ab}(\mathcal{A})$.*

Moreover, in this case F preserves pure-injectivity, and F is fully faithful if and only if f induces an equivalence $\mathrm{Ab}(\mathcal{B})/(\mathrm{Ker}\, f) \overset{\sim}{\to} \mathrm{Ab}(\mathcal{A})$.

Proof One implication is clear since f^* preserves limits and colimits. Thus we assume that F preserves filtered colimits and products. The functor f is constructed as follows. The restriction $F|_{\mathrm{fp}\,\mathcal{A}} \colon \mathrm{fp}\,\mathcal{A} \to \mathcal{B} \hookrightarrow \mathbf{P}(\mathcal{B})$ extends to a left exact functor $\mathrm{fp}\,\mathbf{P}(\mathcal{A}) \to \mathbf{P}(\mathcal{B})$, and this extends to a filtered colimit preserving functor $\bar{F} \colon \mathbf{P}(\mathcal{A}) \to \mathbf{P}(\mathcal{B})$. Note that $\bar{F}$ extends $\mathcal{A} \overset{F}{\to} \mathcal{B} \hookrightarrow \mathbf{P}(\mathcal{B})$. Also $\bar{F}$ is left exact, since an exact sequence in $\mathbf{P}(\mathcal{A})$ can be written as a filtered colimit of exact sequences in $\mathrm{fp}\,\mathbf{P}(\mathcal{A})$; see Remark 11.1.20. Moreover, $\bar{F}$ preserves products since its restriction to the full subcategory of injective objects preserves products. Thus $\bar{F}$ preserves limits and therefore has a left adjoint $\bar{F}_\lambda \colon \mathbf{P}(\mathcal{B}) \to \mathbf{P}(\mathcal{A})$ by the special adjoint functor theorem [183, Theorem 10.6.5]. Note that $\bar{F}_\lambda$ restricts to a functor $f \colon \mathrm{Ab}(\mathcal{B}) = \mathrm{fp}\,\mathbf{P}(\mathcal{B}) \to \mathrm{fp}\,\mathbf{P}(\mathcal{A}) = \mathrm{Ab}(\mathcal{A})$, since $\bar{F}$ preserves filtered colimits. The functor f induces an adjoint pair $(f_!, f^*) = (\bar{F}_\lambda, \bar{F})$ of functors $\mathbf{P}(\mathcal{B}) \rightleftarrows \mathbf{P}(\mathcal{A})$. In particular $f^*|_\mathcal{A} \cong F$.

It remains to show that f is exact. Observe that a sequence $\eta \colon 0 \to A \to B \to C \to 0$ in $\mathrm{fp}\,\mathbf{P}(\mathcal{A})$ is exact if and only if $\mathrm{Hom}_{\mathbf{P}(\mathcal{A})}(\eta, \bar{X})$ is exact for every $X \in \mathcal{A}$, since every injective object in $\mathbf{P}(\mathcal{A})$ is of the form $\bar{X}$ for some $X \in \mathcal{A}$. Thus if a sequence η in $\mathrm{fp}\,\mathbf{P}(\mathcal{B})$ is exact, then $\mathrm{Hom}_{\mathbf{P}(\mathcal{B})}(\eta, \bar{F}(\bar{X}))$ is exact, and therefore the sequence $\mathrm{Hom}_{\mathbf{P}(\mathcal{A})}(f(\eta), \bar{X})$ is exact. It follows that f is exact.

Having shown that f is exact, it follows that $f_!$ is exact, since every exact sequence in $\mathbf{P}(\mathcal{B})$ can be written as a filtered colimit of exact sequences in $\mathrm{fp}\,\mathbf{P}(\mathcal{B})$. Thus its right adjoint f^* preserves injectivity, and therefore F preserves pure-injectivity because of Lemma 12.1.8.

Next we apply Lemma 11.1.30. Thus $f^* \colon \mathbf{P}(\mathcal{A}) \to \mathbf{P}(\mathcal{B})$ is fully faithful if and only if f induces an equivalence $\mathrm{Ab}(\mathcal{B})/(\mathrm{Ker}\, f) \overset{\sim}{\to} \mathrm{Ab}(\mathcal{A})$. It remains to observe that f^* is fully faithful if and only if its restriction to the full subcategories of injective objects is fully faithful; see Lemma 2.1.11. Here we use again that pure-injectives in $\mathcal{A}$ identify with injectives in $\mathbf{P}(\mathcal{A})$. □

Remark 12.2.10. Suppose $F\colon \mathcal{A} \to \mathcal{B}$ preserves filtered colimits and products. If a subcategory $\mathcal{B}' \subseteq \mathcal{B}$ is definable then $F^{-1}(\mathcal{B}') \subseteq \mathcal{A}$ is definable. On the other hand, if F is fully faithful then F maps definable subcategories to definable subcategories.

A consequence of the theorem is a characterisation of coherent functors, which is a special case of Theorem 2.5.26.

Corollary 12.2.11. *A functor $F\colon \mathcal{A} \to \mathrm{Ab}$ preserves filtered colimits and products if and only if it admits a presentation*

$$\mathrm{Hom}_{\mathcal{A}}(D, -) \longrightarrow \mathrm{Hom}_{\mathcal{A}}(C, -) \longrightarrow F \longrightarrow 0$$

which is given by a morphism $C \to D$ in $\mathrm{fp}\,\mathcal{A}$.

Proof One direction is clear. Thus we assume that $F \cong f^*$ for some exact functor $f\colon \mathrm{Ab}(\mathrm{Ab}) \to \mathrm{Ab}(\mathcal{A})$. Then $f(\mathrm{Hom}_{\mathbb{Z}}(\mathbb{Z}, -))$ is an object in $\mathrm{Fp}(\mathrm{fp}\,\mathcal{A}, \mathrm{Ab})$ which yields the presentation of F. □

Next we consider locally finitely presented categories $\mathcal{A}_i$ for $i = 1, 2$ such that each $\mathcal{C}_i := \mathrm{fp}\,\mathcal{A}_i$ is abelian. Fix an exact functor $f\colon \mathcal{C}_1 \to \mathcal{C}_2$. This induces an adjoint pair $(f_!, f^*)$ of functors $\mathcal{A}_1 \rightleftarrows \mathcal{A}_2$ and also an exact functor $\bar{f}\colon \mathrm{Ab}(\mathcal{A}_1) \to \mathrm{Ab}(\mathcal{A}_2)$. We collect these functors in the following commutative diagram, where all vertical downward functors are exact.

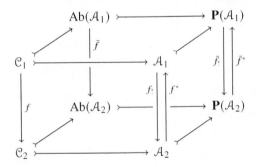

Lemma 12.2.12. *Suppose $f\colon \mathcal{C}_1 \to \mathcal{C}_2$ induces an equivalence $\mathcal{C}_1/(\mathrm{Ker}\,f) \xrightarrow{\sim} \mathcal{C}_2$. Then $\bar{f}$ also induces an equivalence $\mathrm{Ab}(\mathcal{A}_1)/(\mathrm{Ker}\,\bar{f}) \xrightarrow{\sim} \mathrm{Ab}(\mathcal{A}_2)$, and both f^* and $\bar{f}^*$ are fully faithful.*

Proof We apply Lemma 11.1.30. If $f\colon \mathcal{C}_1 \to \mathcal{C}_2$ induces an equivalence $\mathcal{C}_1/(\mathrm{Ker}\,f) \xrightarrow{\sim} \mathcal{C}_2$ then f^* is fully faithful, and therefore also $\bar{f}^*$ is fully faithful.

Restricting the left adjoint $\bar{f}_!$ to subcategories of finitely presented objects, it follows that $\mathrm{Ab}(\mathcal{A}_1)/(\mathrm{Ker}\,\bar{f}) \xrightarrow{\sim} \mathrm{Ab}(\mathcal{A}_2)$. □

12.3 Indecomposable Pure-Injective Objects

In this section we focus on properties of indecomposable pure-injective objects in locally finitely presented categories. In particular, we investigate when objects decompose into indecomposable pure-injectives.

We keep our set-up and fix a locally finitely presented category $\mathcal{A}$. We set $\mathcal{C} = \mathrm{fp}\,\mathcal{A}$ and $\mathrm{Ab}(\mathcal{A}) = \mathrm{Fp}(\mathcal{C}, \mathrm{Ab})^{\mathrm{op}}$, so that objects $X \in \mathcal{A}$ identify with exact functors $\bar{X} \colon \mathrm{Ab}(\mathcal{A})^{\mathrm{op}} \to \mathrm{Ab}$. We set

$$\mathrm{Ab}(X) := \mathrm{Ab}(\mathcal{A})/\mathrm{Ker}\,\bar{X}.$$

Subgroups of Finite Definition

Fix an object $X \in \mathcal{A}$. We consider the exact functor

$$\mathrm{Ab}(\mathcal{A}) \twoheadrightarrow \mathrm{Ab}(X) \longrightarrow \mathrm{Mod}\,\mathrm{End}(X), \quad F \mapsto \bar{F}(X) = \bar{X}(F)$$

and study its image. This leads to the notion of a subgroup of finite definition. In fact, for each object in $\mathrm{fp}\,\mathcal{A}$ the collection of these subgroups forms a lattice which provides a useful invariant of X.

Given a morphism $\phi \colon C \to C'$ in $\mathrm{fp}\,\mathcal{A}$, we denote by X_ϕ the image of the induced map $\mathrm{Hom}(C', X) \to \mathrm{Hom}(C, X)$ and call it a *subgroup of finite definition* of $\mathrm{Hom}(C, X)$. Thus

$$\bar{X}(\mathrm{Im}\,\mathrm{Hom}(\phi, -)) = X_\phi = \mathrm{Im}\,\mathrm{Hom}(\phi, X).$$

Note that any subgroup X_ϕ of finite definition is an $\mathrm{End}(X)$-submodule.

Lemma 12.3.1. *The subgroups of finite definition of* $\mathrm{Hom}(C, X)$ *are closed under finite sums and intersections. Thus they form a lattice, which is anti-isomorphic to the lattice of subobjects of* $\mathrm{Hom}(C, -)$ *in* $\mathrm{Ab}(X)$.

Proof Given morphisms $\phi_i \colon C \to C_i$ $(i = 1, 2)$ in $\mathcal{C}$, the pushout is given by an exact sequence $C \xrightarrow{\phi_1 + \phi_2} C_1 \oplus C_2 \to C' \to 0$. Then

$$X_{\phi_1} + X_{\phi_2} = X_{\phi_1 + \phi_2} \quad \text{and} \quad X_{\phi_1} \cap X_{\phi_2} = X_\phi$$

for $\phi \colon C \to C_i \to C'$. Also observe that $X_{\phi_2} \subseteq X_{\phi_1}$ if $X_{\phi_2} = X_{\psi\phi_1}$ for some $\psi \colon C_1 \to C'$.

Any subobject of $\mathrm{Hom}(C, -)$ in $\mathrm{Fp}(\mathcal{C}, \mathrm{Ab})$ is of the form $F = \mathrm{Im}\,\mathrm{Hom}(\phi, -)$

for some morphism $\phi \colon C \to C'$. The assignment $F \mapsto \bar{X}(F)$ induces an inclusion preserving map from the lattice of subobjects of $\mathrm{Hom}(C, -)$ in $\mathrm{Fp}(\mathcal{C}, \mathrm{Ab})$ to the lattice of subgroups of finite definition of $\mathrm{Hom}(C, X)$. Clearly, this is surjective, and for $F' \subseteq F$ we have $F' = F$ in $\mathrm{Ab}(X)$ if and only if $\bar{X}(F') = \bar{X}(F)$. Finally note that $X_{\phi'} \subseteq X_\phi$ implies $\mathrm{Im}\,\mathrm{Hom}(\phi', -) \subseteq \mathrm{Im}\,\mathrm{Hom}(\phi, -)$, since we may assume $\phi' = \psi\phi$ for some ψ. $\square$

Lemma 12.3.2. *Given a pure-injective object X in $\mathcal{A}$, every cyclic $\mathrm{End}(X)$-submodule of $\mathrm{Hom}(C, X)$ is the intersection of subgroups of finite definition.*

Proof We use the embedding $\mathcal{A} \to \mathbf{P}(\mathcal{A})$ that takes X to $\bar{X}$. Let $\phi \colon C \to X$ be a morphism and write $\mathrm{Ker}\,\bar{\phi} = \sum_i K_i$ as a sum of finiteley generated subobjects in $\mathbf{P}(\mathcal{A})$. For each i choose a morphism $\phi_i \colon C \to C_i$ with $\mathrm{Ker}\,\bar{\phi}_i = K_i$. Then $\phi \in \bigcap_i X_{\phi_i}$. On the other hand, every morphism $C \to X$ in $\bigcap_i X_{\phi_i}$ necessarily factors through ϕ since X is pure-injective. $\square$

Σ-Pure-Injectivity

For a definable subcategory $\mathcal{B} \subseteq \mathcal{A}$ the abelian category $\mathrm{Ab}(\mathcal{B})$ is an important invariant. We illustrate this by the following result.

Theorem 12.3.3. *Let $\mathcal{A}$ be a locally finitely presented category. For a definable subcategory $\mathcal{B}$ of $\mathcal{A}$ the following are equivalent.*

(1) *Every object in $\mathcal{B}$ is pure-injective.*
(2) *Every object in $\mathcal{B}$ decomposes into a coproduct of indecomposable objects with local endomorphism rings.*
(3) *Every object in $\mathrm{Ab}(\mathcal{B})$ is noetherian.*

Proof We begin with some preparations. Identify $\mathcal{B}$ with $\mathrm{Ex}(\mathrm{Ab}(\mathcal{B})^{\mathrm{op}}, \mathrm{Ab})$; see Corollary 12.2.3. Thus we identify an object $X \in \mathcal{B}$ with the exact functor $\bar{X} \colon \mathrm{Ab}(\mathcal{A})^{\mathrm{op}} \twoheadrightarrow \mathrm{Ab}(\mathcal{B})^{\mathrm{op}} \xrightarrow{\bar{X}_{\mathcal{B}}} \mathrm{Ab}$. Set $\mathbf{P}(\mathcal{B}) = \mathrm{Lex}(\mathrm{Ab}(\mathcal{B})^{\mathrm{op}}, \mathrm{Ab})$ and note that $\mathrm{fp}\,\mathbf{P}(\mathcal{B})$ identifies with $\mathrm{Ab}(\mathcal{B})$ by Theorem 11.1.15. Now $X \in \mathcal{B}$ is pure-injective if and only if $\bar{X}$ is injective in $\mathbf{P}(\mathcal{A})$ if and only if $\bar{X}_{\mathcal{B}}$ is injective in $\mathbf{P}(\mathcal{B})$; see Lemma 12.1.8 and Proposition 11.1.31 plus the subsequent remark.

  (1) $\Leftrightarrow$ (3): Apply Corollary 11.2.15, which says that all functors of the form $\bar{X}_{\mathcal{B}}$ are injective if and only if all objects in $\mathrm{Ab}(\mathcal{B})$ are noetherian.

  (2) $\Leftrightarrow$ (3): Observe that all objects in $\mathrm{Ab}(\mathcal{B})$ are noetherian if and only if $\mathbf{P}(\mathcal{B})$ is locally noetherian; see Proposition 11.2.5. Now apply Theorem 11.2.12. $\square$

An object X in $\mathcal{A}$ is called Σ-*pure-injective* if every coproduct of copies of X

is pure-injective. A Σ-pure-injective object admits a decomposition into indecomposable objects. In fact, there is a host of useful properties that characterise Σ-pure-injectivity.

Theorem 12.3.4. *For an object X in $\mathcal{A}$ the following are equivalent.*

(1) *The object X is Σ-pure-injective.*

(2) *Every object in $\mathrm{Ab}(X)$ is noetherian.*

(3) *The object X is pure-injective and the direct summands of products of copies of X form a definable subcategory.*

(4) *The canonical monomorphism $X^{(\mathbb{N})} \to X^{\mathbb{N}}$ splits.*

(5) *Every product of copies of X decomposes into a coproduct of indecomposable objects with local endomorphism rings.*

(6) *There exists an object Y such that every product of copies of X is a pure subobject of a coproduct of copies of Y.*

(7) *The subgroups of finite definition of $\mathrm{Hom}(C, X)$ satisfy the descending chain condition for every $C \in \mathrm{fp}\,\mathcal{A}$.*

Proof We apply Theorem 12.3.3 by taking for $\mathcal{B}$ the smallest definable subcategory of $\mathcal{A}$ containing X. In particular, we have $\mathrm{Ab}(X) = \mathrm{Ab}(\mathcal{B})$. As in the proof of Theorem 12.3.3, we consider $\mathbf{P}(\mathcal{B})$ and characterise the fact that it is locally noetherian.

(1) $\Rightarrow$ (2): We adapt the proof of Theorem 11.2.12. The pure-injectivity of all coproducts of copies of X implies that all coproducts of copies of $\bar{X}$ are injective in $\mathbf{P}(\mathcal{B})$. It follows that all objects in $\mathrm{Ab}(X)$ are noetherian.

(2) $\Rightarrow$ (3): If follows from Theorem 12.3.3 that X is pure-injective. In fact, the object $\bar{X}$ in $\mathbf{P}(\mathcal{B})$ is an injective cogenerator. Thus each $Y \in \mathcal{B}$ is a pure subobject of some product of copies of X. The pure monomorphism splits since Y is pure-injective, again by Theorem 12.3.3.

(3) $\Rightarrow$ (1): All objects in $\mathcal{B}$ are pure-injective. Thus all coproducts of copies of X are pure-injective.

(2) $\Leftrightarrow$ (4) $\Leftrightarrow$ (5) $\Leftrightarrow$ (6): This follows from Proposition 11.2.16 applied to $\bar{X}$ in $\mathbf{P}(\mathcal{B})$. The assumption on $\bar{X}$ in this proposition is satisfied by Lemma 11.1.26.

(2) $\Leftrightarrow$ (7): Every object in $\mathrm{Ab}(X)$ is noetherian if and only if $\mathrm{Hom}(C, -)$ is noetherian in $\mathrm{Ab}(X)$ for all $C \in \mathrm{fp}\,\mathcal{A}$. Now apply Lemma 12.3.1. $\square$

Given an object X in $\mathcal{A}$, every subgroup of finite definition of $\mathrm{Hom}(C, X)$ is an $\mathrm{End}(X)$-submodule. Therefore X is Σ-pure-injective, provided $\mathrm{Hom}(C, X)$ is an artinian module over $\mathrm{End}(X)$ for all $C \in \mathrm{fp}\,\mathcal{A}$. We note the following partial converse.

Lemma 12.3.5. *Let $X \in \mathcal{A}$ be Σ-pure-injective. Then every finitely generated* $\text{End}(X)$*-submodule of* $\text{Hom}(C, X)$ *is a subgroup of finite definition.*

Proof It suffices to show this for cyclic submodules since subgroups of finite definition are closed under finite sums by Lemma 12.3.1. But for cyclic submodules this follows from Lemma 12.3.2 since subgroups of finite definition of $\text{Hom}(C, X)$ satisfy the descending chain condition by Theorem 12.3.4. □

Example 12.3.6. Let Q be a quiver and k a commutative ring. If X is a k-linear representation such that X_i is an artinian k-module for each vertex $i \in Q_0$, then X is Σ-pure-injective.

Proof For a finitely presented representation C, we have an epimorphism $\bigoplus_{j=1}^{n} P(i_j) \to C$ where $P(i_1), \ldots, P(i_n)$ is a finite number of standard projectives corresponding to vertices $i_j \in Q_0$. We have $\text{Hom}(P(i), X) \cong X_i$ for each $i \in Q_0$. Thus $\text{Hom}(C, X)$ identifies with an $\text{End}(X)$-submodule of $\bigoplus_{j=1}^{n} X_{i_j}$ and is therefore artinian, so satisfies the descending chain condition for subgroups of finite definition. □

Product-Complete Objects

We consider a particular class of Σ-pure-injective objects. For an object X let $\text{Add}\, X$ denote the full subcategory consisting of all direct summands of coproducts of copies of X. Analogously, let $\text{Prod}\, X$ denote the full subcategory consisting of all direct summands of products of copies of X.

An object satisfying the equivalent conditions of the following proposition is called *product-complete*.

Proposition 12.3.7. *Let $\mathcal{A}$ be a locally finitely presented category. For an object X the following are equivalent.*

(1) $\text{Prod}\, X = \text{Add}\, X$.
(2) $\text{Add}\, X$ *is a definable subcategory of $\mathcal{A}$.*
(3) X *is Σ-pure-injective and the indecomposable direct summands of X form a Ziegler closed set.*

Proof (1) $\Rightarrow$ (2): It follows from Theorem 12.3.4 that X is Σ-pure-injective. The same result implies that $\text{Add}\, X$ is a definable subcategory.

  (2) $\Rightarrow$ (3): As before, Theorem 12.3.4 implies that X is Σ-pure-injective. The indecomposable objects in $\text{Add}\, X$ form a Ziegler closed set by Theorem 12.2.2.

  (3) $\Rightarrow$ (1): The definable subcategory generated by X equals $\text{Prod}\, X$ by Theorem 12.3.4, since X is Σ-pure-injective. Since all objects in $\text{Prod}\, X$ decompose into indecomposable objects, it follows that $\text{Prod}\, X = \text{Add}\, X$. □

Prüfer Objects

We consider a class of Σ-pure-injective objects which generalises the notion of a Prüfer module over a Dedekind domain.

Let k be a commutative noetherian ring and $\mathcal{A}$ a k-linear locally finitely presented category. An object $X \in \mathcal{A}$ is called a *Prüfer object* if there is an endomorphism $\phi \colon X \to X$ such that for each $C \in \mathrm{fp}\,\mathcal{A}$

(Pr1) each morphism $C \to X$ is annihilated by some power of ϕ, and

(Pr2) the kernel of $\mathrm{Hom}(C, X) \xrightarrow{\phi \circ -} \mathrm{Hom}(C, X)$ is a finite length k-module.

Example 12.3.8. Let $A = k$ be a Dedekind domain and $\mathfrak{p}$ a maximal ideal. Then the Prüfer module

$$A_{\mathfrak{p}^\infty} = \bigcup_{n \geq 0} A/\mathfrak{p}^n = E(A/\mathfrak{p})$$

is a Prüfer object in $\mathrm{Mod}\,A$, because the canonical morphism $A/\mathfrak{p}^2 \twoheadrightarrow A/\mathfrak{p}$ extends to an epimorphism $\phi \colon A_{\mathfrak{p}^\infty} \twoheadrightarrow A_{\mathfrak{p}^\infty}$ with $\mathrm{Ker}\,\phi^n = A/\mathfrak{p}^n$.

Proposition 12.3.9. *A Prüfer object is Σ-pure-injective.*

Proof Let X be a Prüfer object with endomorphism $\phi \colon X \to X$. We show that $\mathrm{Hom}(C, X)$ is an artinian $\mathrm{End}(X)$-module for each $C \in \mathrm{fp}\,\mathcal{A}$. Then the subgroups of finite definition of $\mathrm{Hom}(C, X)$ satisfy the descending chain condition, and therefore the assertion follows from Theorem 12.3.4.

Consider the polynomial ring $k[t]$ in one variable and the homomorphism $k[t] \to \mathrm{End}(X)$ given by $t \mapsto \phi$. Fix $C \in \mathrm{fp}\,\mathcal{A}$ and set $C_n = \mathrm{Ker}\,\mathrm{Hom}(C, \phi^n)$ for $n \geq 0$. An induction shows C_n has finite length as a k-module, since it fits into an exact sequence $0 \to C_{n-1} \to C_n \to C_1$. Also, the socle of the $k[t]$-module $\mathrm{Hom}(C, X)$ has finite length because it is annihilated by t and therefore contained in C_1. It remains to apply the lemma below.   □

Lemma 12.3.10. *Let A be a commutative noetherian ring. Then an A-module M is artinian if and only if M is a directed union of finite length submodules and* $\mathrm{soc}\,M$ *has finite length.*

Proof Suppose that M is artinian. If $M = \bigcup_i M_i$ is written as a directed union of finitely generated submodules, then each M_i has finite length. The module $\mathrm{soc}\,M$ is semisimple and artinian, and therefore has finite length.

For the other implication consider an injective envelope $\mathrm{soc}\,M \to E(\mathrm{soc}\,M)$ which extends to a morphism $\alpha \colon M \to E(\mathrm{soc}\,M)$. We claim that $\mathrm{Ker}\,\alpha = 0$. Otherwise $\mathrm{Ker}\,\alpha$ has a simple submodule, because it is a directed union of finite length submodules. This is impossible, and therefore α is a monomorphism.

It remains to observe that $E(\text{soc}\, M)$ is artinian, since it is a finite direct sum of modules of the form $E(A/\mathfrak{p})$ for some maximal ideal $\mathfrak{p} \in \text{Spec}\, A$, cf. Lemma 2.4.19. $\qquad\square$

Suppose that $\mathcal{A}$ is abelian. Then a Prüfer object with endomorphism $\phi \colon X \to X$ is given by a sequence of extensions

$$
\begin{array}{ccccccccc}
0 & \longrightarrow & X_1 & \longrightarrow & X_n & \xrightarrow{\;\phi_n\;} & X_{n-1} & \longrightarrow & 0 \\
& & \| & & \downarrow & & \downarrow & & \\
0 & \longrightarrow & X_1 & \longrightarrow & X_{n+1} & \xrightarrow{\;\phi_{n+1}\;} & X_n & \longrightarrow & 0
\end{array}
$$

where $X_n = \text{Ker}\, \phi^n$ and the vertical morphisms are the inclusions. In particular, $X = \bigcup_{n \geq 0} X_n$ and $\phi = \bigcup_{n \geq 0} \phi_n$.

Compactness

Recall that a topological space T is *quasi-compact* if for any family $(V_i)_{i \in I}$ of closed subsets $\bigcap_{i \in I} V_i = \varnothing$ implies $\bigcap_{i \in J} V_i = \varnothing$ for some finite subset $J \subseteq I$.

The correspondence in Theorem 12.2.2 provides an inclusion reversing isomorphism between the lattice of closed subsets of $\text{Ind}\,\mathcal{A}$ and the lattice of Serre subcategories of $\text{Ab}(\mathcal{A})$. This yields a criterion for when the space $\text{Ind}\,\mathcal{A}$ is quasi-compact.

An abelian category $\mathcal{C}$ is *finitely generated* if there exists an object $X \in \mathcal{C}$ such that $\mathcal{C}$ equals the smallest Serre subcategory containing X.

Proposition 12.3.11. *Let $\mathcal{A}$ be a locally finitely presented category. The space $\text{Ind}\,\mathcal{A}$ is quasi-compact if and only if the abelian category $\text{Ab}(\mathcal{A})$ is finitely generated.*

Proof Combine the correspondence in Theorem 12.2.2 with Lemma 12.1.16, using the bijection $\text{Ind}\,\mathcal{A} \xrightarrow{\sim} \text{Sp}\,\mathbf{P}(\mathcal{A})$. $\qquad\square$

Corollary 12.3.12. *Suppose there exists an object $G \in \text{fp}\,\mathcal{A}$ such that every object in $\text{fp}\,\mathcal{A}$ is a quotient of G^n for some integer $n \geq 1$. Then $\text{Ind}\,\mathcal{A}$ is quasi-compact.*

Proof The abelian category $\text{Ab}(\mathcal{A})$ is generated by $\text{Hom}_{\mathcal{A}}(G, -)$, since each object in $\text{Fp}(\text{fp}\,\mathcal{A}, \text{Ab})$ is a quotient of some representable functor $\text{Hom}_{\mathcal{A}}(C, -)$ which embeds into $\text{Hom}_{\mathcal{A}}(G^n, -)$ when $G^n \twoheadrightarrow C$. $\qquad\square$

Left Almost Split Morphisms

A morphism $\phi\colon X \to Y$ is *left almost split* if it is not a split monomorphism, and if every morphism $X \to X'$ which is not a split monomorphism factors through ϕ.

We fix a locally finitely presented abelian category $\mathcal{A}$. In the following we use freely the fact that the functor $\mathrm{ev}\colon \mathcal{A} \to \mathbf{P}(\mathcal{A})$ identifies the pure-injective objects in $\mathcal{A}$ with the injective objects in the purity category $\mathbf{P}(\mathcal{A})$; see Lemma 12.1.8.

Theorem 12.3.13. *For an indecomposable pure-injective object $X \in \mathcal{A}$ the following are equivalent.*

(1) *X is the source of a left almost split morphism in $\mathcal{A}$.*
(2) *$\bar{X}$ is an injective envelope of a simple object in $\mathbf{P}(\mathcal{A})$.*
(3) *If X is isomorphic to a direct summand of a product $\prod_{i \in I} Y_i$ of indecomposable objects in $\mathcal{A}$, then $X \cong Y_i$ for some $i \in I$.*

Proof (1) $\Rightarrow$ (2): Let $\phi\colon X \to Y$ be left almost split. Choose a finitely generated subobject $0 \neq C \subseteq \mathrm{Ker}\,\bar{\phi}$, a maximal subobject $U \subseteq C$, and an injective envelope $C/U \to \bar{X}'$. Then the induced morphism $C \to \bar{X}'$ factors through $C \hookrightarrow \bar{X}$ via a morphism $\alpha\colon X \to X'$. We claim that α is a split monomorphism. Otherwise it factors through ϕ, which is impossible since $\bar{\phi}(C) = 0$. Thus α is an isomorphism, and $\bar{X}$ is an injective envelope of a simple object in $\mathbf{P}(\mathcal{A})$.

(2) $\Rightarrow$ (1): Let $S \hookrightarrow \bar{X}$ be an injective envelope of a simple object S in $\mathbf{P}(\mathcal{A})$. We choose a left $\mathcal{A}$-approximation $\bar{X}/S \to \bar{Y}$ which yields a morphism $\phi\colon X \to Y$. This is possible by Proposition 11.1.27, because we identify $\mathcal{A}$ with the full subcategory of exact functors in $\mathbf{P}(\mathcal{A})$. We claim that ϕ is left almost split. Clearly, ϕ is not a split monomorphism since $\bar{\phi}$ is not a monomorphism. Let $\alpha\colon X \to X'$ be a morphism which is not a split monomorphism. Thus $\bar{\alpha}$ is not a monomorphism, and therefore $\bar{\alpha}(S) = 0$. Thus $\bar{\alpha}$ factors through $\bar{X} \to \bar{X}/S$, and therefore through the approximation $\bar{X}/S \to \bar{Y}$ via a morphism $\beta\colon Y \to X'$. Thus $\alpha = \beta\phi$.

(2) $\Rightarrow$ (3): Let $S \hookrightarrow \bar{X}$ be an injective envelope of a simple object S in $\mathbf{P}(\mathcal{A})$. If X is isomorphic to a direct summand of a product $\prod_{i \in I} Y_i$ of indecomposable objects in $\mathcal{A}$, then $\mathrm{Hom}(S, \bar{Y}_i) \neq 0$ for some $i \in I$. This yields a monomorphism $S \to \bar{X} \to \prod_{i \in I} \bar{Y}_i \to \bar{Y}_i$, and therefore $X \to Y_i$ is a pure monomorphism, which splits since X is pure-injective. Thus $X \cong Y_i$ since Y_i is indecomposable.

(3) $\Rightarrow$ (2): Let $\mathcal{U} = \mathrm{Ind}\,\mathcal{A} \setminus \{X\}$ and consider the canonical morphisms

$$\phi\colon X \longrightarrow \prod_{Y \in \mathcal{U}} Y^{\mathrm{Hom}(X,Y)}.$$

The condition (3) implies $\operatorname{Ker} \bar{\phi} \neq 0$. Choose a finitely generated non-zero subobject $U \subseteq \operatorname{Ker} \bar{\phi}$ and a maximal subobject $V \subseteq U$ in $\mathbf{P}(\mathcal{A})$. Set $S = U/V$. The composite $U \twoheadrightarrow S \rightarrowtail E(S)$ extends to a morphism $\bar{X} \to E(S)$ which does not factor through $\bar{\phi}$. Thus $\bar{X} \cong E(S)$ by the construction of ϕ. □

Corollary 12.3.14. *The subset* $\mathcal{U} \subseteq \operatorname{Ind} \mathcal{A}$ *of indecomposables which are the source of a left almost split morphism is dense.*

Proof Set $U = \prod_{X \in \mathcal{U}} X$. This is an injective cogenerator of $\mathbf{P}(\mathcal{A})$ by the above theorem, and therefore $^{\perp}\mathcal{U} = \{0\}$. Thus $\mathcal{U}$ is dense by Theorem 12.2.2. □

Call a point $X \in \operatorname{Ind} \mathcal{A}$ *isolated* if the set $\{X\}$ is open.

Corollary 12.3.15. *If* $X \in \operatorname{Ind} \mathcal{A}$ *is isolated, then X is the source of a left almost split morphism. The converse holds when X is finitely presented.*

Proof Set $\mathcal{U} = \operatorname{Ind} \mathcal{A} \setminus \{X\}$. Then $^{\perp}\mathcal{U} \subseteq \operatorname{Ab}(\mathcal{A})$ is a proper Serre subcategory by Theorem 12.2.2 when X is isolated. Choose $C \in \operatorname{Ab}(\mathcal{A}) \setminus ^{\perp}\mathcal{U}$ and a maximal subobject $U \subseteq C$. Then $\bar{X}$ is an injective envelope of C/U, since $\operatorname{Hom}_{\mathbf{P}(\mathcal{A})}(C, \bar{Y}) = 0$ for all $Y \in \mathcal{U}$.

Now suppose that $S \hookrightarrow \bar{X}$ is an injective envelope of a simple object S in $\mathbf{P}(\mathcal{A})$. If X is finitely presented, then S is finitely presented in $\mathbf{P}(\mathcal{A})$. Thus $\{S\}^{\perp} = \mathcal{U}$ is closed. □

Fp-Injective Objects

Let $\mathcal{A}$ be a locally finitely presented abelian category and set $\mathcal{C} = \operatorname{fp} \mathcal{A}$. An object $X \in \mathcal{A}$ satisfying $\operatorname{Ext}^1_{\mathcal{A}}(-, X)|_{\operatorname{fp} \mathcal{A}} = 0$ is called *fp-injective*.

Lemma 12.3.16. *An fp-injective object is injective if and only if it is pure-injective.*

Proof If X is fp-injective then any exact sequence $0 \to X \to Y \to Z \to 0$ is pure-exact. Thus X is injective if and only if X is pure-injective. □

Now suppose that $\mathcal{C} = \operatorname{fp} \mathcal{A}$ is an abelian category. We write $\operatorname{Eff}(\mathcal{C}, \operatorname{Ab})$ for the Serre subcategory of $\operatorname{Fp}(\mathcal{C}, \operatorname{Ab})$ given by all functors F with presentation

$$0 \longrightarrow \operatorname{Hom}_{\mathcal{C}}(C, -) \longrightarrow \operatorname{Hom}_{\mathcal{C}}(B, -) \longrightarrow \operatorname{Hom}_{\mathcal{C}}(A, -) \longrightarrow F \longrightarrow 0$$

coming from an exact sequence $0 \to A \to B \to C \to 0$ in $\mathcal{C}$.

Proposition 12.3.17. *We have*

$$\operatorname{Eff}(\mathcal{C}, \operatorname{Ab})^{\perp} = \{X \in \mathcal{A} \mid X \text{ is fp-injective}\}.$$

Therefore the fp-injective objects form a definable subcategory and

$$\mathrm{Eff}(\mathcal{C}, \mathrm{Ab})^{\perp} \cap \mathrm{Ind}\,\mathcal{A} = \mathrm{Inj}\,\mathcal{A} \cap \mathrm{Ind}\,\mathcal{A}.$$

Proof Identifying $\mathcal{A} = \mathrm{Lex}(\mathcal{C}^{\mathrm{op}}, \mathrm{Ab})$, the first assertion follows as a reformulation of Lemma 11.1.26. The subcategory $\mathrm{Eff}(\mathcal{C}, \mathrm{Ab})^{\perp}$ is definable by definition, and the last equality then follows from the fact that pure-injectivity and injectivity coincide for fp-injective objects, by Lemma 12.3.16. □

Corollary 12.3.18. *A locally finitely presented abelian category is locally noetherian if and only if the injective objects form a definable subcategory.*

Proof When $\mathcal{A}$ is locally noetherian then every fp-injective object is injective; this follows from Baer's criterion. Thus the injective objects form a definable subcategory. Conversely, if the injectives form a definable subcategory, then they are closed under coproducts and therefore $\mathcal{A}$ is locally noetherian, by Theorem 11.2.12. □

12.4 Pure-Injective Modules

Let Λ be a ring. We consider the category of Λ-modules and set $\mathcal{A} = \mathrm{Mod}\,\Lambda$. Note that $\mathcal{A}$ is locally finitely presented with $\mathrm{fp}\,\mathcal{A} = \mathrm{mod}\,\Lambda$. In this section we give an explicit description of the embedding $\mathcal{A} \to \mathbf{P}(\mathcal{A})$ into the purity category, and we identify $\mathrm{Ab}(\mathcal{A})$ with the free abelian category over Λ. Also, we consider the set of indecomposable pure-injectives $\mathrm{Ind}\,\Lambda := \mathrm{Ind}(\mathrm{Mod}\,\Lambda)$, which is called the *Ziegler spectrum* of Λ.

The Free Abelian Category

The *free abelian category* over Λ is by definition

$$\mathrm{Ab}(\Lambda) := \mathrm{Fp}(\mathrm{mod}\,\Lambda, \mathrm{Ab})^{\mathrm{op}} = \left(\mathrm{mod}\left((\mathrm{mod}\,\Lambda)^{\mathrm{op}}\right)\right)^{\mathrm{op}}.$$

We identify Λ with the representable functor $\mathrm{Hom}_{\Lambda}(\Lambda, -)$ in $\mathrm{Ab}(\Lambda)$. The following universal property of $\mathrm{Ab}(\Lambda)$ justifies its name.

Proposition 12.4.1. *For a ring Λ the category $\mathrm{Ab}(\Lambda)$ is abelian. Given an object X in an abelian category $\mathcal{A}$ and a ring homomorphism $\phi\colon \Lambda \to \mathrm{End}_{\mathcal{A}}(X)$, there exists a unique (up to isomorphism) exact functor $\mathrm{Ab}(\Lambda) \to \mathcal{A}$ sending Λ to X and inducing the homomorphism ϕ.*

Proof The category $\mathrm{Ab}(\Lambda)$ is abelian by Lemma 2.1.6 since $\mathrm{mod}\,\Lambda$ has co-kernels. A homomorphism $\phi\colon \Lambda \to \mathrm{End}_{\mathcal{A}}(X)$ extends uniquely to an additive functor $\phi_0\colon \mathrm{proj}\,\Lambda \to \mathcal{A}$, and therefore uniquely to a right exact functor $\phi_1\colon \mathrm{mod}\,\Lambda \to \mathcal{A}$, by Lemma 2.1.7. Then ϕ_1 extends uniquely to a left exact functor $\phi_2\colon \mathrm{Ab}(\Lambda) \to \mathcal{A}$, again by Lemma 2.1.7. The functor ϕ_2 is exact by Lemma 2.1.8. Clearly, ϕ_2 agrees on Λ with ϕ and is uniquely determined, up to isomorphism. □

The universal property of $\mathrm{Ab}(\Lambda)$ yields an equivalence

$$\mathrm{Ab}(\Lambda)^{\mathrm{op}} \xrightarrow{\sim} \mathrm{Ab}(\Lambda^{\mathrm{op}})$$

extending the identity $\Lambda^{\mathrm{op}} \to \Lambda^{\mathrm{op}}$. We give an explicit description. To this end define for $F \in \mathrm{Ab}(\Lambda^{\mathrm{op}})$

$$F^{\vee}\colon \mathrm{mod}\,\Lambda \longrightarrow \mathrm{Ab}, \quad X \mapsto \mathrm{Hom}(F, X \otimes_{\Lambda} -).$$

Then we have for $X \in \mathrm{mod}\,\Lambda$ and $Y \in \mathrm{mod}(\Lambda^{\mathrm{op}})$

$$(X \otimes_{\Lambda} -)^{\vee} = \mathrm{Hom}_{\Lambda}(X, -) \qquad \text{and} \qquad (\mathrm{Hom}_{\Lambda^{\mathrm{op}}}(Y, -))^{\vee} = - \otimes_{\Lambda} Y.$$

Lemma 12.4.2. *The assignment $F \mapsto F^{\vee}$ yields mutually inverse equivalences between $\mathrm{Ab}(\Lambda)^{\mathrm{op}}$ and $\mathrm{Ab}(\Lambda^{\mathrm{op}})$.*

Proof Given $F \in \mathrm{Ab}(\Lambda)$ and $G \in \mathrm{Ab}(\Lambda^{\mathrm{op}})$, we have

$$\mathrm{Hom}(F, G^{\vee}) \cong \mathrm{Hom}(G, F^{\vee}).$$

This is clear for $F = \mathrm{Hom}_{\Lambda}(X, -)$ and follows for arbitrary F by exactness, since a presentation

$$\mathrm{Hom}_{\Lambda}(X_1, -) \longrightarrow \mathrm{Hom}_{\Lambda}(X_0, -) \longrightarrow F \longrightarrow 0$$

yields an exact sequence

$$0 \longrightarrow F^{\vee} \longrightarrow X_0 \otimes_{\Lambda} - \longrightarrow X_1 \otimes_{\Lambda} -$$

in $\mathrm{Ab}(\Lambda^{\mathrm{op}})$. Thus $F^{\vee\vee} \cong F$ since this holds for all representable functors. □

For $\mathcal{A} = \mathrm{Mod}\,\Lambda$ we have by definition $\mathrm{Ab}(\mathcal{A}) = \mathrm{Ab}(\Lambda)$. The following gives an explicit description of the purity category $\mathbf{P}(\mathcal{A})$.

Proposition 12.4.3. *The assignment $F \mapsto \mathrm{Hom}((-)^{\vee}, F)$ induces an equivalence*

$$\mathrm{Add}(\mathrm{mod}(\Lambda^{\mathrm{op}}), \mathrm{Ab}) \xrightarrow{\sim} \mathrm{Lex}(\mathrm{Fp}(\mathrm{mod}\,\Lambda, \mathrm{Ab}), \mathrm{Ab}) = \mathbf{P}(\mathcal{A}). \qquad (12.4.4)$$

Proof Both categories are locally finitely presented. We have

$$\mathrm{fp\,Add}(\mathrm{mod}(\Lambda^{\mathrm{op}}), \mathrm{Ab}) = \mathrm{Fp}(\mathrm{mod}(\Lambda^{\mathrm{op}}), \mathrm{Ab}) = \mathrm{Ab}(\Lambda^{\mathrm{op}})^{\mathrm{op}}$$

and

$$\mathrm{fp}\,\mathbf{P}(\mathcal{A}) = \mathrm{Fp}(\mathrm{mod}\,\Lambda, \mathrm{Ab})^{\mathrm{op}} = \mathrm{Ab}(\Lambda);$$

see Proposition 11.1.9 and Theorem 11.1.15. Thus the assertion follows from the second part of Theorem 11.1.15 and the equivalence $\mathrm{Ab}(\Lambda)^{\mathrm{op}} \xrightarrow{\sim} \mathrm{Ab}(\Lambda^{\mathrm{op}})$. □

Corollary 12.4.5. *The embedding* $\mathrm{ev} \colon \mathcal{A} \to \mathbf{P}(\mathcal{A})$ *identifies with the functor*

$$\mathcal{A} \longrightarrow \mathrm{Add}(\mathrm{mod}(\Lambda^{\mathrm{op}}), \mathrm{Ab}), \quad X \mapsto X \otimes_\Lambda -$$

via the equivalence (12.4.4). *In particular,* Λ-*modules identify with exact functors* $\mathrm{Ab}(\Lambda)^{\mathrm{op}} \to \mathrm{Ab}$, *by sending a* Λ-*module* X *to the functor*

$$\mathrm{Ab}(\Lambda) \ni F \longmapsto \mathrm{Hom}(F^\vee, X \otimes_\Lambda -).$$

Proof We need to check that there is a natural isomorphism

$$\mathrm{ev}(X) = \bar{X} \cong \mathrm{Hom}((-)^\vee, X \otimes_\Lambda -)$$

for every Λ-module X. For $F = \mathrm{Hom}_\Lambda(C, -)$ we have

$$\bar{X}(F) = \mathrm{Hom}_\Lambda(C, X) \cong \mathrm{Hom}(C \otimes_\Lambda -, X \otimes_\Lambda -) \cong \mathrm{Hom}(F^\vee, X \otimes_\Lambda -).$$

The functors $\bar{X}$ and $\mathrm{Hom}((-)^\vee, X \otimes_\Lambda -)$ are both exact; so we have the isomorphism for all $F \in \mathrm{Ab}(\mathcal{A})$.

The second assertion is an immediate consequence. □

Corollary 12.4.6. *A sequence* $0 \to X \to Y \to Z \to 0$ *of* Λ-*modules is pure-exact if and only if the induced sequence*

$$0 \longrightarrow X \otimes_\Lambda C \longrightarrow Y \otimes_\Lambda C \longrightarrow Z \otimes_\Lambda C \longrightarrow 0$$

is exact for every Λ^{op}-*module* C.

Proof Combine Lemma 12.1.6 and Corollary 12.4.5. □

A Criterion for Pure-Injectivity

Let k be a commutative ring and Λ a k-algebra. We fix a minimal injective cogenerator E over k and set $D(X) := \mathrm{Hom}_k(X, E)$ for every k-module X. This induces Matlis duality between right and left Λ-modules.

A Λ-module Q is *pure-injective* if Q is a pure-injective object in $\mathrm{Mod}\,\Lambda$, so every pure monomorphism $X \to Y$ induces a surjective map $\mathrm{Hom}_\Lambda(Y, Q) \to \mathrm{Hom}_\Lambda(X, Q)$.

Proposition 12.4.7. *For a Λ-module X the following are equivalent.*

(1) *The module X is pure-injective.*
(2) *The natural morphism $\phi_X\colon X \to D^2(X)$ given by $\phi_X(x)(\alpha) = \alpha(x)$ for $x \in X$ and $\alpha \in D(X)$ is a split monomorphism.*
(3) *There is a bimodule $_\Lambda Y_\Gamma$ for some ring Γ and an injective Γ-module I such that X is isomorphic to a direct summand of $\mathrm{Hom}_\Gamma(Y, I)$.*

Proof (1) $\Rightarrow$ (2): The composite $D(X) \xrightarrow{\phi_{D(X)}} D^3(X) \xrightarrow{D(\phi_X)} D(X)$ is the identity. Thus $D(\phi_X)$ is a split epimorphism, and therefore

$$D(\phi_X \otimes_\Lambda C) \cong \mathrm{Hom}_\Lambda(C, D(\phi_X))$$

is an epimorphism for every left Λ-module C. It follows that $\phi_X \otimes_\Lambda C$ is a monomorphism, and therefore ϕ_X is a pure monomorphism by Corollary 12.4.6. We conclude that ϕ_X splits when X is pure-injective.

(2) $\Rightarrow$ (3): Take $_\Lambda Y_\Gamma = {}_\Lambda D(X)_k$. Then X is isomorphic to a direct summand of $\mathrm{Hom}_k(Y, E) = D^2(X)$.

(3) $\Rightarrow$ (1): The functor

$$\mathrm{Hom}_\Lambda(-, \mathrm{Hom}_\Gamma(Y, I)) \cong \mathrm{Hom}_\Gamma(- \otimes_\Lambda Y, I)$$

sends pure-exact sequences to exact sequences, by the description of pure-exact sequences in Corollary 12.4.6. Thus $\mathrm{Hom}_\Gamma(Y, I)$ is pure-injective. □

Corollary 12.4.8. *Every Λ-module X admits a pure monomorphism into a pure-injective module of the form $X \to \prod_{i \in I} D(Y_i)$ that is given by a family of finitely presented Λ^{op}-modules $(X_i)_{i \in I}$.*

Proof Choose an epimorphism $\coprod_{i \in I} Y_i \to D(X)$ and take the composite $X \to D^2(X) \to \prod_{i \in I} D(Y_i)$. □

Example 12.4.9. Let $_\Gamma X_\Lambda$ be a bimodule and suppose that X is artinian over Γ. Then X is a Σ-pure-injective Λ-module, because the descending chain condition for subgroups of finite definition is satisfied; see Theorem 12.3.4.

Duality

There is no global duality between right and left Λ-modules, but there is a bijective correspondence between specific classes of modules. This is based on the equivalence $\mathrm{Ab}(\Lambda)^{\mathrm{op}} \xrightarrow{\sim} \mathrm{Ab}(\Lambda^{\mathrm{op}})$ given by $F \mapsto F^\vee$, which induces a bijection between Serre subcategories. Applying the bijective correspondence between Serre subcategories of $\mathrm{Ab}(\Lambda)$ and definable subcategories of $\mathrm{Mod}\,\Lambda$ from Theorem 12.2.2, we obtain for definable subcategories a bijection

$$\mathrm{Mod}\,\Lambda \supseteq \mathcal{X} \longmapsto ((^{\perp}\mathcal{X})^{\vee})^{\perp} \subseteq \mathrm{Mod}\,\Lambda^{\mathrm{op}}.$$

For a Λ-module X we consider the Serre subcategory

$$^{\perp}X := \{F \in \mathrm{Ab}(\Lambda) \mid \mathrm{Hom}(F^{\vee}, X \otimes_{\Lambda} -) = 0\}$$

and observe that $(^{\perp}X)^{\perp} \subseteq \mathrm{Mod}\,\Lambda$ is the smallest definable subcategory containing X. We call a pair of Λ-modules $(X_{\Lambda}, {}_{\Lambda}Y)$ a *dual pair* if the following equivalent conditions are satisfied:

$$(^{\perp}X)^{\vee} = {}^{\perp}Y \qquad \Longleftrightarrow \qquad {}^{\perp}X = (^{\perp}Y)^{\vee}.$$

Proposition 12.4.10. *Let $_{\Gamma}X_{\Lambda}$ be a Γ-Λ-bimodule and fix an injective cogenerator $I \in \mathrm{Mod}\,\Gamma$. Then $(X, \mathrm{Hom}_{\Gamma}(X, I))$ is a dual pair of Λ-modules.*

Proof Choose $F \in \mathrm{Ab}(\Lambda)$ with presentation

$$\mathrm{Hom}_{\Lambda}(C', -) \longrightarrow \mathrm{Hom}_{\Lambda}(C, -) \longrightarrow F \longrightarrow 0$$

given by a morphism in $\phi\colon C \to C'$ in $\mathrm{mod}\,\Lambda$. Then $F^{\vee} \in \mathrm{Ab}(\Lambda^{\mathrm{op}})$ has the presentation

$$0 \longrightarrow F^{\vee} \longrightarrow C \otimes_{\Lambda} - \longrightarrow C' \otimes_{\Lambda} -$$

and we have

$$
\begin{aligned}
F \in {}^{\perp}X &\Longleftrightarrow \mathrm{Hom}_{\Lambda}(\phi, X) \text{ is an epimorphism}\\
&\Longleftrightarrow \mathrm{Hom}_{\Gamma}(\mathrm{Hom}_{\Lambda}(\phi, X), I) \text{ is a monomorphism}\\
&\Longleftrightarrow \phi \otimes_{\Lambda} \mathrm{Hom}_{\Gamma}(X, I) \text{ is a monomorphism}\\
&\Longleftrightarrow F^{\vee} \in {}^{\perp}\mathrm{Hom}_{\Gamma}(X, I)\\
&\Longleftrightarrow F \in (^{\perp}\mathrm{Hom}_{\Gamma}(X, I))^{\vee}.
\end{aligned}
$$
$\square$

Example 12.4.11. Let Λ be a k-algebra over a commutative ring k. Then a Λ-module X together with its Matlis dual $D(X)$ form a dual pair $(X, D(X))$.

Pure-Semisimplicity

A ring Λ is called *right pure-semisimple* when every pure-exact sequence of Λ-modules is split exact.

Proposition 12.4.12. *For a ring Λ the following are equivalent.*

(1) *The ring Λ is right pure-semisimple.*

(2) *Every Λ-module is pure-injective.*

(3) *Every Λ-module decomposes into a coproduct of indecomposable modules with local endomorphism rings.*

(4) *Every object in $\text{Ab}(\Lambda)$ is noetherian.*

Proof Apply Theorem 12.3.3. □

Modules of Finite Projective Dimension

Let Λ be a ring. We consider for fixed $n \in \mathbb{N}$ the full subcategory of Λ-modules X such that the projective dimension proj.dim X is bounded by n.

Proposition 12.4.13. *Let Λ be a ring that is right perfect and left coherent. Then for each $n \in \mathbb{N}$ the Λ-modules of projective dimension at most n form a definable subcategory of* Mod Λ.

Recall that Λ is *right perfect* if every flat module is projective. The ring Λ is *right coherent* if the category of finitely presented Λ-modules is abelian. For example, every right artinian ring is right perfect and right coherent.

Proof Because Λ is right perfect, we have proj.dim $X \leq n$ if and only if $\text{Tor}^\Lambda_{n+1}(X, -) = 0$. We can test the vanishing of $\text{Tor}^\Lambda_{n+1}(X, -)$ on finitely presented left modules, since $\text{Tor}^\Lambda_{n+1}(X, -)$ preserves filtered colimits and every module is a filtered colimit of finitely presented modules. Because Λ is left coherent, a finitely presented left Λ-module Y admits a projective resolution

$$\cdots \longrightarrow P_2 \longrightarrow P_1 \longrightarrow P_0 \longrightarrow Y \longrightarrow 0$$

such that each P_i is finitely generated. It follows that $\text{Tor}^\Lambda_{n+1}(-, Y)$ preserves products, since each functor $- \otimes_\Lambda P_i$ preserves products and taking products of abelian groups is exact. In particular

$$\bigcap_{Y \in \text{mod}(\Lambda^{\text{op}})} \text{Ker} \, \text{Tor}^\Lambda_{n+1}(-, Y)$$

is a definable subcategory of Mod Λ, which equals the subcategory of modules of projective dimension at most n. □

A consequence of Theorem 12.2.2 is then the fact that we may test on Ind Λ the *finitistic dimension* of Λ, that is, the supremum of all finite projective dimensions proj.dim X.

The Ziegler Spectrum of an Artin Algebra

Let Λ be an Artin algebra over a commutative artinian ring k. We write $D = \mathrm{Hom}_k(-, E)$ for the Matlis duality over k given by a minimal injective cogenerator E.

We consider $\mathcal{A} = \mathrm{Mod}\,\Lambda$ and identify $\mathbf{P}(\mathcal{A}) = \mathrm{Add}(\mathrm{mod}(\Lambda^{\mathrm{op}}), \mathrm{Ab})$. Finitely presented and finite length modules over Λ coincide because Λ is artinian. We find some further finiteness conditions that are equivalent.

Proposition 12.4.14. *The assignment $X \mapsto \mathrm{soc}(X \otimes_\Lambda -)$ induces a bijection between*

- *the isomorphism classes of indecomposable finite length Λ-modules, and*
- *the isomorphism classes of simple objects in $\mathrm{Add}(\mathrm{mod}(\Lambda^{\mathrm{op}}), \mathrm{Ab})$.*

Proof Write $(\mathrm{mod}\,\Lambda, \mathrm{mod}\,k)$ for the category of k-linear functors $\mathrm{mod}\,\Lambda \to \mathrm{mod}\,k$ and observe that $F \mapsto D(F) := D \circ F$ induces an equivalence

$$(\mathrm{mod}\,\Lambda, \mathrm{mod}\,k)^{\mathrm{op}} \xrightarrow{\;\sim\;} ((\mathrm{mod}\,\Lambda)^{\mathrm{op}}, \mathrm{mod}\,k).$$

Next observe that every simple functor in $((\mathrm{mod}\,\Lambda)^{\mathrm{op}}, \mathrm{mod}\,k)$ is of the form

$$S_Y = \mathrm{Hom}_\Lambda(-, Y)/\mathrm{Rad}_\Lambda(-, Y)$$

for some indecomposable finitely presented Λ-module Y since $\mathrm{End}_\Lambda(Y)$ is local.

Let $X \in \mathrm{mod}\,\Lambda$ be indecomposable. The functor

$$D(X \otimes_\Lambda -) \cong \mathrm{Hom}_{\Lambda^{\mathrm{op}}}(-, D(X))$$

has a unique simple quotient $S_{D(X)}$. Thus $X \otimes_\Lambda -$ has $D(S_{D(X)})$ as a unique simple subobject in $\mathbf{P}(\mathcal{A})$, and this implies

$$\mathrm{soc}(X \otimes_\Lambda -) \cong D(S_{D(X)}).$$

Let $S \in \mathbf{P}(\mathcal{A})$ be simple. Then $S = D(S_Y)$ for some indecomposable $Y \in \mathrm{mod}\,\Lambda^{\mathrm{op}}$. We have $D\,\mathrm{Hom}_{\Lambda^{\mathrm{op}}}(-, Y) \cong D(Y) \otimes_\Lambda -$, and this implies

$$\mathrm{soc}(D(Y) \otimes_\Lambda -) \cong S. \qquad \square$$

Theorem 12.4.15. *For an indecomposable pure-injective Λ-module X the following are equivalent.*

(1) *X is finitely presented.*
(2) *X is the source of a left almost split morphism.*
(3) *X is isolated.*

Proof We use the embedding $\mathcal{A} \to \mathbf{P}(\mathcal{A})$ which sends Y to $\bar{Y} = Y \otimes_{\Lambda} -$; see Corollary 12.4.5.

(1) $\Leftrightarrow$ (2): The module X is finitely presented if and only if $\bar{X}$ is the injective envelope of a simple object in $\mathbf{P}(\mathcal{A})$, by Proposition 12.4.14, and this happens if and only if X is the source of a left almost split morphism, by Theorem 12.3.13.

(2) $\Leftrightarrow$ (3): Apply Corollary 12.3.15, using also the first part of the proof. $\square$

Corollary 12.4.16. *An Artin algebra of infinite representation type has an indecomposable pure-injective module of infinite length.*

Proof The space $\operatorname{Ind}\Lambda$ is quasi-compact by Corollary 12.3.12; so it cannot consist of infinitely many isolated points. $\square$

The Zariski Spectrum

Let Λ be a commutative noetherian ring. We consider the *Zariski spectrum* $\operatorname{Spec}\Lambda$ consisting of all prime ideals, where a subset is *Zariski closed* if it is of the form

$$\mathcal{V}(\mathfrak{a}) = \{\mathfrak{p} \in \operatorname{Spec}\Lambda \mid \mathfrak{a} \subseteq \mathfrak{p}\}$$

for some ideal $\mathfrak{a}$ of Λ. Recall that the assignment $\mathfrak{p} \mapsto E(\Lambda/\mathfrak{p})$ yields a bijection

$$\Phi \colon \operatorname{Spec}\Lambda \xrightarrow{\sim} \operatorname{Sp}(\operatorname{Mod}\Lambda)$$

onto the spectrum consisting of a representative set of the isomorphism classes of indecomposable injective Λ-modules (Corollary 2.4.15).

Let us compare via Φ the Zariski topology on $\operatorname{Spec}\Lambda$ with the topology on $\operatorname{Sp}(\operatorname{Mod}\Lambda)$ which is defined in Lemma 12.1.12.

Proposition 12.4.17. *For a subset $\mathcal{V} \subseteq \operatorname{Spec}\Lambda$ the following conditions are equivalent.*

(1) $\Phi(\mathcal{V})$ *is closed.*
(2) $\Phi(\mathcal{V})$ *is closed under products. If $X \subseteq \prod_{i \in I} Y_i$ for some indecomposable injective module X and a family of modules $Y_i \in \Phi(\mathcal{V})$, then $X \in \Phi(\mathcal{V})$.*
(3) $(\operatorname{Spec}\Lambda) \setminus \mathcal{V}$ *is specialisation closed.*
(4) $\mathcal{V} = \bigcap_{i \in I} \mathcal{U}_i$ *for a family of Zariski open subsets $\mathcal{U}_i \subseteq \operatorname{Spec}\Lambda$.*

Proof (1) $\Leftrightarrow$ (2): This follows from Lemma 12.1.14.

(1) $\Leftrightarrow$ (3): By definition, $\Phi(\mathcal{V})$ is closed if it is of the form $\mathcal{C}^{\perp}$ for some Serre subcategory $\mathcal{C} \subseteq \operatorname{mod}\Lambda$. Such a Serre subcategory corresponds to a specialisation closed subset of $\operatorname{Spec}\Lambda$ via $\mathcal{C} \mapsto \operatorname{Supp}\mathcal{C}$; see Proposition 2.4.8. Using the theory of associated primes, the map Φ identifies $(\operatorname{Spec}\Lambda) \setminus (\operatorname{Supp}\mathcal{C})$ with $\mathcal{C}^{\perp}$ by Corollary 2.4.16.

(3) $\Leftrightarrow$ (4): A subset of Spec Λ is specialisation closed if and only if it is the union of Zariski closed subsets. $\square$

Corollary 12.4.18. *The assignment* $\mathfrak{p} \mapsto E(\Lambda/\mathfrak{p})$ *gives a map* $\Phi\colon \operatorname{Spec}\Lambda \to$ Ind Λ *that identifies* Spec Λ *with a Ziegler closed subset of* Ind Λ. *For a subset* $\mathcal{V} \subseteq \operatorname{Spec}\Lambda$ *the following conditions are equivalent.*

(1) $\Phi(\mathcal{V})$ *is Ziegler closed.*
(2) $\Phi(\mathcal{V})$ *is closed under products. If* $X \subseteq \prod_{i \in I} Y_i$ *for some indecomposable injective module* X *and a family of modules* $Y_i \in \Phi(\mathcal{V})$, *then* $X \in \Phi(\mathcal{V})$.
(3) $(\operatorname{Spec}\Lambda) \setminus \mathcal{V}$ *is specialisation closed.*
(4) $\mathcal{V} = \bigcap_{i \in I} \mathcal{U}_i$ *for a family of Zariski open subsets* $\mathcal{U}_i \subseteq \operatorname{Spec}\Lambda$.

Proof The injective Λ-modules form a definable subcategory of Mod Λ since Λ is noetherian, by Corollary 12.3.18. Clearly, every injective module is pure-injective, and it follows that Φ identifies Spec Λ with a Ziegler closed subset of Ind Λ.

The second part of the assertion follows from Proposition 12.4.17 since the topology on Sp(Mod Λ) is the restriction of the Ziegler topology on Ind Λ. $\square$

Remark 12.4.19. For a commutative noetherian ring Λ, Zariski and Ziegler topology on Spec Λ are related via *Hochster duality* as follows. The prime ideal spectrum of any commutative ring with its Zariski topology is a *spectral space*. For a spectral space there is a dual topology with closed subsets of the form $\bigcap_{i \in I} \mathcal{U}_i$ for any family of quasi-compact and open subsets $\mathcal{U}_i$ (the Ziegler closed subsets). The dual space is again spectral, and its Hochster dual topology coincides with the original Zariski topology of Spec Λ.

Injective Cohomology Representations

Let G be a finite group and k a field. We consider modules over the group algebra kG and note that kG is a self-injective algebra. The *group cohomology*

$$R := H^*(G, k) := \operatorname{Ext}^*_{kG}(k, k)$$

is by definition the Ext-algebra of the trivial representation k; it is a graded commutative and noetherian k-algebra by a theorem of Golod, Venkov and Evens [29, Corollary 3.10.2]. We consider only graded modules over R. Let R^+ denote the unique maximal ideal consisting of positive degree elements and call an R-module *torsion* if each element is annihilated by some power of R^+. The torsion modules form a localising subcategory which is denoted by $\operatorname{Mod}_0 R$.

We extend the functor $H^*(G,-) = \text{Ext}^*_{kG}(k,-)$ from kG-modules to the category $\mathbf{K}(\text{Inj}\,kG)$ of complexes of injective kG-modules. Set

$$\text{Hom}^*_{\mathbf{K}(kG)}(X,Y) = \bigoplus_{n\in\mathbb{Z}} \text{Hom}_{\mathbf{K}(kG)}(X,\Sigma^n Y)$$

for each pair of complexes X,Y and

$$H^*(G,X) = \text{Hom}^*_{\mathbf{K}(kG)}(ik,X)$$

where ik denotes an injective resolution of the trival representation. Note that $\text{End}^*_{\mathbf{K}(kG)}(ik) \cong R$. The functor $H^*(G,-)$ is cohomological and preserves coproducts; it induces the following commutative diagram.

$$
\begin{array}{ccccc}
\text{Loc}(kG) & \rightarrowtail & \mathbf{K}(\text{Inj}\,kG) & \twoheadrightarrow & \mathbf{K}_{\text{ac}}(\text{Inj}\,kG) \\
\downarrow & & \downarrow{\scriptstyle H^*(G,-)} & & \downarrow \\
\text{Mod}_0\,R & \rightarrowtail & \text{Mod}\,R & \twoheadrightarrow & \text{Mod}\,R/\text{Mod}_0\,R
\end{array}
$$

The upper row of the diagram is taken from Proposition 9.1.10, where $\text{Loc}(kG)$ denotes the localising subcategory generated by kG, viewed as a complex concentrated in degree zero, and keeping in mind that kG is self-injective. Note that $H^*(G,X)$ is torsion for each $X \in \text{Loc}(kG)$, since $H^*(G,kG)$ is torsion (in fact just k in degree zero). Thus the diagram does commute.

We wish to explain that the functor $H^*(G,-)$ admits a partial adjoint when we restrict to injective R-modules.

Given a pair of R-modules M,N we write

$$\text{Hom}^*_R(M,N) = \bigoplus_{n\in\mathbb{Z}} \text{Hom}^n_R(M,N)$$

for the graded abelian group of R-linear morphisms $\phi\colon M \to N$ satisfying $\phi(M^i) \subseteq M^{i+n}$ for $\phi \in \text{Hom}^n_R(M,N)$ and $i,n \in \mathbb{Z}$. The module M is called *torsion free* if $\text{Hom}^*_R(-,M)$ vanishes on $\text{Mod}_0\,R$. Recall that the category of injective R-modules is closed under coproducts since R is noetherian.

Also, we use the triangle equivalence $Z^0\colon \mathbf{K}_{\text{ac}}(\text{Inj}\,kG) \xrightarrow{\sim} \text{StMod}\,kG$ and view it as an identification (Proposition 4.4.18). A quasi-inverse maps a kG-module X to a complete resolution tX.

Proposition 12.4.20. *There is a fully faithful functor $T\colon \text{Inj}\,R \to \mathbf{K}(\text{Inj}\,kG)$ with a natural isomorphism*

$$\text{Hom}^*_R(H^*(G,-),I) \cong \text{Hom}^*_{\mathbf{K}(kG)}(-,T(I))$$

for each $I \in \text{Inj}\,R$. The functor preserves products and coproducts. Moreover,

T restricts to a fully faithful functor

$$\{I \in \operatorname{Inj} R \mid I \text{ torsion free}\} \longrightarrow \mathbf{K}_{\mathrm{ac}}(\operatorname{Inj} kG) \xrightarrow{\ \sim\ } \operatorname{StMod} kG$$

and $T(I)$ is a Σ-pure-injective kG-module for each torsion free $I \in \operatorname{Inj} R$.

Proof The triangulated category $\mathbf{K}(\operatorname{Inj} kG)$ is compactly generated (Proposition 9.3.12) and therefore every cohomological functor $\mathbf{K}(\operatorname{Inj} kG)^{\mathrm{op}} \to \operatorname{Ab}$ that preserves coproducts is representable, by Brown's representability theorem (Theorem 3.4.16). This yields for each $I \in \operatorname{Inj} R$ an object $T(I)$ that represents $\operatorname{Hom}_R^*(H^*(G, -), I)$. Given $J \in \operatorname{Inj} R$ we compute

$$
\begin{aligned}
\operatorname{Hom}_{\mathbf{K}(kG)}^*(T(I), T(J)) &\cong \operatorname{Hom}_R^*(H^*(G, T(I)), J) \\
&\cong \operatorname{Hom}_R^*(\operatorname{Hom}_{\mathbf{K}(kG)}^*(ik, T(I)), J) \\
&\cong \operatorname{Hom}_R^*(\operatorname{Hom}_R^*(H^*(G, ik), I), J) \\
&\cong \operatorname{Hom}_R^*(\operatorname{Hom}_R^*(R, I), J) \\
&\cong \operatorname{Hom}_R^*(I, J).
\end{aligned}
$$

Thus T is fully faithful. Clearly, T preserves products. To show that it preserves coproducts consider for any family (I_α) of injective R-modules the canonical morphism $\phi \colon \coprod T(I_\alpha) \to T(\coprod I_\alpha)$. Apply $\operatorname{Hom}_{\mathbf{K}(kG)}^*(iX, -)$ where iX is the injective resolution of a finitely generated kG-module X. Then $\operatorname{Hom}_{\mathbf{K}(kG)}^*(iX, \phi)$ is an isomorphism since iX is compact, and it follows that ϕ is an isomorphism since the objects of the form iX generate $\mathbf{K}(\operatorname{Inj} kG)$ (Proposition 9.3.12). If $I \in \operatorname{Inj} R$ is torsion free, then

$$H^*T(I) \cong \operatorname{Hom}_{\mathbf{K}(kG)}^*(kG, T(I)) \cong \operatorname{Hom}_R^*(H^*(G, kG), I) = 0,$$

and therefore $T(I)$ is acyclic.

Next we use the identification $\mathbf{K}_{\mathrm{ac}}(\operatorname{Inj} kG) \xrightarrow{\sim} \operatorname{StMod} kG$ and show that $T(I)$ is a Σ-pure-injective kG-module for each torsion free $I \in \operatorname{Inj} R$. We apply the criterion of Theorem 12.3.4 and show that the canonical monomorphism $T(I)^{(\mathbb{N})} \to T(I)^{\mathbb{N}}$ splits in $\operatorname{Mod} kG$. Clearly, the canonical monomorphism $I^{(\mathbb{N})} \to I^{\mathbb{N}}$ splits since R is noetherian. The functor T preserves products and coproducts. Thus $T(I)^{(\mathbb{N})} \to T(I)^{\mathbb{N}}$ splits in $\mathbf{K}_{\mathrm{ac}}(\operatorname{Inj} kG)$ and therefore also in $\operatorname{StMod} kG$. It remains to apply the lemma below. $\square$

Lemma 12.4.21. *Let $\mathcal{A}$ be a Frobenius category. Then a monomorphism in $\mathcal{A}$ splits if and only if it splits in $\operatorname{St}\mathcal{A}$.* $\square$

Let $\mathfrak{p}$ be a non-maximal prime ideal in R and $n \in \mathbb{Z}$. Then $I_\mathfrak{p} = E(R/\mathfrak{p})$ and its twist $I_\mathfrak{p}(n)$ are indecomposable injective R-modules. We may assume that the corresponding kG-module $T(I_{\mathfrak{p}}(n))$ is indecomposable, by removing all non-zero injective summands.

For a kG-module X one defines its *Tate cohomology*

$$\hat{H}^n(G, X) := \widehat{\operatorname{Ext}}^n_{kG}(k, X) := H^n \operatorname{Hom}_{kG}(k, tX) \qquad (n \in \mathbb{Z})$$

and more generally

$$\widehat{\operatorname{Ext}}^n_{kG}(-, X) := H^n \operatorname{Hom}_{kG}(-, tX) \qquad (n \in \mathbb{Z})$$

where tX denotes a complete resolution of X (cf. Lemma 4.4.19). Note that $\hat{H}^*(G, X)$ is naturally an R-module via restriction along

$$\operatorname{Ext}^*_{kG}(k, k) \longrightarrow \widehat{\operatorname{Ext}}^*_{kG}(k, k).$$

Corollary 12.4.22. *Let $I \in \operatorname{Inj} R$ be torsion free. The kG-module $T(I)$ satisfies $\hat{H}^*(G, T(I)) \cong I$ and is uniquely determined (up to isomorphism in $\operatorname{StMod} kG$) by the isomorphism*

$$\operatorname{Hom}^*_R(\hat{H}^*(G, -), I) \cong \widehat{\operatorname{Ext}}^*_{kG}(-, T(I)).$$

Moreover, after removing all non-zero injective summands, $T(I)$ admits a unique decomposition into indecomposable modules of the form $T(I_{\mathfrak{p}}(n))$, with $\mathfrak{p}$ a prime ideal in R and $n \in \mathbb{Z}$.

Proof The first part follows from the defining isomorphism for $T(I)$. More precisely, taking a complete resolution tk of k we have

$$I \cong \operatorname{Hom}^*_R(H^*(G, tk), I) \cong \operatorname{Hom}^*_{\mathbf{K}(kG)}(tk, T(I)) \cong \hat{H}^*(G, T(I)),$$

where the first isomorphism is induced by $R = H^*(G, ik) \to H^*(G, tk)$ since $H^*(G, pk)$ is torsion, the second isomorphism defines $T(I)$, and the third isomorphism is from Lemma 4.4.19. Similarly, we have for $X \in \operatorname{StMod} kG$

$$\operatorname{Hom}^*_R(\hat{H}^*(G, X), I) \cong \operatorname{Hom}^*_R(H^*(G, tX), I)$$
$$\cong \operatorname{Hom}^*_{\mathbf{K}(kG)}(tX, T(I))$$
$$\cong \widehat{\operatorname{Ext}}^*_{kG}(X, T(I)),$$

and the uniqueness of $T(I)$ then follows from Yoneda's lemma.

The module $T(I)$ is Σ-pure-injective and therefore decomposes uniquely into indecomposables, by Theorem 12.3.4. Then one uses the description of the indecomposable injective R-modules via $\operatorname{Spec} R$ (Corollary 2.4.15). $\square$

Let us denote by $\operatorname{Proj} R$ the set of all homogeneous prime ideals of R except the maximal ideal consisting of positive degree elements.

Corollary 12.4.23. *Taking a prime ideal* $\mathfrak{p}$ *to* $T(I_{\mathfrak{p}})$ *yields an injective map*

$$\operatorname{Proj} H^*(G, k) \longrightarrow \operatorname{Ind} kG. \qquad \square$$

Corollary 12.4.24. *The modules of the form* $T(I)$ *(* $I \in \operatorname{Inj} H^*(G, k)$ *torsion free) form a definable subcategory of* $\operatorname{Mod} kG$.

Proof The torsion free injective $H^*(G, k)$-modules form a definable subcategory, because they are closed under products, coproducts, and direct summands, keeping in mind that $H^*(G, k)$ is noetherian (Corollary 12.3.18). Then this category equals $\operatorname{Add} I_0 = \operatorname{Prod} I_0$ for some product-complete module I_0. It follows that $T(I_0)$ is a product-complete kG-module, since T preserves products and coproducts. Thus the image of the functor T is definable, by Proposition 12.3.7. $\qquad \square$

Notes

The notion of a pure subgroup (Servanzuntergruppe) of an abelian group was introduced by Prüfer [163]. For modules over arbitrary rings the concept of purity is due to Cohn [53]. Pure-injective modules are also known as algebraically compact modules [119, 200]. It was shown by Kiełpiński [124] and independently by Stenström [196] that every module admits a pure-injective envelope. For the characterisation of Σ-pure-injective modules, see Gruson and Jensen [99], Zimmermann [204], and Zimmermann-Huisgen [205], building on work of Chase [49, 50], but also Garavaglia [86] in a model theoretic setting. The space of indecomposable pure-injective modules is known as the Ziegler spectrum because it was introduced by Ziegler in his work on the model theory of modules [203]. For an Artin algebra of infinite representation type the existence of a large indecomposable module was established by Auslander [10]. Product-complete modules were introduced in joint work with Saorín [134].

The study of pure-exactness via the embedding of a module category into a bigger Grothendieck category (the purity category) goes back to work of Gruson and Jensen [100]; see also Simson [192]. The systematic treatment of purity for locally finitely presented categories is due to Crawley-Boevey [58]. Definable subcategories were introduced by Crawley-Boevey [59] for module categories, and more generally for locally finitely presented categories in [126]. The related notion of an elementary subcategory and its connection with Ziegler closed subsets appear already in [203]. For the correspondence between definable subcategories and Ziegler closed subsets, see Herzog [111] and [125].

The free abelian category on a category was introduced by Freyd [75]. The characterisation of pure-injective modules via duality is taken from Auslander's work [11], where it arises from the description of morphisms determined by objects. Almost split morphisms were introduced by Auslander and Reiten [15]; for the connection with simple functors, see [11]. The characterisation of indecomposable pure-injectives which are the source of a left almost split morphism combines results from [10] and [57].

Any ring of finite representation type is known to be right and left pure-semisimple, by a result of Ringel and Tachikawa [175]. In fact, pure-semisimple rings were introduced by Simson, and no ring is known which is right pure-semisimple but not of finite representation type [191, 193].

For commutative rings the connection between the Zariski spectrum and the Ziegler spectrum was clarified by Prest [160] in terms of Hochster's duality for spectral spaces [113].

The construction of pure-injective modules for group algebras via group cohomology is taken from work with Benson [32]. For example these modules play a role in the study of local duality for representations of finite groups, and more generally of finite group schemes [31].

For more material and further references about infinite length modules and purity, see [133, 162].

13

Endofiniteness

Contents

13.1	**Endofinite Objects and Subadditive Functions**	**415**
	Subadditive Functions	415
	Endofinite Functors	416
	Endofinite Objects	418
	Finite Type	421
	Properties of Endofinite Objects	423
	Uniserial Categories	424
13.2	**Endofinite Modules**	**426**
	Properties of Endofinite Modules	426
	Examples of Endofinite Modules	427
	Finite Representation Type	429
	Noetherian Algebras	429
	Quasi-Frobenius Rings	430
	The Space of Indecomposables	431
	Duality	432
Notes		**433**

In this chapter we study a particular finiteness condition for objects in a locally finitely presented category. An object X is called endofinite if the morphisms from any finitely presented object form a finite length module over $\mathrm{End}(X)$. For example, a ring is of finite representation type if and only if all its modules are endofinite. We present a remarkable classification of endofinite objects in purely numerical terms, using subadditive functions on finitely presented objects. A basic idea is to identify an object X with an exact functor $\bar{X} : \mathcal{C} \to \mathrm{Ab}$ for an appropriate abelian category $\mathcal{C}$. Then X is endofinite if and only if the quotient $\mathcal{C}/\mathrm{Ker}\,\bar{X}$ is a length category. Thus the study of endofinite objects is equivalent to the study of exact functors from length categories to abelian groups. Of particular interest are indecomposable endofinite objects that are not finitely presented; often they represent families of finitely presented objects.

We illustrate this by looking at endofinite modules over Artin algebras. Other interesting examples arise from locally finitely presented categories such that the finitely presented objects form a uniserial category.

13.1 Endofinite Objects and Subadditive Functions

Let $\mathcal{A}$ be a locally finitely presented category and let $\mathrm{fp}\,\mathcal{A}$ be the full subcategory of finitely presented objects. In this section we study the notion of endofiniteness. An object $X \in \mathcal{A}$ is called *endofinite* if

$$\ell_{\mathrm{End}_{\mathcal{A}}(X)}(\mathrm{Hom}_{\mathcal{A}}(C, X)) < \infty \quad \text{for all} \quad C \in \mathrm{fp}\,\mathcal{A}.$$

We begin with a discussion of subadditive functions which are defined on additive categories with cokernels.

Subadditive Functions

Let $\mathcal{C}$ be an additive category with cokernels. A *subadditive function* $\chi \colon \mathcal{C} \to \mathbb{N}$ assigns to each object in $\mathcal{C}$ a non-negative integer such that

(SF1) $\chi(X \oplus Y) = \chi(X) + \chi(Y)$ for all $X, Y \in \mathcal{C}$, and
(SF2) $\chi(X) + \chi(Z) \geq \chi(Y)$ for each exact sequence $X \to Y \to Z \to 0$ in $\mathcal{C}$.

If the category $\mathcal{C}$ is abelian, then an *additive function* $\chi \colon \mathcal{C} \to \mathbb{N}$ assigns to each object in $\mathcal{C}$ a non-negative integer such that $\chi(X) + \chi(Z) = \chi(Y)$ for each exact sequence $0 \to X \to Y \to Z \to 0$.

The sum $\chi_1 + \chi_2$ of (sub)additive functions χ_1 and χ_2 is again (sub)additive. More generally, if $(\chi_i)_{i \in I}$ is a family of (sub)additive functions and if for any X in $\mathcal{C}$ the set $\{i \in I \mid \chi_i(X) \neq 0\}$ is finite, then we can define the *locally finite sum* $\sum_{i \in I} \chi_i$.

A (sub)additive function $\chi \neq 0$ is *irreducible* if χ cannot be written as a sum of two non-zero (sub)additive functions.

We give a quick proof of the following result using the localisation theory for abelian categories.

Lemma 13.1.1. *Let $\mathcal{C}$ be an abelian category. Every additive function $\mathcal{C} \to \mathbb{N}$ can be written uniquely as a locally finite sum of irreducible additive functions.*

Proof Fix an additive function $\chi \colon \mathcal{C} \to \mathbb{N}$. The objects X satisfying $\chi(X) = 0$ form a Serre subcategory of $\mathcal{C}$ which we denote by $\mathcal{S}_{\chi}$. The quotient category $\mathcal{C}/\mathcal{S}_{\chi}$ is an abelian length category since the length of each object X is bounded by $\chi(X)$. Let $\mathrm{Sp}\,\chi$ (the *spectrum* of χ) denote a representative set of simple objects in $\mathcal{C}/\mathcal{S}_\chi$. For each S in $\text{Sp}\,\chi$ let $\chi_S\colon \mathcal{C} \to \mathbb{N}$ denote the map sending X to the multiplicity of S in a composition series of X in $\mathcal{C}/\mathcal{S}_\chi$. Clearly, χ_S is irreducible and we have the expression

$$\chi = \sum_{S \in \text{Sp}\,\chi} \chi(S)\chi_S \qquad (13.1.2)$$

which is unique by the Jordan–Hölder theorem. □

Lemma 13.1.3. *Let $\mathcal{C}$ be an additive category with cokernels and let $h\colon \mathcal{C} \to \text{Fp}(\mathcal{C}, \text{Ab})$ be the Yoneda embedding. Then the assignment $\chi \mapsto \chi \circ h$ induces an additive bijection between*

(1) *additive functions $\text{Fp}(\mathcal{C}, \text{Ab}) \to \mathbb{N}$, and*
(2) *subadditive functions $\mathcal{C} \to \mathbb{N}$.*

Proof The inverse map sends $\chi\colon \mathcal{C} \to \mathbb{N}$ to the map $\hat{\chi}$ that takes $F \in \text{Fp}(\mathcal{C}, \text{Ab})$ with presentation

$$0 \to \text{Hom}_{\mathcal{C}}(Z, -) \to \text{Hom}_{\mathcal{C}}(Y, -) \to \text{Hom}_{\mathcal{C}}(X, -) \to F \to 0$$

to $\hat{\chi}(F) = \chi(X) - \chi(Y) + \chi(Z)$. □

Corollary 13.1.4. *Let $\mathcal{C}$ be an additive category with cokernels. Every subadditive function $\mathcal{C} \to \mathbb{N}$ can be written uniquely as a locally finite sum of irreducible subadditive functions.* □

Endofinite Functors

Let $\mathcal{C}$ be an essentially small abelian category. An exact functor $F\colon \mathcal{C}^{\text{op}} \to \text{Ab}$ is called *endofinite* if $F(X)$ has finite length as $\text{End}(F)$-module for each object X. An endofinite exact functor F induces an additive function

$$\chi_F\colon \mathcal{C} \longrightarrow \mathbb{N}, \qquad X \mapsto \ell_{\text{End}(F)}(F(X)).$$

We need the following elementary lemma.

Lemma 13.1.5. *Let $\mathcal{A}$ be an abelian category and E an injective object. Then*

$$\ell_{\text{End}_{\mathcal{A}}(E)}(\text{Hom}_{\mathcal{A}}(X, E)) \le \ell_{\mathcal{A}}(X) \quad \text{for} \quad X \in \mathcal{A};$$

equality holds provided that $\text{Hom}_{\mathcal{A}}(Y, E) = 0$ implies $Y = 0$ for all $Y \in \mathcal{A}$.

Proof Observe that $\ell_{\text{End}_{\mathcal{A}}(E)}(\text{Hom}_{\mathcal{A}}(X, E)) \le 1$ when X is simple. Now use induction on $\ell_{\mathcal{A}}(X)$. □

Lemma 13.1.6. *Let* $F: \mathcal{C}^{op} \to \text{Ab}$ *be an exact functor and let* $\mathcal{D} = \mathcal{C}/\mathcal{S}_F$, *where* $\mathcal{S}_F$ *denotes the Serre subcategory of objects* X *satisfying* $F(X) = 0$. *Then*

$$\ell_{\text{End}(F)}(F(X)) \geq \ell_{\mathcal{D}}(X) \quad \text{for} \quad X \in \mathcal{C};$$

equality holds when all objects in $\mathcal{D}$ *have finite length.*

Proof The first assertion is clear by induction on $\ell_{\mathcal{D}}(X)$. Now suppose that all objects in $\mathcal{D}$ have finite length. We consider the abelian category $\text{Lex}(\mathcal{D}^{op}, \text{Ab})$ of left exact functors $\mathcal{D}^{op} \to \text{Ab}$. The Yoneda functor

$$\mathcal{D} \longrightarrow \text{Lex}(\mathcal{D}^{op}, \text{Ab}), \quad X \mapsto h_X = \text{Hom}_{\mathcal{D}}(-, X) \qquad (13.1.7)$$

identifies $\mathcal{D}$ with the full subcategory of finite length objects. We write F as the composite $\mathcal{C}^{op} \twoheadrightarrow \mathcal{D}^{op} \xrightarrow{\bar{F}} \text{Ab}$ and note that $\text{End}(\bar{F}) \cong \text{End}(F)$. Then $\bar{F}$ is an injective object in $\text{Lex}(\mathcal{D}^{op}, \text{Ab})$ by Corollary 11.2.15. Using Lemma 13.1.5 we compute

$$\ell_{\text{End}(F)}(F(X)) = \ell_{\text{End}(\bar{F})}(\text{Hom}(h_X, \bar{F})) = \ell(h_X) = \ell_{\mathcal{D}}(X). \qquad \square$$

Proposition 13.1.8. *The assignment* $F \mapsto \chi_F$ *induces a bijection between the isomorphism classes of indecomposable endofinite exact functors* $\mathcal{C}^{op} \to \text{Ab}$ *and the irreducible additive functions* $\mathcal{C} \to \mathbb{N}$.

Proof We construct the inverse map. Let $\chi: \mathcal{C} \to \mathbb{N}$ be an irreducible additive function. Following the proof of Lemma 13.1.1, we consider the Serre subcategory $\mathcal{S}_\chi$ of $\mathcal{C}$ consisting of the objects X satisfying $\chi(X) = 0$. The quotient category $\mathcal{D} = \mathcal{C}/\mathcal{S}_\chi$ is a length category, and $\chi(X)$ equals the length of X in $\mathcal{D}$ for each object X, since χ is irreducible. Now consider the abelian category $\text{Lex}(\mathcal{D}^{op}, \text{Ab})$ of left exact functors $\mathcal{D}^{op} \to \text{Ab}$. The Yoneda functor (13.1.7) identifies $\mathcal{D}$ with the full subcategory of finite length objects. There is a unique simple object in $\text{Lex}(\mathcal{D}^{op}, \text{Ab})$ since χ is irreducible, and we denote by F an injective envelope. It follows that F is indecomposable, and the injectivity implies that F is exact. For each X in $\mathcal{D}$ we have

$$\ell_{\text{End}(F)}(F(X)) = \ell_{\text{End}(F)}(\text{Hom}(h_X, F)) = \ell_{\mathcal{D}}(X) = \chi(X)$$

by Lemma 13.1.5.

Let $F': \mathcal{C}^{op} \to \text{Ab}$ be the composite of F with the quotient functor $\mathcal{C} \to \mathcal{D}$ and observe that $\text{End}(F') \cong \text{End}(F)$. Then F' has the desired properties: it is indecomposable endofinite exact and $\chi_{F'} = \chi$.

It remains to show for an indecomposable endofinite exact functor $F: \mathcal{C}^{op} \to \text{Ab}$ that the function χ_F is irreducible. Set $\mathcal{D} = \mathcal{C}/\mathcal{S}_{\chi_F}$ and view F as an exact functor $\mathcal{D}^{op} \to \text{Ab}$. Note that $\text{Hom}(h_S, F) = F(S) \neq 0$ for each simple object

S in $\mathcal{D}$. The indecomposability of F implies that all simple objects in $\mathcal{D}$ are isomorphic, and the equation (13.1.2) then implies that χ is irreducible since for each simple object S

$$\chi_F(S) = \ell_{\text{End}(F)}(F(S)) = \ell_{\mathcal{D}}(S) = 1$$

by Lemma 13.1.6.   □

Endofinite Objects

Let $\mathcal{A}$ be a locally finitely presented category. We recall the embedding $\mathcal{A} \to \mathbf{P}(\mathcal{A})$ into the purity category that identifies an object $X \in \mathcal{A}$ with the exact functor $\bar{X} \colon \text{Ab}(\mathcal{A})^{\text{op}} \to \text{Ab}$. This yields the abelian category $\text{Ab}(X) = \text{Ab}(\mathcal{A})/\mathcal{S}_X$, where $\mathcal{S}_X = \text{Ker}\,\bar{X}$; it is a useful invariant, because endofiniteness of X is controlled by $\text{Ab}(X)$.

Proposition 13.1.9. *An object X in $\mathcal{A}$ is endofinite if and only if every object in* $\text{Ab}(X)$ *has finite length. In that case X is Σ-pure-injective and decomposes into a coproduct of indecomposable endofinite objects with local endomorphism rings.*

Proof We identify X with the exact functor $\bar{X} \colon \text{Ab}(\mathcal{A})^{\text{op}} \twoheadrightarrow \text{Ab}(X)^{\text{op}} \to \text{Ab}$. Using the Yoneda embedding

$$\text{fp}\,\mathcal{A} \longrightarrow \text{Ab}(\mathcal{A}), \quad C \mapsto h_C = \text{Hom}_{\text{fp}\,\mathcal{A}}(C, -)$$

we have $\text{Hom}_{\mathcal{A}}(C, X) = \bar{X}(h_C)$. Thus X is endofinite if and only if $\bar{X}$ is an endofinite functor. It follows from Lemma 13.1.6 that X is endofinite if and only if every object in $\text{Ab}(X)$ has finite length. The second part of the assertion then follows from Theorem 12.3.4.   □

Corollary 13.1.10. *An object X in $\mathcal{A}$ is endofinite if and only if the subgroups of finite definition of* $\text{Hom}(C, X)$ *satisfy the ascending and descending chain conditions for every $C \in \text{fp}\,\mathcal{A}$. In this case $\text{End}(X)$-submodules and subgroups of finite definition of* $\text{Hom}(C, X)$ *coincide.*

Proof The first part follows from the above proposition since the lattice of subgroups of finite definition of $\text{Hom}(C, X)$ identifies with the lattice of subobjects of $\text{Hom}(C, -)$ in $\text{Ab}(X)$ by Lemma 12.3.1. For the second part, see Lemma 12.3.5.   □

An endofinite object X gives rise to a subadditive function χ_X by setting

$$\chi_X(C) = \ell_{\text{End}_{\mathcal{A}}(X)}(\text{Hom}_{\mathcal{A}}(C, X)) \quad \text{for} \quad C \in \text{fp}\,\mathcal{A}.$$

Note that $\chi_X(C) = \ell_{\text{Ab}(X)}(h_C)$ by Lemma 13.1.6.

Theorem 13.1.11. *Let $\mathcal{A}$ be a locally finitely presented category.*

(1) *Any subadditive function* $\mathrm{fp}\,\mathcal{A} \to \mathbb{N}$ *can be written uniquely as a locally finite sum of irreducible subadditive functions.*

(2) *The assignment* $X \mapsto \chi_X$ *induces a bijection between the isomorphism classes of indecomposable endofinite objects in* $\mathcal{A}$ *and the irreducible subadditive functions* $\mathrm{fp}\,\mathcal{A} \to \mathbb{N}$.

(3) *Let* $X \in \mathcal{A}$ *be endofinite and* $(X_i)_{i \in I}$ *a representative set of indecomposable direct summands of X. Then* $\chi_X = \sum_{i \in I} \chi_{X_i}$.

Proof (1) This is Corollary 13.1.4.

(2) Following the proof of Proposition 13.1.9, endofinite objects in $\mathcal{A}$ correspond to endofinite exact functors $\mathrm{Ab}(\mathcal{A})^{\mathrm{op}} \to \mathrm{Ab}$. Thus the bijective correspondence $X \mapsto \chi_X$ between endofinite objects and subadditive functions follows from Proposition 13.1.8.

(3) We identify X with the induced exact functor $\mathrm{Ab}(X)^{\mathrm{op}} \to \mathrm{Ab}$. Then $\chi_X(C) = \ell_{\mathrm{Ab}(X)}(h_C)$ for all $C \in \mathrm{fp}\,\mathcal{A}$, by Lemma 13.1.6. Let $\mathrm{Sp}\,\chi_X$ denote a representative set of simple objects in $\mathrm{Ab}(X)$. For $S \in \mathrm{Sp}\,\chi_X$, consider the irreducible subadditive function $\chi_S \colon \mathrm{fp}\,\mathcal{A} \to \mathbb{N}$ that maps C to the multiplicity of S in a composition series of h_C in $\mathrm{Ab}(X)$. Then we have $\chi_S = \chi_{X_i}$ for a unique $i \in I$ by the first part of this proof, and therefore $\chi_X = \sum_S \chi_S = \sum_i \chi_{X_i}$ by the identity (13.1.2). $\square$

Remark 13.1.12. Let $X \in \mathcal{A}$ be endofinite. Then the isomorphism classes of indecomposable direct summands of X are in canonical bijection to the isomorphism classes of simple objects in $\mathrm{Ab}(X)$. If X_i corresponds to $S_i \in \mathrm{Ab}(X)$, then $\mathrm{End}(X_i)/J(\mathrm{End}(X_i)) \cong \mathrm{End}(S_i)$.

Proof The first assertion follows from the proof of part (3) of Theorem 13.1.11. Thus an indecomposable summand X_i of X identifies with an injective envelope of a simple object in $\mathrm{Lex}(\mathrm{Ab}(X)^{\mathrm{op}}, \mathrm{Ab})$; see also Proposition 13.1.8 and its proof. For an abelian category and a simple object T with injective envelope $E = E(T)$, we have $\mathrm{soc}\,E = T$ and the assignment $\phi \mapsto \phi|_{\mathrm{soc}\,E}$ yields a surjective homomorphism $\mathrm{End}(E) \to \mathrm{End}(T)$ with kernel $J(\mathrm{End}(E))$. $\square$

For an object X in $\mathcal{A}$ let $\mathrm{Add}\,X$ denote the full subcategory formed by all coproducts of copies of X and their direct summands.

For subadditive functions χ', χ we write $\chi' \leq \chi$ if $\chi - \chi'$ is a subadditive function.

Corollary 13.1.13. *Let* $X \in \mathcal{A}$ *be an endofinite object. Then* $\mathrm{Add}\,X$ *is a definable subcategory of* $\mathcal{A}$, *consisting of all endofinite objects* $Y \in \mathcal{A}$ *such that* $\chi_Y \leq \chi_X$.

Proof Recall that $\text{Ab}(X) = \text{Ab}(\mathcal{A})/\mathcal{S}_X$, where $\mathcal{S}_X = \text{Ker}\,\bar{X}$. Let $\mathcal{B}$ denote the smallest definable subcategory of $\mathcal{A}$ containing X; it identifies with the category $\text{Ex}(\text{Ab}(X)^{\text{op}}, \text{Ab})$ by Corollary 12.2.3. All objects in $\mathcal{B}$ are endofinite by Proposition 13.1.9. Let $(X_i)_{i \in I}$ be a representative set of indecomposable direct summands of X. Then it follows from Theorem 13.1.11 and its proof that the objects in $\mathcal{B}$ are precisely those of the form $Y = \coprod_{i \in I} Y_i$ with Y_i a coproduct of copies of X_i. In particular $\chi_Y \leq \chi_X$. Conversely, if $\chi_Y \leq \chi_X$ for some object $Y \in \mathcal{A}$, then $\mathcal{S}_X \subseteq \mathcal{S}_Y$. Thus $\bar{Y}\colon \text{Ab}(\mathcal{A})^{\text{op}} \to \text{Ab}$ factors through $\text{Ab}(\mathcal{A})^{\text{op}} \twoheadrightarrow \text{Ab}(X)^{\text{op}}$ and therefore $Y \in \mathcal{B}$. $\qquad\square$

Corollary 13.1.14. *For an indecomposable object $X \in \mathcal{A}$ the following are equivalent.*

(1) *The object X is endofinite.*
(2) *The coproducts of copies of X form a definable subcategory.*
(3) *Every product of copies of X is a coproduct of copies of X.*

Proof (1) $\Rightarrow$ (2): Apply Corollary 13.1.13.

(2) $\Rightarrow$ (3): This is clear, since a definable subcategory is closed under products.

(3) $\Rightarrow$ (1): The object X is Σ-pure-injective and every object in $\text{Ab}(X)$ is noetherian, by Theorem 12.3.4. In fact, this result also implies that $\{X\}$ is a Ziegler closed subset of $\text{Ind}\,\mathcal{A}$, by Theorem 12.2.2. Let $\text{Ab}(X)_0$ denote the full subcategory of finite length objects. If this is a proper subcategory, then $\text{Ab}(X)_0$ corresponds to a proper Ziegler closed subset of $\{X\}$, by Theorem 12.2.2. This is impossible, and therefore all objects in $\text{Ab}(X)$ have finite length. Thus X is endofinite by Proposition 13.1.9. $\qquad\square$

Corollary 13.1.15. *Let $X_1, \ldots, X_n$ be endofinite objects in $\mathcal{A}$. Then $X_1 \oplus \cdots \oplus X_n$ is endofinite.*

Proof Set $X = X_1 \oplus \cdots \oplus X_n$. Then we have ${}^{\perp}X = \bigcap_i {}^{\perp}X_i$ in $\text{Fp}(\text{fp}\,\mathcal{A}, \text{Ab})$. Thus if $\text{Ab}(X_i)$ is a length category for each i, then $\text{Ab}(X)$ is a length category, by the lemma below. Now the assertion is clear from the characterisation of endofiniteness of X via $\text{Ab}(X)$ in Proposition 13.1.9. $\qquad\square$

Lemma 13.1.16. *Let $\mathcal{C}$ be an abelian category and $\mathcal{C}_1, \ldots, \mathcal{C}_n$ Serre subcategories of $\mathcal{C}$. If each localisation $\mathcal{C}/\mathcal{C}_i$ is a length category, then $\mathcal{C}/\bigcap_i \mathcal{C}_i$ is a length category.*

Proof The product $\prod_i \mathcal{C}/\mathcal{C}_i$ is a length category since $\ell(X) = \sum_i \ell(X_i)$ for each object $X = (X_i)$. The kernel of the canonical functor $\mathcal{C} \to \prod_i \mathcal{C}/\mathcal{C}_i$ equals $\bigcap_i \mathcal{C}_i$. This yields a faithful and exact functor $\mathcal{C}/\bigcap_i \mathcal{C}_i \to \prod_i \mathcal{C}/\mathcal{C}_i$. Clearly, the length of each object in $\mathcal{C}/\bigcap_i \mathcal{C}_i$ is bounded by the length of its image in $\prod_i \mathcal{C}/\mathcal{C}_i$.   □

Finite Type

Let $\mathcal{A}$ be a locally finitely presented category. We wish to characterise the fact that all objects in $\mathcal{A}$ are endofinite. This requires a study of representable functors of finite length and we begin with some preparations.

Let $\mathcal{C}$ be an additive category. Let us call $\mathcal{C}$ *left Hom-finite* if for all objects X, Y in $\mathcal{C}$ the $\operatorname{End}(Y)$-module $\operatorname{Hom}(X, Y)$ has finite length. Clearly, this property implies that $\mathcal{C}$ is a Krull–Schmidt category, assuming that $\mathcal{C}$ is idempotent complete.

Lemma 13.1.17. *Let $\mathcal{C}$ be a Krull–Schmidt category. Then the following are equivalent.*

(1) *The category $\mathcal{C}$ is left Hom-finite.*
(2) *For all indecomposable objects X, Y the $\operatorname{End}(Y)$-module $\operatorname{Hom}(X, Y)$ has finite length.*
(3) *Every object in $\mathcal{C}$ has a left artinian endomorphism ring.*

Proof (1) $\Leftrightarrow$ (2): Fix a pair of objects X, Y in $\mathcal{C}$. First observe that for any decomposition $X = \bigoplus_i X_i$ we have

$$\ell_{\operatorname{End}(Y)}(\operatorname{Hom}(X, Y)) = \sum_i \ell_{\operatorname{End}(Y)}(\operatorname{Hom}(X_i, Y)).$$

Now fix a decomposition $Y = \bigoplus_j Y_j^{n_j}$, $n_j > 0$ such that the Y_j are indecomposable and pairwise non-isomorphic. Set $Y' = \bigoplus_j Y_j$. Then

$$\ell_{\operatorname{End}(Y)}(\operatorname{Hom}(X, Y)) = \ell_{\operatorname{End}(Y')}(\operatorname{Hom}(X, Y'))$$
$$= \sum_j \ell_{\operatorname{End}(Y_j)}(\operatorname{Hom}(X, Y_j))$$

since

$$\operatorname{End}(Y')/J(\operatorname{End}(Y')) \cong \prod_j \operatorname{End}(Y_j)/J(\operatorname{End}(Y_j)).$$

Now the assertion follows.

(1) $\Leftrightarrow$ (3): One direction is clear. So fix objects $X, Y \in \mathcal{C}$ and suppose that $\Lambda = \operatorname{End}(X \oplus Y)$ is left artinian. Thus $\ell(_\Lambda \Lambda)$ is finite. Let $e \in \Lambda$ be the idempotent given by projecting onto Y. Observe that $\ell_{e \Lambda e}(eM) \leq \ell_\Lambda(M)$ for every left Λ-module M. Now $\operatorname{Hom}(X, Y)$ is a direct summand of $e\Lambda = \operatorname{Hom}(X \oplus Y, Y)$ and has therefore finite length over $e\Lambda e = \operatorname{End}(Y)$.   □

Now suppose that $\mathcal{C}$ is a Krull–Schmidt category. Let ind $\mathcal{C}$ denote a representative set of the isoclasses of indecomposable objects. For an additive functor $F\colon \mathcal{C} \to \mathrm{Ab}$ we define its *support*

$$\mathrm{Supp}(F) = \{X \in \mathrm{ind}\,\mathcal{C} \mid FX \neq 0\}$$

and let $\ell(F)$ denote the composition length of F in $\mathrm{Add}(\mathcal{C}, \mathrm{Ab})$.

Lemma 13.1.18. *For an additive functor $F\colon \mathcal{C} \to \mathrm{Ab}$ we have*

$$\ell(F) = \sum_{X \in \mathrm{ind}\,\mathcal{C}} \ell_{\mathrm{End}(X)}(FX).$$

Proof   The assignment

$$F \mapsto \tilde{\ell}(F) := \sum_{X \in \mathrm{ind}\,\mathcal{C}} \ell_{\mathrm{End}(X)}(FX)$$

satisfies $\tilde{\ell}(F) = \tilde{\ell}(F') + \tilde{\ell}(F'')$ for every exact sequence $0 \to F' \to F \to F'' \to 0$, and $\tilde{\ell}(F) \neq 0$ for every $F \neq 0$. Thus $\ell(F) = \infty$ implies $\tilde{\ell}(F) = \infty$.

Now suppose that $\ell(F) < \infty$. If F is a simple functor and $FX \neq 0$ for some $X \in \mathrm{ind}\,\mathcal{C}$, then we have $\mathrm{Supp}(F) = \{X\}$ and $\ell(F) = 1 = \ell_{\mathrm{End}(X)}(FX)$. From this the assertion follows by induction on $\ell(F)$. $\square$

Remark 13.1.19.   Let $\mathcal{F} = \mathrm{Fp}(\mathcal{C}, \mathrm{Ab})$ be abelian and $F \in \mathcal{F}$. Consider the embedding $\mathcal{F} \to \bar{\mathcal{F}} = \mathrm{Add}(\mathcal{C}, \mathrm{Ab})$. Then $\ell_{\mathcal{F}}(F) = \ell_{\bar{\mathcal{F}}}(F)$ since the embedding is right exact and every simple object in $\mathcal{F}$ is simple in $\bar{\mathcal{F}}$.

Theorem 13.1.20. *For a locally finitely presented category $\mathcal{A}$ the following are equivalent.*

(1) *Every object in $\mathcal{A}$ is endofinite.*
(2) *The abelian category $\mathrm{Ab}(\mathcal{A})$ is a length category.*
(3) *For all $C \in \mathrm{fp}\,\mathcal{A}$ the functor $\mathrm{Hom}(C, -)\colon \mathrm{fp}\,\mathcal{A} \to \mathrm{Ab}$ has finite length.*
(4) *For all $C \in \mathrm{fp}\,\mathcal{A}$ the endomorphism ring $\mathrm{End}(C)$ is left artinian and there are, up to isomorphism, only finitely many indecomposable objects $D \in \mathrm{fp}\,\mathcal{A}$ such that $\mathrm{Hom}(C, D) \neq 0$.*

In this case each object in $\mathcal{A}$ decomposes into a coproduct of indecomposable finitely presented objects.

Proof   (1) $\Leftrightarrow$ (2): For every object $X \in \mathcal{A}$ we have the quotient $\mathrm{Ab}(\mathcal{A}) \twoheadrightarrow \mathrm{Ab}(X)$. Note that $\mathrm{Ab}(\mathcal{A}) \xrightarrow{\sim} \mathrm{Ab}(U)$ for $U = \prod_{X \in \mathrm{Ind}\,\mathcal{A}} X$; see Theorem 12.2.2. Now apply Proposition 13.1.9. Thus when U is endofinite, then $\mathrm{Ab}(\mathcal{A})$ is a length category. On the other hand, when $\mathrm{Ab}(\mathcal{A})$ is a length category, then $\mathrm{Ab}(X)$ is a length category for every $X \in \mathcal{A}$.

(2) $\Leftrightarrow$ (3): Clearly, $\mathrm{Ab}(\mathcal{A})$ is a length category if and only if for each $C \in \mathrm{fp}\,\mathcal{A}$ the representable functor $\mathrm{Hom}(C, -)$ has finite length, keeping in mind Remark 13.1.19.

(3) $\Leftrightarrow$ (4): We apply Lemma 13.1.17 and Lemma 13.1.18. Then each representable functor $\mathrm{fp}\,\mathcal{A} \to \mathrm{Ab}$ has finite length if and only if $\mathrm{fp}\,\mathcal{A}$ is left Hom-finite and the support of each representable functor is finite.

To prove the final assertion, observe that each endofinite object decomposes into a coproduct of indecomposables by Proposition 13.1.9. Thus it remains to show that each indecomposable $X \in \mathcal{A}$ is finitely presented. Let $X = \mathrm{colim}\, X_i$ be written as a filtered colimit of objects in $\mathrm{fp}\,\mathcal{A}$. There is a unique simple object $S \in \mathrm{fp}\,\mathbf{P}(\mathcal{A})$ such that $E(S) = \bar{X}$. The inclusion $S \to \bar{X}$ factors though $\bar{X}_i$, and we may assume that X_i is indecomposable. Thus $\bar{X}_i \cong E(S)$, and therefore $X \cong X_i$. $\qquad \square$

Example 13.1.21. Let Λ denote the Kronecker algebra and consider the subcategory $\mathcal{J} \subseteq \mathrm{mod}\,\Lambda$ of postinjective Kronecker representations (finite direct sums of indecomposable representations with dimension vector $(n + 1, n)$). Then $\vec{\mathcal{J}}$ satisfies the conditions of the above theorem.

Properties of Endofinite Objects

Let $\mathcal{A}$ be a locally finitely presented category. The indecomposable endofinite objects may be viewed as points of $\mathrm{Ind}\,\mathcal{A}$. These give rise to discrete subsets.

Proposition 13.1.22. *Let $X \in \mathcal{A}$ be an endofinite object. Then each subset $\mathcal{U} \subseteq \mathrm{Add}\, X \cap \mathrm{Ind}\,\mathcal{A}$ is a closed subset of $\mathrm{Ind}\,\mathcal{A}$.*

Proof It follows from Corollary 13.1.13 that the coproducts of copies of objects in $\mathcal{U}$ form a definable subcategory of $\mathcal{A}$. Thus $\mathcal{U}$ is a closed subset of $\mathrm{Ind}\,\mathcal{A}$ by Theorem 12.2.2. $\qquad \square$

We have the following immediate consequence.

Corollary 13.1.23. *Suppose that $\mathrm{Ind}\,\mathcal{A}$ is quasi-compact. If X is an endofinite object in $\mathcal{A}$, then the number of isomorphism classes of indecomposable direct summands of X is finite.*

Proof The indecomposable direct summands of X form a discrete space by the above proposition. This space is necessarily finite if it is quasi-compact. $\qquad \square$

Next we study the endomorphism ring of an endofinite object. Recall that for an object C in an abelian category the height $\mathrm{ht}(C)$ is bounded by its composition length $\ell(C)$.

Proposition 13.1.24. *Let X be an endofinite object and J the Jacobson radical of its endomorphism ring. Then $\bigcap_{n \geq 0} J^n = 0$. Moreover, for $n \geq 0$ we have*

$$J^n = 0 \quad \Longleftrightarrow \quad \mathrm{ht}(F) \leq n \text{ for all } F \in \mathrm{Ab}(X)$$
$$\Longleftrightarrow \quad \mathrm{ht}(\mathrm{Hom}(C, X)) \leq n \text{ for all } C \in \mathrm{fp}\, \mathcal{A}.$$

Proof As before we identify X with an exact functor $\bar{X}: \mathrm{Ab}(X)^{\mathrm{op}} \to \mathrm{Ab}$ in $\mathrm{Lex}(\mathrm{Ab}(X)^{\mathrm{op}}, \mathrm{Ab})$. This is a locally finite Grothendieck category and then the assertion follows from Proposition 11.2.8 and Remark 11.2.9. □

Corollary 13.1.25. *Suppose there exists an object $G \in \mathrm{fp}\, \mathcal{A}$ such that every object in $\mathrm{fp}\, \mathcal{A}$ is a quotient of G^n for some integer $n \geq 1$. Then for every endofinite object in $\mathcal{A}$ the Jacobson radical of its endomorphism ring is nilpotent.*

Proof Let $X \in \mathcal{A}$ be endofinite. For $C \in \mathrm{fp}\, \mathcal{A}$ an epimorphism $G^n \to C$ induces a monomorphism $\mathrm{Hom}(C, X) \to \mathrm{Hom}(G^n, X)$. Thus $\mathrm{ht}(\mathrm{Hom}(C, X)) \leq \mathrm{ht}(\mathrm{Hom}(G, X))$. □

Uniserial Categories

Let $\mathcal{A}$ be a locally finitely presented abelian category and set $\mathcal{C} = \mathrm{fp}\, \mathcal{A}$. Our aim is to describe all indecomposable objects in $\mathcal{A}$ when $\mathcal{C}$ is an abelian category which is uniserial. In particular, we see that all indecomposables are endofinite.

Fix an object $X \in \mathcal{A}$. The composition length of X is denoted by $\ell(X)$, and the height $\mathrm{ht}(X)$ is the smallest $n \geq 0$ such that $\mathrm{soc}^n(X) = X$.

Now suppose that every object in $\mathrm{fp}\, \mathcal{A}$ has finite length. Let $X = \mathrm{colim}\, X_i$ be written as a filtered colimit of finitely presented objects. Then $\mathrm{soc}^n(X) = \mathrm{colim}\,\mathrm{soc}^n(X_i)$ for all $n \geq 0$, and therefore $X = \bigcup_{n \geq 0} \mathrm{soc}^n(X)$.

Recall that $\mathcal{C}$ is *uniserial* if $\mathcal{C}$ is a length category and each indecomposable object has a unique composition series.

Lemma 13.1.26. *Let $\mathcal{C}$ be a length category. Then $\mathcal{C}$ uniserial if and only if $\mathrm{ht}(X) = \ell(X)$ for every indecomposable $X \in \mathcal{C}$.*

Proof Let $X \in \mathcal{C}$ be indecomposable. If $\mathrm{ht}(X) = \ell(X)$, then the socle series of X is the unique composition series of X.

Now assume $\mathrm{ht}(X) \neq \ell(X)$. Then there exists some $n \geq 0$ such that

$$\mathrm{soc}^{n+1}(X)/\mathrm{soc}^n(X) = S_1 \oplus \cdots \oplus S_r$$

with all S_i simple and $r > 1$. Choose n minimal and let $\mathrm{soc}^n(X) \subseteq U_i \subseteq X$ be given by $U_i/\mathrm{soc}^n(X) = S_i$. Then we have at least r different composition series

$$0 = \mathrm{soc}^0(X) \subseteq \mathrm{soc}^1(X) \subseteq \cdots \subseteq \mathrm{soc}^n(X) \subseteq U_i \subseteq \cdots \subseteq \mathrm{soc}^{n+1}(X) \subseteq \cdots$$

of X.   □

Lemma 13.1.27. *Let $\mathcal{C}$ be a uniserial category. Then $\mathcal{C}$ is left Hom-finite.*

Proof We apply Lemma 13.1.17 and need to show for all X, Y in $\mathcal{C}$ that the $\mathrm{End}(Y)$-module $\mathrm{Hom}(X, Y)$ has finite length, and it suffices to assume that Y is indecomposable. We claim that

$$\ell_{\mathrm{End}(Y)}(\mathrm{Hom}(X, Y)) \leq \ell(X).$$

Using induction on $\ell(X)$ the claim reduces to the case that X is simple. So let $S = \mathrm{soc}(Y)$ and write $E = E(Y)$ for its injective envelope. Note that $\mathrm{soc}^n(E) = Y$ for $n = \ell(Y)$, by Lemma 13.1.26. Thus any endomorphism $E \to E$ restricts to a morphism $Y \to Y$. Write $i : S \to Y$ for the inclusion. Then any morphism $j : S \to Y$ induces an endomorphism $f : E \to E$ such that $f|_Y \circ i = j$. Thus the $\mathrm{End}(Y)$-module $\mathrm{Hom}(S, Y)$ is cyclic, and it is annihilated by the radical of $\mathrm{End}(Y)$. Therefore $\mathrm{Hom}(S, Y)$ is simple.   □

Theorem 13.1.28. *Let $\mathcal{A}$ be a locally finitely presented category and suppose that $\mathrm{fp}\,\mathcal{A}$ is a uniserial category. Then every non-zero object in $\mathcal{A}$ has an indecomposable direct summand that is finitely presented or injective.*

Proof From Lemma 13.1.26 it follows that for every indecomposable injective object E in $\mathcal{A}$ the subobjects form a chain

$$0 = E_0 \subseteq E_1 \subseteq E_2 \subseteq \cdots$$

with $E_n = \mathrm{soc}^n(E)$ in $\mathcal{C}$ for all $n \geq 0$ and $E = \bigcup_{n \geq 0} E_n$. Note that $E = E_{\ell(E)}$ when $\ell(E) < \infty$.

Fix $X \neq 0$ in $\mathcal{A}$ and choose a simple subobject $S \subseteq X$. Let $U \subseteq X$ be a maximal subobject containing S such that $S \subseteq U$ is essential; this exists by Zorn's lemma. Then U is injective or belongs to $\mathcal{C}$. In the first case we are done. So assume $U \in \mathcal{C}$. We claim that the inclusion $U \to X$ is a pure monomorphism. To see this, choose a morphism $C \to X/U$ with $C \in \mathcal{C}$. This yields the following commutative diagram with exact rows.

$$
\begin{array}{ccccccccc}
0 & \longrightarrow & U & \longrightarrow & V & \longrightarrow & C & \longrightarrow & 0 \\
 & & \| & & \downarrow & & \downarrow & & \\
0 & \longrightarrow & U & \longrightarrow & X & \longrightarrow & X/U & \longrightarrow & 0
\end{array}
$$

Write $V = \bigoplus V_i$ as a direct sum of indecomposable objects. Then there exists an index i such that the composite $S \hookrightarrow U \to V_i \to X$ is non-zero. Thus $S \to V_i$ is essential and $V_i \to X$ is a monomorphism. It follows from the maximality of U that $U \to V_i$ is an isomorphism. Therefore the top row splits, and this yields the claim. It remains to observe that every object in $\mathcal{C}$ is pure-injective since $\mathcal{C}$ is left Hom-finite; see Proposition 13.1.9 and Lemma 13.1.27.   □

Corollary 13.1.29. *Every indecomposable object in $\mathcal{A}$ is endofinite.*   □

Corollary 13.1.30. *Every indecomposable object in $\mathcal{A}$ is the source of a left almost split morphism.*

Proof Let X be indecomposable. If X is injective, then $X \twoheadrightarrow X/\mathrm{soc}(X)$ is left almost split. If X is not injective, then X is finitely presented and there exists a monomorphism $X \to X'$ such that X' is indecomposable and $\ell(X') = \ell(X) + 1$. It is easily checked that $X \to X' \oplus X/\mathrm{soc}(X)$ is left almost split.   □

13.2 Endofinite Modules

Let Λ be a ring. We consider the category of Λ-modules and set $\mathcal{A} = \mathrm{Mod}\,\Lambda$. Note that $\mathcal{A}$ is locally finitely presented with $\mathrm{fp}\,\mathcal{A} = \mathrm{mod}\,\Lambda$. In this section we apply the general theory of endofinite objects and study the Λ-modules which are endofinite.

Properties of Endofinite Modules

For a Λ-module X with $\Gamma = \mathrm{End}_\Lambda(X)$ we denote by $\ell_\Lambda(X)$ its composition length and call $\mathrm{endol}_\Lambda(X) := \ell_\Gamma(X)$ the *endolength* of X. Clearly, $\mathrm{endol}_\Lambda(X) < \infty$ if and only if X is an endofinite object in $\mathrm{Mod}\,\Lambda$, since for any epimorphism $\Lambda^n \twoheadrightarrow C$ we have

$$\ell_\Gamma(\mathrm{Hom}_\Lambda(C, X)) \leq n \cdot \mathrm{endol}_\Lambda(X).$$

Given a bimodule $_\Sigma X_\Lambda$ we have $\mathrm{endol}_\Lambda(X) \leq \ell_\Sigma(X)$. Thus when Λ is a k-algebra over some commutative ring k, then a Λ-module of finite length over k is endofinite. In particular, when Λ is an Artin k-algebra we have

$$\mathrm{endol}_\Lambda(X) \leq \ell_k(X) \leq \ell_\Lambda(X) \cdot \ell_k(\Lambda/J(\Lambda)).$$

The following summarises the basic properties of endofinite modules.

Proposition 13.2.1. *Endofinite modules are Σ-pure-injective. The class of endofinite modules is closed under finite direct sums, and arbitrary products or coproducts of copies of one module. If $X' \subseteq X$ is a pure submodule of an endofinite module X, then X' is a direct summand and $\mathrm{endol}(X') \leq \mathrm{endol}(X)$.*

Proof Endofinite modules are Σ-pure-injective by Proposition 13.1.9. Being closed under finite direct sums follows from Corollary 13.1.15. Being closed under arbitrary (co)products of copies of one module follows from Corollary 13.1.13. The same result shows that endofinite modules are closed under pure submodules, because a definable subcategory is closed under pure subobjects, by Theorem 12.2.5. Also, we have for a pure submodule $X' \subseteq X$

$$\mathrm{endol}(X') = \chi_{X'}(\Lambda) \le \chi_X(\Lambda) = \mathrm{endol}(X)$$

by Corollary 13.1.13. $\square$

The next result establishes the decomposition of endofinite modules into indecomposables. For a cardinal α let $X^{(\alpha)}$ denote a coproduct of α copies of X.

Proposition 13.2.2. *A Λ-module X is endofinite if and only if there are pairwise non-isomorphic indecomposable endofinite Λ-modules $X_1, \ldots, X_n$ and cardinals $\alpha_i > 0$ such that $X = \bigoplus_{i=1}^{n} X_i^{(\alpha_i)}$. In that case*

$$\mathrm{endol}(X) = \sum_{i=1}^{n} \mathrm{endol}(X_i).$$

Proof Let X be endofinite. Then X decomposes into a coproduct of indecomposable modules because X is Σ-pure-injective; see Theorem 12.3.4. The number of isomorphism classes of indecomposable summands in such a decomposition is finite by Corollary 13.1.23, since $\mathrm{Ind}\,\Lambda$ is quasi-compact; see Corollary 12.3.12. The summation formula for $\mathrm{endol}_\Lambda(X)$ then follows from Theorem 13.1.11.

Now let $X_1, \ldots, X_n$ be indecomposable modules which are endofinite. Then $\bigoplus_{i=1}^{n} X_i^{(\alpha_i)}$ is endofinite for any choice of cardinals α_i, by Proposition 13.2.1. $\square$

Examples of Endofinite Modules

We collect various examples of endofinite modules.

Example 13.2.3. A ring Λ is right artinian if and only if every injective Λ-module is endofinite.

Proof Let I be an injective Λ-module and set $\Gamma = \mathrm{End}_\Lambda(I)$. Given a Λ-module X, we have $\ell_\Gamma(\mathrm{Hom}_\Lambda(X, I)) \le 1$ when X is simple, and therefore by induction

$$\ell_\Gamma(\mathrm{Hom}_\Lambda(X, I)) \le \ell_\Lambda(X),$$

with equality when I is an injective cogenerator. Now the assertion follows by setting $X = \Lambda$. □

Example 13.2.4. Let Λ be a commutative noetherian ring and $\mathfrak{p} \in \operatorname{Spec} \Lambda$. Then the injective envelope $E(\Lambda/\mathfrak{p})$ is endofinite if and only if $\mathfrak{p}$ is minimal.

Proof All injective Λ-modules decompose into indecomposables because Λ is noetherian. Also, we use the bijection $\mathfrak{q} \mapsto E(\Lambda/\mathfrak{q})$ between $\operatorname{Spec} \Lambda$ and the set of isoclasses of indecomposable injective Λ-modules (Corollary 2.4.15).

Now fix $\mathfrak{p} \in \operatorname{Spec} \Lambda$ and consider the specialisation closed set $\mathcal{V} = \{\mathfrak{q} \in \operatorname{Spec} \Lambda \mid \mathfrak{q} \not\subseteq \mathfrak{p}\}$. This yields a split torsion pair

$$(\{X \in \operatorname{Inj} A \mid \operatorname{Ass} X \subseteq \mathcal{V}\}, \{X \in \operatorname{Inj} A \mid \operatorname{Ass} X \cap \mathcal{V} = \varnothing\})$$

(Corollary 2.4.16). Then $\{X \in \operatorname{Inj} A \mid \operatorname{Ass} X \cap \mathcal{V} = \varnothing\}$ is closed under products and consists of coproducts of copies of $E(\Lambda/\mathfrak{p})$ if and only if $\mathfrak{p}$ is minimal. From this the assertion follows because of the characterisation of indecomposable endofinite objects in Corollary 13.1.14. □

The next example takes up the construction of pure-injective modules over group algebras from Proposition 12.4.20.

Example 13.2.5. Let G be a finite group and k a field. If $\mathfrak{p}$ is a prime ideal in $R = H^*(G, k)$ and $I_{\mathfrak{p}} = E(R/\mathfrak{p})$ the corresponding indecomposable injective R-module, then we may assume the corresponding kG-module $T(I_{\mathfrak{p}})$ to be indecomposable, by removing all non-zero injective summands.

If $\mathfrak{p}$ is a minimal prime, then $T(I_{\mathfrak{p}})$ is endofinite. This follows from the above Example 13.2.4 and the characterisation of indecomposable endofinite objects in Corollary 13.1.14, because the functor T preserves products and coproducts.

Example 13.2.6. The indecomposable endofinite Λ-modules X with $\operatorname{End}_\Lambda(X)$ a division ring correspond bijectively to a ring epimorphism $\phi\colon \Lambda \to \Gamma$ such that Γ is simple artinian. The correspondence sends X to $\Lambda \to \operatorname{End}_{\operatorname{End}_\Lambda(X)}(X)$ and ϕ to the restriction of the simple Γ-module.

Example 13.2.7. For a commutative ring Λ the modules of endolength one correspond bijectively to prime ideals, by taking $\mathfrak{p} \in \operatorname{Spec} \Lambda$ to the quotient field $Q(\Lambda/\mathfrak{p})$.

The following is an analogue of Example 12.3.6.

Example 13.2.8. Let Q be a quiver and k a commutative ring. If X is a k-linear representation such that X_i is a finite length k-module for each vertex $i \in Q_0$, then X is endofinite.

Example 13.2.9. Let Λ be a ring and X an indecomposable Σ-pure injective Λ-module. Then the Ziegler closure of $\{X\}$ contains an endofinite module; see Proposition 14.1.19.

Finite Representation Type

A right artinian ring Λ is said to be of *finite representation type* if the number of isomorphism classes of finitely presented indecomposable Λ-modules is finite.

Theorem 13.2.10. *For a ring Λ the following conditions are equivalent.*

(1) *The ring Λ is right artinian and of finite representation type.*
(2) *Every Λ-module is endofinite.*
(3) *Every object in $\mathrm{Ab}(\Lambda)$ is of finite length.*

In this case each Λ-module decomposes into a coproduct of finitely presented indecomposables.

Proof Most implications as well as the final assertion are direct consequences of Theorem 13.1.20.

(1) $\Rightarrow$ (3): See [9, Theorem 3.1].
(2) $\Leftrightarrow$ (3): See Theorem 13.1.20.
(2) $\Rightarrow$ (1): The ring Λ is right artinian, thanks to Example 13.2.3. From Theorem 13.1.20 it follows that for all $C \in \mathrm{mod}\,\Lambda$ there are, up to isomorphism, only finitely many indecomposable objects $D \in \mathrm{mod}\,\Lambda$ such that $\mathrm{Hom}(C, D) \neq 0$. Taking $C = \Lambda_\Lambda$, it follows that Λ is of finite representation type. $\square$

Note that the finite representation type is left-right symmetric because of the duality $\mathrm{Ab}(\Lambda)^{\mathrm{op}} \xrightarrow{\sim} \mathrm{Ab}(\Lambda^{\mathrm{op}})$.

Noetherian Algebras

Let Λ be a noetherian k-algebra over a commutative ring k. In this case there is a natural finiteness condition for an endofinite Λ-module.

Lemma 13.2.11. *An endofinite Λ-module is noetherian if and only if it is artinian.*

Proof Let X be an endofinite Λ-module and set $\Gamma = \mathrm{End}_\Lambda(X)$. Suppose first that X_Λ is noetherian. Then Γ is a noetherian k-algebra and therefore any finite length Γ-module is of finite length over k. It follows that X is of finite length over Λ.

Suppose that X_Λ is artinian. Then the socle series of X_Λ is finite since each term is a Γ-submodule of X. It remains to note that a semisimple artinian module has finite length. $\qquad\qquad\square$

The following is an analogue of Theorem 12.4.15 for Artin algebras.

Theorem 13.2.12. *For an indecomposable endofinite Λ-module X the following are equivalent.*

(1) *X is finitely presented.*
(2) *X is of finite length.*
(3) *X is the source of a left almost split morphism.*
(4) *X is isolated.*

Proof (1) $\Rightarrow$ (2): Follows from Lemma 13.2.11.

(2) $\Rightarrow$ (3): Adapt the proof of Theorem 12.4.15.

(3) $\Rightarrow$ (2): It follows from Theorem 12.3.13 that $\bar{X}$ is the injective envelope of a simple object $C \in \mathbf{P}(\operatorname{Mod}\Lambda)$. The Λ-module X is a direct summand of $\prod_{i\in I} DY_i$ for some collection of finitely presented left Λ-modules Y_i, where $D = \operatorname{Hom}_k(-,E)$ denotes the Matlis duality for k-modules; see Corollary 12.4.8. We have $\operatorname{Hom}(C, \bar{X}_{i_0}) \neq 0$ for some $i_0 \in I$ where $X_i = DY_i$. Then it follows that X is a direct summand of DY_{i_0}. We have $E = \coprod_S E(S)$, where S runs though the simple k-modules, and therefore $DY = \coprod_s \operatorname{Hom}_k(Y, E(S))$ when Y is finitely generated. It follows as before that X is a direct summand of $\operatorname{Hom}_k(Y_{i_0}, E(S_0))$ for some simple S_0 and therefore artinian over k, since the k-module $E(S_0)$ is artinian; cf. Lemma 2.4.19. Then the above Lemma 13.2.11 implies that X_Λ is of finite length.

(2) $\Rightarrow$ (1): Clear.

(3) $\Leftrightarrow$ (4): Apply Corollary 12.3.15. $\qquad\qquad\square$

The above theorem suggests that we can single out the indecomposable endofinite modules which have infinite length; these are called *generic*.

Quasi-Frobenius Rings

Right artinian rings are characterised by the fact that all projective and all injective modules are endofinite. The following theorem describes the right artinian rings such that both classes of modules coincide.

Theorem 13.2.13. *For a ring Λ the following conditions are equivalent.*

(1) *Projective and injective Λ-modules coincide.*
(2) *The category $\operatorname{Mod}\Lambda$ of Λ-modules is a Frobenius category.*
(3) *The ring Λ is right noetherian and the module Λ_Λ is injective.*

(4) *The ring Λ is right artinian and* mod Λ *is a Frobenius category.*

Proof (1) $\Rightarrow$ (2): This is clear, since Mod Λ is a category with enough projective and enough injective objects

(2) $\Rightarrow$ (3): The category of injective modules is closed under coproducts, and therefore Λ is a right noetherian ring (Theorem 11.2.12).

(3) $\Rightarrow$ (4): The module Λ_Λ decomposes into finitely many indecomposables with local endomorphism rings. Let $P_1, \ldots, P_n$ represent the isomorphism classes of indecomposable summands, and set $S_i = P_i/\mathrm{rad}\, P_i$. Then $S_1, \ldots, S_n$ represent the isomorphism classes of simple Λ-modules, and $P_1, \ldots, P_n$ represent the isomorphism classes of their injective envelopes. Here one uses that the P_i are injective. For any i, any product of copies of P_i decomposes into indecomposables, and the socle of each summand is isomorphic to soc P_i. Thus the characterisation of indecomposable endofinite objects in Corollary 13.1.14 implies that P_i is endofinite. But then Λ is endofinite, and therefore all injective Λ-modules are endofinite. This implies that Λ is right artinian, by Example 13.2.3. It is clear that mod Λ has enough projective and enough injective objects. Moreover, the indecomposable projectives and injectives coincide, and therefore all projectives and injectives in mod Λ coincide, by the Krull–Remak–Schmidt theorem. Thus mod Λ is a Frobenius category.

(4) $\Rightarrow$ (1): Since Λ is right artinian, all projective and all injective Λ-modules are direct sums of indecomposable modules. These indecomposables belong to mod Λ, and therefore projective and injective Λ-modules coincide. $\quad\square$

The Space of Indecomposables

We study indecomposable endofinite Λ-modules as points of the spectrum Ind Λ. We use the embedding Mod $\Lambda \to \mathbf{P}(\mathrm{Mod}\,\Lambda)$ and identify each Λ-module X with the corresponding exact functor $\bar{X} \colon \mathrm{Ab}(\Lambda)^{\mathrm{op}} \to \mathrm{Ab}$; see Corollary 12.4.5.

Lemma 13.2.14. *The endolength of a Λ-module X equals the length of the object $F = \mathrm{Hom}_\Lambda(\Lambda, -)$ in $\mathrm{Ab}(X)$. For an integer $n \geq 0$, we have* $\mathrm{endol}(X) \leq n$ *if and only if for every chain of subobjects*

$$0 = F_{n+1} \subseteq \cdots \subseteq F_1 \subseteq F_0 = \mathrm{Hom}_\Lambda(\Lambda, -)$$

in $\mathrm{Ab}(\Lambda)$ *we have* $\bar{X}(F_i/F_{i+1}) = 0$ *for some* $0 \leq i \leq n$.

Proof Observe that $\bar{X}(F) = X$. Thus the proof of Proposition 13.1.8 shows

$$\mathrm{endol}(X) = \ell_{\mathrm{End}(\bar{X})}(\bar{X}(F)) = \ell_{\mathrm{Ab}(X)}(F).$$

Every chain of subobjects

$$0 = F_{n+1} \subseteq \cdots \subseteq F_1 \subseteq F_0 = F$$

in $\mathrm{Ab}(X)$ is isomorphic to the image of a chain of subobjects

$$0 = F_{n+1} \subseteq \cdots \subseteq F_1 \subseteq F_0 = F$$

in $\mathrm{Ab}(\Lambda)$ under the canonical functor $\mathrm{Ab}(\Lambda) \twoheadrightarrow \mathrm{Ab}(X)$. Clearly, $\ell_{\mathrm{Ab}(X)}(F) \leq n$ if and only if for every such chain $F_i/F_{i+1} = 0$ in $\mathrm{Ab}(X)$ for some $0 \leq i \leq n$. It remains to observe that $F_i/F_{i+1} = 0$ in $\mathrm{Ab}(X)$ if and only if $\bar{X}(F_i/F_{i+1}) = 0$.   □

Proposition 13.2.15. *Let Λ be a ring. Then $\mathcal{U}_n = \{X \in \mathrm{Ind}\,\Lambda \mid \mathrm{endol}(X) \leq n\}$ is a closed subset of* $\mathrm{Ind}\,\Lambda$ *for every integer $n \geq 0$.*

Proof   For a chain $\phi = (F_i)_{0 \leq i \leq n+1}$ of subobjects

$$0 = F_{n+1} \subseteq \cdots \subseteq F_1 \subseteq F_0 = \mathrm{Hom}_\Lambda(\Lambda, -)$$

in $\mathrm{Ab}(\Lambda)$ we set $\mathcal{U}_{\phi,i} = \{X \in \mathrm{Ind}\,\Lambda \mid \bar{X}(F_i/F_{i+1}) = 0\}$ and $\mathcal{U}_\phi = \bigcup_{i=0}^n \mathcal{U}_{\phi,i}$. These are closed subsets of $\mathrm{Ind}\,\Lambda$, and therefore the intersection $\mathcal{U} = \bigcap_\phi \mathcal{U}_\phi$ is also closed, where ϕ runs through all chains $\phi = (F_i)_{0 \leq i \leq n+1}$. It remains to observe that $\mathcal{U} = \mathcal{U}_n$ by the preceding lemma.   □

A compactness argument provides for Artin algebras the existence of inde-composable endofinite modules which are of infinite length. Such modules are called *generic*.

Corollary 13.2.16. *Let Λ be an Artin algebra and set* $\mathrm{ind}\,\Lambda = \mathrm{Ind}\,\Lambda \cap \mathrm{mod}\,\Lambda$. *If a subset of $\{X \in \mathrm{ind}\,\Lambda \mid \mathrm{endol}(X) \leq n\}$ is infinite for some fixed $n \in \mathbb{N}$, then its closure contains a point $Y \in \mathrm{Ind}\,\Lambda \setminus \mathrm{ind}\,\Lambda$ with $\mathrm{endol}(Y) \leq n$.*

Proof   The space $\mathcal{U}_n$ is quasi-compact since $\mathrm{Ind}\,\Lambda$ is quasi-compact; see Corollary 12.3.12. On the other hand, $\mathcal{U}_n \cap \mathrm{ind}\,\Lambda$ is discrete by Theorem 12.4.15. Thus $\mathcal{U}_n \subseteq \mathrm{ind}\,\Lambda$ implies that $\mathcal{U}_n$ is finite.   □

Generic modules also arise from Prüfer modules; see Example 14.1.21.

Duality

Let Λ be a ring. There is a bijective correspondence between indecomposable endofinite right and left Λ-modules. Recall that a Λ-module X determines the following Serre subcategory of $\mathrm{Ab}(\Lambda)$

$$^\perp X = \{F \in \mathrm{Ab}(\Lambda) \mid \mathrm{Hom}(F^\vee, X \otimes_\Lambda -) = 0\}.$$

For a module X we set $\Delta(X) = \mathrm{End}(X)/J(\mathrm{End}(X))$.

Lemma 13.2.17. *Let $S \subseteq \mathrm{Ab}(\Lambda)$ be a Serre subcategory such that $\mathrm{Ab}(\Lambda)/S$ is a length category with a unique simple object S. Then there is up to isomorphism a unique indecomposable Λ-module X with $^{\perp}X = S$. Moreover, $\mathrm{endol}(X)$ equals the length of $\mathrm{Hom}_{\Lambda}(\Lambda, -)$ in $\mathrm{Ab}(\Lambda)/S$, and $\Delta(X) \cong \mathrm{End}(S)$.*

Proof Let $S^{\perp}$ denote the definable subcategory corresponding to S. Then $X \in S^{\perp}$ is endofinite if $\mathrm{Ab}(\Lambda)/S$ is a length category, by Proposition 13.1.9. The assertion about $\mathrm{endol}(X)$ then follows from Lemma 13.2.14, and for the rest see Remark 13.1.12. □

Theorem 13.2.18. *Let Λ be a ring. There is a bijection $X \mapsto DX$ between the isomorphism classes of indecomposable endofinite right and left Λ-modules. It is determined by any of the following conditions.*

(1) $^{\perp}DX = (^{\perp}X)^{\vee}$.
(2) *Let Γ be a ring such that $_{\Gamma}X_{\Lambda}$ is a bimodule. If I is an injective Γ-module, then $\mathrm{Hom}_{\Gamma}(X, I)$ is a coproduct of copies of DX.*

Moreover, we have $D^2X \cong X$, $\mathrm{endol}(DX) = \mathrm{endol}(X)$, and $\Delta(DX) \cong \Delta(X)^{\mathrm{op}}$.

Proof We apply the equivalence $\mathrm{Ab}(\Lambda)^{\mathrm{op}} \xrightarrow{\sim} \mathrm{Ab}(\Lambda^{\mathrm{op}})$ and combine this with Lemma 13.2.17. Observe for a Serre subcategory $S \subseteq \mathrm{Ab}(\Lambda)$ that we have an induced equivalence

$$(\mathrm{Ab}(\Lambda)/S)^{\mathrm{op}} \xrightarrow{\sim} \mathrm{Ab}(\Lambda^{\mathrm{op}})/S^{\vee}.$$

If $S = {}^{\perp}X$ for a Λ-module X, then X is endofinite if and only if $\mathrm{Ab}(\Lambda)/S$ is a length category, by Proposition 13.1.9. If X is indecomposable and endofinite, then DX is given by $S^{\perp}$, or as a direct summand of $\mathrm{Hom}_{\Gamma}(X, I)$, see Proposition 12.4.10. Clearly, $D^2X \cong X$ since $S^{\vee\vee} = S$, and the rest follows from Lemma 13.2.17. □

Example 13.2.19. Let Λ be a noetherian k-algebra and $X \mapsto \mathrm{Hom}_k(X, E)$ the Matlis duality over k. For an indecomposable endofinite Λ-module X of finite length we have $DX = \mathrm{Hom}_k(X, E)$.

Notes

Modules of finite endolength were introduced by Crawley-Boevey [56, 57]. Of particular interest are generic modules, that is, the indecomposable endofinite modules that are not of finite length; they can be used to describe the representation type of an algebra, because generic modules parametrise families of finite dimensional representations [56]. In the more general context of locally finitely presented categories, endofinite objects were introduced and studied in [58]. This work contains the classification in terms of subadditive functions. The existence of generic modules over Artin algebras is closely related to the 'strongly unbounded representation type'. This yields a link to the second Brauer–Thrall conjecture [117] which is explained in [57]. The existence proof given here uses the compactness of the Ziegler spectrum and follows Herzog [111]. The characterisation of indecomposable endofinite modules of finite length over noetherian algebras is taken from [57]. The duality between indecomposable endofinite right and left modules is due to Herzog [110]. The study of quasi-Frobenius and self-injective rings goes back to Eilenberg and Nakayama [73], generalising work of Brauer and Nesbitt for finite dimensional algebras [40].

14

Krull–Gabriel Dimension

Contents

14.1	**The Krull–Gabriel Filtration**	**435**
	Filtrations of Abelian Categories	436
	The Lattice of Subobjects	438
	Example: Commutative Noetherian Rings	441
	Pure-Injective Objects	442
	Endofinite Objects	443
14.2	**Examples of Krull–Gabriel Filtrations**	**444**
	Uniserial Categories	444
	Dedekind Domains	445
	The Projective Line	449
	The Kronecker Quiver	452
	Injective Cohomology Representations	455
Notes		**456**

The Krull–Gabriel dimension is an invariant which is defined for any essentially small abelian category. The definition is based on a filtration which can be used to describe injective objects of a locally finitely presented Grothendieck category. In fact, we use this technique to classify all pure-injective objects for some interesting examples, including sheaves on the projective line and representations of the Kronecker quiver.

14.1 The Krull–Gabriel Filtration

In this section we introduce the Krull–Gabriel filtration of an essentially small abelian category. This filtration is then used to classify all injective objects in a locally finitely presented Grothendieck category, provided its category of finitely presented objects is abelian and the Krull–Gabriel dimension is defined.

This is done in two steps: first the indecomposable objects, and then the general case. There is an analogue of this filtration for modular lattices, which we apply to the lattice of subobjects of an object of an abelian category. From the classification of injective objects we deduce the classification of pure-injective objects of a locally finitely presented category. In particular, the last layer of the Krull–Gabriel filtration produces endofinite objects.

Filtrations of Abelian Categories

We begin with a brief discussion of ordinals. For historical reasons let -1 denote the ordinal possessing no element, and $n - 1$ denotes the ordinal possessing a finite number of n elements. The sum of ordinals α and β (first α and then β) is denoted by $\alpha \perp \beta$. For example, we have $\alpha \perp \beta = \alpha + \beta + 1$ when α and β are finite.

Let $\mathcal{C}$ be an essentially small abelian category. We set $\mathcal{A} = \text{Lex}(\mathcal{C}^{\text{op}}, \text{Ab})$ and recall that $\mathcal{A}$ is a locally finitely presented Grothendieck category with $\mathcal{C} \xrightarrow{\sim} \text{fp} \, \mathcal{A}$.

The *Krull–Gabriel filtration* of $\mathcal{C}$ is given by a sequence of Serre subcategories which is indexed by the ordinals $\alpha \geq -1$.

- $\mathcal{C}_{-1}$ is the full subcategory containing only the zero objects.
- $\mathcal{C}_\alpha$ is the full subcategory of objects that become of finite length in $\mathcal{C}/\mathcal{C}_\beta$, if $\alpha = \beta + 1$.
- $\mathcal{C}_\alpha = \bigcup_{\gamma < \alpha} \mathcal{C}_\gamma$, if α is a limit ordinal.

We set $\mathcal{C}_\infty = \bigcup_\alpha \mathcal{C}_\alpha$. If $\mathcal{C}_\infty = \mathcal{C}$, then the smallest ordinal α such that $\mathcal{C} = \mathcal{C}_\alpha$ is called the *Krull–Gabriel dimension* and is denoted KG.dim $\mathcal{C}$. We write KG.dim $\mathcal{C} < \infty$ in this case, and we say that the Krull–Gabriel dimension is defined.

We collect some elementary properties of this dimension.

Lemma 14.1.1. *Let $\mathcal{C}' \subseteq \mathcal{C}$ be a Serre subcategory and set $\mathcal{C}'' = \mathcal{C}/\mathcal{C}'$. Then*

$$\sup(\text{KG.dim} \, \mathcal{C}', \text{KG.dim} \, \mathcal{C}'') \leq \text{KG.dim} \, \mathcal{C} \leq \text{KG.dim} \, \mathcal{C}' \perp \text{KG.dim} \, \mathcal{C}''.$$

Proof First observe that $\mathcal{C}'_\alpha = \mathcal{C}_\alpha \cap \mathcal{C}'$ for every ordinal α; see Lemma 14.1.10 for a detailed proof. Also, the canonical functor $\mathcal{C} \twoheadrightarrow \mathcal{C}''$ maps $\mathcal{C}_\alpha$ into $\mathcal{C}''_\alpha$. Thus $\mathcal{C}_\alpha = \mathcal{C}$ implies $\mathcal{C}'_\alpha = \mathcal{C}'$ and $\mathcal{C}''_\alpha = \mathcal{C}''$. This yields the first inequality. For the second one suppose that $\mathcal{C}' \subseteq \mathcal{C}_\alpha$ for some ordinal α. Then $\mathcal{C} \twoheadrightarrow \mathcal{C}/\mathcal{C}_\alpha$ induces a functor $\mathcal{C}/\mathcal{C}' \twoheadrightarrow \mathcal{C}/\mathcal{C}_\alpha$ which maps $(\mathcal{C}/\mathcal{C}')_\beta$ into $(\mathcal{C}/\mathcal{C}_\alpha)_\beta$ for every ordinal β. The functor $\mathcal{C} \twoheadrightarrow \mathcal{C}/\mathcal{C}_\alpha$ identifies $\mathcal{C}_{\alpha \perp \beta}$ with $(\mathcal{C}/\mathcal{C}_\alpha)_\beta$. Thus $(\mathcal{C}/\mathcal{C}')_\beta = \mathcal{C}/\mathcal{C}'$ implies $\mathcal{C}_{\alpha \perp \beta} = \mathcal{C}$. $\square$

Lemma 14.1.2. *Let* $\mathcal{C}$ *be finitely generated as an abelian category and suppose* KG.dim $\mathcal{C} < \infty$. *Then* KG.dim $\mathcal{C} = \beta + 1$ *for some ordinal* β.

Proof Suppose that X generates $\mathcal{C}$ as an abelian category, i.e. there are no proper Serre subcategories containing X. Let α be a limit ordinal. If $\mathcal{C} = \bigcup_{\gamma < \alpha} \mathcal{C}_\gamma$, then $X \in \mathcal{C}_\gamma$ for some $\gamma < \alpha$ and therefore KG.dim $\mathcal{A} < \alpha$. $\square$

Next we consider indecomposable injective objects in $\mathcal{A}$. Recall that $\mathrm{Sp}\,\mathcal{A}$ denotes a representative set of the isomorphism classes of indecomposable injective objects. For each ordinal α, let $\mathrm{Sp}_\alpha\,\mathcal{A}$ denote the set of functors $F \in \mathrm{Sp}\,\mathcal{A}$ such that $F(\mathcal{C}_\alpha) = 0$ and $F(X) \neq 0$ for some object X which is simple in $\mathcal{C}/\mathcal{C}_\alpha$.

Lemma 14.1.3. *The relation* $F(X) \neq 0$ *yields a bijection between the isomorphism classes of simple objects* $X \in \mathcal{C}/\mathcal{C}_\alpha$ *and the elements* $F \in \mathrm{Sp}_\alpha\,\mathcal{A}$.

Proof An object $F \in \mathrm{Sp}\,\mathcal{A}$ is an exact functor, and $F(\mathcal{C}_\alpha) = 0$ implies that F can be viewed as an object in $\mathrm{Lex}((\mathcal{C}/\mathcal{C}_\alpha)^{\mathrm{op}}, \mathrm{Ab})$. If $X \in \mathcal{C}/\mathcal{C}_\alpha$ is simple, then the corresponding representable functor is simple in $\mathrm{Lex}((\mathcal{C}/\mathcal{C}_\alpha)^{\mathrm{op}}, \mathrm{Ab})$, and $F(X) \neq 0$ implies that F is an injective envelope; see Proposition 11.1.31 and the subsequent remark. It remains to observe that non-isomorphic simples have non-isomorphic injective envelopes. $\square$

Proposition 14.1.4. *Suppose that* KG.dim $\mathcal{C} = \kappa$. *Then* $\mathrm{Sp}\,\mathcal{A}$ *equals the disjoint union* $\bigsqcup_{\alpha < \kappa} \mathrm{Sp}_\alpha\,\mathcal{A}$. *Moreover, for each ordinal* $\alpha < \kappa$ *there is a bijection between* $\mathrm{Sp}_\alpha\,\mathcal{A}$ *and the set of isomorphism classes of simple objects in* $\mathcal{C}/\mathcal{C}_\alpha$.

Proof Let $F \in \mathrm{Sp}\,\mathcal{A}$ and choose β minimal such that $F(\mathcal{C}_\beta) \neq 0$. Then β is not a limit ordinal, so of the form $\beta = \alpha + 1$, because otherwise any $0 \neq X \in \mathcal{C}_\beta$ belongs to $\mathcal{C}_\gamma$ for some $\gamma < \beta$. Thus there is some object X which is simple in $\mathcal{C}/\mathcal{C}_\alpha$ such that $F(X) \neq 0$. Therefore $F \in \mathrm{Sp}_\alpha\,\mathcal{A}$. The second assertion follows from the preceding lemma. $\square$

If the Krull–Gabriel dimension of $\mathcal{C} = \mathrm{fp}\,\mathcal{A}$ is defined, then one obtains a classification of all injective objects in $\mathcal{A}$.

Proposition 14.1.5. *Suppose that* KG.dim $\mathrm{fp}\,\mathcal{A} = \kappa$. *Then every injective object* X *in* $\mathcal{A}$ *is the injective envelope of a coproduct of indecomposable injectives* $\bigsqcup_{i \in I} X_i$. *The isomorphism classes of the* X_i *and their multiplicities are uniquely determined by* X.

Proof First observe that every non-zero injective object X admits an indecomposable summand. To this end let α be an ordinal that is minimal such that X admits a non-zero morphism $\phi \colon C \to X$ with $C \in (\mathrm{fp}\,\mathcal{A})_\alpha$. Clearly,

$\alpha = \beta + 1$. Then C has finite length in $\operatorname{fp} \mathcal{A}/(\operatorname{fp} \mathcal{A})_\beta$. Choose a composition series $0 = C_0 \subseteq C_1 \subseteq \cdots \subseteq C_n = C$ and let r be minimal such that $\phi(C_r) \neq 0$ but $\phi(C_{r-1}) = 0$. Set $C' = C_r/C_{r-1}$ and $D = \operatorname{Im} \phi$. Because C' is simple in $\operatorname{fp} \mathcal{A}/(\operatorname{fp} \mathcal{A})_\beta$, it follows that D is *uniform*, so $D \neq 0$ and any pair of non-zero subobjects has a non-zero intersection. Thus the injective envelope of D yields an indecomposable direct summand of X.

Given a non-zero injective object X, we use Zorn's lemma and obtain a maximal family of indecomposable injective subobjects $(X_i)_{i \in I}$ such that the sum $X' = \sum_{i \in I} X_i$ is direct. This yields a decomposition $X = E(X') \oplus X''$, and $X'' = 0$ by the first part of the proof.

We sketch the argument for the uniqueness statement. One passes to the *spectral category* of $\mathcal{A}$, which is by definition the localisation $\mathcal{A}[\operatorname{Ess}^{-1}]$ with respect to the class Ess of essential monomorphisms (cf. Proposition 2.5.9). This is a split exact Grothendieck category, so each object decomposes into simple objects, and the canonical functor $\mathcal{A} \to \mathcal{A}[\operatorname{Ess}^{-1}]$ identifies each object in $\mathcal{A}$ with its injective envelope. Now one applies the Krull–Remak–Schmidt–Azumaya theorem (Theorem 2.5.8).   □

Let us discuss a variation of the Kull-Gabriel filtration which yields a possible refinement. Fix again an essentially small abelian category $\mathcal{C}$ and consider an ascending chain of Serre subcategories $\mathcal{C}^\alpha \subseteq \mathcal{C}$ which is indexed by the ordinals $\alpha \leq \kappa$:

$$0 = \mathcal{C}^{-1} \subseteq \mathcal{C}^0 \subseteq \mathcal{C}^1 \subseteq \cdots \subseteq \mathcal{C}^\kappa = \mathcal{C}.$$

Suppose that $\mathcal{C}^{\alpha+1}/\mathcal{C}^\alpha$ is a length category for all $\alpha < \kappa$, and $\mathcal{C}^\alpha = \bigcup_{\beta<\alpha} \mathcal{C}^\beta$ for any limit ordinal $\alpha \leq \kappa$. As before, define $\operatorname{Sp}^\alpha \mathcal{A}$ to be the set of objects $F \in \operatorname{Sp} \mathcal{A}$ such that $F(\mathcal{C}^\alpha) = 0$ and $F(X) \neq 0$ for some object X which is simple in $\mathcal{C}^{\alpha+1}/\mathcal{C}^\alpha$.

Proposition 14.1.6. *We have* KG.dim $\mathcal{C} \leq \kappa$, *and* $\operatorname{Sp} \mathcal{A}$ *equals the disjoint union* $\bigsqcup_{\alpha<\kappa} \operatorname{Sp}^\alpha \mathcal{A}$. *Moreover, for each* $\alpha < \kappa$ *there is a bijection between* $\operatorname{Sp}^\alpha \mathcal{A}$ *and the set of isomorphism classes of simple objects in* $\mathcal{C}^{\alpha+1}/\mathcal{C}^\alpha$.

Proof It follows by induction from Lemma 14.1.1 that KG.dim $\mathcal{C} \leq \kappa$. The description of $\operatorname{Sp} \mathcal{A}$ is analogous to Proposition 14.1.4, and the proof is essentially the same.   □

The Lattice of Subobjects

Let $\mathcal{C}$ be an essentially small abelian category. For an object $X \in \mathcal{C}$ we denote by $\mathbf{L}(X)$ its lattice of subobjects and note that this lattice is modular. The Krull–

Gabriel filtration of $\mathcal{C}$ is reflected by a cofiltration of $\mathbf{L}(X)$ for each object $X \in \mathcal{C}$. The description of this cofiltration requires the following definition.

Let L be a modular lattice. Recall that a lattice has *finite length* if there is a finite chain $0 = x_0 \leq x_1 \leq \cdots \leq x_n = 1$ which cannot be refined. An equivalent condition is that L satisfies both chain conditions. Given elements $x, y \in L$ we write $x \sim y$ if the interval $[x \wedge y, x \vee y]$ has finite length. Then $\sim$ defines a *congruence relation* on L. This means the set of equivalence classes $L/\sim$ carries again the structure of a modular lattice and the canonical map $L \to L/\sim$ is a lattice homomorphism.

Let us define a cofiltration of L which is indexed by the ordinals $\alpha \geq -1$:

- $L_{-1} = L$,

- $L_\alpha = L_\beta/\sim$ when $\alpha = \beta + 1$,

- $L_\alpha = \mathrm{colim}_{\gamma < \alpha} L_\gamma$ when α is a limit ordinal.

We denote by L_∞ the colimit of this cofiltration. If $L_\infty = 0$ then the smallest ordinal α such that $L_\alpha = 0$ is called the *minimal dimension* (or simply *m-dimension*) of L and is denoted m.dim L. We write m.dim $L < \infty$ in this case, and m.dim $L = \infty$ when $L_\infty \neq 0$.

Remark 14.1.7. If m.dim $L < \infty$, then m.dim $L = \beta + 1$ for some ordinal β. To see this consider the set I_α of elements $x \in L$ that $L \to L_\alpha$ maps to 0. Note that $I_\alpha = \bigcup_{\gamma < \alpha} I_\gamma$ when α is a limit ordinal. Moreover, $L_\alpha = 0$ if and only if $1 \in I_\alpha$.

We collect some basic properties of the m-dimension and begin with an elementary observation.

Lemma 14.1.8. *Let $F \colon \mathcal{C} \to \mathcal{D}$ be an exact functor between abelian categories and $X \in \mathcal{C}$ an object. Then F induces a lattice homomorphism $\mathbf{L}(X) \to \mathbf{L}(FX)$. This map is surjective when F is a quotient functor.*

Proof Given subobjects $U, V \subseteq X$, then $F(U \cap V) = F(U) \cap F(V)$ and $F(U + V) = F(U) + F(V)$ since F is exact. Thus $\mathbf{L}(X) \to \mathbf{L}(FX)$ is a homomorphism. When F is a quotient functor then we can apply the lemma below. $\qquad\square$

Lemma 14.1.9. *Let $\mathcal{C}$ be an abelian category and $Q \colon \mathcal{C} \to \mathcal{C}/\mathcal{B}$ the quotient functor given by a Serre subcategory $\mathcal{B} \subseteq \mathcal{C}$. Then for any exact sequence $0 \to X \to Y \to Z \to 0$ in $\mathcal{C}/\mathcal{B}$ there is an exact sequence $0 \to X' \xrightarrow{\phi} Y \xrightarrow{\psi} Z' \to 0$*

in $\mathcal{C}$ *inducing the following commutative diagram.*

$$
\begin{array}{ccccccccc}
0 & \longrightarrow & X & \longrightarrow & Y & \longrightarrow & Z & \longrightarrow & 0 \\
 & & \downarrow{\scriptstyle\wr} & & \| & & \downarrow{\scriptstyle\wr} & & \\
0 & \longrightarrow & X' & \xrightarrow{Q(\phi)} & Y & \xrightarrow{Q(\psi)} & Z' & \longrightarrow & 0
\end{array}
$$

Proof We consider the morphism $X \to Y$ in $\mathcal{C}/\mathcal{B}$. Then there are subobjects $X_1 \subseteq X$ and $Y_1 \subseteq Y$ in $\mathcal{C}$ such that X/X_1 and Y_1 belong to $\mathcal{B}$, plus a morphism $\phi_1 \colon X_1 \to Y/Y_1$ in $\mathcal{C}$ inducing the following commutative square (Lemma 2.2.4).

$$
\begin{array}{ccc}
X & \longrightarrow & Y \\
\uparrow & & \downarrow \\
X_1 & \xrightarrow{Q(\phi_1)} & Y/Y_1
\end{array}
$$

We form in $\mathcal{C}$ the following pullback.

$$
\begin{array}{ccc}
X_2 & \xrightarrow{\phi_2} & Y \\
\downarrow & & \downarrow \\
X_1 & \xrightarrow{\phi_1} & Y/Y_1
\end{array}
$$

Now choose for $\phi \colon X' \to Y$ the inclusion $\operatorname{Im}\phi_2 \to Y$ and for $\psi \colon Y \to Z'$ the cokernel of ϕ. □

The following lemma describes the lattice of subobjects corresponding to the Krull–Gabriel filtration of an abelian category.

Lemma 14.1.10. *Let $\mathcal{C}$ be an essentially small abelian category and $q_\alpha \colon \mathcal{C} \to \mathcal{C}/\mathcal{C}_\alpha$ the quotient functor for an ordinal α. Given an object $X \in \mathcal{C}$, the canonical map $\mathbf{L}(X) \to \mathbf{L}(q_\alpha X)$ induces an isomorphism $\mathbf{L}(X)_\alpha \xrightarrow{\sim} \mathbf{L}(q_\alpha X)$. Therefore $X \in \mathcal{C}_\alpha$ if and only if $\mathbf{L}(X)_\alpha = 0$, and $X \in \mathcal{C}_\infty$ if and only if $\mathbf{L}(X)_\infty = 0$.*

Proof It follows from Lemma 14.1.8 that the map $\mathbf{L}(X) \to \mathbf{L}(q_\alpha X)$ is surjective. We compare this map with the canonical map $\mathbf{L}(X) \to \mathbf{L}(X)_\alpha$, and it suffices to show that both maps identify the same elements in $\mathbf{L}(X)$. This is done by induction. For the step $\beta \mapsto \beta + 1$ suppose $\mathbf{L}(X)_\beta \xrightarrow{\sim} \mathbf{L}(q_\beta X)$ and consider subobjects $U \subseteq V \subseteq X$. Then the map $\mathbf{L}(q_\beta X) \to \mathbf{L}(q_{\beta+1} X)$ identifies U and V if and only if V/U has finite length in $\mathcal{C}/\mathcal{C}_\beta$ if and only if $\mathbf{L}(X)_\beta \to \mathbf{L}(X)_{\beta+1}$ identifies U and V. This yields the isomorphism $\mathbf{L}(X)_\alpha \xrightarrow{\sim} \mathbf{L}(q_\alpha X)$. In particular, $X \in \mathcal{C}_\alpha$ if and only if $\mathbf{L}(X)_\alpha = 0$. □

A *dense chain* in a lattice is a sublattice $C \neq 0$ having the property that for every pair $x < y$ in C there is $z \in C$ with $x < z < y$. Note that having a dense chain is equivalent to having a sublattice isomorphic to

$$D = \{p \cdot 2^{-n} \in [0,1] \cap \mathbb{Q} \mid p, n \in \mathbb{N}\}.$$

Lemma 14.1.11. *There is a dense chain in L if and only if* m.dim $L = \infty$.

Proof Let $\pi \colon L \to L_\infty$ be the canonical map. If $L_\infty \neq 0$, then L_∞ is a dense chain in L_∞. Thus we can construct inductively a dense chain isomorphic to D in L, since for any pair $x < y$ in L with $\pi(X) < \pi(y)$ there is some $z \in L$ with $\pi(X) < \pi(z) < \pi(y)$, and therefore $x < z' < y$ for $z' = (x \vee z) \wedge y$. Now suppose there is a dense chain in L, say between x and y. Using induction one shows that $\pi(x) \neq \pi(y)$, and therefore m.dim $L = \infty$. □

We record a useful consequence.

Proposition 14.1.12. *Let C be an essentially small abelian category. Then C_∞ contains all noetherian and all artinian objects.*

Proof Let X be noetherian or artinian. Then $\mathbf{L}(X)$ does not contain a dense chain, and therefore m.dim $\mathbf{L}(X) < \infty$ by Lemma 14.1.11. Thus $X \in C_\infty$ by Lemma 14.1.10. □

Corollary 14.1.13. KG.dim $C < \infty$ *when all objects in C are noetherian.*

Example: Commutative Noetherian Rings

Let Λ be a commutative noetherian ring. We compute the Krull–Gabriel dimension of the abelian category mod Λ. This justifies the terminology, because this dimension coincides with the Krull dimension of the ring Λ.

Let $X \in C$ be an object of an abelian category C and suppose $X \in C_\alpha$ for some ordinal α. Then the smallest ordinal α with this property is called the *Krull–Gabriel dimension* and is denoted KG.dim X.

Proposition 14.1.14. *Let Λ be a commutative noetherian ring. For $\mathfrak{p} \in$ Spec Λ and $n \in \mathbb{N}$ we have* KG.dim $\Lambda/\mathfrak{p} \leq n$ *if and only if every proper chain*

$$\mathfrak{p} = \mathfrak{p}_0 \subset \mathfrak{p}_1 \subset \mathfrak{p}_2 \subset \cdots \subset \mathfrak{p}_r$$

in Spec Λ *has length $r \leq n$. Therefore* KG.dim(mod Λ) $\leq n$ *if and only if every proper chain of prime ideals has length at most n.*

Proof We use the correspondence between Serre subcategories $C \subseteq$ mod Λ and specialisation closed subsets of Spec Λ, which takes C to Supp C (Proposition 2.4.8). Then the assertion follows easily by an induction on n from the lemma below. □

Lemma 14.1.15. *Let $\mathcal{C} \subseteq \mathrm{mod}\,\Lambda$ be a Serre subcategory and $\mathcal{V} = \mathrm{Supp}\,\mathcal{C}$. Then the object $\Lambda/\mathfrak{p}$ is simple in $(\mathrm{mod}\,\Lambda)/\mathcal{C}$ if and only if $\mathfrak{p} \notin \mathcal{V}$ and $\mathfrak{q} \in \mathcal{V}$ for all $\mathfrak{p} \subset \mathfrak{q}$.*

Proof We may assume that $\Lambda/\mathfrak{p}$ is not in $\mathcal{C}$. Then it follows from Lemma 14.1.9 that $\Lambda/\mathfrak{p}$ is simple in $(\mathrm{mod}\,\Lambda)/\mathcal{C}$ if and only if for every proper epimorphism $\Lambda/\mathfrak{p} \to \Lambda/\mathfrak{a}$ in $\mathrm{mod}\,\Lambda$ we have $\Lambda/\mathfrak{a} \in \mathcal{C}$. Recall that $\mathrm{Supp}\,\Lambda/\mathfrak{a} = \mathcal{V}(\mathfrak{a})$ for every ideal $\mathfrak{a}$ (Lemma 2.4.1). Thus $\Lambda/\mathfrak{a} \in \mathcal{C}$ for every $\mathfrak{p} \subset \mathfrak{a}$ if and only if $\mathcal{V}(\mathfrak{a}) \subseteq \mathcal{V}$ for every $\mathfrak{p} \subset \mathfrak{a}$ if and only if $\mathfrak{q} \in \mathcal{V}$ for all $\mathfrak{p} \subset \mathfrak{q}$. □

Pure-Injective Objects

Let $\mathcal{A}$ be a locally finitely presented category. The Krull–Gabriel filtration of $\mathrm{Ab}(\mathcal{A})$ provides a method of classifying the pure-injective objects of $\mathcal{A}$. As before, this is done in two steps: first the indecomposable objects, and then the general case.

We use the embedding $\mathrm{ev}\colon \mathcal{A} \to \mathbf{P}(\mathcal{A})$ into the purity category, which identifies $\mathrm{Ab}(\mathcal{A}) = \mathrm{fp}\,\mathbf{P}(\mathcal{A})$ and $\mathrm{Ind}\,\mathcal{A} = \mathrm{Sp}\,\mathbf{P}(\mathcal{A})$; see Lemma 12.1.4 and Lemma 12.1.17. For each ordinal α set $\mathrm{Ind}_\alpha\,\mathcal{A} = \mathrm{Sp}_\alpha\,\mathbf{P}(\mathcal{A})$.

The following is a direct consequence of Proposition 14.1.4.

Corollary 14.1.16. *Suppose that $\mathrm{KG.dim}\,\mathrm{Ab}(\mathcal{A}) = \kappa$. Then $\mathrm{Ind}\,\mathcal{A}$ equals the disjoint union $\bigsqcup_{\alpha<\kappa} \mathrm{Ind}_\alpha\,\mathcal{A}$. Moreover, for each ordinal $\alpha < \kappa$ there is a bijection between $\mathrm{Ind}_\alpha\,\mathcal{A}$ and the set of isomorphism classes of simple objects in $\mathrm{Ab}(\mathcal{A})/\mathrm{Ab}(\mathcal{A})_\alpha$.* □

If the Krull–Gabriel dimension is defined, then one obtains a classification of all pure-injective objects. This follows from Proposition 14.1.5, since pure-injective objects in $\mathcal{A}$ identify with injective objects in $\mathbf{P}(\mathcal{A})$ by Lemma 12.1.8.

Corollary 14.1.17. *Suppose that $\mathrm{KG.dim}\,\mathrm{Ab}(\mathcal{A}) = \kappa$. Then every pure-injective object X in $\mathcal{A}$ is the pure-injective envelope of a coproduct of indecomposable pure-injectives $\bigsqcup_{i\in I} X_i$. The isomorphism classes of the X_i and their multiplicities are uniquely determined by X.* □

The Krull–Gabriel filtration provides a useful method of classifying objects, even when $\mathrm{KG.dim}\,\mathrm{Ab}(\mathcal{A}) = \infty$. Given a class of objects $\mathcal{X} \subseteq \mathcal{A}$, we may consider the dimension $\mathrm{KG.dim}\,\mathrm{Ab}(\mathcal{X})$ of the corresponding abelian category $\mathrm{Ab}(\mathcal{X})$. For example, an object $X \in \mathcal{A}$ is endofinite if and only if $\mathrm{KG.dim}\,\mathrm{Ab}(X) \le 0$, by Proposition 13.1.9. Also, $\mathrm{KG.dim}\,\mathrm{Ab}(X) < \infty$ when X is Σ-pure-injective, by Theorem 12.3.4 and Corollary 14.1.13.

Endofinite Objects

Let $\mathcal{A}$ be a locally finitely presented category. Then the Krull–Gabriel filtration produces endofinite objects, provided the dimension is defined and is not a limit ordinal.

Proposition 14.1.18. *Suppose that* KG.dim $\mathrm{Ab}(\mathcal{A}) = \kappa + 1$. *Then the objects in* $\mathrm{Ind}_\kappa \, \mathcal{A}$ *are endofinite.*

Proof Let $X \in \mathrm{Ind}_\kappa \, \mathcal{A}$. Then the abelian category $\mathrm{Ab}(X)$ is by definition a quotient of $\mathrm{Ab}(\mathcal{A})/\mathrm{Ab}(\mathcal{A})_\kappa$, which is a length category by assumption. Thus $\mathrm{Ab}(X)$ is a length category, and therefore X is endofinite by Proposition 13.1.9.
□

Proposition 14.1.19. *Suppose that* $\mathrm{Ind}\,\mathcal{A}$ *is quasi-compact and let* $X \neq 0$ *be a* Σ-*pure-injective object in* $\mathcal{A}$. *Then the definable subcategory generated by* X *contains an indecomposable endofinite object.*

Proof The objects in $\mathrm{Ab}(X)$ are noetherian since X is Σ-pure-injective; see Theorem 12.3.4. Thus KG.dim $\mathrm{Ab}(X) < \infty$ by Proposition 14.1.12. Because $\mathrm{Ind}\,\mathcal{A}$ is quasi-compact, the abelian category $\mathrm{Ab}(X)$ is finitely generated by Proposition 12.3.11. Thus KG.dim $\mathrm{Ab}(X) = \kappa + 1$ for some ordinal κ by Lemma 14.1.2. Consider the quotient functor

$$Q \colon \mathrm{Ab}(\mathcal{A}) \twoheadrightarrow \mathrm{Ab}(X) \twoheadrightarrow \mathrm{Ab}(X)/\mathrm{Ab}(X)_\kappa$$

and choose any object $Y \in \mathrm{Ind}\,\mathcal{A}$ such that $\bar{Y}$ factors through Q. Then $\mathrm{Ab}(Y)$ is a quotient of $\mathrm{Ab}(X)/\mathrm{Ab}(X)_\kappa$ and therefore a length category. Thus Y is endofinite by Proposition 13.1.9. Also, Y belongs to the definable subcategory generated by X since $\mathrm{Ab}(Y)$ is a quotient of $\mathrm{Ab}(X)$.
□

Remark 14.1.20. When X is Σ-pure-injective, then any object in the definable closure of X is actually a direct summand of a product of copies of X, by Theorem 12.3.4.

Interesting examples arise from Prüfer modules over Artin algebras, which are Σ-pure-injective by Proposition 12.3.9. With an additional assumption there are no finite length indecomposable direct summands.

Example 14.1.21. Let Λ be an Artin algebra and X a Prüfer module, given by an endomorphism $\phi \colon X \to X$ such that $X = \bigcup_{n \geq 0} \mathrm{Ker}\,\phi^n$ with $\mathrm{Ker}\,\phi$ of finite length. Suppose that each inclusion $\mathrm{Ker}\,\phi^n \to \mathrm{Ker}\,\phi^{n+1}$ is a radical morphism. Then there is a generic module in the definable subcategory generated by X.

Proof We can apply Proposition 14.1.19 since $\mathrm{Ind}\,\Lambda$ is quasi-compact. The assumption on the inclusions $\operatorname{Ker} \phi^n \to \operatorname{Ker} \phi^{n+1}$ implies that X has no inde-composable direct summand of finite length. It follows from Theorem 12.4.15 that there is no indecomposable module of finite length in the definable sub-category generated by X.   □

14.2 Examples of Krull–Gabriel Filtrations

In this section we compute the Krull–Gabriel filtration for several examples. In each case we obtain as a consequence an explicit classification of all pure-injective objects.

Uniserial Categories

Let $\mathcal{A}$ be a locally finitely presented abelian category and set $\mathcal{C} = \operatorname{fp}\mathcal{A}$. Let us compute the Krull–Gabriel filtration of $\operatorname{Ab}(\mathcal{A}) = \operatorname{Fp}(\mathcal{C}, \operatorname{Ab})^{\operatorname{op}}$ when $\mathcal{C}$ is uniserial.

We write $\operatorname{Fp}_0(\mathcal{C}, \operatorname{Ab})$ for the Serre subcategory consisting of all finite length objects in $\operatorname{Fp}(\mathcal{C}, \operatorname{Ab})$, and $\operatorname{Eff}(\mathcal{C}, \operatorname{Ab})$ denotes the Serre subcategory of efface-able functors in $\operatorname{Fp}(\mathcal{C}, \operatorname{Ab})$ given by all functors F with presentation

$$0 \longrightarrow \operatorname{Hom}(Z, -) \longrightarrow \operatorname{Hom}(Y, -) \longrightarrow \operatorname{Hom}(X, -) \longrightarrow F \longrightarrow 0$$

coming from an exact sequence $0 \to X \to Y \to Z \to 0$ in $\mathcal{C}$.

Proposition 14.2.1. *Let* $\mathcal{C} = \operatorname{fp}\mathcal{A}$ *be uniserial. Then we have*

$$\operatorname{Fp}_0(\mathcal{C}, \operatorname{Ab})^\perp = \{X \in \operatorname{Ind}\mathcal{A} \mid X \notin \operatorname{fp}\mathcal{A}\} \quad and \quad \operatorname{Eff}(\mathcal{C}, \operatorname{Ab})^\perp = \operatorname{Ind}\mathcal{A} \cap \operatorname{Inj}\mathcal{A}.$$

In particular, $\operatorname{Eff}(\mathcal{C}, \operatorname{Ab}) \subseteq \operatorname{Fp}_0(\mathcal{C}, \operatorname{Ab})$.

Proof Let $X \in \operatorname{Ind}\mathcal{A}$. Then X is the source of a left almost split morphism by Corollary 13.1.30, and therefore $\bar{X}$ is an injective envelope of a simple object S in $\mathbf{P}(\mathcal{A})$ by Theorem 12.3.13. Note that S is finitely presented when X is finitely presented. Thus $X \in \operatorname{fp}\mathcal{A}$ implies $X \notin \operatorname{Fp}_0(\mathcal{C}, \operatorname{Ab})^\perp$. On the other hand, when F in $\operatorname{Fp}(\mathcal{C}, \operatorname{Ab})$ is simple and the quotient of $\operatorname{Hom}(Y, -)$, then we may choose Y indecomposable, and it is pure-injective since Y is endofinite by Corollary 13.1.29. Thus $\bar{Y}$ is an injective envelope of F in $\mathbf{P}(\mathcal{A})$. It follows that $X \notin \operatorname{Fp}_0(\mathcal{C}, \operatorname{Ab})^\perp$ implies $X \in \operatorname{fp}\mathcal{A}$.

The identity $\operatorname{Eff}(\mathcal{C}, \operatorname{Ab})^\perp = \operatorname{Ind}\mathcal{A} \cap \operatorname{Inj}\mathcal{A}$ is Proposition 12.3.17.

The inclusion $\operatorname{Eff}(\mathcal{C}, \operatorname{Ab}) \subseteq \operatorname{Fp}_0(\mathcal{C}, \operatorname{Ab})$ is clear since $X \in \operatorname{Ind}\mathcal{A}$ and $X \notin \operatorname{fp}\mathcal{A}$ implies $X \in \operatorname{Inj}\mathcal{A}$, by Theorem 13.1.28.   □

We recall from Proposition 2.3.3 the equivalence

$$\text{Fp}(\mathcal{C}, \text{Ab})/\text{Eff}(\mathcal{C}, \text{Ab}) \xrightarrow{\sim} \mathcal{C}^{\text{op}}$$

and record some consequences.

Corollary 14.2.2. *We have* $\text{KG.dim Ab}(\mathcal{A}) \le 1$. □

Corollary 14.2.3. *We have* $\text{Fp}_0(\mathcal{C}, \text{Ab}) = \text{Eff}(\mathcal{C}, \text{Ab})$ *if and only if all injective objects in* $\mathcal{C}$ *are zero.* □

Dedekind Domains

Let A be a *Dedekind domain*, that is, a commutative hereditary integral domain. For simplicity we assume that A is not a field. We write $\text{mod}_0 A$ for the category of finite length A-modules, $Q(A)$ for the quotient field, and $\text{Max } A$ for the set of maximal ideals. Note that

$$\text{mod}_0 A = \coprod_{\mathfrak{p} \in \text{Max } A} \mathcal{T}_\mathfrak{p} \quad \text{and} \quad (\text{mod } A)/(\text{mod}_0 A) \xrightarrow{\sim} \text{mod } Q(A),$$

where $\mathcal{T}_\mathfrak{p}$ denotes the uniserial category of finite length $\mathfrak{p}$-torsion modules. Recall that a module is $\mathfrak{p}$-*torsion* if each element is annihilated by some power of $\mathfrak{p}$.

Let us consider the functor $\pi \colon \text{Fp}(\text{mod } A, \text{Ab}) \to (\text{mod } A)^{\text{op}}$ given by

$$\text{Coker Hom}(\phi, -) \longmapsto \text{Ker } \phi \qquad (\phi \text{ a morphism in } \text{mod } A).$$

We write $\text{Eff}(\text{mod } A, \text{Ab})$ for the Serre subcategory of $\text{Fp}(\text{mod } A, \text{Ab})$ given by all functors F with presentation

$$0 \longrightarrow \text{Hom}(Z, -) \longrightarrow \text{Hom}(Y, -) \longrightarrow \text{Hom}(X, -) \longrightarrow F \longrightarrow 0$$

coming from an exact sequence $0 \to X \to Y \to Z \to 0$ in $\text{mod } A$. Clearly, $\text{Ker } \pi = \text{Eff}(\text{mod } A, \text{Ab})$. Furthermore, let $\text{Fp}(\text{mod } A, \text{Ab})'$ denote the kernel of the composite

$$\text{Fp}(\text{mod } A, \text{Ab}) \xrightarrow{\ \pi\ } (\text{mod } A)^{\text{op}} \xrightarrow{\ -\otimes Q(A)\ } (\text{mod } Q(A))^{\text{op}}.$$

Lemma 14.2.4. *The functor* π *induces an equivalence*

$$\text{Fp}(\text{mod } A, \text{Ab})/\text{Eff}(\text{mod } A, \text{Ab}) \xrightarrow{\sim} (\text{mod } A)^{\text{op}}$$

which yields further equivalences

$$\text{Fp}(\text{mod } A, \text{Ab})'/\text{Eff}(\text{mod } A, \text{Ab}) \xrightarrow{\sim} (\text{mod}_0 A)^{\text{op}}$$

and

$$\mathrm{Fp}(\mathrm{mod}\,A, \mathrm{Ab})/\mathrm{Fp}(\mathrm{mod}\,A, \mathrm{Ab})' \xrightarrow{\sim} (\mathrm{mod}\,Q(A))^{\mathrm{op}}.$$

Proof The functor π is a left adjoint of $(\mathrm{mod}\,A)^{\mathrm{op}} \to \mathrm{Fp}(\mathrm{mod}\,A, \mathrm{Ab})$ given by $X \mapsto \mathrm{Hom}(X, -)$. Thus the first equivalence follows from Proposition 2.3.3. The other equivalences are then consequences which are derived from the equivalence $\frac{\mathrm{mod}\,A}{\mathrm{mod}_0\,A} \xrightarrow{\sim} \mathrm{mod}\,Q(A)$; see Proposition 2.2.8. $\qquad \square$

Next consider the functor

$$\tau\colon \mathrm{Eff}(\mathrm{mod}\,A, \mathrm{Ab}) \longrightarrow \mathrm{mod}\,A, \qquad F \mapsto F(A).$$

Lemma 14.2.5. *The functor τ induces an equivalence*

$$\mathrm{Eff}(\mathrm{mod}\,A, \mathrm{Ab})/(\mathrm{Ker}\,\tau) \xrightarrow{\sim} \mathrm{mod}_0\,A,$$

and the inclusion $\mathrm{mod}_0\,A \to \mathrm{mod}\,A$ *induces an equivalence*

$$\mathrm{Eff}(\mathrm{mod}_0\,A, \mathrm{Ab}) \xrightarrow{\sim} \mathrm{Ker}\,\tau.$$

Proof A torsion module X with projective presentation $0 \to P_1 \to P_0 \to X \to 0$ induces an exact sequence

$$0 \to \mathrm{Hom}(X, -) \to \mathrm{Hom}(P_0, -) \to \mathrm{Hom}(P_1, -) \to \mathrm{Ext}^1(X, -) \to 0$$

in $\mathrm{Eff}(\mathrm{mod}\,A, \mathrm{Ab})$, and evaluating at A yields an isomorphism $\mathrm{Ext}^1(X, A) \cong \mathrm{Tr}\,X$. Thus the functor τ is a right adjoint of $\sigma\colon \mathrm{mod}_0\,A \to \mathrm{Eff}(\mathrm{mod}\,A, \mathrm{Ab})$ given by $X \mapsto \mathrm{Ext}^1(\mathrm{Tr}\,X, -)$ and satisfying $\tau \circ \sigma \cong \mathrm{id}$. Now the first equivalence follows from Proposition 2.2.11.

The second equivalence is easily checked; it uses that $\mathrm{Hom}(X, A) = 0$ for all $X \in \mathrm{mod}_0\,A$, and a quasi-inverse is given by $F \mapsto F|_{\mathrm{mod}_0\,A}$. $\qquad \square$

For an abelian category $\mathcal{A}$ let $\mathcal{A}_0$ denote the full subcategory consisting of all objects of finite length.

Lemma 14.2.6. *We have* $\mathrm{Eff}(\mathrm{mod}_0\,A, \mathrm{Ab}) = \mathrm{Fp}_0(\mathrm{mod}_0\,A, \mathrm{Ab})$. *In particular, the assignment*

$$A/\mathfrak{p}^n \longmapsto \mathrm{Hom}(A/\mathfrak{p}^n, -)/\mathrm{Rad}(A/\mathfrak{p}^n, -)$$

identifies the indecomposable objects in $\mathrm{mod}_0\,A$ *with the simple objects of the category* $\mathrm{Eff}(\mathrm{mod}_0\,A, \mathrm{Ab})$.

Proof The category $\mathrm{mod}_0\,A$ is uniserial and has no non-zero injective objects. Thus finite length functors and effaceable functors in $\mathrm{Fp}(\mathrm{mod}_0\,A, \mathrm{Ab})$ coincide. This follows from Corollary 14.2.3.

Each indecomposable object in $\mathrm{mod}_0\,A$ is the source of an almost split morphism $A/\mathfrak{p}^n \to A/\mathfrak{p}^{n-1} \oplus A/\mathfrak{p}^{n+1}$, and this provides in $\mathrm{Fp}(\mathrm{mod}_0 A, \mathrm{Ab})$ the presentation of a simple functor

$$\mathrm{Hom}(A/\mathfrak{p}^{n-1}, -) \oplus \mathrm{Hom}(A/\mathfrak{p}^{n+1}, -) \longrightarrow \mathrm{Hom}(A/\mathfrak{p}^n, -) \longrightarrow S_{A/\mathfrak{p}^n} \longrightarrow 0.$$

Clearly, any simple functor S is of this form, because we have $S(X) \neq 0$ for some indecomposable object X. $\qquad\square$

For $\mathfrak{p} \in \mathrm{Max}\, A$ set

$$A_{\mathfrak{p}^\infty} = \mathrm{colim}\, A/\mathfrak{p}^n \qquad \text{and} \qquad \hat{A}_{\mathfrak{p}} = \lim A/\mathfrak{p}^n.$$

Note that $A_{\mathfrak{p}^\infty}$ is an injective envelope of $A/\mathfrak{p}$, while

$$\hat{A}_{\mathfrak{p}} \cong \lim \mathrm{Hom}(A/\mathfrak{p}^n, A_{\mathfrak{p}^\infty}) \cong \mathrm{Hom}(\mathrm{colim}\, A/\mathfrak{p}^n, A_{\mathfrak{p}^\infty}) \cong \mathrm{Hom}(A_{\mathfrak{p}^\infty}, A_{\mathfrak{p}^\infty}).$$

In particular, both modules are indecomposable and pure-injective. The module $A_{\mathfrak{p}^\infty}$ is called a *Prüfer module*, while $\hat{A}_{\mathfrak{p}}$ is called *adic*.

Theorem 14.2.7. *The abelian category* $\mathrm{Ab}(A) = \mathrm{Fp}(\mathrm{mod}\, A, \mathrm{Ab})$ *admits a filtration*

$$\{0\} \subseteq \mathrm{Eff}(\mathrm{mod}_0 A, \mathrm{Ab}) \subseteq \mathrm{Eff}(\mathrm{mod}\, A, \mathrm{Ab}) \subseteq \mathrm{Fp}(\mathrm{mod}\, A, \mathrm{Ab})' \subseteq \mathrm{Ab}(A)$$

such that each quotient is a length category. This provides a complete list of indecomposable pure-injective A-modules, by taking the injective envelope of a functor that is simple in one of the quotient categories.

(1) *The simples in* $\mathrm{Eff}(\mathrm{mod}_0 A, \mathrm{Ab})$ *correspond to* $A/\mathfrak{p}^n$, $\mathfrak{p} \in \mathrm{Max}\, A$, $n \geq 1$.
(2) *The simples in* $\frac{\mathrm{Eff}(\mathrm{mod}\, A, \mathrm{Ab})}{\mathrm{Eff}(\mathrm{mod}_0 A, \mathrm{Ab})}$ *correspond to* $\hat{A}_{\mathfrak{p}}$, $\mathfrak{p} \in \mathrm{Max}\, A$.
(3) *The simples in* $\frac{\mathrm{Fp}(\mathrm{mod}\, A, \mathrm{Ab})'}{\mathrm{Eff}(\mathrm{mod}\, A, \mathrm{Ab})}$ *correspond to* $A_{\mathfrak{p}^\infty}$, $\mathfrak{p} \in \mathrm{Max}\, A$.
(4) *The simple in* $\frac{\mathrm{Fp}(\mathrm{mod}\, A, \mathrm{Ab})}{\mathrm{Fp}(\mathrm{mod}\, A, \mathrm{Ab})'}$ *corresponds to* $Q(A)$.

The modules $A/\mathfrak{p}^n$ *and* $Q(A)$ *are endofinite. The modules* $A_{\mathfrak{p}^\infty}$ *are* Σ-*pure-injective.*

Proof The filtration of $\mathrm{Fp}(\mathrm{mod}\, A, \mathrm{Ab})$ follows from the series of the above lemmas. This yields a classification of all indecomposable pure-injective A-modules, using the analogue of Corollary 14.1.16 which is based on the Krull–Gabriel filtration; see also Proposition 14.1.6.

Lemma 14.2.6 takes care of the first layer of the filtration. Lemma 14.2.4 describes the last two layers of the filtration, which yield the indecomposable injective A-modules by Proposition 12.3.17. The remaining layer is given by Lemma 14.2.5 and yields the remaining indecomposable pure-injectives.

It is clear that $A/\mathfrak{p}^n$ has endolength n, while $Q(A)$ has endolength 1. The module $A_{\mathfrak{p}^\infty}$ is actually injective, and therefore Σ-injective since the ring A is noetherian; see Theorem 11.2.12. □

Corollary 14.2.8. *We have* KG.dim Ab$(A) = 2$ *and obtain the following filtration of* Ind A:

$$\mathrm{Ind}_{-1}\, A = \{A/\mathfrak{p}^n \mid \mathfrak{p} \in \mathrm{Max}\, A,\, n \geq 1\}$$
$$\mathrm{Ind}_0\, A = \{A_{\mathfrak{p}^\infty} \mid \mathfrak{p} \in \mathrm{Max}\, A\} \cup \{\hat{A}_\mathfrak{p} \mid \mathfrak{p} \in \mathrm{Max}\, A\}$$
$$\mathrm{Ind}_1\, A = \{Q(A)\}.$$

Proof For the category of finite length objects in Ab$(A) = \mathrm{Fp}(\mathrm{mod}\, A, \mathrm{Ab})$ we have

$$\mathrm{Eff}(\mathrm{mod}_0\, A, \mathrm{Ab}) = \mathrm{Fp}_0(\mathrm{mod}_0\, A, \mathrm{Ab}) = \mathrm{Ab}(A)_0$$

and obtain the following commutative diagram where all functors are exact.

$$
\begin{array}{ccccc}
\mathrm{Eff}(\mathrm{mod}_0\, A, \mathrm{Ab}) & \rightarrowtail & \mathrm{Fp}(\mathrm{mod}\, A, \mathrm{Ab}) & \twoheadrightarrow & \mathrm{Ab}(A)/\mathrm{Ab}(A)_0 \\
\downarrow & & \| & & \downarrow \\
\mathrm{Eff}(\mathrm{mod}\, A, \mathrm{Ab}) & \rightarrowtail & \mathrm{Fp}(\mathrm{mod}\, A, \mathrm{Ab}) & \twoheadrightarrow & (\mathrm{mod}\, A)^{\mathrm{op}} \\
\downarrow & & \| & & \downarrow \\
\mathrm{Fp}(\mathrm{mod}\, A, \mathrm{Ab})' & \rightarrowtail & \mathrm{Fp}(\mathrm{mod}\, A, \mathrm{Ab}) & \twoheadrightarrow & (\mathrm{mod}\, Q(A))^{\mathrm{op}}
\end{array}
$$

The assignment

$$(X, Y) \longmapsto \mathrm{Ext}^1(\mathrm{Tr}\, X, -) \oplus \mathrm{Hom}(Y, -)$$

provides an equivalence

$$(\mathrm{mod}_0\, A) \times (\mathrm{mod}_0\, A)^{\mathrm{op}} \overset{\sim}{\longrightarrow} \mathrm{Ab}(A)_1/\mathrm{Ab}(A)_0.$$

Thus Ab$(A)_1 = \mathrm{Fp}(\mathrm{mod}\, A, \mathrm{Ab})'$ and we obtain the Krull–Gabriel filtration

$$\{0\} \subseteq \mathrm{Eff}(\mathrm{mod}_0\, A, \mathrm{Ab}) \subseteq \mathrm{Fp}(\mathrm{mod}\, A, \mathrm{Ab})' \subseteq \mathrm{Ab}(A).$$

It follows that KG.dim Ab$(A) = 2$. □

A subset $\mathcal{U} \subseteq \mathrm{Ind}\, A$ is Ziegler closed if and only if the following conditions are satisfied.

(1) If $\mathfrak{p} \in \mathrm{Max}\, A$ and $\{n \geq 1 \mid A/\mathfrak{p}^n \in \mathcal{U}\}$ is infinite, then $A_{\mathfrak{p}^\infty}, \hat{A}_\mathfrak{p} \in \mathcal{U}$.
(2) If $\mathcal{U}$ contains infinitely many finite length modules or a module that is not of finite length, then $Q(A) \in \mathcal{U}$.

The Projective Line

Let k be a field and $\mathbb{P}^1_k$ the projective line over k. We view $\mathbb{P}^1_k$ as a scheme and consider the category $\operatorname{Qcoh} \mathbb{P}^1_k$ of quasi-coherent sheaves on $\mathbb{P}^1_k$. This is a locally finitely presented category and the subcategory of finitely presented objects identifies with $\operatorname{coh} \mathbb{P}^1_k$. We denote by $\operatorname{Ind} \mathbb{P}^1_k$ the set $\operatorname{Ind}(\operatorname{Qcoh} \mathbb{P}^1_k)$ of indecomposable pure-injectives.

Recall that we have the following pullback of abelian categories

$$
\begin{array}{ccc}
\operatorname{coh} \mathbb{P}^1_k & \longrightarrow & \operatorname{mod} k[y] \\
\downarrow & & \downarrow \\
\operatorname{mod} k[y^{-1}] & \longrightarrow & \operatorname{mod} k[y, y^{-1}]
\end{array}
$$

which extends to a pullback of Grothendieck categories.

$$
\begin{array}{ccc}
\operatorname{Qcoh} \mathbb{P}^1_k & \longrightarrow & \operatorname{Mod} k[y] \\
\downarrow & & \downarrow \\
\operatorname{Mod} k[y^{-1}] & \longrightarrow & \operatorname{Mod} k[y, y^{-1}]
\end{array}
$$

This reflects the covering $\mathbb{P}^1_k = U' \cup U''$ where we identify

$$U' = \operatorname{Spec} k[y] \qquad U'' = \operatorname{Spec} k[y^{-1}] \qquad U' \cap U'' = \operatorname{Spec} k[y, y^{-1}].$$

We use this covering to describe $\operatorname{Ind} \mathbb{P}^1_k$, though it does not extend to a full covering of $\operatorname{Ind} \mathbb{P}^1_k$. To be more precise, the functors $\operatorname{Qcoh} \mathbb{P}^1_k \to \operatorname{Mod} k[y]$ and $\operatorname{Qcoh} \mathbb{P}^1_k \to \operatorname{Mod} k[y^{-1}]$ admit fully faithful right adjoints, which identify $\operatorname{Mod} k[y]$ and $\operatorname{Mod} k[y^{-1}]$ with definable subcategories of $\operatorname{Qcoh} \mathbb{P}^1_k$; this follows from Theorem 12.2.9. In particular, we have embeddings $\operatorname{Ind} k[y] \subseteq \operatorname{Ind} \mathbb{P}^1_k$ and $\operatorname{Ind} k[y^{-1}] \subseteq \operatorname{Ind} \mathbb{P}^1_k$ with

$$\operatorname{Ind} k[y] \cap \operatorname{Ind} k[y^{-1}] = \operatorname{Ind} k[y, y^{-1}].$$

However, $\operatorname{Ind} \mathbb{P}^1_k \neq \operatorname{Ind} k[y] \cup \operatorname{Ind} k[y^{-1}]$.

The classification of indecomposable pure-injective modules over the Dedekind domain $A = k[y]$ from Theorem 14.2.7 provides a description of most objects in $\operatorname{Ind} \mathbb{P}^1_k$. For instance, the inclusion $\operatorname{Ind} A \to \operatorname{Ind} \mathbb{P}^1_k$ extends the inclusion $\operatorname{Spec} A \to \mathbb{P}^1_k$ and is given by

$$A/\mathfrak{p}^n \mapsto \mathscr{O}_{\mathfrak{p}^n} \qquad A_{\mathfrak{p}^\infty} \mapsto \mathscr{O}_{\mathfrak{p}^\infty} := \operatorname{colim} \mathscr{O}_{\mathfrak{p}^n} \qquad \hat{A}_{\mathfrak{p}} \mapsto \hat{\mathscr{O}}_{\mathfrak{p}} := \lim \mathscr{O}_{\mathfrak{p}^n}.$$

Also, $Q(A) \mapsto \mathscr{Q}$, with $\mathscr{Q}$ the sheaf of rational functions.

Theorem 14.2.9. *The following is, up to isomorphism, a complete list of indecomposable pure-injective quasi-coherent sheaves on $\mathbb{P}^1_k$.*

(1) *For each $n \in \mathbb{Z}$, the sheaf $\mathcal{O}(n)$.*
(2) *For each closed point $\mathfrak{p} \in \mathbb{P}^1_k$ and $r \geq 1$, the sheaf $\mathcal{O}_{\mathfrak{p}^r}$.*
(3) *For each closed point $\mathfrak{p} \in \mathbb{P}^1_k$, the sheaves $\mathcal{O}_{\mathfrak{p}^\infty}$ and $\hat{\mathcal{O}}_{\mathfrak{p}}$.*
(4) *The sheaf of rational functions $\mathscr{Q}$.*

The coherent sheaves and $\mathscr{Q}$ are endofinite objects. The sheaves $\mathcal{O}_{\mathfrak{p}^\infty}$ are Σ-pure-injective objects.

Note that $\mathcal{O}_{\mathfrak{p}^\infty}$ is the injective envelope of $\mathcal{O}_{\mathfrak{p}}$. Also, the sheaf of rational functions $\mathscr{Q}$ is an injective object. This follows from the analogous fact in $\operatorname{Mod} A$ for $A = k[y]$, using that the functor $\operatorname{Mod} A \to \operatorname{Qcoh} \mathbb{P}^1_k$ preserves injectivity.

The proof of the above theorem amounts to an analysis of the abelian category

$$\operatorname{Ab}(\mathbb{P}^1_k) := \operatorname{Ab}(\operatorname{Qcoh} \mathbb{P}^1_k) = \operatorname{Fp}(\mathcal{C}, \operatorname{Ab})^{\operatorname{op}} \quad \text{with} \quad \mathcal{C} = \operatorname{coh} \mathbb{P}^1_k,$$

and we begin with some preparations.

Let $\mathcal{C}_0 := \operatorname{coh}_0 \mathbb{P}^1_k$ denote the category of torsion sheaves, which equals the category of finite length objects in $\mathcal{C}$, and set

$$\mathcal{C}_+ := \{Y \in \mathcal{C} \mid \operatorname{Hom}(X, Y) = 0 \text{ for all } X \in \mathcal{C}_0\}.$$

Lemma 14.2.10. *$(\mathcal{C}_0, \mathcal{C}_+)$ is a split torsion pair for $\mathcal{C}$.*

Proof Fix an object $X \in \mathcal{C}$. Since X is noetherian, there exists a maximal subobject X_0 of finite length. Then every finite length subobject is necessarily contained in X_0, so it is unique. Next, if $\phi \colon S \to X/X_0$ is non-zero for some simple S, then ϕ is injective and we can form the pullback to obtain a larger finite length subobject of X, a contradiction. Thus $X/X_0 \in \mathcal{C}_+$, and therefore $(\mathcal{C}_0, \mathcal{C}_+)$ is a torsion pair.

We know from the classification of objects in $\mathcal{C}$ that $\mathcal{C} = \mathcal{C}_0 \vee \mathcal{C}_+$. Thus the torsion pair $(\mathcal{C}_0, \mathcal{C}_+)$ is split, so X_0 is a direct summand of X. Alternatively, one uses Serre duality (Example 6.5.4). $\qquad \square$

Next we consider the subcategories $\vec{\mathcal{C}}_0$ and $\vec{\mathcal{C}}_+$ of $\operatorname{Qcoh} \mathbb{P}^1_k$ which are obtained by closing under filtered colimits. Observe that $\vec{\mathcal{C}}_0$ and $\vec{\mathcal{C}}_+$ are locally finitely presented categories, with $\operatorname{fp} \vec{\mathcal{C}}_0 = \mathcal{C}_0$ and $\operatorname{fp} \vec{\mathcal{C}}_+ = \mathcal{C}_+$.

Proposition 14.2.11. *We have*

$$\operatorname{Ind} \mathbb{P}^1_k = \operatorname{Ind} \vec{\mathcal{C}}_0 \sqcup \operatorname{Ind} \vec{\mathcal{C}}_+.$$

Proof The subcategory $\mathcal{C}_+ \subseteq \mathcal{C}$ is covariantly finite because the inclusion admits a left adjoint. From this it follows that $\vec{\mathcal{C}}_+$ is a definable subcategory; see Example 12.2.6. Therefore $\operatorname{Ind} \mathbb{P}^1_k \cap \vec{\mathcal{C}}_+ = \operatorname{Ind} \vec{\mathcal{C}}_+$.

The category $\mathcal{C}_0$ is uniserial, and therefore all indecomposable objects in $\vec{\mathcal{C}}_0$ are pure-injective, by Theorem 13.1.28. Thus $\operatorname{Ind}\mathbb{P}^1_k \cap \vec{\mathcal{C}}_0 = \operatorname{Ind}\vec{\mathcal{C}}_0$.

Any object $X \in \operatorname{Ind}\mathbb{P}^1_k$ fits into a pure-exact sequence $0 \to X_0 \to X \to X_+ \to 0$ with $X_0 \in \vec{\mathcal{C}}_0$ and $X_+ \in \vec{\mathcal{C}}_+$; see Example 12.1.10. If $X_0 \neq 0$, then X_0 has an indecomposable direct summand which is pure-injective, again by Theorem 13.1.28. Thus $X_0 = X$. □

Proposition 14.2.12. *We have* $\operatorname{KG.dim}(\operatorname{grmod} k[x,y]) = 2$, *with filtration*

$$0 \subseteq \operatorname{grmod}_0 k[x,y] \subseteq \operatorname{grmod}_1 k[x,y] \subseteq \operatorname{grmod} k[x,y]$$

and subquotients $\operatorname{grmod}_0 k[x,y] \xrightarrow{\sim} \operatorname{grmod} k$,

$$\frac{\operatorname{grmod}_1 k[x,y]}{\operatorname{grmod}_0 k[x,y]} \xrightarrow{\sim} \operatorname{coh}_0 \mathbb{P}^1_k \quad and \quad \frac{\operatorname{grmod} k[x,y]}{\operatorname{grmod}_1 k[x,y]} \xrightarrow{\sim} \operatorname{mod} k(t).$$

Proof Proposition 5.1.6 yields equivalences

$$\frac{\operatorname{grmod} k[x,y]}{\operatorname{grmod}_0 k[x,y]} \xrightarrow{\sim} \operatorname{coh}\mathbb{P}^1_k \quad and \quad \frac{\operatorname{coh}\mathbb{P}^1_k}{\operatorname{coh}_0 \mathbb{P}^1_k} \xrightarrow{\sim} \operatorname{mod} k(t).$$

From this the assertion follows. □

Proof of Theorem 14.2.9 First observe that all objects from the list are pure-injective. In fact, each object $X \in \mathcal{C}$ is endofinite since $\operatorname{Hom}(C, X)$ has finite length over k and therefore also over $\operatorname{End}(X)$ for all $C \in \mathcal{C}$. The inclusion $\operatorname{Ind} k[y] \to \operatorname{Ind}\mathbb{P}^1_k$ preserves endofiniteness and Σ-pure-injectivity, since these properties are preserved by any inclusion of a definable subcategory. Thus $\mathcal{Q}$ is endofinite and each $\mathcal{O}_{\mathrm{p}^\infty}$ is Σ-pure-injective, by Theorem 14.2.7.

It remains to show that the list is complete. We argue via a filtration of $\operatorname{Ab}(\mathbb{P}^1_k)$. The split torsion pair $(\mathcal{C}_0, \mathcal{C}_+)$ yields a sequence of additive functors

$$\mathcal{C}_+ \xrightarrow{\ i\ } \mathcal{C} \xrightarrow{\ p\ } \mathcal{C}/\mathcal{C}_+ \xrightarrow{\sim} \mathcal{C}_0$$

which induces a diagram of exact functors

$$\operatorname{Fp}(\mathcal{C}_0, \operatorname{Ab}) \xrightarrow{\ p^*\ } \operatorname{Fp}(\mathcal{C}, \operatorname{Ab}) \xrightarrow{\ i^*\ } \operatorname{Fp}(\mathcal{C}_+, \operatorname{Ab})$$

by Proposition 2.2.20 and Example 2.2.21. Now observe that the canonical functor $\operatorname{grmod} k[x,y] \twoheadrightarrow \operatorname{coh}\mathbb{P}^1_k$ induces an equivalence $\operatorname{grproj} k[x,y] \xrightarrow{\sim} \mathcal{C}_+$ when restricted to the subcategory of projective modules. Thus we have filtrations

$$0 \subseteq \operatorname{Eff}(\mathcal{C}_0, \operatorname{Ab}) \subseteq \operatorname{Fp}(\mathcal{C}_0, \operatorname{Ab})$$

and

$$0 \subseteq \operatorname{grmod}_0 k[x,y] \subseteq \operatorname{grmod}_1 k[x,y] \subseteq \operatorname{grmod} k[x,y] \cong \operatorname{Fp}(\mathcal{C}_+, \operatorname{Ab})$$

such that each subquotient is a length category, by Corollary 14.2.3 and Proposition 14.2.12.

From these filtrations we obtain a classification of all indecomposable pure-injective objects, using the analogue of Corollary 14.1.16 which is based on the Krull–Gabriel filtration; see also Proposition 14.1.6. To be more precise, the filtration of $\mathrm{Fp}(\mathcal{C}_0, \mathrm{Ab})$ yields the objects in $\mathrm{Ind}\,\vec{\mathcal{C}}_0$, while the filtration of $\mathrm{Fp}(\mathcal{C}_+, \mathrm{Ab})$ yields the objects in $\mathrm{Ind}\,\vec{\mathcal{C}}_+$, keeping in mind the decomposition

$$\mathrm{Ind}\,\mathbb{P}_k^1 = \mathrm{Ind}\,\vec{\mathcal{C}}_0 \sqcup \mathrm{Ind}\,\vec{\mathcal{C}}_+$$

from Proposition 14.2.11. We have $\mathcal{O}_{\mathfrak{p}^r}$ and $\mathcal{O}_{\mathfrak{p}^\infty}$ in $\vec{\mathcal{C}}_0$, with the sheaves $\mathcal{O}_{\mathfrak{p}^r}$ corresponding to the simple objects in $\mathrm{Fp}(\mathcal{C}_0, \mathrm{Ab})$. On the other hand, the sheaves $\mathcal{O}(n)$ correspond to the simple objects in $\mathrm{Fp}(\mathcal{C}_+, \mathrm{Ab})$, the sheaves $\hat{\mathcal{O}}_{\mathfrak{p}}$ correspond to the simple objects in the next layer of $\mathrm{Fp}(\mathcal{C}_+, \mathrm{Ab})$, while the sheaf of rational functions $\mathcal{Q}$ arises from the last layer of $\mathrm{Fp}(\mathcal{C}_+, \mathrm{Ab})$. □

Remark 14.2.13. We have $\mathrm{KG.dim}\,\mathrm{Ab}(\mathbb{P}_k^1) = 2$ and obtain the following filtration of $\mathrm{Ind}\,\mathbb{P}_k^1$:

$$\mathrm{Ind}_{-1}\,\mathbb{P}_k^1 = \{\mathcal{O}(n) \mid n \in \mathbb{Z}\} \cup \{\mathcal{O}_{\mathfrak{p}^r} \mid \mathfrak{p} \in \mathbb{P}_k^1, r \geq 1\}$$
$$\mathrm{Ind}_0\,\mathbb{P}_k^1 = \{\mathcal{O}_{\mathfrak{p}^\infty} \mid \mathfrak{p} \in \mathbb{P}_k^1\} \cup \{\hat{\mathcal{O}}_{\mathfrak{p}} \mid \mathfrak{p} \in \mathbb{P}_k^1\}$$
$$\mathrm{Ind}_1\,\mathbb{P}_k^1 = \{\mathcal{Q}\}.$$

This is the analogue of Corollary 14.2.8 with a similar proof.

The Kronecker Quiver

We consider the following Kronecker quiver

$$\circ \rightrightarrows \circ$$

and fix a field k. A representation (V, W, ϕ, ψ) consists of a pair of vector spaces together with a pair of linear maps between them

$$V \overset{\phi}{\underset{\psi}{\rightrightarrows}} W.$$

The representations of the Kronecker quiver identify with modules over its path algebra, which is the Kronecker algebra.

The sheaf $\mathcal{T} = \mathcal{O} \oplus \mathcal{O}(1)$ is a tilting object of $\mathrm{coh}\,\mathbb{P}_k^1$ and its endomorphism algebra $\Lambda = \mathrm{End}(\mathcal{T})$ identifies with the Kronecker algebra, because of (5.1.7) and (5.1.8). Thus the functor $\mathrm{Hom}(\mathcal{T}, -)$ induces a triangle equivalence

$$\mathbf{D}^b(\mathrm{coh}\,\mathbb{P}_k^1) \overset{\sim}{\longrightarrow} \mathbf{D}^b(\mathrm{mod}\,\Lambda).$$

This follows from Theorem 5.1.2. In fact, the tilting object $\mathcal{T}$ induces a split torsion pair $(\mathcal{T}, \mathcal{F})$ for $\operatorname{coh} \mathbb{P}_k^1$, and we have

$$\mathcal{T} = (\operatorname{coh}_0 \mathbb{P}_k^1) \vee (\operatorname{add}\{\mathcal{O}(n) \mid n \geq 0\}) \qquad \text{and} \qquad \mathcal{F} = \operatorname{add}\{\mathcal{O}(n) \mid n < 0\}.$$

On the other hand, there is a split torsion pair $(\mathcal{U}, \mathcal{V})$ for $\operatorname{mod} \Lambda$ with equivalences

$$\operatorname{Hom}(\mathcal{T}, -): \mathcal{T} \xrightarrow{\sim} \mathcal{V} \qquad \text{and} \qquad \operatorname{Ext}^1(\mathcal{T}, -): \mathcal{F} \xrightarrow{\sim} \mathcal{U}.$$

An explicit description is given in Proposition 5.1.17. In particular, we have

$$\mathcal{V} = (\operatorname{reg} \Lambda) \vee (\operatorname{add}\{P_n \mid n \geq 0\}) \qquad \text{and} \qquad \mathcal{U} = \operatorname{add}\{I_n \mid n \geq 0\}.$$

The following diagram illustrates the tilting from $\operatorname{coh} \mathbb{P}_k^1$ to $\operatorname{mod} \Lambda$.

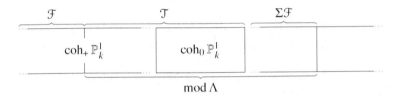

Next we extend the torsion pairs by taking for each subcategory the closure under filtered colimits; see Example 12.1.10. Observe that $\vec{\mathcal{F}}$ and $\vec{\mathcal{U}}$ are locally finitely presented categories such that each object decomposes into a coproduct of finitely presented objects, and such that each object is pure-injective and pure-projective; see Theorem 13.1.20 and Example 13.1.21.

The following is the analogue of Proposition 14.2.11.

Proposition 14.2.14. *We have*

$$\operatorname{Ind} \mathbb{P}^1 = \operatorname{Ind} \vec{\mathcal{T}} \sqcup \operatorname{Ind} \vec{\mathcal{F}} \qquad and \qquad \operatorname{Ind} \Lambda = \operatorname{Ind} \vec{\mathcal{U}} \sqcup \operatorname{Ind} \vec{\mathcal{V}}.$$

Proof The proof is essentially the same as that for Proposition 14.2.11. That each indecomposable pure-injective object over $\mathbb{P}_k^1$ or Λ belongs to one of the subcategories uses the fact that the objects in $\vec{\mathcal{F}}$ and $\vec{\mathcal{U}}$ are pure-injective and pure-projective so that both torsion pairs yield split exact sequences. $\square$

The functor $\operatorname{Hom}(\mathcal{T}, -)$ preserves filtered colimits and extends therefore to an equivalence $\vec{\mathcal{T}} \xrightarrow{\sim} \vec{\mathcal{V}}$. Analogously, $\operatorname{Ext}^1(\mathcal{T}, -)$ yields an equivalence $\vec{\mathcal{F}} \xrightarrow{\sim} \vec{\mathcal{U}}$. Combining these equivalences with Proposition 5.1.17 and Proposition 14.2.14 gives the following bijection $\operatorname{Ind} \mathbb{P}^1 \xrightarrow{\sim} \operatorname{Ind} \Lambda$:

$$\mathcal{O}_{\mathfrak{p}^n} \mapsto R_{\mathfrak{p}^n} \qquad \mathcal{O}_{\mathfrak{p}^\infty} \mapsto R_{\mathfrak{p}^\infty} \qquad \hat{\mathcal{O}}_{\mathfrak{p}} \mapsto \hat{R}_{\mathfrak{p}} \qquad \mathcal{Q} \mapsto Q$$

and

$$\mathcal{O}(n) \mapsto P_n \quad (n \geq 0) \qquad \mathcal{O}(n) \mapsto I_{-n+1} \quad (n < 0).$$

In fact, we use that $\mathcal{O}_{\mathfrak{p}^\infty} = \operatorname{colim} \mathcal{O}_{\mathfrak{p}^n}$, so

$$R_{\mathfrak{p}^\infty} := \operatorname{colim} R_{\mathfrak{p}^n} \cong \operatorname{colim} \operatorname{Hom}(\mathcal{T}, \mathcal{O}_{\mathfrak{p}^n}) \cong \operatorname{Hom}(\mathcal{T}, \mathcal{O}_{\mathfrak{p}^\infty}).$$

Analogously, $\hat{\mathcal{O}}_{\mathfrak{p}} = \lim \mathcal{O}_{\mathfrak{p}^n}$, so

$$\hat{R}_{\mathfrak{p}} := \lim R_{\mathfrak{p}^n} \cong \lim \operatorname{Hom}(\mathcal{T}, \mathcal{O}_{\mathfrak{p}^n}) \cong \operatorname{Hom}(\mathcal{T}, \hat{\mathcal{O}}_{\mathfrak{p}}).$$

Finally, the distinguished sheaf $\mathcal{Q}$ (indecomposable endofinite but not finitely presented) corresponding to the generic point of $\mathbb{P}^1_k$ is mapped to the distinguished module Q, which is indecomposable endofinite but not finitely presented.

The following theorem summarises the description of Ind Λ.

Theorem 14.2.15. *The following is, up to isomorphism, a complete list of indecomposable pure-injective Λ-modules.*

(1) *For each $n \geq 0$, the modules P_n and I_n.*
(2) *For each closed $\mathfrak{p} \in \mathbb{P}^1_k$ and $r \geq 1$, the module $R_{\mathfrak{p}^r}$.*
(3) *For each closed $\mathfrak{p} \in \mathbb{P}^1_k$, the Prüfer module $R_{\mathfrak{p}^\infty}$ and the adic module $\hat{R}_{\mathfrak{p}}$.*
(4) *The generic module Q.*

The finitely presented modules and Q are endofinite. The modules $R_{\mathfrak{p}^\infty}$ are Σ-pure-injective.   □

Remark 14.2.16. We have KG.dim Ab$(\Lambda) = 2$ and obtain the following filtration of Ind Λ:

$$\operatorname{Ind}_{-1} \Lambda = \operatorname{Ind} \Lambda \cap \operatorname{mod} \Lambda$$
$$\operatorname{Ind}_0 \Lambda = \{R_{\mathfrak{p}^\infty} \mid \mathfrak{p} \in \mathbb{P}^1_k\} \cup \{\hat{R}_{\mathfrak{p}} \mid \mathfrak{p} \in \mathbb{P}^1_k\}$$
$$\operatorname{Ind}_1 \Lambda = \{Q\}.$$

This is the analogue of Corollary 14.2.8 with a similar proof.

A subset $\mathcal{U} \subseteq \operatorname{Ind} \Lambda$ is Ziegler closed if and only if the following conditions are satisfied.

(1) $R_{\mathfrak{p}^\infty} \in \mathcal{U}$ provided $\operatorname{Hom}(R_{\mathfrak{p}}, X) \neq 0$ for infinitely many $X \in \mathcal{U} \cap \operatorname{mod} \Lambda$.
(2) $\hat{R}_{\mathfrak{p}} \in \mathcal{U}$ provided $\operatorname{Hom}(X, R_{\mathfrak{p}}) \neq 0$ for infinitely many $X \in \mathcal{U} \cap \operatorname{mod} \Lambda$.
(3) $Q \in \mathcal{U}$ provided $\mathcal{U}$ contains infinitely many finite length modules or a module that is not of finite length.

Injective Cohomology Representations

Let G be a finite group and k a field of characteristic $p > 0$. We recall the functor $T \colon \operatorname{Inj} H^*(G, k) \to \operatorname{StMod} kG$; it identifies the torsion free injective $H^*(G, k)$-modules with a definable subcategory of $\operatorname{Mod} kG$ which we denote by $\mathcal{T}(G, k)$ (Corollary 12.4.24).

The *p-rank* of a finite group G is the largest integer n such that G has an elementary abelian subgroup of order p^n. We note that the Krull dimension of $H^*(G, k)$ equals the p-rank of G by a theorem of Quillen [29, Theorem 5.3.8].

Proposition 14.2.17. $\operatorname{KG.dim} \operatorname{Ab}(\mathcal{T}(G, k)) + 1$ *equals the p-rank of* G.

Proof Set $R = H^*(G, k)$. We consider the category of graded R-modules and the definable subcategory $\operatorname{Inj} R$ of injective R-modules. It follows from Proposition 14.1.14 that the Krull dimension of $H^*(G, k)$ equals

$$\operatorname{KG.dim} \operatorname{Ab}(\operatorname{Inj} R) = \operatorname{KG.dim} \operatorname{Ab}(\operatorname{grmod} R).$$

For the definable subcategory $\operatorname{Inj}_+ R$ of torsion free modules we have

$$\operatorname{KG.dim} \operatorname{Ab}(\operatorname{Inj}_+ R) = \operatorname{KG.dim} \operatorname{Ab}((\operatorname{grmod} R)/(\operatorname{grmod}_0 R))$$
$$= \operatorname{KG.dim} \operatorname{Ab}(\operatorname{grmod} R) - 1.$$

We claim that

$$\operatorname{KG.dim} \operatorname{Ab}(\operatorname{Inj}_+ R) = \operatorname{KG.dim} \operatorname{Ab}(\mathcal{T}(G, k)),$$

and then the proof is complete because of Quillen's result. The claim follows from an iterated application of the lemma below, because the functor T preserves products and coproducts, so it identifies isolated points. □

Let $\mathcal{A}$ be a locally finitely presented category and $\mathcal{B} \subseteq \mathcal{A}$ a definable subcategory. We consider the Krull–Gabriel filtration of $\operatorname{Ab}(\mathcal{B})$ and this yields subsets $\operatorname{Ind}_\alpha \mathcal{B} \subseteq \operatorname{Ind} \mathcal{B} := \mathcal{B} \cap \operatorname{Ind} \mathcal{A}$ for each ordinal $\alpha \geq -1$.

Lemma 14.2.18. *Suppose that* $\operatorname{Ab}(\mathcal{B})$ *is noetherian. Then for* $\alpha \geq -1$ *and* $X \in \mathcal{U} := \bigsqcup_{\beta \geq \alpha} \operatorname{Ind}_\beta \mathcal{B}$ *the following are equivalent.*

(1) X *is isolated, so* $\{X\} \subseteq \mathcal{U}$ *is open.*
(2) $X \in \operatorname{Ind}_\alpha \mathcal{B}$.
(3) *If* X *is isomorphic to a direct summand of a product* $\prod_{i \in I} Y_i$ *of indecomposable objects in* $\mathcal{U}$, *then* $X \cong Y_i$ *for some* $i \in I$.

Proof Apply Theorem 12.3.13. The assumption on $\operatorname{Ab}(\mathcal{B})$ to be noetherian is needed; it implies that each simple object in $\mathbf{P}(\mathcal{B})$ is finitely presented. □

Example 14.2.19. Let $G = \mathbb{Z}/2 \times \mathbb{Z}/2$ and k be a field of characteristic two. We denote by Λ the path algebra of the Kronecker quiver. Then (5.1.23) provides a functor $\operatorname{Mod} \Lambda \to \operatorname{Mod} kG$ which identifies the Prüfer modules $R_{\mathfrak{p}^\infty}$ ($\mathfrak{p} \in \mathbb{P}^1_k$) and the generic module Q with the indecomposable objects in $\mathcal{T}(G, k)$.

Notes

The method of classifying the pure-injective objects via the Krull–Gabriel filtration is taken from Jensen and Lenzing [118], which is modelled after [79]. Note that the corresponding dimensions may differ, depending on the use of *all* simple objects versus the use of the *finitely presented* simple objects in a locally finitely presented Grothendieck category.

The general decomposition theory of injective objects in Grothendieck categories is based on the spectral category in the sense of Gabriel and Oberst [82].

The classification of pure-injectives works well for modules over Dedekind domains, and in particular for abelian groups; for the classical approach see Kaplansky [119] and Fuchs [77].

The striking parallel between abelian groups and modules over tame hereditary algebras was pointed out by Ringel [171]. The Krull–Gabriel dimension for a tame hereditary algebra is equal to two by work of Geigle [87], and for the Ziegler topology we refer to [161, 174]. The computation of the Krull–Gabriel filtration for uniserial categories is taken from joint work with Vossieck [135].

References

[1] J. Adámek and J. Rosický, *Locally presentable and accessible categories*, London Mathematical Society Lecture Note Series, 189, Cambridge University Press, Cambridge, 1994.

[2] K. Akin and D. A. Buchsbaum, Characteristic-free representation theory of the general linear group. II. Homological considerations, Adv. Math. **72** (1988), no. 2, 171–210.

[3] K. Akin, D. A. Buchsbaum and J. Weyman, Schur functors and Schur complexes, Adv. Math. **44** (1982), no. 3, 207–278.

[4] L. Alonso Tarrío, A. Jeremías López and M. J. Souto Salorio, Localization in categories of complexes and unbounded resolutions, Can. J. Math. **52** (2000), no. 2, 225–247.

[5] L. Angeleri Hügel, D. Happel and H. Krause (eds.), *Handbook of tilting theory*, London Mathematical Society Lecture Note Series, 332, Cambridge University Press, Cambridge, 2007.

[6] I. Assem and A. Skowroński, Iterated tilted algebras of type $\tilde{A}_n$, Math. Z. **195** (1987), no. 2, 269–290.

[7] M. Auslander, Coherent functors, in *Proc. Conf. categorical algebra (La Jolla, Calif., 1965)*, 189–231, Springer, New York, 1966.

[8] M. Auslander, Comments on the functor Ext, Topology **8** (1969), 151–166.

[9] M. Auslander, Representation theory of Artin algebras II, Commun. Algebra **1** (1974), 269–310.

[10] M. Auslander, Large modules over Artin algebras, in *Algebra, topology, and category theory (a collection of papers in honor of Samuel Eilenberg)*, 1–17, Academic Press, New York, 1976.

[11] M. Auslander, Functors and morphisms determined by objects, in *Representation theory of algebras (Proc. Conf., Temple Univ., Philadelphia, Pa., 1976)*, 1–244, Lecture Notes in Pure and Applied Mathematics, 37, Dekker, New York, 1978.

[12] M. Auslander and D. A. Buchsbaum, Homological dimension in noetherian rings II, Trans. Am. Math. Soc. **88** (1958), 194–206.

[13] M. Auslander and R.-O. Buchweitz, The homological theory of maximal Cohen–Macaulay approximations, Mém. Soc. Math. Fr. **38** (1989), 5–37.

[14] M. Auslander, M. I. Platzeck and I. Reiten, Coxeter functors without diagrams, Trans. Am. Math. Soc. **250** (1979), 1–46.

[15] M. Auslander and I. Reiten, Representation theory of Artin algebras. III. Almost split sequences, Commun. Algebra **3** (1975), 239–294.

[16] M. Auslander and I. Reiten, Applications of contravariantly finite subcategories, Adv. Math. **86** (1991), no. 1, 111–152.

[17] M. Auslander and I. Reiten, Cohen–Macaulay and Gorenstein Artin algebras, in *Representation theory of finite groups and finite-dimensional algebras (Bielefeld, 1991)*, 221–245, Progress in Mathematics, 95, Birkhäuser, Basel, 1991.

[18] M. Auslander, I. Reiten and S. O. Smalø, *Representation theory of Artin algebras*, corrected reprint of the 1995 original, Cambridge Studies in Advanced Mathematics, 36, Cambridge University Press, Cambridge, 1997.

[19] G. Azumaya, Corrections and supplemetaries to my paper concerning Krull–Remak–Schmidt's theorem, Nagoya Math. J. **1** (1950), 117–124.

[20] R. Baer, Erweiterung von Gruppen und ihre Isomorphismen, Math. Z. **38** (1934), no. 1, 375–416.

[21] R. Baer, Abelian groups that are direct summands of every containing abelian group, Bull. Am. Math. Soc. **46** (1940), 800–806.

[22] H. Bass, On the ubiquity of Gorenstein rings, Math. Z. **82** (1963), 8–28.

[23] H. Bass and M. P. Murthy, Grothendieck groups and Picard groups of abelian group rings, Ann. Math. (Ser. 2) **86** (1967), 16–73.

[24] P. Baumann and C. Kassel, The Hall algebra of the category of coherent sheaves on the projective line, J. Reine Angew. Math. **533** (2001), 207–233.

[25] A. A. Beilinson, Coherent sheaves on $\mathbf{P}^n$ and problems in linear algebra, Funktsional. Anal. Prilozhen. **12** (1978), no. 3, 68–69.

[26] A. A. Beilinson, J. Bernšteĭn and P. Deligne, Faisceaux pervers, in *Analysis and topology on singular spaces, I (Luminy, 1981)*, 5–171, Astérisque, 100, Societé Mathématique de France, Paris, 1982.

[27] A. Beilinson, V. Ginzburg and W. Soergel, Koszul duality patterns in representation theory, J. Am. Math. Soc. **9** (1996), no. 2, 473–527.

[28] A. Beligiannis and I. Reiten, *Homological and homotopical aspects of torsion theories*, Memoirs of the American Mathematical Society, 188, American Mathematical Society, Providence, RI, 2007, no. 883.

[29] D. J. Benson, *Representations and cohomology II*, Cambridge Studies in Advanced Mathematics, 31, Cambridge University Press, Cambridge, 1991.

[30] D. J. Benson, *Representations of elementary abelian p-groups and vector bundles*, Cambridge Tracts in Mathematics, 208, Cambridge University Press, Cambridge, 2017.

[31] D. Benson, S. B. Iyengar, H. Krause and J. Pevtsova, Local duality for representations of finite group schemes, Compos. Math. **155** (2019), no. 2, 424–453.

[32] D. Benson and H. Krause, Pure injectives and the spectrum of the cohomology ring of a finite group, J. Reine Angew. Math. **542** (2002), 23–51.

[33] G. M. Bergman, Coproducts and some universal ring constructions, Trans. Am. Math. Soc. **200** (1974), 33–88.

[34] I. N. Bernšteĭn, I. M. Gel'fand and S. I. Gel'fand, A certain category of g-modules, Funktsional. Anal. Prilozhen. **10** (1976), no. 2, 1–8.

[35] I. N. Bernšteĭn, I. M. Gel'fand and S. I. Gel'fand, Algebraic vector bundles on $\mathbf{P}^n$ and problems of linear algebra, Funktsional. Anal. Prilozhen. **12** (1978), no. 3, 66–67.

[36] I. N. Bernšteĭn, I. M. Gel'fand and V. A. Ponomarev, Coxeter functors, and Gabriel's theorem, Usp. Mat. Nauk **28** (1973), no. 2(170), 19–33.

[37] M. Bökstedt and A. Neeman, Homotopy limits in triangulated categories, Compos. Math. **86** (1993), no. 2, 209–234.

[38] A. I. Bondal and M. M. Kapranov, Representable functors, Serre functors, and mutations, Izv. Akad. Nauk SSSR Ser. Mat. **53** (1989), no. 6, 1183–1205, 1337; translation in Math. USSR-Izv. **35** (1990), no. 3, 519–541.

[39] N. Bourbaki, *Éléments de mathématique: Algèbre*, Chapters 4–7, Lecture Notes in Mathematics, 864, Masson, Paris, 1981.

[40] R. Brauer and C. Nesbitt, On the regular representations of algebras, Proc. Natl. Acad. Sci. U.S.A. **23** (1937), 236–240.

[41] S. Brenner and M. C. R. Butler, Generalizations of the Bernšteĭn Gel'fand Ponomarev reflection functors, in *Representation theory, II (Proc. Second Int. Conf., Carleton Univ., Ottawa, Ont., 1979)*, 103–169, Lecture Notes in Mathematics, 832, Springer, Berlin, 1980.

[42] E. H. Brown, Jr., Cohomology theories, Ann. Math. (Ser. 2) **75** (1962), 467–484.

[43] A. B. Buan and H. Krause, Tilting and cotilting for quivers and type $\tilde{A}_n$, J. Pure Appl. Algebra **190** (2004), 1–21.

[44] R.-O. Buchweitz, Maximal Cohen–Macaulay modules and Tate-cohomology over Gorenstein rings, Universität Hannover, 1986.

[45] H. Cartan, *Algèbres d'Eilenberg–MacLane*, Exposés 2–11, Sém. H. Cartan, Éc. Normale Sup. (1954–1955), Sécrétariat Mathématique, Paris, 1956.

[46] H. Cartan and S. Eilenberg, *Homological algebra*, Princeton University Press, Princeton, NJ, 1956.

[47] R. W. Carter and G. Lusztig, On the modular representations of the general linear and symmetric groups, Math. Z. **136** (1974), 193–242.

[48] A.-L. Cauchy, Mémoire sur les fonctions alternées et sur les sommes alternées, Exercices d'analyse et de physique mathématique, ii (1841), 151–159; Œuvres complètes, 2ème série xii, 173–182, Gauthier-Villars, Paris, 1916.

[49] S. U. Chase, Direct products of modules, Trans. Am. Math. Soc. **97** (1960), 457–473.

[50] S. U. Chase, On direct sums and products of modules, Pacific J. Math. **12** (1962), 847–854.

[51] T. Church, J. S. Ellenberg, B. Farb and R. Nagpal, FI-modules over Noetherian rings, Geom. Topol. **18** (2014), no. 5, 2951–2984.

[52] E. Cline, B. Parshall and L. Scott, Finite-dimensional algebras and highest weight categories, J. Reine Angew. Math. **391** (1988), 85–99.

[53] P. M. Cohn, On the free product of associative rings, Math. Z. **71** (1959), 380–398.

[54] P. M. Cohn, *Free rings and their relations*, Academic Press, London, 1971.

[55] L. Corry, *Modern algebra and the rise of mathematical structures*, second edition, Birkhäuser, Basel, 2004.

[56] W. Crawley-Boevey, Tame algebras and generic modules, Proc. London Math. Soc. (Ser. 3) **63** (1991), no. 2, 241–265.

[57] W. Crawley-Boevey, Modules of finite length over their endomorphism rings, in *Representations of algebras and related topics (Tsukuba, 1990)*, 127–184, London Mathematical Society Lecture Note Series, 168, Cambridge University Press, Cambridge, 1992.

[58] W. Crawley-Boevey, Locally finitely presented additive categories, Commun. Algebra **22** (1994), no. 5, 1641–1674.

[59] W. Crawley-Boevey, Infinite-dimensional modules in the representation theory of finite-dimensional algebras, in *Algebras and modules, I (Trondheim, 1996)*, 29–54, CMS Conference Proceedings, 23, American Mathematical Society, Providence, RI, 1998.

[60] W. Crawley-Boevey, Classification of modules for infinite-dimensional string algebras, Trans. Am. Math. Soc. **370** (2018), no. 5, 3289–3313.

[61] C. de Concini, D. Eisenbud and C. Procesi, Young diagrams and determinantal varieties, Invent. Math. **56** (1980), no. 2, 129–165.

[62] P. Deligne, Cohomologie à supports propres, in *SGA 4, Théorie des Topos et Cohomologie Etale des Schémas*, vol. 3, 250–480, Lecture Notes in Mathematics, 305, Springer, Heidelberg, 1973.

[63] A. Djament, La propriété noethérienne pour les foncteurs entre espaces vectoriels [d'après A. Putman, S. Sam et A. Snowden], Astérisque, 380, Séminaire Bourbaki, Vol. 2014/2015 (2016), Exp. No. 1090, 35–60.

[64] V. Dlab and C. M. Ringel, Quasi-hereditary algebras, Illinois J. Math. **33** (1989), no. 2, 280–291.

[65] V. Dlab and C. M. Ringel, The module theoretical approach to quasi-hereditary algebras, in *Representations of algebras and related topics (Kyoto, 1990)*, 200–224, London Mathematical Society Lecture Note Series, 168, Cambridge University Press, Cambridge, 1992.

[66] S. Donkin, A filtration for rational modules, Math. Z. **177** (1981), no. 1, 1–8.

[67] S. Donkin, On Schur algebras and related algebras I, J. Algebra **104** (1986), no. 2, 310–328.

[68] S. Donkin, On tilting modules for algebraic groups, Math. Z. **212** (1993), no. 1, 39–60.

[69] P. Doubilet, G.-C. Rota and J. Stein, On the foundations of combinatorial theory IX. Combinatorial methods in invariant theory, Stud. Appl. Math. **53** (1974), 185–216.

[70] B. Eckmann and A. Schopf, Über injektive Moduln, Arch. Math. (Basel) **4** (1953), 75–78.

[71] S. Eilenberg, Homological dimension and syzygies, Ann. Math. (Ser. 2) **64** (1956), 328–336.

[72] S. Eilenberg and S. MacLane, Group extensions and homology, Ann. Math. (Ser. 2) **43** (1942), 757–831.

[73] S. Eilenberg and T. Nakayama, On the dimension of modules and algebras II. Frobenius algebras and quasi-Frobenius rings, Nagoya Math. J. **9** (1955), 1–16.

[74] J. Franke, On the Brown representability theorem for triangulated categories, Topology **40** (2001), no. 4, 667–680.

[75] P. Freyd, Representations in abelian categories, in *Proc. Conf. categorical algebra (La Jolla, Calif., 1965)*, 95–120, Springer, New York, 1966.

[76] E. M. Friedlander and A. Suslin, Cohomology of finite group schemes over a field, Invent. Math. **127** (1997), no. 2, 209–270.

[77] L. Fuchs, *Infinite abelian groups*, vol. I, Pure and Applied Mathematics, 36, Academic Press, New York, 1970.

[78] W. Fulton, *Young tableaux*, London Mathematical Society Student Texts, 35, Cambridge University Press, Cambridge, 1997.

[79] P. Gabriel, Des catégories abéliennes, Bull. Soc. Math. Fr. **90** (1962), 323–448.

[80] P. Gabriel, Auslander–Reiten sequences and representation-finite algebras, in *Representation theory, I (Proc. Workshop, Carleton Univ., Ottawa, Ont., 1979)*, 1–71, Lecture Notes in Mathematics, 831, Springer, Berlin, 1980.

[81] P. Gabriel, *Un jeu? Les nombres de Catalan*, UniZürich, Mitteilungsblatt des Rektorates **6** (1981), 4–5.

[82] P. Gabriel and U. Oberst, Spektralkategorien und reguläre Ringe im von-Neumannschen Sinn, Math. Z. **92** (1966), 389–395.

[83] P. Gabriel and A. V. Roiter, *Representations of finite-dimensional algebras* [with a chapter by B. Keller], Encyclopaedia of Mathematical Sciences, 73, Algebra, VIII, Springer, Berlin, 1992.

[84] P. Gabriel and F. Ulmer, *Lokal präsentierbare Kategorien*, Lecture Notes in Mathematics, 221, Springer, Berlin, 1971.

[85] P. Gabriel and M. Zisman, *Calculus of fractions and homotopy theory*, Springer, New York, 1967.

[86] S. Garavaglia, Decomposition of totally transcendental modules, J. Symbolic Logic **45** (1980), no. 1, 155–164.

[87] W. Geigle, The Krull–Gabriel dimension of the representation theory of a tame hereditary Artin algebra and applications to the structure of exact sequences, Manuscripta Math. **54** (1985), no. 1–2, 83–106.

[88] W. Geigle and H. Lenzing, A class of weighted projective curves arising in representation theory of finite-dimensional algebras, in *Singularities, representation of algebras, and vector bundles (Lambrecht, 1985)*, 265–297, Lecture Notes in Mathematics, 1273, Springer, Berlin, 1987.

[89] W. Geigle and H. Lenzing, Perpendicular categories with applications to representations and sheaves, J. Algebra **144** (1991), no. 2, 273–343.

[90] Ch. Geiß and I. Reiten, Gentle algebras are Gorenstein, in *Representations of algebras and related topics*, 129–133, Fields Institute Communications, 45, American Mathematical Society, Providence, RI, 2005.

[91] E. L. Green and D. Zacharia, The cohomology ring of a monomial algebra, Manuscripta Math. **85** (1994), no. 1, 11–23.

[92] J. A. Green, *Polynomial representations of* GL_n, Lecture Notes in Mathematics, 830, Springer, Berlin, 1980.

[93] J. A. Green, Combinatorics and the Schur algebra, J. Pure Appl. Algebra **88** (1993), no. 1–3, 89–106.

[94] A. Grothendieck, Sur quelques points d'algèbre homologique, Tôhoku Math. J. (2) **9** (1957), 119–221.

[95] A. Grothendieck, The cohomology theory of abstract algebraic varieties, in *Proc. Int. Congress of mathematicians (Edinburgh, 1958)*, 103–118, Cambridge University Press, New York, 1960.

[96] A. Grothendieck, Groupes de classes des categories abéliennes et triangulées, complexes parfaits, in *Cohomologie l-adique et fonctions L*, 351–371, Lecture Notes in Mathematics, 589, Springer, Berlin, 1977.

[97] A. Grothendieck and J. A. Dieudonné, *Eléments de géométrie algébrique. I*, Grundlehren der Mathematischen Wissenschaften, 166, Springer, Berlin, 1971.

[98] A. Grothendieck and J. L. Verdier, Préfaisceaux, in *SGA 4, Théorie des Topos et Cohomologie Etale des Schémas*, vol1, *Théorie des Topos*, 1–217, Lecture Notes in Mathematics, 269, Springer, Heidelberg, 1972–1973.

[99] L. Gruson and C. U. Jensen, Deux applications de la notion de *L*-dimension, C. R. Acad. Sci. Paris Sér. A–B **282** (1976), no. 1, A23–A24.

[100] L. Gruson and C. U. Jensen, Dimensions cohomologiques reliées aux foncteurs $\lim^{(i)}$, in *Séminaire d'Algèbre Paul Dubreil et Marie-Paule Malliavin (Paris, 1980)*, 234–294, Lecture Notes in Mathematics, 867, Springer, Berlin, 1981.

[101] D. Happel, On the derived category of a finite-dimensional algebra, Comment. Math. Helv. **62** (1987), no. 3, 339–389.

[102] D. Happel, *Triangulated categories in the representation theory of finite-dimensional algebras*, London Mathematical Society Lecture Note Series, 119, Cambridge University Press, Cambridge, 1988.

[103] D. Happel, Auslander–Reiten triangles in derived categories of finite-dimensional algebras, Proc. Am. Math. Soc. **112** (1991), no. 3, 641–648.

[104] D. Happel, A characterization of hereditary categories with tilting object, Invent. Math. **144** (2001), no. 2, 381–398.

[105] D. Happel, I. Reiten and S. O. Smalø, *Tilting in abelian categories and quasitilted algebras*, Memoirs of the American Mathematical Society, 120, American Mathematical Society, Providence, RI, 1996, no. 575.

[106] R. Hartshorne, *Residues and duality*, Lecture notes of a seminar on the work of A. Grothendieck, given at Harvard 1963/64 [with an appendix by P. Deligne], Lecture Notes in Mathematics, 20, Springer, Berlin, 1966.

[107] A. Heller, Homological algebra in abelian categories, Ann. Math. (Ser. 2) **68** (1958), 484–525.

[108] A. Heller, The loop-space functor in homological algebra, Trans. Am. Math. Soc. **96** (1960), 382–394.

[109] H.-W. Henn, J. Lannes and L. Schwartz, The categories of unstable modules and unstable algebras over the Steenrod algebra modulo nilpotent objects, Am. J. Math. **115** (1993), no. 5, 1053–1106.

[110] I. Herzog, Elementary duality of modules, Trans. Am. Math. Soc. **340** (1993), no. 1, 37–69.

[111] I. Herzog, The Ziegler spectrum of a locally coherent Grothendieck category, Proc. London Math. Soc. (Ser. 3) **74** (1997), no. 3, 503–558.

[112] D. Hilbert, Über die Theorie der algebraischen Formen, Math. Ann. **36** (1890), no. 4, 473–534.

[113] M. Hochster, Prime ideal structure in commutative rings, Trans. Am. Math. Soc. **142** (1969), 43–60.

[114] D. Hughes and J. Waschbüsch, Trivial extensions of tilted algebras, Proc. London Math. Soc. (Ser. 3) **46** (1983), no. 2, 347–364.

[115] J. E. Humphreys, *Representations of semisimple Lie algebras in the BGG category $\mathcal{O}$*, Graduate Studies in Mathematics, 94, American Mathematical Society, Providence, RI, 2008.

[116] Y. Iwanaga, On rings with finite self-injective dimension, Commun. Algebra **7** (1979), no. 4, 393–414.

[117] J. P. Jans, On the indecomposable representations of algebras, Ann. Math. (Ser. 2) **66** (1957), 418–429.

|118| C. U. Jensen and H. Lenzing, *Model-theoretic algebra with particular emphasis on fields, rings, modules*, Gordon and Breach Science Publishers, New York, 1989.

|119| I. Kaplansky, *Infinite abelian groups*, University of Michigan Press, Ann Arbor, MI, 1954, revised 1969.

|120| B. Keller, Chain complexes and stable categories, Manuscripta Math. **67** (1990), 379–417.

|121| B. Keller, Deriving DG categories, Ann. Sci. Éc. Norm. Supér. (Sér. 4) **27** (1994), no. 1, 63–102.

|122| B. Keller and H. Krause, Tilting preserves finite global dimension, C. R. Math. Acad. Sci. Paris **358** (2020), no. 5, 563–571.

|123| B. Keller and D. Vossieck, Sous les catégories dérivées, C. R. Acad. Sci. Paris Sér. I Math. **305** (1987), no. 6, 225–228.

|124| R. Kiełpiński, On Γ-pure injective modules, Bull. Acad. Polon. Sci. Sér. Sci. Math. Astronom. Phys. **15** (1967), 127–131.

|125| H. Krause, The spectrum of a locally coherent category, J. Pure Appl. Algebra **114** (1997), no. 3, 259–271.

|126| H. Krause, Exactly definable categories, J. Algebra **201** (1998), no. 2, 456–492.

|127| H. Krause, A Brown representability theorem via coherent functors, Topology **41** (2002), no. 4, 853–861.

|128| H. Krause, Coherent functors and covariantly finite subcategories, Algebras Represent. Theory **6** (2003), no. 5, 475–499.

|129| H. Krause, The stable derived category of a Noetherian scheme, Compos. Math. **141** (2005), no. 5, 1128–1162.

|130| H. Krause, Koszul, Ringel and Serre duality for strict polynomial functors, Compos. Math. **149** (2013), no. 6, 996–1018.

|131| H. Krause, Krull–Schmidt categories and projective covers, Expo. Math. **33** (2015), no. 4, 535–549.

|132| H. Krause, Completing perfect complexes |with appendices by T. Barthel and B. Keller|, Math. Z. **296** (2020), 1387–1427.

|133| H. Krause and C. M. Ringel, *Infinite length modules*, Trends in Mathematics, Birkhäuser, Basel, 2000.

|134| H. Krause and M. Saorín, On minimal approximations of modules, in *Trends in the representation theory of finite-dimensional algebras (Seattle, WA, 1997)*, 227–236, Contemporary Mathematics, 229, American Mathematical Society, Providence, RI, 1998.

|135| H. Krause and D. Vossieck, Length categories of infinite height, in *Geometric and topological aspects of the representation theory of finite groups*, 213–234, Springer Proceedings in Mathematics and Statistics, 242, Springer, Cham, 2018.

|136| T. Y. Lam, *A first course in noncommutative rings*, Graduate Texts in Mathematics, 131, Springer, New York, 1991.

|137| H. Lenzing, Über die Funktoren $\mathrm{Ext}^1(\cdot, E)$ und $\mathrm{Tor}_1(\cdot, E)$, Dissertation, FU Berlin, 1964.

|138| H. Lenzing, Endlich präsentierbare Moduln, Arch. Math. (Basel) **20** (1969), 262–266.

[139] H. Lenzing, Auslander's work on Artin algebras, in *Algebras and modules, I (Trondheim, 1996)*, 83–105, CMS Conference Proceedings, 23, American Mathematical Society, Providence, RI, 1998.

[140] I. G. Macdonald, *Symmetric functions and Hall polynomials*, second edition, Oxford Mathematical Monographs, Oxford University Press, New York, 1995.

[141] S. Mac Lane, *Homology*, Die Grundlehren der mathematischen Wissenschaften, 114, Academic Press, New York; Springer, Berlin, 1963.

[142] S. Mac Lane, *Categories for the working mathematician*, second edition, Graduate Texts in Mathematics, 5, Springer, New York, 1998.

[143] E. Matlis, Injective modules over Noetherian rings, Pacific J. Math. **8** (1958), 511–528.

[144] J. Milnor, On axiomatic homology theory, Pacific J. Math. **12** (1962), 337–341.

[145] B. Mitchell, Rings with several objects, Adv. Math. **8** (1972), 1–161.

[146] M. Nagata, *Local rings*, Interscience Tracts in Pure and Applied Mathematics, 13, Interscience Publishers, New York, 1962.

[147] A. Neeman, The derived category of an exact category, J. Algebra **135** (1990), no. 2, 388–394.

[148] A. Neeman, The Grothendieck duality theorem via Bousfield's techniques and Brown representability, J. Am. Math. Soc. **9** (1996), no. 1, 205–236.

[149] A. Neeman, Brown representability for the dual, Invent. Math. **133** (1998), no. 1, 97–105.

[150] A. Neeman, *Triangulated categories*, Annals of Mathematics Studies, 148, Princeton University Press, Princeton, NJ, 2001.

[151] O. Ore, Linear equations in non-commutative fields, Ann. Math. (Ser. 2) **32** (1931), no. 3, 463–477.

[152] D. O. Orlov, Triangulated categories of singularities and D-branes in Landau–Ginzburg models, Proc. Steklov Inst. Math. **2004**, no. 3(246), 227–248; translated from Tr. Mat. Inst. Steklova **246** (2004), Algebr. Geom. Metody, Svyazi Prilozh., 240–262.

[153] D. O. Orlov, Formal completions and idempotent completions of triangulated categories of singularities, Adv. Math. **226** (2011), no. 1, 206–217.

[154] B. L. Osofsky, Homological dimension and cardinality, Trans. Am. Math. Soc. **151** (1970), 641–649.

[155] Z. Papp, On algebraically closed modules, Publ. Math. Debrecen **6** (1959), 311–327.

[156] B. Pareigis, *Categories and functors*, translated from the German, Pure and Applied Mathematics, 39, Academic Press, New York, 1970.

[157] B. J. Parshall, Finite-dimensional algebras and algebraic groups, in *Classical groups and related topics (Beijing, 1987)*, 97–114, Contemporary Mathematics, 82, American Mathematical Society, Providence, RI,1989.

[158] B. J. Parshall and L. L. Scott, Derived categories, quasi-hereditary algebras, and algebraic groups, in *Proc. Ottawa—Moosonee Workshop in algebra (Ottawa, 1987)*, 1–104, Carleton Mathematical Lecture Notes, 3, Carleton University, Ottawa, Ont., 1988.

[159] N. Popescu and P. Gabriel, Caractérisation des catégories abéliennes avec générateurs et limites inductives exactes, C. R. Acad. Sci. Paris **258** (1964), 4188–4190.

|160| M. Prest, Remarks on elementary duality, Ann. Pure Appl. Logic **62** (1993), no. 2, 185–205.

|161| M. Prest, Ziegler spectra of tame hereditary algebras, J. Algebra **207** (1998), no. 1, 146–164.

|162| M. Prest, *Purity, spectra and localisation*, Encyclopedia of Mathematics and its Applications, 121, Cambridge University Press, Cambridge, 2009.

|163| H. Prüfer, Untersuchungen über die Zerlegbarkeit der abzählbaren primären Abelschen Gruppen, Math. Z. **17** (1923), no. 1, 35–61.

|164| D. Puppe, On the structure of stable homotopy theory, in *Colloquium on algebraic topology*, 65–71, Aarhus Universitet Matematisk Institut, Aarhus, 1962.

|165| D. Quillen, Higher algebraic K-theory. I, in *Algebraic K-theory, I: Higher K-theories (Proc. Conf., Battelle Memorial Inst., Seattle, Wash., 1972)*, 85–147, Lecture Notes in Mathematics, 341, Springer, Berlin, 1973.

|166| A. Ranicki, *Non-commutative localization in algebra and topology*, Proc. Workshop, Edinburgh, April 29–30, 2002, London Mathematical Society Lecture Note Series, 330, Cambridge University Press, Cambridge, 2006.

|167| D. C. Ravenel, Localization with respect to certain periodic homology theories, Am. J. Math. **106** (1984), no. 2, 351–414.

|168| I. Reiten and M. Van den Bergh, Noetherian hereditary abelian categories satisfying Serre duality, J. Am. Math. Soc. **15** (2002), no. 2, 295–366.

|169| G. Richter, Noetherian semigroup rings with several objects, in *Group and semigroup rings (Johannesburg, 1985)*, 231–246, North-Holland Mathematical Studies, 126, North-Holland, Amsterdam, 1986.

|170| J. Rickard, Morita theory for derived categories, J. London Math. Soc. (Ser. 2) **39** (1989), no. 3, 436–456.

|171| C. M. Ringel, Infinite-dimensional representations of finite-dimensional hereditary algebras, in *Symposia Mathematica*, vol. XXIII *(Conf. abelian groups and their relationship to the theory of modules, INDAM, Rome, 1977)*, 321–412, Academic Press, London, 1979.

|172| C. M. Ringel, The canonical algebras, in *Topics in algebra, Part 1 (Warsaw, 1988)*, 407–432, Banach Center Publications, 26, Part 1, PWN, Warsaw, 1990.

|173| C. M. Ringel, The category of modules with good filtrations over a quasi-hereditary algebra has almost split sequences, Math. Z. **208** (1991), no. 2, 209–223.

|174| C. M. Ringel, The Ziegler spectrum of a tame hereditary algebra, Colloq. Math. **76** (1998), no. 1, 105–115.

|175| C. M. Ringel and H. Tachikawa, QF – 3 rings, J. Reine Angew. Math. **272** (1974), 49–72.

|176| N. Roby, Lois polynomes et lois formelles en théorie des modules, Ann. Sci. Éc. Norm. Supér. (Sér. 3) **80** (1963), 213–348.

|177| J.-E. Roos, Sur la décomposition bornée des objets injectifs dans les catégories de Grothendieck, C. R. Acad. Sci. Paris Sér. A-B **266** (1968), A449–A452.

|178| J. E. Roos, Locally Noetherian categories and generalized strictly linearly compact rings. Applications, in *Category theory, homology theory and their applications, II (Battelle Institute Conf., Seattle, Wash., 1968, Vol. Two)*, 197–277, Springer, Berlin, 1969.

[179] L. Salce, Cotorsion theories for abelian groups, in *Symposia Mathematica*, vol. XXIII *(Conf. abelian groups and their relationship to the theory of modules, INDAM, Rome, 1977)*, 11–32, Academic Press, London, 1979.

[180] S. V. Sam and A. Snowden, Gröbner methods for representations of combinatorial categories, J. Am. Math. Soc. **30** (2017), no. 1, 159–203.

[181] M. Schlichting, Negative K-theory of derived categories, Math. Z. **253** (2006), no. 1, 97–134.

[182] A. H. Schofield, *Representation of rings over skew fields*, London Mathematical Society Lecture Note Series, 92, Cambridge University Press, Cambridge, 1985.

[183] H. Schubert, *Categories* [translated from the German by Eva Gray], Springer, New York, 1972.

[184] I. Schur, Über eine Klasse von Matrizen, die sich einer gegebenen Matrix zuordnen lassen. Dissertation, Berlin, 1901. In I. Schur, Gesammelte Abhandlungen I, 1–70, Springer, Berlin, 1973.

[185] I. Schur, Über die rationalen Darstellungen der allgemeinen linearen Gruppe, Sitzungsber. Preuß. Akad. Wiss., Phys.-Math. Kl. (1927), 58–75. In I. Schur, Gesammelte Abhandlungen III, 68–85, Springer, Berlin, 1973.

[186] S. Schwede, Algebraic versus topological triangulated categories, in *Triangulated categories*, 389–407, London Mathematical Society Lecture Note Series, 375, Cambridge University Press, Cambridge, 2010.

[187] L. L. Scott, Simulating algebraic geometry with algebra. I. The algebraic theory of derived categories, in *The Arcata Conf. on representations of finite groups (Arcata, Calif., 1986)*, 271–281, Proceedings of Symposia in Pure Mathematics, 47, Part 2, American Mathematical Society, Providence, RI, 1987.

[188] J.-P. Serre, Faisceaux algébriques cohérents, Ann. Math. (Ser. 2) **61** (1955), 197–278.

[189] J.-P. Serre, Cohomologie et géométrie algébrique, in *Proc. Int. Congress of mathematicians, 1954, Amsterdam*, vol. III, 515–520, Erven P. Noordhoff, Groningen, 1956.

[190] J.-P. Serre, Sur les modules projectifs, in *Séminaire P. Dubreil, M.-L. Dubreil-Jacotin et C. Pisot, 14ième année: 1960/61. Algèbre et théorie des nombres. Fasc. 1*, 1–16, Faculté des Sciences de Paris, Secrétariat mathématique, Paris, 1963.

[191] D. Simson, Pure semisimple categories and rings of finite representation type, J. Algebra **48** (1977), no. 2, 290–296.

[192] D. Simson, On pure semi-simple Grothendieck categories. I, Fund. Math. **100** (1978), no. 3, 211–222.

[193] D. Simson, On right pure semisimple hereditary rings and an Artin problem, J. Pure Appl. Algebra **104** (1995), no. 3, 313–332.

[194] N. Spaltenstein, Resolutions of unbounded complexes, Compos. Math. **65** (1988), no. 2, 121–154.

[195] R. P. Stanley, *Enumerative combinatorics*, vol. 2, Cambridge Studies in Advanced Mathematics, 62, Cambridge University Press, Cambridge, 1999.

[196] B. T. Stenström, Pure submodules, Ark. Mat. **7** (1967), 159–171.

[197] B. Stenström, *Rings of quotients*, Springer, New York, 1975.

[198] B. Totaro, Projective resolutions of representations of GL(n), J. Reine Angew. Math. 482 (1997) 1–13.

[199] J.-L. Verdier, *Des catégories dérivées des catégories abéliennes*, Astérisque, 239, Société Mathématique de France, 1996.

[200] R. B. Warfield, Jr., Purity and algebraic compactness for modules, Pacific J. Math. **28** (1969), 699–719.

[201] N. Yoneda, On the homology theory of modules, J. Fac. Sci. Univ. Tokyo Sect. I **7** (1954), 193–227.

[202] A. Zaks, Injective dimension of semi-primary rings, J. Algebra **13** (1969), 73–86.

[203] M. Ziegler, Model theory of modules, Ann. Pure Appl. Logic **26** (1984), no. 2, 149–213.

[204] W. Zimmermann, Rein injektive direkte Summen von Moduln, Commun. Algebra **5** (1977), no. 10, 1083–1117.

[205] B. Zimmermann-Huisgen, Rings whose right modules are direct sums of indecomposable modules, Proc. Am. Math. Soc. **77** (1979), no. 2, 191–197.

Notation

Abbreviations

Ab	category of abelian groups, xv
Ab($\mathcal{A}$)	abelian category associated with a locally finitely presented category $\mathcal{A}$, 384
Ab(X)	abelian category associated with an object X of a locally finitely presented category, 392
Ab(Λ)	free abelian category over a ring Λ, 400
Ac($\mathcal{A}$)	acyclic complexes in an exact category $\mathcal{A}$, 107
add($\mathcal{X}$)	closure of $\mathcal{X}$ under finite direct sums and summands, xxi
Add($\mathcal{X}$)	closure of $\mathcal{X}$ under all coproducts and direct summands, xxiii
Add($\mathcal{C}$, Ab)	additive functors $\mathcal{C} \to$ Ab, 345
Ann(X)	annihilator of a module X, 48
Ass(X)	associated prime ideals of a module X, 50
C($\mathcal{A}$)	complexes in an additive category $\mathcal{A}$, 102
C^b($\mathcal{A}$)	bounded complexes in $\mathcal{A}$, 109
C$^+$($\mathcal{A}$)	bounded below complexes in $\mathcal{A}$, 109
C$^-$($\mathcal{A}$)	bounded above complexes in $\mathcal{A}$, 109
card(X)	cardinality of a set X, xv
coh($\mathbb{X}$)	coherent sheaves on a scheme $\mathbb{X}$, 344
Coker(ϕ)	cokernel of a morphism ϕ, xxii
colim(F)	colimit of a functor F, xix
Cone(ϕ)	cone of a morphism ϕ, 76
Cores($\mathcal{C}$)	objects that admit a finite coresolution in $\mathcal{C}$, 208
D($\mathcal{A}$)	derived category of an exact category $\mathcal{A}$, 107
D^b($\mathcal{A}$)	derived category of bounded complexes in $\mathcal{A}$, 109
D$^+$($\mathcal{A}$)	derived category of bounded below complexes in $\mathcal{A}$, 109

$\mathbf{D}^-(\mathcal{A})$	derived category of bounded above complexes in $\mathcal{A}$, 109
$\mathbf{D}(\mathcal{A}, \mathcal{A}_0)$	derived category of a Frobenius pair $(\mathcal{A}, \mathcal{A}_0)$, 88
$\mathbf{D}(A)$	derived category of a dg algebra A, 299
$\mathbf{D}^{\mathrm{perf}}(A)$	perfect complexes of a ring or dg algebra A, 167, 300
$\mathbf{D}_{\mathrm{sg}}(A)$	singularity category of a ring A, 182
$D(X)$	Matlis dual of a module X, xxxi
Δ_i, ∇_i	(co)standard module over a quasi-hereditary algebra, 232
$E(X)$	injective envelope of an object X, xxviii
$\mathrm{eff}(\mathcal{A})$	effaceable functors $\mathcal{A}^{\mathrm{op}} \to \mathrm{Ab}$, 42
$\mathrm{Eff}(\mathcal{A})$	locally effaceable functors $\mathcal{A}^{\mathrm{op}} \to \mathrm{Ab}$, 44
$\mathrm{Eff}(\mathcal{A}, \mathrm{Ab})$	effaceable functors $\mathcal{A} \to \mathrm{Ab}$, 399
$\mathrm{End}(X)$	endomorphisms of an object X, xvii
$\mathcal{E}nd(X)$	endomorphism dg algebra of X, 299
$\mathrm{endol}(X)$	endolength of a module X, 426
Ess	essential monomorphisms in an abelian category, 59
$\mathrm{Ex}(\mathcal{C}, \mathrm{Ab})$	exact functors $\mathcal{C} \to \mathrm{Ab}$, 353
$\mathrm{Ext}^n(X, Y)$	degree n extensions between objects X, Y, xxv
$\widehat{\mathrm{Ext}}^n(X, Y)$	degree n Tate extensions between objects X, Y, 143
$\mathcal{F}(\Lambda)$	finitely generated free modules over a ring Λ, 375
$\mathrm{Filt}(\mathcal{X})$	extension closed subcategory generated by a class $\mathcal{X}$, xxvi
$\mathrm{fp}(\mathcal{A})$	finitely presented objects of a category $\mathcal{A}$, 343
$\mathrm{Fp}(\mathcal{C}, \mathrm{Ab})$	finitely presented functors $\mathcal{C} \to \mathrm{Ab}$, 346
$\mathrm{Fun}(\mathcal{C}, \mathcal{D})$	functors $\mathcal{C} \to \mathcal{D}$, xviii
$\mathrm{GI}(X)$	Gorenstein injective approximation of a module X, 193
$\mathrm{Ginj}(\Lambda)$	Gorenstein injective modules over a ring Λ, 192
$\underline{\mathrm{Ginj}}(\Lambda)$	stable category of Gorenstein injective modules, 193
$\mathrm{GL}_n(\Lambda)$	general linear group of $n \times n$ matrices over a ring Λ, 375
$\mathrm{gl.dim}(\mathcal{A})$	global dimension of an exact category $\mathcal{A}$, xxix
$\mathrm{Gor.dim}(\Lambda)$	dimension of a Gorenstein ring Λ, 184
$\mathrm{GP}(X)$	Gorenstein projective approximation of a module X, 193
$\mathrm{Gproj}(\Lambda)$	Gorenstein projective modules over a ring Λ, 179
$\underline{\mathrm{Gproj}}(\Lambda)$	stable category of Gorenstein projective modules, 182
$\mathrm{GrMod}(\Lambda)$	graded modules over a graded ring Λ, 54
$\mathrm{grmod}(\Lambda)$	finitely presented graded modules, 54
$\underline{\mathrm{grmod}}(\Lambda)$	projectively stable module category, 54
$\mathrm{grproj}(\Lambda)$	finitely generated projective graded modules, 54
Γ	category of finite sets, 373
Γ_{inj}	category of finite sets with injective morphisms, 375
Γ_{os}	category of finite sets with ordered surjections, 373

Γ_{sur}	category of finite sets with surjective morphisms, 373	
$\Gamma^*(V)$	algebra of symmetric tensors of a module V, 244	
$\Gamma^d \mathcal{P}_k$	category of symmetric tensors over k, 244	
h_λ	complete symmetric function for a partition λ, 243	
$H^n(X)$	cohomology of degree n of a complex X, 104	
$H^n(G, X)$	cohomology of degree n of a group G with coefficients in a module X, 408	
$\hat{H}^n(G, X)$	Tate cohomology of degree n of a group G with coefficients in a module X, 411	
$\text{hocolim}(X_n)$	homotopy colimit of a sequence $(X_n \to X_{n+1})_{n \in \mathbb{N}}$, 90	
$\text{holim}(X_n)$	homotopy limit of a sequence $(X_{n+1} \to X_n)_{n \in \mathbb{N}}$, 90	
$\text{Hom}(X, Y)$	set (or complex) of morphisms $X \to Y$, xvii, 129	
$\overline{\text{Hom}}(X, Y)$	stable morphisms modulo injectives, 190	
$\underline{\text{Hom}}(X, Y)$	stable morphisms modulo projectives, 83, 190	
$\mathcal{H}om(\mathcal{C}, \mathcal{D})$	functors $\mathcal{C} \to \mathcal{D}$, xviii	
$\mathcal{H}om(X, Y)$	dg module of morphisms $X \to Y$, 298	
$\text{ht}(X)$	height of an object X, xxiv	
$i(X)$	injective resolution of an object X, 112	
$\mathbf{i}(X)$	K-injective resolution of a complex X, 123	
id_X	identity morphism of an object X, xvii	
$\text{id}_{\mathcal{C}}$	identity functor of a category $\mathcal{C}$, xvii	
$\text{Im}(\phi)$	image of a morphism ϕ, xxii	
$\text{Im}(F)$	essential image of a functor F, xviii	
$\text{ind}(\mathcal{A})$	indecomposable objects of a Krull–Schmidt category $\mathcal{A}$, 422	
$\text{Ind}(\mathcal{A})$	indecomposable pure-injective objects of a locally finitely presented category $\mathcal{A}$, 384	
$\text{Ind}(\Lambda)$	indecomposable pure-injective modules over a ring Λ, 400	
$\text{Inj}(\mathcal{A})$	injective objects of an exact category $\mathcal{A}$, xxviii	
$\text{Inj}(\Lambda)$	injective modules over a ring Λ, 23	
$\text{inj}(\Lambda)$	finitely presented injective modules, 195	
$\text{inj.dim}(X)$	injective dimension of an object X, xxix	
$J(\Lambda)$	Jacobson radical of a ring Λ, xxiv	
$K_{\lambda\mu}$	Kostka number for partitions λ, μ, 242	
$K_0(\mathcal{C})$	Grothendieck group of an exact or triangulated category $\mathcal{C}$, xxx, 110	
$K_0(\Lambda)$	Grothendieck group of a ring Λ, xxx	
$\mathbf{K}(\mathcal{A})$	homotopy category of complexes in an additive category $\mathcal{A}$, 103	

$\mathbf{K}^b(\mathcal{A})$	homotopy category of bounded complexes in $\mathcal{A}$, 109
$\mathbf{K}^+(\mathcal{A})$	homotopy category of bounded below complexes in $\mathcal{A}$, 109
$\mathbf{K}^-(\mathcal{A})$	homotopy category of bounded above complexes in $\mathcal{A}$, 109
$\mathbf{K}^{+,b}(\mathcal{C})$	homotopy category of bounded below complexes with bounded cohomology, 114
$\mathbf{K}^{-,b}(\mathcal{C})$	homotopy category of bounded above complexes with bounded cohomology, 114
$\mathbf{K}_{ac}(\mathcal{P})$	acyclic complexes of projectives in a Frobenius category, 142
$\mathbf{K}_{inj}(\mathcal{A})$	K-injective complexes in an exact category $\mathcal{A}$, 122
$\mathbf{K}_{proj}(\mathcal{A})$	K-projective complexes in an exact category $\mathcal{A}$, 122
$\mathrm{Ker}(\phi)$	kernel of a morphism ϕ, xxii
$\mathrm{Ker}(F)$	kernel of a functor F, xxi
$\mathrm{KG.dim}(\mathcal{A})$	Krull–Gabriel dimension of an abelian category $\mathcal{A}$, 436
$\mathrm{KG.dim}(X)$	Krull–Gabriel dimension of an object X, 441
$\mathbf{L}F$	left derived functor of a functor F, 128
$\mathbf{L}(X)$	lattice of subobjects of an object X, xxxi
$\ell(X)$	composition length of an object X, xxiv
$\mathrm{Lex}(\mathcal{A})$	left exact functors $\mathcal{A}^{op} \to \mathrm{Ab}$ for an exact category $\mathcal{A}$, 44
$\mathrm{Lex}(\mathcal{C}^{op}, \mathrm{Ab})$	left exact functors $\mathcal{C}^{op} \to \mathrm{Ab}$ for an additive category $\mathcal{C}$ with cokernels, 349
$\lim(F)$	limit of a functor F, xix
$\mathrm{Loc}(\mathcal{X})$	localising subcategory generated by a class $\mathcal{X}$, 92
$\Lambda^*(V)$	exterior algebra of a module V, 246
$\Lambda(n,d)$	compositions of d into n parts, 241
$M_n(\Lambda)$	semigroup of $n \times n$ matrices over a ring Λ, 375
$\mathrm{Max}(\Lambda)$	maximal ideals of a commutative ring Λ, 445
$\mathrm{m.dim}(L)$	m-dimension of a lattice L, 439
$\mathrm{Mod}(\mathcal{C})$	additive functors $\mathcal{C}^{op} \to \mathrm{Ab}$, 16
$\mathrm{mod}(\mathcal{C})$	finitely presented functors $\mathcal{C}^{op} \to \mathrm{Ab}$, 17
$\mathrm{mod}_\alpha(\mathcal{C})$	α-presentable functors $\mathcal{C}^{op} \to \mathrm{Ab}$, 65
$\mathrm{Mod}(\Lambda)$	modules over a ring Λ, xv
$\underline{\mathrm{mod}}(\Lambda)$	finitely presented modules, xv
$\overline{\mathrm{mod}}(\Lambda)$	injectively stable module category, 191
$\underline{\mathrm{mod}}(\Lambda)$	projectively stable module category, 43
$\mathrm{Mor}(\mathcal{C})$	morphisms of a category $\mathcal{C}$, xvii
$\mathbb{N}$	set of non-negative integers, xvi
$\bar{\mathbb{N}}, \vec{\mathbb{N}}$	category of non-negative integers, 367
$\mathrm{noeth}(\mathcal{A})$	noetherian objects of an abelian category $\mathcal{A}$, 37

ν	Nakayama functor, 195
$\mathcal{O}_{\mathbb{X}}$	structure sheaf of a scheme $\mathbb{X}$, 152
$\mathrm{Ob}(\mathcal{C})$	objects of a category $\mathcal{C}$, xvii
$\Omega(X)$	syzygy of a module X, xxix
$\Omega_{\mathbb{X}/k}$	sheaf of differential forms of $\mathbb{X}$ over k, 318
$\mathbb{P}^n_k$	projective n-space over a field k, 329
$\mathbf{P}(\mathcal{A})$	purity category of a locally finitely presented category $\mathcal{A}$, 379
$\mathcal{P}(\Lambda)$	modules that admit a finite resolution in proj Λ, 218
$\mathcal{P}_k$	finitely generated projective k-modules, 243
$p(X)$	projective resolution of an object X, 112
$\mathbf{p}(X)$	K-projective resolution of a complex X, 123
$\mathrm{pcoh}(\Lambda)$	pseudo-coherent modules over a ring Λ, 171
$\mathrm{Ph}(X,Y)$	phantom morphisms $X \to Y$, 167
$\mathrm{Pol}^d \mathcal{P}_k$	strict polynomial functors of degree d over k, 251
$\mathrm{pol}^d \mathcal{P}_k$	finite strict polynomial functors of degree d over k, 251
$\mathrm{Prod}(X)$	closure of X under all products and direct summands, 395
$\mathrm{Proj}(\mathcal{A})$	projective objects of an exact category $\mathcal{A}$, xxviii
$\mathrm{Proj}(\Lambda)$	projective modules over a ring Λ, 23
$\mathrm{proj}(\Lambda)$	finitely generated projective modules, xv
$\mathrm{proj.dim}(X)$	projective dimension of an object X, xxviii
$\mathrm{Qcoh}(\mathbb{X})$	quasi-coherent sheaves on a scheme $\mathbb{X}$, 344
Qis	quasi-isomorphisms of complexes in an exact category, 107
$\mathrm{rad}(X)$	radical of an object X, xxiv
$\mathrm{Rad}(X,Y)$	group of radical morphisms $X \to Y$, xxiv
$\mathrm{rank}_k(X)$	rank of a free k-module X, xxiv
$\mathrm{reg}(\Lambda)$	regular modules over an Artin algebra Λ, 158
$\mathrm{rep}(\Gamma,k)$	finite k-linear representations of Γ, 285
$\mathrm{Rep}(\Gamma,k)$	k-linear representations of Γ, 348
$\mathrm{Res}(\mathcal{C})$	objects that admit a finite resolution in $\mathcal{C}$, 177
$\mathbf{R}F$	right derived functor of a functor F, 128
$\mathrm{RHom}(X,Y)$	derived hom of complexes X and Y, 130
$\mathrm{Rlim}\, F$	right derived limit of a functor F, 321
s_λ	Schur function for a partition λ, 243
$S^*(V)$	symmetric algebra of a module V, 244
Sch^λ	Schur functor for a partition λ, 279
Set	category of sets, xv
$\mathrm{sgn}(\sigma)$	signum of a permutation σ, 279
$\mathrm{Sh}(X)$	sheaves on a topological space X, 35
$\mathrm{soc}(X)$	socle of an object X, xxiv

$\mathrm{span}_k(X)$	k-linear span of a set X, 265
$\mathrm{Sp}(\mathcal{A})$	spectrum of a Grothendieck category $\mathcal{A}$, 37
$\mathrm{Spec}(A)$	prime ideal spectrum of a commutative ring A, 48
$S_k(n,d)$	Schur algebra over k given by parameters n, d, 245
$S(\mathbf{p}, \lambda)$	coordinate algebra of a weighted projective line, 334
$\mathrm{Sq}(\mathbf{p}, \lambda)$	squid algebra given by a weighted projective line, 336
$\mathrm{St}(\mathcal{A})$	injectively stable category of an exact category $\mathcal{A}$, 28, 83
$\mathrm{StMod}(\Lambda)$	stable module category of a quasi-Frobenius ring Λ, 88
$\mathrm{Sub}(X)$	poset of subobjects of an object X, 370
$\mathrm{Supp}(F)$	support of a functor F, 422
$\mathrm{Supp}(\mathscr{F})$	support of a sheaf $\mathscr{F}$, 153
$\mathrm{Supp}(X)$	support of a module or complex X, 48, 162, 133
SW_{f}	Spanier–Whitehead category of finite CW-complexes, 308
$\mathfrak{S}_d$	symmetric group, 244
$\mathfrak{S}_\lambda$	Young subgroup of the symmetric group, 247
$\Sigma(X)$	suspension or shift of an object X, 73
$\sigma_{\leq n}X, \sigma_{\geq n}X$	brutal truncations of a complex X, 111
$t(X)$	complete resolution of an object X, 143
$T(A)$	trivial extension algebra of an Artin algebra A, 164
$T^*(V)$	tensor algebra of a module V, 246
$\mathrm{Thick}(\mathcal{X})$	thick subcategory generated by a class of objects $\mathcal{X}$, xxvii, 77
$\mathrm{top}(X)$	top of an object X, xxiv
$\mathrm{Tor}_n^\Lambda(X, Y)$	Tor group of degree n of modules X and Y, 130
$\mathrm{Tr}(X)$	transpose of a finitely presented module X, 191
$\tau_{\leq n}X, \tau_{\geq n}X$	soft truncations of a complex X, 111
$\mathcal{V}(\mathfrak{a})$	prime ideals containing an ideal $\mathfrak{a}$, 48
$\mathrm{vect}(\mathbb{X})$	vector bundles on a scheme $\mathbb{X}$, 317
Weyl^λ	Weyl functor for a partition λ, 279
$\mathrm{w.dim}(X)$	weak dimension of a module X, 179
$\mathbb{Z}$	set of integers, xvi
$Z(\mathcal{C})$	centre of a preadditive category $\mathcal{C}$, xxx, 347
$Z(\Lambda)$	centre of a ring Λ, xxx
$Z^n(X)$	cocycles of degree n of a complex X, 104

Constructions

A^{op}	opposite of a ring A, xv
A_Σ	universal localisation of a ring A with respect to Σ, 46
$A^!$	quadratic dual of an algebra A, 331

$\mathcal{A}^\alpha$	α-presentable objects of a cocomplete category $\mathcal{A}$, 61
$\mathcal{A}_1 \times_{\mathcal{A}} \mathcal{A}_2$	pullback of abelian categories, 41
$\mathcal{C}^{\mathcal{J}}$	diagrams of type $\mathcal{J}$ in a category $\mathcal{C}$, xix
$\mathcal{C}^{\mathrm{op}}$	opposite of a category $\mathcal{C}$, xvii
$\mathcal{C}^2$	morphisms in a category $\mathcal{C}$, xvii
$\mathcal{C}(x, y)$	morphisms $x \to y$ in a category $\mathcal{C}$, 369
$\mathcal{C}(x)$	morphisms terminating at x in a category $\mathcal{C}$, 369
$\vec{\mathcal{C}}$	closure of $\mathcal{C}$ under filtered colimits, 345
$\coprod \mathcal{C}$	closure of $\mathcal{C}$ under all coproducts, 97
$(\mathcal{C}_\alpha)_\alpha$	Krull–Gabriel filtration of an abelian category $\mathcal{C}$, 436
$\mathcal{C}/\mathcal{D}$	quotient of an additive, abelian, or triangulated category $\mathcal{C}$ with respect to a subcategory $\mathcal{D} \subseteq \mathcal{C}$, 29, 30, 78
$\mathcal{C}/F$	slice category of $\mathcal{C}$ over a functor F, 345
$\mathcal{C}/X$	slice category of $\mathcal{C}$ over an object X, 343
$\mathcal{C}[S^{-1}]$	localisation of a category $\mathcal{C}$ with respect to S, 3
$S^{-1}\mathcal{C}$	category of left fractions of $\mathcal{C}$ with respect to S, 10
$\mathcal{C} = \coprod_{i \in I} \mathcal{C}_i$	orthogonal decomposition of an additive category $\mathcal{C}$, xxi
$\mathcal{C} = \bigvee_{i \in I} \mathcal{C}_i$	direct decomposition of an additive category $\mathcal{C}$, xxi
$\mathcal{C}^\perp, {}^\perp\mathcal{C}$	perpendicular categories in an abelian, triangulated, exact, or locally finitely presented category, 30, 77, 176, 385
F_λ, F_ρ	left and right adjoint of a functor F, xix
F^λ	λ-fold tensor product of a graded functor F^*, 254
$F^\vee$	dual of a functor F, 43, 401
F°	dual of a functor F, 243
$k\mathcal{C}$	linearisation of a category $\mathcal{C}$ over a commutative ring, 348
kG	group algebra of a group G over a commutative ring, 348
kQ	path category of a quiver Q over a commutative ring, 348
$k[X]$	free module with basis X over a commutative ring, 348
k_{sgn}	sign representation of the symmetric group, 275
$(L_\alpha)_\alpha$	minimal cofiltration of a modular lattice L, 439
$\lambda \vdash d$	partition of an integer d, 241
$S^\perp$	class of S-local objects with respect to morphisms in S, 4
X^*	dual of a module X, 180
$X^\vee$	dual of a module X, 243
X^G	invariants of a module X with G-action, 244
X_G	coinvariants of a module X with G-action, 244
$X(V)_\lambda$	weight space of $X(V)$ for a composition λ, 260
X_ϕ	subgroup of finite definition of $\mathrm{Hom}(C, X)$ with respect to $\phi: C \to C'$, 363, 392

$X^{(I)}$, $X[I]$	coproduct of copies of X indexed by a set I, xxiii
$X \otimes_\Lambda Y$	tensor product of (complexes of) modules X and Y, xxxiii, 129
$X \otimes_\Lambda^L Y$	derived tensor product of complexes X and Y, 130
$\mathfrak{X} * \mathfrak{Y}$	extensions in a triangulated category of objects in $\mathfrak{X}, \mathfrak{Y}$, 97

Arrows

$X \rightarrowtail Y$	monomorphism, xvii
$X \twoheadrightarrow Y$	epimorphism, xvii
$X \xrightarrow{\sim} Y$	isomorphism, xvii
$\mathcal{C} \rightarrowtail \mathcal{D}$	fully faithful functor, xviii
$\mathcal{C} \twoheadrightarrow \mathcal{D}$	quotient functor, xviii
$\mathcal{C} \xrightarrow{\sim} \mathcal{D}$	equivalence, xviii
$\mathcal{C} \rightleftarrows \mathcal{D}$	adjoint pair of functors, xviii

Index

algebra
  Artin, xxx
  Beilinson, 330
  canonical, 335
  dg, 298
  differential graded, 298
  exterior, 246, 332
  gentle, 185
  integral group, 184
  Koszul, 330
  Kronecker, 157
  noetherian, xxx
  quadratic, 330
  quasi-hereditary, 232
  Schur, 229, 245
  split quasi-hereditary, 285
  squid, 336
  symmetric, 196, 254, 332
  tensor, 246, 331
  trivial extension, 164
antisymmetriser, 274
approximation
  left, 212, 352
  right, 38, 212
arrow, xxxi
  start of, xxxi
  terminus of, xxxi
artinian conjecture, 376
Auslander–Reiten formula, 192

Baer's criterion, 362
Brown representability theorem, 92, 98

calculus
  of left fractions, 10
  of right fractions, 10
cardinality, xv

Catalan number, 225
category, xvii
  α-filtered, 60
  abelian, xxii, 16
  abelian quotient, 30
  additive, xx, 15
  additive quotient, 29
  algebraic triangulated, 85, 300
  BGG category $\mathcal{O}$, 292
  cocomplete, xxiii
  compactly generated triangulated, 98, 305
  connected, xxi
  derived, 88, 107, 299
  dg, 300
  differential graded, 300
  directed, xx
  essentially small, xvii
  exact, xxvi, 21
  exact products, 123
  filtered, xix, 11, 342
  finitely generated abelian, 383, 397
  free abelian, 400
  Frobenius, xxx, 83, 182
  Gröbner, 372
  Grothendieck, xxiii
  having enough injectives, xxviii, 23
  having enough projectives, xxviii, 23
  having injective envelopes, 27
  having projective covers, 27
  hereditary, xxix, 139
  homotopy, 103, 299
  idempotent complete, xxv
  injectively stable, 28, 191
  k-linear, xxx, 347
  Krull–Schmidt, xxvi
  left Hom-finite, 421

length, xxiv
locally α-presentable, 61
locally finitely presented, xxiii, 343
locally presentable, 61
locally small, xv
of morphisms, xvii, 20
opposite, xvii
path, 348
perfectly generated triangulated, 92
perpendicular, 176
preadditive, 347
projectively stable, 43, 191
purity, 379
singularity, 182
slice, 170, 343
small, xvii
Spanier–Whitehead, 308
spectral, 59, 438
split exact, xxix
stabilised derived, 182
stable, 83, 182
suspended, 73
symmetric tensors, 229, 244
triangulated, 73
uniserial, 424
with exact (co)products, 108
Cauchy filtration
exterior powers, 283
symmetric powers, 271
symmetric tensors, 267
Cauchy identity, 292
centre, xxx, 347
Chase's lemma, 362
class, xv
cocycle, 104
cogenerating set
perfectly, 96
cogenerator
injective, xxviii
cohomology
of a complex, 104
Tate, 143, 411
coideal
of a poset, 285
coinvariants, 244
cokernel, xxii
colimit, xix
α-filtered, 60
α-small, 60
directed, xx, 342
filtered, xx, 342

homotopy, 89
complex, 102
acyclic, 105, 106
bounded, 109, 196, 209
cochain, 102
contractible, 102
dualising, 317, 318
homologically perfect, 312
homotopically minimal, 131
homotopy injective, 122
homotopy projective, 122
K-injective, 122
K-projective, 122
Koszul, 331
perfect, 167, 189
quasi-isomorphic to its cohomology, 139
shifted, 103
split, 105
totally acyclic, 180
composition, 241
cone
mapping, 104
of a morphism, 76, 83
content, 242
coresolution
finite $\mathcal{C}$-, 208
cotorsion pair, 176, 211
counit, xviii
cover
projective, 25

decomposition of a category
direct, xxi
orthogonal, xxi
Dedekind domain, 445
degree
cohomology, 104
complex, 102
homogeneous element, 53
dense chain, 440
diagram, xix
differential, 102, 298, 300
dimension
finitistic, 222, 405
flat, 179
global, xxix
injective, xxix
Krull–Gabriel, of a category, 436
Krull–Gabriel, of an object, 441
minimal, 439
of a Gorenstein ring, 179
projective, xxviii

weak, 179
direct summand, xxi
dual
  quadratic, 331
  Ringel, 241
dual pair, 404

endolength of a module, 426
envelope
  injective, xxviii, 25, 35
  pure-injective, 381
epimorphism, xvii
  homological, 137
  of rings, 45
equivalence
  derived, 311
  homotopy, 103
  of categories, xviii
  of extensions, 21, 116
  of fractions, 11
  of monomorphisms, xxii
  of tilting objects, 220
  stable, of objects, 28
  triangle, 74
  up to direct summands, 301
  up to permutation, 241
essential image, xviii
extension
  exact category, xxv, 22, 116
  triangulated category, 97
  universal, 236
extension of scalars, xxxiv
exterior power, 246

filling, 242
  strictly increasing each column, 242
  weakly increasing each row, 242
five lemma, xxix
fraction
  left, 10
  right, 10
Frobenius pair, 88
function
  additive, 415
  complete symmetric, 243
  irreducible (sub)additive, 415
  Schur, 243
  subadditive, 415
functor, xvii
  α-coherent, 67
  additive, xxi, 16
  adjoint, xviii
  cofinal, xx, 344

coherent, 67
cohomological, 75
contravariantly finite, 372
covariantly finite, 372
derived Nakayama, 196, 316
diagonal, xix
effaceable, 42
endofinite, 416
essentially surjective, xviii
exact, xxv, 22, 74
faithful, xviii
FI-module, 375
finitely presented, 17, 378
full, xviii
graded, 254
k-linear, 347
left derived, 128
left exact, 44, 349
localisation, xix, 7
locally effaceable, 43
Nakayama, 164, 195, 277
noetherian, 370
quotient, xviii
representable, 17
right derived, 127, 128
Schur, 257, 280
Serre, 189
strict polynomial, 230, 251
support, 422
triangle, 74
weakly left exact, 17
Weyl, 280
Yoneda, 6, 17

Gabriel–Popescu theorem, 56
general linear group, 250
generating set, 356
  compactly, 98
  perfectly, 92
generator, xxiii
  projective, xxviii
good filtration, 293
grading group, 53
Grothendieck category
  locally finite, 357, 360
  locally finitely generated, 357
  locally finitely presented, 121, 357
  locally noetherian, 37, 325, 357, 359, 369
Grothendieck group
  of a ring, xxx
  of a triangulated category, 110
  of an exact category, xxx, 109

group algebra, 348
group cohomology, 408
height, xxiv
Hilbert's basis theorem, 368
Hochster duality, 408
homogeneous element, 53
   even, 54
horseshoe lemma, xxix
ideal, xxi
   associated prime, 50
   of a poset, 285, 368
   of null-homotopic morphisms, 102
   prime, 48
image
   of functor, xviii
   of morphism, xxii
invariants, 244
isomorphism, xvii
Jacobson radical, xxiv
k-algebra, xxx
k-linearisation, 348
kernel
   of functor, xxi, 16
   of morphism, xxii
   weak, 17
Kostka number, 242
Krull–Gabriel filtration, 436
Krull–Remak–Schmidt–Azumaya theorem, 59
lattice, xxxi
   congruence relation, 439
   finite length, 439
   modular, xxxi
   Tamari, 225
Leibniz rule, 298
length, xxiv
limit, xix
   direct, xx
   homotopy, 90
localisation
   of a category, 3
   universal, 46
   Verdier, 78
Loewy length, xxiv
Matlis duality, xxxi
minimal decomposition, 27, 131
module, xv
   acyclic dg, 299
   adic, 447
   basic tilting, 226
   bimodule, xv

characteristic tilting, 240, 289
   costandard, 232, 288
   dg, 298, 300
   differential, 185
   generic, 430, 432
   Gorenstein injective, 192
   Gorenstein projective, 179, 192
   graded, 54
   K-projective dg, 303
   maximal Cohen–Macaulay, 179
   p-torsion, 445
   perfect dg, 300
   Prüfer, 447
   pseudo-coherent, 171
   pure-injective, 402
   Schur, 279
   stable, 161
   standard, 232, 285
   tilting, 219
   torsion, 408
   torsion free, 409
   twisted, 54, 152
   Weyl, 279
monomorphism, xvii
morphism, xvii
   admissible epimorphism, xxvi, 21
   admissible monomorphism, xxvi, 21
   comparison, 267
   connecting, xxv
   essential epi, 25
   essential mono, xxviii, 25
   left almost split, 398
   left minimal, 212
   null-homotopic, 20, 102, 299
   phantom, 167
   pure mono, 380
   radical, xxiv
   right minimal, 212
   standard, 261, 280
Noether isomorphism, xxxi
object, xvii
   α-presentable, 61
   artinian, xxiv
   closed, 4
   compact, 96, 167, 304, 309, 362
   costandard, 271
   discrete, 133
   endofinite, 415
   finite length, xxiv, 360
   finitely generated, 357
   finitely presented, xxiii, 121, 343, 369

fp-injective, 399
homologically finite, 77, 138, 309
indecomposable, xxiv, 58
initial, 314
injective, xxviii, 23
local, 4
noetherian, xxiv, 359
orthogonal, 4
Prüfer, 396
product-complete, 395
projective, xxviii, 23
pure-injective, 380
semisimple, xxiii
Σ-pure-injective, 393
simple, xxiii
small, 96
standard, 264
super-decomposable, 58
tilting, 147, 215, 307, 308
uniform, 58, 438
octahedral axiom, 74
ordered surjection, 373

p-rank, 455
partial order
  admissible, 370
  dominance, 242
  lexicographic, 242
partition, 241
  conjugate, 241
path, xxxi
  differential, 186
  length, xxxi
  maximal differential, 186
point, 150
  closed, 150, 156
  generic, 150
  isolated, 399
  rational, 150
polynomial map, 250
poset
  noetherian, 368
  strongly noetherian, 368
primitive cycle, 186
projective presentation, 23
pullback of abelian categories, 41

quasi-compact space, 383, 397
quasi-isomorphism, 105, 107, 299, 306
quiver, xxxi
  Kronecker, xxxii, 155
  opposite, 336
  underlying diagram, 224

quotient, xxii
radical, xxiv, 160
recollement, 34
regular cardinal, 60
representation
  generic, 375
  group, 348
  k-linear, 348
  postinjective, 156
  preprojective, 156
  quiver, xxxii, 156, 348
  regular, 156, 158
  separated, 160
  sign, 275
resolution
  complete, 142, 301
  finite $\mathcal{C}$-, 177, 208
  injective, 112
  K-injective, 122
  K-projective, 122, 303
  projective, 113
restriction of scalars, xxxiv
ring
  associated graded, 52
  complete intersection, 183
  finite, 375
  finite representation type, 429
  Gorenstein, 179, 183
  graded, 53
  graded commutative, 54
  graded noetherian, 54
  hypersurface, 183
  Iwanaga-Gorenstein, 179
  local, xxiv, 59
  opposite, xv
  quasi-Frobenius, xxx, 87, 184
  right coherent, 167, 313, 405
  right hereditary, xxix
  right perfect, 405
  right pure-semisimple, 404
  right self-injective, 87
  self-injective, 88
  semiperfect, xxvi
  semiprimary, 37
  semisimple, xxix
  with several objects, 376
scheme, 344
  Gorenstein, 318
  projective, 317
sections of a sheaf, 151
sequence

admissible, xxvi, 21
approximation, 176
bad, 374
colocalisation, 7
cone, 83
exact, xxv, xxvi, 21
localisation, 7
long exact, xxv
mapping cone, 104
minimal, 374
pure-exact, 380
short exact, xxv
set, xv
sheaf
locally free, 153
of differential forms, 204, 318
structure, 152
torsion, 153
twisted, 152
shift of a category, 73
shuffle product, 246
snake lemma, xxix
socle, xxiv, 154
source, 224
specialisation closed, 48, 162
spectral space, 408
spectrum
of a commutative ring, 48
of a Grothendieck category, 37, 381
of an additive function, 415
of indecomposable pure-injectives, 384
Zariski, 407
Ziegler, 384, 400
stable under base change, 263
stalk of a sheaf, 153
standard basis theorem, 270
straightening, 264
subcategory
cofinal, xx, 344
contravariantly finite, 38, 212
coresolving, 212
covariantly finite, 212, 352
definable, 385
extension closed, xxvi, 22
full exact, 22
left cofinal, 12, 119
localising, 33, 92
resolving, 211
right cofinal, 119
self-orthogonal, 208
Serre, xxvii, 30

thick, xxvii, 76, 138, 208
triangulated, 76
subgroup of finite definition, 392
subobject, xxii
sum
Baer, 22
direct, xxi
locally finite, 415
support
of a complex, 133
of a module, 48, 162
of a sheaf, 153
suspension of a category, 73
symmetric group, 244
symmetric power, 244
symmetric tensors, 244
comultiplication map, 248
diagonal map, 248
multiplication map, 246
symmetriser, 274
syzygy, xxix, 180
t-structure, 319
canonical, 320
tensor product, 68, 252
top, xxiv
torsion pair, xxvii, 147
split, xxviii
transpose, 191
triangle, 73
exact, 73
standard, 84
truncation
brutal, 111
exact, 111
soft, 111
two out of three property, xxvii, 88, 138, 208
type of a diagram, xix
unit, xviii
vector bundle, 153
vertex, xxxi
Wakamatsu's lemma, 212
weight sequence, 334
weight space decomposition, 260
weighted projective line, 334
Yoneda's lemma, 17
Young diagram, 242
Young subgroup, 247
Young tableau, 242
Zariski closed, 48, 407
Ziegler closed, 385